MANUAL
OF
MINERALOGY

MANUAL OF MINERALOGY

(*after James D. Dana*)

19th Edition

CORNELIUS S. HURLBUT, JR.

Harvard University

CORNELIS KLEIN

Indiana University

John Wiley & Sons

NEW YORK   SANTA BARBARA   LONDON   SYDNEY   TORONTO

The *Manual of Mineralogy* was written by James D. Dana in 1848 and
revised by him in 1857, 1878 as the *Manual of Mineralogy and Lithol-
ogy,* and in 1887 as the *Manual of Mineralogy and Petrography.* An edition number was given to some, but not all reprintings of each revision.
For example, the 1887 revision was reprinted in 1891 as the 10th Edition but in 1893, 1895, and 1900 as the 12th Edition.* Subsequent revisions as *Dana's Manual of Mineralogy* have each been given an edition
number as follows: *13*, 1912 and *14*, 1929 by William E. Ford. *15*, 1941;
16, 1952; *17*, 1959; and *18*, 1971 by C. S. Hurlbut, Jr.

* The information regarding revisions by James D. Dana was supplied by Clifford
J. Awald, Buffalo Museum of Science.

Library of Congress Cataloging in Publication Data

Dana, James Dwight, 1813–1895.
  Manual of mineralogy (after James D. Dana).

  Includes bibliographies.
  1. Mineralogy. I. Hurlbut, Cornelius Searle,
1906–   II. Klein, Cornelis, 1937–   III. Title.
QE372.D2 1977   549   77-1131
ISBN 0-471-42226-6

Printed in the United States of America

10 9 8 7 6 5 4 3 2 1

A Mineral

A mineral is a wondrous thing,
At least it is to me,
For in its ordered structure
Lies a world of mystery.

The secrets that it has withheld
For countless ages past
And clung to most tenaciously
Are being learned at last.

Each year using new techniques
Or with a new device,
We make our knowledge more complete,
Our data more precise.

But let us not in trying to solve
A mineral mystery
Forget that minerals are a part
Of natural history.

Nor in our quest for more detail
When probing an unknown,
Forget that every mineral
Has a beauty of its own.

With progress in technology
Each year sees new machines
That try to copy nature
By sophisticated means.

But for all these modern methods
We can not yet compete
With the world of ordered beauty
That lies beneath our feet.

C.S.H.

PREFACE

This book, like the preceding edition known as *Dana's Manual of Mineralogy,* is designed for the beginning course in mineralogy and as a permanent mineralogic reference. It is intended for students who will do further work in mineralogy as well as for those for whom this will be their only exposure to the subject. In the nineteenth edition we have expanded the coverage of many mineralogic, and to some extent introductory petrologic, concepts; this expanded coverage should make it easier for students who continue their training in mineralogy and petrology to use effectively more advanced books, references, and journal publications.

The general organization of the present edition, in 12 chapters, allows instructors to select those subjects they consider most essential. It will be impossible in many courses, especially those that are only one semester or one quarter in length, to cover all of the material presented. The order of the various chapters is reasonably logical except for the location of Chapter 11, which deals with introductory aspects of petrology. The material in this

chapter should go hand in hand with much of Systematic Mineralogy (Chapters 7, 8, 9, and 10), especially Chapter 10 on Silicates. In our treatment of the various subjects grouped under mineralogy we discuss basic concepts before describing mineral groups and species.

In Chapter 2, Crystallography, the first portion is devoted to various aspects of the symmetry of ordered patterns. Although this discussion is not essential to the subsequent treatment of crystal morphology, it is most helpful in the understanding of the external forms of naturally occurring crystals. In our treatment of crystal morphology we proceed from the crystal system of lowest symmetry to that of the highest symmetry. This arrangement conforms to our discussion of the symmetries of ordered patterns. Within each crystal system, however, the crystal classes are arranged in order of decreasing symmetry. Although this appears illogical with respect to the increasing symmetry of the crystal systems, this sequence of discussion of crystal classes is least cumbersome because the lower symmetry classes can be described as special cases of the highest symmetry class. Moreover, if time does not permit development of all crystal classes, it is those of highest symmetry in each crystal system that are usually discussed.

Chapters 3 (X-ray Crystallography) and 6 (Optical Properties of Crystals) have changed little from the eighteenth edition, except for a short description of the determination of crystal structures by X-ray techniques. Although the contents of these two chapters may be used in only a few introductory mineralogy courses, they are included because in the subsequent treatment on Systematic Mineralogy each mineral species description includes a listing of diagnostic X-ray and optical parameters as well as a discussion of its crystal structure. In other words, if students wish to find out, on their own, by what techniques such information is obtained they can locate it in these two chapters.

Chapter 4, on the crystal chemistry, crystal structure, and chemical composition of minerals, covers bonding forces in crystals, concepts such as polymorphism, polytypism, solid solution and exsolution, the recalculation of mineral analyses, and the graphical representation of mineral compositions. Much of the material in this chapter is essential to a full understanding of the discussion of mineral species under Systematic Mineralogy. This chapter also contains a condensed discussion of laboratory chemical and blowpipe tests for elements found in common minerals.

Chapters 7, 8, 9 and 10 deal with the systematic descriptions of about 200 minerals. The description of each mineral group or major mineral series is preceded by a synopsis of some of the highlights of its chemistry, crystal chemistry, and structure. Each of the approximately 200 descriptions

also contains a new section on *Composition and Structure*. The mineral structures are illustrated by some of the best available polyhedral and ball-and-stick representations. In these four chapters use is made of stability diagrams and, under the heading of *Occurrence*, petrologic terms such as facies and assemblage are often used; these concepts are covered in Chapter 11.

Chapter 11 presents a brief and elementary introduction to petrology providing a link between standard mineralogy and petrology books.

At the end of most chapters are selected references which should guide the student to the most relevant literature in the subject. The Determinative Tables (Chapter 12) and Mineral Index, both of which were up-dated wherever necessary, are especially useful in the laboratory for study and identification of unknown minerals.

In short, in writing the nineteenth edition of the *Manual of Mineralogy* our aim was for an introductory mineralogy book that would serve two purposes: (1) to provide the student with an understanding of basic concepts in crystallography, crystal chemistry, chemistry, and introductory petrology—concepts essential to the understanding of the genesis of minerals and rocks; and (2) to provide a reference for quick and unambiguous identification of the common minerals in the field and in the laboratory.

We are grateful to John C. Drake of the University of Vermont for his careful review of the manuscript and to Robert P. Wintsch of Indiana University for his comments on Chapter 11. We thank Mrs. Thea Brown-Fritzinger for her enthusiastic and extraordinarily skillful typing.

Cornelius S. Hurlbut, Jr.
Cornelis Klein

CONTENTS

1
INTRODUCTION

Mineralogy is the study of naturally occurring, crystalline substances—minerals. Everyone has a certain familiarity with minerals for they are present in the rocks of the mountains, the sand of the sea beach, and the soil of the garden. Less familiar, but also composed of minerals, are meteorites and the lunar surface material. A knowledge of what minerals are, how they were formed, and where they occur is basic to an understanding of the materials largely responsible for our present technologic culture. For all inorganic articles of commerce, if not minerals themselves, are mineral in origin.

DEFINITION OF MINERAL

Although it is difficult to formulate a succinct definition for the word mineral, the following is generally accepted:
*A mineral is a naturally occurring homogeneous solid with a definite (but generally not fixed) chemi-*cal *composition and an ordered atomic arrangement. It is usually formed by inorganic processes.*

A step-by-step analysis of this defintion will aid in its understanding. The qualification *naturally occurring* distinguishes between substances formed by natural processes and those made in the laboratory. Industrial and research laboratories routinely produce synthetic equivalents of many naturally occurring materials including valuable gemstones such as emeralds, rubies, and diamonds. Since the beginning of the twentieth century, mineralogic studies have relied heavily on the results from synthetic systems in which the products are given the names of their naturally occurring counterparts. Such practice is generally accepted although at variance with the strict interpretation of *naturally occurring*. In this book *mineral* means a naturally occurring substance and its name will be qualified by *synthetic* if purposely produced by laboratory techniques. One might now ask how to refer to $CaCO_3$ (calcite) which sometimes forms in concentric layers in city water mains. The material is precipitated from water by natural processes but in a man-made system. Most mineralogists would refer to it by its mineral name, calcite, since humanity's part in its formation was inadvertent.

The definition further states that a mineral is a *homogeneous solid*. This means that it consists of a single, solid substance that cannot be physically subdivided into simpler chemical compounds. The determination of homogeneity is difficult because it is related to the scale on which it is defined. For example, a specimen that appears homogeneous to the naked eye may prove to be inhomogeneous, made up of several materials, when viewed with a microscope at high magnification. The qualification *solid* excludes gases and liquids. Thus H_2O as ice in a glacier is a mineral, but water is not. Likewise liquid mercury, found in some mercury deposits, must be excluded by a strict interpretation of the definition. However, in a classification of natural materials such substances that otherwise are like minerals in chemistry and occurrence are called *mineraloids* and fall in the domain of the mineralogist.

The statement that a mineral has a *definite chemical composition* implies it can be expressed by a specific chemical formula. For example, the chemical composition of quartz is expressed as SiO_2. Because quartz contains no chemical elements other than silicon and oxygen, its formula is definite. Quartz is, therefore, often referred to as a pure substance. Most minerals, however, do not have such well-defined compositions. Dolomite, $CaMg(CO_3)_2$, is not always a pure Ca-Mg-carbonate. It may contain considerable amounts of Fe and Mn in place of Mg. Because these amounts vary, the composition of dolomite is said to range between certain limits and is, therefore, *not fixed*. Such a compositional range may be expressed by a formula with the same atomic (or more realistically, ionic) ratios as pure $CaMg(CO_3)_2$ in which $Ca:Mg:CO_3 = 1:1:2$. This leads to a more general expression for dolomite as: $Ca(Mg, Fe, Mn)(CO_3)_2$.

An *ordered atomic arrangement* indicates an internal structural framework of atoms (or ions) arranged in a regular geometric pattern. Since this is the criterion of a crystalline solid, minerals are crystalline. Solids, such as glass, that lack an ordered atomic arrangement are called *amorphous*. Several natural solids are amorphous. They, with the liquids water and mercury which also lack internal order, are classified as mineraloids.

According to the traditional definition, a mineral is *formed by inorganic processes*. We prefer to preface this statement with *usually* and thus include in the realm of mineralogy the few organically produced compounds that answer all the other requirements of a mineral. The outstanding example is the calcium carbonate of mollusk shells. The shell of the oyster and the pearl that may be within it are composed in large part of aragonite, identical to the inorganically formed mineral. Other examples are elemental sulfur formed by bacterial action and iron oxide precipitated by iron bacteria. But petroleum and coal, frequently referred to as mineral fuels, are excluded; for, although naturally formed, they have neither a definite chemical composition nor an ordered atomic arrangement. However, in places coal beds have been subjected to high temperatures which have driven off the volatile hydrocarbons and crystallized the remaining carbon. This residue is the mineral, graphite.

HISTORY OF MINERALOGY

Although it is impossible in a few paragraphs to trace systematically the development of mineralogy, some of the highlights of its development can be singled out. The emergence of mineralogy as a science is relatively recent but the practice of mineralogical arts is as old as human civilization. Natural pigments made of red hematite and black manganese oxide were used in cave paintings by early humans, and flint tools were prized possessions during the Stone Age. Tomb paintings in the Nile Valley executed nearly 5000 years ago show busy artificers weighing malachite and precious metals, smelting mineral ores, and making delicate gems of lapis lazuli and emerald. As the Stone Age gave way to the Bronze Age, other minerals were sought from which metals could be extracted.

FIG. 1.1. Prospecting with a forked stick (A) and trenching (B) in the fifteenth century (from: Agricola, *De Re Metallica*, translated into English, Dover Publications, New York, N.Y.).

We are indebted to the Greek philosopher, Theophrastus (372–287 B.C.) for the first written work on minerals and to Pliny, who 400 years later recorded the mineralogical thought of his time. During the following 1300 years, the few works that were published on minerals contained much lore and fable with little factual information. If one were to select a single event signalling the emergence of mineralogy as a science, it would be the publication in 1556 of *De Re Metallica* by the German physician, Georgius Agricola. This work gives a detailed account of the mining practices of the time and includes the first factual account of minerals. The book was translated into English from the Latin in 1912* by the former President of the United States, Herbert Hoover, and his wife, Lou Henry Hoover. In

* Published in 1950 by Dover Publications, Inc., New York.

1669 an important contribution was made to crystallography by Nicholas Steno through his study of quartz crystals. He noted that despite their differences in origin, size, or habit, the angles between corresponding faces were constant. More than a century passed before the next major contributions were made. In 1780 Carangeot invented a device (contact goniometer) for the measurement of interfacial crystal angles. In 1783 Romé de I'Lsle made angular measurements on crystals confirming Steno's work and formulated the law of the consistency of interfacial angle. The following year, 1784, René J. Haüy showed that crystals were built by stacking together tiny identical building blocks, which he called integral molecules. The concept of integral molecules survives almost in its original sense in the unit cells of modern crystallography. Later (1801) Haüy, through his study of hundreds of crystals,

NICOLAVS STENONIVS

FIG. 1.2. Portrait of Niels Stensen (Latinized to Nicolaus Steno). Steno was born in Copenhagen, Denmark, in 1638 and died in 1686 (from: G. Scherz, *Steno, Geological Papers*, Odense Univ. Press, 1969).

developed the theory of rational indices for crystal faces.

In the early nineteenth century rapid advances were made in the field of mineralogy. In 1809 Wollaston invented the reflecting goniometer, which permitted highly accurate and precise measurements of the positions of crystal faces. Where the contact goniometer had provided the necessary data for studies on crystal symmetry, the reflecting goniometer would provide extensive, highly accurate measurements on naturally occurring and artificial crystals. These data made crystallography an exact science. Between 1779 and 1848 Berzelius, a Swedish chemist, and his students studied the chemistry of minerals and developed the principles of our present chemical classification of minerals.

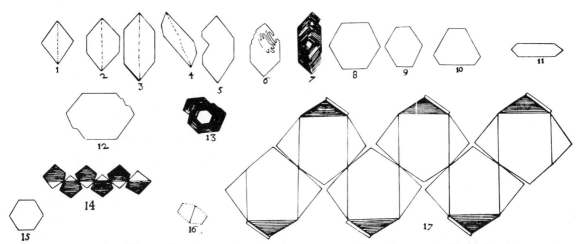

FIG. 1.3. Steno's drawings of various quartz and hematite crystals, illustrating the constancy of angles among crystals of different habits (from: J. J. Schafkranovski, *Die Kristallographischen Entdeckungen N. Stenens,* in *Steno as Geologist,* Odense Univ. Press, 1971).

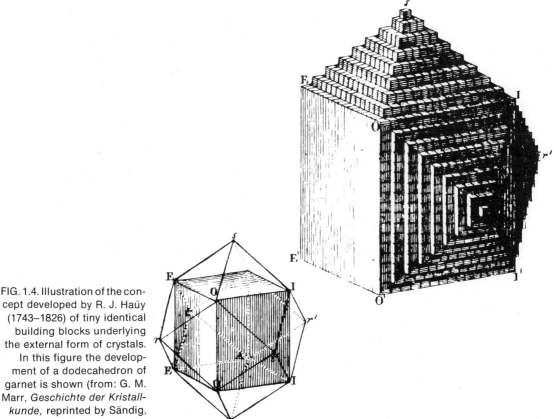

FIG. 1.4. Illustration of the concept developed by R. J. Haüy (1743–1826) of tiny identical building blocks underlying the external form of crystals. In this figure the development of a dodecahedron of garnet is shown (from: G. M. Marr, *Geschichte der Kristallkunde,* reprinted by Sändig, Walluf, West Germany, 1970).

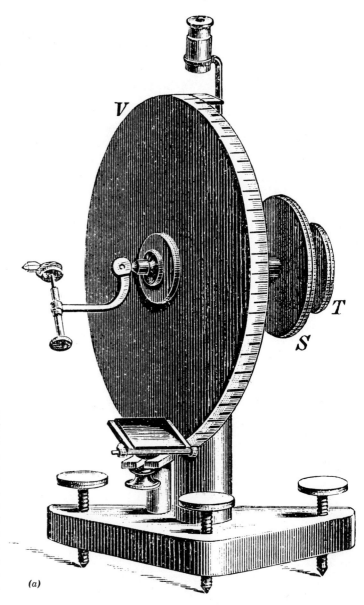

FIG. 1.5. (*a*) Earliest reflecting goniometer as invented by W. H. Wollaston in 1809 (from: G. Tschermak and F. Becke, *Lehrbuch der Mineralogie*, 1921, Hölder-Pichler-Tempsky, Vienna). (*b*) Two-circle reflecting goniometer as developed in the latter part of the nineteenth century (from: P. Groth, *Physikalische Krystallographie*, 1895, Leipzig). Compare with Fig. 2.40.

(*a*)

In 1815, the French naturalist Cordier, whose legacy to mineralogy is honored in the name of the mineral *cordierite*, turned his microscope on crushed mineral fragments immersed in water. He thereby initiated the "immersion method" which others, later in the century, developed into an important technique for the study of the optical properties of mineral fragments. The usefulness of the microscope in the study of minerals was greatly enhanced by the invention in 1828 by the Scotsman, William Nicol, of a polarizing device that permitted the systematic study of the behavior of light in crystalline substances. The polarizing microscope became, and still is, a powerful determinative tool in mineralogical studies. In the latter part of the nineteenth century Fedorov, Schoenflies, and Barlow,

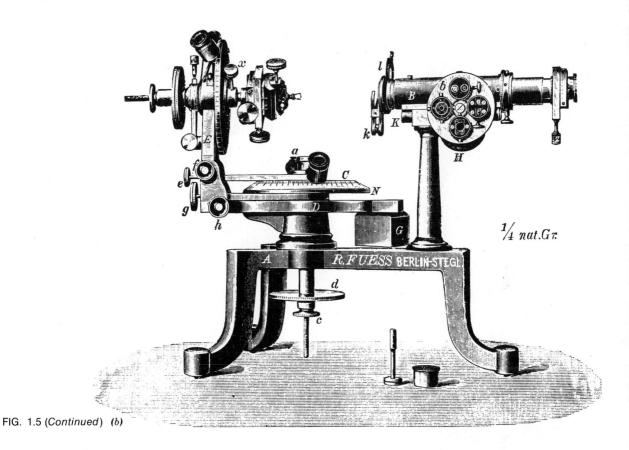

FIG. 1.5 *(Continued)* *(b)*

working independently, almost simultaneously developed theories for the internal symmetry and order within crystals which became the foundations for later work in X-ray crystallography.

The most far-reaching discovery of the twentieth century must be attributed to Max von Laue of the University of Munich. In 1912 in an experiment performed by Friedrich and Knipping at the suggestion of von Laue, it was demonstrated that crystals could diffract X-rays. Thus was proved for the first time the regular and ordered arrangement of atoms in crystalline material. Almost immediately X-ray diffraction became a powerful method for the study of minerals and all other crystalline substances, and in 1914 the earliest crystal structure determinations were published by W. H. Bragg and W. L. Bragg in England. Modern X-ray diffraction equipment with on-line, dedicated computers has made possible the relatively rapid determination of highly complex crystal structures. The advent of the electron microprobe in the early 1960s, for the study of the chemistry of minerals on a microscale has provided yet another powerful tool that is now routinely used for the study of the chemistry of minerals, synthetic compounds, and glasses.

The field of mineralogy now encompasses a wide area of study which includes X-ray crystallography, experimental mineralogy, petrology (the study of rocks), and aspects of metallurgy, crystal physics, and ceramics.

ECONOMIC IMPORTANCE OF MINERALS

Since before historic time, minerals have played a major role in humanity's way of life and standard of living. With each successive century they have be-

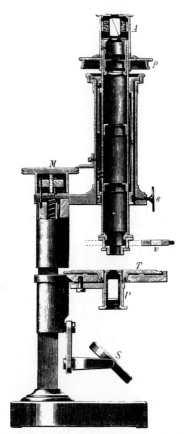

FIG. 1.6. Polarizing microscope as available in the mid-nineteenth century (from: G. Tschermak and F. Becke, *Lehrbuch der Mineralogie*, 1921, Hölder-Pichler-Tempsky, Vienna). Compare with Fig. 6.10.

come increasingly important, and today we depend on them in countless ways—from the construction of skyscrapers to the manufacture of televisions. Modern civilization depends on and necessitates the prodigious use of minerals. A few minerals such as talc, asbestos, and sulfur are used essentially as they come from the ground, but most are first processed to obtain a useable material. Some of the more familiar of these products are: bricks, glass, cement, plaster, and a score of metals ranging from iron to gold. Metallic ores and industrial minerals are mined on every continent wherever specific minerals are sufficiently concentrated to be economically extracted.

The location of mineable metal and industrial mineral deposits, and the study of the origin, size, and ore grade of these deposits is the domain of economic geologists. But a knowledge of the chemistry, occurrence, and physical properties of minerals is basic to pursuits in economic geology.

NAMING OF MINERALS

Minerals are most commonly classified on the basis of the presence of a major chemical component (an anion or anionic complex) into oxides, sulfides, silicates, carbonates, phosphates, and so forth. This is especially convenient because most minerals contain only one major anion. However, the naming of minerals is not based on such a logical chemical scheme.

The careful description and identification of minerals often requires highly specialized techniques such as chemical analysis and measurement of physical properties, among which are the specific gravity, optical properties, and X-ray parameters which relate to the atomic structure of minerals. However, the names of minerals are not arrived at in an analogous scientific manner. Minerals may be given names on the basis of some physical property or chemical aspect, or they may be named after a locality, a public figure, a mineralogist, or almost any other subject considered appropriate. Some examples of mineral names and their derivations are as follows:

Albite ($NaAlSi_3O_8$) from the Latin, *albus* (white), in allusion to its color.

Rhodonite ($MnSiO_3$) from the Greek, *rhodon* (a rose), in allusion to its characteristically pink color.

Chromite ($FeCr_2O_4$) because of the presence of a large amount of chromium in the mineral.

Magnetite (Fe_3O_4) because of its magnetic properties.

Franklinite ($ZnFe_2O_4$) after a locality, Franklin, New Jersey, where it occurs as the dominant zinc mineral.

Sillimanite (Al_2SiO_5) after Professor Benjamin Silliman of Yale University (1779–1864).

An international committee, the Commission on New Minerals and New Mineral Names of the

FIG. 1.7 (*a*) Portrait of Sir William Henry Bragg (1862–1942) and (*b*) of his son Sir William Lawrence Bragg (1890–1971). Father and son received the Nobel Prize for Physics in 1915. Both men are eminently known for their researches in the field of crystal structure by X-ray methods (*a* from Godfrey Argent, London, photograph by Walter Stoneman; *b* from Times Newspapers, Ltd., London).

International Mineralogical Association, now reviews all new mineral descriptions and judges the appropriateness of new mineral names as well as the scientific characterization of newly discovered mineral species.

REFERENCES AND LITERATURE OF MINERALOGY

The first comprehensive book on mineralogy in English, *A System of Mineralogy,* was written by James D. Dana in 1837. Since then, through subsequent revisions, it has remained a standard refer-

ence work. The last complete edition (the sixth) was published in 1892 with supplements in 1899, 1909, and 1915. Parts of a seventh edition known as *Dana's System of Mineralogy,* have appeared as three separate volumes in 1944, 1951, and 1962. The first two volumes treat the nonsilicate minerals and volume three deals with silica (quartz and its polymorphs). Additional volumes on silicates are in preparation. A more recent reference is the five-volume work, *Rock-Forming Minerals,* by W. A. Deer, R. A. Howie, and J. Zussman. The treatment of the physical properties of all minerals in *Dana's System* is exhaustive. The coverage in *Rock-Forming Minerals,* however, is more topical and expansive in the areas of chemistry, structure, and

FIG. 1.8. Computer automated electron microprobe, Autoprobe, manufactured by Etec Corporation (courtesy: Etec Corporation, Hayward, California).

experimental studies, but is essentially confined to the rock-making minerals. Another very useful, up-to-date reference on the chemistry and nomenclature of minerals is *Mineralogische Tabellen* (in German) by H. Strunz.

The diverse literature of mineralogy is found in papers published in scientific journals all over the world. The most widely circulated mineralogical journals in the English-speaking world are the *American Mineralogist,* published by the Mineralogical Society of America, and the *Mineralogical Magazine,* published by the Mineralogical Society of Great Britain. *The Mineralogical Record,* first published in 1970, devotes itself to mineralogical subjects that are often of more general appeal to the hobbyists or mineral collectors than those found in the two above mentioned journals.

References and Suggested Reading

Standard Mineralogic Reference Works
Dana, J. D., *A System of Mineralogy,* 7th ed., vol. 1, 1944; vol. 2, 1951; vol. 3, 1962. John Wiley & Sons, New York, Rewritten by C. Palache, H. Berman, and C. Frondel.

Deer, W. A., Howie, R. A., and Zussman, J., 1962, *Rock-Forming Minerals,* 5 vols. John Wiley & Sons, New York.

Strunz, H., 1970, *Mineralogische Tabellen,* 5th ed. Akademische Verlagsgesellschaft, Leipzig, 621 pp.

Historical Accounts
Ford, W. E., 1918, The growth of mineralogy from 1818 to 1918, *Amer. J. Sci.,* v. 46, pp. 240–254.

Knopf, A., 1941, Petrology. *Geol. Soc. Amer.,* 50th anniversary vol., Geology 1888–1938, pp. 333–364.

Kraus, E. H., 1941, Mineralogy. *Geol. Soc. Amer.,* 50th anniversary vol., Geology 1888–1938, pp. 307–332.

Pirsson, L. V., 1918, The rise of petrology as a science. *Amer. J. Sci.,* v. 46, pp. 222–240.

Economic Geology
American Institute of Mining, Metallurgical and Petroleum Engineers, 1960, *Industrial Minerals and Rocks,* 934 pp.

Bateman, A. M., 1950, *Economic Mineral Deposits,* 2nd ed. John Wiley & Sons, New York, 916 pp.

Park, C. F., Jr., 1975, *Earthbound: Minerals, Energy and*

Man's Future. Freeman and Cooper, San Francisco, 279 pp.

Park, C. F., Jr., 1968, *Affluence in Jeopardy; Minerals and the Political Economy*. Freeman and Cooper, San Francisco, 368 pp.

Park, C. F., Jr. and MacDiarmid, R. A., 1975, *Ore Deposits*, 3rd ed. W. H. Freeman, San Francisco, 530 pp.

Skinner, B. J., 1976, *Earth Resources*, 2nd ed. Prentice-Hall, Englewood Cliffs, N.J., 151 pp.

Voskuil, W. H., 1955, *Minerals in World Industry*. McGraw-Hill, New York, 324 pp.

2
CRYSTALLOGRAPHY

Minerals, with few exceptions, possess the internal, ordered arrangement that is characteristic of crystalline solids. When conditions are favorable, they may be bounded by smooth plane surfaces and assume regular geometric forms known as *crystals*. Today, most material scientists use the term crystal to describe any solid with an ordered internal structure, regardless of whether it possesses external faces. Because bounding faces are mostly an accident of growth and their absence in no way changes the fundamental properties of a crystal, this usage is reasonable. Thus we may frame a broader definition of a crystal as a *homogeneous solid possessing long-range, three-dimensional internal order*. The study of these crystalline solids and the laws that govern their growth, external shape, and internal structure is called *crystallography*. Although crystallography was originally developed as a branch of mineralogy, today it has become a separate science that deals not only with minerals but with all crystalline matter.

This chapter presents the essential principles

and facts of crystallography most useful in introductory mineralogy. The discussion is concerned with some aspects of the *internal structure* of crystals, as well as their *external geometry*, or *morphology*.

In this book, the general term *crystalline* is used to denote the ordered arrangement of atoms in the crystal structure. The term *crystal*, without a modifier, is generally used in the traditional sense of a regular geometric solid bounded by smooth plane surfaces. *Crystal* is also used in its broader sense with modifiers indicating perfection of development. Thus, a crystalline solid with well-formed faces is *euhedral*; if it has imperfectly developed faces, it is *subhedral*; and without faces, *anhedral*.

Crystalline substances may occur in such fine-grained aggregates that their crystalline nature (or crystallinity) can be determined only with the aid of a microscope. These are designated as *microcrystalline*. If the aggregates are so fine that the individual crystallites cannot be resolved with the microscope but can be detected by X-ray diffraction techniques they are referred to as *cryptocrystal-*

line. Although most substances, both natural and synthetic, are crystalline, some lack any ordered internal atomic arrangement and are called *amorphous*. Naturally occurring amorphous substances are designated as *mineraloids*.

CRYSTALLIZATION

Crystals are formed from solutions, melts, and vapors. The atoms in these disordered states have a random distribution but with changing temperature, pressure, and concentration they may join in an ordered arrangement characteristic of the crystalline state.

As an example of crystallization from solution consider sodium chloride dissolved in water. If the water is allowed to evaporate, the solution contains more and more Na^+ and Cl^- per unit volume. Ultimately, the point is reached when the remaining water can no longer retain all the salt in solution, and solid salt begins to precipitate. If the evaporation of the water is very slow, the ions of sodium and chlorine will group themselves together to form one or a few crystals with characteristic shapes and often with a common orientation. If evaporation is rapid, many centers of crystallization will be set up usually resulting in many small, randomly oriented crystals.

Crystals also form from solution by lowering the temperature or pressure. Hot water will dissolve slightly more salt, for instance, than cold water; and, if a hot solution is allowed to cool, a point will be reached where the solution becomes sufficiently concentrated that salt will crystallize. Again, the higher the pressure, the more salt water can hold in solution. Thus, with lowering the pressure of a saturated solution, supersaturation will result and crystals will form. Therefore, in general, crystals may form from a solution by evaporation of the solvent, by lowering the temperature, or by decreasing the pressure.

A crystal is formed from a melt in much the same way as from a solution. The most familiar example of crystallization from fusion is the formation of ice crystals when water freezes. Although it is not ordinarily considered in this way water is fused ice.

When the temperature is lowered sufficiently, the H_2O molecules, which were free to move in any direction in the liquid state, become fixed and arrange themselves in a definite order to build up a solid, crystalline mass. The formation of igneous rocks from molten magmas, though more complicated, is similar to the freezing of water. In the magma there are ions of many elements in an uncombined state. As the magma cools the various ions are attracted to one another to form crystal nuclei of the different minerals. Crystallization proceeds with the addition of more ions to the crystal nuclei to form the mineral grains of the resulting rock.

Although crystallization from a vapor is less common than from a solution or a melt, the underlying principles are much the same. As the vapor is cooled, the dissociated atoms or molecules are brought closer together, eventually locking themselves into a crystalline solid. The most familiar example of this mode of crystallization is the formation of snowflakes from air laden with water vapor.

INTERNAL ORDER IN CRYSTALS

The *three-dimensional internal order* of a crystal can be considered as a repetition of a *motif* (a unit of pattern) in such a way that the environment of and around each repeated motif is identical. A simple and ordered arrangement of a motif in two dimensions is shown in Fig. 2.1 with a comma as the motif. In real crystals the motifs may be molecules such as H_2O, anionic groups such as $(SiO_4)^{-4}$ or $(PO_4)^{-3}$, ions such as Ca^{2+}, Mg^{2+}, Fe^{2+}, atoms such as Cu, or combinations of anionic groups, ions and/or atoms.

FIG. 2.1. Two-dimensional order. The commas are the motif.

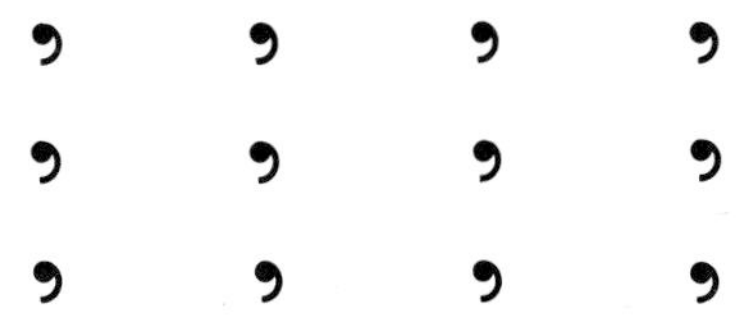

Ordered Patterns and Their Properties

The ordered patterns that characterize crystalline materials represent a lower energy state than random patterns. Intuitively, a brick wall constructed with carefully positioned bricks in an ordered arrangement provides a more stable and less energetic configuration than a brick wall made with a random arrangement of similar bricks (see Fig. 2.2a). The bricks in these walls can be considered as the motif and can be replaced by a comma as in Fig. 2.2b. This shows how an ordered pattern is generated by a motif repeated in a regular sequence of new locations. Any motion that brings the original motif into coincidence with the same motif elsewhere in the pattern is referred to as an *operation*. A homogeneous pattern, therefore, can be generated from a single motif by a set of geometrical operations (homogeneous meaning that each motif unit bears a similar relationship to the rest of the pattern). Many wallpaper patterns, for example, are based on a two-dimensional pattern in which the motif (sprays of flowers, dots, figurines) is arranged in a regular geometrical pattern. More abstract, and less symmetrical arrangements, as in many contemporary wallpapers may have a much less clearly defined pattern, if any at all.

The types of geometric operations by which homogeneous patterns can be generated are: translation, rotation, and by an operation that combines *translation* and *rotation*.

Translation in one dimension alone generates a linear pattern at intervals equal to the translation distance, t, as in Fig. 2.3a.

Rotation alone, through an angle (α), about an imaginary axis, generates a sequence of the motif along a circle. The orientation of the successive units differ by the angle α. In Fig. 2.3b the angle α of 90° generates a pattern with four motifs.

Translation and *rotation* together generate a regular pattern as long as the axis of rotation is parallel with the translation direction. This combination of two operations results in a *screw motion* (see Fig. 2.3c), which will be discussed further under the heading of "Space Groups."

The above types of operations produce patterns in which the original motif and those generated

FIG. 2.3. Generation of patterns by translation (a), rotation (b), and a combination of translation and rotation (c).

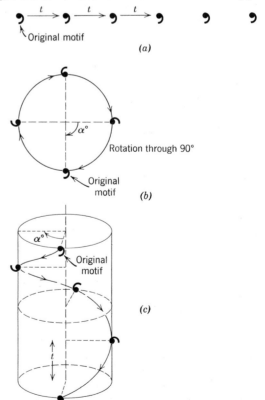

FIG. 2.2(a). Two-dimensional order and random arrangement of bricks in a brick wall. (b) The brick patterns are represented by commas as motifs.

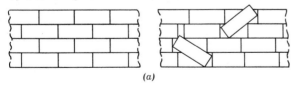

(a)

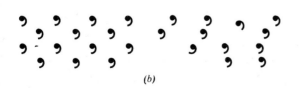

(b)

from it are identical in orientation with respect to each other. In other words, the original motif and the newly generated ones can be superimposed by the operations already listed. This generates *congruent patterns*. Additional operations, as provided by a *miror plane* (or *mirror*) or a *center of symmetry* (or *center of inversion, i*), produce patterns in which the original motif is related to the newly generated motif by a mirror image or an inversion. The relationship between the motifs in such patterns is described as *enantiomorphous*, meaning that the motifs are related by a mirror or inversion and that

they can not be superimposed on each other. Illustrations of such relations are given Fig. 2.4a and b.

A combination of a *mirror and a translation* produces a pattern with a *glide reflection* or *glide plane*, which will be discussed further in the section "Space Groups." Such a pattern is illustrated in Fig. 2.4c using human footprints. A set of footprints of a robbin, on the other hand, as in Fig. 2.4d shows only a mirror reflection (without a translation component). It should be noted that rotation about an axis, reflection by a mirror, and inversion about a point are referred to as *symmetry operations*. The

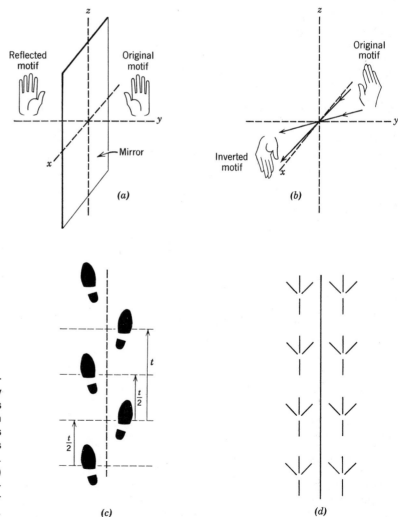

FIG. 2.4(*a*). Right- and left-handed motifs related by mirror reflection. (*b*) Motifs related by inversion through a center. (*c*) Human tracks showing relation of motifs (footprints) by a glide plane. Glide component = $t/2$. (*d*) Robbin tracks showing relationship of motifs by a mirror reflection.

geometrical locus that aids in the visualization of the symmetry of an ordered arrangement is known as a *symmetry element*. Rotation axes, mirror planes, and centers of symmetry are examples of such symmetry elements (see also p. 38).

A *rotation axis* is an imaginary line about which a motif unit may be rotated and repeat itself in appearance once or several times during a complete rotation (see Fig. 2.3b).

A *mirror plane (m)* is an imaginary plane that reflects a motif into a mirror image thereof (see Figs. 2.4a and d).

A *center of symmetry* (or *inversion center, i*) relates a motif unit in a pattern to an equivalent motif unit by inversion through a point (see Fig. 2.4b).

Rotation

The ordered arrangement in a pattern may be related to rotation through an angle (α) as in Fig. 2.3b. Not just any angular rotation will produce an ordered pattern, however. The types of rotation found in the internal order of crystals, and also expressed in their external shape (morphology), are *1-fold* $(\alpha = 360°)$, *2-fold* $(\alpha = 180°)$, *3-fold* $(\alpha = 120°)$, *4-fold* $(\alpha = 90°)$, and *6-fold* $(\alpha = 60°)$. A 5-fold axis is not possible. This becomes clear when one tries to completely cover a plane surface with a five-sided motif such as a pentagon, without mismatches and gaps, Fig. 2.5a. On the other hand, Fig. 2.5b shows how hexagons can completely cover a surface.

The reason for the lack of a 5-fold axis and other rotations than those listed above can also be derived by geometric considerations of an ordered two-dimensional arrangement. Fig. 2.6 illustrates the geometric restrictions on rotation axes in ordered arrangements that also contain translation. If the motif units, represented by large nodes in Fig. 2.6, are part of an ordered arrangement, then the distances AB and BC must be equal. If the motif at B contains a rotation axis with the axis $\perp$ to the plane of the figure, then the translations require similar axes at A and C. Furthermore, if points D, E, F, and G are related to B by a rotation, then $BC = BD = BE = BF = BG = t$. This also means that the distance ED, which lies on a line parallel to AC, must be equal to AB or a multiple thereof. In other words, $ED = u = mt$ where $m = $ integer. If the rotation by which A, F, G, C, D, and E are related is through an angle α, the following geometric relations hold: $\cos \alpha = x/t$ and also $x = \frac{1}{2}ED = \frac{1}{2}u$; $\therefore \cos \alpha = \frac{1}{2}u/t = u/2t$; $\therefore 2t \cos \alpha = u$. Combining $u = mt$ and $u = 2t \cos \alpha$ gives $mt = 2t \cos \alpha$, or $\cos \alpha = m/2$, where m is an integer. This leads to restrictions on the solutions possible for the angle of rotation α. For $m = 2$, $m/2 = 1$, $\alpha = 0°$ or $360°$; $m = 1$, $m/2 = 1/2$, $\alpha = 60°$; $m = 0$, $m/2 = 0$, $\alpha = 90°$; $m = -1$, $m/2 = -1/2$, $\alpha = 120°$; and for $m = -2$, $m/2 = -1$, $\alpha = 180°$. Any other integral values of m produce solutions of $\cos \alpha$ greater or less than ± 1 which are impossible. Other rotation angles produce non integer values of m. For example, a 5-fold rotation axis would require an angle of rotation of $72°$. This leads to a value for the $\cos 72° = 0.30902$. Such a number cannot equal $m/2$ in which m must be an integer. Therefore, a 5-fold rotation axis is not possible.

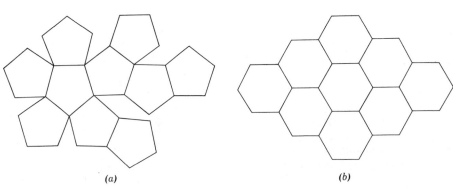

FIG. 2.5(a). Arrangement of pentagons, with 5-fold symmetry axes perpendicular to the page, leads to gaps in the pattern. (b) Arrangement of hexagons with 6-fold axes perpendicular to the page, as in honeycombs.

(a)

(b)

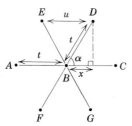

FIG. 2.6. Motifs separated by translation (t) and a possible axis of rotation, perpendicular to the page, at each of the motif units. One axis of rotation at motif B is shown.

The possible rotation axes are portrayed in Fig. 2.7 with the graphic symbols used to represent them. The number of duplications of the motif during a 360° rotation gives the rotation axis its name. For example, two equivalent units per 360° rotation are related by a *2-fold rotation axis*.

Rotation with Inversion

In addition to the symmetrical order generated by rotation axes, there are 1-, 2-, 3-, 4-, and 6-fold rota-

tions that can be combined with inversion and are known as *rotoinversion axes*. The rotation axes have been illustrated by motifs that lie in the same plane, as in Fig. 2.7. In combining rotation with inversion it is easier to observe the order of a pattern in three dimensions.

Figure 2.8*a* illustrates the combination of symmetry operations in a 1-fold rotoinversion. This is known as a 1-fold rotoinversion axis, and is symbolized as $\bar{1}$ (read: bar one). The original motif is rotated 360°, so that it returns to its original position and is then inverted through a center. This combination of operations produces the same result as does the presence of a *center of symmetry*. The $\bar{1}$ operation is therefore also referred to as a *center of symmetry*, or *i* (for inversion). The right-hand portion of Fig. 2.8*a* illustrates how the three-dimentional arrangement of the motif commas would appear projected on the equatorial plane of the globe in Fig. 2.8*a*. The rotoinversion operations for $\bar{2}$, $\bar{3}$, $\bar{4}$, and $\bar{6}$ (read: bar two, bar three, etc.) are also shown in Fig. 2.8. The $\bar{2}$ operation is equivalent to the operation

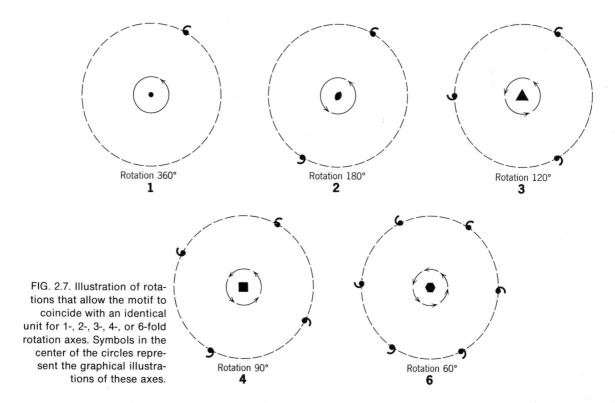

FIG. 2.7. Illustration of rotations that allow the motif to coincide with an identical unit for 1-, 2-, 3-, 4-, or 6-fold rotation axes. Symbols in the center of the circles represent the graphical illustrations of these axes.

Rotation 360° **1**

Rotation 180° **2**

Rotation 120° **3**

Rotation 90° **4**

Rotation 60° **6**

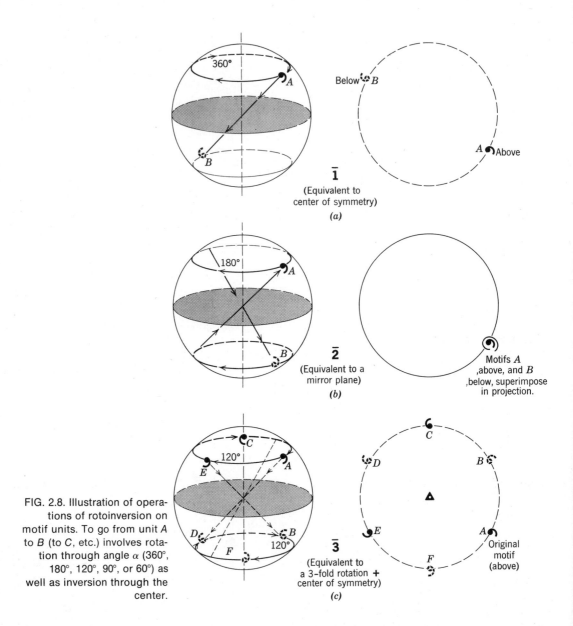

FIG. 2.8. Illustration of operations of rotoinversion on motif units. To go from unit A to B (to C, etc.) involves rotation through angle α (360°, 180°, 120°, 90°, or 60°) as well as inversion through the center.

$\overline{1}$
(Equivalent to center of symmetry)
(a)

$\overline{2}$
(Equivalent to a mirror plane)
(b)

$\overline{3}$
(Equivalent to a 3-fold rotation + center of symmetry)
(c)

of a mirror plane coincident with the equatorial plane of the globe (Fig. 2.8b). The $\overline{3}$ operation is the equivalent of a 3-fold rotation axis and inversion (i), which is the same as a 3-fold rotation and a center of symmetry. The $\overline{4}$ operation is not resolvable into other operations and as such is unique. The $\overline{6}$ operation is equivalent to a 3-fold axis of rotation with a mirror plane perpendicular to the rotation axis.

It should be noted that the original motif unit (denoted A in all illustrations of Fig. 2.8) has an *enantiomorphic** relationship with the second motif unit (denoted B), because of the inversion. The third motif unit, however (denoted C), is congruent (similar) with the original motif unit (A).

---

* Two enantiomorphic motifs are related by mirror reflection or inversion.

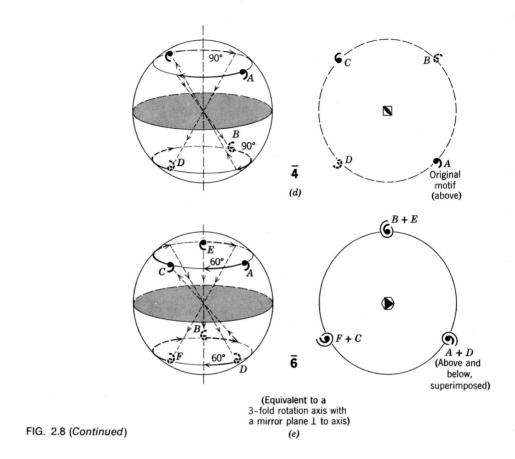

$\overline{4}$

(d)

Original
motif
(above)

$\overline{6}$

(Equivalent to a
3–fold rotation axis with
a mirror plane ⊥ to axis)

(e)

FIG. 2.8 (*Continued*)

Combinations of Rotations

Until now we have considered patterns generated by a single axis of rotation or rotoinversion. We can, however, combine various axes of rotation and generate regular three-dimensional patterns. For example, we might combine a 4-fold rotation axis (4) perpendicular to the plane of the page with a 2-fold rotation axis (2) in the plane of the page. Another example would be a combination of a 6-fold rotation axis (6) perpendicular to the plane of the page with a 2-fold axis (2) in the plane of the page. Both examples are illustrated in Fig. 2.9. The 4- and 6-fold axes in both symmetry combinations are at point *A*, the center of the circle. The 2-fold axis is to the right of *A*, along the east-west direction. The presence of the 4- and 6-fold axes will generate three and five more 2-fold axes, respectively. These generated 2-fold axes are dashed. Although we have generated

three and five 2-fold axis extensions, they constitute only two 2-fold axis directions at 90° to each other in Fig. 2.9a and three 2-fold axis directions at 120° to each other in Fig. 2.9b. Let us now enter a comma (marked by *B* in the drawings) above the page in a position slightly north of the original 2-fold axis. This comma is marked (+) indicating that it lies above the page, in the positive direction of the *z* axis. The original 2-fold axis (in the east-west direction) will generate another comma from the one given at *B*, namely on the south side of the 2-fold axis, and below the page. This generated comma is accompanied by a minus (−) sign, indicating its position below the page. The 4- and 6-fold axes will generate three, and five additional motif pairs, respectively, as shown by dashed commas. If now we carefully observe the arrangement of all of the commas it becomes clear that we have generated

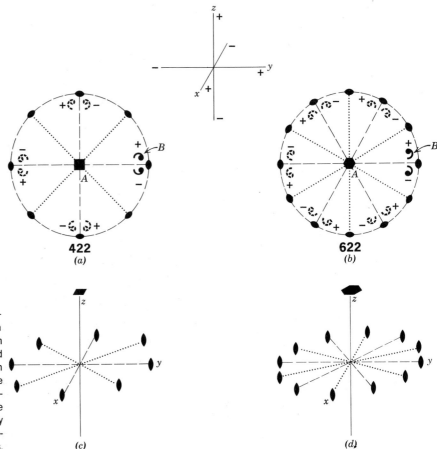

FIG. 2.9. Combination of rotation axes and the generation of ordered patterns from such combinations. (a) and (b) Plan views of the location of the symmetry axes and the motif units. (c) and (d) Three-dimensional sketches of the distribution of the symmetry axes. See text for development of these figures.

yet another set of 2-fold axes. These axes are dotted and at 45° to the original 2-fold axes in Fig. 2.9a and at 30° to the original 2-fold axes in Fig. 2.9b. The total symmetry in Fig. 2.9a, therefore, consists of a 4-fold rotation axis perpendicular to the page and two sets of 2-fold axes, the original set in the E–W, N–S directions, and the second at 45° thereto. The total symmetry in Fig. 2.9b consists of a 6-fold rotation axis perpendicular to the page and six 2-fold rotation axes in the plane of the page. The 2-fold axes are at 60° from each other. These types of combinations of axes can be represented by a sequence of digits for the type of rotation axes involved. For the examples in Fig. 2.9 this would result in 422 and 622, respectively. Three-dimensional representa-

tions of the locations of the rotation axes in the 422 and 622 combinations are given in Figs. 2.9c and d.

Other possible combinations of rotational symmetry elements are 222, 32, 23, and 432. Note that in two sequences (32 and 23) there are only two rotational symmetries indicated rather than three as in the other combinations. Figure 2.10 shows that combining a 3-fold axis with a 2-fold axis in a plane perpendicular to it generates no symmetry axes in addition to three 2-folds. The 432 symmetry axis combination is one of high symmetry with the location of the axes in specialized positions. Figure 2.11 shows the location of such axes with reference to a cube-like outline. The 4-fold axes are perpendicular

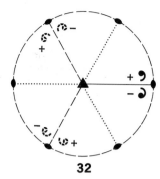

32

FIG. 2.10. Combination of a 3-fold rotation axis perpendicular to the page and a 2-fold rotation axis lying (E–W) in the page. The original motif units are on the right hand side. Generated motif units and symmetry axes are indicated by dashes and dots. The resultant symmetry sequence is 32, not 322. Compare with Fig. 2.9.

to the cube faces, the 3-fold axes are at the corners of the cube, and the 2-fold axes are located at the centers of the edges of the cube.

Combinations of Rotation Axes and Mirrors

In the previous section we discussed the possibility of combining rotational symmetries as in the sequences 622, 422, 32, etc. Let us consider some ex-

FIG. 2.11. The location of symmetry axes in 432 with respect to a cube-like outline.

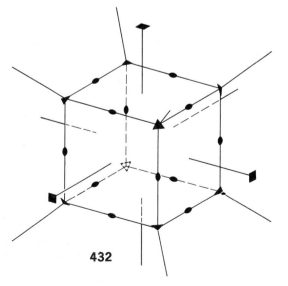

432

amples of combinations of rotation axes and mirror planes. In Fig. 2.12a a 4-fold rotation axis is combined with a mirror plane perpendicular to the axis and Fig. 2.12b shows a 6-fold axis with a mirror perpendicular to it. In Fig. 2.12a the arrangement of the motif is compatible with the 4-fold rotation axis and is shown by commas above the mirror plane. These are reflected by the mirror plane giving rise to another set of four commas below as shown by the dashed commas. Normally the symmetry elements and motif units are shown in a two-dimensional projection as on the right side of Fig. 2.12a. The motif units above the mirror as well as those below the mirror are projected onto the mirror itself. This causes the commas from above and below to coincide. In order to distinguish motif units that lie above the plane of projection (the mirror plane, in this case) from those below the plane, motif units above the plane are generally shown as solid dots and those below the plane as small open circles. When this convention is used in the projection of Fig. 2.12a it results in a 4-fold rotation axis surrounded by four motif units (dots) above, and four identical motif units (circles) below the mirror plane. Note that the mirror plane is conventionally shown by a solid circle. This type of combination of symmetry elements is represented by $4/m$ (read: four over m). The symmerty combination in Fig. 2.12b is represented by $6/m$. Other similar combinations are $2/m$ and $3/m$.

In a previous section we derived several combinations of rotation axes such as 622, 422, 222. If we add mirror planes perpendicular to each of the rotation axes the following symmetry combinations result: $6/m \, 2/m \, 2/m$, $4/m \, 2/m \, 2/m$, $2/m \, 2/m \, 2/m$. In Figs. 2.13a, b, and c illustrations are given for the combinations 422, $4/m \, 2/m \, 2/m$, and 4mm (compare with Fig. 2.9). The motif units are shown as solid dots and equivalent open circles indicating their position above and below the plane of projection, respectively. In 422 we have four motif units above the plane and four below, making a total of eight symmetrically related motifs. In $4/m \, 2/m \, 2/m$ we have a mirror plane perpendicular to the 4-fold axis (shown in Fig. 2.13b as a solid circle) and we have mirror planes perpendicular to each of the four

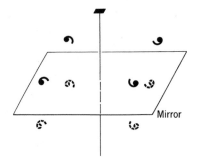

$$\frac{4}{m}$$

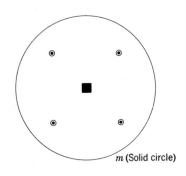

m (Solid circle)

(a)

FIG. 2.12(*a*). Combination of a 4-fold symmetry axis and a mirror plane perpendicular to it. The motif units that can be represented by commas are more conventionally shown by solid dots and small open circles in order to differentiate motif units above and below the mirror plane, respectively. (*b*) Combination of a 6-fold rotation axis and a mirror plane perpendicular to it.

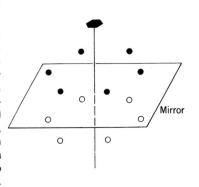

$$\frac{6}{m}$$

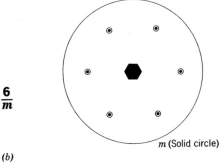

m (Solid circle)

(b)

FIG. 2.13(*a*). Combination of a 4-fold rotation axis and four 2-fold rotation axes (see also Fig. 2.9). Motif units above the page are shown as solid dots, those below the page as open circles. (*b*) Combination of a 4-fold rotation axis, four 2-fold rotation axes, and mirror planes perpendicular to each of the axes. (*c*) Combination of a 4-fold rotation axis and four mirror planes parallel to the 4-fold axis.

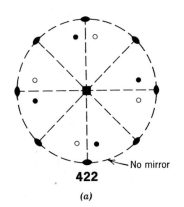

No mirror

422

(a)

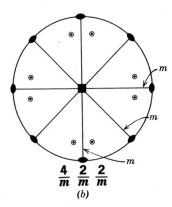

m

m

m

m

$$\frac{4}{m}\,\frac{2}{m}\,\frac{2}{m}$$

(b)

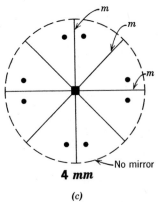

m

m

m

No mirror

4 *mm*

(c)

2-fold rotation axes that lie in the equatorial plane. The trace of the mirror plane perpendicular to the E–W axis coincides with the N–S axis, and so forth. The $4/m$ $2/m$ $2/m$ symmetry combination, therefore, contains four vertical mirror planes in addition to the one horizontal mirror perpendicular to the 4-fold axis. In Figs. 2.13b and c the traces of the mirrors are shown, by convention, as solid lines. In Fig. 2.13b we generate eight motif units above the equatorial mirror and eight below, making a total of 16 symmetrically related units. Yet another possible combination of 4-fold rotation and mirror planes is expressed by $4mm$. This symbolism denotes one 4-fold rotation axis and four mirror planes that intersect in the 4-fold axis and whose traces in the equatorial plane lie N–S, E–W, and at 45° to these directions. In other words, instead of the four 2-fold axes shown in the projection of 422 (see Fig. 2.13a) we now have traces of four mirror planes in the same locations (see Fig. 2.13c). The total number of motif units related by the combined symmetry elements in this case is eight, all of which lie on one side of the projection (above the page as shown in Fig. 2.13c). Other similar combinations are $6mm$, $3m$, and $2mm$.

Resume of Symmetry Operations Without Translation

In the preceding discussion we have introduced several symmetry elements: rotation axes (1, 2, 3, 4,

and 6), rotoinversion axes ($\bar{1}$, $\bar{2}$, $\bar{3}$, $\bar{4}$, and $\bar{6}$), a center of symmetry (i), and mirror planes (m). We have also discussed some of the combinations of rotation axes such as 622, 422, 222, and of rotation axes and mirror planes such as $6/m$ $2/m$ $2/m$, $4/m$ $2/m$ $2/m$, and $4mm$. The number of possible symmetry combinations is not unlimited; indeed the total number of nonidentical symmetry elements and combinations of symmetry elements is only 32. In Table 2.1 they are arranged in sequence from the lowest rotational symmetry (1) to the highest rotational symmetry (6). The 32 possible elements and combinations of elements are identical to the 32 possible *crystal classes* to which crystals can be assigned on the basis of their morphology (see pages 34–39) or internal atomic arrangement.

In the above discussion we have illustrated a number of combinations of symmetry elements but we have not rigorously derived the possible 32 nonidentical symmetry elements or combinations of symmetry elements, which are also known as the 32 *point groups*. The word *point* indicates that the symmetry operations leave one particular point of the pattern unmoved. The word *group* relates to the mathematical theory of groups which allows for a systematic derivation of all the possible and nonidentical symmetry combinations (see, for example, D. E. Sands, *Introduction to Crystallography*, 1975).

Until now we have concerned ourselves with

Table 2.1.
THIRTY-TWO POSSIBLE NON-IDENTICAL SYMMETRY ELEMENTS AND COMBINATIONS OF SYMMETRY ELEMENTS

	Increasing Rotational Symmetry →				
Rotation axis only	1	2	3	4	6
Rotoinversion axis only	$\bar{1}(=i)$	$\bar{2}(=m)$	$\bar{3}$	$\bar{4}$	$\bar{6}(=3/m)$
Combination of rotation axes		222	32	422	622
One rotation with perpendicular mirror		2/m	$3/m(=\bar{6})$	4/m	6/m
One rotation with parallel mirrors		2mm	3m	4mm	6mm
Rotoinversion with rotation and mirror			$\bar{3}\ 2/m$	$\bar{4}\ 2m$	$\bar{6}\ 2m$
Three rotation axes and perpendicular mirrors		2/m 2/m 2/m		4/m 2/m 2/m	6/m 2/m 2/m
Additional symmetry combinations present in isometric patterns		23	$2/m\ \bar{3}$	432 $\bar{4}\ 3\ m$ (see Fig. 2.11)	$4/m\ \bar{3}\ 2/m$

Table 2.2.

THIRTY-TWO CRYSTAL CLASSES (see also Table 2.3)

Crystal System	Symmetry of Crystal Classes
Triclinic	1 and $\bar{1}$
Monoclinic	2, m, and 2/m
Orthorhombic	222, 2mm, and 2/m 2/m 2/m
Tetragonal	4, $\bar{4}$, 4/m, 422, 4mm, $\bar{4}$ 2m, and 4/m 2/m 2/m
Hexagonal (rhombohedral division)	3, $\bar{3}$, 32, 3m, and $\bar{3}$ 2/m
Hexagonal (hexagonal division)	6, $\bar{6}$, 6/m, 622, 6mm, $\bar{6}$2m, and 6/m 2/m 2/m
Isometric	23, 2/m $\bar{3}$, 432, $\bar{4}$ 3/m, and 4/m $\bar{3}$ 2/m

some of the symmetry elements that give rise to the ordered, internal atomic arrangement of crystals. A crystal, however, under favorable circumstances of growth will develop smooth plane, external surfaces (faces) that may assume regular geometric forms which are the expression of its regular atomic internal arrangement. In crystals with well-developed faces one can recognize the elements of symmetry such as rotation axes, rotoinversion axes, a center of symmetry, and mirror planes. A systematic study of the external forms of crystals leads to 32 possible symmetries or symmetry combinations, which are the same 32 as the point groups noted above. Certain of the 32 *crystal classes* have symmetry characteristics in common with others, permitting them to be grouped together in one of six *crystal systems*. Table 2.2 shows the conventional arrangement of crystal systems and classes. Compare this with Table 2.1 and note that the symmetry elements and combinations are the same in both tables, but that their groupings are somewhat different.

Translation Lattices

Although we have mentioned the existence of screw axes and glide planes in relation to rotation axes and mirror planes we have not yet looked into the translational periodicity of an ordered two- or three-dimensional array of points or motif units. As stated earlier, a crystal is a homogeneous solid possessing long-range *three-dimensional internal order*. In addition to the translational order in screw axis and glide plane operations (see Space Groups), there is the more basic translational order present in a sequence of motifs that are arranged in space in a homogeneous pattern (homogeneous meaning that each motif unit bears a similar relationship to the rest of the pattern). Fig. 2.1 illustrates a two-dimensional array of motifs (commas). The order in such an array can be expressed in terms of two translations (t_1 and t_2) at 90° to each other (see Fig. 2.14a). A somewhat less symmetric pattern of identically arranged motifs is shown in Fig. 2.14b where the t_1 direction is the same as in Fig. 2.14a but the t_2

FIG. 2.14(*a*). A 2-dimensional pattern with translation components t_1 and t_2 at 90° to each other. (*b*) A two-dimensional pattern with translation components t_1 and t_2 at <90° to each other. (*c*) A three-dimensional pattern with translation components t_1, t_2, and t_3. None of the translation components make 90° with each other.

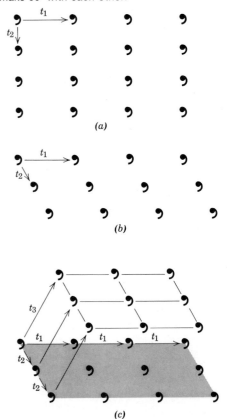

translation is at an angle $< 90°$ to the t_1 translation. These two illustrations can be thought of as infinite strings of units along the t_1 direction, which have been repeated by parallel and identical infinite strings along a translation direction t_2. The translations, as marked by t_1 and t_2, are vectors.

A three-dimensional ordered pattern can be obtained by adding yet another translation component (t_3) that does not lie in the plane of t_1 and t_2 (see Fig. 2.14c). This results in a pattern that is infinite in all directions in three-dimensional space. In crystals such a pattern is not exactly infinite, although it is generally considered so. The magnitude of the translations in inorganic crystals are on the order of 1 Ångstrom* (Å) or 10^{-8} cm. This means that a dimension of 1 cm in a crystal would contain approximately 100 million translations; indeed this can be considered infinite!

It is often convenient to ignore the actual shape of the motif units in a pattern and to concentrate only on the geometry of the repetitions in space. If the motif (commas in Fig. 2.14) is replaced by points we have a regular pattern of points which is referred to as a *lattice*. A *lattice* is, therefore, an imaginative pattern of points in which every point has an environment that is identical to that of any other point in the pattern. A lattice has no specific origin as it can be shifted parallel to itself.

A *row* represents equally spaced points along a line. A *plane lattice* or *net* represents a regular arrangement of points in two-dimensions and a *space lattice* depicts the distribution of equivalent points in three-dimensions. There are only five types of plane lattices, which are illustrated in Fig. 2.15. The various types differ from each other in equality or inequality of the periodicities and variations in the angle (γ) subtended between the translation directions. The most general two-dimensional lattice is a parallelogram net as shown in Fig. 2.15a. A rectangular lattice, as shown in Fig. 2.15b, contains a 90° angle between the translation directions. In a diamond lattice (see Fig. 2.15c) the translation periodicities are equal but the angle between them

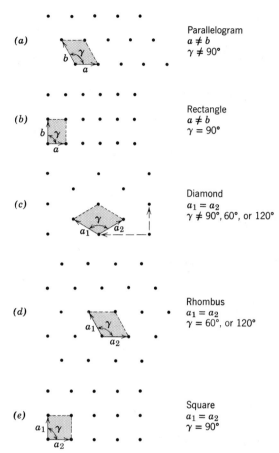

FIG. 2.15. Five types of plane lattices or nets. The translation distances are represented by a, b, a_1, and a_2. The angle between the translation vectors is γ. The symbol $\neq$ implies nonequality.

must not be 90°, 60°, or 120°. In this case a different, orthogonal set of lattice directions could be chosen but note that this would lead to lattice points in the centers of the net as well as at the corners. A special case of the diamond lattice occurs when the angle between t_1 and $t_2 = 120°$ and a rhombic lattice is produced (Fig. 2.15d). A square lattice contains two equal translation vectors at 90° to each other (Fig. 2.15e). The shaded areas in each of the lattices are outlined by the two translation vectors and represent the smallest geometric unit of the lattice. A regular repeat of these units will build up the point patterns (lattices) illustrated. In Fig. 2.15c a second choice of unit is indicated by the dashed arrows.

---

* Named after the Swedish physicist K. Ångstrom, 1814–1874.

This second choice will contain a lattice point at its center, as well as its corners.

It should be noted that the various plane lattices in Fig. 2.15 are compatible with different symmetry elements or combinations of symmetry elements. For example, the lattice in Fig. 2.15a is compatible with 2-fold rotation axes perpendicular to the page; the lattice in Fig. 2.15b is compatible with 2-fold rotation axes perpendicular to the page and mirror planes perpendicular to each other which intersect in the 2-fold axes, resulting in *2mm* symmetry; the lattice in Fig. 2.15d can contain 3- or 6-fold rotation axes and the lattice in Fig. 2.15e is compatible with 4-fold rotation axes perpendicular to the page and mirror planes at 90° as well as 45° to each other, which intersect in the 4-fold rotation axis.

Daily we see plane patterns such as in brick walls, tiled floors, wallpaper designs, and in prints in curtains, dresses, and neckties. Fig. 2.16 illustrates how a wallpaper designer has used the concepts of symmetry in repetition of a motif. Each motif in this design contains a 2-fold axis perpendicular to the

page and two mirror planes perpendicular to each other. The pattern of the motifs can be described by an oblique, *primitive unit,* outlined by the a_1 and b axes, but it can also be represented by a *multiple, centered unit* with the two axes a_2 and b at 90° to each other. The choice of unit in the wallpaper pattern may be arbitrary, but, in the three-dimensional patterns displayed by crystal structures, the choice is generally made according to the following rules:
1. The edges of the unit cell should, if possible, coincide with symmetry axes of the lattice,
2. The edges should be related to each other by the symmetry of the lattice, and
3. The smallest possible cell in accordance with (1) and (2) should be chosen.
If the above rules are applied to the wallpaper pattern in Fig. 2.16 the centered unit would be appropriate.

Three-dimensional lattices can be constructed by adding one additional translation direction (vector) to the plane lattices of Fig. 2.15; this third vector must not lie in the plane of the two-dimensional nets. The unit-cells of the resultant space-lattices (as based on the primitive nets in Fig. 2.15) are all parallelepipeds with lattice points only at their corners. A *unit cell* is a small parallelipiped built up on three translation directions selected as unit translations. Although a unit cell is imaginary it has an actual shape and a definite volume. *Unit cells* with lattice points only at the corners are referred to as *primitive* whereas those with additional lattice points are known as *nonprimitive.* All the plane nets in Fig. 2.15 are primitive except for the second orthogonal choice of cell in Fig. 2.15c which is nonprimitive. Fig. 2.17 illustrates the primitive and nonprimitive unit cells in some three-dimensional lattices. A space lattice, as in Fig. 2.17, can be expressed in terms of three vectorial directions, t_1, t_2, and t_3; however, conventionally these directions are described as a, b, and c. The lengths along the a, b, and c axes are known as the *unit cell dimensions* (a, b, c) and are expressed in Ångstrom units. If the unit cell is nonprimitive, the centering may occur on a pair of opposite faces of the unit and is called *A-*, *B-*, or *C*-centered depending on whether the centering takes place along the direc-

FIG. 2.16. Choice of alternate unit cells in a wallpaper design. The motif of the wallpaper contains symmetry 2mm.

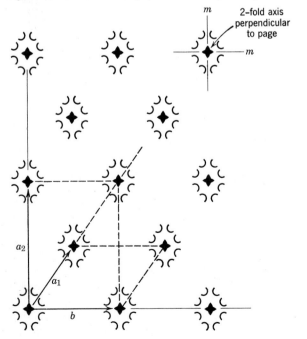

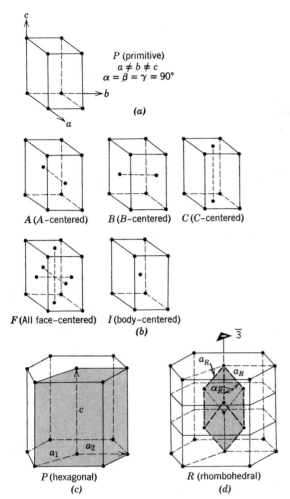

P (primitive)
$a \neq b \neq c$
$\alpha = \beta = \gamma = 90°$

(a)

A (A-centered) B (B-centered) C (C-centered)

F (All face-centered) I (body-centered)

(b)

P (hexagonal)

(c)

R (rhombohedral)

(d)

FIG. 2.17(a). A primitive unit cell with orthogonal axes, a, b, and c and with the lengths along these axes unequal. (b) Possible centerings in the unit cell outlined in a. (c) A primitive unit cell outlined by two equal axes, a_1 and a_2, and a third, c, perpendicular to them. This type of lattice occurs in crystals known as hexagonal. (d) A rhombohedral unit cell outlined by translation direction a_R (the edge of the rhombohedron) and α_R (the angle between two edges). The edges are symmetrical with respect to a $\bar{3}$ axis along the c-direction.

in Fig. 2.17 as well as the two choices of unit cells in space lattices which are derived from the rhombic plane pattern. The space lattice in Fig. 2.17c is based on two equal translation directions (a_1 and a_2) which make an angle of 120° with each other. The unit cell in Fig. 2.17d is known as a rhombohedral unit cell in which the translation directions are a_R and the angle between the edges of the unit-cell is α_R. As we will see in a subsequent discussion on the morphology of crystals the lattice in Fig. 2.17c is that of hexagonal crystals and the lattice in Fig. 2.17d is that of rhombohedral crystals which are classified in the rhombohedral division of the hexagonal system.

Clearly, any regular three-dimensional array of points can be outlined by a primitive lattice. However it is frequently highly desirable and appropriate to choose a nonprimitive unit cell. In Table 2.2 we outlined the 32 nonidentical symmetry elements or combinations of symmetry elements in terms of the crystal classes and crystal systems. The types of space lattices compatible with these 32 point groups are known as the 14 *Bravais lattices*, which are shown in Fig. 2.18. These lattice types are unique as was shown by A. Bravais (1811–1863) after whom they are named.

In Fig. 2.18 the lattice types are arranged in terms of crystal systems. The names of the crystal systems reflect the characteristic symmetries of the lattice types.

For example, in the triclinic system, which includes the symmetries 1 and $\bar{1}$, the unit cell compatible with these symmetries will have very few, if any, constraints. The unit cell shape compatible with triclinic symmetry, therefore, is one of low symmetry. The isometric system however, contains very high symmetry ($4/m\ \bar{3}\ 2/m$, 432, $\bar{4}3m$, $2/m\ \bar{3}$, and 23) and will require a unit cell that reflects the highest symmetrical constraints. It should be noted that a primitive lattice type is compatible with all the crystal systems. Centered lattices occur in five of the crystal systems. It should also be noted that only one face-centered type of lattice (namely C) is shown in Fig. 2.18. If the lattice had been chosen in such a way as to be A-centered or B-centered, rather than C-centered, this would not in-

tion of the a, b, or c axis. The centering may also occur on all faces of the unit cell and is then referred to as F (for face-centered), or it may be present in the center of the unit cell and is referred to as I (body-centered, from the German word "*innenzentrierte*"). These various types of unit cells are shown

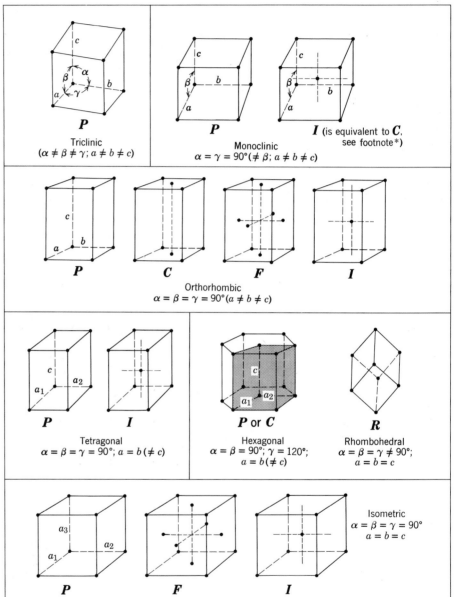

FIG. 2.18. The 14 unique types of space lattices, known as the Bravais lattices. Axial lengths are indicated by a, b, and c and axial angles by α, β, and γ. Each lattice type has its own constraints regarding lengths of edges a, b, and c and angles between edges, α, β, and γ. In the notations the nonequivalence of angles or edges that usually exist but are not mandatory are set off by parentheses.

*In the monoclinic system the unit cell can be described by a body centered or a C-face centered cell by a change in choice of the length of the a axis and the angle β. Vectorially these relations are: $a_I = c_C + a_C$; $b_I = b_C$; $c_I = -c_C$; and $a_I \sin \beta_I = a_C \sin \beta_C$. Subscripts I and C refer to the unit cell types.

troduce a new category of lattice types. The A-, B-, and C-lattices are symmetrically identical and can be converted to each other by an appropriate exchange of the crystallographic axes.

The shape and size of unit cells of minerals are determined by X-ray diffraction techniques (see Chapter 3). However, very recently, high resolution transmission electron microscopy has allowed the direct observation of projected images of crystal structures on photographic plates. Such a structure image (after Buseck and Iijima, 1974) is shown in Fig. 2.19 for the mineral cordierite. The dark parts of

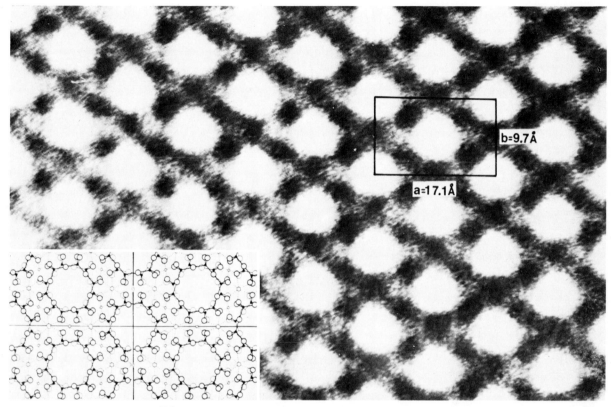

FIG. 2.19. High magnification structure image of an *a-b* section through the mineral cordierite. An orthorhombic unit cell is outlined and distances are given in Ångstroms. The insert shows the idealized structure of cordierite, as determined by X-ray diffraction techniques. The scales of the idealized structure and the electron transmission image are identical. (From Buseck and Iijima, *American Mineralogist*, vol. 59, pp. 1–22, 1974.)

the photograph outline the projected image of the structure and the superimposed lines outline an orthorhombic unit cell for cordierite.

Space Groups

We have discussed the translation-free symmetry elements (rotation, rotoinversion, mirror, and center of symmetry), and elements of pure translation as in lattices. Consideration of the translation-free symmetry elements led to the 32 *point groups* or *crystal classes*. Consideration of the types of unit cells and lattice types compatible with the 32 point groups led to the 14 *Bravais lattice types*. If we wish to con-

sider what combinations of symmetry elements (translational as well as translation-free) are possible with the space lattices we arrive at the concept of *space groups*. Space groups, therefore, represent the various ways in which motifs (such as atoms in crystals) can be arranged in space in a homogeneous array (homogeneous meaning that each motif unit bears a similar relationship to the rest of the pattern). Space groups reflect the lattice type, the point group symmetry, as well as the presence of additional translational symmetry elements as represented by *screw axes* and *glide planes*.

A combination of a rotation axis with a translation (*t*) parallel to the axis is known as a screw axis.

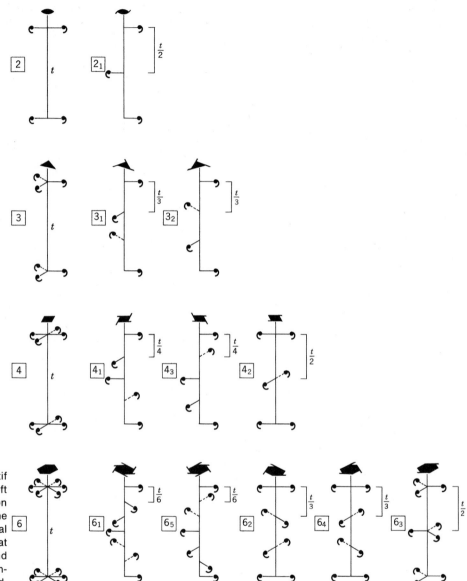

FIG. 2.20. Repetition of motif units by screw axes. The left column represents rotation axes and the columns to the right represent isogonal screw axes. The symbols at the top of the rotation and screw axes are internationally accepted.

The 2-, 3-, 4-, and 6-fold rotation axes can all be combined with such a translation. A 1-fold rotation axis combined with a translation is equivalent to a translation only. These screw axis operations affect motif units in three dimensions and their operations are graphically illustrated and compared with rotation axes in Fig. 2.20. The path along which the motif units are generated from the original unit is helical (as in a screw, see Fig. 2.3c) and the orientation of each unit differs from the neighboring motif unit by the angle of rotation (α). As with any screw motion, *screw axes are right- or left-handed*. A right-handed screw may be defined as one that advances away from the observer when rotated clockwise. The screw axis symbols consist of the symbols for rotation axes (2, 3, 4, and 6) followed by a subscript that represents the fraction of the translation (*t*) inherent in the operation of the screw axis. For example 2_1 means that $\frac{1}{2}t$ (obtained by placing the subscript over the main axis symbol, as in a fraction) is the translation involved. For a 3-fold rotation there are two possible screw axes, namely 3_1 and 3_2. The translation component in both screw axes is $\frac{1}{3}t$ but a convention allows for the distinction between the directions of the screw. When the ratio of the subscript to the number of the rotation axis (as $\frac{1}{3}$ for 3_1) $< \frac{1}{2}$ the screw is right-handed. When this ratio $> \frac{1}{2}$, it is left-handed (as in 3_2), and when the ratio $= \frac{1}{2}$, the screw is considered neutral in direction (see Fig. 2.20). In other words 3_1 and 3_2 are an *enantiomorphous pair of screw axes,* with 3_1 right-handed and 3_2 left-handed. Similarly the following pairs are enantiomorphous: 4_1 and 4_3; 6_1 and 6_5; and 6_2 and 6_4. Another way of representing the operation of two enantiomorphous pairs of screw axes is shown in

Fig. 2.21. Here the motif units are projected onto the plane of the page (from below the page) and the resulting representation is very similar to the illustrations in Figs. 2.7 and 2.8 for rotation and rotoinversion axes, respectively. The fractions next to the motif units in Fig. 2.10 represent the distance the units lie below the surface of the page. The fractions that represent t/n, are preceded by a minus sign ($-$) in order to indicate that the motif units lie below the page (in the negative direction of z in an x, y, z coordinate system).

Screw axes are said to be *isogonal* (from the Greek meaning "same angle") with the appropriate rotation axes. This means that 3_1 and 3_2 screw axes rotate the motif through the same angle as the 3-fold rotation axis (120° in both cases).

In addition to the repetition of a motif by a mirror reflection (see Fig. 2.4a) one can generate a regular pattern by a combination of a mirror reflec-

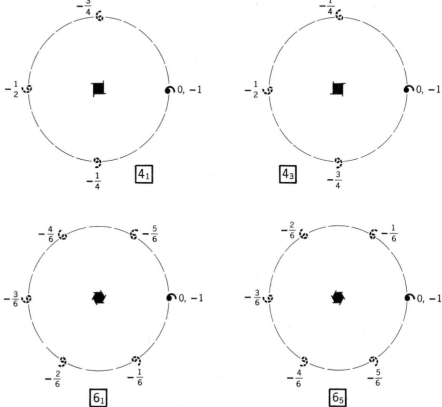

FIG. 2.21. Generation of pattern of motif units by two enantiomorphic pairs of screw axes (4_1 and 4_3; 6_1 and 6_5). The fractions next to the motif units represent the distance of the motif unit below the page. In this illustration the motif units have been projected upward from below the page.

tion and a translation. This operation is referred to as a *glide plane* or *glide reflection*. A glide reflection causes a motif unit to be reflected across a mirror and to be translated parallel to the mirror. Fig. 2.22 shows how motif units are related to a glide plane that has a translation component of $t/2$.

Specific glide directions can be identified in two- and three-dimensional patterns and expressed in terms of a set of axes such as x, y, and z. However, crystallographers refer the internal order as well as external morphology of crystals to three axes, a, b, and c. The c axis is vertical and the a and b axes lie in a plane that does not contain c. If the glide component $(t/2)$ in a three-dimensional ordered arrangement is parallel to the a axis, it is referred to as an *a glide* and is represented by the symbol a. Similarly if the glide component $(t/2)$ is parallel to the b or c axes, the glide is referred to as a *b or c glide*, respectively. If the glide component involves two axial directions but is parallel to the third, and if the translation component can be represented by $a/2 + b/2$, $a/2 + c/2$ or $b/2 + c/2$, it is referred to as a *diagonal glide*, and is represented by the symbol n. If the glide component can be represented by $a/4 + b/4$, $b/4 + c/4$, $a/4 + c/4$, or $a/4 + b/4 + c/4$ it is known as a *diamond glide* and is symbolized by the letter d.

FIG. 2.22(a). A three-dimensional view of the functions of a glide plane with the translation component of the glide = $t/2$. (b) A two-dimensional view of the glide plane shown in Fig. 2.22a.

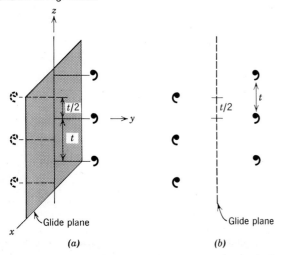

(a) (b)

There are 230 different ways in which symmetry elements (including screw axes and glide planes) can be combined with space lattices. As the derivation of all of the 230 space groups is a lengthy assignment, we will only illustrate some of the concepts related to space groups and will derive examples of a small number of the lower symmetry space groups. As was stated earlier, the presence of glide planes, or screw axes, or periods of translation (as in a lattice) cannot be detected morphologically because the translations are on the order of 1 to 10 Å and are unobservable by the naked eye. The translation-free symmetry combinations are point groups, whereas space groups define the symmetry and translations in space (on an atomic level). If we were to ignore the translation components in the 230 space groups we would end up with the 32 point groups.

Space groups have the following characteristics: (1) *They are isogonal with one of the 32 point groups.* Isogonal implies that rotation and screw axes with the same rotational repeat have the same rotational angle (e.g., 60° in a 6-fold rotation or 6-fold screw axis). This means that the screw axes 6_1, 6_2, 6_3, 6_4, and 6_5 are isogonal with the rotation axis 6. In other words, the point group is the translation-free residue of a family of possible isogonal space groups. (2) *They are based on one of the 14 Bravais lattices which is compatible with the specific point group.*

A specific point group designation consists of a series of symmetry elements, as in $2/m\ 2/m\ 2/m$. For each of the specific point group symmetry elements there is a possible space group element. In other words, instead of a mirror plane perpendicular to the first 2-fold rotation axis and a mirror plane perpendicular to the second 2-fold rotation axis, there may be glide planes in their places that would be symbolized as $2/b\ 2/a\ 2/m$. The space group symbol is furthermore preceded by a symbol that designates the general lattice type (P, A, B, C, I, F, or R). The full symbol for a space group that is isogonal with $2/m\ 2/m\ 2/m$ might be $I\ 2/b\ 2/a\ 2/m$. Another example of a space group isogonal with $2/m\ 2/m\ 2/m$ would be $P\ 2_1/b\ 2_1/c\ 2_1/a$ and so on.

Let us now illustrate the derivation of some space groups in the triclinic and monoclinic

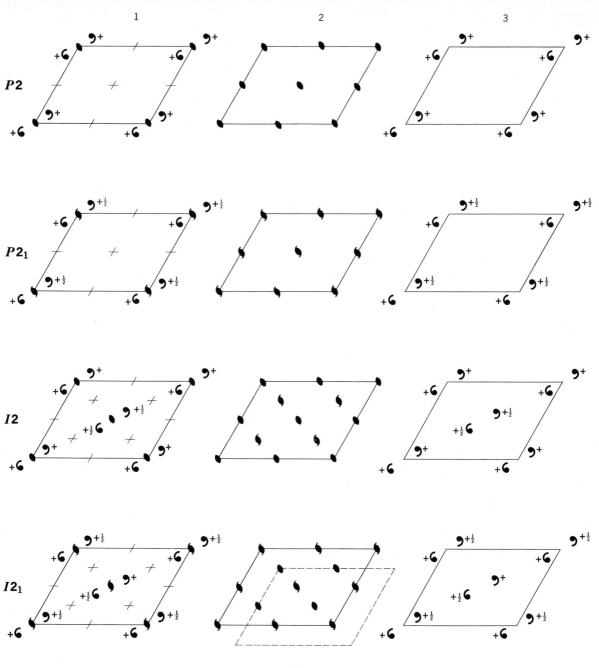

P2

P2₁

I2

I2₁

1   2   3

$\mathbf{9}^+$ ══ Motif unit lies some distance above the page

$\mathbf{9}^{+\frac{1}{2}}$ ══ Motif unit lies an additional $\frac{1}{2}$ of the unit cell edge above $\mathbf{9}^+$

FIG. 2.23. Derivation of the monoclinic space groups P2, P2₁, I2, and I2₁. Column 1 depicts only the lattice type, the motif units and the symmetry element specified for each lattice point (corners and centers). In the illustrations in column 1 it should become clear that additional symmetry elements are inherent in the unit cell at the points indicated by +. Column 2 depicts all the symmetry elements inherent in the pattern. Column 3 shows the arrangement of motif units only. For space group I2₁ it becomes clear that the symmetry contained in the unit cell is equivalent to the symmetry in the unit cell for I2. Only a shift in origin of the unit cell is needed to obtain a symmetry pattern identical to that for I2 (column 2). The dashed unit cell indicates the actual shift necessary. In other words I2 and I2₁ are identical and therefore only one of the two arrangements is unique.

systems. In the triclinic system only two possible space groups can occur, namely $P1$ and $P\bar{1}$. Here we have combined the two possible point groups in the triclinic system, 1 and $\bar{1}$, (see Table 2.1) and the only possible lattice type, P (see Fig. 2.18). In the monoclinic system we have three possible point groups (2, m, and $2/m$) and two possible lattice types (P and I) to consider. Let us restrict ourselves to a consideration of one of the point groups (2) and the two possible lattice types (P and I; I in the monoclinic system can be transferred into A, B, or C; see footnote to Fig. 2.18). The four possible space group notations are: $P2$, $P2_1$, $I2$, and $I2_1$. Fig. 2.23 illustrates the arrangement of motif units (commas) about the 2-fold rotation and 2-fold screw axes in relation to a monoclinic unit cell. Of the four possible space groups only three turn out to be unique because $I2$ and $I2_1$ are equivalent in their arrangement of symmetry elements except for a change in the location of the origin chosen for the lattice. The three unique space groups for the monoclinic point group 2 are therefore: $P2$, $P2_1$, and $I2$ (which is equivalent to $B2$). A more complete discussion of the derivation of some additional space groups can be found in M. J. Buerger, *Elementary Crystallography*, 1956. A complete listing of all possible space groups and their graphical representation is given in the *International Tables for X-ray Crystallography, vol. 1, 1969*.

It should be noted that the symbols we have used throughout our discussion for the symmetry elements are *international symbols* referred to as the *Hermann-Mauguin symbols* after their inventors. For space groups, however, another type notation may be encountered in the literature, which is referred to as the *Schoenflies notation*. The Schoenflies notation does not logically follow from the point group symbols and is not used here. References at the end of this chapter will allow you to find the Schoenflies system.

One aspect of the Hermann-Mauguin notation that may need some clarification is the use by crystallographers of what is known as *abbreviated symbols*. In our discussion we have always used the complete symbol such as $2/m\ 2/m\ 2/m$ and $4/m\ 2/m\ 2/m$. The abbreviated symbolism for these two point groups would be mmm and $4/mmm$, respectively. The reason for the abbreviation for $2/m\ 2/m\ 2/m$ is the understanding that the three mutually perpendicular mirror planes intersect each other along 2-fold axes. Similar reasoning explains the abbreviation $4/mmm$. In much of this text, for reasons of clarity, we will use only complete point group symbols. Abbreviated symbols are generally used in the literature for space group notation. For example, a mineral with point group $4/m\ \bar{3}\ 2/m$ may have a space group that is reported as $Fm3m$, because the presence of 4-fold and 2-fold rotation axes is implied. Such abbreviated symbols for the space group notation will be used in the four chapters on Systematic Mineralogy.

CRYSTAL MORPHOLOGY

As crystals are formed by the repetition in three dimensions of a unit of structure, the limiting surfaces, which are known as the faces of a crystal, depend in part on the shape of the unit. They also depend on the conditions in which the crystal grows. These conditions include all the external influences such as temperature, pressure, nature of solution, direction of movement of the solution, and availability of open space for free growth. The angular relationships, size, and shape of faces on a crystal are aspects of *crystal morphology*.

If a cubic unit cell is repeated in three dimensions to build up a crystal with n units along each edge, a larger cube will result containing n^3 units. With a similar orderly repeat mechanism, different shapes may result, as shown in Fig. 2.24 for malformed cubes, octahedron, and dodecahedron. The octahedral and dodecahedral forms are common in many crystals but as the unit cell dimensions are on the Ångstrom level the steps are invisible to the eye and the resulting faces appear as smooth plane surfaces.

With a given internal structure, a limited number of planes bound a crystal, and only a comparatively few are common. In determining the types of crystal faces that may develop on a crystal we must also consider the internal lattice. Faces are

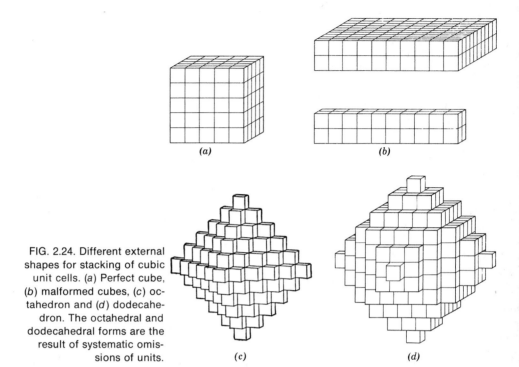

FIG. 2.24. Different external shapes for stacking of cubic unit cells. (a) Perfect cube, (b) malformed cubes, (c) octahedron and (d) dodecahedron. The octahedral and dodecahedral forms are the result of systematic omissions of units.

most likely to form on crystals parallel to lattice planes that have a high density of *lattice points* (or *nodes*). The frequency with which a given face is observed is roughly proportional to the number of nodes it intersects in the lattice; the larger the number of nodes, the more common the face, as is illustrated in Fig. 2.25. This rule, known as the *law of Bravais*, is generally confirmed by observations. Although there are exceptions to the law, as pointed out by Donnay and Harker in 1937, it is usually possible to choose the lattice in such a way that the rule holds true.

Because crystal faces have a direct relationship to the internal structure, it follows that the faces have a definite relationship to each other. This fact was observed in 1669 by Nicolaus Steno who pointed out that the angles between corresponding faces on crystals of quartz are always the same. This observation is generalized today as Steno's law of the constancy of interfacial angles which states: *the angles between equivalent faces of crystals of the same substance, measured at the same temperature, are constant* (see Fig. 1.3). For this reason crystal

morphology is frequently a valuable tool in mineral identification. A mineral may be found in crystals of widely varying shapes and sizes, but the angles between pairs of corresponding faces are always the same.

FIG. 2.25. This figure represents one layer of *lattice points* in a cubic lattice. Several lines are possible through the network that include a greater or lesser number of *lattice points* (or *nodes*). These lines represent the traces of possible crystal planes. It is found that the planes with the highest density of lattice points are the most common, such as *AB* and *AC*.

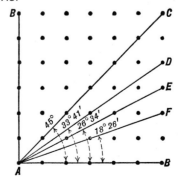

Although crystals possess a regular, orderly internal structure, different planes and directions within them have differences in atomic environments. Consider Fig. 2.26 which illustrates the packing of ions in sodium chloride (NaCl), the mineral halite. Any plane parallel to the front face of the cube is composed of half Na^+ and half Cl^- ions. On the other hand, cutting the corner of the cube, planes containing only Na^+ ions alternate with planes containing only Cl^- ions.

These different atomic arrangements along different crystal planes or directions give rise to *vectorial properties*. Because the magnitude of the property is dependent on direction, it differs with changing crystallographic direction. Some of the vectorial properties of crystals are hardness, conductivity for heat and electricity, thermal expansion, speed of light, growth rate, solution rate, diffraction of X-rays, and many others.

Of these properties, some vary continuously with direction within the crystal. Hardness, electrical and heat conductivity, thermal expansion, and the speed of light in crystals are all examples of such *continuous vectorial properties*.

The *hardness* of some crystals varies so greatly with crystallographic direction that the difference may be detected by simple scratch tests. Thus, kyanite, a mineral that characteristically forms elongate bladed crystals, may be scratched with an ordinary pocketknife in a direction parallel to the elongation, but cannot be scratched by the knife perpendicular to the elongation. Some directions in a diamond crystal are much harder than others. When diamond dust is used for cutting or grinding, a certain percentage of the grains always presents the hardest surface and, hence, the dust is capable of cutting along planes in the crystal of lesser hardness. If a perfect sphere cut from a crystal is placed in a cylinder with abrasive and tumbled for a long time, the softer protions of the crystal wear away most rapidly. The nonspherical solid resulting serves as a hardness model for the substance being tested.

The directional character of *electrical conductivity* is of great importance in the manufacture of silicon and germanium diodes, tiny bits of silicon and germanium crystals used to rectify alternating current. In order to obtain the optimum rectifying effect, the small bit of semimetal must be oriented crystallographically, as the conduction of electricity through such crystals varies greatly with orientation.

Ball bearings of synthetic ruby sound very attractive, as the great hardness of ruby would cut down wear and give long life to the bearing. However, when heated, ruby expands vectorially, and ruby ball bearings would rapidly become nonspherical with the rise of temperature from friction during operation. However, since the *thermal expansion* figure of ruby is an ellipsoid of revolution with a circular cross section, roller bearings are practical. Most minerals have unequal coefficients of thermal expansion in different directions, leading to poor resistance to thermal shock and easy cracking with heating or cooling. Quartz glass, which has an irregular internal structure as compared with quartz crystal, is more resistant to thermal shock than the mineral.

The *velocity of light* in all transparent crystals,

FIG. 2.26. Packing model of halite, NaCl, with cubo-octahedral outline. Na^+ small, Cl^- large. Note that the cube faces have sheets of equal numbers of Na^+ and Cl^- ions whereas the octahedral planes (at the corners of the cube) consist of alternating sheets of Na^+ and sheets of Cl^- ions. Na^+ and Cl^- ions are both surrounded by six closest neighboring ions (six coordination) in a face-centered lattice. This type of structure is also found in PbS, galena, MgO, and many other *AX* compounds.

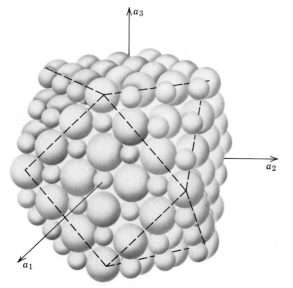

except those that are isotropic (see Chapter 6) varies continuously with crystallographic direction. Of all the vectorial properties of crystals, the optical parameters are most easily determined quantitatively and are expressed as the index of refraction (n), the reciprocal of the velocity of light in the crystal relative to the velocity of light in air or vacuum.

Discontinuous vectorial properties, on the other hand, pertain only to certain definite planes or directions within the crystal. There are no intermediate values of such properties connected with intermediate crystallographic directions. An example of such a property is *rate of growth*. The rate of growth of a plane in a crystal is intimately connected with the density of lattice points in the plane. We saw that a plane such as *AB* in Fig. 2.25 has a much greater density of nodes than plane *AD*, *AE*, or *AF*. Calculations of the energy involved indicate that the energy of particles in a plane such as *AB*, in which there is a high density of nodes, is less than the energy of particles in less densely populated planes such as *AF*. Hence, the plane *AB* will be the most stable, because in the process of crystallization the configuration of lowest energy is that of maximum stability. Planes *AF*, *AD*, *AE*, etc., will however grow faster than *AB* because fewer particles need to be added per unit area. In the growth of a crystal from a nucleus, the early forms that appear will be those of relatively high energy and rapid growth. Continued addition of material to these planes will build them out while the less rapidly growing planes lag behind. Thus, the edges and corners of a cube may be built out by addition of material to the planes cutting off the corners and edges, while little material is added to the cube faces. As growth progresses, the rapidly growing faces disappear, literally growing themselves out of existence, building the slower-growing, more stable forms in the process. After this stage is complete, growth is much slower, as addition is now entirely to the slower-growing, lowest-energy form. Thus, crystals themselves, if taken at various stages of their development, serve as models of the rate of growth for the compound being studied.

The *rate of solution* of a crystal in a chemical solvent is likewise a discontinuous vectorial process, and solution of a crystal or of any fragment of a single crystal may yield a more or less definite solution polyhedron. A clearer illustration of the vectorial nature of the rate of solution is afforded by etch pits. If a crystal is briefly treated with a chemical solvent that attacks it, the faces are etched or pitted. The shape of these pits is regular and depends on the structure of the crystal, the face being attacked, and the nature of the solvent. Valuable information about the internal geometry of arrangement of crystals may be obtained from a study of such etch pits.

Cleavage may be thought of as a discontinuous vectorial property and, like crystal form, reflects the internal structure, as cleavage always takes place along those planes across which there exist the weakest bonding forces. Those planes are generally the most widely spaced and the most densely populated.

Crystal Symmetry

The external shape of well-formed crystal reflects the presence or absence of the previously discussed translation-free symmetry elements. These are rotation axes, rotoinversion axes, center of symmetry, and mirror planes. The presence of these symmetry elements can be detected, in a well-formed crystal, by the angular arrangement of the bounding faces and sometimes by their size and shape. In poorly developed or malformed crystals the symmetry is generally not obvious, but can be derived from careful measurement of the angular relation of the bounding faces (see page 49). The recognition of symmetry elements in crystals, or wooden models that display the morphology of perfect crystals, is very similar to the recognition of symmetry in ordered patterns as discussed earlier (see pages 14–24). However, it may be useful to review some aspects of symmetry elements in relation to crystal morphology.

An axis of rotation is an imaginary line through a crystal about which the crystal may be rotated and repeat itself in appearance 1, 2, 3, 4, or 6 times during a complete rotation (see Fig. 2.7). Fig. 2.27a depicts a 6-fold rotation axis. When rotated about this axis the crystal repeats itself each 60° or six times in a 360° rotation.

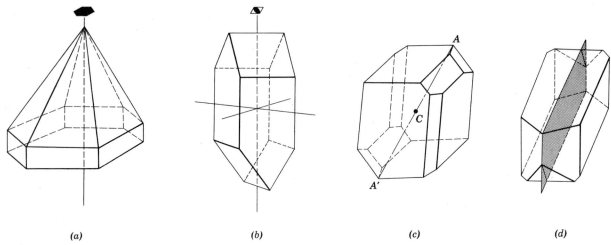

(a)   (b)   (c)   (d)

FIG. 2.27. Translation-free symmetry elements as expressed by the morphology of crystals. (a) 6-fold axis of rotation (6); (b) 4-fold axis of rotoinversion ($\bar{4}$); (c) Center of symmetry (i); (d) Mirror plane (m).

An axis of rotoinversion combines rotation about an axis with inversion (see Fig. 2.8). As discussed earlier, $\bar{1}$ is equivalent to a center of symmetry (or center of inversion, *i*), $\bar{2}$ is equivalent to a mirror plane, $\bar{3}$ is equivalent to a 3-fold rotation plus a center of symmetry, $\bar{4}$ is unique, and $\bar{6}$ is equivalent to a 3-fold rotation axis with a mirror plane perpendicular to the axis. Fig. 2.27b illustrates the morphological expression of a 4-fold rotoinversion axis.

A center of symmetry is present in a crystal if an imaginary line can be passed from any point on its surface through its center and a similiar point is found on the line at an equal distance beyond the center. This is equivalent to $\bar{1}$, or inversion. A center is represented by *i*. Fig. 2.27c illustrates a center of symmetry in a crystal.

A mirror plane is an imaginary plane that divides a crystal into halves, each of which, in a perfectly developed crystal, is the mirror image of the other. Fig. 2.27d illustrates the nature and position of a mirror, also called a *symmetry plane*.

In our earlier discussion of symmetry elements in patterns we noted the existence of 32 point groups or crystal classes (see Tables 2.1 and 2.2). These point groups, uniquely defined by their symmetry, have been given names derived from the general form in each crystal class (see page 43).

This scheme of nomenclature, as proposed by Groth in 1895, is used in Table 2.3. In the sections on mineral descriptions the crystal class is designated by the international (Hermann-Mauguin) notation and the name is omitted. As can be seen in Tables 2.2 and 2.3 certain groups of crystal classes have common symmetry characteristics. These groups of crystal classes are known as *crystal systems*. There are six such systems, with the hexagonal system having two subdivisions (hexagonal and rhombohedral) on the basis of the presence of a 6-fold or a 3-fold symmetry axis.

Crystallographic Axes

In the description of crystals it is convenient to refer, after the methods of analytic geometry, the external forms or internal symmetry to a set of three reference axes. These imaginary reference lines are known as the *crystallographic axes* and are generally taken parallel to the intersection edges of major crystal faces. Such axes are in most instances fixed by the symmetry and coincide with symmetry axes or with normals to symmetry planes. For some crystals there may be more than one choice of crystallographic axes when the selection is made by morphology alone. Ideally the axes should be paral-

Table 2.3.
THE THIRTY-TWO
CRYSTAL CLASSES
AND THEIR
SYMMETRY

Crystal System	Crystal Class	Name	Symmetry Content
Triclinic	1	Pedial	none
	$\bar{1}$	Pinacoidal	i
Monoclinic	2	Sphenoidal	$1A_2$
	m	Domatic	$1m$
	$2/m$	Prismatic	$i, 1A_2, 1m$
Orthorhombic	222	Rhombic-disphenoidal	$3A_2$
	$mm2$	Rhombic-pyramidal	$1A_2, 2m$
	$2/m\ 2/m\ 2/m$	Rhombic-dipyramidal	$i, 3A_2, 3m$
Tetragonal	4	Tetragonal-pyramidal	$1A_4$
	$\bar{4}$	Tetragonal-disphenoidal	$1\bar{A}_4$
	$4/m$	Tetragonal-dipyramidal	$i, 1A_4, 1m$
	422	Tetragonal-trapezohedral	$1A_4, 4A_2$
	$4mm$	Ditetragonal-pyramidal	$1A_4, 4m$
	$\bar{4}\ 2m$	Tetragonal-scalenohedral	$1\bar{A}_4, 2A_2, 2m$
	$4/m\ 2/m\ 2/m$	Ditetragonal-dipyramidal	$i, 1A_4, 4A_2, 5m$
Hexagonal Rhombohedral division	3	Trigonal-pyramidal	$1A_3$
	$\bar{3}$	Rhombohedral	$1\bar{A}_3(= i + 1A_3)$
	32	Trigonal-trapezohedral	$1A_3, 3A_2$
	$3m$	Ditrigonal-pyramidal	$1A_3, 3m$
	$\bar{3}\ 2/m$	Hexagonal-scalenohedral	$1\bar{A}_3, 3A_2, 3m$ $(1\bar{A}_3 = i + 1A_3)$
Hexagonal	6	Hexagonal-pyramidal	$1A_6$
	$\bar{6}$	Trigonal-dipyramidal	$1\bar{A}_6(= 1A_3 + m)$
Hexagonal division	$6/m$	Hexagonal-dipyramidal	$i, 1A_6, 1m$
	622	Hexagonal-trapezohedral	$1A_6, 6A_2$
	$6mm$	Dihexagonal-pyramidal	$1A_6, 6m$
	$\bar{6}m2$	Ditrigonal-dipyramidal	$1\bar{A}_6, 3A_2, 3m$ $(1\bar{A}_6 = 1A_3 + m)$
	$6/m\ 2/m\ 2/m$	Dihexagonal-dipyramidal	$i, 1A_6, 6A_2, 7m$
Isometric	23	Tetartoidal	$3A_2, 4A_3$
	$2/m\ \bar{3}$	Diploidal	$3A_2, 3m, 4\bar{A}_3$ $(1\bar{A}_3 = 1A_3 + i)$
	432	Gyroidal	$3A_4, 4A_3, 6A_2$
	$\bar{4}\ 3m$	Hextetrahedral	$3A_4, 4A_3, 6m$
	$4/m\ \bar{3}\ 2/m$	Hexoctahedral	$3A_4, 4\bar{A}_3, 6A_2, 9m$ $(1\bar{A}_3 = 1A_3 + i)$

lel to and their lengths proportional to the edges of the unit cell.

All crystals, with the exception of those belonging to the hexagonal system (see page 77), are referred to three crystallographic axes designated as $a, b,$ and c. In the general case (triclinic system) all the axes are of different lengths and at oblique angles to each other. In the orthorhombic system the three axes are also of different lengths but are mutually perpendicular. This set of axes, illustrated in Fig. 2.28a, is conventionally oriented as follows: the *a* axis is horizontal and in a front-back position; the *b* axis is horizontal and in a left-right position; the *c* axis is vertical. The ends of each axis are designated plus or minus; the front end of *a*, the right-hand end of *b*, and the upper end of *c* are positive; the opposite ends are negative. The general case (triclinic) is illustrated in Fig. 2.28b. The angles between the positive ends of the axes are conventionally designated by the Greek letters α, β, and γ. The α angle is enclosed between axial directions *b* and *c*, the β angle between *a* and *c*, and the γ angle between *a* and *b*. In summary (see also Fig. 2.18) the six crystal systems are referred to the following axial directions and axial angles:

Triclinic. Three unequal axes all intersecting at oblique angles.

Monoclinic. Three unequal axes, two of which are inclined to each other at an oblique angle and the third perpendicular to the plane of the other two.

Orthorhombic. Three mutually perpendicular axes all of different lengths.

Tetragonal. Three mutually perpendicular axes, two of which (the horizontal axes) are of equal length (a_1 and a_2), but the vertical axis is shorter or longer than the other two.

Hexagonal. Referred to four crystallographic axes; three equal horizontal axes (a_1, a_2, and a_3) intersect at angles of 120°, the fourth (vertical) is of different length and perpendicular to the plane of the other three.

Isometric. Three mutually perpendicular axes of equal lengths (a_1, a_2, and a_3).

Axial Ratios

In all the crystal systems, with the exception of the isometric, there are crystallographic axes differing in length. If it were possible to isolate a unit cell and measure the dimensions along the edges, which are parallel to the crystallographic axes, we would be able to write ratios between the edge lengths. X-ray crystallographers cannot isolate the cell, but they can measure accurately the cell dimensions in Ångstrom units (1 Å = 10^{-8} cm). Thus for the orthorhombic mineral sulfur the unit cell dimensions are given as: $a = 10.47$ Å, $b = 12.87$ Å, and $c = 24.49$ Å (see Fig. 2.29). Using the length of the *b* axis as a unit of measure we can determine the size of *a* and *b* relative to this. Thus $a/b:b/b:c/b = X:1:Y$. In the case of sulfur we can write the ratios $a:b:c = 0.813:1:1.903$. The ratios thus express the relative, not the absolute, lengths of the cell edges that correspond to the crystallographic axes.

We saw earlier (page 35) that the number of crystal planes appearing as faces is limited and that the angular relation between them is dependent on the unit cell, that is, the length of its edges and the angles between the edges. Long before X-ray investigations made it possible to determine the absolute dimensions of the unit cell, this relation of crystal morphology to internal structure was recognized, and axial ratios were calculated. By measuring the interfacial angles on a crystal and making appropriate calculations, it is possible to arrive at axial ratios that express the relative lengths of the crystallographic axes. It is interesting to see how closely the axial ratios calculated from unit cell dimensions compare with the older ratios derived from morphological measurements. For example, from mor-

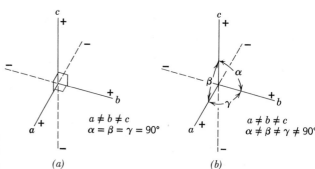

FIG. 2.28. Crystal axes in the orthorhombic (a) and triclinic (b) systems.

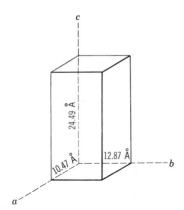

FIG. 2.29. The orthorhombic unit cell of sulfur.

phological measurements and calculations the ratios for sulfur were reported in 1869 as: $a:b:c = 0.8131:1:1.9034$; from unit cell measurements by X-ray techniques in 1960 they are: $a:b:c = 0.8135:1:1.9029$.

Parameters

Crystal faces are defined by indicating their intercepts on the crystallographic axes. Thus, in describing a crystal face it is necessary to determine whether it is parallel to two axes and intersects the third, is parallel to one axis and intersects the other

two, or intersects all three. In addition, one must determine at what *relative* distance the face intersects the different axes. In defining the position of a crystal face one must remember that crystal faces are parallel to a family of possible lattice planes. In Fig. 2.30a we have an a–b plane in an orthorhombic net based on the unit cell dimensions of the mineral sulfur.

The lattice plane AA is parallel to the b and c axes and intersects the a axis at one length (taken as unit length) along the a axis. The intercepts for this plane would be: $1a, \infty b, \infty c$. Similarly the plane A'A', which is parallel to AA but intersects the a axis at two unit lengths, would have intercepts: $2a, \infty b, \infty c$. The plane BB', which is parallel to the a and c axes and intersects the b axis at unit distance, has intercepts: $\infty a, 1b, \infty c$. The plane AB intersects both horizontal axes (a and b) at unit distance but is parallel to c, which leads to the intercepts: $1a, 1b, \infty c$. A plane that intersects all three axes at unit distances would have intercepts $1a, 1b,$ and $1c$. Figure 2.30b shows the development of crystal faces that are parallel to the lattice planes shown in Fig. 2.30a. It must be remembered that the parameters (intercepts) shown on the faces are strictly relative values and do not indicate any actual cutting lengths.

When parameters are assigned to the faces of a crystal, without any knowledge of its unit cell di-

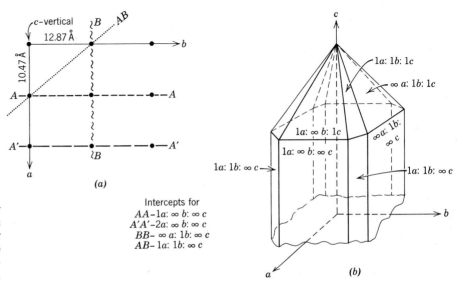

FIG. 2.30(a). Intercepts for planes in an orthorhombic lattice. (b) Intercepts for some crystal faces on the upper half of an orthorhombic crystal.

Intercepts for
AA–$1a: \infty b: \infty c$
A'A'–$2a: \infty b: \infty c$
BB–$\infty a: 1b: \infty c$
AB–$1a: 1b: \infty c$

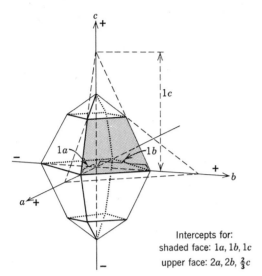

Intercepts for:
shaded face: $1a, 1b, 1c$
upper face: $2a, 2b, \frac{2}{3}c$

FIG. 2.31. Relative intersections of faces on an orthorhombic crystal, all of which cut three crystallographic axes.

mensions, one face that cuts all three axes is arbitrarily assigned the unit parameters $1a$, $1b$, and $1c$. This face, which is referred to as the *unit face*, is generally the largest in case there are several faces that cut all three axes. Figure 2.31 shows an orthorhombic crystal that consists of faces, all of which cut all three crystallographic axes. The largest face (shaded) that intersects all three crystallographic axes at their positive ends is taken as the unit face. Its intercepts are $1a$, $1b$, and $1c$ as shown. The intercepts of the smaller face above it can now be estimated by extending the edges of this face in the directions of the a and b axes. The intercepts for the upper face become $2a$, $2b$, $\frac{2}{3}c$, with respect to the unit face. These intercepts can be divided by the common factor 2 resulting in $1a$, $1b$, $\frac{1}{3}c$. This example illustrates that the parameters $1a$ and $1b$ do not represent actual cutting distances but express only relative values. The parameters of a face have no relation to its size, for a face may be moved parallel to itself for any distance without changing the relative values of its intersections with the crystallographic axes.

Miller Indices

Various methods of notation have been devised to express the intercepts of crystal faces upon the crystal axes. The most universally employed is the system of indices proposed by W. H. Miller which has many advantages over the system of parameters discussed above.

The *Miller indices* of a face consist of a series of whole numbers that have been derived from the parameters by their inversions and, if necessary, the subsequent clearing of fractions. The indices of a face are always given so that the three numbers (four in the hexagonal system) refer to the a, b, and c axes, respectively, and therefore the letters that indicate the different axes are omitted. Like the parameters, the Miller indices express a ratio, but for the sake of brevity the ratio sign is also omitted. For the two upper faces in Fig. 2.31 which cut the positive segments of the crystallographic axes, the parameters are $1a$, $1b$, $1c$, and $2a$, $2b$, $\frac{2}{3}c$, respectively. Inverting these parameters leads to: $\frac{1}{1} \frac{1}{1} \frac{1}{1}$ and $\frac{1}{2} \frac{1}{2} \frac{3}{2}$ respectively. For the unit-face this gives a Miller index of (111) and clearing the fractions for $\frac{1}{2} \frac{1}{2} \frac{3}{2}$ by multiplying all by 2, leads to a Miller index of (113) for the other face. The Miller index (111) is read as "one-one-one." Further examples of conversions of parameters to Miller indices for an isometric lattice and crystal forms are given in Fig. 2.32. Commas are used in Miller indices only when two-digit numbers appear as in (1, 14, 3). For faces that intersect negative ends of crystallographic axes, a line is placed over the appropriate number, as shown in Fig. 2.33. For example, $(1\,\bar{1}\,1)$ reads "one, minus one, one." It shoud be noted that indices given above for specific faces are placed in parentheses. This is to distinguish them from similar symbols used to designate crystal forms (page 44) and axial directions (page 48).

It is sometimes convenient when the exact intercepts are unknown to use a general symbol $(h\,k\,l)$ for the Miller indices; here h, k, and l are, respectively, the reciprocals of rational but undefined intercepts along the a, b, and c axes. The symbol $(h\,k\,l)$ would indicate that a face cuts all three crystallographic axes without reflecting relative units along the

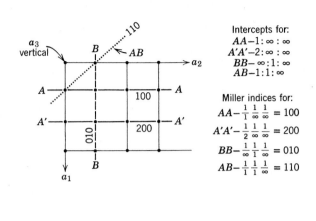

Intercepts for:
$AA - 1 : \infty : \infty$
$A'A' - 2 : \infty : \infty$
$BB - \infty : 1 : \infty$
$AB - 1 : 1 : \infty$

Miller indices for:
$AA - \frac{1}{1} \frac{1}{\infty} \frac{1}{\infty} = 100$

$A'A' - \frac{1}{2} \frac{1}{\infty} \frac{1}{\infty} = 200$

$BB - \frac{1}{\infty} \frac{1}{1} \frac{1}{\infty} = 010$

$AB - \frac{1}{1} \frac{1}{1} \frac{1}{\infty} = 110$

(a)

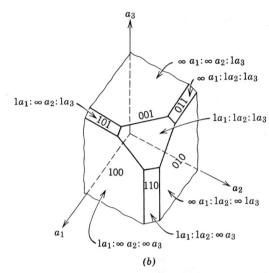

(b)

FIG. 2.32(a). Intercepts and Miller indices for planes in an isometric lattice. (b) Intercepts and Miller indices of some faces modifying the corner of a cube.

axes. If a face is parallel to one of the crystallographic axes and intersects the other two, the general symbols would be written as $(0\ k\ l)$, $(h\ 0\ l)$, and $(h\ k\ 0)$. A face parallel to two of the axes may be considered to intersect the third at unity, and the indices would, therefore, be: (100), (010), and (001), as well as the negative equivalents such as ($\bar{1}$00), (0$\bar{1}$0), and (00$\bar{1}$).

Early in the study of crystals it was discovered that for given faces the indices could always be ex-

pressed by simple whole numbers or zero. This is known as the *law of rational indices*.

Form

In its most familiar meaning the term *form* is used to indicate general outward appearance. In crystallography, external shape is denoted by the word *habit*, whereas *form* is used in a special and restricted sense. Thus a *form* consists of a group of crystal faces, all of which have the same relation to the elements of symmetry and display the same chemical and physical properties because all are underlain by like atoms in the same geometrical arrangement. The relationship between form and the elements of symmetry of a crystal is an important one to understand. For example, as in Fig. 2.34 where we have symmetry elements that belong to two crystal classes, $\bar{1}$ (in the triclinic system) and $4/m\ \bar{3}\ 2/m$ (in the isometric system) we may wish to develop the full form for the unit face (111). In the case of the $\bar{1}$ symmetry (which is equivalent to a center of symmetry) we develop an additional face by inverting through the origin of the three crystallographic axes. This additional face will have indices ($\bar{1}\ \bar{1}\ \bar{1}$). The

FIG. 2.33. Miller indices with respect to positive and negative ends of crystallographic axes.

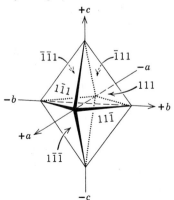

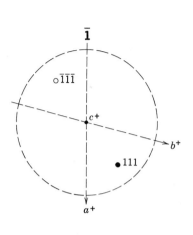

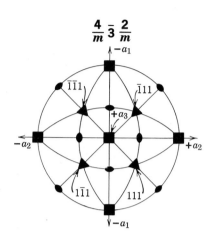

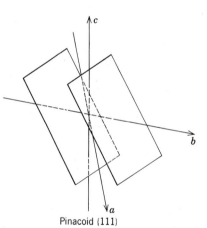

Pinacoid (111)

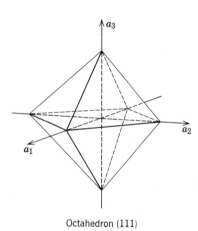

Octahedron (111)

FIG. 2.34. Development of a form from one face with Miller index (111) in the $\bar{1}$ and the $4m\,\bar{3}\,2/m$ crystal classes. In the case of $\bar{1}$ symmetry only a two-faced form, a pinacoid, develops. In the illustration of the isometric class the pole to the (111) face coincides with the position of the 3-fold rotoinversion axis. Faces at the top of the crystal are generally shown by dots, those at the bottom by open circles. Because of the complexity of the isometric figure only the positions of the four top faces of the octahedron are shown by Miller indices; there exists a similar set of four faces underneath.

form in the case of $\bar{1}$ consists, therefore, of two parallel faces only, and is known as a pinacoid. In the case of $4/m\,\bar{3}\,2/m$ the symmetry elements will generate seven additional faces for (111) with indices $(\bar{1}11)$, $(1\bar{1}1)$, $(11\bar{1})$, $(\bar{1}\bar{1}1)$, $(1\bar{1}\bar{1})$, $(\bar{1}1\bar{1})$, and $(\bar{1}\bar{1}\bar{1})$. This form is known as an octahedron. It should be clear, therefore, that the number of faces that belongs to a form is determined by the symmetry of the crystal class.

Although faces of a form may be of different sizes and shapes because of malformation of the crystal, the similarity is frequently evidenced by natural striations, etchings, or growths shown in Fig. 2.35. On some crystals the differences between faces of different forms can be seen only after etch-

ing with acid. In Figs. 2.35a and b there are three forms, each of which has a different physical appearance from the others, and in Fig. 2.35c there are two forms.

In our discussion of Miller indices we said that a crystal face may be designated by a symbol enclosed in parentheses as (hkl), (010), or (111). Miller indices also may be used as form symbols and are then enclosed in braces as {hkl}, {010}. Thus in Fig. 2.34 (111) refers to a specific face; whereas {111} embraces all faces of the form. In choosing a form symbol it is desirable to select, if possible, the face symbol with positive digits; {111} rather than {$1\bar{1}\bar{1}$}, {010} rather than {$0\bar{1}0$}.

In each crystal class there is a form, the faces of

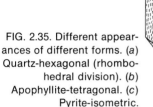

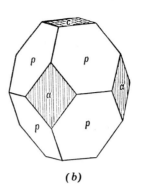

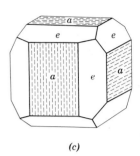

FIG. 2.35. Different appearances of different forms. (a) Quartz-hexagonal (rhombohedral division). (b) Apophyllite-tetragonal. (c) Pyrite-isometric.

(a) *(b)* *(c)*

which intersect all of the crystallographic axes at different lengths; this is the *general form, {hkl}*. All other forms that may be present are *special forms*. In the orthorhombic, monoclinic, and triclinic crystal systems {111} is a general form, for the unit length along each of the axes is different. In the crystal systems of higher symmetry in which the unit distances along two or more of the crystallographic axes are the same, a general form must intersect the like axes at different multiples of the unit length. Thus {121} is a general form in the tetragonal system but a special form in the isometric system and {123} is a general form in the isometric system. The concept of a *general form* can also be related to the symmetry elements of a specific crystal class. An *(hkl)* face will not be parallel or perpendicular to a single symmetry element regardless of the crystal class. A *special form,* however, consists of faces that are parallel or perpendicular to any of the symmetry elements in the crystal class. For most crystal classes the general form contains a larger number of faces than any of the special forms of that same class.

In Fig. 2.34 we developed a two-faced form and an eight-faced form. The two-faced form {111} for crystal class $\overline{1}$ is referred to as an *open form*, as it consists of two parallel faces that do not enclose space (see also Fig. 2.36b). The eight-faced form {111} in crystal class $4/m\,\overline{3}\,2/m$ is a *closed form* as the eight faces together enclose space (see Figs. 2.33 and 2.34).

A crystal will often display several forms in combination with one another but may have only one, provided that it is a closed form. Because any combination of forms must enclose space, a minimum of two open forms is necessary. The two may exist together or be in combination with other closed or open forms. Altogether 48 different types of crystal forms can be distinguished by the angular relations of their faces. Thirty-two of these are represented by the general forms of the 32 crystal classes; 10 are special closed forms of the isometric system; and 6 are special open forms (prisms) of the hexagonal and tetragonal systems. The various forms will be discussed under the symmetry class or classes in which they are found. Each form of the isometric system has a special name, but the same general names are used for the forms present in other systems. They are as follows:

Pedion. A single face comprising a form (Fig. 2.36a).

Pinacoid. An open form made up of two parallel faces (Fig. 2.36b).

Dome. An open form consisting of two nonparallel faces symmetrical with respect to a mirror plane (m). See Fig. 2.36c.

Sphenoid. Two nonparallel faces symmetrical with respect to a 2-fold rotation axis (see Fig. 2.36d).

Disphenoid. A four-faced, closed form consisting of two upper sphenoid faces that alternate with two lower sphenoid faces (see Fig. 2.36e).

Prism. An open form composed of 3, 4, 6, 8, or 12 faces, all of which are parallel to the same axis. Except for certain prisms in the monoclinic system, the axis is one of the principal crystallographic axes (see Figs. 2.36f to k).

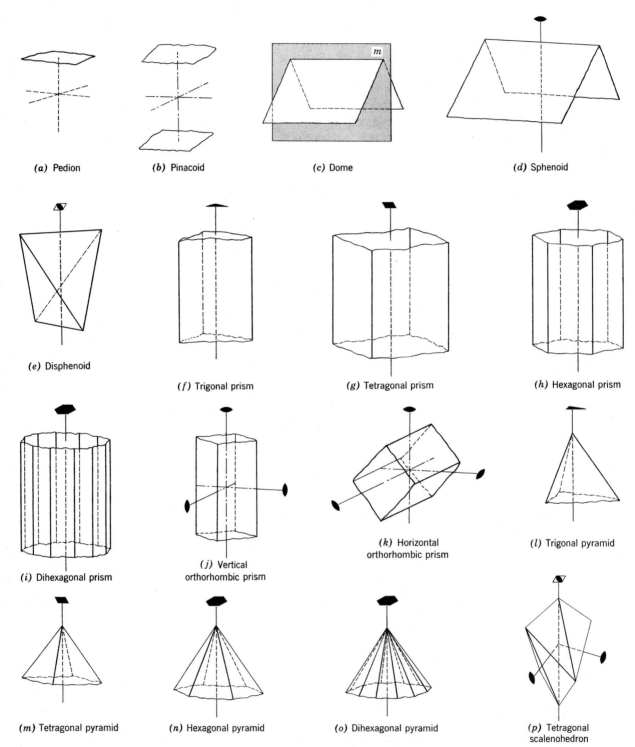

(a) Pedion   (b) Pinacoid   (c) Dome   (d) Sphenoid

(e) Disphenoid   (f) Trigonal prism   (g) Tetragonal prism   (h) Hexagonal prism

(i) Dihexagonal prism   (j) Vertical orthorhombic prism   (k) Horizontal orthorhombic prism   (l) Trigonal pyramid

(m) Tetragonal pyramid   (n) Hexagonal pyramid   (o) Dihexagonal pyramid   (p) Tetragonal scalenohedron

FIG. 2.36. General forms, and some of their symmetry elements.

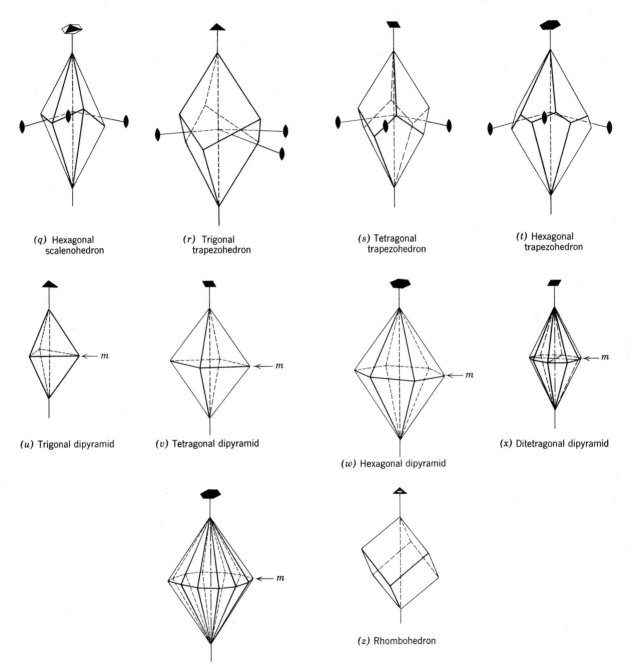

(q) Hexagonal scalenohedron

(r) Trigonal trapezohedron

(s) Tetragonal trapezohedron

(t) Hexagonal trapezohedron

(u) Trigonal dipyramid

(v) Tetragonal dipyramid

(w) Hexagonal dipyramid

(x) Ditetragonal dipyramid

FIG. 2.36 (Continued) (y) Dihexagonal dipyramid

(z) Rhombohedron

Pyramid. An open form composed of 3, 4, 6, 8, or 12 nonparallel faces that meet at a point (see Figs. 2.36*l* to *o*).

Scalenohedron. Eight-faced (tetragonal, Fig. 2.36*p*) or 12-faced (hexagonal, Fig. 2.36*q*), closed forms with faces grouped in symmetrical pairs. In the tetragonal scalenohedron pairs of faces above are related by an axis of 4-fold rotoinversion to pairs of faces below. The 12 faces of the hexagonal scalenohedron display three pairs of faces above and three pairs of faces below in alternating positions. The pairs are related by an axis of 3-fold rotoinversion. In perfectly developed crystals, each face is a scalene triangle.

Trapezohedron. Closed forms that are 6-, 8-, or 12-faced with 3, 4, or 6 faces above offset from 3, 4, or 6 faces below (Fig. 2.36*r* to *t*). These forms are the result of a 3-, 4-, or 6-fold axis combined with perpendicular 2-fold axes. In addition there is an isometric trapezohedron, a 24-faced form. In well-developed trapezohedrons each face is a trapezium.

Dipyramid. Closed forms having 6-, 8-, 12-, 16-, or 24-faces (Figs. 2.36*u* to *y*). The dipyramids can be considered as formed from pyramids by reflection across a horizontal mirror plane.

Rhombohedron. A closed form composed of six faces of which three faces at the top alternate with three faces at the bottom, the two sets of faces being offset by 60° (see Fig. 2.36*z*). Rhombohedrons are found only on crystals of the rhombohedral division of the hexagonal system.

In crystal drawings it is convenient to indicate the faces of a form by the same letter. The choice of which letter shall be assigned to a given form rests largely with the person who first describes the crystal. However, there are certain simple forms that, by convention, usually receive the same letter. Thus the three pinacoids that cut the *a*, *b*, and *c* axes are lettered respectively *a*, *b*, and *c*, the letter *m* is usually given to {110} and *p* to {111} (Fig. 2.37).

Zones

One of the conspicuous features on many crystals is the arrangement of a group of faces in such a manner that their intersection edges are mutually

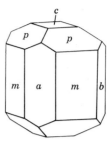

FIG. 2.37. Conventional lettering on crystal drawings.

parallel. Considered collectively, these faces form a *zone*. A line through the center of the crystal that parallels the lines of face intersections is called the *zone axis*. In Fig. 2.38 the faces *m'*, *a*, *m*, and *b* are in one zone, and *b*, *r*, *c*, and *r'* in another. The lines given as [001] and [100] are the zone axes.

A zone is indicated by a symbol similar to that for Miller indices of faces, the generalized expression of which is [*uvw*]. Any two nonparallel faces determine a zone. Assume one wishes to determine the zone axis of two such faces, (*hkl*) and (*pqr*). The method usually used is to write twice the symbol of one face and directly beneath, twice the symbol of the other face. The first and last digits of each line are disregarded and the remaining numbers, joined by sloping arrows, multiplied. In each set the product of 2 is subtracted from the product of 1 as:

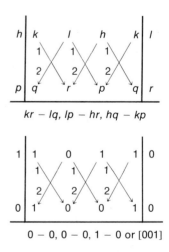

$$kr - lq, \; lp - hr, \; hq - kp$$

$$0 - 0, \; 0 - 0, \; 1 - 0 \text{ or } [001]$$

As a specific example assume *m*, Fig. 2.37 is (*hkl*) with index (110) and *b* is (010), the zone axis is thus

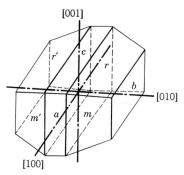

FIG. 2.38. Crystal zones and zone axes.

[001] as shown above. It should be noted that the zone symbols are enclosed in brackets, as [uvw], to distinguish them from face and form symbols.

Crystal Habit

The term *habit* is used to denote the general shapes of crystals as cubic, octahedral, prismatic. Because habit is controlled by the environment in which crystals grow, it may vary with locality. At one place it may be equant and at others tabular or fibrous. Only rarely do crystals present an ideal geometrical shape. But even in malformed crystals evidence of the symmetry is present in the physical appearance of the faces and in the symmetrical arrangement of

the interfacial angles. Given in Fig. 2.39 are various crystal forms, first ideally developed and then malformed.

MEASUREMENT OF CRYSTAL ANGLES

As stated by Steno's Law the angles between equivalent faces of the same substance are constant, regardless of whether the crystal is misshapen or ideal. The angles that are measured between the normals to the crystal faces characterize a crystal and should be measured carefully. A graphical representation of the distribution of the angles, and the normals to the crystal faces, may define the symmetry of the crystal and therefore its crystal class. Angles are measured with *goniometers,* and for accurate work a *reflecting goniometer* is used. A crystal mounted on this instrument can be rotated about a zone axis and will reflect a collimated beam of light from its faces through a telescope to the eye. The angle through which a crystal must be turned in order to throw successive beams of light from two adjacent faces into the telescope determines the angle between the faces. Fig. 2.40 illustrates a modern, two-circle reflecting goniometer.

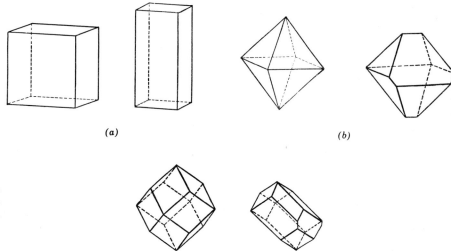

FIG. 2.39. Examples of some crystal habits in the isometric system. (a) Cube and malformed cube. (b) Octahedron and malformed octahedron. (c) Dodecahedron and malformed dodecahedron.

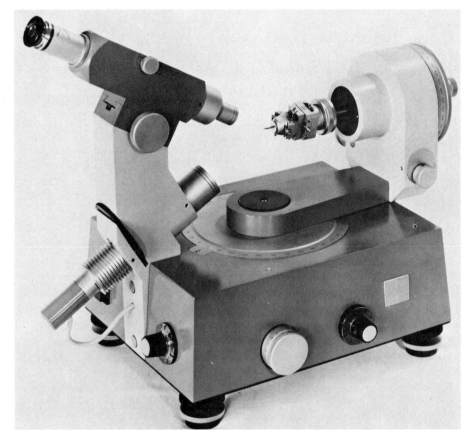

FIG. 2.40. Two-circle reflecting goniometer with crystal mounted at the right, a telescope at the upper left and a light source at the lower left (courtesy: Huber Diffraktionstechnik, X-ray Crystallography Equipment, Rimstig, West Germany). Compare with Fig. 1.5*b*.

FIG. 2.41(*a*). Contact goniometer. (*b*) Schematic enlargement of *a* showing the measurement of the internal angle α.

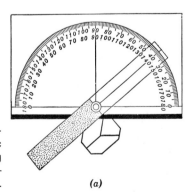

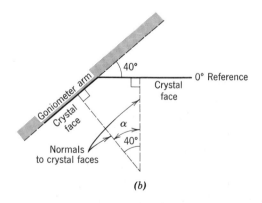

(*a*)

(*b*)

A simpler instrument used for approximate work and with larger crystals is known as a *contact goniometer* (Fig. 2.41a). In using a contact goniometer it is imperative that the plane determined by the two arms of the goniometer be exactly at right angles to the edge between the measured faces. It becomes clear from the construction in Fig. 2.41b that the angle between the poles to the crystal faces, the *internal angle,* is measured. Thus in Fig. 2.41a the angle should be read as 40°, not as 140°.

CRYSTAL PROJECTIONS

A crystal projection is a means of representing the three-dimensional crystal on a two-dimensional plane surface. Different projections are used for different purposes, but each is made according to some definite rule so that the projection bears a known and reproducible relationship to the crystal. The crystal drawings in this book are known as *clinographic projections* and are a type of perspective drawing which yields a portrait-like picture of the crystal in two dimensions. This is the best means of conveying the appearance of a crystal and generally serves much better for the purpose than a photograph.

Spherical Projection

Because the actual size and shape of the different faces on a crystal are chiefly the result of accidents of growth, we wish to minimize this aspect of crystals in projecting, but at the same time emphasize the angular relation of the faces. We can do this using the *spherical projection.* Although the spherical projection is itself three dimensional, an understanding of it is necessary, for it serves as the basis for other crystal projections. We can envisage the construction of such a projection in the following way (see Fig. 2.43). Imagine a hollow model of a crystal containing a bright point source of light. Now let us place this model within a large hollow sphere of translucent material in such a way that the light source is at the center of the sphere. If we now make a pinhole in each of the faces so that the ray of light emerging from the hole is perpendicular to the

face, these rays of light will fall on the inner surface of the sphere and make bright spots. The whole situation resembles a planetarium in which the crystal model with its internal light source and pinholes is the projector and the translucent sphere is the dome. If we now mark on the sphere the position of each spot of light, we may remove the model and have a permanent record of its faces. Each of the crystal faces is represented on the sphere by a point called the face *pole.* This is the spherical projection.

The position of each pole and thus its angular relationship to other poles can be fixed by angular coordinates on the sphere. This is done in a manner similar to the location of points on the earth's surface by means of longitude and latitude. For example, the angular coordinates 74°00' west longitude, 40°45' north latitude, locate a point in New York City. This means that the angle, measured at the center of the earth, between the plane of the equator and a line from the center of the earth through that point in New York is exactly 40°45'; and the angle between the Greenwich meridian and the meridian passing through the point in New York, measured west in the plane of the equator, is exactly 74°00'. These relations are shown in Fig. 2.42.

A similar system may be used to locate the poles of faces on the spherical projection of a crystal. There is one major difference between locating points on a spherical projection and locating points on the earth's surface. On the earth, the latitude is measured in degrees north or south from the equator, whereas the angle used in the spherical projection is the colatitude, or *polar angle,* which is measured in degrees from the north pole. The north pole of a crystal projection is thus at a colatitude of 0°, the equator at 90°. The colatitude of New York City is 49°15'. This angle is designated in crystallography by the Greek letter ρ (rho).

The "crystal longitude" of the pole of a face on the spherical projection is measured, just as are longitudes on earth, in degrees up to 180°, clockwise and counterclockwise from a starting meridian analogous to the Greenwich meridian of geography. To locate this starting meridian, the crystal is oriented in conventional manner with the (010) face to the right of the crystal. The meridian passing through the pole of this face is taken as zero. Thus, to determine the crystal longitude of any crystal face, a meridian is passed through the pole of that face, and the angle between it and the zero meridian is measured in the plane of the equator. This angle is designated by the Greek letter ϕ (phi).

If any plane is passed through a sphere, it will intersect the surface of the sphere in a circle. The circles of maximum diameter are those formed by planes passing through the center and having a diameter equal to the diameter of the sphere. These are called *great circles*. All other circles formed by passing planes through the sphere are *small circles*. The meridians on the earth are great circles, as is the equator, whereas the parallels of latitude are small circles.

The spherical projection of a crystal brings out interesting zonal relationships, for the poles of all the faces in a zone lie along a great circle of the projection. In Fig. 2.43 (spherical projection) faces (001), (101), (100), (10$\overline{1}$), and (00$\overline{1}$) lie in a zone with the zone axis [010]. Because the great circle along which the poles of these faces lie passes through the north and south poles of the projection, it is called a *vertical great circle*. The zone axis is always perpendicular to the plane containing the face poles; thus all vertical circles have horizontal zone axes.

Stereographic Projection

It is important in reducing the spherical projection of a crystal to a plane surface that the angular relation of the faces be preserved in such a way as to reveal the true symmetry. This can best be done with the *stereographic projection*.

The stereographic projection is a representation in a plane of half of the spherical projection, usually the northern hemisphere. The plane of the projection is the equatorial plane of the sphere, and the *primitive* circle (the circle outlining the projection) is the equator itself. If one were to view the poles of crystal faces located in the northern hemisphere of the spherical projection with the eye at the south pole, the intersection of the lines of sight with the equatorial plane would be corresponding points representing the poles on the stereographic projection. We can thus construct a stereographic pro-

jection by drawing lines from the south pole to the face poles in the northern hemisphere. The points corresponding to these poles on the stereographic projection are located where these lines intersect the equatorial plane. The relationship of the two projections is illustrated in Fig. 2.44.

In practice one plots poles directly on the stereographic projection. It is, therefore, necessary to determine stereographic distances in relation to angles of the spherical projection. Figure 2.45 shows a vertical section through the spherical projection of a crystal in the plane of the "zero meridian," that is, the plane containing the pole of (010). N and S are, respectively, the north and south poles of the sphere of projection, O is the center of the projected crystal. Consider the face (011). OD is the perpendicular to the face (011), and D is the pole of this face on the spherical projection. The line from the south pole, SD, intersects the trace of the plane of the equator, FG, in the point D', the stereographic pole of (011). The angle NOD will be recognized as the angle ρ (rho). In order to plot D' directly on the stereographic projection, it is necessary to determine the distance OD' in terms of angle ρ. Because $\triangle SOD$ is isosceles, $\angle ODS = \angle OSD$. $\angle ODS + \angle OSD = \angle NOD = \rho$. Therefore, $\angle OSD = \rho/2$. $OS = r$, the radius of the primitive of the projection.

$$\tan \rho/2 = OD'/r \quad \text{or} \quad OD' = r \tan \rho/2$$

To sum up, in order to find the stereographically projected distance from the center of the projection of the pole of any face, find the natural tangent of one-half of ρ of that face and multiply by the radius of the projection. The distance so obtained will be in whatever units are used to measure the radius of the primitive circle of the projection.

In addition to determining the distance a pole should lie from the center of the projection, it is also necessary to determine its "longitude" or ϕ (phi) angle. Because the angle is measured in the plane of the equator, which is also the plane of the stereographic projection, it may be laid off directly on the primitive circle by means of a circular protractor. It is first necessary to fix the "zero meridian" by making a point on the primitive circle to represent the pole of (010). A straight line drawn through this

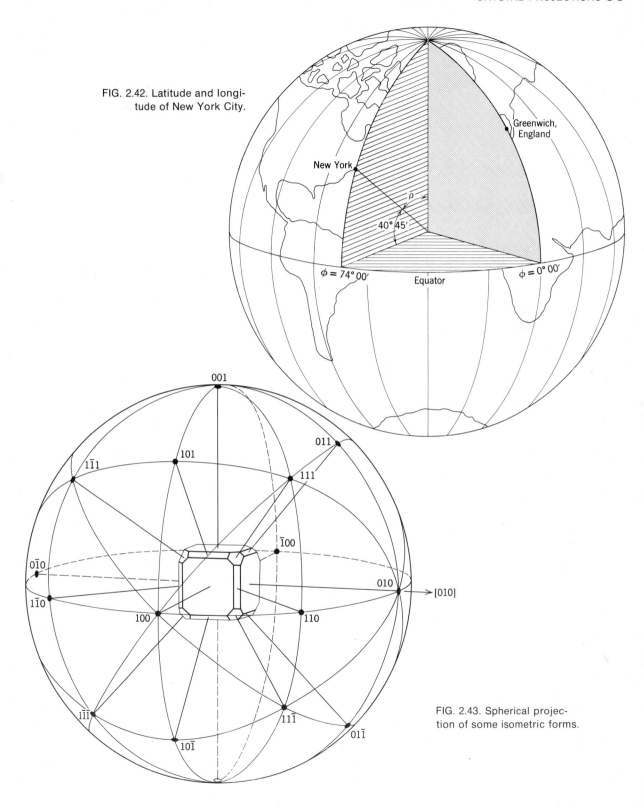

FIG. 2.42. Latitude and longitude of New York City.

Greenwich, England

New York

ρ

40° 45′

$\phi = 74° 00′$ Equator $\phi = 0° 00′$

001

011

101

1$\bar{1}$1

111

$\bar{1}$00

0$\bar{1}$0

010   →[010]

1$\bar{1}$0

100   110

1$\bar{1}\bar{1}$ 11$\bar{1}$ 01$\bar{1}$

10$\bar{1}$

FIG. 2.43. Spherical projection of some isometric forms.

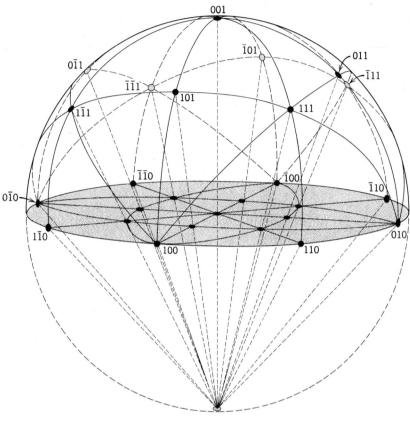

FIG. 2.44. Relation of spherical and stereographic projections (after E. E. Wahlstrom, Optical Crystallography, John Wiley & Sons, New York, 1951).

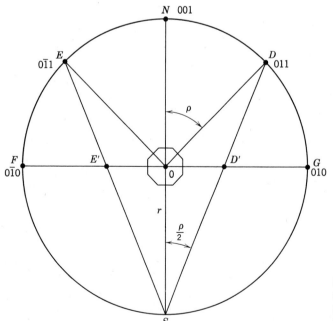

FIG. 2.45. Section through sphere of projection showing relation of spherical to stereographic poles.

point and the center of the projection is the zero meridian. With the protractor edge along this line and the center point at the center of the projection, the ϕ angle can be marked off. On a construction line from the center of the projection through this point lie all possible face poles having the specified ϕ angle. Positive ϕ angles are laid off clockwise from (010); negative ϕ angles are laid off counterclockwise, as shown in Fig. 2.46.

In order to plot the pole of the face having this given ϕ value, it is necessary to find the natural tangent of one-half of ρ, to multiply by the radius of the projection, and to lay off the resulting distance along the ϕ line. Although any projection radius may be chosen, one of 10 cm is usually used. This is large enough to give accuracy, but not be unwieldy, and at the same time simplifies the calculation. With a 10-cm radius, it is only necessary to look up the

natural tangent, move the decimal point one place to the right, and plot the result as centimeters from the center of the projection.

When the poles of crystal faces are plotted stereographically as explained above, their symmetry of arrangement should be apparent (see Fig. 2.47). We have seen (page 53) that a great circle in the spherical projection is the locus of poles of faces lying in a crystal zone. When projected stereographically vertical great circles become diameters of the projection; all other great circles project as circular arcs that subtend a diameter. The limiting case of such great circles is the primitive of the projection, which is itself a great circle common to both the spherical and stereographic projections. The poles of vertical crystal faces lie on the primitive and thus are projected without angular distortion.

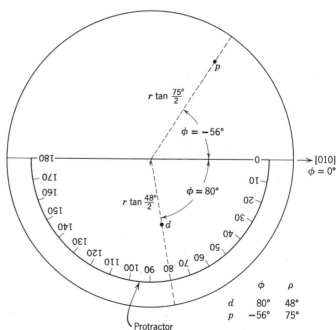

FIG. 2.46. Stereographic projection of crystal faces.

Stereographic Net

Both the measurement and plotting of angles on the stereographic projection are greatly facilitated by means of a stereographic net. A net with radius of 5 cm is shown in Fig. 2.48 and a net with radius of 10 cm is given on the inside of the back cover. This type of net is also called the Wulff net, named after G. V. Wulff, Russian crystallographer (1863–1925). Both great and small circles are drawn on the net at intervals of 1° or 2°. In Fig. 2.48 the intervals are 2°. A projection made on tracing paper may be placed over the net and the angles read directly. The ϕ angles are determined where a straight line from the center of the projection through the face pole intersects the primitive. To determine the ρ angle, the projection must be rotated about the center until the face pole lies on one of the vertical great circles. The angle can then be read directly from the net.

If the ϕ and ρ angles are known, the stereogram can be constructed by reversing the process. First, one should make a mark on the primitive at $\phi = 0°$ so that it is always possible to return the projection

to its initial setting. The pole of a face is located as follows: (1) Mark a point on the primitive at the ϕ angle; (2) rotate the projection until this point is at the end of a vertical great circle; (3) mark off the ρ angle along this line at the proper distance from the center. This is the face pole.

Instead of measuring ϕ and ρ angles the beginning student usually measures interfacial angles that can be plotted easily with the help of the stereographic net. As an example consider Fig. 2.49a, a crystal drawing of the orthorhombic mineral, anglesite. The starting point, as in all projections, is (010), face b. The pole of this face should be placed on the primitive at 0° (Fig. 2.49b). The interfacial angles $b \wedge n = 32\frac{1}{2}°$, and $b \wedge m = 52°$ can be measured and plotted as ϕ angles on the primitive. Face c is (001); it makes an angle of 90° with b and its pole should be placed at the center of the projection. Face o is in zone with c and b and thus has a ϕ angle of 0°. Its ρ angle, $c \wedge o = 52°$, can be measured directly and plotted along the vertical great circle. Face d lies in a vertical zone at 90° to the zone c, o,

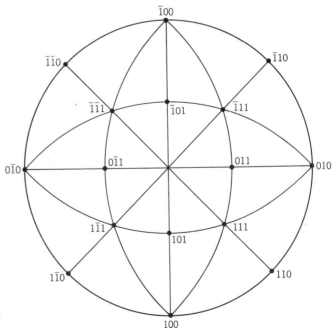

FIG. 2.47. Stereographic projection of some isometric crystal faces.

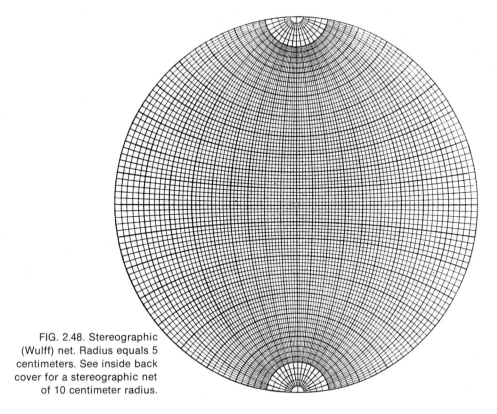

FIG. 2.48. Stereographic (Wulff) net. Radius equals 5 centimeters. See inside back cover for a stereographic net of 10 centimeter radius.

b. It thus has $\phi = 90°$ and $c \wedge d = 39\frac{1}{2}°$ which can be plotted along the vertical great circle of the net. The pole of face *y* cannot be plotted directly, but the angles $b \wedge y = 50°$ and $c \wedge y = 57°$ can be measured. To locate this pole, the projection is rotated 90° so that *b* lies along the radii of the small

circles of the net, and a tracing of the 50° circle is made. This small circle is the locus of all poles 50° from *b*. The projection is again rotated until this tracing of the circle intersects a vertical great circle at 57° ($c \wedge y$). This is the pole of *y*.

To check this position, measure the angle *m* ∧

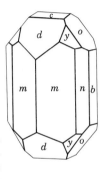

FIG. 2.49. A crystal of the orthorhombic mineral anglesite (*a*) and the stereographic projection of some of the faces on this mineral (*b*).

(*a*)

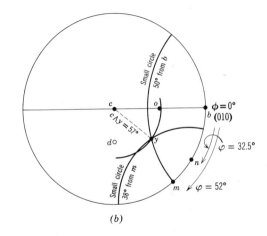

(*b*)

FIG. 2.50. Crystal with symmetry $2/m\overline{3}$. The stereogram shows a mirror plane at right angles to each of the 2-fold rotation axes, and four 3-fold axes of rotoinversion. The solid primitive circle denotes a horizontal mirror plane; faces at the bottom of the crystal directly below those at the top.

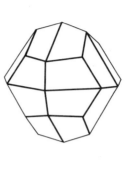

(a)

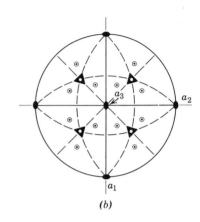

(b)

$y = 38°$. Now with the pole m placed along the radii of the small circles, trace the 38° small circle. It should also intersect at the common point. Having located pole y, its ϕ ($32\frac{1}{2}°$) and ρ (57°) angles can be read directly from the stereographic net.

THE THIRTY-TWO CRYSTAL CLASSES

In the following section the 32 classes listed in Table 2.3 are described under the crystal systems in which they are grouped. The crystal systems will be treated in order of increasing symmetry. Within each system, however, the classes will be discussed in order of decreasing symmetry. The symmetry of each class is given by the Hermann-Mauguin notation but it is also shown by means of stereograms, giving projections of all the faces of the general forms. These are the forms from which the classes derive their names. In the stereograms it is necessary to show faces in the southern hemisphere as well as in the northern hemisphere in order to give completely the symmetry of the class. This is done by superimposing stereographic projections of the two hemispheres with the poles in the northern hemisphere represented by solid points and those in the southern hemisphere by circles. Thus, if two poles lie directly one above the other on the sphere, they will be represented by a solid point surrounded by a circle. A vertical face is represented by a single point on the primitive, for, although such a pole

would appear on projections of both top and bottom of the crystal, it would represent but one face.

Fig. 2.50a is a drawing of a crystal with a horizontal symmetry plane.* The stereogram of this crystal, Fig. 2.50b, consequently, has for all faces solid points circled to indicate corresponding faces at the top and bottom of the crystal. Figure 2.51a is a drawing of a crystal lacking a horizontal mirror plane. Its stereogram, Fig. 2.51b has 12 solid points as poles of faces in the northern hemisphere and an additional 12, independent circles as poles of faces in the southern hemisphere.

Triclinic System
Crystallographic Axes. In the triclinic system the crystal forms are referred to three crystallographic

* A solid primitive circle in the projection indicates a horizontal mirror plane; a broken primitive circle indicates the lack of a horizontal mirror plane. A solid line drawn as a great circle (vertical or otherwise) indicates a mirror plane. Broken straight lines indicate the positions of crystal axes or symmetry axes; crystal axes are labeled a, b, c, whereas symmetry axes are so indicated by symbols at the ends of these lines. Symmetry axes of rotation are denoted with solid symbols as follows: ● 2-fold axis, ▲ 3-fold axis, ■ 4-fold axis, ⬢ 6-fold axis, Axes of rotoinversion are denoted with the following symbols: △ 3-fold, ▨ 4-fold, ⬡6-fold. These are the standard symbols used in the *International Tables for X-ray Crystallography*, The Kynoch Press, Birmingham, England, 1969. A center of symmetry is shown in writing only by the letter i for inversion, which is equivalent to $\overline{1}$. If a symmetry center occurs at the center of the sphere of projection, its presence cannot be shown properly by a symbol on a stereogram, but is detected from the arrangement of poles of equivalent faces.

FIG. 2.51. Crystal with symmetry $\bar{4}3m$. Stereogram shows three 4-fold axes of rotoinversion, six mirror planes, and four 3-fold rotation axes. The broken primitive circle indicates lack of a horizontal mirror plane and faces at the top of the crystal do not lie above those at the bottom.

(a)

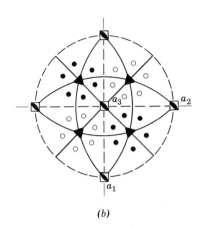

(b)

axes of unequal lengths that make oblique angles with each other (Fig. 2.52). The three rules to follow in orienting a triclinic crystal and thus in determining the position of the crystallographic axes are: (1) The most pronounced zone should be vertical. The axis of this zone then becomes c. (2) $\{001\}$ should slope forward and to the right. (3) Two forms in the vertical zone should be selected: one as $\{100\}$ the other as $\{010\}$. The directions of the a and b axes are determined by the intersections of $\{010\}$ and $\{100\}$, respectively, with $\{001\}$. The b axis should be longer than the a axis. In reporting on the crystallography of a new triclinic mineral or one that has not been recorded in the literature, the convention should be followed that $c < a < b$. The relative lengths of the three axes and the angles between them can be established only with difficulty and must be calculated for each mineral from appropriate measurements. The angles between the positive ends of b and c, c and a, and a and b are designated respectively as α, β, and γ (see Fig. 2.52). For example, the crystal constants of the tri-

clinic mineral axinite are as follows $a:b:c = 0.972:1:0.778$; $\alpha = 102°41'$, $\beta = 98°09'$, $\gamma = 88°08'$.

$\bar{1}$—Pinacoidal Class

Symmetry—*i.* The symmetry consists of a 1-fold axis of rotoinversion which is equivalent to a center of symmetry, or inversion (*i*). Figure 2.53 illustrates a triclinic pinacoid and its stereogram.

Forms. All the forms are pinacoids and thus consist of two similar and parallel faces.* Once a crystal is oriented, the Miller indices of a crystal face establish its position. However, names are given to the various pinacoids which, in a general way, designate their relation to the crystallographic axes. In addition to the front, side, and basal pinacoids, there are first-, second-, third-, and fourth-order pinacoids.

1. *Front, Side, and Basal Pinacoids.* Each of these pinacoids intersects one crystallographic axis and is parallel to the other two. The front or a pinacoid, $\{100\}$, intersects the a axis and is parallel to the other two; the side or b pinacoid, $\{010\}$, intersects the b axis; the basal or c pinacoid, $\{001\}$, intersects the c axis (Fig. 2.54).

---

* In the description of forms on the following pages the geometrically perfect model of the unmodified form is considered in each case. It should be kept in mind that in nature this ideal is rarely obtained, and that crystals not only are frequently malformed but also are usually bounded by a combination of forms.

FIG. 2.52. Triclinic crystal axes.

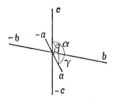

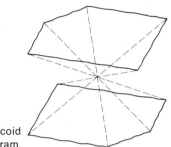 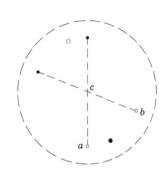

FIG. 2.53. Triclinic pinacoid and stereogram.

2. *First-, Second-, and Third-Order Pina-coids.* For each of these designations there is a positive and negative form whose faces are parallel to one axis and intersect the other two as follows:

First-order: parallel to *a*; (0*kl*) positive, (0$\bar{k}l$) negative.

Second-order: parallel to *b*; (*h*0*l*) positive, ($\bar{h}$0*l*) negative.

Third-order: parallel to *c*; (*hk*0) positive, (*h$\bar{k}$*0) negative.

3. *Fourth-Order Pinacoids,* {*hkl*} positive right, {*h$\bar{k}$l*} positive left, {$\bar{h}$kl} negative right, {$\bar{h}k\bar{l}$} negative left. Each of these two-faced forms can exist independently of the others.

Various first-, second-, third-, and fourth-order

pinacoids may be present depending on the axial intercepts.

Among the minerals that crystallize in the pinacoidal class are:

amblygonite   polyhalite
chalcanthite   rhodonite
microcline   turquoise
pectolite   ulexite
plagioclase feldspars   wollastonite

Of the above mentioned minerals microcline, rhodonite and chalcanthite are fairly common as well-formed crystals (see Fig. 2.55).

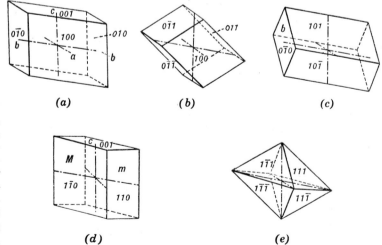

(a)   *(b)*   *(c)*

FIG. 2.54. Triclinic pinacoids. (a) Front {100}, side {010}, and base {001}. (b) First-order {011} positive, {0$\bar{1}$1} negative. (c) Second-order {101} positive, {10$\bar{1}$} negative. (d) Third-order {110} positive, {1$\bar{1}$0} negative. (e) Fourth-order (four different forms).

(d)   *(e)*

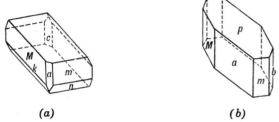

FIG. 2.55. Triclinic crystals. (a) Rhodonite. (b) Chalcanthite.

1—Pedial Class

Symmetry. There is merely a 1-fold rotation axis, which is equivalent to no symmetry. Fig. 2.56 illustrates a triclinic pedion and its stereogram.

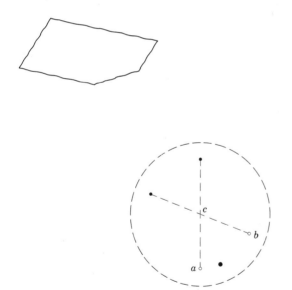

FIG. 2.56. Triclinic pedion and stereogram.

Forms. The general form {*hkl*} as well as all other forms are pedions and thus each face stands by itself.

Axinite was originally classified as 1 but has been found to have symmetry $\bar{1}$.

Monoclinic System

Crystallographic Axes. Monoclinic crystals are referred to three axes of unequal lengths. The only restrictions in the angular relations are that $a \wedge b$ (γ) and $c \wedge b$ (α) = 90°. For most crystals the angle between $+a$ and $+c$ is greater than 90° but in rare instances it too may equal 90°. In such cases the monoclinic symmetry is not apparent from the morphology. The 2-fold rotation axis or the direction perpendicular to the mirror plane is usually taken as the b axis; the a axis is inclined downward toward the front; and c is vertical. This orientation, known as the "second setting," is traditional for mineralogists.[*]

The crystal constants of monoclinic crystals include the axial ratios expressed as $a:b:c$ with b taken as unity, and the angle β. Calculations of axial ratios that in the orthogonal systems are simple and straightforward, become involved and tedious in the inclined systems. For the formulas used in their calculations, one is referred to more advanced books on crystallography. Figure 2.57 represents the crystallographic axes of the monoclinic mineral orthoclase, the axial constants of which are expressed as $a:b:c = 0.658:1:0.553; \beta = 116°01'$.

Although the direction of the b axis is fixed by symmetry, the directions that serve as the a and c axes are matters of choice and depend on crystal

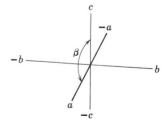

FIG. 2.57. Monoclinic crystal axes.

* Some crystallographers orient monoclinic crystals according to the "first setting" in which the 2-fold axis or the normal to the mirror plane is chosen as the c axis instead of the b axis. Here, all monoclinic crystals are referred to the "second setting."

habit and cleavage. If crystals show an elongated development (prismatic habit) parallel to a direction in the *a–c* plane, that direction often serves as the *c* axis. Further, if there is a prominent sloping plane or planes, such as planes *c* or *r* in the drawings of Fig. 2.65 the *a* axis may be taken as parallel to these. It is quite possible that there may be two, or even more, choices that are equally good, but in the description of a new mineral it is conventional to orient the crystals so that $c < a$.

Cleavage is also an important factor in orienting a monoclinic crystal. If there is a good pinacoidal cleavage parallel to the *b* axis, as in orthoclase, it is usually taken as the basal cleavage. If there are two equivalent cleavage directions, as in the amphiboles and pyroxenes, they are usually taken to be vertical prismatic cleavages.

2/m—Prismatic Class

Symmetry—*i*, $1A_2$, 1*m*. The *b* axis is an axis of 2-fold rotation, and the *a* and *c* axes lie in a mirror plane (Fig. 2.58). The stereogram, Fig. 2.59, shows the symmetry of the prism (fourth-order). Because the *a* axis slopes down and to the front, it does not lie in the equatorial plane, and the positive end intersects the sphere of projection in the southern hemisphere.

Forms. There are only two types of forms in the prismatic class of the monoclinic system—pinacoids and prisms.

1. *Pinacoids.* (See Fig. 2.60)

Front or *a* pinacoid, {100}. Intersects *a*, parallels *b* and *c*.

Side or *b* pinacoid, {010}. Intersects *b*, parallels *a* and *c*.

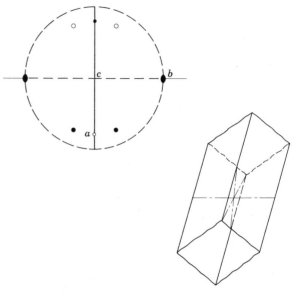

FIG. 2.59. Monoclinic prism and stereogram.

Basal or *c* pincaoid, {001}. Intersects *c*, parallels *a* and *b*.

Second-order pinacoids: {*h0l*} positive, {$\bar{h}0l$} negative.

Because the opposite ends of the *a* axis are not interchangeable, there is no second-order prism. Instead there are pinacoids: {*h0l*} between {100} and {001} and {$\bar{h}0l$} netween {$\bar{1}00$} and {001}. It should be emphasized that these two forms are independent of each other, and the presence of one does not necessitate the presence of the other (see Figs. 2.61 and 2.62).

FIG. 2.60. Monoclinic pinacoids, front {100}, side {010}, base {001}.

FIG. 2.58. Symmetry elements of prismatic class.

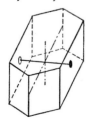

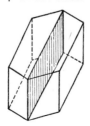

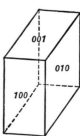

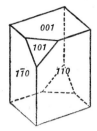 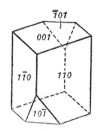

FIG. 2.61. Second-order pinacoids. {101} positive, {$\overline{1}$01} negative with third-order prism {110}, and basal pinacoid {001}.

2. *Prisms.* The four-faced prism is the general form {*hkl*}; the special forms {0*kl*} and {*hk*0} are also prisms.

First-order prism {0*kl*}. Faces intersect the *b* and *c* axis and parallel the *a* axis (see Fig. 2.63).

Third-order or vertical prism {*hk*0}. Faces intersect the *a* and *b* axis and parallel the *c* axis.

Fourth-order prism {*hkl*} *positive,* {$\overline{h}kl$} *negative.* The faces of these two independent forms intersect all three crystallographic axes. If the faces of the prism at the upper end of the crystal intersect the positive end of the *a* axis, the prism is positive; if they intersect the negative end of *a*, the prism is negative (see Fig. 2.64).

The only form in this crystal class that is fixed is the side pinacoid. The other forms may vary with the choice of the *a* and *c* axes. For instance, the front pinacoid, basal pinacoid, and second-order pinacoid may be converted into each other by a rotation about the *b* axis. In the same manner the three prisms can be changed from one position to another.

Characteristic combinations of the forms described above are given in Fig. 2.65.

FIG. 2.62. Positive and negative second-order pinacoids with side pinacoid.

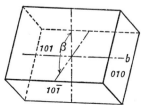

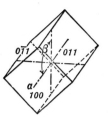

FIG. 2.63. First-order prism {011} and front pinacoid {100}.

Many minerals crystallize in the monoclinic, prismatic class; some of the most common are:

azurite	monazite
borax	muscovite (and
calaverite	other micas)
chlorite	orpiment
colemanite	orthoclase
datolite	realgar
diopside (and	sphene
other pyroxenes)	spodumene
epidote	talc
gypsum	tremolite (and
heulandite	other amphiboles)
kaolinite	wolframite
malachite	

m—Domatic Class

Symmetry—1*m***.** There is one vertical mirror plane (010) that includes the *a* and *c* crystallographic axes. A monoclinic dome and its stereogram are shown in Fig. 2.66.

Forms. The *dome* is a two-faced form symmetrical across a mirror plane in contrast to a sphenoid

FIG. 2.64. Fourth-order prisms. {111} positive, {$\overline{1}$11} negative with third-order prism {110}, and basal pinacoid {001}.

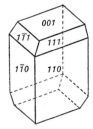 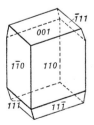

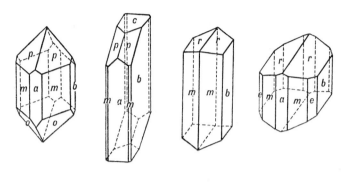

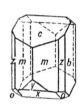

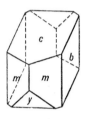

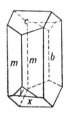

FIG. 2.65. Monoclinic crystals. Forms: a {100}, b {010}, c {001}, m {110}; p {111}, o {$\bar{2}21$}, r {011}, e {120}, x {$\bar{1}01$}, y {$\bar{2}01$}, z {130}.

(Fig. 2.68) which is symmetrical about an axis of 2-fold rotation. The general forms {hkl} and {$\bar{h}kl$}, both of which are domes, each correspond to two of the faces of the fourth-order prism of the prismatic class. The special prisms {$0kl$} and {$hk0$} of class $2/m$ also become domes, {$0kl$}, {$0k\bar{l}$}, and {$hk0$}, {$\bar{h}k0$}. The form {010} is a pinacoid, but all faces lying across the mirror plane as {100}, {$\bar{1}00$}, {001}, {$00\bar{1}$}, and {$h0l$}, {$\bar{h}0l$} are pedions.

The rare minerals hilgardite, $Ca_2ClB_5O_8(OH)_2$ (Fig. 2.67) and clinohedrite, $Ca_2Zn_2(OH)_2Si_2O_7 \cdot H_2O$ crystallize in this class.

2—Sphenoidal Class

Symmetry—$1A_2$. The b crystallographic axis is an axis of 2-fold rotation. Figure 2.68 shows a sphenoid and its stereogram.

Forms. With the absence of an a-c symmetry plane, the b axis is polar, and different forms are present at opposite ends. The {010} pinacoid of class $2/m$ becomes two pedions, {010} and {$0\bar{1}0$}. Likewise the prisms, first-, third-, and fourth-order, each degenerate into a pair of enantiomorphic sphenoids. A *sphenoid* is a two-faced form symmetrical about a 2-fold rotation axis, b, in contrast to a

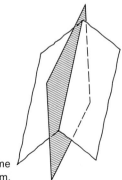

FIG. 2.66. Monoclinic dome and stereogram.

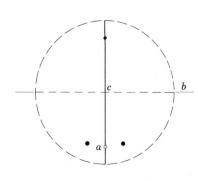

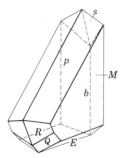

FIG. 2.67. Hilgardite.

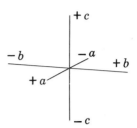

FIG. 2.69. Orthorhombic crystal axes.

dome with symmetry across a plane. The general form is thus the sphenoid {*hkl*} with its enantiomorphic equivalent {*hk̄l*}. The other special prismatic forms of class 2/*m* become sphenoids {0*kl*}, {0*k̄l*}, and {*hk*0}, {*h̄k*0}. The pinacoids {100}, {001}, {*h0l*} and {*h̄0l*} may be present.

Mineral representatives in the sphenoidal class are rare, but chief among them are the members of the halotrichite isostructural group, of which pickeringite, $MgAl_2(SO_4)_4 \cdot 22H_2O$, is the most common member.

FIG. 2.68. Monoclinic sphenoid and stereogram.

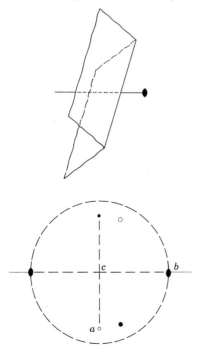

Orthorhombic System

Crystallographic Axes. The crystal forms in the orthorhombic system are referred to three crystallographic axes of unequal length that make angles of 90° with each other. The relative lengths of the axes, or the axial ratios, must be determined for each orthorhombic mineral. In orienting an orthorhombic crystal, the convention is to set the crystal so that $c < a < b$. In the past, however, this convention has not necessarily been observed, and it is customary to conform to the orientation given in the literature. One finds, therefore, that any one of the three axes may have been chosen as *c*. The longer of the other two is then taken as *b* and the shorter as *a*.

In the past, the decision as to which of the three axes should be chosen as the vertical axis rested largely on the crystal habit of the mineral. If its crystals commonly showed an elongation in one direction, this direction was usually chosen as the *c* axis (see Fig. 2.76). If, on the other hand, the crystals showed a prominent pinacoid and therefore were tabular, this pinacoid was usually taken as {001} with *c* normal to it (see Fig. 2.77). Cleavage also aided in orienting orthorhombic crystals. If, as in topaz, there was one pinacoidal cleavage, it was taken as {001}. If, as in barite, there were two equivalent cleavage directions, they were set vertical and their intersection edges determined *c*. After the orientation has been determined, the length of the axis chosen as *b* is taken as unity, and the relative lengths of *a* and *c* are given in terms of it. Figure 2.69 represents the crystallographic axes for the

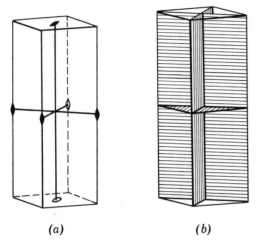

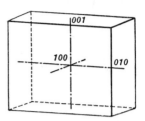

FIG. 2.72. Orthorhombic pinacoids.

(a)   *(b)*

FIG. 2.70. Symmetry elements of rhombic dipyramidal class. (*a*) Rotation axes. (*b*) Mirror planes.

orthorhombic mineral sillimanite with axial ratios $a:b:c = 0.98:1:0.75$.

In the Hermann-Mauguin notation for the orthorhombic system, the symbols refer to the symmetry elements in the order *a, b, c*. For example, in the class *mm2*, the *a* and *b* axes lie in vertical mirror planes and *c* is an axis of 2-fold rotation.

$2/m\ 2/m\ 2/m$—Rhombic-Dipyramidal Class

Symmetry—*i*, 3A_2, 3*m*. The three crystallographic axes are axes of 2-fold rotation and perpendicular to each of them is a mirror plane (Fig. 2.70). A rhombic dipyramid and its stereogram are shown in Fig. 2.71.

Forms. There are three types of forms in the rhombic-dipyramidal class—pinacoids, prisms, and dipyramids.

1. *Pinacoids.* There are three possible pinacoids, the faces of which intersect one crystallographic axis and parallel the other two (see Fig. 2.72). They are:

Front or *a* pinacoid, {100}. Intersects *a*; parallels *b* and *c*.

Side or *b* pinacoid, {010}. Intersects *b*; parallels *a* and *c*.

Basal or *c* pinacoid, {001}. Intersects *c*; parallels *a* and *b*.

2. *Prisms.* Orthorhombic prisms consist of four faces that are parallel to one axis and intersect the other two.

First-order prism, {0*kl*}. Faces parallel *a* (first axis); intersect *b* and *c*.

Second-order prism, {*h0l*}. Faces parallel *b* (second axis); intersect *a* and *c*.

Third-order prism, {*hk0*}. Faces parallel *c* (third axis); intersect *a* and *b*.

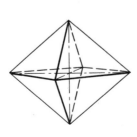

FIG. 2.71. Rhombic dipyramid and stereogram.

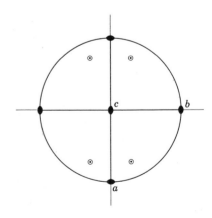

There are various prisms of each order with different axial intercepts. The unit forms are illustrated in Fig. 2.73. Because all prisms intersect two axes and parallel the third, one prism will be transformed into another by a different choice of axes.

3. *Dipyramid* {*hkl*}. An orthorhombic dipyramid has eight triangular faces, each of which intersects all three crystallographic axes. It is the general form from which the orthorhombic-dipyramidal class receives its name. Figure 2.74 represents the unit dipyramid {111}.

Combinations. Practically all orthorhombic crystals consist of combinations of two or more forms. Characteristic combinations of the various forms are given in Figs. 2.75–2.77.

There are many mineral representatives in this class. Among the more common are the following:

andalusite	enstatite (and other
anthophyllite (and other	orthopyroxenes)
orthorhombic amphiboles)	goethite
aragonite (group)	lawsonite
barite (group)	marcasite
brookite	olivine
chrysoberyl	sillimanite
columbite	stibnite
cordierite	sulfur
danburite	topaz

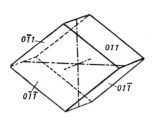

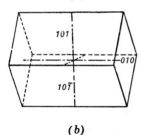

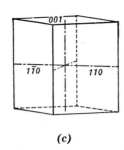

FIG. 2.73. Orthorhombic prisms, unit forms. (*a*) First-order. (*b*) Second-order. (*c*) Third-order.

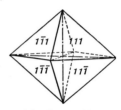

FIG. 2.74. Orthorhombic dipyramid.

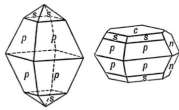

FIG. 2.75. Sulfur crystals. Forms: *p* {111}, *S* {113}, first-order dipyramids. *n* {011}, first-order prism, *c* {001}, basal pinacoid.

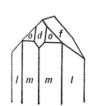 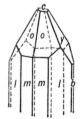

FIG. 2.76. Topaz crystals. Forms: *c* {001}, *b* {010}, pinacoids. *f* {011}, *y* {021}, *d* {101}, *m* {110}, *l* {120}, prisms. *o* {111}, dipyramid.

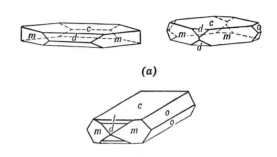

FIG. 2.77. (*a*) Barbite. (*b*) Celestite. Forms: *c* {001}, *o* {011}, *d* {101}, *m* {210}, *l* {104}.

*mm*2—Rhombic-Pyramidal Class

Symmetry—$2m$, $1A_2$. The c crystallographic axis is an axis of 2-fold rotation. Two mirror planes at right angles to each other intersect in this axis. A rhombic pyramid and its stereogram are shown in Fig. 2.78.

Forms. Because of the absence of a horizontal symmetry plane, the forms at the top of the crystal are different from those at the bottom. The rhombic dipyramid thus becomes two rhombic pyramids, $\{hkl\}$ at the top and $\{hk\bar{l}\}$ at the bottom. Likewise the first-order and second-order prisms do not exist. In the place of each there are two domes (two-faced forms). $\{0kl\}$ and $\{0k\bar{l}\}$ are the first-order domes, and $\{h0l\}$ and $\{h0\bar{l}\}$ are second-order domes. In addition to these forms there are also pedions, $\{001\}$ and $\{00\bar{1}\}$, and third-order prisms $\{hk0\}$.

Only a few minerals crystallize in this class; the most common representatives are hemimorphite, $Zn_4Si_2O_7(OH)_2 \cdot H_2O$, Fig. 2.79 and bertrandite, $Be_4Si_2O_7(OH)_2$.

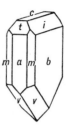

FIG. 2.79. Hemimorphite.

222—Rhombic-Disphenoidal Class

Symmetry—$3A_2$. There are three axes of 2-fold rotation coincident with the crystallographic axes. There are no mirror planes and no center of symmetry. Figure 2.80 illustrates rhombic disphenoids and gives a stereogram of the right disphenoid.

Forms. The rhombic disphenoid is composed of four faces, two in the upper hemisphere and two

FIG. 2.80. Enantiomorphic forms of the rhombic disphenoid, left and right, and a stereogram of the right form.

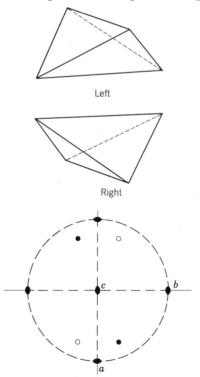

FIG. 2.78. Rhombic pyramid and stereogram.

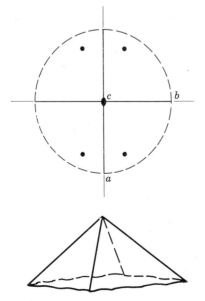

in the lower. It resembles the tetragonal disphenoid, but each face is a scalene triangle; whereas in the tetragonal disphenoid each face is an isosceles triangle. There are two disphenoids. The right $\{hkl\}$ and left $\{h\bar{k}l\}$ are enantiomorphic forms (see Fig. 2.80). The three pinacoids and the three prisms may be present in this class.

Although there are several representative minerals crystallizing in this class, they are all comparatively rare. The most common are epsomite, $MgSO_4 \cdot 7H_2O$, and olivenite, $Cu_2AsO_4(OH)$.

Orthorhombic Axial Ratios

To express the relative lengths of the axes in the orthorhombic system there are two ratios, $a:b$ and $b:c$, where the length of b is taken as unity. They are given as $a:b:c = -:1:-$, and can be calculated using ϕ and ρ angles.

As an example, consider the face of a general form ABC, Fig. 2.81. Assume this to be face (132) of aragonite with $\phi = 28°11'$, $\rho = 50°48'$. OP is the face normal, and OD is normal to AB. Therefore, angle BOD is ϕ and angle COP is ρ. Reducing the indices (132) to intercepts we find $AO = 6a$, $OC = 3c$, and $OB = 2b = 2$, because $b = 1$.

Hence, we may find a (see Fig. 2.81b) by

$$\cot \phi = \frac{6a}{2b} \quad \text{or} \quad a = \frac{b \cot \phi}{3}$$

$$b = 1 \quad a = \frac{1 \times \cot 28°11'}{3} \quad a = 0.6221$$

In triangle COD (Fig. 2.81c), $\tan \rho = 3c/OD$. In triangle BOD, $\cos \phi = OD/2$. Setting both of these expressions equal to OD,

$$OD = \frac{3c}{\tan \rho} = 2 \cos \phi \quad \text{or}$$

$$c = \frac{2 \tan \rho \cos \phi}{3}$$

Substituting ϕ and ρ values of face (132)

$$c = \frac{2 \tan 50°48' \cos 28°11'}{3} = 0.7205$$

Using Miller indices h, k, and l rather than intercepts, a and c relative to b are obtained using the formulas: $a = h/k \cot \phi$ and $c = l/k \tan \rho \cos \phi$. Crystallographic calculations generally involve the variables: (1) axial ratios, (2) indices, and (3) angles ϕ and ρ. When two of these variables are known, the third may be calculated using the above formulas.

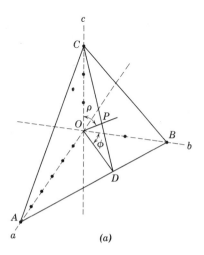

(a)

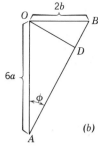

(b)

(c)

FIG. 2.81. Intercepts of face (132).

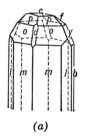 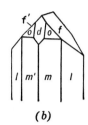

FIG. 2.82. Topaz crystals.

The ϕ and ρ angles may be obtained from interfacial angles. For example (Fig. 2.82a), the angle between b (010) and another face, such as m in the [001] zone is the ϕ angle of m. Other faces, such as p and o, in the same horizontal zone with m, also have $b \wedge m$ as their ϕ angle. When c (001) is present (Fig. 2.82a) the angles of f and o are respectively, $c \wedge f$ and $c \wedge o$. When (010) and (001) are not present (Fig. 2.82b), ϕ and ρ angles must be calculated from interfacial angles. For example, $\phi m = 90° - (m \wedge m')/2$; and $pf = (f \wedge f')/2$.

Tetragonal System

Crystallographic Axes. The forms of the tetragonal system are referred to three crystallographic axes that make right angles with each other. The two horizontal axes, a, are equal in length and interchangeable, but the vertical axis, c, is of a different length. Figure 2.83a represents the crystallographic axes for the tetragonal mineral zircon with c less than a. Figure 2.83b represents the crystallographic axes of the mineral octahedrite with c greater than a. The length of the horizontal axes is taken as unity, and the relative length of the vertical axis is expressed in terms of the horizontal. The axial ratio must be determined for each tetragonal crystal by measuring the interfacial angles and making the proper calculations (see page 76). For zircon, the length of the vertical axis is expressed as $c = 0.901$, for octahedrite as $c = 1.777$. The proper orientation of the crystallographic axes and the method of their notation are shown in Fig. 2.83. When the general form symbols are used, $h < k$.

In the Hermann-Mauguin notation of the sym-

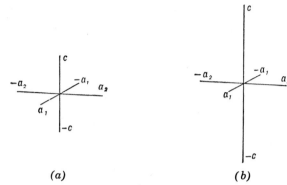

FIG. 2.83. Tetragonal crystal axes.

metry elements in the tetragonal system, the first part of the symbol (consisting of 4 or $\bar{4}$) refers to the c axis, whereas the second or third parts refer to the axial (a_1 and a_2) and diagonal symmetry elements, respectively.

4/m 2/m 2/m—Ditetragonal-Dipyramidal Class

Symmetry—i, $1A_4$, $4A_2$, $5m$. The vertical crystallographic axis is an axis of 4-fold rotation. There are four horizontal axes of 2-fold symmetry, two of which are coincident with the crystallographic axes, and the others at 45° to them. There are five mirror planes, one horizontal and four vertical. One of the horizontal symmetry axes lies in each of the vertical mirror planes. The position of the axes and mirrors is shown in Fig. 2.84. Figure 2.85 illustrates a ditetragonal dipyramid and its stereogram.

Forms. 1. *Basal pinacoid* {001}. A form composed of two horizontal faces. It is shown in combination with various prisms in Fig. 2.86.

FIG. 2.84. Symmetry of ditetragonal dipyramidal class. (a) Axes. (b) Planes.

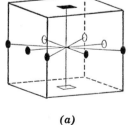 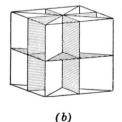

(a)   (b)

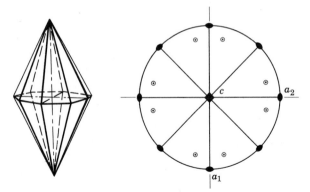

FIG. 2.85. Ditetragonal dipyramid and stereogram.

2. *Prism of First Order* {110}. Consists of four rectangular vertical faces, each of which intersects the two horizontal crystallographic axes equally (Fig. 2.86a).

3. *Prism of Second Order* {010}. Consists of four rectangular vertical faces, each of which intersects one horizontal crystallographic axis and is parallel to the other two axes. The form is represented in Fig. 2.86b.

4. *Ditetragonal Prism* {hk0}. Consists of eight rectangular vertical faces, each of which intersects the two horizontal crystallographic axes unequally. There are various ditetragonal prisms, depending on their differing relations to the horizontal axes. A common form, represented in Fig. 2.86c, has indices {120}.

5. *Dipyramid of First Order* {hhl}. Has eight isosceles triangular faces, each of which intersects all three crystallographic axes, with equal intercepts upon the two horizontal axes. There are various dipyramids of the first order, depending on the inclination of their faces to c. The unit dipyramid {111} (Fig. 2.87a), which intersects all the axes at their unit lengths, is most common. Indices of other dipyramids of the first order are {221}, {331}, {112}, {113}, etc., or, in general, {hhl}.

6. *Dipyramid of Second Order* {0kl}. Composed of eight isosceles triangular faces, each of which intersects one horizontal axis and the vertical axis and is parallel to the second horizontal axis. There are various dipyramids of the second order, with different intersections upon the vertical axis. The most common is the unit dipyramid {011} (Fig. 2.87b). Other dipyramids of the second order have indices {021}, {031}, {012}, {013}, or, in general {0kl}.

The relationship between the dipyramids of the first and second order is similar to that between the prisms of the first and second order.

As a general rule in the absence of other evidence, if one dipyramid is present, it is set as first-order. If two dipyramids of different orders are present, the dominant one is usually set as first order. In the orientation of a crystal, the prisms are subordinate to the dipyramids. Thus, an important prism may be relegated to second order by the presence of a small dipyramid.

7. *Ditetragonal Dipyramid* {hkl}. Composed of 16 triangular faces, each of which intersects all three of the crystallographic axes, cutting the two horizontal axes at different lengths. There are various ditetragonal dipyramids, depending on the

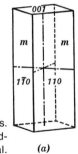

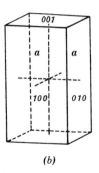

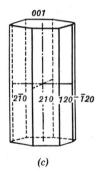

FIG. 2.86. Tetragonal prisms. (a) First-order. (b) Second-order. (c) Ditetragonal.

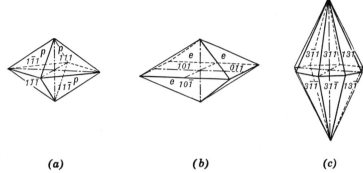

FIG. 2.87. Tetragonal dipyramids. (*a*) First-order. (*b*) Second-order. (*c*) Ditetragonal.

(a)   **(b)**   **(c)**

different intersections on the crystallographic axes. One of the most common is the dipyramid {131} shown in Fig. 2.87*c*.

Several common minerals crystallize in the ditetragonal-dipyramidal class. Major representatives are rutile, anatase, cassiterite, apophyllite, zircon, and idocrase.

Tetragonal Combinations. Characteristic combinations of ditetragonal-dipyramidal forms as found on crystals of different minerals are represented in Fig. 2.88.

$\overline{4}2\,m$—Tetragonal-Scalenohedral Class

Symmetry—$1\overline{A}_4$, $2A_2$, $2m$. The c axis is an axis of 4-fold rotoinversion and the two a axes are axes of 2-fold rotation. At 45° to the a axes are two vertical mirror planes intersecting in the vertical axis (see Fig. 2.89). Figure 2.90 illustrates a tetragonal scalenohedron and its stereogram.

Forms.

1. *Disphenoid* {*hhl*} *positive*, {*hhl*} *negative* are the only important forms in this class. They consist of four isosceles triangular faces that intersect

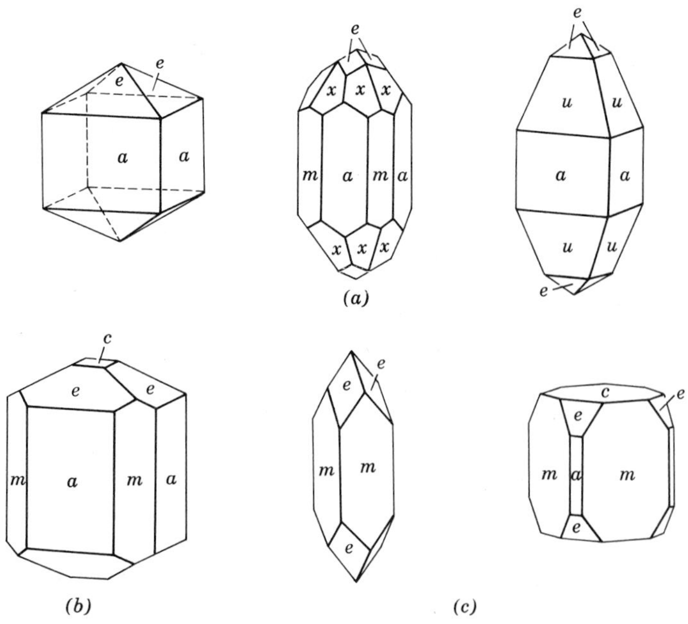

FIG. 2.88. Tetragonal crystals. (*a*) Zircon. (*b*) Idocrase. (*c*) Apophyllite. Forms: *e* {011} and *u* {012}, second-order dipyramids. *c* {001}, basal pinacoid. *a* {010}, second-order prism. *m* {110}, first-order prism. *x* {211}, ditetragonal dipyramid. In this illustration the form designations are based on the knowledge of the orientation of the unit cell. If the forms were named on the basis of morphology first- and second-order designations would be interchanged.

(*a*)

(*b*)   (*c*)

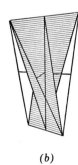

(a)   (b)

FIG. 2.89. Symmetry of tetragonal scalenohedral class. (a) Axes. (b) Planes.

(a)   (b)   (c)

FIG. 2.91. Tetragonal disphenoids. (a) Positive. (b) Negative. (c) Positive and negative.

all three of the crystallographic axes, with equal intercepts on the two horizontal axes. There may be different disphenoids, depending on their varying intersections with the vertical axis. Two different disphenoids are shown in Figs. 2.91a and b. There may also be a combination of a positive and a negative disphenoid as represented in Fig. 2.91c.

The tetragonal disphenoid differs from the tetrahedron in the isometric system in that its vertical crystallographic axis is not the same length as the horizontal axes. The only common mineral in the tetragonal-scalenohedral class is chalcopyrite, crystals of which ordinarily show only the disphenoid {112}. This disphenoid closely resembles a tetrahedron, and it requires accurate measurements to prove its tetragonal character.

2. *Tetragonal Scalenohedron* {hkl}. This form, Fig. 2.90, if it were to occur by itself, is bounded by eight similar scalene triangles. It is a rare form and observed only in combination with others. Other

forms that may be present are: pinacoid, first- and second-order tetragonal prisms, ditetragonal prisms, and second-order tetragonal dipyramids.

Chalcopyrite ($CuFeS_2$) and stannite (Cu_2FeSnS_4) are the only common minerals that crystallize in this class.

4mm—Ditetragonal-Pyramidal Class

Symmetry—$1A_4$, $4m$. The vertical axis is a 4-fold rotation axis, and four mirror planes intersect in this axis. Figure 2.92 illustrates a ditetragonal pyramid and its stereogram.

Forms. The lack of a horizontal symmetry plane gives rise to different forms at the top and bottom of crystals of this class. There are pedions {001} and {00$\bar{1}$}. The first-order {hhl} and second-order {h0l} tetragonal pyramids have corresponding lower forms, {hh$\bar{l}$} and {h0$\bar{l}$}. The ditetragonal pyramid {hkl} is an upper form, whereas {hk$\bar{l}$} is the lower form. The first- and second-order tetragonal prisms as well as the ditetragonal prism may be present.

The rather rare mineral diaboleite,

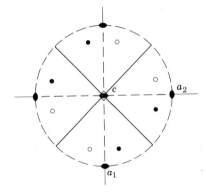

FIG. 2.90. Tetragonal scaleno-hedron and stereogram.

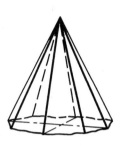
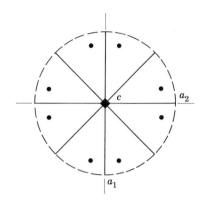

FIG. 2.92. Ditetragonal pyramid and stereogram.

$Pb_2Cu(OH)_4Cl_2$, is the only mineral representative
in this crystal class.

422—Tetragonal-Trapezohedral Class

Symmetry—$1A_4$, $4A_2$. The vertical axis is one of
4-fold rotation and there are four 2-fold axes at right
angles to it. Mirror planes and a center of symmetry
are lacking. The symmetry axes are the same as
those of the ditetragonal-dipyramidal class (Fig.
2.84).

Forms. The tetragonal trapezohedron has eight
faces, corresponding to half the faces of the ditetragonal dipyramid. There are two enantiomorphic
forms, right $\{hkl\}$ and left $\{h\bar{k}l\}$ (Fig. 2.93). Other
forms that may be present are: pinacoid, first- and
second-order tetragonal prisms, ditetragonal prism,
and first- and second-order tetragonal dipyramids.

Phosgenite, $Pb_2CO_3Cl_2$, is the only mineral representative in this class.

4/m—Tetragonal-Dipyramidal Class

Symmetry—i, $1A_4$, $1m$. There is a vertical 4-fold
rotation axis with symmetry plane at right angles.
Figure 2.94 illustrates a tetragonal dipyramid and its
stereogram.

Forms. The tetragonal dipyramid, $\{hkl\}$, is an
eight-faced form having four upper faces directly
above four lower faces. This form by itself appears
to have higher symmetry, and it must be in combination with other forms to reveal the absence of vertical symmetry planes. The basal pinacoid $\{001\}$
and tetragonal prisms $\{hk0\}$ may be present. The tetragonal prism $\{hk0\}$ is equivalent to the four alternate faces of the ditetragonal prism and is present in
those classes of the tetragonal system that have no
vertical mirror planes or horizontal 2-fold rotation
axes.

Mineral representatives in this class are:
scheelite ($CaWO_4$), powellite ($CaMoO_4$), and members of the scapolite series ($Na_4Al_3Si_9O_{24}Cl$ to

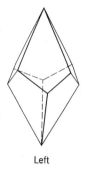
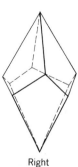
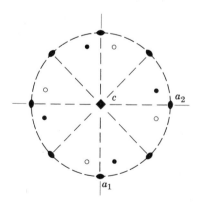

FIG. 2.93. Enantiomorphic
forms of the tetragonal trapezohedron, left and right, and
a stereogram of the right
form.

Left

Right

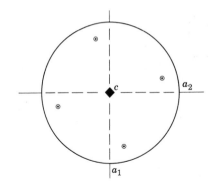

FIG. 2.94. Tetragonal di-
pyramid and stereogram.

$Ca_4Al_6Si_6O_{24}CO_3$). Figure 2.95 illustrates a crystal of scapolite in which the tetragonal dipyramid z reveals the true symmetry of this class.

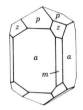

FIG. 2.95. Scapolite.

$\overline{4}$—Tetragonal-Disphenoidal Class

Symmetry—$1\overline{A}_4$. The vertical axis is a 4-fold axis of rotoinversion. There is no other symmetry. Figure 2.96 illustrates a tetragonal disphenoid and its stereogram.

Forms. The tetragonal disphenoid $\{hkl\}$ is a closed form composed of four wedge-shaped faces.

In the absence of other modifying faces, the form appears to have two vertical symmetry planes giving it the symmetry $\overline{4}2m$. The true symmetry is shown only in combination with other forms. The pinacoid and tetragonal prisms may be present.

The only mineral representative in this class is the rare mineral cahnite, $CaB(OH)_4AsO_4$.

4—Tetragonal-Pyramidal Class

Symmetry—$1A_4$. The vertical axis is one of 4-fold rotation. There are no symmetry planes or center. The tetragonal pyramid and its stereogram are shown in Fig. 2.97.

Forms. The tetragonal pyramid is a four-faced form. The upper form $\{hkl\}$ is different from the lower form $\{hk\overline{l}\}$, and each has a right- and left-hand variation. There are thus two enantiomorphic pairs of tetragonal pyramids. Pedions and tetragonal prisms may also be present.

As in some other classes, the true symmetry is

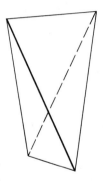

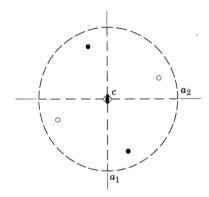

FIG. 2.96. Tetragonal di-
sphenoid and stereogram.

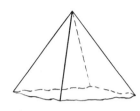

FIG. 2.97. Tetragonal pyramid
and stereogram.

not shown morphologically unless the general form is in combination with other forms. Figure 2.98 is a drawing of wulfenite, $PbMoO_4$. Other mineral representatives are unknown.

FIG. 2.98. Wulfenite.

Tetragonal Axial Ratios

The axial ratio of a tetragonal crystal is expressed as $a:c$, with the length of the two equal a axes taken as unity. It is calculated from ϕ and ρ angles derived from interfacial angles by the general formula

$$c = \frac{l}{k} \tan \rho \cos \phi \qquad \text{where } k \text{ and } l \text{ are Miller}$$

indices.

Tetragonal crystals are so oriented that (010), perpendicular to a_2, has $\phi = 0°$. Thus, for second-order forms, $\{0kl\}$, $\cos \phi = 1$ and the formula becomes: $c = (l/k) \tan \rho$. For the trigonometry involved, consider the calculation of c from angular measurements of face (021), (Fig. 2.99a,b). Tan $\rho_{021} = CO/OA$. $CO = 2c$, $AO = a = 1$. Thus, $c = \tan \rho_{021}/2$. For the unit form $\{011\}$, $c = \tan \rho$.

For first-order tetragonal forms, $\phi = 45°$, $\cos \phi = 0.7071$ and

$$c = \frac{l}{k} \tan \rho \cdot 0.7071.$$

Consider the calculation of c using the angular measurements of face (221), (Fig. 2.99a,c,d). In triangle AOB, $OB = \cos 45°$. In triangle COB, $OC = 2c$, tan $\rho_{221} = OC/OB$. Thus $c = (\tan \rho \cos 45°)/2$.

When (001), face c in Fig. 2.100a, is present, ρ of face p can be measured directly as $c \wedge p$. If (001) is not present (Fig. 2.100b), ρ of face e can be determined as $90° - a(010) \wedge e(011)$ and ρ of face $s = 90° - m (110) \wedge s (111)$. If the pyramidal form and a prism do not lie in a horizontal zone (Fig. 2.100c), measure the interfacial angle over the top

FIG. 2.99. Tetragonal crystal, angular relationships.

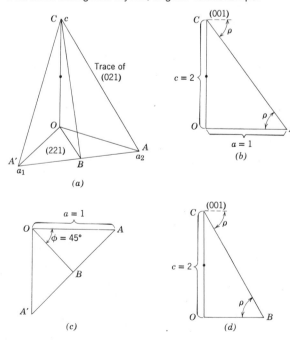

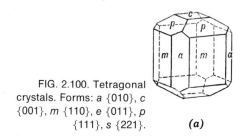

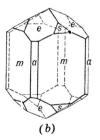

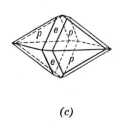

FIG. 2.100. Tetragonal crystals. Forms: a {010}, c {001}, m {110}, e {011}, p {111}, s {221}.

(a)

(b)

(c)

of the crystal as $p \wedge p'$ (where p and p' are faces of the same form differing in ϕ by 180°). ρ of p is one-half this interfacial angle.

Hexagonal System

The 12 crystal classes in the hexagonal system are divided into two groups—the *hexagonal division* and the *rhombohedral division*. The crystal classes in the hexagonal division have a 6-fold axis of either rotation or rotoinversion, whereas the crystal classes in the rhombohedral division have a 3-fold symmetry axis of either rotation or rotoinversion. However, even on well-formed crystals, it may be difficult to determine the crystal class from morphology alone. Rhombohedral crystals may appear hexagonal because of the presence of dominant hexagonal forms; and hexagonal crystals may appear rhombohedral, for a 6-fold axis of rotoinversion ($\bar{6}$) is equivalent to a 3-fold rotation axis with a mirror plane perpendicular to it ($3/m$; see pages 17–19).

Crystallographic Axes. The forms of both hexagonal and rhombohedral divisions are referred to four crystallographic axes as proposed by Bravais.

Three of these, designated a_1, a_2, and a_3, lie in the horizontal plane and are of equal length with angles of 120° between the positive ends; the fourth axis, c, is vertical. When properly oriented, one horizontal crystallographic axis, a_2, is left to right, and the other two make 120° angles on either side of it (Fig. 2.101a). The positive end of a_1 is to the front and left, the positive end of a_2 is to the right, and the positive end of a_3 is to the back and left. Figure 2.101b shows the four axes in clinographic projection. In stating the indices for any face of a hexagonal crystal, four numbers (the Bravais symbol) must be given. The numbers expressing the reciprocals of the intercepts of a face on the axes are given in the order a_1, a_2, a_3, c. Therefore, $(11\bar{2}1)$, which represents the intercepts $3a_1$, $3a_2$, $-3/2a_3$, $3c$, refers to a face that cuts the positive ends of the a_1 and a_2 axes at twice the distance it cuts the negative end of the a_3 axis; it cuts the c axis at the same relative number of units (3) as it cuts the a_1 and a_2 axes. The general Bravais form symbol is $\{hki l\}$ with $h > k$. The third digit of the index is the sum of the first two times -1; or, stated another way, $h + k + i = 0$.

In the Hermann-Mauguin notation the first number refers to the prinicpal axis of symmetry coincident with c. The second and third symbols refer to the axial and intermediate symmetry elements, respectively.

Hexagonal Division

$6/m\ 2/m\ 2/m$—Dihexagonal-Dipyramidal Class

Symmetry—i, $1A_6$, $6A_2$, $7m$. The vertical axis is an axis of 6-fold rotation. There are six horizontal axes of 2-fold rotation, three of them coincident with the

FIG. 2.101. Hexagonal crystal axes.

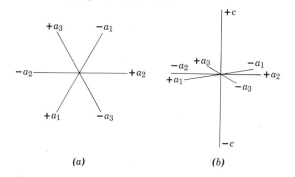

(a)

(b)

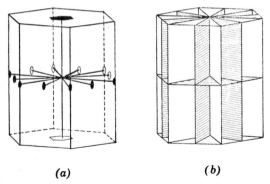

(a)   **(b)**

FIG. 2.102. Symmetry of dihexagonal-dipyramidal class. (a) Axes. (b) Planes.

crystallographic axes and the other three lying midway between them (Fig. 2.102a). There are six vertical mirror planes perpendicular to the 2-fold axes and a horizontal mirror plane (Fig. 2.102b). Figure 2.103 shows a dihexagonal dipyramid and its stereogram.

Forms.

1. *Basal Pinacoid* {0001}. Composed of two horizontal faces. It is shown in combination with various prisms in Fig. 2.104.

2. *Prism of First-Order* {$10\bar{1}0$}. Consists of six vertical faces, each of which intersects two of the horizontal crystallographic axes equally and is parallel to the third (Fig. 2.104a).

3. *Prism of Second-Order* {$11\bar{2}0$}. Six vertical faces, each of which intersects two of the horizontal axes equally and the intermediate horizontal axis at one-half this distance (Fig. 2.104b). The prisms of the first and second order are geometrically identical forms, the distinction between them is only in orientation.

4. *Dihexagonal Prism* {$hki0$}. 12 vertical faces, each of which intersects all three of the horizontal crystallographic axes at different lengths. There are various dihexagonal prisms, depending on their different relations to the horizontal axes. A common dihexagonal prism with indices {$21\bar{3}0$} is shown in Fig. 2.104c.

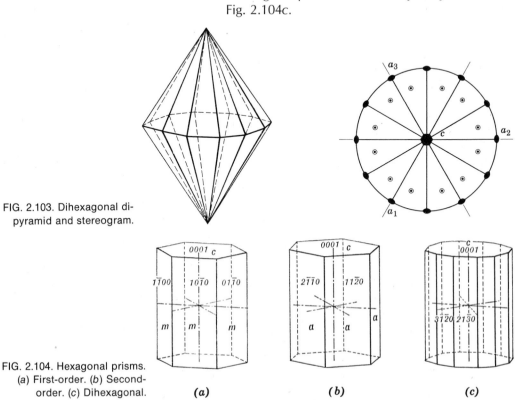

FIG. 2.103. Dihexagonal dipyramid and stereogram.

FIG. 2.104. Hexagonal prisms. (a) First-order. (b) Second-order. (c) Dihexagonal.

(a)   **(b)**   **(c)**

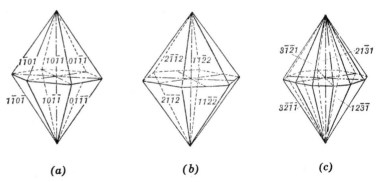

FIG. 2.105. Hexagonal di-
pyramids. (a) First-order.
(b) Second-order.
(c) Dihexagonal.

(a)   **(b)**   **(c)**

5. *Dipyramid of First Order* $\{h\bar{0}hl\}$. Consists of 12 isosceles triangular faces, each of which intersects two horizontal crystallographic axes equally, is parallel to the third, and intersects the vertical axis. Various dipyramids of the first order are possible, depending on the inclination of the faces to the c axis. The unit form has the indices $\{10\bar{1}1\}$ (Fig. 2.105a).

6. *Dipyramid of Second Order* $\{hh\bar{2}hl\}$. Composed of 12 isosceles triangular faces. Each face intersects two of the horizontal axes equally and the third (the intermediate horizontal axis) at one-half this distance; it also intersects the vertical axis. Various dipyramids of the second order are possible, depending on the inclination of the faces to c. A common form (Fig. 2.105b) has the indices $\{11\bar{2}2\}$. The relation between the dipyramids of the first and second order is the same as between the corresponding prisms.

If only one dipyramid is present on a crystal, it is usually set as of the first order. If dipyramids of both orders are present, the dominant one, in the absence of other evidence, is considered the first

order. If a dipyramid is combined with prismatic forms, the orientation of the crystal is usually determined by the dipyramid.

7. *Dihexagonal Dipyramid* $\{hki\bar{l}\}$. Composed of 24 triangular faces. Each face is a scalene triangle that intersects all three of the horizontal axes differently and also intersects the vertical axis. A common form, $\{21\bar{3}1\}$, is shown in Fig. 2.105c.

Combinations of the forms of this class are shown in Fig. 2.106.

Beryl, $Be_3Al_2Si_6O_{18}$, affords the best example of a mineral representative in this class. Other minerals are molybdenite, MoS_2, pyrrhotite, $Fe_{1-x}S$, and niccolite, $NiAs$.

$\bar{6}m2$—Ditrigonal-Dipyramidal Class
Symmetry—$1\bar{A}_6$, $3A_2$, $4m$. The vertical axis is a 6-fold rotoinversion axis which is equivalent to a 3-fold axis of rotation with a horizontal mirror

FIG. 2.106. Combinations of hexagonal forms.

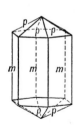
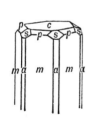
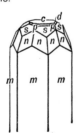

FIG. 2.107. Ditrigonal dipyramid and stereogram.

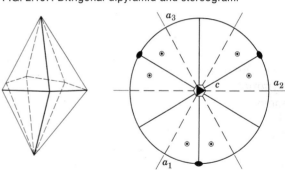

plane. Three mirror planes intersect in the vertical axis, and three horizontal 2-fold axes lie in the vertical mirror planes. A ditrigonal dipyramid and its stereogram are shown in Fig. 2.107.

Forms. The ditrigonal dipyramid {$hki\bar{l}$} is a 12-faced form with six faces at the top of the crystal and six at the bottom. Additional forms that may be present are: pinacoid, trigonal prisms, second-order hexagonal prism, ditrigonal prisms, trigonal dipyramids, and second-order hexagonal dipyramids.

Benitoite, $BaTiSi_3O_9$, is the only mineral that has been described as definitely crystallizing in this class.

6mm—Dihexagonal-Pyramidal Class

Symmetry—$1A_6$, $6m$. There are a vertical 6-fold rotation axis and six vertical mirror planes intersecting in this axis. Figure 2.108 shows a dihexagonal pyramid and its stereogram.

Forms. The forms of the dihexagonal pyramidal class are similar to those of the dihexagonal-dipyramidal class, but, inasmuch as a horizontal mirror plane is lacking, different forms appear at the top and bottom of the crystal. The *dihexagonal pyramid* is thus two forms: {$hki\bar{l}$} upper and {$hki\bar{l}$} lower. The hexagonal-pyramidal forms are: {$h0\bar{h}l$} upper and {$h0\bar{h}l$} lower; and {$hh\bar{2}hl$} upper and {$hh\bar{2}hl$} lower. The pinacoid cannot exist here, but instead there are two pedions {0001} and {$000\bar{1}$}. The first- and second-order hexagonal prisms and the dihexagonal prism may be present.

Wurtzite, ZnS, greenockite, CdS, and zincite, ZnO, are the commonest mineral representatives in this class. Figure 2.109 represents a zincite crystal with a hexagonal prism terminated above by a hexagonal pyramid and below by a pedion.

622—Hexagonal-Trapezohedral Class

Symmetry—$1A_6$, $6A_2$. The symmetry axes are the same as those in the dihexagonal-dipyramidal class (Fig. 2.102a), but there are no mirror planes nor symmetry center.

Forms. The *hexagonal trapezohedrons* {$hki\bar{l}$} *right* and {$ihk\bar{l}$} *left* are enantiomorphic forms, each

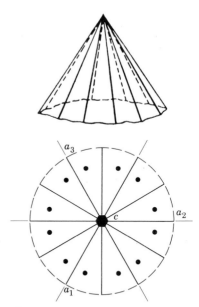

FIG. 2.108. Dihexagonal pyramid and stereogram.

FIG. 2.109. Zincite.

with six trapezium-shaped faces (see Fig. 2.110). Other forms that may be present are the pinacoid, first- and second-order hexagonal prisms, and dipyramids and dihexagonal prisms.

High quartz, SiO_2, and kalsilite, $KAlSiO_4$, are the only mineral representatives in this class.

6/m—Hexagonal-Dipyramidal Class

Symmetry—i, $1A_6$, $1m$. There is a vertical axis of 6-fold rotation, a horizontal mirror plane, and a symmetry center. Figure 2.111 illustrates a hexagonal dipyramid and its stereogram.

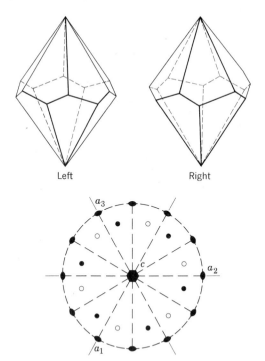

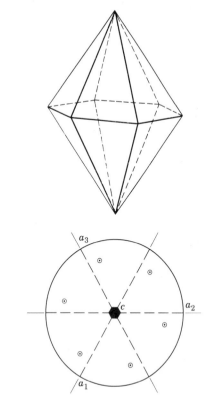

FIG. 2.110. Enantiomorphic, left- and right-handed hexagonal trapezohedrons and a stereogram for a left-handed form.

Forms. The general forms of this class are the dipyramids, {$hk\bar{i}l$} positive, {$hk\bar{i}\bar{l}$} negative. These forms consist of 12 faces, six above and six below, which correspond in position to one-half the faces of a dihexagonal dipyramid. Other forms that may be present are: pinacoid, first- and second-order hexagonal dipyramids and prisms, as well as hexagonal prisms {$hk\bar{i}0$}, {$kh\bar{i}0$}.

The hexagonal-dipyramidal class has as its chief mineral representatives the minerals of the apatite group, $Ca_5(PO_4)_3$ (OH, F, Cl). The dipyramid revealing the symmetry of the class is rarely seen but is illustrated as face μ, Fig. 2.112.

FIG. 2.111. Hexagonal dipyramid and stereogram.

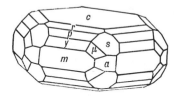

FIG. 2.112. Apatite crystal.

$\bar{6}$—Trigonal-Dipyramidal Class

Symmetry—$1\bar{A}_6(= 1A_3 + 1\,m)$. The vertical axis is a 6-fold axis of rotoinversion ($\bar{6}$) which is equivalent to a 3-fold axis of rotation with a symmetry plane at right angles to it ($3/m$). Figure 2.113 shows a trigonal dipyramid and its stereogram.

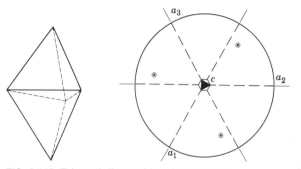

FIG. 2.113. Trigonal dipyramid and stereogram.

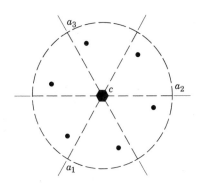

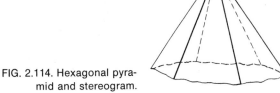

FIG. 2.114. Hexagonal pyramid and stereogram.

Forms. There are four trigonal dipyramids of the general form with six faces each, three above and three below, corresponding to six faces of the dihexagonal dipyramid. The symmetry does not permit hexagonal forms. Thus, instead of first- and second-order hexagonal prisms there are two first-order and two second-order trigonal prisms. Likewise in place of first- and second-order hexagonal dipyramids there are two trigonal dipyramids of each order. The basal pinacoid may be present.

There is no example of an authenticated mineral or other crystalline substance belonging to this crystal class.

6—Hexagonal-Pyramidal Class
Symmetry—$1A_6$. A vertical 6-fold rotation axis is the only symmetry of this class. A hexagonal pyramid and its stereogram are shown in Fig. 2.114.

Forms. The forms of this class are similar to those of the hexagonal-dipyramidal class, but because a horizontal mirror plane is lacking, different forms are present at the top and bottom of the crystal. The hexagonal pyramid is thus four 6-faced forms: upper positive and negative, and lower positive and negative. Likewise the upper first- and second-order hexagonal pyramids at the top of the crystal are different forms from those at the lower end. Present also may be pedions, first- and second-order hexagonal prisms as well as hexagonal prisms, $\{hk\bar{i}0\}$ and $\{kh\bar{i}0\}$.

The form development of crystals is rarely suffi-

cient to enable one without other evidence to place unequivocally a crystal in this class. The mineral nepheline, $(Na, K)AlSiO_4$, is the chief mineral representative.

Rhombohedral Division
The forms of the crystal classes of the rhombohedral division of the hexagonal system are referred to the same four crystallographic axes as the forms in the classes of the hexagonal division. However, some workers have used three axes parallel to the three culminating edges of the unit rhombohedron to describe the forms. All the classes are characterized by a 3-fold axis of either rotation or rotoinversion. The crystals usually show a lower symmetry than those of the hexagonal division, and are generally recognized by a 3-fold distribution of faces at the ends of the principal axis. In the sections on systematic mineralogy the crystal system of minerals crystallizing in the rhombohedral division is designated as *Hexagonal-R*.

$\bar{3}\ 2/m$—Hexagonal-Scalenohedral Class
Symmetry—$1\bar{A}_3$, $3A_2$, $3m$. The vertical crystallographic axis is one of 3-fold rotoinversion and the three horizontal crystallographic axes are axes of 2-fold rotation (see Fig. 2.115a). There are three vertical mirror planes bisecting the angles between the

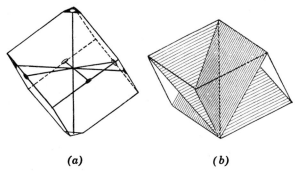

(a)   **(b)**

FIG. 2.115. Symmetry of hexagonal scalenohedral class. (*a*) Aces. (*b*) Planes.

horizontal axes (see Fig. 2.115*b*). Figure 2.116 illustrates a hexagonal scalenohedron and its stereogram.

Forms.

1. *Rhombohedron* {$h0\bar{h}l$} *positive,* {$0h\bar{h}l$} *negative.* The rhombohedron is a form consisting of six rhomb-shaped faces, which correspond in their positions to the alternate faces of a hexagonal dipyramid of the first-order. The relation of these two forms to each other is shown in Fig. 2.117. The rhombohedron may also be thought of as a cube deformed in the direction of one of the axes of 3-fold rotoinversion. The deformation may appear either as an elongation along the rotoinversion axis producing an acute solid angle, or compression along the rotoinversion axis producing an obtuse solid angle. Depending on the angle, the rhombohedron is known as acute or obtuse.

Depending on the orientation the rhombohe-

FIG. 2.116. Hexagonal scalenohedron and stereogram.

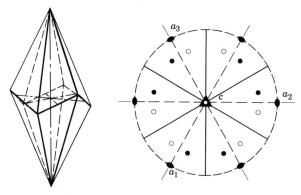

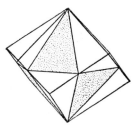

FIG. 2.117. Relation between rhombohedron and first-order hexagonal dipyramid.

dron may be positive or negative (Fig. 2.118). When properly oriented, the positive rhombohedron has one of its faces, and the negative rhombohedron one of its edges, toward the observer. There are various rhombohedrons that differ from each other in the inclination of their faces to the *c* axis. The index symbol of the unit positive rhombohedron is {$10\bar{1}1$} and of the unit negative rhombohedron {$01\bar{1}1$}.

The rhombohedron is such an important form in the hexagonal system that it need not appear externally on a crystal to determine the orientation. The orientation of calcite is determined by the rhombohedral cleavage, and the orientation of corundum is determined by rhombohedral parting. Thus, in calcite the only external rhombohedral form may be negative, and in corundum the rhombohedral parting invariably necessitates orienting the crystal so that the prism is of the second order. However, if but one rhombohedron is present on a crystal, it is oriented, in the absence of other evidence, in the positive position.

2. *Scalenohedron* {$hki\bar{l}$} *positive,* {$kh\bar{i}l$} *negative.* This form consists of 12 scalene triangular faces corresponding in position to alternate *pairs* of faces

FIG. 2.118. Rhombohedrons. (*a*) Positive. (*b*) Negative.

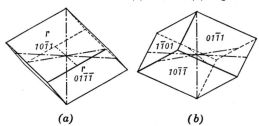

(a)   **(b)**

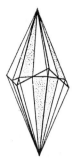

FIG. 2.119. Relation of dihexagonal dipyramid and scaleno-hedron.

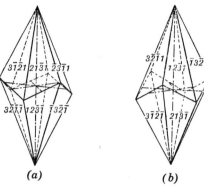

(a)   *(b)*

FIG. 2.120. Hexagonal scalenohedrons. (a) Positive. (b) Negative.

of a dihexagonal dipyramid (Fig. 2.119). The scale-nohedron is differentiated from the dipyramid by the zigzag appearance of the middle edges. It is in the positive position when the angle between the upper and lower faces points down toward the observer (Fig. 2.120a) and in the negative position when the angle points up (Fig. 2.120b).

There are many different scalenohedrons; most frequently seen is {2131}, a common form on cal-cite (Fig. 2.120a).

The rhombohedron and scalenohedron of the

hexagonal-scalenohedral class may combine with forms found in classes of higher hexagonal sym-metry. Thus they may be in combination with the first- and second-order hexagonal prisms, dihex-agonal prisms, second-order dipyramid, and basal pinacoid (Fig. 2.121). (See also Figs. 2.122 and 2.123.)

Several common minerals crystallize in this class. Chief among them is calcite and the other

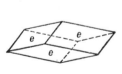

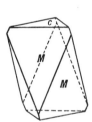

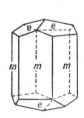

FIG. 2.121. Calcite crystals. Forms: e {01$\bar{1}$2} and f {02$\bar{2}$1}, negative rhombohedrons. r {10$\bar{1}$0} and M {40$\bar{4}$1} positive rhombohedrons. m {10$\bar{1}$0}, first-order prism. c {0001}, basal pinacoid. v {2131}, scalenohedron.

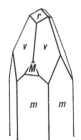

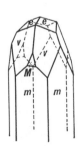

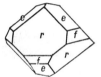

FIG. 2.122. Chabazite.

FIG. 2.123. Corundum.

members of the calcite group. Other minerals are corundum, hematite, brucite, soda niter, arsenic, millerite, antimony, and bismuth.

3m—Ditrigonal-Pyramidal Class

Symmetry—$1A_3$, $3m$. The vertical axis is a 3-fold rotation axis, and there are three mirror planes that intersect in this axis. A ditrigonal pyramid and its stereogram are shown in Fig. 2.124.

 Forms. The forms are similar to those of the hexagonal scalenohedral class but with only half the number of faces. Because of the lack of 2-fold rotation axes, the faces at the top of the crystals belong to different forms from those at the bottom. There are four ditrigonal pyramids, each corresponding to half the faces of either the positive or negative scalenohedron with indices: $\{hki\bar{l}\}$, $\{khi\bar{l}\}$, $\{hki\bar{l}\}$, $\{khi\bar{l}\}$. Other forms that may be present are pedions, second-order hexagonal prisms and pyramids, trigonal pyramids, trigonal prisms, and ditrigonal prisms. There are four trigonal pyramids; two of these are at the top of the crystal, one corresponding to the top three faces of the positive rhombohedron, the other corresponding to the top three faces of the negative rhombohedron. Their respective indices are $\{h0\bar{h}l\}$ and $\{0h\bar{h}l\}$. The two trigonal pyramids at the bottom of the crystal correspond to the bottom faces of the rhombohedrons with indices $\{0h\bar{h}l\}$ and $\{h0\bar{h}l\}$. The trigonal prisms $\{10\bar{1}0\}$ and $\{01\bar{1}0\}$ each corresponds to three faces of the first-order hexagonal prism. There are two ditrigonal prisms,

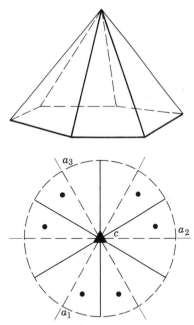

FIG. 2.124. Ditrigonal pyramid and stereogram.

$\{hki\bar{0}\}$ and $\{khi\bar{0}\}$, each corresponding to half the faces of a dihexagonal prism.

 Tourmaline (Fig. 2.125) is the most common mineral crystallizing in this class, but in addition there are pyrargyrite, proustite, and alunite.

32—Trigonal-Trapezohedral Class

Symmetry—$1A_3$, $3A_2$. The vertical crystallographic axis is an axis of 3-fold rotation and the three horizontal crystallographic axes are axes of 2-fold rota-

FIG. 2.125. Tourmaline crystals: Forms: $r = \{10\bar{1}1\}$, o $\{02\bar{2}1\}$, t $\{21\bar{3}1\}$, c $\{0001\}$, a $\{11\bar{2}0\}$, M $\{10\bar{1}0\}$, m $\{01\bar{1}0\}$, e $\{10\bar{1}2\}$, r' $\{01\bar{1}1\}$.

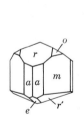
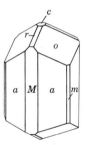
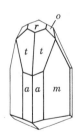

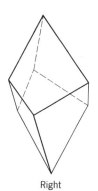
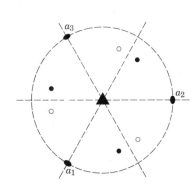

FIG. 2.126. Trigonal trapezohedrons (positive left and
positive right) and a stereogram for the positive right
form.

Left   Right

tion. The symmetry axes are the same as in the scalenohedral class, but planes of symmetry are lacking. Figure 2.126 shows a positive left and a positive right trigonal trapezohedron and a stereogram for the positive right form.

Forms. There are four trigonal trapezohedrons, each made up of six trapezium-shaped faces. These faces correspond in position to one-quarter of the faces of a dihexagonal dipyramid and thus have similar symbols, as follows: positive right $\{hkil\}$, positive left $\{ikhl\}$, negative left $\{khil\}$, negative right $\{kihl\}$. These forms can be grouped into two enantiomorphic pairs each with a *right* and *left* form (see Fig. 2.126). Other forms that may be present are: pinacoid, first-order hexagonal prism, ditrigonal prisms, and rhombohedrons. The second-order hexagonal prism is not present, but instead there are two trigonal prisms $\{11\bar{2}0\}$ and $\{2\bar{1}\bar{1}0\}$. Likewise the second-order hexagonal dipyramid becomes two trigonal dipyramids $\{hh\bar{2}hl\}$ and $\{2h\bar{h}\bar{h}l\}$.

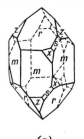

FIG. 2.127. Quartz. (a) Right-hand. (b) Left-hand.

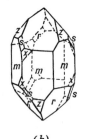

(a)   (b)

Low-temperature quartz is the most common mineral crystallizing in this class, but only rarely can faces of the trigonal trapezohedron be observed. When this form is present, the crystals can be distinguished as right-handed or left-handed (Fig. 2.127), depending on whether, with a prism face fronting the observer, the trigonal trapezohedral faces, x, truncate the edges between prism and the top rhombohedron faces at the right or at the left. The faces marked s are trigonal dipyramids.

Cinnabar, HgS, and the rare mineral berlinite, $AlPO_4$, also crystallize in the trigonal trapezohedral class.

$\bar{3}$—Rhombohedral Class

Symmetry—$1\bar{A}_3$. There is one 3-fold axis of rotoinversion. This is equivalent to a 3-fold rotation axis and a center of symmetry. Figure 2.128 illustrates a rhombohedron and its stereogram.

Forms. As general forms of this class there are four different rhombohedrons, each corresponding to six faces of the dihexagonal-dipyramid. If one of these appeared alone on a crystal, it would have the morphological symmetry of class $\bar{3}2/m$. It is only in combination with other forms that its true symmetry becomes apparent. In addition to the general forms there are the special positive and negative rhombohedrons of the hexagonal-scalenohedral class. The basal pinacoid, first- and second-order hexagonal prisms and the hexagonal prisms $\{hki0\}$, $\{ikh0\}$ may be present.

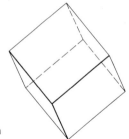
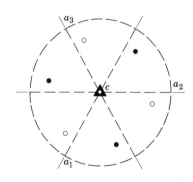

FIG. 2.128. Rhombohedron
and stereogram.

Dolomite, $CaMg(CO_3)_2$, is the most common mineral crystallizing in this class; other representatives are ilmenite, $FeTiO_3$, willemite, Zn_2SiO_4, and phenacite, Be_2SiO_4.

Possibly the mineral gratonite, $Pb_9As_4S_{15}$, belongs in this class; there are no other mineral representatives.

3—Trigonal-Pyramidal Class

Symmetry—$1A_3$. One 3-fold rotation axis is the only symmetry. Figure 2.129 shows a trigonal pyramid and its stereogram.

Forms. There are eight trigonal pyramids of the general form, $\{hkil\}$, four above and four below, each corresponding to three faces of the dihexagonal dipyramid. In addition there are two first-order and two second-order trigonal pyramids above and equivalent, but independent pyramids below. Pedions and several different trigonal prisms may be present. In combination with a pedion a trigonal pyramid appears to have the symmetry of the ditrigonal-pyramidal class ($3m$) with three vertical symmetry planes (Fig. 2.124). It is only when certain trigonal pyramids are in combination with one another that the true symmetry is revealed.

Axial Ratios in the Hexagonal System

With the exception of the hexagonal system, crystals are oriented so that (010) is to the right with $\phi = 0°$ 00'. In the hexagonal system the negative end of the a_3 axis is taken as $\phi = 0°$. According to this convention ϕ of second-order forms is 0°, whereas ϕ of first-order forms is 30°. This apparent inconsistency has definite advantages in working with rhombohedral crystals; it gives positive forms $+\phi$ values and negative forms $-\phi$ values (See Fig. 2.130).

To determine an axial ratio of a crystal, the forms must first be indexed and their ϕ and ρ angles determined. For many crystals ϕ and ρ can be measured directly as interfacial angles. For other crystals it may be necessary to project the measurable interfacial angles and determine ϕ and ρ from the projection (see page 57).

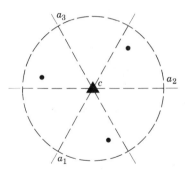

FIG. 2.129. Trigonal pyramid
and stereogram.

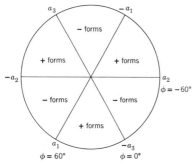

FIG. 2.130. Distribution of rhombohedral forms.

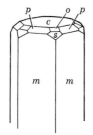

FIG. 2.131. Beryl.

Consider the drawing of the beryl crystal, Fig. 2.131. It is oriented so that m and p are first-order forms and therefore $\phi = 30°$. o and s are second-order forms with $\phi = 0°$. The angles $c \wedge p$, $c \wedge o$, $c \wedge s$ are, respectively, the ρ angles of p, o, and s. If c were not present the ρ angle of p could be determined as the complement of $m \wedge p$. If only faces of form p were present, an interfacial angle could be measured over the top of the crystal between p and p' (p' a face at the back of the crystal 180° removed from p). ρp equals half of this measured angle.

An axial ratio in the hexagonal system expresses the length of c in terms of a as unity or $a:c = 1:?$. For ease of calculations the $-a_3$ axis is taken as unity since it is at the position of $\phi = 0°$. The reciprocal of the intercept on this axis is $i = -(h + k)$.

The formula for determining c from the angles of the general form is:

$$c = \frac{l \tan \rho_{hk\bar{i}l} \cos \phi}{h + k}$$

FIG. 2.132. First- and second- order hexagonal forms.

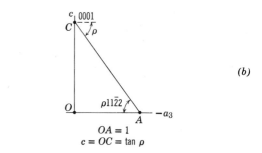

(a)

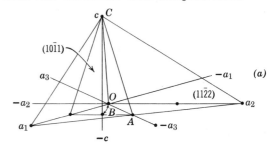

(b)

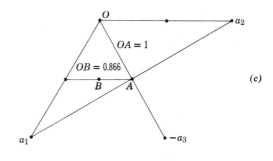

(c)

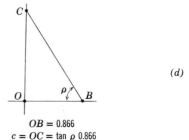

(d)

If a second-order form is used, $\phi = 0°$, and $\cos \phi = 1$; hence $\cos \phi$ disappears from the equation, leaving:

$$c = \frac{l \tan \rho_{hh\bar{2}hl}}{h + k}$$

A face of the second-order form $\{11\bar{2}2\}$ intersects $-a_3$ and c at unit distances but a_1 and a_2 at twice unity (Fig. 2.132a). Thus for this form $l = 2$, $h + k = 2$ and $c = \tan \rho$. For the form $\{11\bar{2}1\}$, $c = \tan \rho/2$.

For a first-order form: $\phi = 30°$, $\cos \phi = 0.8660$ and the equation becomes:

$$c = \frac{l \tan \rho_{ho\bar{h}l} \times 0.8660}{h + k}$$

ISOMETRIC SYSTEM

Crystallographic Axes. The crystal forms of all the classes of the isometric system are referred to three axes of equal length that make right angles with each other. Because the axes are identical, they are interchangeable, and all are designated by the letter a. When properly oriented, one axis, a_1, is horizontal and oriented front to back, a_2 is horizontal and right to left, and a_3 is vertical (see Fig. 2.133).

Form Symbols. Although the symbol of any face of a crystal form might be used as the form symbol, it is conventional, when possible, to use one in which $h, k,$ and l are all positive. In forms that have two or more faces with $h, k,$ and l positive the rule followed is to take the form symbol with $h < k < l$. For example, the form with a face symbol (123) also has faces with symbols (132),

(213), (231), (312), and (321). Following the rule, $\{123\}$ would be taken as the form symbol, for here $h < k < l$.

In giving the angular coordinates of a form, it is customary to give those for only one face; the others can be determined knowing the symmetry. The face for which these coordinates are given is the one with the smallest ϕ and ρ values. This is the face of the form in which $h < k < l$.

$4/m\,\bar{3}\,2/m$—Hexoctahedral Class

Symmetry—$3A_4, 4\bar{A}_3, 6A_2, 9m$. The three crystallographic axes are axes of 4-fold rotation (see Fig. 2.134a). There are also four diagonal axes of 3-fold rotoinversion. These axes emerge in the middle of each of the octants formed by the intersection of the crystallographic axes (Fig. 2.134b). Further, there are six diagonal axes of 2-fold rotation, each of

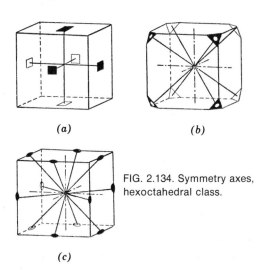

(a)   (b)

FIG. 2.134. Symmetry axes, hexoctahedral class.

(c)

FIG. 2.135. Symmetry planes, hexoctahedral class.

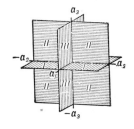

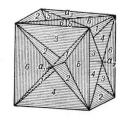

FIG. 2.133. Isometric crystal axes.

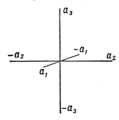

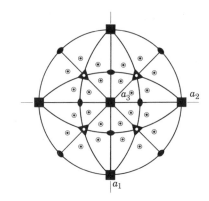

FIG. 2.136. Hexoctahedron
and stereogram.

which bisects one of the angles between two of the crystallographic axes, as illustrated in Fig. 2.134c. There is also a center of symmetry.

This class has nine mirror planes: three of them are known as the axial planes, because each includes two crystallographic axes (Fig. 2.135a); and six are called diagonal planes, because each bisects the angle between two of the axial planes (Fig. 2.135b). This combination of symmetry elements defines the highest symmetry possible in crystals. Every crystal form and every combination of forms that belong to this class must show its complete symmetry. It is important to remember that in this class the three crystallographic axes are axes of 4-fold rotation. Thus one can easily locate the crystallographic axes and properly orient the crystal.

The hexoctahedron, the general form from which the class derives its name, is shown in Fig. 2.136 with a stereogram.

Forms.

1. *Cube* or *Hexahedron* {001}. The cube is composed of six square faces that make 90° angles with each other. Each face intersects one of the crystallographic axes and is parallel to the other two (Fig. 2.137).

2. *Octahedron* {111}. The octahedron is composed of eight equilateral triangular faces, each of which intersects all three of the crystallographic axes equally. Figure 2.138 represents a simple octahedron, and Fig. 2.139 shows combinations of a cube and an octahedron. When in combination the octahedron can be recognized by its eight similar faces, each of which is equally inclined to the three

crystallographic axes. It is to be noted that the faces of an octahedron truncate symmetrically the corners of a cube.

3. *Dodecahedron* or *Rhombic Dodecahedron* {011}. The dodecahedron is composed of 12 rhomb-shaped faces. Each face interesects two of the crystallographic axes equally and is parallel to the third. Figure 2.140 shows a simple dodecahedron. Fig. 2.141 shows combinations of a dodecahedron and a cube, of a dodecahedron and an octahedron, and of a cube, octahedron, and a dodecahedron. Note that the faces of a dodecahedron truncate the edges of both the cube and the octahedron. The dodecahedron is sometimes called the rhombic dodecahedron to distinguish it from the pentagonal dodecahedron and the regular geometrical dodecahedron.

4. *Tetrahexahedron* {0kl}. The tetrahexahedron is composed of 24 isosceles triangular faces, each of which intersects one axis at unity and the second at some multiple and is parallel to the third. There are a number of tetrahexahedrons that differ from each other in respect to the inclination of their faces. The commonest is {012}. The indices of other forms are {013}, {014}, {023}, etc., or in general,

FIG. 2.137. Cube.

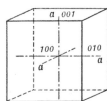

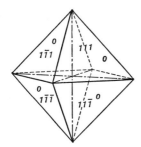

FIG. 2.138. Octahedron.

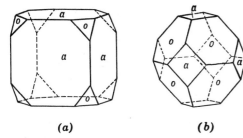

FIG. 2.139. Cube and octahedron.

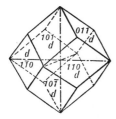

FIG. 2.140. Dodecahedron.

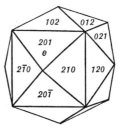

FIG. 2.142. Tetrahexadron.

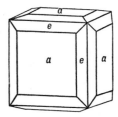

FIG. 2.143. Cube and tetrahexahedron.

{0kl}. It is helpful to note that the tetrahexahedron, as its name indicates, resembles a cube each of the faces of which have been raised to accommodate four others. Figure 2.142 shows a simple tetrahexahedron, and Fig. 2.143 shows a cube with its edges beveled by the faces of a tetrahexahedron.

5. *Trapezohedron* or *Tetragonal Trisoctahedron* {hhl}. The trapezohedron is composed of 24 trapezium-shaped faces, each of which intersects one of the crystallographic axes at unity and the other two at equal multiples. There are various trapezohedrons with their faces having different angles of inclination, but the most common is {112} (Fig. 2.144). This form is sometimes called a tetragonal trisoctahedron to indicate that each of its

FIG. 2.141. Combinations of cube and dodecahedron (a), octahedron and dodecahedron (b), and cube, octahedron, and dodecahedron (c).

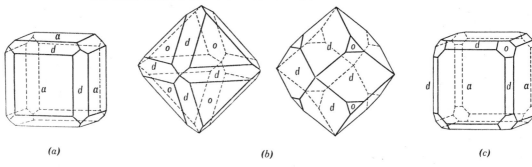

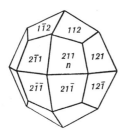

FIG. 2.144. Trapezohedron.

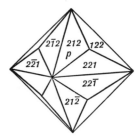

FIG. 2.146. Trisoctahedron.

faces has four edges and to distinguish it from another 24-faced form, the trigonal trisoctahedron.

Figure 2.145 shows the common trapezohedron *n* {112} truncating the edges of the dodecahedron. Both forms by themselves and in combination are common on the mineral garnet.

6. *Trisoctahedron* or *Trigonal Trisoctahedron* {*hll*}. The trisoctahedron is composed of 24 isosceles triangular faces, each of which intersects two of the crystallographic axes at unity and the third axis at some multiple. There are various trisoctahedrons, the faces of which have different inclinations, but most common is {122} (Fig. 2.146). The trisoctahedron, like the trapezohedron, is a form that may be conceived as an octahedron, each face of which has been raised to accommodate three others. It is frequently spoken of as the trigonal trisoctahedron, indicating that its faces have three edges and so differ from those of the trapezohedron. Figure 2.147 shows a combination of an octahedron and trisoctahedron.

7. *Hexoctahedron* {*hkl*}. The hexoctahedron is composed of 48 triangular faces, each of which intersects all three crystallographic axes at different lengths. There are several hexoctahedrons that have varying ratios of axial intercepts. A common hexoc-

tahedron has indices {123}. Other hexoctahedrons have indices {124}, {135}, etc., or, in general, {*hkl*}. Figure 2.148 shows a simple hexoctahedron; Fig. 2.149 shows a combination of dodecahedron and hexoctahedron (a) and a combination of a dodecahedron, trapezohedron, and hexoctahedron (b).

Determination of Indices of Forms. In determining the forms present on any crystal of the hexoctahedral class, it is first necessary to locate the crystallographic axes (axes of 4-fold symmetry). Once the crystal has been oriented by these axes, the faces of the cube, dodecahedron, and octahedron are easily recognized, because they intersect respectively one, two, and three axes at unit distances. The indices can be quickly obtained for faces of other forms that truncate symmetrically the edges between known faces. The algebraic sums of the *h*, *k*, and *l* indices of two faces give the indices of the face symmetrically truncating the edge between them. Thus, in Fig. 2.150 the algebraic sum of the two dodecahedron faces (101) and (011) is (112), or the indices of a face of a trapezohedron.

Occurrence of Isometric Forms of the Hexoctahedral Class. The cube, octahedron, and dodecahedron are the commonest isometric forms. The trapezohedron is also frequently observed as the

FIG. 2.145. Dodecahedron and trapezohedron.

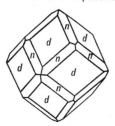

FIG. 2.147. Octahedron and trisoctahedron.

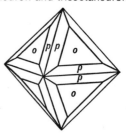

FIG. 2.148. Hexoctahedron.

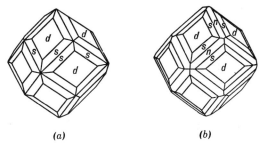

(a)　　　　　　　　　　(b)

FIG. 2.149(a). Dodecahedron and hexoctahedron. (b) Dodecahedron, trapezohedron, and hexoctahedron.

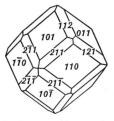

FIG. 2.150. Dodecahedron and trapezohedron.

only form on a few minerals. The other forms, the tetrahexahedron, trisoctahedron, and hexoctahedron, are rare and are ordinarily observed only as small truncations in combinations.

A large group of minerals crystallize in the hexoctahedral class. Some of the most common are:

analcime	galena	leucite
cerargyrite	garnet	silver
copper	gold	spinel group
cuprite	halite	sylvite
fluorite	lazurite	uraninite

$\bar{4}3m$—Hextetrahedral Class

Symmetry—$3\bar{A}_4$, $4A_3$, $6m$. The three crystallographic axes are axes of 4-fold rotoinversion. The four diagonal axes are axes of 3-fold rotation (Fig. 2.151a). There are six diagonal mirror planes (Fig. 2.151b), the same planes shown in Fig. 2.135b for the hexoctahedral class. The general form, the hextetrahedron, is illustrated in Fig. 2.152.

Forms.

1. *Tetrahedron* {111} *positive*, {1$\bar{1}$1} *negative*. The tetrahedron is composed of four equilateral triangular faces, each of which intersects all the crystallographic axes at equal lengths. It can be considered as derived from the octahedron of the hexoctahedral class by the omission of the alternate faces and the extension of the others, as shown in Fig. 2.153. This form, shown also in Fig. 2.154a is known as the positive tetrahedron {111}. If the other four faces of the octahedron had been extended, the tetrahedron resulting would have had a

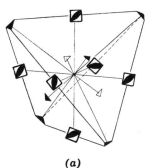

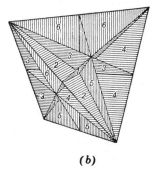

(a)　　　　　　　　　　(b)

FIG. 2.151. Symmetry of hextetrahedral class. (a) Axes. (b) Planes.

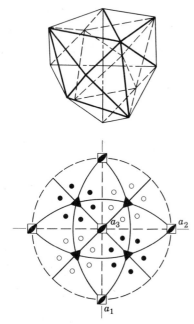

FIG. 2.152. Hextetrahedron and stereogram.

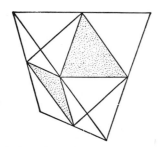

FIG. 2.153. Relation of octahedron and tetrahedron.

2. *Tristetrahedron* $\{hhl\}$ *positive,* $\{h\bar{h}l\}$ *negative.* These forms have 12 faces that correspond to one-half of the faces of a trapezohedron (Fig. 2.155) taken in alternating groups of three above and three below. The positive form may be made negative by a rotation of 90° about the vertical axis.

3. *Deltoid Dodecahedron* $\{hll\}$ *positive,* $\{h\bar{l}l\}$ *negative.* This is a 12-faced form in which the faces correspond to one-half of those of the trisoctahedron (in the hexoctahedral class) taken in alternate groups of three above and three below (Fig. 2.156).

4. *Hextetrahedron* $\{hkl\}$ *positive,* $\{h\bar{k}l\}$ *negative.* The hextetrahedron (Fig. 2.157) has 24 faces that correspond to one-half of the faces of the hexoctahedron taken in groups of six above and six below.

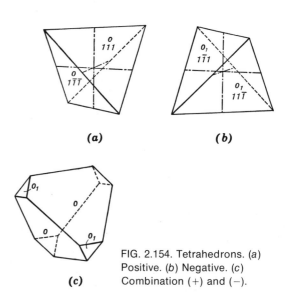

(a)   (b)

(c)

FIG. 2.154. Tetrahedrons. (a) Positive. (b) Negative. (c) Combination (+) and (−).

different orientation, as shown in Fig. 2.154b. This is the negative tetrahedron, $\{1\bar{1}1\}$. The positive and negative tetrahedrons are geometrically identical. The existence of both must be recognized for they may occur together as shown in Fig. 2.154c. If the positive and negative tetrahedron are equally developed on the same crystal, the combination could not be distinguished from an octahedron unless, as often happens, the faces of the two forms showed different lusters, etchings, or striations that would serve to differentiate them.

FIG. 2.155. Tristetrahedron.

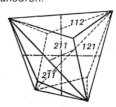

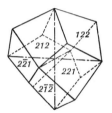

FIG. 2.156. Deltoid dodecahedron.

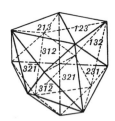

FIG. 2.157. Positive hextetrahedron.

The cube, dodecahedron, and tetrahexahedron are also present on crystals of the hextetrahedral class. Figure 2.158 shows combinations of cube and tetrahedron (a), tetrahedron and dodecahedron (b),

FIG. 2.158. Combination of cube and tetrahedron (a), of tetrahedron and dodecahedron (b), and of dodecahedron, cube, and tetrahedron (c).

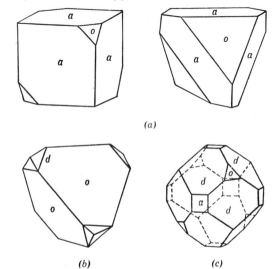

(a)

(b)   (c)

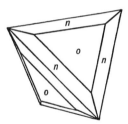

FIG. 2.159. Tetrahedron and tristetrahedron.

and combinations of cube, dodecahedron, and tetrahedron (c). Figure 2.159 shows a combination of tetrahedron and tristetrahedron.

Members of the tetrahedrite-tennantite series, $(Cu, Fe, Zn, Ag)Sb_4S_{13}$ to $(Cu, Fe, Zn, Ag)_{12}As_4S_{13}$, are the only common minerals that ordinarily show distinct hextetrahedral forms. Sphalerite, ZnS, occasionally exhibits them, but commonly its crystals are complex and malformed.

432—Gyroidal Class

Symmetry—$3A_4, 4A_3, 6A_2$. The gyroidal class has all the symmetry axes of the hexoctahedral class but lacks mirror planes and a center of symmetry (Fig. 2.160).

Forms. The *gyroid* $\{hkl\}$ right, $\{khl\}$ left. These two forms each have 24 faces and are enantiomorphic with a left-handed and a right-handed form (see Fig. 2.160). All the forms of the hexoctahedral class, with the exception of the hexoctahedron, can be present in the gyroidal class.

For many years cuprite was considered gyroidal, but more recent studies have shown that it is probably hexoctahedral. With the elimination of cuprite, no known mineral crystallizes in the gyroidal class.

2/m$\bar{3}$—Diploidal Class

Symmetry—$3A_2, 4\bar{A}_3, 3m$. The three crystallographic axes are axes of 2-fold rotation; the four diag-

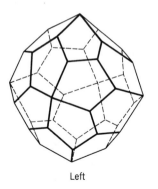

Left

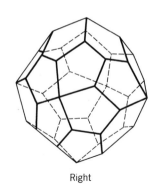

Right

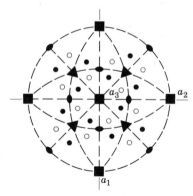

FIG. 2.160. Enantiomorphic forms of the gyroid, with left and right forms, and a stereogram for the left-handed form. ·

onal axes, each of which emerges in the middle of an octant, are axes of 3-fold rotoinversion; the three axial planes are mirror planes (Figs. 2.161a and b). Figure 2.162 illustrates a positive diploid and its stereogram.

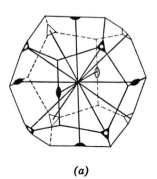

(a)

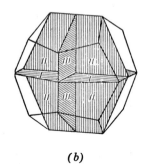

(b)

FIG. 2.161. Symmetry of diploid class. (a) Axis. (b) Planes.

Forms.

1. *Pyritohedron or Pentagonal Dodecahedron* $\{h0l\}$ *positive,* $\{0kl\}$ *negative.* This form consists of 12 pentagonal faces, each of which intersects one crystallographic axis at unity, intersects the second axis at some multiple of unity, and is parallel to the third. A rotation of 90° about a crystallographic axis brings the positive pyritohedron into the negative position. There are a number of pyriotchedrons that differ from each other with respect to the inclination of their faces. The most common positive pyritohedron has indices $\{102\}$ (Fig. 2.163a). Figure 2.163b shows the corresponding negative pyritohedron.

2. *Diploid* $\{hkl\}$ *positive,* $\{khl\}$ *negative.* The diploid is a rare form composed of 24 faces that correspond to one-half of the faces of a hexoctahedron. The diploid may be pictured as having two faces

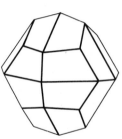

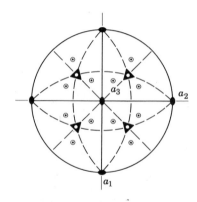

FIG. 2.162. Diploid and stereogram.

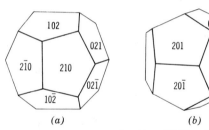

FIG. 2.163. Pyritohedron. (a) Positive. (b) Negative.

built up on each face of the pyritohedron. As in the case of the pyritohedron, a rotation of 90° about one of the crystallographic axes brings the positive diploid into the negative position.

In addition to the pyritohedron and the diploid, there may be present the cube, dodecahedron, octahedron, trapezohedron, and trisoctahedron (see hexoctahedral class). On some crystals these forms may appear alone and so perfectly developed that they cannot be distinguished from the forms of the hexoctahedral class. This is often true of octahedrons and cubes of pyrite. Usually, however, they will show by the presence of striation lines or etching figures that they conform to the symmetry of the diploidal class. This is shown in Fig. 2.164 which

FIG. 2.164. Striated cube.

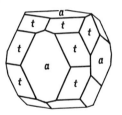

FIG. 2.166. Diploid and cube.

represents a cube of pyrite with characteristic striations showing the lower symmetry. Figure 2.165 shows combinations of the pyritohedron with forms of the hexoctahedral class. Figure 2.166 represents a combination of cube and diploid {124}.

The chief mineral of the diploidal class is pyrite, FeS_2; other rarer minerals of this class are skutterudite, chloanthite, gersdorffite, and sperrylite.

23—Tetartoidal Class

Symmetry—$3A_2$, $4A_3$. The three crystallographic axes are axes of 2-fold rotation and the four diagonal axes are axes of 3-fold symmetry. Figure 2.167 shows drawings of the positive left and positive right tetartoids and a stereogram of a positive right form.

Forms. There are four separate forms of the tetartoid. These are: positive right {hkl}, positive left {khl}, negative right {$k\bar{h}l$}, negative left {$h\bar{k}l$}. They comprise two enantiomorphic pairs, positive right and left, and negative right and left. Other forms that may be present are the cube, dodecahedron (see hexoctahedral class), pyritohedron (see di-

FIG. 2.165. Combinations of pyritohedron with forms of the hexoctahedral class. (a) Cube and pyritohedron. (b) Octahedron and pyritohedron. (c) Pyritohedron, cube, and octahedron.

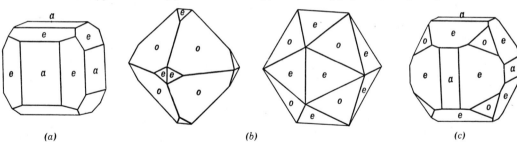

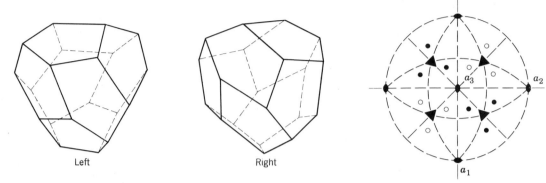

Left   Right

FIG. 2.167. Enantiomorphic forms of the tetartoid. Positive left and positive right and a stereogram of the positive right form.

ploidal class), tetrahedron and deltoid dodecahedron (see hextetrahedral class).

Cobaltite, (Co, Fe)AsS, and ullmanite, NiSbS, are the most common mineral representatives crystallizing in the tetartoidal class.

Characteristics of Isometric Crystals

Four 3-fold symmetry axes are common to all isometric crystals. Symmetrically developed crystals are equidimensional in the three directions of the crystallographic axes. The crystals commonly show faces that are squares, equilateral triangles, or these figures with truncated corners. All forms are *closed forms*. Thus, crystals are characterized by a large number of similar faces; the smallest number of any form of the hexoctahedral class is six.

Some important *interfacial angles* of the isometric system that may aid in the recognition of the commoner forms are as follows:

Cube (100) $\wedge$ cube (010) = 90°00′
Octahedron (111) $\wedge$ octahedron ($\overline{1}$11) = 70°32′
Dodecahedron (011) $\wedge$ dodecahedron (101) = 60°00′
Cube (100) $\wedge$ octahedron (111) = 54°44′
Cube (100) $\wedge$ dodecahedron (110) = 45°00′
Octahedron (111) $\wedge$ dodecahedron (110) = 35°16′

TWIN CRYSTALS

All but a very small percentage of minerals occur as random aggregates of mineral grains in the rocks of the earth's crust. These grains generally are anhedral (lacking external faces) but, being crystalline, they possess an internal order evidenced by cleavage, optical properties, and X-ray diffraction. It is the rare mineral that is found bounded by well-developed crystal faces.

Under certain conditions two or more mineral grains or crystals form a rational, symmetrical intergrowth. Such crystallographically controlled intergrowths are called *twins* or *twinned crystals*. The lattice directions of one unit in a twin bear a definite crystallographic relation to the lattice directions of the other. Figure 2.168 illustrates the possible twinned relationship versus the untwinned pattern in an orthorhombic lattice. The operations that may relate a crystal to its twinned counterpart are: (1) Reflection by a mirror plane (*twin plane*). (2) Rotation about a crystal direction common to both (*twin axis*). Although there are exceptions, the angular rotation is usually 180°. (3) Inversion about a point (*twin center*). Twinning is defined by a *twin law*, which states whether there is a center, an axis, or a plane of twinning, and gives the crystallographic orientation for the axis or plane.

The surface on which two individuals are united is known as the *composition surface*. If this

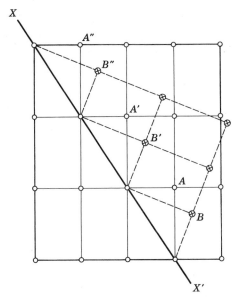

FIG. 2.168. Cross-section through an orthorhombic lattice showing a twinned relationship (shown by *x* in nodes) about the direction *XX'*. Along this direction atoms (or lattice nodes) are compatible with the regular as well as the twinned relationship. The twin may be viewed as the result of a 180° rotation about *XX'*, or of a mirror along *XX'*.

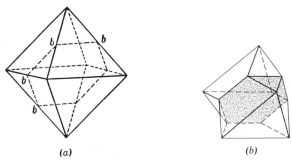

FIG. 2.169(*a*). Octahedron with possible twin plane *b-b*. (*b*) Octahedron twinned on {111}, spinel twin.

Repeated or *multiple twins* are made up of three or more parts twinned according to the same law. If all the successive composition surfaces are parallel, the resulting group is a *polysynthetic twin* (Figs. 2.171 and 2.172). If successive composition planes are not parallel, a *cyclic* twin results (Fig. 2.173). When a large number of individuals in a polysynthetic twin are closely spaced, crystal faces or cleavages crossing the composition planes show striations owing to the reversed position of adjacent individuals.

Twinning in the lower symmetry groups generally produces a resulting aggregate symmetry higher

surface is a plane, it is called the *composition plane*. The composition plane is commonly, but not invariably, the twin plane. However, if the twin law can be defined only by a twin plane, the twin plane is always parallel to a possible crystal face but *never parallel to a plane of symmetry*. The twin axis is a zone axis or a direction perpendicular to a rational lattice plane; but it can *never be an axis of even rotation* (2-, 4-, 6-fold) if the rotation involved is 180°. In some crystals a 90° rotation about a 2-fold axis can be considered a twin operation.

Twin crystals are usually designated as either *contact twins* or *penetration twins*. Contact twins have a definite composition surface separating the two individuals, and the twin is defined by a twin plane such as {111} in Figs. 2.169*a* and *b*. Penetration twins are made up of interpenetrating individuals having an irregular composition surface, and the twin law is usually defined by a twin axis (see Figs. 2.170*a*, *b* and *c*).

FIG. 2.170. Penetration twins. (*a*) Two interpenetrating fluorite cubes twinned on [111] as twin axis. (*b*) Two pyritohedral crystals (in pyrite) forming an Iron Cross, with twin axis [001].(*c*) Orthoclase exhibiting the Carlsbad twin law in which two interpenetrating crystals are twinned by 180° rotation about the *c* axis.

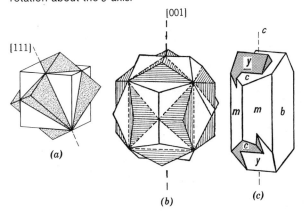

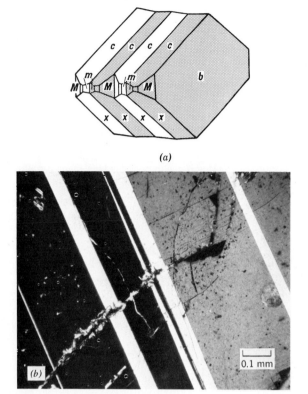

(a)

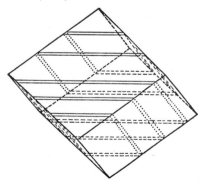

(b)

FIG. 2.171(a). Albite twinned on {010}. (b) The same polysynthetic twinning as in a but as seen in a polarizing microscope. The dark and light lamellae in albite are related by reflection on {010}.

FIG. 2.172. Calcite polysynthetically twinned on negative rhombohedron {1012}.

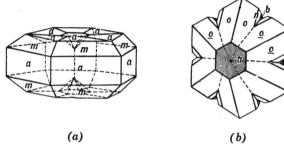

(a)   (b)

FIG. 2.173. Cyclic twinning. (a) Rutile. (b) Chrysoberyl.

than that of each individual because the twin planes, or twin axes, are added symmetry elements.

Origin of Twinning

The various mechanisms by which twins are formed have been discussed by M. J. Buerger, 1945, in terms of *growth twins, transformation twins* and *gliding* (or *deformation*) *twins*.

Growth twins are the result of an emplacement of atoms, or ions (or groups of atoms or ions) on the outside of a growing crystal in such a way that the regular arrangement of the original crystal structure (and, therefore, its lattice) is interrupted. For example, in Fig. 2.168, the line XX' may be considered as a side view of an external face of a growing crystal. An ion, or atom (or group of ions), would have the choice of attaching itself at structural sites represented by lattice nodes A, A', A'', or B, B', B''. In the former case the original structure is continued without interruption but in the latter a twinned relationship results. *Growth twinning* therefore reflects "accidents" during free growth ("nucleation errors") and can be considered as *primary twinning*. Growth twins can be recognized in mineral specimens by the presence of a single, often somewhat irregular boundary between the twinned units or by the interpenetration of two twinned units. Most of the crystal drawings with twins in this section represent growth twins.

Transformation twins occur in crystals after their formation and represent *secondary twinning*. Transformation twinning may result when a crystalline substance that formed at a relatively high temperature is cooled and subsequently rearranges its

High quartz $P6_2 22$

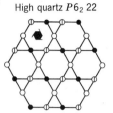

Low quartz $P3_1 21$

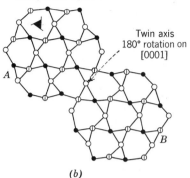

Twin axis
180° rotation on
[0001]

A

B

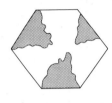

$\bullet = 0, 1$ ⎱ Heights
$\oplus = \frac{1}{3}$ ⎰ above
$\circ = \frac{2}{3}$ ⎰ page

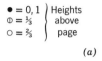

(*a*)   (*b*)   (*c*)

FIG. 2.174(*a*). The distribution of silicon atoms in high quartz. The atom positions given in terms of thirds above the page. (*b*) A part of the low quartz structure (*A*) twinned with respect to another part (*B*) by a rotation axis of 180° perpendicular to the page. Note the presence of 3-fold screw axes instead of a 6-fold screw axes as in high quartz. (*c*) The three-fold distribution of sectors of low quartz twinned according to the Dauphiné Law. The composition surfaces are generally irregular. This type of twin is visible only on sawn and etched basal sections.

structure to one of a different symmetry from that stable at higher temperatures. For example, high quartz with space group $P6_222$, if cooled below 573°C, transforms to low quartz with space group $P3_221$. Figure 2.174*a* illustrates the high quartz structure whereas Fig. 2.174*b* shows the structure of low quartz in which part *A* is related to part *B* by a 180° rotation (see also Fig. 4.21). In the transition to the low quartz structure the original high quartz structure has a choice of two orientations, related by 180° rotation, for the trigonal structural arrangement of low quartz (see Fig. 2.174*b*). The relationship of these two orientations is known as a *Dauphiné twin* and is expressed as a 180° rotation about [0001]. If the twinned crystal of low quartz is heated above 573°C the Dauphiné twinning will disappear and untwinned high quartz will form spontaneously. Fig. 2.174*c* shows a basal section of low quartz with the 3-fold distribution of the twinned orientations of low quartz in the crystal.

Another example of transformation twinning is provided by $KAlSi_3O_8$, which occurs in three different structural forms: high temperature sanidine (space group $C\,2/m$), lower temperature orthoclase (space group $C\,2/m$) and lowest temperature microcline (space group $C\bar{1}$; see also polymorphism, page 153). Microcline invariably shows the microscopic twin pattern shown in Fig. 2.175 which is known as *microcline* or "tartan" *twinning*. The cross-hatched pattern, which is usually visible only between crossed polarizers, consists of two types of

FIG. 2.175. Photomicrograph of transformation twinning in microcline. The specimen is viewed under a microscope with crossed polarizers. The section of the photograph is approximately parallel to (001). The twin laws represented are albite with twin and composition plane (010), and pericline with twin axis direction [010].

0.1 mm

twins, twinned according to the *albite* and *pericline* laws. This combination of twinned relations is a key feature in the symmetry change from monoclinic (as in orthoclase) to triclinic (as in microcline). In this transition from higher temperature the mirror plane and the 2-fold rotation axis are lost, leading to nucleation of triclinic "domains" which are related by twinning.

Gliding (or *deformation*) *twins* result when a crystalline substance is deformed by the application of a mechanical stress. Deformation twinning is another type of *secondary twinning*. If the stress produces slippage of the atoms on a very small scale, twin crystals may result (Fig. 2.176). If the movement of the atoms is large, slip or gliding may occur without twinning, which may finally lead to the rupture of the crystal. Deformation twinning is very common in metals and is frequently present in metamorphosed limestones as shown by the presence of polysynthetically twinned calcite grains (see Fig. 2.172). Similarly, plagioclase feldspars from metamorphic terranes may show deformation twinning.

Deformation twins in clacite may be obtained by carefully applying pressure with a knife edge (or razor blade) to the edge of a calcite rhombohedron (see Fig. 2.183c). The twinning in this case is produced by slippage of $\{01\bar{1}2\}$ layers in the calcite structure.

Common Twin Laws

Triclinic System. The feldspars best illustrate the twinning in the triclinic system. They are almost uni-

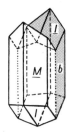

FIG. 2.177. Gypsum. Twin plane {100}.

versally twinned according to the *albite law,* with the {010} twin plane, as shown in Figs. 2.171 a and b. Another important type of twinning in triclinic feldspar is according to the *pericline law,* with {010} the twin axis. When, as occurs frequently in microcline, albite and pericline twins are closely interwoven, a typical cross-hatched, or "tartan" pattern can be seen under a polarizing microscope (Fig. 2.175). In addition, triclinic feldspars twin according to the same laws as monoclinic feldspars (see below).

Monoclinic System. In the monoclinic system twinning on {100} and {001} is most common. Figure 2.177 of gypsum illustrates twinning with {100} the twin plane and Fig. 2.178a shows a *Manebach twin* of orthoclase in which {001} is the twin plane. Orthoclase also forms penetration twins according to the *Carlsbad law,* in which the *c* crystallographic axis is the twin axis, and the individuals are united on an irregular surface roughly parallel to {010} (see Fig. 2.170c). The *Baveno twin* also found in orthoclase has {021} as the twin plane (Fig. 2.178b).

Orthorhombic System. In the orthorhombic system the twin plane is most commonly parallel to a prism face. The contact twin of aragonite (Fig.

FIG. 2.178. Manebach (a) and Baveno (b) twins in feldspar.

FIG. 2.176(a) and (b). Deformation twinning in an oblique lattice due to the application of mechanical stress as indicated by the arrows. Note that the amount of movement of the first layer above and parallel with the twin plane in b is less than that of the successive layers further removed from the twin plane.

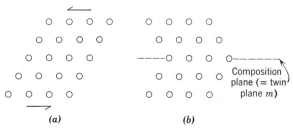

(a)   (b)

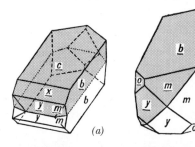

(a)   (b)

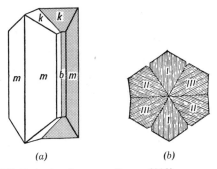

FIG. 2.179. Twinning in aragonite on {110}.

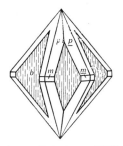

FIG. 2.180. Cerrusite cyclic twin on {110}.

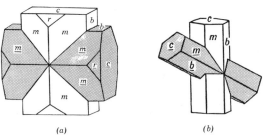

FIG. 2.181. Staurolite twins. (a) {031} Twin plane. (b) {231} Twin plane.

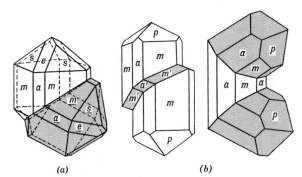

FIG. 2.182. Cassiterite and rutile twinned on {011}. (a) Cassiterite. (b) Rutile.

2.179a), the cyclic twin of the same mineral, in cross section (Fig. 2.179b) and the cyclic twin of cerussite (Fig. 2.180) are all twinned on {110}. The pseudohexagonal appearance of Figs. 2.179b and 2.180 results from the fact that $(110) \wedge (1\bar{1}0)$ is nearly 60°.

The mineral staurolite, which is monoclinic with a β angle of 90°, is pseudoorthorhombic and morphologically appears orthorhombic. It is commonly found in two types of penetration twins. In one with {031} as twin plane a right angle cross results (Fig. 2.181), in the other with twin plane {231}, a 60° cross is formed.

Tetragonal System. The most common type of twin in the tetragonal system has {011} as the twin plane. Crystals of cassiterite and rutile, twinned according to this law, are shown in Fig. 2.182.

Hexagonal System. In the rhombohedral division of the hexagonal system twins are common but in the hexagonal division, twins are rare and unimportant. The rhombohedral carbonates, especially

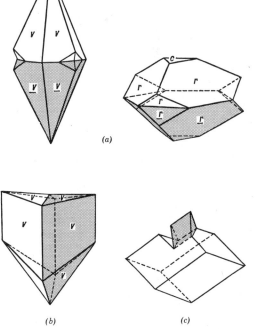

FIG. 2.183. Twinning in calcite. (a) Twinned on {0001}. (b) Twinned on {01$\bar{1}$2}. (c) Artificial twinning on {01$\bar{1}$2}.

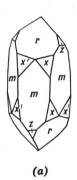
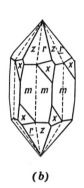
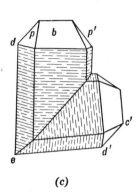

FIG. 2.184. Twinning in quartz. (a) Brazil twin. (b) Dauphiné twin. (c) Japanese twin. For Dauphiné twin see also Fig. 2.174.

(a)   (b)   (c)

calcite, serve as excellent illustrations of three twin laws. The twin plane may be {0001}, with c the twin axis (Fig. 2.183a), or it may be the positive rhombohedron {10$\bar{1}$1}. But twinning on the negative rhombohedron {01$\bar{1}$2} is most common and may yield contact twins (Fig. 2.183a) or polysynthetic twins as the result of pressure (Fig. 2.172). The ease of twinning according to this law can be demonstrated by the artificial twinning of a cleavage fragment of Iceland spar by the pressure of a knife blade (Fig. 2.183c).

In the trigonal trapezohedral class, quartz shows several types of twinning. Figure 2.184a illustrates the *Brazil law* with the twin plane perpendicular to an a axis. Here, right- and left-hand individuals have formed a penetration twin. Figure 2.184b shows a *Dauphiné twin,* a penetration twin with c the twin axis. Such twins are composed either of two right- or left-hand individuals (see also Fig. 2.174). Figure 2.184c illustrates the *Japanese law* with the twin plane {11$\bar{2}$2}. The reentrant angles usually present on twinned crystals do not show on either Brazil or Dauphiné twins.

Isometric System. In the hexoctahedral class of the isometric system the twin axis, with a few rare exceptions, is a 3-fold symmetry axis, and the twin plane is thus parallel to a face of the octahedron. Figure 2.169 shows an octahedron with plane *bb* a possible twin plane, as well as an octahedron

twinned according to this law, forming a contact twin. This type of twin is especially common in gem spinel and hence is called a *spinel twin.* Figure 2.170a shows two cubes forming a penetration twin with the 3-fold rotoinversion axis [111] the twin axis.

In the diploidal class, two pyritohedrons may form a penetration twin (Fig. 2.170b) with a 90° rotation about the twin axis [001]. The twin plane is {011}. This twin is known as the *iron cross.*

References and Suggested Reading

Bloss, F. D., 1971, *Crystallography and Crystal Chemistry,* Holt, Rinehart and Winston, New York, 545 pp.

Buerger, M. J., 1945, The genesis of twin crystals. *Amer. Mineralogist,* v. 30, pp. 469–482.

———, 1956, *Elementary Crystallography,* John Wiley & Sons, New York, 528 pp.

———, 1971, *Introduction to Crystal Geometry,* McGraw-Hill Book Co., New York, 204 pp.

International Tables for X-ray Crystallography, 1969, vol. 1, Symmetry Groups, International Union of Crystallography, The Kynoch Press, Birmingham, England, 558 pp.

Phillips, F. C., 1971, *An Introduction to Crystallography,* John Wiley & Sons, New York, 351 pp.

Sands, D. E., 1975, *Introduction to Crystallography,* W. A. Benjamin, Inc., New York, 166 pp.

3
X-RAY CRYSTALLOGRAPHY

The application of X-rays to the study of crystals was the greatest single impetus ever given to crystallography. Prior to 1912, crystallographers had correctly deduced from cleavage, optical properties, and the regularity of external form that crystals had an orderly structure; but their thinking concerning the geometry of crystal structures had only the force of a hypothesis. X-rays have made it possible not only to measure the distance between successive atomic planes but also to locate the positions of the various atoms or ions within the crystal, thus enabling determinations of crystal structures.

X-rays were discovered accidently by Wilhelm Conrad Roentgen in 1895 while he was experimenting with the production of cathode rays in sealed discharge tubes wrapped in black paper. The electron stream in the discharge tube, striking the glass of the tube, produced low-intensity X-radiation which caused some fluorescent material nearby to glow in the dark. Roentgen correctly inferred that a new type of radiation was being produced, fittingly called ''X-radiation'' because of the many mysteries connected with it. Roentgen was unsuccessful in his efforts to measure the wavelength of X-rays, and it was this unsolved problem that led to the discovery of the diffraction of X-rays by crystals.

The fact that most substances are more or less transparent to X-rays brought about their almost immediate use in hospitals for medical purposes in the location of fractures, foreign bodies, and diseased tissue in much the same manner in which they are used today.

It was not until 1912, 17 years after the discovery of X-radiation, that, at the suggestion of Max von Laue, X-rays were used to study crystals. The original experiments were carried out at the University of Munich where von Laue was a lecturer in Professor Sommerfeld's department. Sommerfeld was interested in the nature and excitation of X-rays and Laue in interference phenomena. Also at the University of Munich was Paul Heinrich Groth, a leading crystallographer. With the gathering together of such a group of distinguished scientists having these special interests, the stage was set for the momentous discovery.

In 1912, Paul Ewald was working under Sommerfeld's direction on a doctoral dissertation involving the scattering of light waves on passing through a crystal. In thinking about this problem, von Laue raised the question: what would be the effect if it were possible to use electromagnetic waves having essentially the same wavelength as the interatomic distances in crystals? Would a crystal act as a three-dimensional diffraction grating forming spectra that could be recorded? If so, it would be possible to measure precisely the wavelength of the X-rays employed, assuming interatomic spacings of the crystal; or assuming a wavelength for the X-rays, measure the interplanar spacings of the crystal. Methods for making such an experiment were discussed, and Freidrich and Knipping, two research students, agreed to carry them out. Several experiments in which copper sulfate was used as the object crystal were unsuccessful. Finally they passed a narrow beam of X-rays through a cleavage plate of sphalerite, ZnS, allowing the beam to fall on a photographic plate. The developed plate showed a large number of small spots arranged in a geometrical pattern around the large spot produced by the direct X-ray beam. This pattern was shown to be identical with that predictable from the diffraction of X-rays by a regular arrangement of scattering points within the crystal. Thus, a single experiment demonstrated the regular, orderly arrangement of atomic particles within crystals and the agreement as to order of magnitude of the wavelength of X-rays with the interplanar spacing of crystals. Although largely replaced by more powerful means of X-ray investigation, this technique, the Laue method, is still in use.

Within the next few years, great strides were made as a result of the work of the English physicists, William Henry Bragg and William Lawrence Bragg, father and son. In 1914 the first structure of a compound, halite, NaCl, was worked out by the Braggs; and in the years following many more structures were solved by them. The Braggs also greatly simplified von Laue's mathematical generalizations as to the geometry of X-ray diffraction and popularized the results of their research through their well-written books on the subject.

X-RAY SPECTRA

Electromagnetic waves form a continuous series varying in wavelength from long radio waves with wavelengths of the order of thousands of meters to cosmic radiation whose wavelengths are of the order of 10^{-12} meters (a millionth of a millionth of a meter!). All forms of electromagnetic radiation have certain properties in common such as propagation along straight lines at a speed of 300,000 km per second in a vacuum, reflection, refraction according to Snell's law, diffraction at edges and by slits or gratings, and a relation between energy and wavelength given by Planck's law: $e = h\nu = hc/\lambda$, where e is energy, ν frequency, c velocity of propagation, λ wavelength, and h Planck's constant. Thus, the shorter the wavelength the greater its energy and the greater its powers of penetration. X-rays occupy only a small portion of the spectrum, with wavelengths varying between slightly more than 100 Å and 0.02 Å (see Fig. 3.1). X-rays used in the investigation of crystals have wavelengths of the order of 1 Å. Visible light has wavelengths between 7200 and 4000 Å, more than 1000 times as great, and hence is less penetrating and energetic than X-radiation.

When electrons moving at high velocity strike

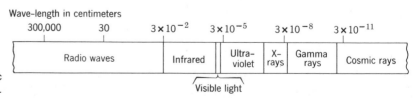

FIG. 3.1. The electromagnetic spectrum.

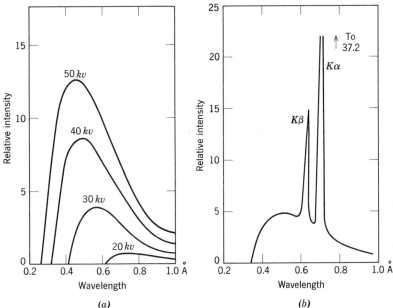

FIG. 3.2. X-ray spectra. (*a*) Distribution of intensity with wavelength in the continuous X-ray spectrum of tungsten at various voltages. (*b*) Intensity curve showing characteristic wavelengths superimposed on the continuous X-ray spectrum of molybdenum. (After Ulrey, *Phys. Rev.*, 11, 401.)

the atoms of any element, as is the case in an X-ray tube where electrons bombard a target material, X-rays are produced. These X-rays result in two types of X-ray spectra—*continuous* and *characteristic* (Fig. 3.2).

In a modern X-ray tube there is nearly a complete vacuum. The tube is fitted with a tungsten filament as a cathode that provides the source of electrons. The anode consists of one of a number of metals such as Mo, Cu, or Fe, and acts as the target for the electrons. When the filament is heated by passage of a current, electrons are emitted which are accelerated toward the target anode by a high voltage applied across the tube. X-rays are generated when the electrons impact on the target (anode). The nature of the X-rays depends on the metal of the target and the applied voltage. No X-rays are produced until the voltage reaches a miminum value dependent on the target material. At that point a *continuous spectrum* is generated. On increasing potential, the intensity of all wavelengths increases, and the value of the minimum wavelength becomes progressively less (Fig. 3.2a). The *continuous spectrum*, also referred to as *white radiation,* is caused by the stepwise loss of energy of bombarding elec-

trons in series of encounters with atoms of the target material. When an electron is instantaneously stopped it loses all of its energy at once and the X-ray radiation emitted is that of the shortest wavelength (see Fig. 3.2a). The stepwise energy losses from a stream of electrons produce a continuous range of wavelengths that can be plotted as a smooth function of intensity against wavelength (Fig. 3.2a). The curve begins at the short wavelength limit, rises to a maximum and extends toward infinity at very low intensity levels.

If the voltage across the X-ray tube is increased to a critical level, which is dependent on the element of the target, there becomes superimposed on the continuous spectrum a *line spectrum of characteristic radiation* peculiar to the target material. This characteristic radiation, many times more intense than the continuous spectrum, consists of several isolated wavelengths as shown in Fig. 3.2b by β and α peaks.

The characteristic X-ray spectrum is produced when the bombarding electrons have sufficient energy to dislodge electrons from the inner electron shells in the atoms of the target material. When these inner electrons are expelled they leave vacancies that are filled by electrons from surrounding electron shells. The electron transitions, from outer to inner shells, are accompanied by the emission of X-radiation with specific wavelengths. Electron transitions from the *L*- to the *K*-shell produce *Kα* radiation, and those from the *M*- to the *K*-shell cause *Kβ* radiation (see Fig. 3.2*b*). The *Kβ* peak can be eliminated by an appropriate filter yielding essentially a single wavelength, which by analogy to monochromatic light is called *monochromatic X-radiation*. The *Kα* radiation consists of two peaks *Kα₁* and *Kα₂* which are very close together in wavelength.

The wavelengths of the characteristic X-radiation emitted by various metals have been accurately determined. The *Kα* wavelengths (weighted averages of *Kα₁* and *Kα₂*) of the most commonly used are:

	Å		Å
Molybdenum	0.7107	Cobalt	1.7902
Copper	1.5418	Iron	1.9373
		Chromium	2.2909

DIFFRACTION EFFECTS AND THE BRAGG EQUATION

Crystals consist of an ordered three-dimensional structure with characteristic periodicities, or *identity periods* along the crystallographic axes. When an X-ray beam strikes such a three-dimensional arrangement, it causes electrons in its path to vibrate with a frequency of the incident X-radiation. These vibrating electrons absorb some of the X-ray energy and, acting as a source of new wave fronts, emit (scatter) this energy as X-radiation of the same frequency and wavelength. In general the scattered waves interfere destructively, but in some specific directions they reinforce one another producing a cooperative scattering effect known as *diffraction*.

In a row of regularly spaced atoms that is bombarded by X-rays every atom can be considered the center of radiating, spherical wave shells (Fig. 3.3). When the scattered waves interfere constructively, they produce wave fronts that are in phase, and diffraction will occur. Figure 3.4 illustrates that rays 1

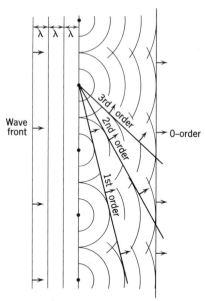

FIG. 3.3. Scattering of X-rays by a row of equally spaced, identical atoms.

and 2 will be in phase only when distance *AB* represents an integral number of wavelengths, in other words when $AB = n\lambda = c \cos \phi$ (where *n* represents whole numbers such as 0, 1, 2, 3, . . . , *n*). For a specific value of *nλ*, *φ* is constant, and the locus of all possible diffracted rays will be represented by a cone with the row of scattering points as the central axis. Because the scattered rays will be in

FIG. 3.4. Conditions for X-ray diffraction from a row of atoms.

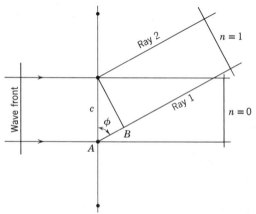

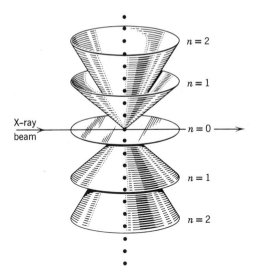

FIG. 3.5. Diffraction cones from row of atoms.

phase for the same angle ϕ on the other side of the incident beam, there will be another similar but inverted cone on that side (see Fig. 3.5). The two cones with $n = 1$ would have ϕ (as in Fig. 3.4) as the angle between the cone axis and the outer surface of the cone. When $n = 0$, the cone becomes a plane that includes the incident beam. The greater the value of n the larger the value of cos ϕ, and hence the smaller the angle ϕ and the narrower the cone. All have the same axis however, and all have their vertices at the same point, the intersection of the incident beam and the row of atoms.

In a three-dimensional lattice there are three axial directions, each with its characteristic periodicity of scattering points and each capable of generating its own series of nested cones with characteristic apical angles. Diffraction cones from any three noncoplanar rows of scattering atoms may or may not intersect each other, but only when all three intersect in a common line is there a diffracted beam (Fig. 3.6). This direction (shown by the arrow in Fig. 3.6) represents the direction of the diffracted beam which can be recorded on a film or registered electronically. The geometry of the three intersecting cones in Fig. 3.6 can be expressed by three independent equations (the Laue equations) in which the three cone angles (ϕ_1, ϕ_2, and ϕ_3) define a common direction along the path of the arrow (the

common intersection of the three cones). In order to produce a diffraction effect (a spot on a photographic plate or film) these three geometric equations must be simultaneously satisfied. The three equations are named after Max von Laue who originally formulated them.

Shortly after the publication of these equations, W. L. Bragg, working on X-ray diffraction in England pointed out that, although X-rays are indeed diffracted by crystals, the diffracted X-rays act as though they were reflected from planes within the crystal. However, unlike the reflection of light, X-rays are not "reflected" continuously from a given crystal plane. Using a given wavelength, λ, Bragg showed that a "reflection" takes place from a family of parallel planes only under certain conditions. These conditions must satisfy the equation: $n\lambda = 2d$ sin θ where n is an integer (1, 2, 3, . . . , n), λ the wavelength, d the distance between successive parallel planes, and θ the angle of incidence and "reflection" of the X-ray beam from the given atomic plane. This equation, known as the *Bragg law*, expresses in a simpler manner the simultaneous fulfillment of the three Laue equations.

We have seen that the faces most likely to appear on crystals are those parallel to atomic planes having the greatest density of lattice nodes. Parallel

FIG. 3.6. Diffraction cones from three noncoplanar rows of scattering atoms, intersecting in a common line.

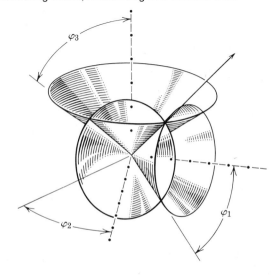

to each face is a family of equispaced identical planes. When an X-ray beam strikes a crystal it penetrates it, and the resulting diffraction effect is not from a single plane but from an almost infinite number of parallel planes, each contributing a small bit to the total diffraction maximum. In order that the diffraction effect ("reflection") be of sufficient intensity to be recorded, the individual "reflections" must be in phase with one another. The following conditions necessary for reinforcement were demonstrated by W. L. Bragg.

In Fig. 3.7 the lines p, p_1, and p_2 represent the trace of a family of atomic planes with spacing d. X-rays striking any of these planes by itself would be reflected at the incident angle θ, whatever the value of θ. However, to reinforce one another in order to give a "reflection" that can be recorded, these "reflected" rays must be in phase. The path of the waves along DEF "reflected" at E is longer than the path of the waves along ABC "reflected" at B. If the two sets of waves are to be in phase, the path difference of ABC and DEF must be a whole number of wave lengths ($n\lambda$). In Fig. 3.7 BG and BH are drawn perpendicular to AB and BC, respectively, so that $AB = DG$ and $BC = HF$. To satisfy the condition that the two waves be in phase $GE + EH$ must be equal to an integral number of wavelengths. BE is perpendicular to the lines p and p_1 and is equal to the interplanar spacing d. In $\triangle GBE$, $d \sin \theta = GE$; and, in $\triangle HBE$, $d \sin \theta = EH$. Thus for in phase "reflection" $GE + EH = 2d \sin \theta = n\lambda$.

This, $n\lambda = 2d \sin \theta$, is the Bragg equation. For a given interplanar spacing (d) and given λ, "reflections" (diffraction maxima) occur only at those angles of θ that satisfy the equation. Suppose, for example, a monochromatic X-ray beam is parallel to a cleavage plate of halite and the plate is supported in such a way that it can be rotated about an axis at right angles to the X-ray beam. As the halite is slowly rotated there is no "reflection" until the incident beam makes an angle θ that satisfies the Bragg equation, with $n = 1$. On continued rotation there are further "reflections" only when the equation is satisfied at certain θ angles with $n = 2, 3, 4, 5$, etc. These are known as the first-, second-, third-order, etc., "reflections." These "reflections" are in actuality the diffraction effects that occur when the three diffraction cones about three noncoplanar rows of atoms intersect in a common direction (see Fig. 3.6).

Laue Method

In the Laue method a single crystal that remains stationary is used. A photographic plate or flat film, enclosed in a light-tight envelope is placed a known distance, usually 5 cm, from the crystal. A beam of white X-radiation is passed through the crystal at right angles to the photographic plate. The direct beam causes a darkening at the center of the photograph so that a small disc of lead is usually placed in front of the film to intercept and absorb it. The angle of incidence, θ, between the X-ray beam and the various atomic planes with their given interplanar spacings (d) within the crystal is fixed. However, because X-rays of a large range of wavelengths are present, the Bragg law, $n\lambda = 2d \sin \theta$, can be satisfied by each family of atomic planes provided ($2d \sin \theta$)/n is within the range of wavelengths. Around the central point of a Laue photograph are arranged

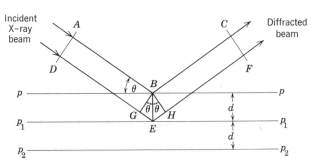

FIG. 3.7. Geometry of X-ray "reflection."

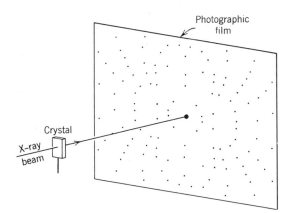

FIG. 3.8. Obtaining a Laue photograph with a stationary crystal.

diffraction spots, each one resulting from diffractions from a certain given series of atomic planes (Fig. 3.8).

The Laue method, although of great historical interest, has largely been replaced by other more powerful methods of X-ray crystal analysis. Its use today is primarily for determining symmetry. If a crystal is so oriented that the X-ray beam is parallel to a symmetry element, the arrangement of spots on the photograph reveals this symmetry. A Laue photograph of a mineral taken with the X-ray beam parallel to the 2-fold axis of a monoclinic crystal will show a 2-fold arrangement of spots; if the beam is parallel to a symmetry plane, the photograph shows a line of symmetry. A photograph of an orthorhombic crystal (symmetry $2/m\ 2/m\ 2/m$) with the beam parallel to a crystallographic axis shows a 2-fold arrangement as well as two lines of symmetry. Figure 3.9 shows a 4-fold arrangement of spots as given by idocrase with the X-ray beam parallel to the 4-fold rotation axis.

Unfortunately, X-rays do not differentiate between opposite ends of a polar axis and thus diffraction effects always introduce a center of symmetry.

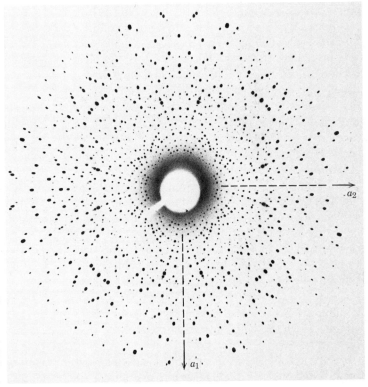

FIG. 3.9. Laue photograph of idocrase with point group symmetry $4/m\ 2/m\ 2/m$. The photograph was taken along the 4-fold rotation axis (c axis) of idocrase, thus revealing 4-fold symmetry and mirrors in the arrangement of diffraction spots.

Rotation, Weissenberg, and Precession Methods

In the rotation method and all refinements of it, as in the Laue method, a single crystal is used. The crystal must be oriented so that it can be rotated around one of the principal crystallographic axes. If crystal faces are present orientation is effected most easily by means of an optical goniometer; without crystal faces orientation is obtained by X-ray orientation techniques. The camera is a cylinder of known diameter, coaxial with the axis of rotation of the crystal, and a photographic film, protected from the light by an envelope, is wrapped around the inside of the cylindrical camera. A beam of monochromatic X-rays enters the camera through a slit and strikes the crystal.

Under these conditions, with a stationary crystal, any diffraction effects that occur would be fortuitous. However, as the crystal is slowly rotated, various families of atomic planes will be brought into position, so that for them θ has a value that will, with known wavelength, λ, satisfy $n\lambda = 2d \sin \theta$. A given family of planes gives rise to separate diffraction maxima when $n = 1, 2, 3, 4$, etc.

In taking a rotation photograph, the crystal is rotated about one of the principal lattice rows, usually a crystallographic axis. This lattice row is at right angles to the incident X-ray beam. Consequently, whatever diffracted beams arise must lie along cones having axes in common with the axis of rotation of the crystal. This axis is also the axis of the cylindrical film, and therefore the cones intersect the film in a series of circles (Fig. 3.10). When the film is developed and laid flat, the circles appear as straight parallel lines (Fig. 3.11). Each of these is a *layer line*, which corresponds to a cone of diffracted rays for which n has some integral value. Thus, the layer line including the incident beam is called the 0-layer, the first layer is that for which $n = 1$, the next $n = 2$, and so on. The layer lines are not continuous because diffraction spots appear only where all three diffraction cones intersect (see Fig. 3.6).

The separation of the layer lines is determined by the cone angles, which are in turn dependent on

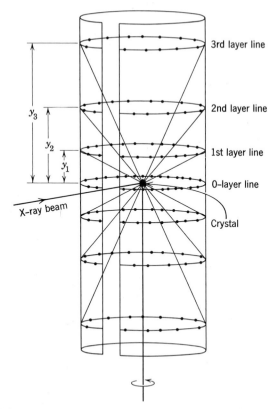

FIG. 3.10. Intersection of diffraction cones with cylindrical film.

the spacing in the lattice row about which the crystal was rotated. Therefore, if we know: (1) the diameter of the cylindrical film ($2r$), (2) the wavelength of the X-rays (λ), (3) the distance from the 0-layer to the nth layer of the film (y_n), it is possible to determine the spacing, or *identity period* (I), along the rotation axis of the crystal from the following relations:

$$\frac{y_n}{r} = \tan \nu \quad \text{(Fig. 3.12a)}$$

$$I = \frac{n\lambda}{\sin \nu} \quad \text{(Fig. 3.12b)}$$

If rotation photographs are taken with the crystal rotating about each of the three crystallographic axes, it is possible to determine the unit cell dimensions. The identity periods determined by rotating the crystal successively about a, b, and c are

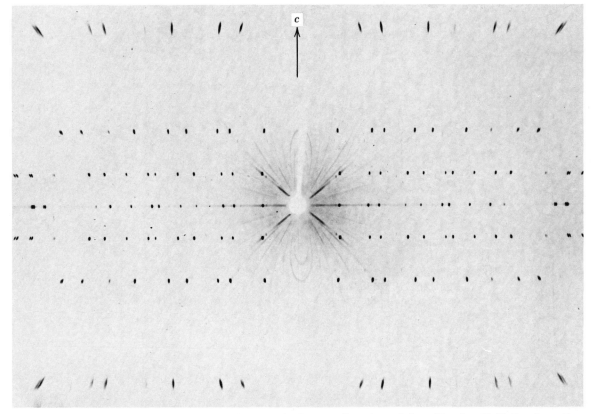

FIG. 3.11. Rotation photograph of quartz. Rotation axis is *c*. (Courtesy C. W. Burnham, Harvard University)

the edges of the unit cell. This is true for a crystal of any symmetry. However, in the isometric system one photograph to determine *a* suffices; in the tetragonal and hexagonal systems, two photographs are necessary, one about the *c* axis and one about an *a* axis; in the orthorhombic system three are needed, about the *a*, *b*, and *c* axes and in the monoclinic and triclinic crystal systems interaxial angles as well as the lengths of cell edges must be known to completely define the unit cell.

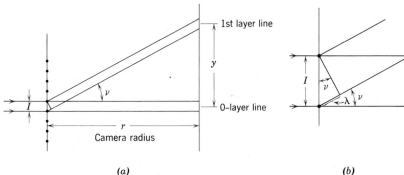

FIG. 3.12. Geometry for calculation of identity period, *I*.

(a)

(b)

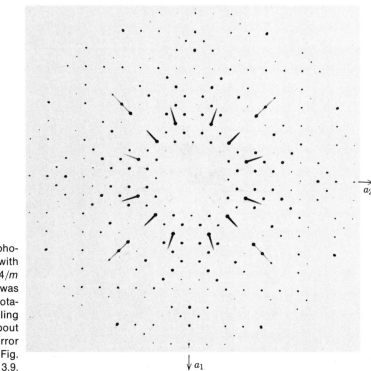

FIG. 3.13. Precession photograph of idocrase, with point group symmetry $4/m$ $2/m$ $2/m$. The photo was taken along the 4-fold rotation (c) axis, thus revealing the 4-fold symmetry about the axis, as well as mirror planes. Compare with Fig. 3.9.

It is frequently desirable to identify specific spots on the film with the atomic planes that gave rise to them. This process of correlation called *indexing* cannot be done with a rotation photograph because the crystal orientation is not completely known. However, several modifications of the rotation method have been devised to permit complete indexing of reflections. Those most commonly used are the *Weissenberg* and *precession* methods. In the Weissenberg method, the camera is translated back and forth during rotation of the crystal, and the spots of a layer line are spread out in festoons on the film. Although it is a laborious process, each spot can be indexed. The precession method, devised by M. J. Buerger, yields the same information in a more straightforward way (Fig. 3.13). In this method a flat film and crystal both move with a complex gyratory motion, mechanically compensating for the distortions produced by the Weissenberg method. Thus a great amount of data is provided in a readily usable form. Further consideration of these methods is beyond the scope of this discussion and the reader is referred to the references at the end of this chapter. However, we have seen that valuable information can be obtained from a rotation photograph without identifying individual reflections.

THE DETERMINATION OF CRYSTAL STRUCTURES

The orderly arrangement of atoms in a crystal is known as the *crystal structure*. It provides information on the location of all the atoms, bond positions and bond types, space group symmetry, and the chemical content of the unit cell.

The first step in the determination of the structure of a crystal is the determination of its space group and unit cell size. This information is obtained by a combination of photographic techniques using, for example, the rotation method and the precession camera as described earlier.

Next the intensities of the X-ray diffraction maxima must be determined because these intensities can be related to the electron densities of the atoms. Because atoms are positions of relatively high electron density the electron density in a crystal varies greatly, with high values over atom centers, lower values over chemical bonds and very low values elsewhere in the structure. Because atoms with high atomic numbers have high electron densities, their positions are especially easily recognized in electron density maps. When the electron density has been mapped in the unit cell and after each recognizable high on the map has been identified with atom and bond positions, the atomic structure can finally be derived.

The intensity of scattering of atoms in the crystal based on the intensity measurements of the reflections can be expressed mathematically by what is known as a three-dimensional Fourier series (see Figs. 3.14a and b). The results of a Fourier series must be complemented by additional information such as that provided by the space group.

The solution of a crystal structure is not a straightforward matter. Many structures are solved by trial-and-error techniques. A preliminary structure is modeled on the basis of knowledge of the chemical composition, the unit cell size and space group, the configuration of some chemical groupings present, and possibly on the basis of some significant physical properties. This preliminary model of the structure is then mathematically compared with the data obtained for the substance

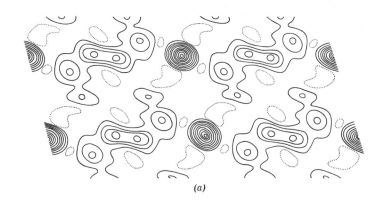

(a)

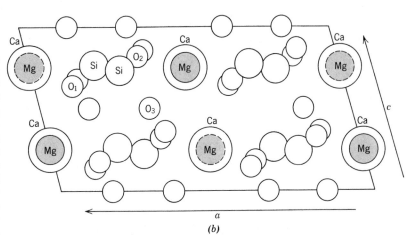

FIG. 3.14(a). The summation of Fourier series for diopside CaMgSi$_2$O$_6$ projected on (010). The distribution of scattering matter is indicated by contour lines drawn through points of equal density in the projection. (b) The atomic positions of diopside (with space group C 2/c) projected on (010), as derived from the distribution in a (redrawn after W. L. Bragg, Zeitschrift für Kristallographie, vol. 70, p. 488, 1929).

(b)

under investigation. When a very close agreement is reached between the model and the measured data the structure is concluded to have been determined.

The intensity values of X-ray diffraction maxima for use in structure determinations were, until recently, obtained from the careful intensity measurements of diffraction spots on photographic films or plates. With the advent of highly sensitive and reproducible X-ray detectors as well as computer programmed motors for simultaneous crystal and X-ray detector orientation, X-ray intensities are now most generally measured on computer-controlled single crystal diffractometers.

POWDER METHOD

The relative rarity of well-formed crystals and the difficulty of making the precise orientation required by single crystal methods led to the discovery of the *powder method* of X-ray investigation. For the powder method, the specimen is ground as finely as possible and bonded together by an amorphous material, such as flexible collodion, into a needle-like spindle 0.2 to 0.3 mm in diameter. This spindle, the *powder mount*, consists ideally of crystalline particles in completely random orientation. To ensure randomness of orientation of these tiny particles with respect to the impinging X-ray beam, the mount is generally rotated in the path of the beam during the exposure.

When the beam of monochromatic X-rays strikes the mount, all possible diffractions take place simultaneously. If the orientation of the crystalline particles in the mount is truly random, for each family of atomic planes with its characteristic interplanar spacing (d), there are many particles whose orientation is such that they make the proper θ angle with the incident beam to satisfy the Bragg law, $n\lambda = 2d \sin \theta$. The diffraction maxima from a given set of planes form cones with the incident beam as axis and the internal angle 4θ. Any set of atomic planes yields a series of nested cones corresponding to "reflections" of the first, second, third, and higher orders ($n = 1, 2, 3, \ldots$). Different families of planes with different interplanar spacings will satisfy the Bragg law at appropriate values of θ for different integral values of n, thus giving rise to separate sets of nested cones of "reflected" rays.

If the rays forming these cones are premitted to fall on a flat photographic plate at right angles to the incident beam, a series of concentric circles will result (Fig. 3.15). However, only "reflections" with small values of the angle 2θ can be recorded in this manner.

In order to record diffraction effects of 2θ up to 180°, the powder camera is used. This is a flat disc-shaped box with an adjustable pin at the center for attachment of the mount. The cylindrical wall of the camera is pierced diametrically for a demountable slit system and opposing beam catcher. The light-tight lid may be removed to insert and remove the film, which, for the most popular type of camera, is a narrow strip about 14 in. long and 1 in. wide. Two holes are punched in the film and so located that, when the film is fitted snugly to the inner curve of the camera, the collimating tube and beam catcher pass through the holes. This type of mounting is called the Straumanis method (Fig. 3.16).

A narrow beam of monochromatic X-rays is allowed to pass through the collimating slit and fall on the spindle-shaped mount, which is carefully centered on the short axis of the camera so that the mount remains in the X-ray beam as it rotates during the exposure. The undeviated beam passes through and around the mount and enters the lead-lined beam catcher. Under these conditions, the film intercepts the cones of diffracted rays along curved lines (Fig. 3.17). Because the axes of the cones coin-

FIG. 3.15. X-ray diffraction from powder mount recorded on a flat plate.

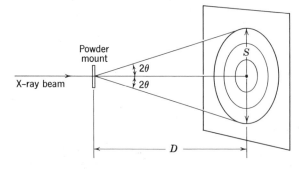

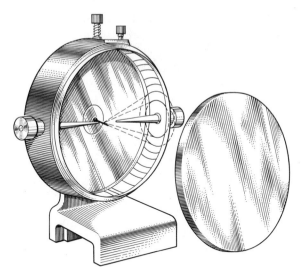

FIG. 3.16. Powder diffraction camera.

cide with the X-ray beam, for each cone there will be two curved lines on the film symmetrically disposed on each side of the slit through which the X-rays leave the camera. The angular distance between these two arcs is 4θ.

When the film is developed and laid flat, the arcs are seen to have their centers at the two holes in the film. Cones of low θ have their center at the exit hole through which the beam catcher passes through the film. Going out from this point, the arcs are of increasing radius until at $2\theta = 90°$ they are straight lines. Cones made by diffraction maxima with $2\theta > 90°$ curve in the opposite direction and are concentric about the hole through which the X-rays enter the camera. These are known as *back reflections*.

If a flat film is used and the distance D from the

FIG. 3.17. X-ray diffraction from powder mount recorded on cylindrical film.

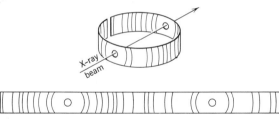

specimen to the film is known, it is possible to calculate θ by measuring S, the diameter of the rings. It may be seen in Fig. 3.15 that $\tan 2\theta = S/2D$. When a cylindrical camera is used, the distance S is measured with the film laid flat. Under these conditions, $S = R \times 4\theta$ in radians* or $\theta = S/4R$ in radians where R is the radius of the camera and S is measured in the same units as R. Most powder cameras are constructed with a radius such that S, measured in millimeters on the flat film can be converted easily to θ. For example, when the radius of the camera is 57.3 mm, the circumference is 360 mm. Using such a camera, each millimeter measured along the film is equal to 1°. Hence, a distance S of 60 mm measured on the film is equal to $60° = 4\theta$, and θ is accordingly equal to 15°.

It is not possible to measure the symmetrical distance S for values of θ much in excess of 40° ($S = 160$ mm) on two-hole films of the Straumanis type. In order to obtain θ for lines of θ higher than about 40°, the center about which the low lines are concentric must first be found by measurement of a number of such lines. Then the distance $S/2$ may be measured. It must be borne in mind that, if the camera radius is not 57.3°, a correction factor must be applied. Thus, if a camera having a radius of 28.65 mm is used, the correction factor is 57.3/28.65 = 2, and all S values measured must be divided by 2 to obtain θ in degrees.

The Straumanis method, the type of mounting described with two holes in the film, is now widely used. In older type powder cameras the X-ray beam entered between the ends of the cylindrical film and left through a centered hole. During the process of developing, the film usually shrinks. With the Straumanis camera, the effective diameter of the film can be measured and compared with the true diameter. This provides a shrinkage factor that can be applied to film measurements in making calculations.

When the diffraction angle θ corresponding to a given line on a powder photograph has been determined, it is possible to calculate the interplanar spacing of the family of atomic planes that gave rise to this line by use of the Bragg equation $n\lambda = 2d$

* One radian = 57.3°

$\sin\theta$, or $d = n\lambda/(2\sin\theta)$. As it is usually impossible to tell the order of a given reflection, n in the formula above is generally given the value 1, and d is determined in every case as though the line was produced by a first-order "reflection." For substances crystallizing in the isometric, and tetragonal systems, it is easy to index the lines of a powder photograph and thus determine cell dimensions. For crystals of the other systems, indexing is more difficult. However, for these crystals, if the unit cell dimensions are known, indexing is performed routinely with high-speed computer techniques.

The powder method finds its chief use in mineralogy as a determinative tool. One can use it for this purpose without knowing anything of the crystal structure or symmetry. Every crystalline substance produces its own powder pattern, which, because it is dependent on the internal structure, is characteristic of that substance. The powder photograph is often spoken of as the "fingerprint" of a mineral, because it differs from the powder pattern of every other mineral. Thus, if an unknown mineral is suspected of being the same as a known mineral, powder photographs of both are taken. If the photographs correspond exactly line for line, the two minerals are identical. Many organizations maintain extensive files of standard photographs of known minerals, and by comparison, an unknown mineral may be identified readily if there is some indication as to its probable nature.

However, one frequently has no clue to the identity of the unknown mineral, and a systematic comparison with the thousands of photographs in a reference file would be too time consuming. When this happens, the investigator turns to the card or microfiche file of X-ray diffraction data prepared by the Joint Committee on Powder Diffraction Standards (JCPDS) (Fig. 3.18). The interplanar spacings for thousands of crystalline substances are recorded on cards or on microfiches. To use these systems, the investigator must calculate the interplanar spacings for the most prominent lines in the powder pattern of the unknown substance and estimate the relative intensities of the lines on a scale where the strongest line is taken as 100. Then he may seek a corresponding series of d's in the JCPDS file cards, which are arranged in order of d of the most intense line. Because many substances have intense lines corresponding to the same d value and because many factors may operate to change the relative intensity of the lines in a powder pattern, all substances are cross-indexed for their second and third most intense lines. After the most likely "suspects" have been selected from the file, comparison of the weaker "reflections," which are also listed on the JCPDS card or microfiche, will speedily identify the substances in most cases. In this way a completely unknown substance may generally be identified in a short time on a very small volume of sample.

5-0490 MINOR CORRECTION

d	3.34	4.26	1.82	4.26	SiO₂	
I/I₁	100	35	17	35	SILICON IV OXIDE	ALPHA QUARTZ

Rad. CuKα₁ λ 1.5405 Filter Ni	d Å	I/I₁	hkl	d Å	I/I₁	hkl
Dia. Cut off Coll.	4.26	35	100	1.228	2	220
I/I₁ G.C. Diffractcmeter d corr. abs.?	3.343	100	101	1.1997	5	213
Ref. Swanson and Fuyat, NBS Circular 539, Vol. III (1953)	2.458	12	110	1.1973	2	221
	2.282	12	102	1.1838	4	114
Sys. Hexagonal S.G. D_3^4 - $P3_121$	2.237	6	111	1.1802	4	310
a₀ 4.913 b₀ c₀ 5.405 A C1.10	2.128	9	200	1.1530	2	311
α β γ Z 3	1.980	6	201	1.1408	<1	204
Ref. Ibid.	1.817	17	112	1.1144	<1	303
	1.801	<1	003	1.0816	4	312
	1.672	7	202	1.0636	1	400
εa nωβ 1.544ε γ 1.553 Sign +	1.659	3	103	1.0477	2	105
2V Dₓ2.647 mp Color	1.608	<1	210	1.0437	2	401
Ref. Ibid.	1.541	15	211	1.0346	2	214
	1.453	3	113	1.0149	2	223
Mineral from Lake Toxaway, N.C. Spect. Anal.:	1.418	<1	300	0.9896	2	402,115
<0.01% Al; <0.001% Ca,Cu,Fe,Mg.	1.382	7	212	.9872	2	313
X-ray pattern at 25°C.	1.375	11	203	.9781	<1	304
	1.372	9	301	.9762	1	320
	1.288	3	104	.9607	2	321
Replaces 1-0649, 2-0458,2-0459, 2-0471, 3-0419, 3-0427, 3-0444	1.256	4	302	.9285	<1	410

FIG. 3.18. JCPDS card for quartz. The interplanar spacings (*d*), their relative intensities and Miller indices are given. At the top of card are given the three strongest lines and their relative intensities. The fourth *d* is of the greatest spacing.

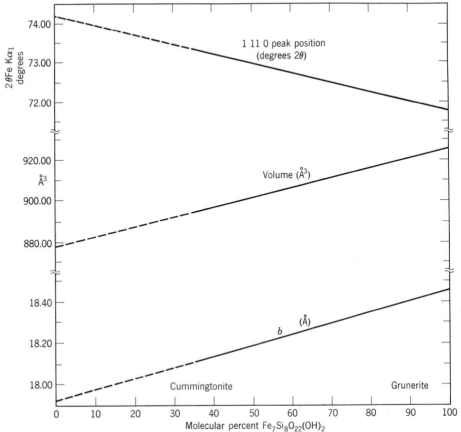

FIG. 3.19. Variation of *a*, *b*, and *V* of the unit cell and the position of the (1,11,0) peak as a function of composition in the monoclinic cummingtonite-grunerite series with composition ranging from $Fe_2Mg_5Si_8O_{22}(OH)_2$ to $Fe_7Si_8O_{22}(OH)_2$. (After Klein and Waldbaum, *Jour. Geol.*, Univ. Chicago Press, 1967.)

The powder method is of wider usefulness, however, and there are several other applications in which it is of great value. Variations in chemical composition of a known substance involve the substitution of atoms, generally of a somewhat different size, in specific sites in a crystal structure. As a result of this substitution the unit cell dimensions and hence the interplanar spacings are slightly changed, and the positions of the lines on the powder photograph corresponding to these interplanar spacings are shifted accordingly. By measuring these small shifts in position of the lines in powder patterns of substances of known structure, changes in chemical composition may often be accurately detected. Figure 3.19 illustrates a variation diagram that correlates unit cell dimensions and changes in the position of a specific diffraction maximum (1,11,0) with composition in the cummingtonite-grunerite series.

Further, the relative proportions of two or more known minerals in a mixture are often conveniently determined by the comparison of the intensities of the same lines in photographs of prepared control samples of known composition.

X-Ray Powder Diffractometer

The usefulness of the powder method has been greatly increased and its field of application extended by the introduction of the X-ray *powder diffractometer*. This powerful research tool uses essentially monochromatic X-radiation and a finely powdered sample, as does the powder film method, but records the information as to the "reflections" present as an inked trace on a printed strip chart.

The automated equipment supplied by one manufacturer is shown in Fig. 3-20.

The sample is prepared for powder diffractometer analysis by grinding to a fine powder, which is then spread uniformly over the surface of a glass slide, using a small amount of adhesive binder. The instrument is so constructed that this slide, when clamped in place, rotates in the path of a collimated X-ray beam while an X-ray detector, mounted on an arm, rotates about it to pick up the diffracted X-ray signals.

When the instrument is set at zero position, the X-ray beam is parallel to the slide and passes directly into the X-ray detector. The slide mount and the counter are driven by a motor through separate gear trains so that, while the slide and specimen rotate through the angle θ, the detector rotates through 2θ.

If the specimen has been properly prepared, there will be thousands of tiny crystalline particles on the slide in random orientation. As in powder photography, all possible "reflections" from atomic planes take place simultaneously. However, instead of recording all of them on a film at one time, the X-ray detector maintains the appropriate geometrical relationship to receive each diffraction maximum separately.

In operation, the sample, the X-ray detector and the paper drive of the strip chart recorder are set in motion simultaneously. If an atomic plane has an interplaner spacing (d) such that a reflection occurs at $\theta = 20°$, there is no evidence of this reflection until the counting tube has been rotated through 2θ, or $40°$. At this point the diffracted beam enters the X-ray detector causing it to respond. The pulse thus generated is amplified and causes a deflection on the recording strip chart. Thus, as the X-ray detector scans, the strip chart records as peaks the diffraction maxima from the specimen. The angle 2θ at which the diffraction occurred may be read directly from the position of the peak on the strip chart. The heights of the peaks are directly proportional to the intensities of the diffraction effects.

FIG. 3.20. Automated X-ray powder diffractometer. (Courtesy of Philips Electronics Co. Inc., Mt. Vernon, N.Y.)

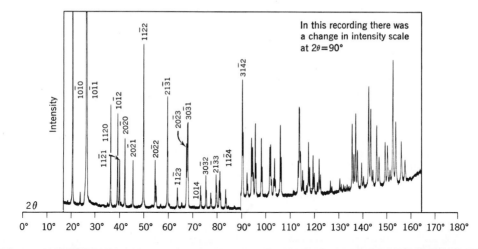

FIG. 3.21. Comparison of diffractometer record and powder film of quartz. On the diffractometer recording are given the Miller indices of the crystal planes that gave rise to the various low angle diffraction peaks. (Courtesy of Philips Electronic Instruments, Inc., Mount Vernon, N.Y.)

The chart on which the record is drawn is divided in tenths of inches and moves at a constant speed, such as e.g., 0.5 in. per minute. At this chart speed and a scanning speed of the X-ray detector of 1° per minute, 0.5 in. on the chart is equivalent to a 2θ of 1°. The positions of the peaks on the chart can be read directly and interplanar spacing giving rise to them can be determined by use of the equation $n\lambda = 2d \sin \theta$.

Although the powder diffractometer yields data similar to those derived from a powder photograph, it has definite advantages. The powder film method requires several hours of exposure plus the time necessary for developing, fixing, washing, and drying the film; a diffractometer run can be made in 1 hour. It is frequently difficult to estimate the intensity of the lines on a powder photograph, whereas the height of peaks on a diffractometer chart can be determined graphically with good precision. The powder photograph must be carefully measured to obtain 2θ values, whereas 2θ can be read directly from the diffractometer strip chart. Figure 3-21 compares a powder diffractometer strip chart with a powder photograph of the same mineral.

References and Suggested Reading

Azaroff, L. V., 1968, *Elements of X-ray Crystallography,* McGraw-Hill Book Co., New York, 610 pp.

—— and M. J. Buerger, 1958, *The Powder Method in X-ray Crystallography,* McGraw-Hill Book Co., New York, 342 pp.

Bragg, W. L., 1949, *The Crystalline State: A General Survey.* G. Bell and Sons Ltd., London, 352 pp.

Klug, H. P. and L. E. Alexander, 1954, *X-Ray Diffraction Procedures for Polycrystalline and Amorphous Materials.* John Wiley & Sons, New York, 716 pp.

Nuffield, E. W., 1966, *X-Ray Diffraction Methods.* John Wiley & Sons, New York, 409 pp.

4
CRYSTAL CHEMISTRY, CRYSTAL STRUCTURE, AND CHEMICAL COMPOSITION OF MINERALS

The chemical composition of a mineral is of fundamental importance, for many of its properties are in a great measure dependent on it. However, such properties depend not only on the chemical composition but also on the geometrical arrangement of the constituent atoms or ions and the nature of the electrical forces that bind them together. Thus for an understanding of minerals one must consider their structure as well as their chemistry.

Before we discuss the chemistry and structure of minerals and mineral groups, it is instructive to look briefly at some aspects of the chemistry of the earth's crust. Patterns of mineral distribution and assessments of our mineral wealth are directly related to the abundance and distribution of elements in the earth's crust. Although a wealth of chemical and mineralogical data are available for the earth's crust, considerable and very significant mineralogic information has also been obtained from extraterrestrial samples as provided by meteorites and lunar materials.

CHEMICAL COMPOSITION OF THE EARTH CRUST

Geophysical investigations indicate a division of the earth into a crust, mantle, and core (Fig. 4.1). The crust is approximately 36 km thick under the continents and under the oceans its thickness ranges from 10 to 13 km. The boundary between the crust and the underlying upper mantle is referred to as the *Mohorovičić discontinuity*. The upper part of the crust, which consists of the materials immediately underfoot, is composed of a relatively large percentage of sedimentary rocks and unconsolidated materials. However, this sedimentary cover forms but a thin veneer on an underlying basement of igneous and metamorphic rocks. Clarke and Washington (1924) estimate that the upper 10 miles of the crust consists of 95% igneous rocks (or their metamorphic equivalents), 4% shale, 0.75% sandstone, and 0.25% limestone. The average composition of igneous rocks, therefore, would very closely approxi-

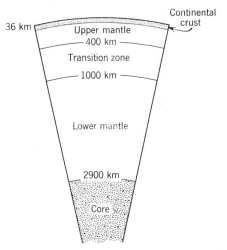

FIG. 4.1. The internal structure of the earth (after Mason, 1966).

mate the average crustal composition. To this effect, Clarke and Washington compiled 5159 "superior" analyses of igneous rocks, the average of which (Table 4.1) represents the average composition of the continental crust. It turns out that this average composition is intermediate between that of granite and basalt (or its coarser grained equivalents, diabase and gabbro) which are the two most common igneous rock types (see also Table 4.1). If the sampling by Clarke and Washington had included an appropriate number of basalts from the deep ocean basins, their average would have been more representative of the average crust than of just the average continental crust. Consideration of the extensive basalts on the ocean floor would lower the average Si, K, and Na values but would raise the Fe, Mg, and Ca contents of the average analysis.

It should be noted that eight elements make up approximately 99 weight percent of the earth crust. Some interesting features result when their weight percent values are recalculated in terms of atomic percent and finally in terms of volume percent (see Table 4.2). The last column in Table 4.2 shows that oxygen constitutes more than 90% of the total volume of the crust. In other words, the earth's crust, on an atomic scale, consists essentially of a packing of oxygen anions with interstitial metal ions, chiefly Si. The listing of the elements in Table 4.2 indicates

Table 4.1

AVERAGE AMOUNTS OF THE ELEMENTS IN CRUSTAL ROCKS, IN WEIGHT PERCENT FOR THE COMMON ELEMENTS (AS INDICATED BY %) AND IN PARTS PER MILLION FOR THE LESS ABUNDANT ELEMENTS[a]

Atomic Number	Element	Crustal Average	Granite (G-1)	Diabase (W-1)
1	H	0.14%	0.04%	0.06%
3	Li	20	24	12
4	Be	2.8	3	0.8
5	B	10	2	17
6	C	200	200	100
7	N	20	8	14
8	O	46.60%	48.50%	44.90%
9	F	625	700	250
11	Na	2.83%	2.46%	1.54%
12	Mg	2.09%	0.24%	3.99%
13	Al	8.13%	7.43%	7.86%
14	Si	27.72%	33.96%	24.61%
15	P	0.10%	0.04%	0.06%
16	S	260	175	135
17	Cl	130	50	
19	K	2.59%	4.51%	0.53%
20	Ca	3.63%	0.99%	7.83%
21	Sc	22	3	34
22	Ti	0.44%	0.15%	0.64%
23	V	135	16	240
24	Cr	100	22	120
25	Mn	0.09%	0.02%	0.13%
26	Fe	5.00%	1.37%	7.76%
27	Co	25	2.4	50
28	Ni	75	2	78
29	Cu	55	13	110
30	Zn	70	45	82
31	Ga	15	18	16
32	Ge	1.5	1.0	1.6
33	As	1.8	0.8	2.2
34	Se	0.05		
35	Br	2.5	0.5	0.5
37	Rb	90	220	22
38	Sr	375	250	180
39	Y	33	13	25
40	Zr	165	210	100
41	Nb	20	20	10
42	Mo	1.5	7	0.05
44	Ru	0.01		
45	Rh	0.005		
46	Pd	0.01	0.01	0.02
47	Ag	0.07	0.04	0.06

(continued on next page)

Table 4.1 (continued)

Atomic Number	Element	Crustal Average	Granite (G-1)	Diabase (W-1)
48	Cd	0.2	0.06	0.3
49	In	0.1	0.03	0.08
50	Sn	2	4	3
51	Sb	0.2	0.4	1.1
52	Te	0.01		
53	I	0.5		
55	Cs	3	1.5	1.1
56	Ba	0.04%	0.12%	0.02%
57	La	30	120	30
58	Ce	60	230	30
59	Pr	8.2	20	2
60	Nd	28	55	15
62	Sm	6.0	11	5
63	Eu	1.2	1.0	1.1
64	Gd	5.4	5	4
65	Tb	0.9	1.1	0.6
66	Dy	3.0	2	4
67	Ho	1.2	0.5	1.3
68	Er	2.8	2	3
69	Tm	0.5	0.2	0.3
70	Yb	3.4	1	3
71	Lu	0.5	0.1	0.3
72	Hf	3	5.2	1.5
73	Ta	2	1.6	0.7
74	W	1.5	0.4	0.45
75	Re	0.001	0.0006	0.0004
76	Os	0.005	0.001	0.004
77	Ir	0.001	0.006	
78	Pt	0.01	0.008	0.009
79	Au	0.004	0.002	0.005
80	Hg	0.08	0.2	0.2
81	Tl	0.5	1.3	0.13
82	Pb	13	49	8
83	Bi	0.2	0.1	0.2
90	Th	7.2	52	2.4
92	U	1.8	3.7	0.52

[a] From *Principles of Geochemistry* by Brian Mason. Copyright 1966 by John Wiley & Sons, Inc.

that oxygen-containing minerals such as silicates, oxides, and carbonates must be the most abundant types of minerals in the earth's crust. Thus the minerals referred to as the "rock-forming" minerals are, with few exceptions, members of these groups.

It is noteworthy that many elements important to our economy have very low values for their average abundances in the crust (see Table 4.1). For

Table 4.2

THE EIGHT MOST COMMON CHEMICAL ELEMENTS IN THE CRUST[a]

	Weight Percent	Atom Percent (wt% divided by atomic wt.)	Ionic Radius (Å)	Volume Percent
O	46.60	62.55	1.40	93.77
Si	27.72	21.22	0.42	0.86
Al	8.13	6.47	0.51	0.47
Fe	5.00	1.92	0.74	0.43
Mg	2.09	1.84	0.66	0.29
Ca	3.63	1.94	0.99	1.03
Na	2.83	2.64	0.97	1.32
K	2.59	1.42	1.33	1.83

[a] From *Principles of Geochemistry* by Brian Mason. Copyright 1966 by John Wiley & Sons, Inc.

example, Cu (atomic number, Z, $= 29$) $= 55$ ppm, Pb ($Z = 82$) $= 13$ ppm, Hg ($Z = 80$) $= 0.08$ ppm. On the other hand, the less commonly used element Zr ($Z = 40$) is more abundant (165 ppm) than Cu. Similarly, Ga ($Z = 31$) is about as abundant (15 ppm) as Hg. Clearly, in order to produce metals needed for our economy one must locate areas of high concentrations in order to make the mining profitable. Copper is not extracted from rocks of average composition, but from *ore deposits* in which copper has been concentrated and is present in specific copper minerals.

Some elements, for example Rb ($Z = 37$) are dispersed throughout common minerals and are never concentrated. Rb does not form specific Rb compounds, but is housed in K-rich minerals, and, therefore, is an example of what is called a *dispersed element*. Other elements are strongly concentrated in specific minerals; for example, Zr in zircon ($ZrSiO_4$) and Ti in rutile (TiO_2) and ilmenite ($FeTiO_3$).

Although we cannot directly determine the average composition of the earth as a whole, assumptions and calculations can be made on the basis of meteorite samples. These are generally concluded to represent materials analogous to materials within the earth. The composition of iron meteorites (mainly FeNi alloy) is believed to be very similar to the composition of the earth's core. Meteorites with about 50% metal and 50% silicate are probably representative of the lower mantle, and stony (silicate) meteorites with little metal are probably similar to materials in the upper mantle and lower crust. On the basis of the known average compositions of these types of meteorites combined with the known volumes of the core, mantle, and crust, one can arrive at an average estimate for the composition of the earth as a whole. Mason (1966) made such a calculation with the following results: Fe, 34.63%; O, 29.53%; Si, 15.20%; Mg, 12.70%; Ni, 2.39%; S, 1.93%; Ca, 1.13%; Al, 1.09%; and seven other elements (Na, K, Cr, Co, Mn, P, and Ti) each in amounts from 0.1 to 1%. Although all of the above elements in the calculated average earth abundance (except for Ni and S) are also major elements in the average crustal abundance listing (Tables 4.1 and 4.2) they appear in a different order because of the consideration of materials from the Fe-Ni-S-rich and O- and H_2O-poor mantle and core.

CRYSTAL CHEMISTRY

In this section the general principles relating the chemistry of minerals to their structure, crystallography, and physical properties are discussed. It was first recognized in the eighteenth century that a relation exists between chemical composition and crystal morphology. Since the discovery of X-rays, X-ray crystallographic investigations (see Fig. 3.14) have greatly developed our understanding of the relationship of chemistry and structure. As a result of common interests among crystallographers and chemists an interdisciplinary science, *crystal chemistry*, came into being. The goal of this science is the elucidation of the relationship between chemical composition, internal structure, and physical properties of crystalline matter. In mineralogy it serves as a unifying thread upon which often apparently unrelated facts of descriptive mineralogy may be strung. Before considering the crystal chemical and structural aspects of minerals, it is necessary to consider some of the elementary concepts of atoms and ions and their bonding forces in crystalline materials.

Atoms, Ions, and the Periodic Table

Atoms are the smallest subdivision of matter that retain the characteristics of the elements. They consist of a very small, massive nucleus composed of protons and neutrons surrounded by a much larger region thinly populated by electrons. Although atoms are so small that it is impossible to see them even with the highest magnification of the electron microscope, their sizes have been measured and are generally expressed as atomic radii in Ångstrom units. For example, the smallest atom, hydrogen, has a radius of only 0.46 Å, whereas the largest, cesium, has a radius of 2.62 Å (see Table 4.3).

We are indebted to the Danish physicist Niels Bohr for the most widely accepted picture of the atom. In 1912 he developed the concept of the "planetary" atom in which the electrons are visualized as circling the nucleus in "orbits" or energy levels at distances from the nucleus depending on the energies of the electrons. According to this mechanical model, the atom can be considered as a minute solar system*). At the center, corresponding to the sun, is the *nucleus* which, except in the hydrogen atom, is made up of *protons* and *neutrons*. The hydrogen *nucleus* is made up of a single proton. Each proton carries a unit charge of positive electricity; the neutron, as the name implies, is electrically neutral. Each *electron,* which, like a planet of the solar system, moves in an orbit around the nucleus, carries a charge of negative electricity. Because the atom as a whole is electrically neutral, there must be as many electrons as protons. The weight of the atom is concentrated in the nucleus, for the mass of an electron is only 1/1837 that of the lightest nucleus. Although the electrons and nuclei are both extremely small, the electrons move so rapidly about the nuclei that they give to the atoms relatively large effective diameters—ten to twenty thousand times the diameter of the nucleus. On a statistical basis, electrons are restricted to specific energy levels or orbital shells, which differ from each other by discrete amounts of energy, or *quanta*. The

---

* For a more detailed discussion see one of the Chemistry books listed at the end of this chapter.

Table 4.3
PERIODIC TABLE OF THE CHEMICAL ELEMENTS

Ia	IIa	IIIb	IVb	Vb	VIb	VIIb	
1 H 0.46							
3 Li 1.52 Li⁺ 0.68	**4** Be 1.12 Be²⁺ 0.35						
11 Na 1.86 Na⁺ 0.97	**12** Mg 1.60 Mg²⁺ 0.66						
19 K 2.31 K⁺ 1.33	**20** Ca 1.96 Ca²⁺ 0.99	**21** Sc 1.51 Sc³⁺ 0.81	**22** Ti 1.46 Ti³⁺ 0.76 Ti⁴⁺ 0.68	**23** V²⁺ 0.88 V³⁺ 0.74 V⁴⁺ 0.63 V⁵⁺ 0.59	**24** Cr 1.25 Cr³⁺ 0.63 Cr⁶⁺ 0.52	**25** Mn 1.12 Mn²⁺ 0.80 Mn³⁺ 0.66 Mn⁴⁺ 0.60 Mn⁷⁺ 0.46	**26** Fe 1.24 Fe²⁺ 0.74 Fe³⁺ 0.64
37 Rb 2.43 Rb⁺ 1.47	**38** Sr 2.15 Sr²⁺ 1.12	**39** Y 1.81 Y³⁺ 0.92	**40** Zr 1.56 Zr⁴⁺ 0.79	**41** Nb 1.43 Nb⁴⁺ 0.74 Nb⁵⁺ 0.69	**42** Mo 1.36 Mo⁴⁺ 0.70 Mo⁶⁺ 0.62	**43** Tc Tc⁷⁺ 0.56	**44** Ru 1.33 Ru⁴⁺ 0.67
55 Cs 2.62 Cs⁺ 1.67	**56** Ba 2.17 Ba²⁺ 1.34	**57** La 1.86 *) La³⁺ 1.14	**72** Hf 1.58 Hf⁴⁺ 0.78	**73** Ta 1.43 Ta⁵⁺ 0.68	**74** W 1.36 W⁴⁺ 0.70 W⁶⁺ 0.62	**75** Re Re⁴⁺ 0.72 Re⁷⁺ 0.56	**76** Os 1.35 Os⁶⁺ 0.69
87 Fr Fr⁺ 1.80	**88** Ra Ra²⁺ 1.43	**89** Ac *) Ac³⁺ 1.18 *)					

Lanthanides *)	**58** Ce 1.82 Ce³⁺ 1.07 Ce⁴⁺ 0.94	**59** Pr 1.81 Pr³⁺ 1.06 Pr⁴⁺ 0.92	**60** Nd 1.80 Nd³⁺ 1.04	**61** Pm	**62** Sm Sm³⁺ 1.00
Actinides *) *)	**90** Th 1.80 Th⁴⁺ 1.02	**91** Pa Pa³⁺ 1.13 Pa⁴⁺ 0.98 Pa⁵⁺ 0.89	**92** U 1.38 U⁴⁺ 0.97 U⁶⁺ 0.80	**93** Np Np³⁺ 1.10 Np⁴⁺ 0.95 Np⁷⁺ 0.71	**94** Pu Pu³⁺ 1.08 Pu⁴⁺ 0.93

concept that atomic particles can exist only with certain energy configurations was developed by the German physicist Max Planck and constitutes the basis for quantum theory. According to this theory, energy exists on an atomic scale only as discrete bundles and not as an infinitely divisible spectrum. Therefore, the electrons surrounding the nucleus can occupy only specific energy levels which differ by discrete numbers of quanta. The further development of wave mechanics does not allow us to picture an electron as a physical particle moving in an orbit of a definite geometrical form; instead, one can speak only of a probability distribution of elec-

tron density. Although in reality electrons exhibit wavelike as well as particlelike properties, in our succeeding discussion it will suffice to consider electrons simply as negatively charged particles.

The innermost *shell,* closest to the nucleus, is designated the K-shell ($n = 1$) and contains a maximum of two electrons. Outer orbitals representing higher energy levels are designated as L ($n = 2$), M ($n = 3$), N ($n = 4$), and so forth. The L-shell can accommodate 8 electrons, the M-shell 18, and the N-shell 32. The higher shells are not completely filled. Furthermore, *subshells* of different energies occur within a specific shell. Such subshells, listed

WITH ATOMIC AND IONIC RADII

	IIIa	IVa	Va	VIa	VIIa	VIIIa
						2 He 1.78
	5 B 0.97; B^{3+} 0.23	**6** C 0.77; C^{4+} 0.16	**7** N 0.71; N^{3+} 0.16; N^{5+} 0.13	**8** O^{2-} 1.40; O 0.60; O^{6+} 0.10	**9** F^- 1.33; F; F^{7+} 0.08	**10** Ne 1.60
	13 Al 1.43; Al^{3+} 0.51	**14** Si^{4-} 1.98; Si 1.17; Si^{4+} 0.39	**15** P 1.10; P^{3+} 0.44; P^{5+} 0.35	**16** S^{2-} 1.74; S 1.04; S^{4+} 0.37; S^{6+} 0.30	**17** Cl^- 1.81; Cl 1.07; Cl^{5+} 0.34; Cl^{7+} 0.27	**18** Ar 1.91

VIIIb		Ib	IIb	IIIa	IVa	Va	VIa	VIIa	VIIIa
27 Co 1.25; Co^{2+} 0.72; Co^{3+} 0.63	**28** Ni 1.24; Ni^{2+} 0.69	**29** Cu 1.28; Cu^+ 0.96; Cu^{2+} 0.72	**30** Zn 1.33; Zn^{2+} 0.74	**31** Ga 1.22; Ga^{3+} 0.62	**32** Ge 1.22; Ge^{2+} 0.73; Ge^{4+} 0.53	**33** As 1.25; As^{3+} 0.58; As^{5+} 0.46	**34** Se^{2-} 1.93; Se 1.16; Se^{4+} 0.50; Se^{6+} 0.42	**35** Br^- 1.96; Br 1.19; Br^{5+} 0.47; Br^{7+} 0.39	**36** Kr 2.01
45 Rh 1.34; Rh^{3+} 0.68	**46** Pd 1.37; Pd^{2+} 0.80; Pd^{4+} 0.65	**47** Ag 1.44; Ag^+ 1.26; Ag^{2+} 0.89	**48** Cd 1.49; Cd^{2+} 0.97	**49** In 1.62; In^{3+} 0.81	**50** Sn^{4-} 2.15; Sn 1.40; Sn^{2+} 0.93; Sn^{4+} 0.71	**51** Sb 1.45; Sb^{3+} 0.76; Sb^{5+} 0.62	**52** Te^{2-} 2.11; Te 1.43; Te^{4+} 0.70; Te^{6+} 0.56	**53** I^- 2.20; I 1.36; I^{5+} 0.62; I^{7+} 0.50	**54** Xe 2.20
77 Ir 1.35; Ir^{4+} 0.68	**78** Pt 1.38; Pt^{2+} 0.80; Pt^{4+} 0.65	**79** Au 1.44; Au^+ 1.37; Au^{3+} 0.85	**80** Hg 1.50; Hg^{2+} 1.10	**81** Tl 1.70; Tl^+ 1.47; Tl^{3+} 0.95	**82** Pb 1.75; Pb^{2+} 1.20; Pb^{4+} 0.84	**83** Bi 1.55; Bi^{3+} 0.96; Bi^{5+} 0.74	**84** Po; Po^{6+} 0.67	**85** At; At^{7+} 0.62	**86** Rn

63	64	65	66	67	68	69	70	71	
Eu; Eu^{3+} 0.98	Gd; Gd^{3+} 0.97	Tb; Tb^{3+} 0.93; Tb^{4+} 0.81	Dy; Dy^{3+} 0.92	Ho; Ho^{3+} 0.91	Er 1.86; Er^{3+} 0.89	Tm; Tm^{3+} 0.87	Yb; Yb^{3+} 0.86	Lu; Lu^{3+} 0.85	
95 Am; Am^{3+} 1.07; Am^{4+} 0.92	**96** Cm	**97** Bk	**98** Cf	**99** Es	**100** Fm	**101** Md	**102** No	**103** Lw	

in order of increasing energy, are *s*, *p*, *d*, and *f*. In some instances, a higher subshell (e.g., 5*s* in the O-shell) will be filled before electrons enter the 4*f* subshell of the N-shell, provided 5*s* has a lower energy than 4*f*. The electron configurations of the elements are presented in Table 4.4. Although the shape of shells and subshells are assumed to be spherical in the Bohr atom, the electron densities for specific subshells deviate greatly from the spherical model.

The simplest atom is that of hydrogen in which the nucleus has one electron moving around it, as is shown diagramatically (in a Bohr atom represen-

tation) in Fig. 4.2. Atoms of the other natural elements have from two electrons (helium) to 92 electrons (uranium) moving in orbits about their nuclei.

The fundamental difference between atoms of the different elements lies in the electrical charge of the nucleus. This positive charge is the same as the number of protons, and this number, equal to the number of electrons in an uncharged atom, is called the *atomic number* (= Z). The sum of the protons and neutrons determines the *characteristic mass,* or mass number of an element. Atoms of the same element but with differing numbers of neutrons are called *isotopes*. For example, oxygen (Z = 8) has

Table 4.4

ELECTRON CONFIGURATIONS OF THE ATOMS

Element	K	L		M			N				O					P			Q
	1s	2s	2p	3s	3p	3d	4s	4p	4d	4f	5s	5p	5d	5f	5g	6s	6p	6d	7s
1. H	1																		
2. He	2																		
3. Li	2	1																	
4. Be	2	2																	
5. B	2	2	1																
6. C	2	2	2																
7. N	2	2	3																
8. O	2	2	4																
9. F	2	2	5																
10. Ne	2	2	6																
11. Na	2	2	6	1															
12. Mg	2	2	6	2															
13. Al	2	2	6	2	1														
14. Si	2	2	6	2	2														
15. P	2	2	6	2	3														
16. S	2	2	6	2	4														
17. Cl	2	2	6	2	5														
18. A	2	2	6	2	6														
19. K	2	2	6	2	6		1												
20. Ca	2	2	6	2	6		2												
21. Sc	2	2	6	2	6	1	2												
22. Ti	2	2	6	2	6	2	2												
23. V	2	2	6	2	6	3	2												
24. Cr	2	2	6	2	6	5	1												
25. Mn	2	2	6	2	6	5	2												
26. Fe	2	2	6	2	6	6	2												
27. Co	2	2	6	2	6	7	2												
28. Ni	2	2	6	2	6	8	2												
29. Cu	2	2	6	2	6	10	1												
30. Zn	2	2	6	2	6	10	2												
31. Ga	2	2	6	2	6	10	2	1											
32. Ge	2	2	6	2	6	10	2	2											
33. As	2	2	6	2	6	10	2	3											
34. Se	2	2	6	2	6	10	2	4											
35. Br	2	2	6	2	6	10	2	5											
36. Kr	2	2	6	2	6	10	2	6											
37. Rb	2	2	6	2	6	10	2	6			1								
38. Sr	2	2	6	2	6	10	2	6			2								
39. Y	2	2	6	2	6	10	2	6	1		2								
40. Zr	2	2	6	2	6	10	2	6	2		2								
41. Nb	2	2	6	2	6	10	2	6	4		1								
42. Mo	2	2	6	2	6	10	2	6	5		1								
43. Tc	2	2	6	2	6	10	2	6	5		2								
44. Ru	2	2	6	2	6	10	2	6	7		1								
45. Rh	2	2	6	2	6	10	2	6	8		1								
46. Pd	2	2	6	2	6	10	2	6	10										
47. Ag	2	2	6	2	6	10	2	6	10		1								
48. Cd	2	2	6	2	6	10	2	6	10		2								
49. In	2	2	6	2	6	10	2	6	10		2	1							
50. Sn	2	2	6	2	6	10	2	6	10		2	2							

Table 4.4 (continued)

Element	K	L		M			N				O					P			Q
	1s	2s	2p	3s	3p	3d	4s	4p	4d	4f	5s	5p	5d	5f	5g	6s	6p	6d	7s
51. Sb	2	2	6	2	6	10	2	6	10		2	3							
52. Te	2	2	6	2	6	10	2	6	10		2	4							
53. I	2	2	6	2	6	10	2	6	10		2	5							
54. Xe	2	2	6	2	6	10	2	6	10		2	6							
55. Cs	2	2	6	2	6	10	2	6	10		2	6				1			
56. Ba	2	2	6	2	6	10	2	6	10		2	6				2			
57. La	2	2	6	2	6	10	2	6	10		2	6	1			2			
*58. Ce	2	2	6	2	6	10	2	6	10	1	2	6	1			2			
*59. Pr	2	2	6	2	6	10	2	6	10	2	2	6	1			2			
*60. Nd	2	2	6	2	6	10	2	6	10	3	2	6	1			2			
*61. Pm	2	2	6	2	6	10	2	6	10	4	2	6	1			2			
*62. Sm	2	2	6	2	6	10	2	6	10	5	2	6	1			2			
*63. Eu	2	2	6	2	6	10	2	6	10	6	2	6	1			2			
*64. Gd	2	2	6	2	6	10	2	6	10	7	2	6	1			2			
*65. Tb	2	2	6	2	6	10	2	6	10	8	2	6	1			2			
*66. Dy	2	2	6	2	6	10	2	6	10	9	2	6	1			2			
*67. Ho	2	2	6	2	6	10	2	6	10	10	2	6	1			2			
*68. Er	2	2	6	2	6	10	2	6	10	11	2	6	1			2			
*69. Tm	2	2	6	2	6	10	2	6	10	12	2	6	1			2			
*70. Yb	2	2	6	2	6	10	2	6	10	13	2	6	1			2			
*71. Lu	2	2	6	2	6	10	2	6	10	14	2	6	1			2			
72. Hf	2	2	6	2	6	10	2	6	10	14	2	6	2			2			
73. Ta	2	2	6	2	6	10	2	6	10	14	2	6	3			2			
74. W	2	2	6	2	6	10	2	6	10	14	2	6	4			2			
75. Re	2	2	6	2	6	10	2	6	10	14	2	6	5			2			
76. Os	2	2	6	2	6	10	2	6	10	14	2	6	6			2			
77. Ir	2	2	6	2	6	10	2	6	10	14	2	6	7			2			
78. Pt	2	2	6	2	6	10	2	6	10	14	2	6	9			1			
79. Au	2	2	6	2	6	10	2	6	10	14	2	6	10			1			
80. Hg	2	2	6	2	6	10	2	6	10	14	2	6	10			2			
81. Tl	2	2	6	2	6	10	2	6	10	14	2	6	10			2	1		
82. Pb	2	2	6	2	6	10	2	6	10	14	2	6	10			2	2		
83. Bi	2	2	6	2	6	10	2	6	10	14	2	6	10			2	3		
84. Po	2	2	6	2	6	10	2	6	10	14	2	6	10			2	4		
85. At	2	2	6	2	6	10	2	6	10	14	2	6	10			2	5		
86. Rn	2	2	6	2	6	10	2	6	10	14	2	6	10			2	6		
87. Fr	2	2	6	2	6	10	2	6	10	14	2	6	10			2	6		1
88. Ra	2	2	6	2	6	10	2	6	10	14	2	6	10			2	6		2
89. Ac	2	2	6	2	6	10	2	6	10	14	2	6	10			2	6	1	2
*90. Th	2	2	6	2	6	10	2	6	10	14	2	6	10			2	6	2	2
*91. Pa	2	2	6	2	6	10	2	6	10	14	2	6	10	2		2	6	1	2
*92. U	2	2	6	2	6	10	2	6	10	14	2	6	10	3		2	6	1	2
*93. Np	2	2	6	2	6	10	2	6	10	14	2	6	10	4		2	6	1	2
*94. Pu	2	2	6	2	6	10	2	6	10	14	2	6	10	6		2	6		2
*95. Am	2	2	6	2	6	10	2	6	10	14	2	6	10	7		2	6		2
*96. Cm	2	2	6	2	6	10	2	6	10	14	2	6	10	7		2	6	1	2
*97. Bk	2	2	6	2	6	10	2	6	10	14	2	6	10	8		2	6	1	2
*98. Cf	2	2	6	2	6	10	2	6	10	14	2	6	10	10		2	6		2
*99. E	2	2	6	2	6	10	2	6	10	14	2	6	10	11		2	6		2
*100. Fm	2	2	6	2	6	10	2	6	10	14	2	6	10	12		2	6		2
*101. Mv	2	2	6	2	6	10	2	6	10	14	2	6	10	13		2	6		2
*102. No	2	2	6	2	6	10	2	6	10	14	2	6	10	14		2	6		2

* Lanthanide and actinide elements; some configurations uncertain.

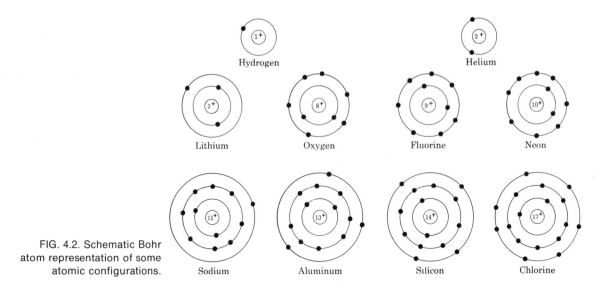

FIG. 4.2. Schematic Bohr atom representation of some atomic configurations.

three isotopes, the commonest of which has a nucleus with 8 protons and 8 neutrons (= 16); this is known as O^{16}. Rarer and heavier isotopes of oxygen carry 8 protons and 9 or 10 neutrons; these are O^{17} and O^{18}, respectively.

The elements in the periodic table (Table 4.3) are arranged according to increasing atomic number (Z). The *atomic weight* of an element is a number expressing its relative weight in terms of the weight of the element oxygen, which is taken as 16. The atomic weight of the elements are listed with the symbols for the elements in Table 4.5.

The characteristics of an element depend to a great extent on the configuration of the electronic structure of its atoms. Table 4.4 lists the occupancy of the electron shells ($n = 1$ to $n = 7$) in atomic structures. A close relationship exists between the electronic structure of an atom, the chemical properties of the element and its place in the periodic table (Table 4.3). In groups Ia, IIa, IIIb, IVb, Vb, VIb, and VIIa the number of electrons in the outermost shell (see also Table 4.4) is the same for the atoms of each element in a group and is equal to the group number. The atoms of the alkali-metals, Group Ia, have one such electron and the atoms of the halogens, Group VIIa, have seven such electrons. It is the number of these outer shell electrons or *valence electrons* that largely determines the chemical prop-

erties of an element. With the exception of helium, with only one shell occupied by two electrons, the other inert gases of Group VIII have atoms with eight electrons in their outermost shell.

Changes in ionic sizes are also regularly reflected by the arrangement of the elements in the periodic table. For elements of the same group the ionic radii increase as the atomic number increases. For example, in group IIa the smallest ion is Be^{2+} with a radius of 0.35 Å and the largest ion is Ra^{2+} with a radius of 1.43 Å. An apparent contradiction to this rule is provided by the rare earth elements. The trivalent ions of these elements decrease in radius with increasing atomic number, from La^{3+} (Z = 57) with a radius of 1.14 Å to Lu^{3+} (Z = 71) with a radius of 0.85 Å. This feature, known as the *lanthanide contraction,* is the result of building up inner electron shells instead of the addition of a new shell (see Table 4.4). As a result of the increasing nuclear charge, an increased attraction is exerted on the outer electrons, which causes an effective decrease in ionic radius.

For positive ions with the same electronic structure, the radii decrease with increasing charge. For example, the ionic radii of elements in the second horizontal row, all of which have 2 electrons in the first shell and 8 in the second shell, decrease from Na^+ with a radius of 0.97 Å to Cl^{7+} with a radius of

Table 4.5
ATOMIC WEIGHTS, 1971

	Symbol	Atomic Weight		Symbol	Atomic Weight
Aluminum	Al	26.9815	Neodymium	Nd	144.24
Antimony	Sb	121.75	Neon	Ne	20.179
Argon	A	39.948	Nickel	Ni	58.71
Arsenic	As	74.9216	Niobium	Nb	92.906
Barium	Ba	137.34	Nitrogen	N	14.0067
Beryllium	Be	9.0122	Osmium	Os	190.2
Bismuth	Bi	208.981	Oxygen	O	15.9994
Boron	B	10.811	Palladium	Pd	106.4
Bromine	Br	79.904	Phosphorus	P	30.9738
Cadmium	Cd	112.40	Platinum	Pt	195.09
Calcium	Ca	40.08	Potassium	K	39.102
Carbon	C	12.0111	Praseodymium	Pr	140.907
Cerium	Ce	140.12	Protactinium	Pa	231.036
Cesium	Cs	132.905	Radium	Ra	226.025
Chlorine	Cl	35.453	Radon	Rn	222.
Chromium	Cr	51.996	Rhenium	Re	186.2
Cobalt	Co	58.9332	Rhodium	Rh	102.905
Copper	Cu	63.546	Rubidium	Rb	85.467
Dysprosium	Dy	162.50	Ruthenium	Ru	101.07
Erbium	Er	167.26	Samarium	Sm	150.4
Europium	Eu	151.96	Scandium	Sc	44.956
Fluorine	F	18.9984	Selenium	Se	78.96
Gadolinium	Gd	157.25	Silicon	Si	28.086
Gallium	Ga	69.72	Silver	Ag	107.868
Germanium	Ge	72.59	Sodium	Na	22.9898
Gold	Au	196.967	Strontium	Sr	87.62
Hafnium	Hf	178.49	Sulfur	S	32.064
Helium	He	4.0026	Tantalum	Ta	180.947
Holmium	Ho	164.930	Tellurium	Te	127.60
Hydrogen	H	1.0080	Terbium	Tb	158.9254
Indium	In	114.82	Thallium	Tl	204.37
Iodine	I	126.9045	Thorium	Th	232.038
Iridium	Ir	192.2	Thulium	Tm	168.934
Iron	Fe	55.847	Tin	Sn	118.69
Krypton	Kr	83.80	Titanium	Ti	47.90
Lanthanum	La	138.91	Tungsten	W	183.85
Lead	Pb	207.19	Uranium	U	238.029
Lithium	Li	6.941	Vanadium	V	50.9414
Lutetium	Lu	174.97	Xenon	Xe	131.30
Magnesium	Mg	24.305	Ytterbium	Yb	173.04
Manganese	Mn	54.9380	Yttrium	Y	88.9059
Mercury	Hg	200.59	Zinc	Zn	65.37
Molybdenum	Mo	95.94	Zirconium	Zr	91.22

0.27 Å. The size of ions with similar electron configurations decreases because the increased nuclear charge exerts a greater pull on the electrons, thus decreasing the effective radius of the ion.

For an element that can exist in several valence states (ions of the same element with different charges) the higher the charge on the positive ion, the smaller its radius. For example, $Mn^{2+} = 0.80$ Å, $Mn^{3+} = 0.66$ Å, $Mn^{4+} = 0.60$ Å and $Mn^{5+} = 0.46$ Å. This decrease in size is due to the greater pull exerted by the nucleus on a reduced electron cloud.

Bonding Forces in Crystals

The forces that bind together the atoms (or ions, or ionic groups) of crystalline solids are electrical in nature. Their type and intensity are largely responsible for the physical and chemical properties of minerals. Hardness, cleavage, fusibility, electrical and thermal conductivity, and the coefficient of thermal expansion are directly related to binding forces. In general, the stronger the bond the harder the crystal, the higher its melting point, and the smaller its coefficient of thermal expansion. The great hardness of a diamond is attributed to the very strong electrical forces linking its constituent carbon atoms. The structural patterns of the minerals periclase, MgO, and halite, NaCl, are similar, yet periclase melts at 2800°C, whereas halite melts at 801°C. The greater amount of heat energy required to separate the atoms in periclase indicates it has a stronger electrical bond than halite.

These electrical forces are *chemical bonds* and can be described as belonging to one of four principal *bond types:* ionic, covalent, metallic, and van der Waals'. It should be understood that this classification is one of expediency and that transitions may exist between all types.* The electrical interaction of the ions or atoms constituting the structural units determines the properties of the resulting crystal. It is the similarity in properties among crystals having similar types of electrical interaction that justifies

---

* For further discussion of bonding see one of the chemistry books listed at the end of this chapter.

the use of the classification of bonding mechanisms. Thus, the bonding forces linking the atoms of silicon and oxygen in quartz display in almost equal amount the characteristics of the ionic and the covalent bond. In addition, galena, PbS, displays characteristics of the metallic bond, as expressed by good electrical conductivity, and some of the ionic bond, as shown by excellent cleavage and brittleness. Furthermore, many minerals, such as mica, contain two or more bond types of different character and strength. Such crystals are called *heterodesmic* in contrast to crystals like diamond and halite, in which all bonds are of the same kind—*homodesmic*.

Ionic Bond

A comparison of the chemical activity of elements with the configuration of their outer, or valence, electron shells leads to the conclusion that all atoms have a strong tendency to achieve a stable configuration of the outer shell in which all possible electron sites are filled. In the noble gases, helium, neon, argon, krypton, and xenon, which are almost completely inert, this condition is met (see Table 4.4). We have seen that sodium, for example, has a single valence electron in its outer shell, which it loses readily, leaving the atom with an unbalanced positive charge. Such a charged atom is called an *ion.* Positively charged atoms are *cations,* and an atom bearing a single positive charge, such as sodium, is *monovalent* and is represented by the symbol Na^+. In similar fashion, we will speak of divalent, trivalent, tetravalent, and pentavalent cations (Ca^{2+}, Fe^{3+}, Ti^{4+}, and P^{5+}). On the other hand chlorine and the other elements of the seventh group of the periodic table most easily attain the stable configuration by capturing an electron to fill the single vacancy in their outer valence shells. This produces *monovalent anions* (Cl^-, Br^-, F^-, I^-) with a net unbalanced negative charge. Oxygen, sulfur, and elements of the sixth group form *divalent anions* ($O^=$, $S^=$, $Se^=$, $Te^=$).

In a crystal of NaCl, the Na metal has achieved an ionic state (Na^+) through loss of one electron, and Cl has become ionized through the gain of one electron (Cl⁻). Now Na⁺ has the electron configuration of the noble gas Ne (neon) and Cl⁻ the configuration of A (argon) (see Fig. 4.3a). In other words, in the case of NaCl, one may describe the formation of the ionic bond between Na⁺ and Cl⁻ as the result of the exchange of an electron from the metal to the anion. With the elements present in their ionic states, the attraction between their unlike electrostatic charges holds the ions together in a crystal. This type of bonding mechanism is referred to as *ionic* or *electrostatic*.

A solution of sodium chloride is thought of as containing free ions of Na⁺ and Cl⁻. The incessant colliding of molecules of the solvent with the ions keeps them dissociated as long as the free energy of the particles in the solution is kept high. If, however, the temperature is reduced, or the volume of the solution decreased below a certain critical value, the

FIG. 4.3(*a*). Schematic representation of the electron exchange between Na and Cl to form Na⁺ and Cl⁻ ions. (*b*) Schematic representation of electron sharing between two Cl atoms to form Cl₂.

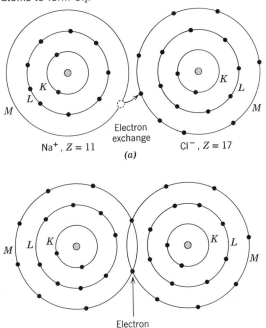

mutual attraction of the opposing electrical charges of Na⁺ and Cl⁻ exceeds the disruptive forces of collision, and the ions lock together to form the atomic beginnings of a crystal. When sufficient numbers of ions attach themselves to the growing crystallite, it settles out of solution as a crystalline precipitate.

A crystal of sodium chloride has characteristic properties by which it may be recognized. The cubic crystal habit, cleavage, specific gravity, index of refraction, and so forth are subject to little variation. These properties in no way resemble those of the shining metal (Na) or the greenish, acrid gas (Cl₂), which are the elemental constituents of the substance. Touching the crystal to the tongue yields the taste of the solution. In other words, the properties conveyed into the crystal by its constituent elements are the properties of the ions, not the elements.

As we have seen in the example of NaCl, the ionic bond confers on crystals in which it is dominant the important property of dissolving in polar solvents, such as water, to yield conducting solutions containing free ions.

Physically, ionic bonded crystals are generally of moderate hardness and specific gravity, have fairly high melting and boiling points, and are very poor conductors of electricity and heat. Because the electrostatic charge constituting the ionic bond is spread over the whole surface of the ion, a cation will tend to surround itself with as many anions as can be fitted around the cation. This means that the ionic bond is of a nondirectional nature, and the symmetry of the resultant crystals is generally high (see Table 4.7).

Ionic Radius

Atoms and ions do not possess definite surfaces and must be thought of as tiny, very dense, and highly charged nuclei surrounded by space that is sparsely inhabited by clouds of electrons whose density varies with distance from the nucleus, falling finally to zero. Hence, the radius of an ion cannot be defined precisely except in terms of its interactions with other ions.

Between any pair of oppositely charged ions there exists an attractive electrostatic force directly proportional to the product of their charges and inversely proportional to the square of the distance between their centers (Coulomb's law). When ions approach each other under the influence of these forces, repulsive forces are set up. These repulsive forces arise from the interaction of the negatively charged electron clouds and from the opposition of the positively charged nuclei and increase rapidly with diminishing internuclear distance. The distance at which these repulsive forces balance the attractive forces is the characteristic interionic spacing for this pair of ions. In the simplest case, when both cations and anions are fairly large and feebly charged and both have numerous symmetrically disposed neighbors of opposite sign, ions may be regarded as spheres in contact. Sodium chloride, in which both cation and anion are monovalent, fairly large, and surrounded by six neighbors of opposite polarity, is a good example. In such crystals, the interionic distance may be regarded as the sum of the radii of the two ions in contact.

The interionic distance may be measured as an interplanar spacing from X-ray diffraction data. Thus, if we assume the radius of one of the ions, the radius of the other may be found. It was in this way that the ionic radii given in Table 4.3 were obtained. The values shown are based in part on a radius of 1.40 Å for the oxygen ion but contain numerous correction factors. These ionic radii apply only to ionic bonded crystals and strictly only to those in which both cations and anions have six closest neighbors. Atoms and ions are not rigid bodies but respond to external electrical forces by dilatation and deformation. A larger number of neighboring ions tends to distend the central ion; a smaller number allows it to collapse a little. Some distortion of shape may accompany the distension of ions. These effects are collectively called *polarization* and are of great importance in crystal structures.

If the bond is not predominantly ionic, the interionic spacing will not agree with the sum of the radii as taken from Table 4.3, even after applying correction for polarization. Thus, the length of the silicon-oxygen bond in silicates is characteristically about 1.6 Å, much less than the sum of the radii taken from the table (1.79 Å). The large discrepancy in the case of the silicon-oxygen bond, which is regarded as about 50% ionic and 50% covalent, may be contrasted with the good agreement of the sum of the tabulated radii with the measured interionic spacing in NaCl. (Sum of the radii from Table 4.3 = 2.78 Å; measured interionic spacing from X-ray data is 2.81 Å.)

As noted earlier (page 130), the size of the ions in the same family in the periodic table increases with atomic number up to the element lanthanum ($Z = 57$). Between elements with $Z = 57$ and $Z = 71$ the lanthanide contraction occurs. As noted also, for any atomic number the size of the ion depends on the state of ionization. Cations tend to be smaller, more rigid, and less extensible than anions, and for a given element the positive ionizations are always markedly smaller than the negative (see, for example S, $Z = 16$ in Table 4.3). Anions are generally larger, more easily polarized, and less sensitive to variations in size with change in valence than cations.

Bond Strength and Physical Properties

The strength of the ionic bond, that is, the amount of energy required to break it, depends on two factors: the center-to-center spacing between the two ions and their total charge. In a crystal with a "pure" ionic bonding mechanism, such as sodium chloride, this law holds rigorously.

The effect of increased interionic distance on the strength of the ionic bond is readily seen in the halides of sodium. Table 4.6a shows the melting points and interionic distances for these compounds. It is apparent that the strength of the bond, as measured by the melting temperature, is inversely proportional to the length of the bond. The melting temperatures of the fluorides of the alkali metals illustrate that it does not matter whether the size of the anion or cation is varied; bond strength varies in inverse proportion to bond length (Table 4.6a). LiF offers an intriguing exception to this generalization, explained by anion-anion repulsion in a structure having a very small cation.

The charge on the coordinated ions has an even more powerful effect on the strength of the bond. Comparison of the absolute values of melting

Table 4.6a

MELTING POINT VS. INTERIONIC DISTANCE IN IONIC-BONDED COMPOUNDS

Compound	Interionic Distance (Å)	M.P. (°C)	Compound	Interionic Distance (Å)	M.P. (°C)
NaF	2.31	988	SrO	2.57	2430
NaCl	2.81	801	BaO	2.76	1923
NaBr	2.98	755			
NaI	3.23	651	LiF	2.01	842
			NaF	2.31	988
MgO	2.10	2800	KF	2.67	846
CaO	2.40	2580	RbF	2.82	775

Data from *Handbook of Chem. and Phys.*, 52nd ed., Chem. Rubber Publishing Co., 1972.

temperature for the alkali-earth oxides, which are divalent compounds, with the absolute values for the monovalent alkali fluorides, in which the interionic spacings are closely comparable, reveals the magnitude of this effect. Although the interionic distance is almost the same for corresponding oxides and fluorides, the bonds uniting the more highly charged ions are obviously much stronger. Table 4.6b illustrates the effect of interionic spacing and charge on hardness. All substances cited as examples in the table have the same structure and may be regarded as ionic bonded.

Covalent Bond

We have seen that ions of chlorine may enter into ionic bonded crystals as stable units of structure because, by taking a free electron from the environment, they achieve a filled outer valence shell. A single atom of chlorine, with a void in its valence shell, is in a highly reactive condition. It seizes upon and combines with almost anything in its neighborhood. Generally, its nearest neighbor is another chlorine atom, and the two unite in such a way that

one electron does double duty in the outer valence shells of both atoms, and both thus achieve the stable inert gas configuration. As a result of this sharing of an electron the two atoms of chlorine are very strongly bound together (see Fig. 4.3b).

This electron-sharing or *covalent* bond is the strongest of the chemical bonds. Minerals so bonded are characterized by general insolubility, great stability, and very high melting and boiling points. They do not yield ions in the dissolved state and are nonconductors of electricity both in the solid state and in solution. Because the electrical forces constituting the bond are sharply localized in the vicinity of the shared electron, the bond is highly directional and the symmetry of the resulting crystals is likely to be lower than where ionic bonding occurs (see Table 4.7). In chlorine, the bonding energy of the atom is entirely consumed in linking to one neighbor, and stable Cl_2 molecules result, which show little tendency to link together. Certain other elements—in general, those near the middle of the periodic table, such as carbon, silicon, aluminum, and sulfur—have two, three, and four vacancies in their outer electron shell, and hence

Table 4.6b

HARDNESS VS. INTERIONIC DISTANCE AND CHARGE IN IONIC-BONDED COMPOUNDS

Compound	Interionic Distance (Å)	Hardness (Mohs)	Compound	Interionic Distance (Å)	Hardness (Mohs)
BeO	1.65	9.0	Na^+F^-	2.31	3.2
MgO	2.10	6.5	$Mg^{2+}O^{-2}$	2.10	6.5
CaO	2.40	4.5	$Sc^{3+}N^{-3}$	2.23	7–8
SrO	2.57	3.5	$Ti^{4+}C^{-4}$	2.23	8–9
BaO	2.76	3.3			

Data from *Crystal Chemistry*, R. C. Evans, Cambridge University Press, 1952.

Table 4.7

EXAMPLES OF PROPERTIES CONFERRED BY THE PRINCIPAL TYPES OF CHEMICAL BOND

Property	Bond Type			
	Ionic (Electrostatic)	Covalent (Electron-shared)	Metallic	Van der Waals (Residual)
Bond Strength	Strong	Very strong	Variable strength, generally moderate	Weak
Mechanical	Hardness moderate to high, depending on interionic distance and charge; brittle	Hardness great Brittle	Hardness low to moderate; gliding common; high plasticity; sectile, ductile, malleable	Crystals soft and somewhat plastic
Electrical	Poor conductors in the solid state; melts and solutions conduct by ion transport	Insulators in solid state and melt	Good conductors; conduction by electron transport	Insulators in both solid and liquid state
Thermal (melting point = m.p.; coefficient of thermal expansion = coef.)	m.p. moderate to high depending on interionic distance and charge; low coef.	m.p. high; low coef.; atoms and molecules in melt	Variable m.p. and coef.; atoms in melt	Low m.p.; high coef.; liquid crystal molecules in melt
Solubility	Soluble in polar solvents to yield solutions containing ions	Very low solubilities	Insoluble, except in acids or alkalis by chemical reaction	Soluble in organic solvents to yield solutions
Structural	Nondirected; gives structures of high coordination and symmetry	Highly directional; gives structures of lower coordination and symmetry	Nondirected; gives structures of very high coordination and symmetry	Nondirected; symmetry low because of shape of molecules
Examples	Halite, NaCl; Calcite, $CaCO_3$; Fluorite, CaF_2; most minerals	Diamond, C; Sphalerite, ZnS; molecules of O_2; organic molecules; graphite (strong bond)	Copper, Cu; Silver, Ag; Gold, Au; Electrum, (Au,Ag); most metals	Iodine, I_2; organic compounds; graphite (weak bond).

not all the bonding energy is consumed in linking to one neighboring atom. Such elements tend, therefore, to unite in covalent bonding with a number of adjacent atoms, forming very stable groups of atoms of rigidly fixed shape and dimensions, which may then link together to form larger aggregates or groups.

Carbon is an outstanding example of such an atom. Carbon atoms have four vacancies in their valence shells which they may fill by electron sharing with four other carbon atoms, forming a very stable, firmly bonded configuration having the shape of a tetrahedron with carbon atoms at the four apices. Every carbon atom is linked in this way to four others to form a continuous network. The energy of the bond is strongly localized in the vicinity of the shared electrons, producing a very rigid and highly polarized structure—that of diamond, the hardest natural substance (see Fig. 7.8a).

Covalent Atomic Radii. In covalent bonded structures, the interatomic distance is generally equal to the arithmetic mean of the interatomic distances in crystals of the elemental substances. Thus, in diamond, the C-C spacing is 1.54 Å; in metallic silicon the Si-Si distance is 2.34 Å. We may therefore suppose that if these atoms unite to form a compound, SiC, the silicon-carbon distance will be near 1.94 Å, the arithmetic mean of the elemental spacings. X-ray measurement determines this spacing in the familiar synthetic abrasive, silicon carbide, as 1.93 Å.

Estimation of the Character of the Bonding Mechanism. It is now generally recognized that there is some electron sharing in most ionic bonded crystals, whereas atoms in covalent bonded substances often display some electrostatic charges. Assessment of the relative proportions of ionic to covalent character is based in part on the polarizing power and polarizability of the ions involved. Compounds of a highly polarizing cation with an easily polarized anion, such as AgI, may show strongly covalent character. In contrast, AgF, because of the lower polarizability of the smaller fluorine ion, is a dominantly ionic bonded compound.

Bonds between elements of the first and seventh groups of the periodic table and between the second and the sixth groups are dominantly ionic. Examples are the alkali halides and the alkali-earth oxides. Bonds between like atoms or atoms close together in the periodic table will be covalent. The silicon-oxygen bond is 50% ionic, the aluminum-oxygen bond 63%, but the boron-oxygen bond only 44% ionic. Linus Pauling has generalized this concept and rendered it quantitative by assigning to each element a numerical value of electronegativity. The greater the difference in electronegativity between any two elements the more ionic the bond between them. (See Table 4.8 and Fig. 4.4.)

Van der Waals' Bond

Returning to the chlorine, Cl_2, molecules, if we take energy from them by cooling the gas, the molecules will ultimately collapse into the close-packed, chaotic liquid state. If still more heat energy is removed, the amplitude of vibrations of the Cl_2 molecules is further reduced, and ultimately the minute, stray electrical fields existing about the essentially

Table 4.8
ELECTRONEGATIVITY OF ELEMENTS[a]

Li	Be	B	C		N	O	F
1.0	1.5	2.0	2.5		3.0	3.5	4.0
Na	Mg	Al	Si		P	S	Cl
0.9	1.2	1.5	1.8		2.1	2.5	3.0
K	Ca	Sc	Ti	Ge	As	Se	Br
0.8	1.0	1.3	1.5	1.8	2.0	2.4	2.8
Rb	Sr	Y	Zr	Sn	Sb	Te	I
0.8	1.0	1.2	1.4	1.8	1.9	2.1	2.5
Cs	Ba						
0.7	0.9						

[a] After L. Pauling, *The Nature of the Chemical Bond*, Cornell University Press, 1960.

electrically neutral atoms will serve to lock the sluggishly moving molecules into the orderly structure of the solid state. This phenomenon of the solidification of chlorine takes place at very low temperatures and warming above $-102°C$ will permit the molecules to break the very weak bonds and return to the disordered state of a liquid.

Neutral molecules such as Cl_2 may develop a small concentration of positive charge at one end, with a corresponding dearth of positive charge

FIG. 4.4. Curve relating the amount of ionic character of a bond A—B to the difference in electronegativity $X_A - X_B$ of the atoms. (After Linus Pauling, *The Nature of the Chemical Bond*, Cornell University Press, Ithaca, 1960.)

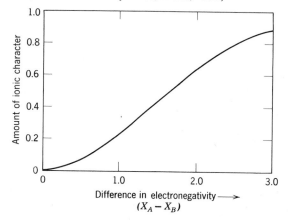

(which results in a small negative charge) at another end. Such molecules would act as very weakly charged dipoles, with unlike charges at opposite ends. In the formation of crystals these molecules are aligned with negative poles against positive poles of neighboring molecules. Such is the mechanism that bonds the Cl_2 molecules in solid Cl_2. This weak bond, which ties neutral molecules and essentially uncharged structure units into a cohesive structure by virtue of small residual charges on their surfaces, is called the *van der Waals'* or residual bond and is one of the weakest of the chemical bonds. Common in organic compounds and solidified gases, it is not often encountered in minerals, but, when it is, it generally defines a zone of ready cleavage and low hardness. An example is the mineral graphite, which consists of covalent bonded sheets of carbon atoms linked only by van der Waals' bonds (see Fig. 4.5a and Table 4.7).

Metallic Bond

Metallic sodium is soft, lustrous, opaque, and sectile and conducts heat and electric current well. X-ray diffraction analysis reveals that it has the regular repetitive pattern of a true crystalline solid. The properties of the metal differ so from those of its salts that we are led to suspect a different mechanism of bonding. Sodium, like all true metals, conducts electricity, which means that electrons are free to move readily through the structure. So prodigal with their electrons are sodium and its close relatives, cesium, rubidium, and potassium, that the impact of the radiant energy of light knocks a considerable number entirely free of the structure. This photoelectric effect, on which such instruments as exposure meters depend, shows that the electrons are very weakly tied into the metal structure. We may thus postulate that the structural units of true metals are really the atomic nuclei bound together by the aggregate electric charge of a cloud of electrons that surrounds the nuclei. Many of the electrons owe no allegiance to any particular nucleus and are free to drift through the structure or even out of it entirely without disrupting the bonding mechanism (see Fig. 4.5b). This type of bond is fittingly called the *metallic bond*. To it metals owe their high plasticity, tenacity, ductility, and conductivity, as well as their generally low hardness, and melting and boiling points. Among minerals, only the native metals display pure metallic bonding. Table 4.7 gives a brief listing of some properties as related to bond type in crystalline materials.

Crystals with More than One Bond Type

Among naturally occurring substances, with their tremendous diversity and complexity, the presence of only one type of bonding is rare, and two or more bond types coexist in most minerals. Where this is

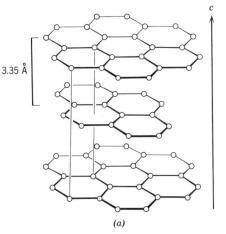

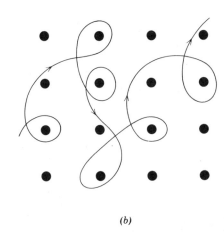

FIG. 4.5(a). Perspective sketch of a graphite structure with covalent bonding between carbon atoms within layers and residual (van der Waals') bonding between layers. Note large separation (3.35 Å) between layers. (b) Schematic representation of a possible electron path between atomic nuclei in a metal.

3.35 Å

(a)

(b)

so, the crystal shares in the properties of the different bond types represented, and often strongly directional properties result. Thus, in the mineral graphite, the cohesion of the thin sheets in which the mineral generally occurs is the result of the strong covalent bonding in the plane of the sheets, whereas the excellent cleavage reflects the weak van der Waals' bond joining the sheets together. The micas, which consist of sheets of strongly bonded silica tetrahedra with a relatively weak ionic bond joining the sheets together through the cations, similarly reflect in their remarkable cleavage the difference in strength of the two bond types. As we shall see, the prismatic habit and cleavage of the pyroxenes and amphiboles, and the chunky, blocky habit and cleavage of the feldspars, may likewise be traced to the influence of relatively weak bonds joining together more strongly bonded structure units having a chain, band, or block shape.

Coordination Principle

When oppositely charged ions unite to form a crystal structure in which the binding forces are dominantly electrostatic, each ion tends to gather to itself, or to *coordinate,* as many ions of opposite sign as size permits. When the atoms are linked by simple electrostatic bonds, they may be regarded as spheres in contact, and the geometry is simple. The coordinated ions always cluster about the central coordinating ion in such a way that their centers lie at the apices of a polyhedron. Thus, in a stable crystal structure, each cation lies at the center of a *coordination polyhedron* of anions. The number of anions in the polyhedron is the *coordination number* (**C.N.**) of the cation with respect to the given anion, and is determined by their relative sizes. Thus, in NaCl each Na^+ has six closest Cl^- neighbors and is said to be in 6 coordination with Cl (**C.N.** 6). In fluorite, CaF_2, each calcium ion is at the center of a coordination polyhedron consisting of eight fluorine ions and hence is in 8 coordination with respect to fluorine (**C.N.** 8).

Anions may also be regarded as occupying the centers of coordination polyhedra formed of cations. In NaCl each chloride ion has six sodium neighbors and hence is in 6 coordination with respect to sodium. Because both sodium and chlorine are in 6 coordination, there must be equal numbers of both, in agreement with the formula, NaCl. On the other hand, examination of the fluorite structure reveals that each fluorine ion has four closest calcium neighbors and hence is in 4 coordination with respect to calcium (**C.N.** 4). Although these four calcium ions do not touch each other, they form a definite coordination polyhedron about the central fluorine ion in such a way that the calcium ions lie at the apices of a regular tetrahedron. Because each calcium ion has eight fluorine neighbors, while each fluorine ion has only four calcium neighbors, it is obvious that there are twice as many fluorine as calcium ions in the structure. This accords with the formula CaF_2, and with the valences for calcium and fluorine.

It is easily seen that the relative sizes of the calcium and fluorine ions would permit a structure containing equal numbers of each with both ions in 8 coordination. The fact that in fluorite only half the possible calcium sites are filled calls attention to an important restriction on crystal structures; viz., *the total numbers of ions of all kinds in any stable crystal structure must be such that the crystal as a whole is electrically neutral.* That is, the total number of positive charges must equal the total number of negative charges; hence, in fluorite there may be only half as many divalent positive calcium ions as there are monovalent negative fluorine ions.

Radius Ratio

Although each ion in a crystal affects every other ion to some extent, the strongest forces exist between ions that are nearest neighbors. These are said to constitute the *first coordination shell*. The geometry of arrangement of this shell and hence the coordination number are dependent on the relative sizes of the coordinated ions. The relative size of ions is generally expressed as a *radius ratio*, $R_X : R_Z$, where R_X is the radius of the cation and R_Z the radius of the anion in Ångstrom units. The radius ratio of sodium and chlorine in halite, NaCl, is therefore

$R_{Na^+} = 0.97 \text{ Å} \qquad R_{Cl^-} = 1.81 \text{ Å}$
$R_{Na^+} : R_{Cl^-} = 0.97/1.81 = 0.54$

The radius ratio of calcium and fluorine in fluorite, CaF_2, is

$R_{Ca^{2+}} = 0.99 \text{ Å} \qquad R_{F^-} = 1.33 \text{ Å}$
$R_{Ca^{2+}} : R_{F^-} = 0.99/1.33 = 0.74$

When two or more cations are present in a structure, coordinated with the same anion, separate radius ratios must be computed for each. Thus, in spinel, $MgAl_2O_4$, both magnesium and aluminum coordinate oxygen anions. Hence,

$R_{Mg^{2+}} = 0.66 \text{ Å} \quad R_{Al^{3+}} = 0.51 \text{ Å} \quad R_{O^{2-}} = 1.40 \text{ Å}$
$R_{Mg^{2+}} : R_{O^{2-}} = 0.66/1.40 = 0.47$
$R_{Al^{3+}} : R_{O^{2-}} = 0.51/1.40 = 0.36$

When coordinating and coordinated ions are the same size, the radius ratio is 1. Trial with a tray of identical spheres, such as pingpong balls, reveals that spherical units may be arranged in three dimensions in either of two ways, called hexagonal closest packing (*HCP*) and cubic closest packing (*CCP*). In either arrangement, each sphere is in contact with its 12 closest neighbors (**C.N.** 12). With reference to Fig. 4.6*a* it is clear that a sphere can be surrounded

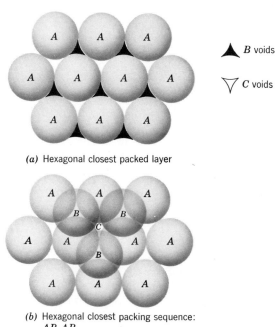

▲ *B* voids

▽ *C* voids

(*a*) Hexagonal closest packed layer

(*b*) Hexagonal closest packing sequence: *AB AB*.....

The plane of this drawing turns out to be perpendicular to the 3-fold axis at the corner of a cube

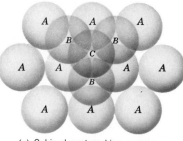

FIG. 4.6. Closest packing of like spheres.

(*c*) Cubic closest packing sequence: *ABC ABC*.....

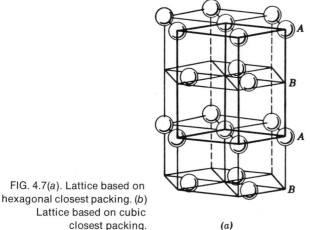

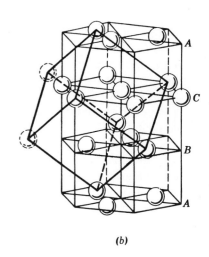

FIG. 4.7(a). Lattice based on hexagonal closest packing. (b) Lattice based on cubic closest packing.

(a)

(b)

by six spheres of equal size, all of which lie in the same plane and touch each other. This is referred to as a *hexagonal closest packed layer* because of the hexagonal arrangement of spheres (marked A in Fig. 4.6a). Between the spheres one can distinguish two types of voids, B and C, on the basis of the orientation of their triangular shapes. In Fig. 4.6b a part of a second hexagonal closest packed layer is superimposed on top of the B voids. This sequence of stacking can be represented by the combination of letters AB. If a third layer is now put on top of the second layer but with the spheres resting in the dimples of the second layer above the A spheres in the first layer, we obtain a sequence ABA, which can be extended upward by another layer of spheres on top of the B voids, giving rise to an indefinite stacking sequence, ABABAB . . . , which is known

as *hexagonal closest packing* (HCP). If, however, we choose to stack the third layer on top of the second layer of the AB sequence in the dimples that directly overlie the C voids in the first layer we form a three-layer sequence ABC. We can continue this type of stacking sequence indefinitely resulting in an ABCABCABC . . . sequence, or *cubic closest packing* (CCP). Figure 4.7 illustrates the hexagonal lattice that underlies the HCP arrangement and the cubic lattice that underlies the CCP arrangement. The space group based on the HCP stacking would be $P\ 6_3/m\ 2/m\ 2/c$, and the space group compatible with CCP stacking is $F\ 4/m\ \bar{3}\ 2/m$ (the faces of the cubic lattice are populated by spheres as expressed by F). Twelve-fold coordination is rare in minerals, with the exception of the native metals; they, and many alloys, have structures based on

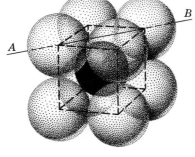

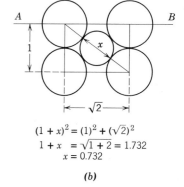

Fig. 4.8(a). Cubic or 8 coordination of Z ions about an X ion. $R_X : R_Z > 0.732$. (b) Limiting conditions for cubic coordination.

(a)

$$(1 + x)^2 = (1)^2 + (\sqrt{2})^2$$
$$1 + x = \sqrt{1 + 2} = 1.732$$
$$x = 0.732$$

(b)

HCP or CCP.

When the coordinating cation is slightly smaller than the anions, 8 coordination results. This is also called *cubic coordination* for the centers of the anions lie at the eight corners of a cube (Fig. 4.8a). If we consider a cubic coordination polyhedron in which the anions touch each other as well as the central cation, we may compute the limiting value of radius ratio for **C.N.** = 8. Allowing the radius of the anion to equal unity, the radius of the cation for this limiting condition must be 0.732 (Fig. 4.8b). Hence, cubic coordination has maximum stability for radius ratios between 0.732 and 1.000.

For values of radius ratio less than 0.732, 8 coordination is not as stable as 6, in which the centers of the coordinated ions lie at the apices of a regular octahedron. Six-fold coordination is accordingly called *octahedral coordination* (Fig. 4.9a). We may, as before, calculate the limiting value of radius ratio for the condition in which the six coordinated anions touch each other and the central cation. The lower limit of radius ratio for stable 6 coordination is found to be 0.414 (Fig. 4.9b). Hence we may expect six to be the common coordination number when the radius ratio lies between 0.732 and 0.414. Na and Cl in halite, Ca and CO_3 in calcite, the *B*-type cations in spinel and many cations in silicates are examples of 6 coordination.

Figure 4.10 illustrates the change from 6 to 8 coordination in the alkali chlorides with increasing

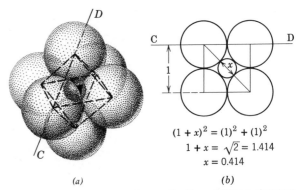

$$(1 + x)^2 = (1)^2 + (1)^2$$
$$1 + x = \sqrt{2} = 1.414$$
$$x = 0.414$$

(a)   *(b)*

FIG. 4.9(a). Octahedral or 6 coordination of *Z* ions about an *X* ion. $R_X : R_Z = 0.732-0.414$. (b) Limiting condition for octahedral coordination.

ionic radius of the cation. It is interesting to note that rubidium may be in both 6 and 8 coordination and thus rubidium chloride is polymorphous.

For values of radius ratio less than 0.414, 6 coordination is not as stable as 4 coordination in which the centers of the coordinated ions lie at the apices of a regular tetrahedron. Four-fold coordination is therefore called *tetrahedral coordination* (Fig. 4.11a). Calculating the limiting value of radius ratio for the condition in which four coordinated anions touch each other and the central ion, we find it to be 0.225 (Fig. 4.11b). Therefore, we may expect 4 to be the common coordination number when the radius ratio lies between 0.414 and 0.225. Tetrahedral coordination is typified by the SiO_4 group in sili-

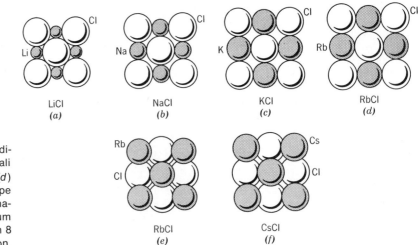

LiCl *(a)*   NaCl *(b)*   KCl *(c)*   RbCl *(d)*

RbCl *(e)*   CsCl *(f)*

FIG. 4.10. Change in coordination from 6 to 8 in alkali chlorides. (a), (b), (c), and (d) have sodium chloride type structures with 6 coordination; (e) and (f) have cesium chloride structures with 8 coordination.

G is location of
center of small ion,
in center of tetrahedron

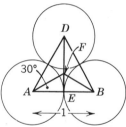

In base triangle

$$\cos 30° = \frac{AE}{AF}$$

$$\therefore AF = \frac{AE}{\cos 30°} = \frac{½}{\cos 30°}$$

$$= ½ \cdot \frac{2}{\sqrt{3}} = \frac{1}{\sqrt{3}}$$

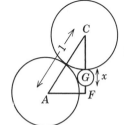

In vertical triangle CAF

$$CF = \sqrt{AC^2 - AF^2} =$$

$$\sqrt{(1)^2 - \left(\frac{1}{\sqrt{3}}\right)^2} = \sqrt{1 - \frac{1}{3}} =$$

$$\frac{\sqrt{2}}{3} = .81649$$

Also $CG = ¾\ CF$, because center of
tetrahedron G is ¼ up from the base.

Furthermore $CG = ½ + ½x$

$$\therefore ½ + ½x = ¾ \cdot .81649 = .6124$$

$$\therefore ½x = .612 - .5 - .1124$$

$$x = 0.225$$

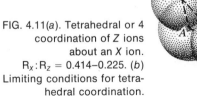

FIG. 4.11(a). Tetrahedral or 4
coordination of Z ions
about an X ion.
$R_X : R_Z = 0.414-0.225$. (b)
Limiting conditions for tetrahedral coordination.

(a)

(b)

cates, by the *A*-type ion in spinel, and by the ZnS structure.

Triangular or *3 coordination* is stable between limits of 0.225 and 0.155 (see Figs. 4.12a and b) and is common in nature in CO_3, NO_3 and BO_3 groups.

Linear or *2 coordination* (see Fig. 4.13) is very rare in ionic bonded crystals. Examples are the uranyl group $(UO_2)^{2+}$, the nitrite group $(NO_2)^=$, and copper with respect to oxygen in cuprite, Cu_2O.

Although rare, examples of 5, 7, 9, and 10 coordination are known. Such coordination numbers are possible only in complex structures and result from the filling of interstices between other coordination polyhedra.

The coordination polyhedra are frequently distorted. The smaller and more strongly polarizing the coordinating cation, or the larger and more polarizable the anion, the greater the distortion and the wider the departure from the theoretical radius ratio limits. Also, if the bonding mechanism is not purely ionic, radius ratio considerations may not be safely used to determine the coordination number.

Obviously, every ion in a crystal structure has some effect on every other ion—it is attractive if the charges are opposite, repulsive if the same. Hence, ions tend to group themselves in crystal structures in such a way that cations are as far apart as possible yet consistent with the coordination of the anions that will result in electrical neutrality. Thus, when cations share anions between them, they do so in such a way as to place themselves as far apart as possible. Hence, the coordination polyhedra

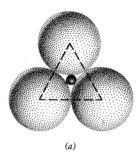

(a)

FIG. 4.12(a). Triangular or 3 coordination of Z ions about an X ion. $R_X : R_Z = 0.225–0.155$. (b) Limiting conditions for triangular coordination.

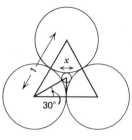

$$\cos 30° = \frac{\frac{1}{2}}{\frac{1}{2} + \frac{1}{2}x}$$

$$\frac{1}{2} + \frac{1}{2}x = \frac{\frac{1}{2}}{\cos 30°} = \frac{\frac{1}{2}}{0.8660} = 0.5774$$

$$\frac{1}{2}x = 0.5774 - 0.50 = 0.0774$$

$$x = 0.155$$

(b)

formed around each are linked more commonly through corners than through edges or faces (Fig. 4.14). Cations tend to share as small a number of anions as possible, and sharing of as many as three or four anions is rare.

Pauling's Rules. Every stable crystal structure bears witness to the operation of some broad generalizations which determine the structure of solid matter. These principles were enunciated in 1929 by Linus Pauling in the form of the following five rules:

Rule 1. A coordination polyhedron of anions is formed about each cation, the cation-anion distance being determined by the radius sum and the coordination number (i.e., number of nearest neighbors) of the cation by the radius ratio.

Rule 2. The electrostatic valency principle. In a stable crystal structure the total strength of the valency bonds which reach an anion from all the neighboring cations is equal to the charge of the anion.

FIG. 4.13. Linear or 2 coordination of Z ions about an X ion. $R_X : R_Z < 0.155$.

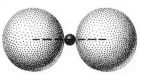

Rule 3. The existence of edges, and particularly of faces, common to two anion polyhedra in a coordinated structure decreases its stability (see Fig. 4.14); this effect is large for cations with high valency and small coordination number and is especially large when the radius ratio approaches the lower limit of stability of the polyhedron.

Rule 4. In a crystal containing different cations, those of high valency and small coordination number tend *not* to share polyhedral elements with each other.

Rule 5. The principle of parsimony. The number of essentially different kinds of constituents in a crystal tends to be small, because, characteristically, there are only a few types of contrasting cation and anion sites. Thus, in structures with very complex compositions, a number of different ions may occupy the same structural position (site). These ions must be considered as a single "constituent," as stated in rule 5.

Rule 2, which concerns the *electrostatic valency principle,* needs further examination. The strength of an electrostatic bond may be defined as an ion's valence charge divided by its coordination number. The resulting number, called the *electrostatic valency* (e.v.) is a measure of the strength of any of the bonds reaching the coordinating ion from its nearest neighbors. For instance, in NaCl the Na$^+$

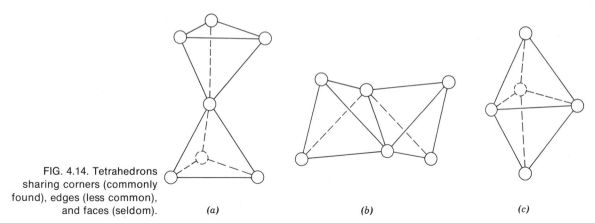

FIG. 4.14. Tetrahedrons sharing corners (commonly found), edges (less common), and faces (seldom).

(a) (b) (c)

is surrounded by six Cl^- neighbors, and each of the bonds reaching Na^+ has a strength (e.v.) of $\frac{1}{6}$. The Cl^- also has six neighbors (Na^+) so that the e.v. for each of the bonds reaching Cl^- is also $\frac{1}{6}$. Crystals in which all bonds are of equal strength are called *isodesmic*.

This generalization is so simple that it seems trivial, but in some cases unexpected results emerge from calculation of electrostatic valencies. For example, minerals of the spinel group have formulas of the type AB_2O_4, where A is a divalent cation such as Mg^{2+} or Fe^{2+} and B is a trivalent cation such as Al^{3+} or Fe^{3+}. Such compounds have been called aluminates and ferrates, by analogy with such compounds as borates and oxalates. This nomenclature suggests that ionic clusters or radicals are present in the structure. X-ray data reveal that the A ions are in 4 coordination whereas the B ions are in 6 coordination. Hence, for the A ions, e.v. $= \frac{2}{4} = \frac{1}{2}$; and, for B ions, e.v. $= \frac{3}{6} = \frac{1}{2}$. All bonds are the same strength and such crystals are isodesmic multiple oxides, not oxysalts.

When small, highly charged cations coordinate larger and less strongly charged anions, compact, firmly bonded groups result, as in the carbonates and nitrates. If the strength of the bonds within such groupings is computed, the numerical value of the e.v. is always greater than one-half the total charge on the anion. This means that in such groups, the anions are more strongly bonded to the central coordinating cation than they can possibly be bonded to any other ion. For example, in the carbonate group, C^{+4} is in 3 coordination with O^{-2} and hence we may compute the e.v. $= \frac{4}{3} = 1\frac{1}{3}$. This is greater than one-half the charge on the oxygen ion, and hence a functional group, or radical, exists, This is the carbonate triangle, the basic structural unit of carbonate minerals (see Fig. 4.17). Another example is the sulfate group. O^{-2} is in 4 coordination with S^{+6}; hence the e.v. $= \frac{6}{4} = 1\frac{1}{2}$. Because this is greater than one-half the charge on the oxygen ion, the sulfate radical forms a tightly knit group, and oxygen is more strongly bonded to sulfur than it can be bound to any other ion in the structure. This is the tetrahedral unit that is the fundamental basis of the structure of all sulfates. Compounds such as sulfates and carbonates are said to be *anisodesmic*.

Of course, it must be understood that if the tightly bonded groups are regarded as single structural units, then in a compound such as $CaCO_3$, calcite, all Ca to $(CO_3)^{-2}$ group bonds are of equal strength and resemble those of an isodesmic crystal. Likewise, in simple sulfates like barite, all Ba to $(SO_4)^{-2}$ group bonds may be equal in strength. The crystals are, however, called anisodesmic because of the presence of the strong $C—O$ and $S—O$ bonds in addition to the weaker Ca to $(CO_3)^{-2}$ group and Ba to $(SO_4)^{-2}$ group bonds.

Logic requires yet another case: that in which the strength of bonds joining the central coordinating cation to its coordinated anions equals exactly half the bonding energy of the anion. In this case each anion may be bonded to some other unit in the structure just as strongly as it is to the coordinating cation. The other unit may be an identical cation, and the anion shared between two cations may enter into the coordination polyhedra of both. Let us consider the case of boron-oxygen groupings. B^{3+} has an ionic radius 0.23 Å. Hence the radius ratio with oxygen is 0.164, indicating triangular coordination, and the e.v. will hence be $\frac{3}{3} = 1$. This is half the bonding strength of the oxygen ion. Consequently, a BO_3 triangle may link to some other ion just as strongly as to the central B^{3+} ion. If this ion is another B^{3+}, two triangles may result, linked through a common oxygen to form a single $(B_2O_5)^{-4}$ group. In similar fashion, BO_3 triangles may join, or *polymerize,* to form chains, sheets, or boxworks by sharing oxygen ions. Such crystals are called *mesodesmic,* and the most important example is the silicates.

All ionic crystals may be classified on the basis of the relative strengths of their bonds into isodesmic, and anisodesmic crystals.

CRYSTAL STRUCTURE

A knowledge of the structure of crystalline materials is of great importance in the understanding of properties such as cleavage, hardness, density, melting point, refractive index, X-ray diffraction pattern, solid solution, and exsolution. Furthermore, our knowledge of atomic and ionic sizes, and of the nature of the chemical bonds in crystals, derives from the precise determinations of crystal structures.

An X-ray crystallographer determines the structure of a crystalline material through the interpretation of the interrelationship of various properties such as external symmetry, X-ray diffraction effects (see Fig. 3.14), density, and chemical composition. The resulting crystal structure provides us with a knowledge of the geometric arrangement of all of the atoms (or ions) in the unit cell, and the bonding and coordination between them. The symmetry of the internal, atomic arrangement of a crystal is expressed by its space group, whereas the morphology of the crystal is reflected in the point group symmetry.

Let us illustrate briefly the chemical and physical interrelationships in the derivation of the simple structure of sodium chloride (NaCl). The external morphology of halite is consistent with $4/m\ \bar{3}\ 2/m$ symmetry. X-ray diffraction data indicate that the unit cell has an *a* axis dimension of 5.64 Å and that the unit cell is not primitive but face-centered (*F*). As no glide planes or screw axes are present (as determined from X-ray single crystal photographs) the space group of halite is $F\ 4/m\ \bar{3}\ 2/m$. This is often abbreviated by crystallographers to *Fm3m*. A chemical analysis yields 39.4 weight percent Na and 60.6 weight percent Cl. Division of these weight percent figures by the appropriate atomic weights yields an atomic ratio for Na:Cl of 1:1, which is expressed as NaCl. The density of halite is 2.165 g/cm³. If the density (*D*) and the volume (*V*) of the unit cell (*a*³) are known then the number of formula weights per unit cell (*Z*) can be calculated from $D = (Z \times M)/(N \times V)$ (see page 189). This number for halite is 4, which means that the unit cell contains 4 NaCl units, or 4Na⁺ and 4Cl⁻ ions. It is now possible to derive the structural arrangement of the ions in NaCl. The Na⁺ (or Cl⁻) ions must be located in a face-centered cubic array. The radius ratio of Na⁺ to Cl⁻ would predict a coordination of 6 about each of the ions. A reasonable and indeed correct interpretation of the crystal structure is given in Fig. 4.15.

Illustration of Crystal Structures

As in the case of illustrations of the external morphology, where clinographic and stereographic projections (see page 51 and page 53) are used to illustrate crystal forms, the three-dimensional arrangement of crystal structures must be represented on a two-dimensional page. A commonly used method is to project the three-dimensional structure onto the page by the use of *fractional cell coordinates,* x, y, and z. For example, as in Fig. 4.16a the location of an atom within a cell with edges a, b, and c can be described in terms of distances from the origin whose components are parallel with the edges. These components are x', y', and

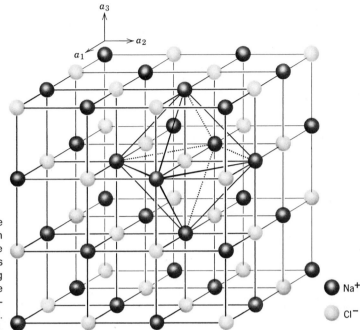

FIG. 4.15. Sodium chloride (NaCl) structure based on face-centered cubic lattice type. Note that each ion is surrounded by 6 neighboring ions of opposite charge (**CN** = 6), outlining octahedral polyhedra (see also Fig. 4.19).

FIG. 4.16(a). Fractional cell coordinates (x, y, z) of an atom in a unit cell as outlined. (b) Atomic arrangement in high tridymite polymorph of SiO_2, projected on (0001). Oxygens occur in the base of the unit cell (0), one quarter $(\frac{1}{4})$, halfway $(\frac{1}{2})$, and three-quarters way up $(\frac{3}{4})$. In this projection some oxygen positions superimpose as shown by the z coordinates, $0, \frac{1}{2}$. Silicon positions, above and below oxygen locations occur at $\frac{1}{16}, \frac{7}{16}, \frac{9}{16}$, and $\frac{15}{16}$th of the z axis.

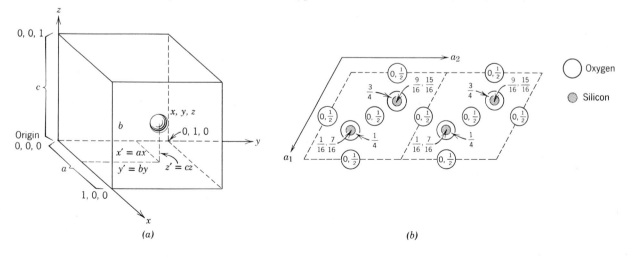

(a)

(b)

z', where $x = x'/a$, $y = y'/b$, and $z = z'/c$; x, y, and z range from 0 to 1. A listing of the fractional coordinates of all atoms (and/or ions) in a unit cell defines the crystal structure and a two-dimensional projection can be constructed. Usually such projections are made onto one of the cell faces, or on planes perpendicular to a specific direction in the crystal (e.g., the crystallographic axes). The fractional heights of atoms above the plane of projection are noted adjacent to them, although values of 0 and 1 corresponding to positions on the bottom and top faces of the unit cell are often omitted. The coordinates 0, 0, 0 specify an atom at the origin of the unit cell. An example of a structure projection on the (0001) plane of high tridymite is shown in Fig. 4.16b. Two unit cells are shown. The x and y coordinates of the oxygen and silicon ions in the structure are shown by their locations in the a_1-a_2 plane. The z coordinates of the ions are shown by fractions indicating their location above the page of the projection. Instead of using fractional coordinates, structure projections may be represented by atomic positions accompanied by numbers ranging from 0 to 100. An atom in the lower face of the unit cell is indicated by 0, one on the top face by 100, and intermediate heights are given accordingly. An

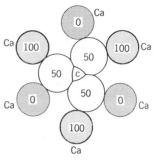

FIG. 4.17. Projection of some of the ion positions in calcite onto (0001). The upper layer of Ca^{2+} ions is shown by the number 100; the Ca^{2+} ions in the lower layer by 0. Note the triangular outline of the CO_3 group.

atom marked 75, for example, is located at three-quarters up the unit cell. An example of such an illustration is given in Fig. 4.17.

For some purposes it may be useful to represent a crystal structure in terms of coordination polyhedra instead of the locations of the atoms or ions. Fig. 4.18 illustrates the structure of high tridymite in terms of the linkage of SiO_4 tetrahedra (compare with Fig. 4.16b). To aid in the visualization of complex crystal structures, crystal structure models can be built or obtained commercially. Such models re-

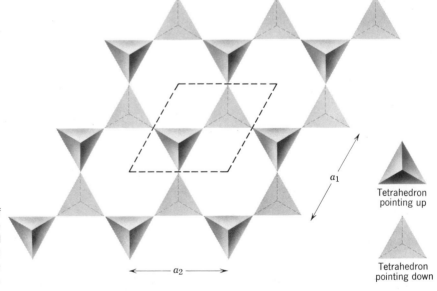

FIG. 4.18. A single sheet of the high tridymite structure (polymorph of SiO_2) projected onto the (0001) plane. Unit cell is outlined. Compare with Fig. 4.16b.

Tetrahedron pointing up

Tetrahedron pointing down

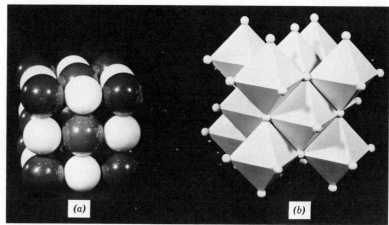

FIG. 4.19(*a*). Close packed model of NaCl structure. Na⁺ ions light; Cl⁻ ions dark. (*b*) Polyhedral representation of octahedral coordination of Cl⁻ about Na⁺ (as well as Na⁺ about Cl⁻) in NaCl.

produce the internal atomic arrangement on an enormously magnified scale (e.g., 1 Å might be represented by 1 cm in a model). Several types of models are illustrated in Figs. 4.19 and 4.20. A *close-packed model* is most realistic in terms of the representation of the filling of space, on an atomic scale (see Fig. 4.19a). It is difficult to visualize the interior of such a model, however. A *polyhedral model** (Figs. 4.19b and 4.20a) represents the coordination polyhedra instead of specific atomic locations. Note in Fig. 4.19 the very different appearances of two models, both of which represent the NaCl structure. A polyhedral model of a complex structure, such as in Fig. 4.20a, is most useful in visualizing the regular repeat units of the structure. This often idealized regularity of the polyhedral arrangement also allows for a relatively easy identification of the symmetry elements of the structure. An *open "ball and stick" model** (Fig. 4.20b) allows the viewing of the coordination inside the model and portrays best the irregularly coordinated environment of ions in complex structures.

Structure Type

Although it seems that uraninite, UO_2, and fluorite, CaF_2, have little in common, their X-ray powder pat-

* Polyhedral models available from Gary R. Gaynor, Minneapolis, Minn. 55406; ball and stick models from Crystal Structures Limited, Bottisham, Cambridge, England.

terns show analogous lines, although different in spacing and intensity. Structure analysis reveals that the U^{4+} ions in uraninite are in 4-fold coordination with respect to oxygen, whereas eight O^{-2} are grouped about each uranium. In fluorite, four Ca^{2+} are grouped about each fluorine, and eight F^{-1} are packed about each calcium. Uraninite and fluorite have structures that are analogous in every respect, although the cell dimensions are different and other properties are, of course, totally dissimilar. These two substances are said to be *isostructural,* or *isotypous,* and belong to the same *structure type.* Crystals in which the centers of the constituent atoms occupy geometrically similar positions, regardless of the size of the atoms or the absolute dimensions of the structure, are said to belong to the same structure type. For example, all crystals in which there are equal numbers of cations and anions in 6 coordination belong to the NaCl structure type. A few of the large number of minerals of diverse composition belonging to this structure type are: KCl, sylvite; MgO, periclase; NiO, bunsenite; PbS, galena; MnS, alabandite; AgCl, cerargyrite; TiN, osbornite.

Of great importance in mineralogy is the concept of the *isostructural group:* a group of minerals related to each other by analogous structures, generally having a common anion, and frequently displaying extensive ionic substitution. Many groups of minerals are isostructural, of which the barite group

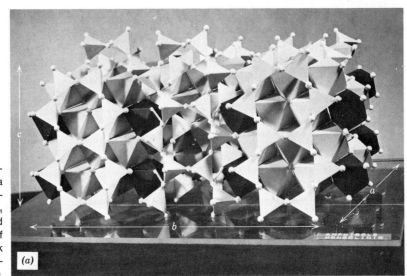

FIG. 4.20(a). Polyhedral structure model representing a monoclinic amphibole. Tetrahedral linking of SiO_4 (Si_4O_{11} chains parallel to c axis) and octahedral coordination of cations between chains. Dark octahedra represent M_4 position in structure.

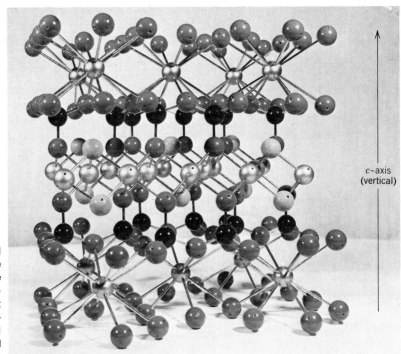

FIG. 4.20(b). Expanded "ball and stick" model of the structure of biotite $K(Mg,Fe)_3(AlSi_3O_{10})(OH)_2$. Top and bottom layers represent K^+ ions in 12 coordination with oxygen. Central layer represents Mg^{2+} and Fe^{2+} in 6 coordination with O^{2-} and $(OH)^-$. (b)

Table 4.9

ARAGONITE GROUP OF ORTHORHOMBIC ISOSTRUCTURAL CARBONATES (SPACE GROUP *Pnam*)

Mineral	Chemical Composition	Cation Size (Å)		Unit Cell Dimensions (Å)			110 ∧ 1$\bar{1}$0
				a	b	c	
Aragonite	CaCO$_3$	Ca^{2+}	0.99	5.72	7.94	4.94	63°48'
Strontianite	SrCO$_3$	Sr^{2+}	1.12	6.08	8.40	5.12	62°41'
Cerussite	PbCO$_3$	Pb^{2+}	1.20	6.13	8.48	5.17	62°46'
Witherite	BaCO$_3$	Ba^{2+}	1.37	6.39	8.83	5.28	62°12'

of sulfates, the calcite group of carbonates, and the aragonite group of carbonates are perhaps the best examples. The extremely close relationship that exists among the members of many groups is illustrated by the aragonite group listed in Table 4.9.

Polymorphism

The ability of a specific chemical substance to crystallize with more than one type of structure is known as *polymorphism*. The various structures of such a chemical element or compound are known as *polymorphic forms,* or *polymorphs*. Examples of some polymorphous minerals are given in Table 4.10. An example of the polymorphism of H$_2$O is shown as a stability diagram in Fig. 4.30.

Three major types of mechanisms are recognized by which one polymorphic form of a substance can change to another. These are: *displacive, reconstructive,* and *order-disorder* polymorphism.

In a *displacive* polymorphic reaction the internal adjustment in going from one form to another is very small and requires little energy. The structure is generally left completely intact and no bonds between ions must be broken; only a slight displacement of atoms (or ions) and readjustment of bond angles ("kinking") between ions is needed. This type of transformation occurs instantaneously and is reversible. An example of such a displacive transformation occurs when the low quartz form of SiO$_2$ is heated to above 573°C (at atmospheric pressure; see also Fig. 10.69) and rearranges its structure to that of high quartz. This difference between the two forms of quartz is expressed by their space groups (low quartz, $P3_121$; high quartz, $P6_222$) and can be portrayed in a basal projection of the SiO$_2$ framework for both forms (see Fig. 4.21). The structural arrangement in the low-temperature form is slightly less symmetric and somewhat more dense than that of the high-temperature form. The transi-

Table 4.10

EXAMPLES OF POLYMORPHOUS MINERALS

Composition	Mineral Name	Crystal System and Space Group	Hardness	Specific Gravity
C	Diamond	Isometric—*Fd3m*	10	3.52
	Graphite	Hexagonal—*P6$_3$/mmc*	1	2.23
FeS$_2$	Pyrite	Isometric—*Pa3*	6	4.99
	Marcasite	Orthorhombic—*Pnnm*	6	4.85
CaCO$_3$	Calcite	Rhombohedral—*R$\bar{3}$c*	3	2.71
	Aragonite	Orthorhombic—*Pnam*	3$\frac{1}{2}$	2.93
SiO$_2$	Low quartz	Rhombohedral—*P3$_1$21*	7	2.65
	High quartz	Hexagonal—*P6$_2$22*		2.53
	High tridymite	Hexagonal—*C6/mmc*	7	2.20
	High cristobalite	Isometric—*P2$_1$3(?)*	6$\frac{1}{2}$	2.20
	Coesite	Monoclinic—*C2/c*	7$\frac{1}{2}$	3.01
	Stishovite	Tetragonal—*P4/mnm*		4.30

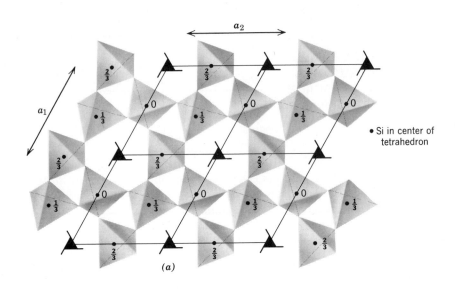

FIG. 4.21(a). Projection of the tetrahedral SiO_2 framework of low quartz onto (0001). Four unit cells and some 3_1 axes are shown. Fractional heights represent locations of centers (Si) of tetrahedra. (b) Projection of the tetrahedral SiO_2 framework of high quartz onto (0001). Four unit cells and 6_2 axes are shown. Fractional heights represent locations of centers (Si) of tetrahedra.

tion can be viewed as the result of "kinking" and "unkinking" of atomic bonds in the structures. Because the high-temperature form of quartz (high quartz) shows a higher symmetry than the low-temperature form twinning ("transformation twinning") may result in the transition from high to low quartz. Such twins, known as Dauphiné twins (see

Fig. 2.174c) represent a megascopic expression of the presence of obversed and reversed units of structure in the low quartz crystal.

In a *reconstructive* polymorphic reaction the internal rearrangement in going from one form to another is extensive. It involves the breaking of atomic bonds and a reassembly of the structural units in a different arrangement. This type of transformation requires a large amount of energy, is nonreversible and very sluggish. An example of such a polymorphic reaction is the change from tridymite (or cristobalite) to low quartz (see Fig. 4.18 for tridymite structure). Cristobalite and tridymite are formed at high temperatures and relatively low pressures, such as in SiO_2-rich lava flows. A high activation energy is needed to reconstruct the cristobalite (or tridymite) SiO_2 network in the arrangement of the low quartz structure. Cristobalite and tridymite are metastable in terms of atmospheric conditions; however, both minerals are abundantly present in very old terrestrial volcanic flows as well as in Precambrian lunar lavas. Such persistence of metastable minerals testifies to the fact that high energy is required to activate the polymorphic transformation. The differences between the polymorphs for C, FeS_2, and $CaCO_3$ (Table 4.10) are such that a major reworking and rearrangement of the structure is needed to go from one structure type to the other; each of these examples represent reconstructive polymorphism (see also graphite-diamond relations and calcite-aragonite discussion on pages 228 and 303, respectively).

Yet another type of polymorphism is referred to as an *order-disorder* transformation. This is most frequently observed in alloys but occurs also in minerals. An alloy of composition AB with 50% of A and 50% of B can exist in various states of disorder of which a totally disordered and a perfectly ordered state are two extreme conditions. In a perfectly ordered state atoms of A are arranged in perfect and regular repeat with respect to B atoms (see Fig. 4.22a). Atom A is always in structural site 1 and atom B in structural site 2. A somewhat disordered state for the atomic arrangement of such an alloy is shown in Fig. 4.22b. Here the distribution is not perfectly ordered and yet also not random. Out of every four no. 1 sites three are occupied by A (on the average) and one is occupied by B (on the average). The opposite holds for site no. 2. In other words, in Fig. 4.22b the probability ratio for site 1 being occupied by atoms of A rather than B is 3:1. A state of total disorder for the alloy AB is shown schematically in Fig. 4.22c. Total disorder, on an atomic scale, implies equal probability of finding A or B

atoms in a specific site in the structure. In other words, the probability of a given atomic site being occupied by one type of atom instead of another equals 1. This means that for a graphical representation, as in Fig. 4.22c, each atomic position is indicated to represent A as well as B occupancy (in a statistical sense).

There is no definite transition point between perfect order and complete disorder. Perfect order is possible only at 0° Kelvin (absolute zero) and with increasing temperature the degree of order generally decreases. At high temperatures, close to but below the melting point of a substance, atoms (or ions) tend to become completely disordered and are indeed ready to break away from the structure.

An example of order-disorder polymorphism in a mineral is shown by potassium feldspar ($KAlSi_3O_8$) in which Al occupies a structural position identical with and replacing Si in the mineral. The high-temperature form, sanidine, with space group $C\ 2/m$, shows a disordered distribution of Al in the SiO_2 network. The low-temperature K-feldspar, microcline with space group $C\overline{1}$, however, shows an ordered distribution of Al in the SiO_2 network (see page 421). Intermediate order (or disorder) states are present between the high-temperature sanidine and the low-temperature microcline.

Polytypism

A special variety of polymorphism, known as *polytypism,* occurs when two polymorphs differ only in the stacking of identical, two-dimensional sheets or layers. As a consequence, the unit cell dimensions parallel to the sheets will be identical in the two polytypes. However, their unit cell lengths perpendicular to the stacking sheets (or layers) will be related to each other as multiples or submultiples. Polytypism is a well-known feature of SiC, ZnS, and the micas. The only difference between sphalerite (ZnS) and its polymorph wurtzite (ZnS) is that the S^{-2} ions in sphalerite are in cubic closest packing whereas in wurtzite they are in hexagonal closest packing. However, wurtzite shows extensive polytypism as reflected in c dimensions of the unit-cell. Examples of wurtzite polytypes and their c dimensions are: *4H*, 12.46 Å; *6H*, 18.73 Å; *8H*, 24.96 Å; and *10H*,

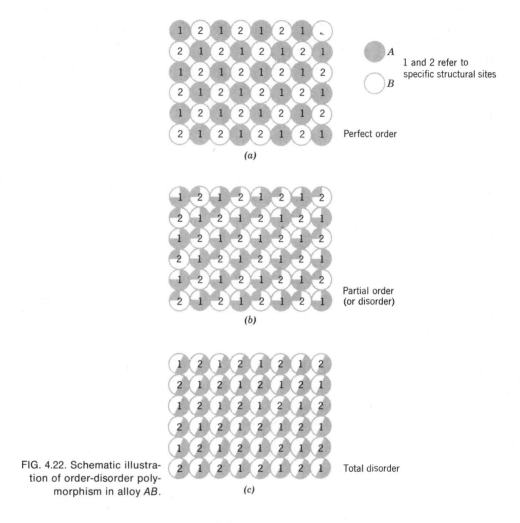

A

B

1 and 2 refer to specific structural sites

Perfect order

(a)

Partial order (or disorder)

(b)

Total disorder

(c)

FIG. 4.22. Schematic illustration of order-disorder polymorphism in alloy *AB*.

31.20 Å (*H* refers to hexagonal cell). The "thickness" of the basic ZnS layer in these polytypes is 3.12 Å. The examples show multiples of this basic unit in their *c* dimension. The polytypism of the micas will be discussed in the section on phyllosilicates (see Fig. 10.63).

Pseudomorphism

The existence of a mineral with the outward crystal form of another mineral species is known as *pseudomorphism*. If a crystal of a mineral is altered so that the internal structure or chemical composition is changed but the external form is preserved, it is

called a *pseudomorph*, or *false form*. The chemical composition and structure of a pseudomorph belong to one mineral species, whereas the crystal forms correspond to another. For example, pyrite, FeS_2, may change to limonite, $FeO \cdot OH \cdot nH_2O$, but preserve all the external features of the pyrite. Such a crystal is described as a pseudomorph of *limonite after pyrite*. Pseudomorphs are usually further defined according to the manner in which they were formed, as by:

1. Substitution. In this type of pseudomorph there is a gradual removal of the original material and a corresponding and simultaneous replacement of it by another with no chemical reaction between the

two. A common example of this is the substitution of silica for the wood fiber to form petrified wood. Another example is quartz, SiO_2, after fluorite, CaF_2.

2. Encrustation. In the formation of this type of pseudomorph a crust of one mineral is deposited over crystals of another. A common example is quartz encrusting cubes of fluorite. The fluorite may later be entirely carried away by solution, but its former presence is indicated by the casts left in the quartz.

3. Alteration. In this type of pseudomorph there has been only a partial addition of new material, or a partial removal of the original material. The change of anhydrite, $CaSO_4$, to gypsum, $CaSO_4 \cdot 2H_2O$, and the change of galena, PbS, to anglesite, $PbSO_4$, are examples of alteration pseudomorphs. A core of the unaltered mineral may be found in such pseudomorphs.

Mineraloids (Noncrystalline Minerals)

The definition of a mineral states that "a mineral . . . has an ordered atomic arrangement." There are, however, a number of noncrystalline, natural solids classed as either *amorphous* or *metamict* minerals.

Amorphous minerals include gel minerals and glasses. Gel minerals (mineraloids) are usually formed under conditions of low temperature and low pressure and commonly originate during weathering processes. They characteristically occur in mammillary, botryoidal, and stalactitic masses. The ability of amorphous materials to adsorb other substances accounts for their often wide variations in chemical composition. A well-known example of a not totally amorphous material is opal. Its chemical composition can be represented as $SiO_2 \cdot nH_2O$ with an average range of H_2O content from 4 to 9 weight percent. Opal was generally considered to be completely without internal structure, however careful X-ray studies show that it contains an ordered arrangement of very small SiO_2 spheres. Another amorphous mineral example is limonite $FeO \cdot OH \cdot nH_2O$.

Other types of naturally occurring amorphous solids are known as *metamict* minerals. These originally formed as crystalline solids but their crystal structure has been destroyed by radiation from radioactive elements. As one expects of amorphous substances, metamict minerals do not diffract X-rays and are optically isotropic. In addition, they have no cleavage but break with a conchoidal fracture. Many metamict minerals are bounded by crystal faces and are thus amorphous pseudomorphs after an earlier crystalline phase. When a metamict mineral is heated its crystal structure may be reconstituted and its density is increased. The recrystallization is frequently accompanied by the internal generation of so much heat that the mineral becomes incandescent.

All metamict minerals are radioactive and it is believed that the structural breakdown results from the bombardment of alpha particles emitted from radioactive uranium or thorium contained in these minerals. However, the process is not well understood for some minerals containing less than 0.5% of these elements may be metamict, whereas others containing high percentages of them may be crystalline.

CHEMICAL COMPOSITION

Chemical analyses of minerals are obtained by a variety of analytical techniques. Prior to 1947 quantitative mineral analyses were obtained mainly by "wet" analytical techniques in which the mineral is dissolved in a solution from which various insoluble compounds are precipitated by reaction. From the weights of the precipitates one obtains a *gravimetric analysis*. Since 1947 various "rapid" analytical schemes have been used, which offer increased speed of determination at the expense of some accuracy. Since 1960 the majority of analyses have been made by instrumental techniques such as optical emission, X-ray fluorescence, atomic absorption spectroscopy, and electron microprobe analysis. Each of these techniques has its own specific sample preparation requirements and fairly well-established error ranges. The results of any analysis procedure are generally represented in a table of weight percentages of the elements or oxide components in the mineral analyzed. "Wet" analytical techniques allow for the quantitative determination of various oxidation states of cations (such as Fe^{2+} vs. Fe^{3+}) and also for the determination of the H_2O content of hydrous minerals. Instrumental methods generally do not provide information as to the oxidation state of elements or the presence or absence of H_2O.

Minerals submitted for quantitative or qualitative chemical analysis must consist of one mineral species only (the one being analyzed for) and must be free of weathering or other alteration products. As it is often very difficult, if not impossible, to separate as much as $\frac{1}{10}$ gram to 1 gram of clean material for analysis, analysis results not infrequently must be accompanied by a statement concerning the amount of impurities. An instrumental method, such as electron probe microanalysis, allows for quantitative analysis, *in situ* (in a polished section), of mineral grains as small as 1 micron in diameter (10^{-3} mm). All of the above mentioned techniques require special analytical skills and often very expensive equipment. Some qualitative analytical tests that can be performed rapidly in the laboratory are described at the end of this chapter under the heading Blowpipe and Chemical Tests.

Recalculation of Analyses

Elements such as gold, arsenic, and sulfur occur in the native state and their formulas are the chemical symbols of the elements. Most minerals, however, are compounds composed of two or many more elements and their formulas, recalculated from the results of quantitative chemical analyses, indicate the atomic proportions of the elements present. Thus, in galena, PbS, there is one atom of sulfur for each atom of lead; and in $CuFeS_2$, there are two atoms of sulfur for each atom of copper and iron. Such specific atomic proportions underlie the definition of a mineral which states that "a mineral . . . has a definite (but generally not fixed) chemical composition." Very few minerals, however, have a truly constant composition; examples of specific and essentially constant compositions are provided by

quartz, SiO_2, and kyanite, Al_2SiO_5. Such minerals are often referred to as *pure substances*. The great majority of minerals show large variations in composition within specific atomic sites of their structures. For example ZnS, sphalerite, shows a large range of Fe content. The Fe substitutes for Zn in the sphalerite structure, and in general terms an Fe-rich (ferroan) sphalerite composition can be expressed as (Zn, Fe)S. Here the total cation-anion ratio is still fixed, 1 : 1, however the Zn and Fe contents are variable.

As noted above, a quantitative chemical analysis provides the basic information for the atomic formula of a mineral. A weight percent analysis in terms of metals, or oxides, gives us a listing of what elements are present and in what quantities, but it gives us no direct information about how the elements (or ions) occur in the structure of the mineral. A quantitative weight percent analysis should add up close to 100%. Slight variations, above or below 100%, occur because of cumulative small errors that are part of the analytical procedure. Let us, as an example, consider the analysis of chalcopyrite in Table 4.11. The percentages in column 1, reported by the chemist, represent the percentages by weight

Table 4.11
CHALCOPYRITE ($CuFeS_2$) RECALCULATIONS

	1 Weight Percent	2 Atomic Weights	3 Atomic Proportions	4 Atomic Ratios	
Cu	34.30	63.54	0.53982	1	
Fe	30.59	55.85	0.54772	1	approx.
S	34.82	32.07	1.08575	2	
Total	99.71				

	Atomic Weights	
Cu	63.54	Cu(in %) $= \dfrac{63.54}{183.53} \times 100\% = 34.62$
Fe	55.85	Fe(in %) $= \dfrac{55.85}{183.53} \times 100\% = 30.43$
S(2×)	(32.07) × 2	S(in %) $= \dfrac{64.14}{183.53} \times 100\% = 34.94$
Total	183.53	Total 99.99

of the various elements. Because the elements have different atomic weights, these percentages do not represent the ratios of the different atoms. To arrive at their relative proportions, the weight percentage in each case is divided by the atomic weight of the element. This gives a series of numbers (column 3), the *atomic proportions,* from which the *atomic ratios* can be quickly derived. In the analysis of chalcopyrite, these ratios are Cu:Fe:S = 1:1:2; thus $CuFeS_2$ is the chemical formula.

The reverse procedure, that of calculating the percentage composition from the formula is shown in the lower half of Table 4.11. The formula of chalcopyrite is $CuFeS_2$, which makes for a gram-formula weight of 183.53. The calculated Cu, Fe, and S weight percentage values are very similar to the measured percentages reported in column 1 of Table 4.11. The calculated values are slightly different from the determined values probably because of a slight experimental error in the analysis technique.

Let us now examine the recalculation of some sphalerite analyses, which show a considerable range in Fe, and of troilite, FeS (Table 4.12). The upper half of the table shows the weight percentages for the various elements and the lower part of the table provides atomic proportions obtained by dividing each weight percentage figure by the appro-

priate atomic weight. One of the lower lines in Table 4.12 shows that the ratio of (Zn + Fe + Mn + Cd):S is constant in the sphalerite analyses, namely 1:1, and that it is 1:1 in the troilite as well. However, the Zn in the sphalerite structure is partially substituted for by variable amounts of Fe, Mn, and Cd. The maximum amount of Fe reported in Table 4.12 is in column 4 which shows 18.25 weight percent Fe. This recalculates to 0.327 atomic proportions out of a total of 1.060 atomic proportions for (Zn + Fe + Mn + Cd). These analyses of naturally occurring sphalerites lead to the conclusion that the representation of sphalerite compositions as ZnS is a great oversimplification. Only analysis 1 (Table 4.12) is pure ZnS and analysis 5 is pure FeS. Analyses 2, 3, and 4 contain variable amounts of Fe, Mn, and Cd. Analysis 4, which contains the largest amounts of elements other than Zn and S, can serve to illustrate some of the useful atomic proportion recalculations. Its total cation content (in atomic percentage) is 1.060. Fe is $0.327/1.060 \times 100\%$ = 30.8%; similarly Mn, Cd, and Zn values are 4.5, 0.2, and 64.4%, respectively. These cation proportion values can be used as subscripts to represent the specific composition of the sample of sphalerite in column 4; this leads to the formulation $(Zn_{64.4}Fe_{30.8}Mn_{4.5}Cd_{.2})S$. The other specific formulas in Table 4.12 were obtained in a similar manner. It

Table 4.12	Weight Percentage[a]				
RECALCULATION					
OF SEVERAL	**1**	**2**	**3**	**4**	**5**
SPHALERITE Fe	0	0.15	7.99	18.25	63.53
ANALYSES AND Mn	0	0	0	2.66	0
Cd	0	0	1.23	0.28	0
OF TROILITE Zn	67.10	66.98	57.38	44.67	0
S	32.90	32.78	32.99	33.57	36.47
Total	100.00	99.91	99.59	99.43	100.00

	Recalculated in Terms of Atomic Proportions				
Fe	0	0.003 ⎫	0.143 ⎫	0.327 ⎫	1.137
Mn	0	0 ⎪	0 ⎪	0.048 ⎪	0
Cd	0	0 ⎬ 1.027	0.011 ⎬ 1.032	0.002 ⎬ 1.060	0
Zn	1.026	1.024 ⎭	0.878 ⎭	0.683 ⎭	0
S	1.026	1.022	1.029	1.047	1.137
(Zn + Fe + Mn + Cd):S	1:1	1:1	1:1	1:1	1:1
Formulas	ZnS	$(Zn_{99.7}Fe_{.3})S$	$(Zn_{85.1}Fe_{13.8}Cd_{1.1})S$	$(Zn_{64.4}Fe_{30.8}Mn_{4.5}Cd_{.2})S$	FeS
Fe:Zn	0:100	0.3:99.7	14.0:86.0	32.4:67.6	100:0

[a] Analyses 1–4 from *Dana's System,* vol. 1, 1944; analysis 5 for pure troilite FeS.

is obvious from the analyses that Zn and Fe are the main variables. Often it is necessary to obtain the cation distribution of such major elements only. For analysis 4 in Table 4.12 we then must eliminate from consideration the quantities of Cd and Mn. The total of Fe + Zn in the atomic proportions column = 1.010, rather than 1.060 for Fe + Mn + Cd + Zn. If now we wish to know the Fe versus Zn percentage in this analysis we factor $0.327/1.010 \times 100\% = 32.4\%$. This value, as well as those similarly obtained for the other analyses, are listed in the last line of Table 4.12. These Fe : Zn ratios show that a selection of sphalerites may show *a compositional range from pure ZnS to $(Zn_{.68}Fe_{.32})S$*. The mineral troilite is found only in meteorites and shows no Zn in its chemical composition; it can therefore be considered as a compound of constant composition, FeS. The large variation of Fe in sphalerites is due to the substitution of Fe for Zn in the same structural site in the sphalerite structure; this is also known as *solid solution* of Fe in ZnS (see page 161). The largest amount of Fe solid solution in the analyses of Table 4.12 is shown by analysis 4, with composition $(Zn_{67.6}Fe_{32.4})S$.

The majority of minerals such as silicates, oxides, carbonates, phosphates, sulfates, etc. are compounds containing large amounts of oxygen. By convention the analyses of such minerals are reported as percentages of oxides, rather than as percentages of metals. Calculations similar to those outlined above are performed with the oxide components to arrive at molecular proportions of the oxides, rather than atomic proportions of the elements. Table 4.13 gives an example of the recalculation of an analysis of gypsum. The analytically determined oxide components in column 1 are divided by the molecular weights of the corresponding oxides (column 2) to arrive at molecular proportions (column 3). From the molecular ratios in column 4 we see that $CaO : SO_3 : H_2O = 1 : 1 : 2$ and the composition can be written as $CaO \cdot SO_3 \cdot 2H_2O$, or as $CaSO_4 \cdot 2H_2O$. The latter formula is preferred because it avoids creating the erroneous impression that the mineral is composed of discrete oxide molecules.

In the lower part of Table 4.13 the chemical analysis of a member of the olivine series is given. The desired result of a recalculation of the analysis again is a statement of the molecular proportions of the oxide components, or of the cation components. The steps in going from columns 1 to 3 are the same as in the gypsum analysis. Column 4 lists the values

Table 4.13
RECALCULATION OF GYPSUM AND OLIVINE ANALYSES

Gypsum	1 Weight Percent	2 Molecular Weights	3 Molecular Proportions	4 Molecular Ratios		
CaO	32.44	56.08	0.57846	1		
SO₃	46.61	80.08	0.58211	1	approx.	
H₂O	20.74	18.0	1.15222	2		
Total	99.79					

Olivine	1 Weight Percent	2 Molecular Weights	3 Molecular Proportions	4 Atomic Proportions Cations	5 Oxygens	6 On Basis of 4 Oxygens	7 Atomic Ratios
SiO₂	34.96	60.09	0.58179	Si 0.5818	1.1636	0.989	1
FeO	36.77	71.85	0.51176	Fe²⁺ 0.5118	0.5118	0.870	
MnO	0.52	70.94	0.00733	Mn 0.0073	0.0073	0.012 } 2.022	2
MgO	27.04	40.31	0.67080	Mg 0.6708	0.6708	1.140	
Total	99.29				2.3535		

Olivine in terms of end member compositions: Mg = 1.140, Fe = 0.870; total: 2.010. Percentage Mg = 56.7; percentage Fe = 43.3; percentage Mg_2SiO_4 = % forsterite = % Fo = 56.7%; percentage Fe_2SiO_4 = % fayalite = % Fa = 43.3%.

for the atomic proportions of the various atoms. The atomic proportions of the metal atoms are the same as the corresponding molecular proportions in column 3, because, for example, one "molecule" of SiO_2 contributes 1 Si and one "molecule" of FeO contributes 1 Fe. However, the number of oxygens contributed by each molecular proportion is not the same as the numbers listed in column 3. For each of the divalent metals there is one oxygen atom, but for each silicon there are two oxygens. In other words, one molecular proportion of FeO contributes 1 oxygen, but one molecular proportion of SiO_2 contributes two oxygens. This is reflected in the numbers of column 5.

The total number of oxygens contributed by the atomic proportions in column 5 is 2.3535. From crystal structure data we know that olivine, with the general formula of $(Mg, Fe)_2 SiO_4$, has four oxygens per formula unit. In order to arrive at the cation proportions in terms of 4 oxygens, we multiply each of the cation numbers in column 4 with the 4/2.3535 ratio. This leads to the numbers in columns 6

and 7. The final formulation for this olivine is $(Mg_{1.12}Fe_{.86}Mn_{.01}) SiO_4$ obtained by assuming 2.0 for total cations, instead of 2.022 as calculated. Such a formula is often expressed in terms of *end member compositions*. In olivine these are Mg_2SiO_4, forsterite, and Fe_2SiO_4, fayalite. The amounts of forsterite (Fo) and fayalite (Fa) in the general formula are directly proportional to the atomic proportions of Mg and Fe, or molecular proportions of MgO and FeO in the analysis. For example, in Table 4.13, this leads to Fo_{57} and Fa_{43}. If the olivine composition ranges mainly between Mg and Fe compositions, it suffices to state the final recalculation as Fo_{57}.

The olivine structure is relatively simple with Mg and Fe substituting for each other in the same structural position (site). Many silicates, however, have various structural sites among which elements can be distributed. An example would be a member of the diopside-hedenbergite series ($CaMgSi_2O_6$-$CaFeSi_2O_6$) in the pyroxenes. Table 4.14 gives a chemical analysis, as well the various recalculation steps for pyroxene similar to that outlined for

Table 4.14 RECALCULATION OF PYROXENE ANALYSIS

	1	2	3	4	5	6		7
	Weight Percent	Molecular Proportions	Cation Proportions	Number of Oxygens	Cations on Basis of 6 Oxygens	Cation Assignment		End Member Recalculation
SiO_2	49.23	0.8193	0.8193	1.6386	1.848	Si	1.848 ⎫	MgO as $MgSiO_3$ (enstatite = En)
TiO_2	2.25	0.0282	0.0282	0.0564	0.064	Ti	0.064 ⎬ 2.0	FeO as $FeSiO_3$ (ferrosilite = Fs)
Al_2O_3	2.01	0.0197	0.0394	0.0591	0.089	Al^{IV}	0.088 ⎭	CaO as $CaSiO_3$ (wollastonite = Wo)
Fe_2O_3	1.95	0.0122	0.0244	0.0366	0.055	Al^{VI}	0.001 ⎫	Using molecular proportions:
FeO	4.53	0.0630	0.0630	0.0630	0.142	Fe^{3+}	0.055 ⎪	MgO = 0.3644
MnO	0.09	0.0013	0.0013	0.0013	0.003	Fe^{2+}	0.142 ⎬ 1.023	FeO = 0.0630
MgO	14.69	0.3644	0.3644	0.3644	0.822	Mn	0.003 ⎪ ≈1	CaO = 0.4321
CaO	24.23	0.4321	0.4321	0.4321	0.974	Mg	0.822 ⎭	Total = 0.8595
Na_2O	0.46	0.0074	0.0148	0.0074	0.033	Ca	0.974 ⎫	%En = 42.39
K_2O	0.15	0.0016	0.0032	0.0016	0.007	Na	0.033 ⎬ 1.014	%Fs = 7.33
				Total O = 2.6605		K	0.007 ⎭ ≈1	%Wo = 50.27
$H_2O(+)$	0.38							
Total	99.89							

Oxygen factor $\dfrac{6}{2.6605} = 2.255215$

olivine. In column 1 of Table 4.14 percentages are reported for FeO as well as Fe_2O_3. In analyses made by instrumental techniques Fe^{3+} and Fe^{2+} cannot be distinguished, and total Fe is then reported as FeO only, or Fe_2O_3 only. As pyroxenes are generally considered to be anhydrous, the small amount of H_2O reported is probably not part of the pyroxene structure and will be neglected in subsequent calculations. Column 2 provides molecular proportions, obtained by dividing the weight percent figures by the appropriate molecular weights. Column 3 lists the cation proportions for each oxide "molecule." Note, for example, that Al_2O_3 contains 2Al, therefore 0.0197 is multiplied by 2. Column 4 lists the number of oxygens contributed by each oxide "molecule." Note that SiO_2 contains two oxygens per "molecule"; the molecular proportion figure (0.8193) is therefore multiplied by 2. The total number for all oxygens contributed by all the oxide "molecules" in column 4 is 2.6605. From crystal structure data we know that a pyroxene of this type has a formula like $Ca(Mg,Fe)(Si,Al)_2O_6$. The formula is, therefore, recalculated on the basis of 6 oxygens and the numbers in column 4 are multiplied by 2.25521 to obtain the figures in column 5. These cations are subsequently assigned to specific atomic sites in the pyroxene structure. If necessary, sufficient Al is added to Si so that the sum Si + Al = 2.0 in column 6; the remainder of the Al is added to the sum of the intermediate size cations. If necessary Ti values may also be added in the sum of Al + Si + (Ti) = 2.0; Ti^{4+}, of small size and high charge, similar to Si^{4+}, may occupy sites which generally house Si. Intermediate size cations (Al, Fe^{3+}, Fe^{2+}, Mn, Mg) are assigned to the $M1$ cation position (see Fig. 10.37) in the structure; these cation numbers together amount to 1.023, in agreement with the general formula $Ca_1(Mg, Fe)_1(Si, Al)_2O_6$. The remaining larger cations (Ca, Na, K) are assigned to the $M2$ position in the pyroxene structure, and their total is 1.014, again approximately 1.

Pyroxene analyses are often recalculated in terms of *end member compositions*. The analysis in Table 4.14 contains major amounts of CaO, MgO and FeO and can, therefore, be recalculated in terms of $CaSiO_3$ (wollastonite), $MgSiO_3$ (enstatite)

and $FeSiO_3$ (ferrosilite) end members. The calculations are given in column 7 showing that the weight percent analysis can be recast as $Wo_{50.3}En_{42.4}Fs_{7.3}$.

Complex hydrous silicates such as amphiboles are recalculated by the same sequence of steps as illustrated for olivine and pyroxene, but the H_2O content is evaluated as (OH) groups in the amphibole structure. Table 4.15 gives an example of a Na-containing actinolite analysis. A general actinolite formula is $Ca_2(Mg, Fe)_5Si_8O_{22}(OH)_2$. Column 1 lists weight percentages for $H_2O(+)$ and $H_2O(-)$. The $H_2O(+)$ is considered part of the actinolite structure but the $H_2O(-)$ is not and is therefore neglected in subsequent calculations. Column 2 lists the molecular proportions, column 3 the cation proportions, and column 4 gives the total contribution of (O, OH) for each of the "molecules" in column 2. The sum in column 4 divided into 24 gives the ratio by which the whole analysis must be multiplied in order to put it on a basis of 24 (O, OH). Column 3 is now multiplied by the factor 8.4803, giving the results in column 5 on the basis of 24 (O, OH). In general it will be found that Si is less than 8.0 and Al is added to bring (Si, Al) to 8.0. The remaining Al, if any, is added to the cation group immediately below consisting of Fe^{3+}, Fe^{2+}, and Mg. The total for this group of intermediate size cations is 5.111, close to 5.0 and the total for the remaining larger size cations in 2.068, close to 2.0. The intermediate size cations enter the $M1$, $M2$, and $M3$ sites in the amphibole structure and the larger cations enter $M4$ (see Fig. 10.48). Mn could be distributed among $M1$, $M2$, and $M3$ as well as $M4$ because its ionic size lies between that of Ca^{2+} and Mg^{2+}. The (OH) content turns out to be about 2. The recalculated analysis now has the general form $(Ca, Na, Mn)_2(Fe, Mg)_5(Si, Al)_8O_{22}(OH)_2$. Instrumental analysis results of hydrous minerals such as amphiboles do not provide information on the oxidation state of iron (Fe^{2+} vs. Fe^{3+}), for example, or on the presence of structural water. Such anhydrous analytical results are frequently recalculated on an anhydrous basis. For the analysis in Table 4.14 such a basis would be 23 oxygens, replacing $2(OH)^-$ by one O^{-2}. Although this is not correct in terms of the structure of amphiboles, it does allow for comparison among an-

Table 4.15

RECALCULATION OF AMPHIBOLE ANALYSIS

	1 Weight Percent	2 Molecular Proportions	3 Number of Cations	4 Total Number of (O, OH)	5 Cations on Basis of 24 (O, OH)	6 Cations on Basis of 23 Oxygens
SiO_2	56.16	0.9346	0.9346	1.8692	7.926 ⎫ 7.958	7.931
Al_2O_3	0.20	0.0019	0.0038	0.0057	0.032 ⎭ ≈8	0.032
Fe_2O_3	1.81	0.0113	0.0226	0.0339	0.192 ⎫	0.192
FeO	6.32	0.0880	0.0880	0.0880	0.746 ⎬ 5.111	0.747
MgO	19.84	0.4921	0.4921	0.4921	4.173 ⎭ ≈5	4.176
MnO	2.30	0.0324	0.0324	0.0324	0.275 ⎫	0.275
CaO	9.34	0.1665	0.1665	0.1665	1.412 ⎬ 2.068	1.413
Na_2O	1.30	0.0210	0.0420	0.0210	0.356 ⎭ ≈2	0.356
K_2O	0.14	0.0015	0.0030	0.0015	0.025	0.025
$H_2O(+)$	2.16	0.1198	0.2396	0.1198	2.032 ≈2	
$H_2O(-)$	0.48	—		—		
Total	100.05			2.8301		

$$\frac{24}{2.8301} = 8.4803; \quad 2.8301 - 0.1198 = 2.7103; \quad \frac{23}{2.7103} = 8.4861$$

alytical results in which H_2O was not determined. In column 6 the actinolite is recalculated on an anhydrous basis and the cation results are similar to, but not identical with, the results in column 5.

Compositional Variation in Minerals

As already noted, it is the exceptional mineral that is a *pure substance,* and most minerals display extensive variation in chemical composition. Sphalerite, Table 4.12, is an example of a mineral displaying extensive variation in chemical composition.

Compositional variation is the result of substitution, in a given structure, of an ion or ionic group for another ion or ionic group. This type of relationship is referred to as *ionic substitution,* or *solid solution.* For example, with reference to the sphalerite analyses in Table 4.12 we may say that analysis 4 contains 32.4 atomic percent Fe in substitution for Zn. Another way of referring to the compositional variation of sphalerite as expressed by the analyses in Table 4.12 is to say that a solid solution series of Fe in ZnS exists from pure ZnS to $(Zn_{.68}Fe_{.32})S$. The types of solid solution can be discussed in terms of

substitutional, interstitial, and *omission* solid solution mechanisms.

Substitutional Solid Solution

This is the most common type of ionic substitution, and it is illustrated by the analyses in Tables 4.12 to 4.15. Fe^{2+} substitutes for Zn^{2+} in the same atomic site of the sphalerite structure; Fe^{2+} replaces Mg^{2+} in a specific site of the olivine structure, and so forth. Several factors influence the extent to which substitution may take place. The most important is the size of the ion. Ions of two elements can readily substitute for each other only if their ionic radii are similar or differ by less than 15%. If the radii of two ions differ by 15 to 30%, substitution is limited or rare; if they differ by more than 30%, little substitution is likely. These limits are only approximate and are based on empirical data derived from the study of a large number of mineral systems displaying ionic substitution. Another factor of great importance is the temperature at which the crystal grew. The higher the temperature, the greater the amount of thermal disorder, and the less stringent the space requirements of the structure. Hence, crystals grown at high temperatures may display extensive ionic substitution that would not have been possible at lower temperatures. An example is the high-temperature form of $KAlSi_3O_8$ (sanidine), which can house much larger amounts of Na (in substitution for K) than can be accommodated in the low-temperature form (microcline). Yet another factor in substitution is the need for neutrality of the structure. If a monovalent ion such as Na^+ is substituted for a divalent ion such as Ca^{2+}, other substitutions are necessary in the structure to maintain electrical neutrality.

The simplest types of ionic substitution are *simple cationic* or *anionic* substitutions. In a compound of A^+X^- type composition, A^+ may be in part or wholly replaced by B^+. In this instance there is no valence change, and such a substitution is illustrated by K in the Na position of NaCl. A simple anionic substitution can be represented in an A^+X^- compound in which part or all of X^- can be replaced by Y^-. An example is the incorporation of Br^- in the structure of KCl in place of Cl^-. An example of a *complete solid solution series* (meaning substitution of one element by another over the total possible compositional range, as defined by two end member compositions) is provided by olivine (Mg, Fe)$_2SiO_4$. Mg^{2+} can be replaced in part, or in total by Fe^{2+}; the *end members* of the olivine series between which there is complete solid solution, are Mg_2SiO_4 (forsterite) and Fe_2SiO_4 (fayalite). Intermediate compositions are defined in terms of Fo_x Fa_y as shown in Table 4.13. Another example of a total solid solution series is given by (Mn, Fe)CO_3, which occurs as a series from $MnCO_3$ (rhodochrosite) to $FeCO_3$ (siderite). An example of a complete anionic series between two compounds is given by KCl and KBr. The size of the two anions are within 8.5% of each other allowing for complete substitution of Cl^- and Br^-, and vice versa.

If in a general composition of $A^{2+}X^{2-}$ a cation B^{3+} substitutes for some of A^{2+}, electrical neutrality can be maintained if an identical amount of A^{2+} is at the same time replaced by cation C^+. This can be represented by $2A^{2+} \rightleftarrows 1B^{3+} + 1C^+$, with identical total electrical charges on both sides of the equation. This type of substitution is known as *coupled substitution*. The plagioclase feldspar series can be represented in terms of two end members

$NaAlSi_3O_8$ (albite) and $CaAl_2Si_2O_8$ (anorthite). The complete solid solution between these two end member compositions illustrates coupled substitution: $Na^+Si^{4+} \rightleftarrows Ca^{2+}Al^{3+}$. This means that for each Ca^{2+} that replaces Na^+ in the feldspar structure, one Si^{4+} is replaced by Al^{3+} in the Si-O framework. The equation shows that the total electrical charges on both sides of the equation are identical; as such the structure remains neutral. An example of only limited coupled solid solution is provided by two pyroxenes, diopside (CaMgSi$_2O_6$) and jadeite (NaAlSi$_2O_6$). The coupled replacement can be represented as: $Ca^{2+}Mg^{2+} \rightleftarrows Na^+Al^{3+}$. Although the pyroxene structure is neutral with either type of cation pair, the atomic sites and coordination polyhedra do not allow for a complete solid solution of this type.

Interstitial Solid Solution

Between atoms or ions, or ionic groups, of a crystal structure *interstices* exist, which normally are considered voids. When ions or atoms are located in these structural voids, we speak of *interstitial substitution* or *interstitial solid solution*. In some crystal structures these voids may be channel-like cavities, as in beryl (Be$_3Al_2Si_6O_{18}$). In this ring silicate, large ions as well as gases occupy the tubular cavities of the superimposed rings (see Fig. 4.23). Considerable amounts of K, Rb, Cs, and H_2O as well as He are reported in beryl analyses, housed interstitially in the hexagonal channels.

Omission Solid Solution

The best known mineralogical example of this type of solid solution is provided by pyrrhotite, $Fe_{(1-x)}S$. In pyrrhotite, sulfur occurs in hexagonal closest packing and iron in 6 coordination with sulfur. If each octahedral site were occupied by Fe^{2+}, the formula of pyrrhotite would be FeS. However, in pyrrhotites there is a variation in the percentage of *vacancies* in the octahedral sites, causing the composition to range from Fe_6S_7 through $Fe_{11}S_{12}$, close to FeS. The formula is generally expressed as $Fe_{(1-x)}S$ with x ranging from 0 to 0.2. Minerals, such as pyrrhotite, in which a particular structural site is incompletely filled are known as *defect structures*. When Fe^{2+} is absent from some of the octahedral sites in

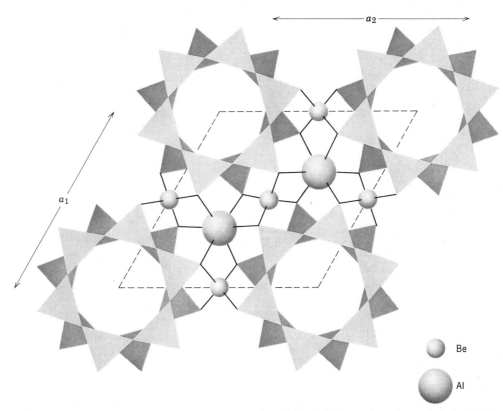

FIG. 4.23. Plan of the hexagonal structure of beryl, $Be_3Al_2Si_6O_{18}$ projected onto the basal plane (0001). The Si_6O_{18} rings at two different heights are shown. The hexagonal channels are the locus for large alkali ions, neutral molecules such as H_2O and of atoms as He. A unit cell is outlined by dashed lines.

pyrrhotite, with the sulfur net completely intact, the structure is not electrically neutral. It is possible that some of the Fe is in the Fe^{3+} state to compensate for the deficiency in Fe^{2+}. If this is so, a neutral pyrrhotite formula can be written as $(Fe^{2+}_{1-3x} Fe^{3+}_{2x} V_x)S_1$. If x in this example is taken as 0.1, this formulation means that 0.1 vacancies (V) and 0.7 Fe^{2+} and 0.2 Fe^{3+} occur in the structure of pyrrhotite.

Graphical Representation of Compositional Variations

Because most minerals show solid solution features ranging from partial to complete, it is frequently useful to show such variations in composition in graphical form. A composition graph can be constructed after we have chosen the chemical compo-

nents pertaining to the mineral compositions to be graphed. The mineral kyanite, Al_2SiO_5, is known to be essentially constant in composition. A chemical analysis of a pure kyanite yields the following data, in terms of weight percent: Al_2O_3, 62.91 and SiO_2, 37.08. Selecting Al_2O_3 and SiO_2 as the two end components of a bar graph, we can position the kyanite composition on it in terms of weight percent of the oxides (see Fig. 4.24a). In order to plot a mineral composition in terms of the weight percent of its components we need to obtain its chemical analysis or we can recalculate the weight percentages of the components from the formula as shown in Table 4.11. However, formulas that provide direct information about atomic proportions of elements or molecular proportions of components can be used directly for graphing of compositions. The formula

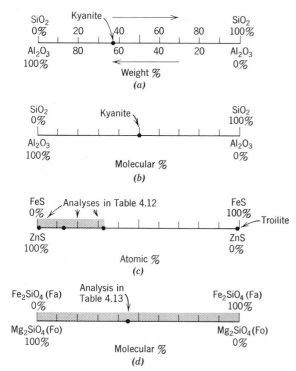

FIG. 4.24. Graphical representations of chemical compositions in terms of bar diagrams. (a) Kyanite (Al_2SiO_5) composition in terms of weight percentage of the oxides. (b) Kyanite in terms of molecular percentage of the oxides. (c) Sphalerite analyses from Table 4.12 in terms of atomic percentages of Zn and Fe. (d) Complete solid solution series of olivine and composition of olivine from Table 4.13 in terms of molecular percentages of forsterite and fayalite.

of kyanite can be written as $1Al_2O_3 + 1SiO_2 = 1Al_2SiO_5$. In other words, the composition of kyanite can be expressed in terms of the molecular proportions of the oxide components Al_2O_3 and SiO_2. This is shown in Fig. 4.24b.

In Table 4.12 we recalculated several sphalerite analyses. These compositions can be shown graphically (see Fig. 4.24c) in terms of two end members ZnS and FeS. Intermediate compositions can be obtained directly from the Zn : Fe ratios in the last line of Table 4.12. After these numbers are plotted, and with the addition of further literature analyses as well, the solid solution extent of FeS in sphalerite can be generalized on the basis of a population of analysis points. Although only very few

points are shown on the bar graph in Fig. 4.24c, it can be stated that sphalerite shows a partial solid solution series from ZnS to at least $Zn_{.676}Fe_{.324}S$. This statement is shown graphically by the shading. Troilite, on the other hand, shows no Zn content and must appear as a single point on the bar graph, indicating a total lack of Zn substitution. The graph shows that between troilite and the most Fe-rich sphalerite plotted there is a region in which homogeneous compositions of Fe-rich sphalerites are not found. In Table 4.13 we recalculated an olivine analysis in terms of molecular percentages of MgO and FeO; such percentages can also be expressed as Mg_2SiO_4(Fo) and Fe_2SiO_4(Fa), the two end member compositions of the olivine solid solution series. Figure 4.24d shows a bar graph depicting the complete solid solution series between Fo and Fa as well as the specific composition of the olivine of Table 4.13. Bar diagrams of this type form the horizontal axis in *variation diagrams* that relate changes in physical properties to variation in composition (see Figs. 3.19 and 5.6). Composition bars also form the horizontal axis in temperature-composition diagrams (see Fig. 4.29).

As we have already seen in Tables 4.13 to 4.15 mineral analyses can be very complex and may show substitution of several elements in a specific atomic site of the mineral structure. In order to portray the variation of at least three components instead of just two as in Fig. 4.24, a *triangular diagram* is frequently used (see Fig. 4.25). Triangular coordinate graph paper is available commercially* with ruling for each 1% division. Let us illustrate the use of such paper with some relatively simple compositions in the pyroxene family of the silicates. Table 4.14 gives a pyroxene analysis which is recalculated in terms of three components, $CaSiO_3$ (wollastonite), $MgSiO_3$ (enstatite), $FeSiO_3$ (ferrosilite). The ratios of these three components are the same as they would be for CaO, MgO, and FeO (these two sets of alternate components are completely interchangeable). In Fig. 4.25 we will use the three mineral components, wollastonite (Wo), enstatite (En) and ferrosilite (Fs), which mark the corners of

* From Keuffel and Esser Co.

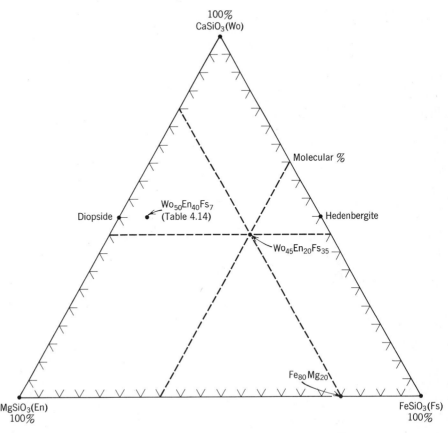

FIG. 4.25. Plotting of chemical compositions on triangular coordinate paper. The specific compositions represented are for the pyroxene family of silicates. See text for explanation of plotting procedure.

the triangular graph paper. Any pyroxene composition that involves only two of the three components can be plotted along an edge of the triangle, identical to the procedure in the bar graphs of Fig. 4.24. For example, diopside ($CaMgSi_2O_6$) can be plotted along the left edge of the triangle: $1CaSiO_3 + 1MgSiO_3 = 1CaMgSi_2O_6$. Similarly, hedenbergite ($CaFeSi_2O_6$) can be plotted along the right edge, according to $1CaSiO_3 + 1FeSiO_3 = 1CaFeSiO_3$. Compositions along the bottom edge can be expressed in general as $(Mg, Fe)SiO_3$. A specific compositional location can be obtained from a formulation $(Fe_{0.80}Mg_{0.20})SiO_3$, which can be restated as 80 molecular percent ferrosilite (Fs) and 20 molecular percent enstatite (En). This composition is shown in Fig. 4.25.

A more general pyroxene composition that shows the presence of all three components may be expressed as $Wo_{45}En_{20}Fs_{35}$. This is shorthand for a pyroxene containing 45 molecular percent CaO(Wo), 20 molecular percent MgO(En), and 35 molecular percent FeO(Fs). In order to locate this composition on the triangular graph paper, note that the join En-Wo represents 0% Fs, similarly the En-Fs join represent 0% Wo, and so forth. The distance between the corner and the opposite side of the triangle is graduated in % lines from 100% at the corner to 0% along the join. For example, the $CaSiO_3$ corner represents 100% $CaSiO_3$ and horizontal lines between this corner and the base of the triangle mark the change from 100% to 0% $CaSiO_3$. If we wish to plot $Wo_{45}En_{20}Fs_{35}$, locate first the Wo_{45} line (marked on Fig. 4.25), then locate the En_{20} line (also marked). Where the two intersect the composition is located, on the Fs_{35} line. The position of the composition, $Wo_{50}En_{43}Fs_7$, from Table 4.14 is also shown in Fig. 4.25.

Like bar graphs, triangular diagrams are frequently used to depict the extent of solid solution in minerals. Figure 4.26 shows the extent of solid solution among minerals in the chemical system SiO_2-MgO-FeO. In this representation we can show the olivine series $(Mg, Fe)_2SiO_4$ and the orthorpyroxene series $(Mg, Fe)SiO_3$, both of which show an essentially complete solid solution between the end member compositions. SiO_2 (quartz and its various polymorphs) shows no substitution in terms of FeO or MgO and is drawn as a point, indicating its constant composition. Forsterite (Mg_2SiO_4) can be written as $2MgO + 1SiO_2 = 1Mg_2SiO_4$; of the three oxide "molecules," one is SiO_2 and the other two are MgO. On the left side of the triangle Fo is therefore located at $\frac{1}{3}$ away from the MgO corner. Enstatite $(MgSiO_3)$ can be written as $1MgO + 1SiO_2 = 1MgSiO_3$, which is located halfway between the MgO and SiO_2 corners.

In addition to depicting graphically the extent of solid solution, triangular diagrams are useful in showing the minerals that make up (coexist in) a specific rock type. In such diagrams, known as *assemblage diagrams,* the minerals that coexist with each other (i.e., touch each other along the borders of their grains) are connected by *tielines.* Such tielines denote the fact that 2, 3, or 4 minerals are found next to each other. In Fig. 4.26 some tielines have been drawn for coexistences of possible olivine and orthopyroxene compositions, as well as for the coexistence of a high-temperature form of SiO_2 (cristobalite) and an Fe-rich orthopyroxene and fayalite. This mineral assemblage could be found in a high-temperature, Fe-rich basalt. Such a three-mineral coexistence is denoted by an *assemblage triangle,* of which each of the three sides are tielines.

Triangular diagrams would tend to limit us to representation of only three components, which may be simple elements, compound oxides, or

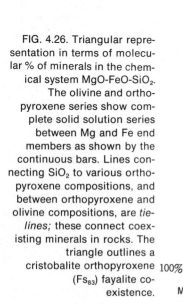

FIG. 4.26. Triangular representation in terms of molecular % of minerals in the chemical system MgO-FeO-SiO₂. The olivine and orthopyroxene series show complete solid solution series between Mg and Fe end members as shown by the continuous bars. Lines connecting SiO₂ to various orthopyroxene compositions, and between orthopyroxene and olivine compositions, are *tielines;* these connect coexisting minerals in rocks. The triangle outlines a cristobalite orthopyroxene (Fs_{83}) fayalite coexistence.

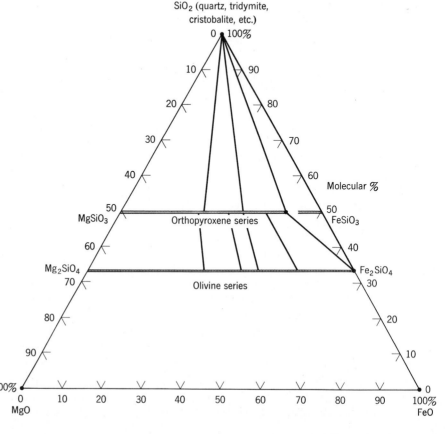

more complex components as expressed by mineral formulas. In order to represent more than three components on one triangle, some components are frequently combined and some may not be considered in the graphical representation. For example, in an attempt to represent carbonate compositions in the system $CaO-MgO-FeO-MnO-CO_2$ we must somehow reduce the possible number of compositional variables. As all carbonates contain CO_2 there will be little gained in using the CO_2 component to show the small CO_2 variations that do exist. We will therefore ignore CO_2 in the graphical representation. We may furthermore decide to combine the FeO and MnO because Fe^{2+} and Mn^{2+} substitute freely for each other in the carbonate structure. This leaves three components, CaO, MgO, and (FeO + MnO). Figure 4.27a shows that a complete solid solution series may exist between magnesite ($MgCO_3$) and siderite + rhodochrosite ($FeCO_3 + MnCO_3$); the completeness of this series is not well documented. It also shows a complete series between dolomite, $CaMg(CO_3)_2$, and ankerite, $CaFe(CO_3)_2 + $ kutnahorite, $CaMn(CO_3)_2$. Calcite ($CaCO_3$) shows a very

limited ionic substitution by Mg, Fe, and Mn. Tielines show coexistences of magnesite-dolomite, calcite-dolomite, and calcite-ankerite-siderite in the triangle. The three-mineral assemblage is not uncommon in banded Precambrian iron-formations. In chemically more complex systems, such as exhibited by most rocks, additional combinations of components are necessary in order to portray the chemistry of the minerals in the rock on a triangular diagram. A frequently used presentation of this type is known as an *ACF diagram* in which $A = Al_2O_3 + Fe_2O_3 - (Na_2O + K_2O)$; $C = CaO$; and $F = MgO + MnO + FeO$, all in molecular proportions. Figure 4.27b shows an example of such a diagram for rocks of medium-grade metamorphic origin. Triangles connect compositions of three coexisting minerals in these rocks.

Exsolution

The term *exsolution* refers to the process whereby an initially homogeneous solid solution separates into two (or possibly more) distinct crystalline min-

FIG. 4.27(a). Compositions of carbonates in the system $CaO-MgO-FeO-MnO-CO_2$. The CO_2 component is not considered in this diagram. (FeO + MnO) is a combined component. A complete solid solution series exists between dolomite and ankerite. A complete series may also exist between magnesite and siderite, but such a series is not well substantiated. Tielines connect possible coexisting carbonate members. The three-mineral triangle portrays the coexistence of calcite-ankerite-siderite. (b) Mineral compositions in metamorphosed carbonate rocks, pelitic schists, and amphibolites in terms of an *ACF* diagram. See text.

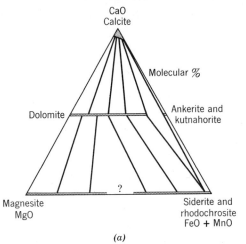

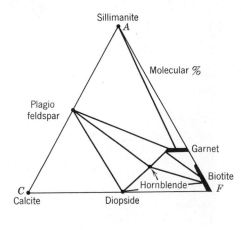

(a)   (b)

erals without the addition or removal of material to or from the system; this means that no change in the bulk composition takes place. Exsolution generally, although not necessarily, occurs on cooling. Exsolution features are best observed under a high-power microscope because the scale of exsolution is generally on about a 1 micron level or smaller (1 μ = 10^{-3} mm). An example of microscopic exsolution in amphiboles is shown in a photomicrograph in Fig. 4.28. Exsolution lamellae which separate from the originally homogeneous host mineral are generally crystallographically oriented as seen in Fig. 4.28. In amphiboles the orientation of exsolved lamellae is generally parallel to {001} and {100}. In alkali (Na-K) feldspars exsolution lamellae can often be seen in handspecimen to be approximately parallel to {100}. Such a coarse-grained intergrowth of exsolution lamellae in alkali feldspars may consist of Na-rich feldspar lamellae in a K-rich host and is

FIG. 4.28. Photomicrograph of exsolution lamellae of hornblende (hbl) in grunerite (gru) and grunerite (gru) lamellae in hornblende (hbl). The lamellae are oriented parallel to (001) and (100), respectively.

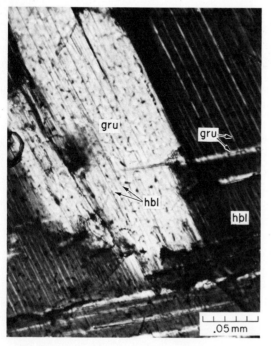

known as *perthite*. If such exsolution is on a scale that it can be resolved only microscopically, it is referred to as *microperthite;* and if X-ray diffraction techniques are needed to resolve extremely fine lamellae, the intergrowth is known as *cryptoperthite*. These types of crystallographically controlled intergrowths resulting from exsolution are common in many mineral systems: alkali feldspars, pyroxenes, amphiboles, and iron oxides, to name a few examples.

The origin of exsolution textures is best illustrated with reference to a schematic temperature-composition diagram in Fig. 4.29. The horizontal axis represents the compositional variation, in terms of molecular percentages, between two possible silicates with very similar, if not identical, structures. The hypothetical silicates are monoclinic and are represented as ASi_xO_y and BSi_xO_y. The ionic sizes of A and B differ by about 25%. If we were to collect from the literature the chemical analyses for all low-temperature geologic occurrences of these two minerals we might conclude that there is very limited solid solution between the two end members as shown graphically by the low-temperature (T_1) composition bars. These same two hypothetical silicates might also be present in high-temperature basalt flows which cooled very rapidly. At such high temperature (T_2) we may find a complete solid solution series (shown by shading in Fig. 4.29). What happens to the extent of solid solution between T_2 and T_1 is outlined by the *miscibility gap* in the diagram. This gap represents a temperature-composition field in which solid solution between the end members decreases gradually from higher to lower T. For specific compositions in this region two minerals coexist rather than one homogeneous mineral as above the gap.

Let us take a specific composition (X_1) in the region above the miscibility gap at T_2. At this temperature, the structure of the silicate is in a high-energy state and will allow for relatively easy accommodation of A and B (although differing by as much as 25% in ionic radius) in the same atomic sites. In other words, the A and B ions randomly occupy the various cationic sites in the structure. On

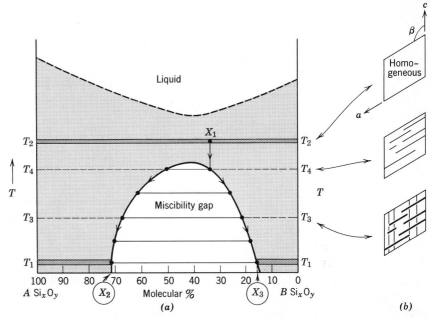

FIG. 4.29(a). Schematic *T-X* (*X* = composition and is expressed as molecular or atomic %) diagram for hypothetical silicates ASi_xO_y and BSi_xO_y. At temperature T_2 a complete solid solution exists between the two end members. At lower temperature a miscibility gap occurs. The arrows along the edge of the gap indicate the change in compositions of the coexisting silicates with decreasing *T*. The process of exsolution is discussed in the text. (*b*) Cross sections of mineral (with composition X_1 in Fig. 4.29*a*) illustrating sequence of appearance of possible exsolution lamellae. (1) homogeneous, at T_2. (2) exsolution lamellae // (001) at T_4. (3) additional set of exsolution lamellae // (100) at T_3, and coarsened lamellae // (001).

very slow cooling of this composition we reach a temperature (T_4) at which the structure is unable to accommodate a completely random distribution of *A* and *B* ions. Stresses in the structure resulting from the difference in size of the cations causes diffusion of A^+ ions to produce silicate regions with a preponderance of A^+; simultaneous B^+ ion diffusion results in silicate regions rich in B^+. This is the beginning of the process known as *unmixing*, which means that an originally homogeneous mineral segregates into two chemically different minerals. The miscibility gap continually widens in lowering the temperature from T_4. Under equilibrium conditions the compositions of lamellae of the exsolution intergrowth which forms as a function of decreasing temperature, are given by points along the edge of the misci-

bility gap; such compositional points are joined by tielines. At the lowest temperature, T_1, we may find a rather coarse-grained, crystallographically well-oriented intergrowth of (*A*, *B*)Si_xO_y lamellae with the specific composition (X_2) in a host of (*B*, *A*)Si_xO_y with specific composition (X_3). Although the segregation of ions within the silicate structure, which began at T_4, is complex on an atomic level, it is due mainly to the inability of a single structure to house ions of disparate sizes in a random distribution when decreasing temperature lowers the energy of the crystal.

Although exsolution intergrowths are common in many mineral groups, the resulting textures are rarely visible in hand specimens. Perthite in the alkali feldspar series is an example of exsolution on a

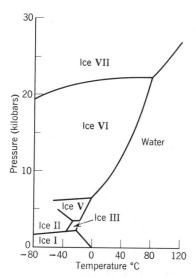

FIG. 4.30. *P-T* diagram for H_2O. Six polymorphic types of ice are indicated by I, II, III, V, VI, and VII (after P. W. Bridgeman, *Jour. Chem. Phys.*, 1937, v. 5, p. 965 and *Phase Diagrams for Ceramists*, copyright 1964 by the American Ceramic Society). One kilobar = 1000 bars; 1 bar = 0.987 atmosphere.

large scale, visible in hand specimen A schematic sequence of cross-sections in Fig. 4.29*b* illustrates what may be seen under the microscope as a result of the processes depicted in Fig. 4.29*a*. An originally homogeneous crystal of composition X_1 develops sets of differently oriented exsolution lamellae between temperatures T_4 and T_3.

These segregation processes all take place within the silicate structure in the solid state. The *T-X* diagram depicting this sequence of events is, therefore, in general referred to as a *subsolidus phase diagram*. Other types of phase diagrams, which are used as illustrations in the subsequent chapters on systematic mineralogy, are *liquidus* and *P-T diagrams*. Liquidus diagrams portray the crystallization of minerals from a liquid in terms of temperature and compositions of liquid and solids; *P-T* diagrams may be used to show polymorphic reaction curves, chemical reaction curves, and so forth. Such diagrams are used to outline *stability fields* of minerals in terms of pressure and temperature (*P-T*). An

example of a *P-T* diagram for H_2O is given in Fig. 4.30. See Chapter 11 for further discussion and examples of phase diagrams.

BLOWPIPE AND CHEMICAL TESTS

Although sophisticated analytical apparatus is now routinely used for the chemical study of minerals, such apparatus is generally not available in a college mineralogy laboratory. A number of quick and easy tests, known as *blowpipe tests,* were developed during the nineteenth century for making qualitative determinations of most of the common elements in minerals. In this section we will mention only a few of the tests, those that are *easy* to perform (with a minimum number of steps) and are *highly diagnostic* for specific elements.*

The ordinary blowpipe consists essentially of a tapering tube ending in a small opening through which air can be forced in a thin stream. When this current of air is directed into a luminous flame, combustion takes place more rapidly and completely, producing a very hot flame.

Figure 4.31*a* represents a common type of blowpipe. The mouthpiece, either *c* or *c'*, is fitted into the upper end of the tube, and air from the lungs forced into it issues from the small opening at the other end. The tip of the blowpipe, *b*, is placed just within a flame rich in carbon such as is obtained from a candle or ordinary household gas. When gas is available, the flame may be produced using a Bunsen burner in which an inner tube has been inserted to cut off the supply of air at the base and thus yield a luminous flame. The upper end of the tube is flattened and cut at an angle (Fig. 4.32). The gas flame should be adjusted to about 1 inch in height. The resulting blowpipe flame should be nonluminous, narrow, sharp-pointed, and clean-cut (Fig. 4.32). If illuminating gas is not available, a candle can be used.

---

* A more complete listing of tests is given in Hurlbut, 1971, *Dana's Manual,* 18th ed., and in Brush and Penfield, 1926; see references.

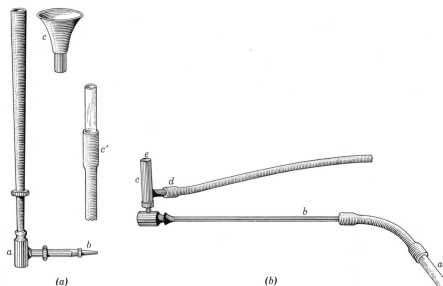

FIG. 4.31(*a*). Common blow-
pipe. (*b*) Improved blowpipe. (*a*) (*b*)

Figure 4.31*b* illustrates a more elaborate blow-pipe, which is constructed with a gas fitting, *c*, so that air forced into the tube through the mouthpiece *a*, and gas admitted through tube *d*, issue together from the opening *e*.

It usually requires some practice before one can produce a steady and continuous blowpipe flame. Some blowpipe tests can be completed before it is necessary to replenish the supply of air in the lungs. Frequently, however, a test takes a longer time than this permits, and an interruption in order to fill the lungs afresh materially interferes with the success of the experiment. Consequently, it is important to be able to maintain a steady stream of air

FIG. 4.32. Blowpipe flame. (*a*) Oxidizing flame. (*b*) Reducing flame. (*c*) Unburned gas. *o* and *r* indicate positions of assay in oxidizing and reducing flames, respectively.

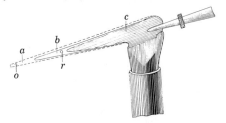

from the blowpipe for a considerable time. This is accomplished by distending the cheeks so as to form a reservoir of air in the mouth. When the supply of air in the lungs must be replenished, the passage from the mouth into the throat is closed by lifting the root of the tongue; thus, while breathing in through the nose, a steady stream of air is also being forced out of the reservoir in the mouth. In this way a constant flame may be obtained.

Fusion by Means of the Blowpipe Flame

The hottest part of a blowpipe flame is just beyond the visible portion, at *a* Fig. 4.32, and may reach a temperature as high as 1500°C, but the temperature varies depending on the type of gas and the mixture of gas and air. The ease with which a mineral fuses is an important aid to its identification. To make this test the *assay*, a small fragment of the mineral, is held in forceps and placed in the hottest portion of the flame. It may melt completely or only the sharp edges may round over. In either case it is said to be fusible.

Minerals can therefore be divided into two

Table 4.16

SCALE OF FUSIBILITY

Number	Mineral	Approximate Fusing Point (°C)	Remarks
1	Stibnite	525	Very easily fusible in the candle flame
2	Chalcopyrite	800	A small fragment will fuse easily in the Bunsen burner flame
3	Garnet (Almandite)	1050	Infusible in the Bunsen burner flame but fuses easily in the blowpipe flame
4	Actinolite	1200	A sharp-pointed splinter fuses with little difficulty in the blowpipe flame
5	Orthoclase	1300	The edges of fragments are rounded with difficulty in the blowpipe flame
6	Bronzite	1400	Practically infusible in the blowpipe flame; only the fine ends of splinters are rounded
7	Quartz	1710	Infusible in the blowpipe flame

classes: fusible and infusible. The fusible minerals can be further classified according to the ease with which they fuse. To assist in this classification, six minerals with different degrees of fusibility have been chosen as a scale of reference. For instance, a mineral has a fusibility of 3 if it fuses as easily as the mineral listed as 3 in the scale. In making such comparative tests, it is necessary to use fragments of uniform size, preferably a splinter not larger than 1 mm in cross-section. The scale of fusibility is given in Table 4.16.

Reducing and Oxidizing Flames

Reduction consists essentially of taking oxygen away from a chemical compound, and oxidation of adding oxygen to it. These two opposite chemical reactions can be accomplished by the flame of a Bunsen burner or of a blowpipe. Cone b, Fig. 4.32 contains CO, carbon monoxide. This is a reducing agent that, because of its strong tendency to take up oxygen to become CO_2, will, if possible, take oxygen away from another substance with which it is in contact. For instance, if a small fragment of ferric oxide, Fe_2O_3, is held in this part of the blowpipe

flame, it will be reduced by the removal of oxygen to ferrous oxide, FeO. In Fig. 4.32 cone b is the reducing part of the blowpipe flame, and if a reduction test is to be performed the mineral fragment is placed at r.

If oxidation is to be accomplished, the mineral must be placed entirely outside of the flame, at o in Fig. 4.32, where the oxygen of the air can have free access to it, but where it is heated by the flame. Under these conditions, if the reaction is possible, oxygen will be added to the mineral and the substance will be oxidized. Pyrite, FeS_2, for instance, if placed in the oxidizing flame, would be converted into ferric oxide, Fe_2O_3, and sulfur dioxide, SO_2.

Charcoal in Blowpiping

Charcoal blocks measuring about $4 \times 1 \times \frac{1}{2}$ in. are used to reduce metals from their minerals and to obtain characteristic oxide coatings on the surface (Fig. 4.33). Although it is impossible to extract the metal from some minerals by blowpipe methods, a few can be reduced merely by heating on charcoal and others can be reduced by means of a flux. A mixture of sodium carbonate and charcoal in equal propor-

FIG. 4.33. Charcoal block with antimony oxide coating.

tions, called the *reducing mixture,* serves as a good flux for most reductions. In all reduction experiments the mineral should be powdered and placed in a small depression in the charcoal. When the flux is used, it should be mixed with the mineral in the proportion, 2 flux to 1 mineral. Table 4.17 describes the metallic globules for several metals.

Sublimates on Charcoal and Plaster

Certain elements yield characteristic sublimates when minerals containing them are placed in a small depression in the charcoal block and heated in the oxidizing flame (see Fig. 4.33). In some cases a plaster of Paris tablet is preferable to charcoal, for the white background brings out the color of subli-

mates poorly displayed on the black charcoal. Table 4.18 lists sublimates of As, Sb, Pb, and Bi on charcoal and of Mo on plaster.

Open-Tube Test

Hard glass tubing about 5 in. long and $\frac{1}{4}$ in. i.d. is used in making what are known as open-tube tests. A small amount of powdered mineral is placed in the tube about an inch from one end. The tube is then inclined at as sharp an angle as possible, with the mineral near the lower end. A Bunsen burner flame is at first placed on the upper part of the tube to convert the inclined tube into a chimney, up which a current of air flows. After a moment the tube is shifted so the flame heats it just above the mineral, or in some cases directly beneath the mineral. Under these conditions in a current of air, the mineral will oxidize if such a reaction is possible. Various oxides may come off as gases and either escape at the end of the tube or be condensed as sublimates on its walls. Open-tube tests are characteristic

Table 4.17

METALLIC GLOBULES REDUCED ON CHARCOAL

Element	Color and Character of Globule	Remarks
Silver Ag	White, soft, no coating, remains bright	Usually necessary to use reducing mixture. To distinguish from other globules, dissolve in nitric acid, add hydrochloric acid to obtain white silver chloride precipitate
Tin Sn	White, soft, becomes dull on cooling. White coating of oxide film	Globules form with difficulty even with reducing mixture. Metallic globule is oxidized in nitric acid to a white hydroxide
Copper Cu	Red, soft, surface black when cold, difficult to fuse	Copper minerals should be roasted to drive off S, As, and Sb before mixing with reducing mixture
Lead Pb	Gray, soft, fusible. Bright in reducing flame, iridescent in oxidizing flame	The incandescent charcoal will reduce the lead. To distinguish from other globules dissolve in nitric acid, and from clear solution precipitate white lead sulfate by addition of sulfuric acid

Note. Easily fusible metallic beads are often obtained on heating metallic compounds containing sulfur, antimony, or arsenic. Such beads are always very brittle and often magnetic. Magnetic masses or globules are obtained when compounds of iron, nickel, and cobalt are heated on charcoal.

Table 4.18

SUBLIMATES ON CHARCOAL AND PLASTER

Element	Composition of Coating	Color and Character of Coating on *Charcoal*	Remarks
As	Arsenious oxide As_2O_3	White and volatile, depositing at some distance from the assay	Usually accompanied by garlic odor
Sb	Antimony oxides Sb_2O_3, Sb_2O_4	White and volatile, depositing close to the assay	Less volatile than As_2O_3
Pb	Lead oxide PbO	Yellow near the mineral and white farther away. Volatile	Coating may be composed of white sulfite and sulfate of lead in addition to PbO
Bi	Bismuth iodide BiI_3	Bright red with yellow ring near assay	This reaction occurs when bismuth minerals are heated with iodide flux

Element	Composition of Coating	Color and Character of Coating on *Plaster*	Remarks
Mo	MoO_3	White in oxidizing flame. Red MoO_2 under assay	Coating turns deep ultramarine blue if touched with reducing flame

OPEN-TUBE

Element	Composition of Oxidation Product	Color and Character of Oxidation Product in the *Open Tube*	Remarks
S	Sulfur dioxide SO_2	SO_2, a colorless gas, issues from the upper end of the tube	The gas has a pungent and irritating odor. Moistened blue litmus paper placed at the upper end of the tube becomes red, owing to the acid reaction of sulfurous acid

for several elements but sulfur is the only major element for which some other test is not equally good. The open-tube test for sulfur in unoxidized sulfur compounds is given in Table 4.18.

Closed-Tube Test

The closed tube made of soft glass should have a length of about $3\frac{1}{2}$ in. and an internal diameter from $\frac{1}{8}$ to $\frac{3}{16}$ in. Two closed tubes can be made by fusing the center of a piece of tubing 7 in. in length

and pulling it apart. The closed-tube test is used to determine what takes place when a mineral is heated in the absence of oxygen. In performing the test the mineral is broken into small fragments or powdered, placed in the closed end of the tube, and heated in the Bunsen burner flame (see Table 4.19).

Flame Test

Some elements may volatilize and impart characteristic colors to the flame when minerals containing

Table 4.19

CLOSED-TUBE TESTS

Element	Substance	Color and Character	Remarks
	H_2O	Colorless liquid, easily volatile	Minerals containing water of crystallization or hydroxyl give, on moderate heating, a deposit of drops of water on the cold upper walls of the tube. The water may be acid from hydrochloric, hydrofluoric, sulfuric, or other volatile acid
S	S	Red when hot, yellow when cold. Volatile	Given only by native sulfur and those sulfides that contain a high percentage of sulfur
Hg	HgS	Black amorphous sublimate	This test given when cinnabar, HgS, is heated alone
Hg	Hg	Gray, metallic globules	Metallic mercury is obtained when cinnabar is mixed with sodium carbonate and heated

Table 4.20

FLAME COLORATIONS

Element	Color of Flame	Remarks
Strontium Sr	Crimson	Strontium minerals that give the flame color also give alkaline residues after being heated
Lithium Li	Crimson	Lithium minerals that give the flame color do not give alkaline residues after being heated (unlike strontium)
Boron B	Yellow-green	Minerals giving a boron flame rarely give alkaline residues after ignition. Many boron minerals will give a green flame only after they have been broken down by sulfuric acid or the "boron flux"
Copper Cu	Emerald green	Obtained from copper oxide
Copper Cu	Azure blue	Obtained from copper chloride. Any copper mineral will give the copper chloride flame after being moistened with hydrochloric acid
Chlorine Cl	Azure blue (copper chloride flame)	If a mineral containing chlorine is mixed with copper oxide and introduced into the flame, the copper chloride flame results
Sodium Na	Intense yellow	A very delicate reaction. The flame should be very strong and persistent to indicate the presence of sodium in the mineral as an essential constituent
Barium Ba	Yellow-green	Minerals that give the barium flame also give alkaline residues after ignition

them are heated. A flame test is made by heating a small fragment of the mineral held in forceps, but a more decisive test is usually obtained when the fine powder of the mineral is introduced into the Bunsen burner flame on a piece of platinum wire. Some minerals contain elements that normally color the flame, but because the compound is nonvolatile they fail to give the characteristic flame until broken down by an acid or a flux.

Frequently a flame color is masked by the presence of a sodium flame. Although a mineral may contain no sodium, it may be coated with dust of a sodium compound ever present in the laboratory, and a yellow flame may result. A filter of blue glass held in front of the flame will completely absorb the yellow sodium flame and allow the characteristic flame colors of other elements to be observed. See Table 4.20 for a listing of some diagnostic flame colorations.

FIG. 4.34. Loop in platinum wire.

before it cools, a small amount of the powdered mineral is introduced into it and is dissolved by further heating.* The color of the resulting bead may depend on whether it was heated in the oxidizing or reducing flame.

Bead Test

Some elements, when dissolved in fluxes, give a characteristic color to the fused mass (see Table 4.21). The fluxes that are most commonly used are borax, $Na_2B_4O_7 \cdot 10H_2O$, and salt of phosphorus, $HNaNH_4PO_4 \cdot 4H_2O$. Sodium carbonate, Na_2CO_3, is used in the test for manganese, giving an opaque bluish-green coloration. The operation is most satisfactorily performed by first fusing the flux on a small loop of platinum wire into the form of a lens-shaped bead. For best results, the loop on the wire should have the shape and size shown in Fig. 4.34. After the flux has been fused into a bead on the wire, but

Spectroscopic Analysis

Optical spectrographic analysis has for years been standard procedure in mineralogical research, but the necessary equipment is costly and can be operated only by the skilled technician. However, several small, relatively low-priced spectroscopes and spectrographs are now available. Because of their simplicity of operation, they can be used as a deter-

* All metallic minerals should be roasted before being introduced into the bead, for it is the oxide of the metal that is desired. It is especially desirable to rid the mineral of arsenic, for even small amounts of this element make the platinum brittle and cause the end of the wire to break off.

Table 4.21
BEAD TESTS[a]

	Borax Bead		Phosphorous Salt Bead	
Oxides of	Oxidizing Flame	Reducing Flame	Oxidizing Flame	Reducing Flame
Chromium	Yellowish green	Fine green	Fine Green	Fine green
Uranium	Yellow	Pale green to nearly colorless	Pale greenish yellow	Fine green
Cobalt	Blue	Blue	Blue	Blue
Manganese	Reddish violet	Colorless	Violet	Colorless

[a] The colors given are those obtained after the bead is cold.

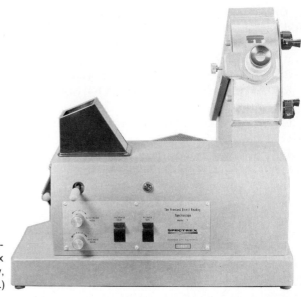

FIG. 4.35. Vreeland spectroscope. (Courtesy of Spectrex Co., Redwood City, California.)

minative tool in the elementary mineralogy laboratory. One such instrument, the Vreeland Spectroscope, is illustrated in Fig. 4.35.

We have seen that when certain elements are volatilized they may impart characteristic colors to a flame. Each of these colors represents a different wavelength in the visible region of the electromagnetic spectrum (see Fig. 3.1). In flame tests the elements are volatilized by the heat of the Bunsen burner, and in a spectroscope the elements are vaporized by the high temperature of a carbon arc.

FIG. 4.36. Sketch of optical path in spectroscope.

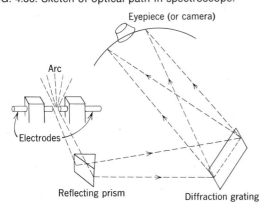

Eyepiece (or camera)

Arc

Electrodes

Reflecting prism

Diffraction grating

When the atoms are in an excited (vapor) state they emit light of definite wavelengths, and the dispersion of this light by a spectroscope produces a spectrum that is *characteristic* of the atom or molecule. The production of the emitted light results from the falling of electrons from outer levels of the electron cloud into inner levels. Such electrons were moved into the outer levels during the excitation by a high-energy source (arc). The energy difference between the two levels (excited vs. lower level) determines the wavelength of the emitted radiation. Because the electronic configuration of the atom is different for each element, every element has a *characteristic spectrum*. Figure 4.36 shows schematically how a spectrum produced in an arc is analyzed in a spectroscope. Part of the light source is dispersed via an optical prism and subsequently by a diffraction grating. This dispersed spectrum is viewed with an eyepiece (spectroscope) or recorded on film (spectrograph). In the Vreeland instrument, while the spectrum from the vaporizing mineral is being observed, standard spectra of many different elements can be brought into the field of view for comparison and the elements composing the mineral can be identified.

1. Dry Reagents

Sodium carbonate (anhydrous) Na_2CO_3. Used as a flux to decompose minerals by fusion on charcoal, and in some bead tests.

Borax, $Na_2B_4O_7 \cdot 10H_2O$. Used for bead tests.

Salt of phosphorus, $HNaNH_4PO \cdot 4H_2O$. Used for bead tests.

Cupric oxide, CuO. Used for making a flame test for chlorine.

Potassium iodide and sulfur in equal parts (bismuth flux) used in a test for bismuth, lead, etc.

Metallic zinc or tin to make reduction tests in HCl solutions.

2. Wet Reagents

Hydrochloric acid, HCl. Used for solution of minerals. The concentrated acid is diluted with three parts of water for ordinary use.

Nitric acid, HNO_3. Commonly used in concentrated form.

Sulfuric acid, H_2SO_4. For normal use the concentrated acid is diluted with four parts of water. *Note.* Dilute by pouring acid into the water, not vice versa.

Ammonium hydroxide, NH_4OH.

Ammonium oxalate, $(NH_4)_2C_2O_4$, for test for calcium.

Sodium phosphate, $Na_2HPO_4 \cdot 12H_2O$. Used as aqueous solution, in test for magnesium.

Ammonium molybdate, $(NH_4)_2MoO_4$, for phosphate test.

Silver nitrate, $AgNO_3$, for test for chlorine.

Cobalt nitrate, $Co(NO_3)_2$, in dilute solution.

Dimethylglyoxime, used in test for nickel.

Hydrogen peroxide, H_2O_2, used in titanium test.

Summary of Tests for Some Elements

A few of the diagnostic blowpipe and chemical tests for the determination of specific elements in minerals are given here. For a more complete treatment the student is referred to Hurlbut, *Dana's Manual,* 18th ed., 1971 and Bush and Penfield, 1926 (see references).

Aluminum. Light-colored infusible aluminum minerals turn blue when moistened with $Co(NO_3)_2$ and ignited. Zinc silicates also turn blue under similar conditions. A flocculent white precipitate of $Al(OH)_3$ forms when an excess of NH_4OH is added to an acid solution of an aluminum-bearing mineral.

Antimony. When an antimony mineral is heated on charcoal, a white sublimate forms near the assay, Table 4.18.

Arsenic. Unoxidized arsenic minerals give a white, volatile coating on charcoal at some distance from the assay, Table 4.18.

Barium. Yellowish-green flame is characteristic of barium, Table 4.20. White $BaSO_4$ is precipitated from an acid solution by the addition of dilute H_2SO_4.

Bismuth. When heated with reducing mixture on charcoal, a bismuth mineral yields a metallic globule and an oxide coating. The metal is easily fusible but only imperfectly malleable and when hammered breaks into grains. When the assay is heated with bismuth flux, the sublimate is yellow near the mineral and red farther away (Table 4.18).

Boron. A yellow green flame is characteristic of boron, Table 4.20.

Calcium. Calcium is precipitated from alkaline solutions as white calcium oxalate by the addition of ammonium oxalate.

Carbon. All carbonate minerals dissolve with effervescence in HCl. Some dissolve only in hot acid; and others (for example, cerussite, $PbCO_3$) form an insoluble chloride which quickly brings the reaction to an end.

Chlorine. A copper chloride flame test is obtained from a mixture of chloride with CuO, Table 4.20. When $AgNO_3$ is added to a dilute HNO_3 solution of a chloride, a white precipitate of AgCl forms. The precipitate is soluble in NH_4OH.

Chromium. The green bead tests for chromium are characteristic, Table 4.21.

Cobalt. Cobalt minerals yield a distinctive dark blue color to the borax bead, Table 4.21.

Copper. When a copper mineral is moistened with HCl and heated, it yields an intense blue flame tinged with green. Copper sulfides should first be roasted before moistening with acid, Table 4.20. Copper can be reduced from some minerals on charcoal, Table 4.17.

Fluorine. Fluorine can be detected by converting it to HF and observing the etching effects on glass as follows: Cover a watch glass with paraffin and then remove the coating in places. Place the powdered mineral in the watch glass with a few drops of concentrated H_2SO_4. If the mineral is a soluble fluoride, HF will be liberated and etch the glass where it is exposed.

Iron. Minerals containing an appreciable amount of iron become magnetic when heated in the reducing flame. Cobalt and nickel minerals react similarly. A reddish-brown flocculent precipitate of $Fe(OH)_3$ forms when NH_4OH is added to an acid solution of an iron mineral.

Lead. Lead minerals when powdered, mixed with reducing mixture, and heated on charcoal yield a metallic globule, Table 4.17. A coating of PbO forms on the charcoal, yellow near the assay and white farther away, Table 4.18.

Lithium. Lithium imparts a strong crimson color to the flame, Table 4.20.

Magnesium. The only satisfactory test for magnesium is to precipitate it as NH_4MgPO_4. This is done by adding $Na_2HPO_4 \cdot 12H_2O$ to a strongly ammoniacal solution of the mineral.

Manganese. Manganese gives a bluish-green color to a Na_2CO_3 bead. This is a more sensitive test than the borax bead test, Table 4.21.

Mercury. A mercury mineral mixed with anhydrous Na_2CO_3 and heated in the closed tube yields a deposit of metallic mercury on the upper part of the tube, Table 4.19.

Molybdenum. A white sublimate of MoO_3 forms when an unoxidized molybdenum mineral is heated by the oxidizing flame on plaster. Touched with the reducing flame, the sublimate turns a deep ultramarine blue, Table 4.18.

Nickel. When an alcohol solution of dimethylglyoxime is added to an ammoniacal solution containing nickel, a scarlet precipitate forms. This is a very sensitive test.

Phosphorus. To test for phosphorus, dissolve the mineral in nitric acid and add a few drops of the solution to an excess of ammonium molybdate solution. A yellow precipitate of ammonium phosphomolybdate indicates phosphorus.

Silicon. When a powdered silicate mineral is heated in a salt of phosphorus bead, the bases are dissolved, leaving the silica as a translucent skeleton.

Silver. When HCl is added to a nitric acid solution of a silver mineral, white, curdy AgCl precipitates. This is a very sensitive test. Many silver minerals can be reduced to silver by heating them with reducing mixture on charcoal.

The resulting globules are white and malleable, Table 4.17.

Sodium. Sodium minerals give a strong, persistent yellow flame, Table 4.20.

Strontium. Strontium minerals give a strong, persistent crimson flame, Table 4.20.

Sulfur. Sulfides heated in the open tube give off SO_2 that may be detected by its pungent odor. Moistened blue litmus paper inserted into the upper end of the tube turns red, Table 4.18. Sulfur in sulfates can be detected as follows: Fuse the mineral with reducing mixture on charcoal, and place the fused mass on a clean silver surface with a drop of water. A dark brown stain of silver sulfide results.

Tin. Fragments of tin minerals placed in dilute HCl with metallic zinc and warmed become coated with metallic tin. Tin can be reduced from its minerals by heating with reducing mixture on charcoal, Table 4.17.

Titanium. To test for titanium, fuse the mineral with anhydrous sodium carbonate, and dissolve the fused mass in equal amounts of concentrated H_2SO_4 and water. The cold solution turns yellow on addition of hydrogen peroxide.

Tungsten. To test for tungsten, powder the mineral and fuse with anhydrous sodium carbonate. When the fused mass is dissolved in HCl, a yellow precipitate, WO_3, forms. The solution turns blue when boiled with grains of metallic tin or zinc.

Uranium. Uranium colors the salt of phosphorus bead yellow-green in the oxidizing flame, Table 4.21. This bead will fluoresce yellow in ultraviolet light.

Zinc. When fused with reducing mixture on charcoal, zinc minerals yield a nonvolatile coating, yellow when hot, white when cold. The coating, moistened with cobalt nitrate and heated, turns green.

References and Suggested Reading

Bragg, W. L. and G. F. Claringbull, 1965, *Crystal Structures of Minerals,* Cornell University Press, Ithaca, 409 pp.

Brush, G. J. and S. L. Penfield, 1926, *The Manual of Determinative Mineralogy with An Introduction to Blowpipe Analysis,* 16th ed. John Wiley & Sons, New York.

Evans, R. C., 1966, *An Introduction to Crystal Chemistry,* 2nd ed. The University Press, Cambridge, 410 pp.

Mason, B., 1966, *Principles of Geochemistry,* 3rd ed. John Wiley & Sons, New York, 329 pp.

Chemistry

Busch, D. H., Schull, H., and Conley, R. T., 1973, *Chemistry*. Allyn and Bacon, Inc., Boston, 929 pp.

Dickerson, R. E. and Geis, I., 1976, *Chemistry, Matter and the Universe*. W. A. Benjamin, Inc., Menlo Park, Cal., 669 pp.

Masterson, W. L. and Slowinski, E. J., 1973, *Chemical Principles*, 3rd ed. W. B. Saunders Co., Philadelphia, PA., 707 pp.

Pauling, L., 1960, *The Nature of the Chemical Bond,* 3rd ed. Cornell University Press, Ithaca, 644 pp.

Sienko, M. J. and Plane, R. A., 1976, *Chemistry,* 5th ed. McGraw Hill Book Co., New York, 624 pp.

Slabaugh, W. H. and Parsons, T. D., 1976, *General Chemistry,* 3rd ed. John Wiley & Sons, New York, 568 pp.

5
PHYSICAL PROPERTIES OF MINERALS

In this chapter we examine those physical properties that can be determined by inspection or by relatively simple tests. Because they are determined in hand specimens, they are important in the rapid recognition of minerals. Other physical properties, such as those determined by X-ray or optical techniques, require special and often sophisticated equipment and may involve elaborate sample preparation. X-ray and optical properties are discussed separately in Chapters 3 and 6. Crystal symmetry and forms (also physical properties) are treated in Chapter 2.

CRYSTAL HABITS AND AGGREGATES

The habit or appearance of single crystals as well as the manner in which crystals grow together in aggregates are of considerable aid in mineral recognition. Terms used to express habit and state of aggregation are given below.

1. Minerals in isolated or distinct crystals may be described as:
    (a) *Acicular*. Slender, needlelike crystals.
    (b) *Capillary and filiform*. Hairlike or threadlike crystals.
    (c) *Bladed*. Elongated crystals flattened like a knife blade.
2. For groups of distinct crystals the following terms are used:
    (a) *Dendritic*. Arborescent, in slender divergent branches, somewhat plantlike.
    (b) *Reticulated*. Latticelike groups of slender crystals.
    (c) *Divergent or radiated*. Radiating crystal groups.
    (d) *Drusy*. A surface covered with a layer of small crystals.
3. Parallel or radiating groups of individual crystals, are described as:
    (a) *Columnar*. Stout, columnlike individuals.
    (b) *Bladed*. An aggregate of many flattened blades.

(c) *Fibrous*. Aggregate of slender fibers, parallel or radiating.

(d) *Stellated*. Radiating individuals forming starlike or circular groups.

(e) *Globular*. Radiating individuals forming small spherical or hemispherical groups.

(f) *Botryoidal*. Globular forms resembling, as the word derived from the Greek implies, a "bunch of grapes."

(g) *Reniform*. Radiating individuals terminating in round kidney-shaped masses (Fig. 5.1).

(h) *Mammillary*. Large rounded masses resembling mammae, formed by radiating individuals.

(i) *Colloform*. Spherical forms composed of radiating individuals without regard to size; this includes botryoidal, reniform, and mammillary.

4. A mineral aggregate composed of scales or lamellae is described as:

(a) *Foliated*. Easily separable into plates or leaves.

(b) *Micaceous*. Similar to foliated, but splits into exceedingly thin sheets, as in the micas.

(c) *Lamellar or tabular*. Flat, platelike individuals superimposed upon and adhering to each other.

(d) *Plumose*. Fine scales with divergent or featherlike structure.

5. A mineral aggregate composed of grains is *granular*.

6. Miscellaneous terms:

(a) *Stalactitic*. Pendent cylinders or cones. Stalactites form by deposition from mineral-bearing waters dripping from the roofs of caverns.

(b) *Concentric*. More or less spherical layers superimposed upon one another about a common center.

(c) *Pisolitic*. Rounded masses about the size of peas.

(d) *Oölitic*. A mineral aggregate formed of small spheres resembling fish roe.

(e) *Banded*. A mineral in narrow bands of different color or texture.

(f) *Massive*. Compact material without form or distinguishing features.

(g) *Amygdaloidal*. A rock such as basalt containing almond-shaped nodules.

FIG. 5.1. Reniform hematite, Cumberland, England.

(h) *Geode.* A rock cavity lined by mineral matter but not wholly filled. Geodes may be banded as in agate, due to successive depositions of material, and the inner surface is frequently covered with projecting crystals.

(i) *Concretion.* Masses formed by deposition of material about a nucleus. Some concretions are roughly spherical whereas others assume a great variety of shapes.

CLEAVAGE, PARTING, AND FRACTURE

Cleavage

A mineral has *cleavage* if it breaks along definite plane surfaces. Cleavage may be perfect as in the micas, or more or less obscure as in beryl and apatite. In some minerals, it is completely absent.

Cleavage depends on crystal structure and takes place parallel to atomic planes. If a family of parallel atomic planes has a weak binding force between them, cleavage is likely to take place along these planes. This weakness may be a result of a weak type of bond, a greater spacing between the planes, or a combination of the two. Graphite has a platelike cleavage. Within the plates there is a strong bond, but across the plates there is a weak bond giving rise to the cleavage. A weak bond is usually accompanied by a large interplanar spacing because the attractive force cannot hold the planes closely together. Diamond has but one bond type, and its excellent cleavage takes place along those atomic planes having the largest interplanar spacing.

Because cleavage is the breaking of a crystal between atomic planes, it is a directional property, and any parallel plane through a crystal is a potential cleavage plane. Moreover, it is always parallel to crystal faces or possible crystal faces (usually those with simplest indices), because both faces and cleavage reflect the same crystal structure.

In describing a cleavage its quality and crystallographic direction should be given. The quality is expressed as perfect, good, fair, etc. The direction is expressed by the name or indices of the form which the cleavage parallels, such as cubic $\{001\}$ (Fig. 5.2), octahedral $\{111\}$, rhombohedral $\{10\bar{1}1\}$, prismatic $\{110\}$, or pinacoidal $\{001\}$. Cleavage is always consistent with the symmetry; thus, if one octahedral cleavage direction is developed, it implies that there must be three other symmetry related directions. If one dodecahedral cleavage direction is present, it likewise implies five other symmetry related directions. Not all minerals show cleavage, and only a comparatively few show it in an eminent

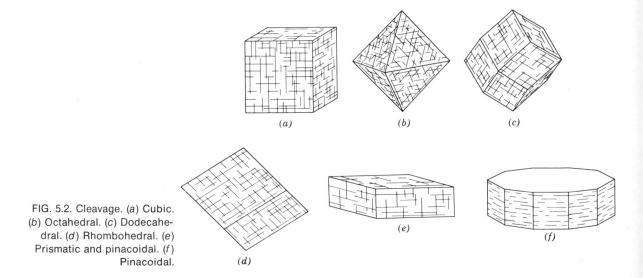

FIG. 5.2. Cleavage. (*a*) Cubic. (*b*) Octahedral. (*c*) Dodecahedral. (*d*) Rhombohedral. (*e*) Prismatic and pinacoidal. (*f*) Pinacoidal.

degree, but in these it serves as an outstanding diagnostic criterion.

Parting

When minerals break along planes of structural weakness, they have parting. The weakness may result from pressure or twinning; and, because it is parallel to rational crystallographic planes, it resembles cleavage. However, parting, unlike cleavage, is not shown by all specimens but only those that are twinned or have been subjected to the proper pressure. Even in these specimens there are a limited number of planes in a given direction along which the mineral will break. For example, twinned crystals part along composition planes but between these planes they fracture irregularly. Familiar examples of parting are found in the octahedral parting of magnetite, the basal parting of pyroxene, and the rhombohedral parting of corundum (see Figs. 5.3a and b).

Fracture

The way a mineral breaks when it does not yield along cleavage or parting surfaces is its *fracture*. Different kinds of fracture are designated as follows:

a. *Conchoidal.* The smooth, curved fracture resembling the interior surface of a shell (see Fig. 5.4). This is most commonly observed in such substances as glass and quartz. b. *Fibrous or splintery.* c.

FIG. 5.4. Conchoidal fracture, obsidian.

Hackly. Jagged fractures with sharp edges. *d. Uneven or irregular.* Fractures producing rough and irregular surfaces.

HARDNESS

The resistance that a smooth surface of a mineral offers to scratching is its *hardness* (designated by **H**). Like the other physical properties of minerals, hardness is dependent on the crystal structure. The stronger the binding force between the atoms, the harder the mineral (see Table 4.6b). The degree of hardness is determined by observing the comparative ease or difficulty with which one mineral is scratched by another, or by a file or knife. The hardness of a mineral might then be said to be its "scratchability." A series of 10 common minerals were chosen by the Austrian mineralogist F. Mohs in 1824 as a scale, by comparison with which the relative hardness of any mineral can be told. The following minerals arranged in order of increasing hardness comprise what is known as the *Mohs scale of hardness*:

1. Talc
2. Gypsum
3. Calcite
4. Fluorite
5. Apatite
6. Orthoclase
7. Quartz
8. Topaz
9. Corundum
10. Diamond

FIG. 5.3(a). Basal parting, pyroxene. (b) Rhombohedral parting, corundum.

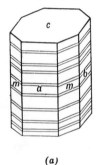
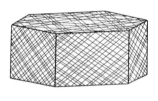

(a)   (b)

The above minerals are arranged in an order of increasing relative hardness. The hardness of these same minerals can be measured by more quantitative techniques than a scratch test, and this leads to an absolute hardness scale as shown in Fig. 5.5. The relative position of the minerals in the Mohs scale is preserved, but corundum, for example, is two times as hard as topaz but four times harder than quartz.

Talc, number 1 in the Mohs scale, has a structure made up of plates so weakly bound to one another that the pressure of the fingers is sufficient to slide one plate over the other. At the other end of the scale is diamond with its constituent carbon atoms so firmly bound to each other that no other mineral can force them apart to cause a scratch.

In order to determine the relative hardness of any mineral in terms of this scale, it is necessary to find which of these minerals it can and which it cannot scratch. In making the determination, the following should be observed: sometimes when one mineral is softer than another, portions of the first will leave a mark on the second that may be mis-

FIG. 5.5. Comparison of Mohs relative hardness scale and absolute measurements of hardness.

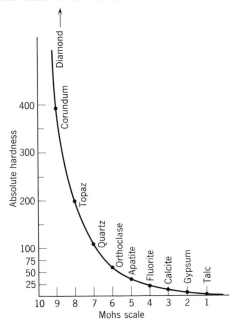

taken for a scratch. Such a mark can be rubbed off, whereas a true scratch will be permanent. The surfaces of some minerals are frequently altered to material that is much softer than the original mineral. A *fresh surface* of the specimen to be tested should therefore be used. The physical nature of a mineral may prevent a correct determination of its hardness. For instance, if a mineral is pulverulent, granular, or splintery, it may be broken down and apparently scratched by a mineral much softer than itself. It is always advisable when making the hardness test to confirm it by reversing the order of procedure; that is, do not only try to scratch mineral A by mineral B, but also try to scratch B by A.

The following materials serve in addition to the above scale: the hardness of the finger nail is a little over 2, a copper coin about 3, the steel of a pocket knife a little over 5, window glass $5\frac{1}{2}$, and the steel of a file $6\frac{1}{2}$. With a little practice, the hardness of minerals under 5 can be quickly estimated by the ease with which they can be scratched with a pocket knife.

Hardness, as we have seen (page 36), is a vectorial property. Thus crystals may show varying degrees of hardness depending on the directions in which they are scratched. The directional hardness differences in most common minerals are so slight that, if they can be detected at all, it is only through the use of delicate instruments. Two exceptions are kyanite and calcite. In kyanite, $H = 5$ parallel to the length, but $H = 7$ across the length of the crystal. The hardness of calcite is 3 on all surfaces except $\{0001\}$. On this form, however, it can be scratched by the fingernail and has a hardness of 2.

TENACITY

The resistance that a mineral offers to breaking, crushing, bending, or tearing—in short, its cohesiveness—is known as tenacity. The following terms are used to describe tenacity in minerals:

1. *Brittle.* A mineral that breaks or powders easily.
2. *Malleable.* A mineral that can be hammered out into thin sheets.

3. *Sectile.* A mineral that can be cut into thin shavings with a knife.
4. *Ductile.* A mineral that can be drawn into wire.
5. *Flexible.* A mineral that bends but does not resume its original shape when the pressure is released.
6. *Elastic.* A mineral that, after being bent, will resume its original position upon the release of the pressure.

SPECIFIC GRAVITY

Specific gravity (**G**) or relative density* is a number that expresses the ratio between the weight of a substance and the weight of an equal volume of water at 4°C. Thus a mineral with a specific gravity of 2, weighs twice as much as the same volume of water. The specific gravity of a mineral is frequently an important aid in its identification, particularly in working with fine crystals or gemstones, when other tests would injure the specimens.

The specific gravity of a crystalline substance depends on (1) the kind of atoms of which it is composed, and (2) the manner in which the atoms are packed together. In isostructural compounds (see page 149), in which the packing is constant, those with elements of higher atomic weight will usually have higher specific gravities. This is well illustrated by the orthorhombic carbonates listed in Table 5.1 in which the chief difference is in the cations.

Table 5.1
SPECIFIC GRAVITY INCREASE WITH INCREASING ATOMIC WEIGHT OF CATION IN ORTHORHOMBIC CARBONATES

Mineral	Composition	Atomic Weight of Cation	Specific Gravity
Aragonite	$CaCO_3$	40.08	2.95
Strontianite	$SrCO_3$	87.62	3.76
Witherite	$BaCO_3$	137.34	4.29
Cerussite	$PbCO_3$	207.19	6.55

* Density and specific gravity are sometimes used interchangeably. However, density requires the citation of units, for example, grams per cubic centimeter or pounds per cubic foot.

In a solid solution series (see page 161), there is a continuous change in specific gravity (or density) with change in chemical composition. For example, the mineral olivine, $(Mg, Fe)_2SiO_4$, is a solid solution series between forsterite, Mg_2SiO_4 (**G** 3.3), and fayalite, Fe_2SiO_4 (**G** 4.4). Thus from determination of specific gravity one can obtain a close approximation of the chemical composition of an Mg-Fe olivine (see Fig. 10.9). A similar relationship in an amphibole series is shown in Fig. 5.6.

The influence of the packing of atoms on specific gravity is well illustrated in polymorphous compounds (see Table 4.10). In these compounds the composition remains constant, but the packing of the atoms varies. The most dramatic example is given by diamond and graphite, both elemental carbon. Diamond with specific gravity 3.5 has a closely packed structure, giving a high density of atoms per unit volume; whereas in graphite, specific gravity 2.23, the carbon atoms are loosely packed.

Average Specific Gravity
Most people from everyday experience have acquired a sense of relative weight even in regard to minerals. For example, ulexite (**G** 1.96) seems light, whereas barite (**G** 4.5) seems heavy for nonmetallic minerals. This means that one has developed an idea of an average specific gravity or a feeling of what a nonmetallic mineral of a given size should weigh. This average specific gravity can be considered to be between 2.65 and 2.75. The reason for

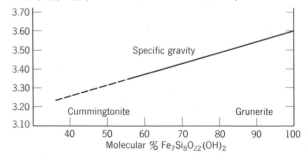

FIG. 5.6. Variation of specific gravity with composition in the monoclinic cummingtonite-grunerite series ranging in composition from $Fe_2Mg_5Si_8O_{22}(OH)_2$ to $Fe_7Si_8O_{22}(OH)_2$ (after Klein, *Amer. Miner*, 1964).

this is that the specific gravities of quartz (**G** 2.65), feldspar (**G** 2.60–2.75), and calcite (**G** 2.72), the most common and abundant nonmetallic minerals, fall mostly within this range. The same sense may be developed in regard to metallic minerals: graphite (**G** 2.23) seems light, whereas silver (**G** 10.5) seems heavy. The average specific gravity for metallic minerals is about 5.0, that of pyrite. Thus, with a little practice, one can, by merely lifting specimens, distinguish minerals that have comparatively small differences in specific gravity.

Determination of Specific Gravity

In order to determine specific gravity accurately, the mineral must be homogeneous and pure, requirements frequently difficult to fulfill. It must also be compact with no cracks or cavities within which bubbles or films of air could be imprisoned. For normal mineralogical work, the specimen should have a volume of about one cubic centimeter. If these conditions cannot be met, a specific gravity determination by any rapid and simple method means little.

The necessary steps in making an ordinary specific gravity determination are, briefly, as follows: the mineral is first weighed in air. Let this weight be represented by W_a. It is then immersed in water and weighed again. Under these conditions it weighs less, since in water it is buoyed up by a force equivalent to the weight of the water displaced. Let the weight in water be represented by W_w. Then $W_a - W_w$ equals the apparent loss of weight in water, or the weight of an equal volume of water. The expression $W_a/(W_a - W_w)$ will therefore yield a number which is the specific gravity.

Jolly Balance. Because specific gravity is merely a ratio, it is not necessary to determine the absolute weight of the specimen but merely values proportional to the weights in air and in water. This can be done by means of a *Jolly balance* (Fig. 5.7),* with which the data for making the calculations are obtained by the stretching of a spiral spring. In using the balance, a fragment is first placed on the upper

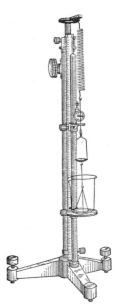

FIG. 5.7. Jolly balance.

scale pan and the elongation of the spring noted. This is proportional to the weight in air (W_a). The fragment is then transferred to the lower pan and immersed in water. The elongation of the spring is now proportional to the weight of the fragment in water (W_w).

A torsion balance was adapted by the late Harry Berman of Harvard University for obtaining specific gravities of small particles weighing less than 25 milligrams (Fig. 5.8).* To the advanced worker interested in accurate determinations this balance is particularly useful, because it is frequently possible to obtain only a tiny mineral fragment free from impurities. In using it, however, one must make a correction for temperature and use a liquid with a low surface tension.

Pycnometer. When a mineral cannot be obtained in a homogenous mass large enough to permit use of one of the balance methods, the specific gravity of a powder or an aggregate of mineral fragments can be accurately obtained by means of a *pycnometer*. The pycnometer is a small bottle (Fig. 5.9) fitted with a ground-glass stopper through

* Manufactured by Eberbach and Son, Ann Arbor, Michigan.

* This balance is distributed through Bethlehem Instrument Company, Bethlehem, Pennsylvania.

FIG. 5.8. Berman balance.

FIG. 5.9. Pycnometer.

which a capillary opening has been drilled. In making a specific gravity determination, the dry bottle with stopper is first weighed empty (P). The mineral fragments are then introduced into the bottle and a second weighing (M) is made. ($M - P$) represents the weight of the sample in air. Subsequently the bottle (containing the mineral sample) is partially filled with distilled water and boiled for a few minutes to drive off any air bubbles. After cooling, the pycnometer is further filled with distilled water and weighed (S), care being taken that the water rises to the top of the capillary opening but that no excess water is present. The last weighing (W) is made after emptying the bottle and refilling with distilled water alone. In this last step, the pycnometer contains more water than in the previous weighing; the volume of water added is equal to the aggregate volume of the grains comprising the sample. The specific gravity can be determined from:

$$G = \frac{(M - P)}{W + (M - P) - S}$$

where $M - P$ = weight of sample,
W = pycnometer + water content,
S = sample + pycnometer + undisplaced water,
$W + (M - P) - S$ = weight of water displaced by sample.

Heavy liquids. Several liquids with relatively high densities are sometimes used in the determination of the specific gravity of minerals. The two liquids most easily used are bromoform (G 2.89) and methylene iodide (G 3.33).* These liquids are miscible with acetone (G 0.79), and thus, by mixing, a solution of any intermediate specific gravity may be obtained. A mineral grain is introduced into the heavy liquid, and the liquid diluted with acetone until the mineral neither rises nor sinks. The specific gravity of the liquid and the mineral are then the same, and that of the liquid may be quickly determined by means of a Westphal balance.

Heavy liquids are frequently used in the separation of grains from mixtures composed of several constituents. For example, a separation of the constituent mineral grains of a sand composed of quartz (G 2.65), tourmaline (G 3.20), and garnet (G 4.25) could be quickly made. In bromoform the quartz would float and the tourmaline and garnet would sink. After removing and washing these "heavy minerals" in acetone, they could be separated from each other in methylene iodide; the tourmaline would float and the garnet would sink.

Calculation of Density. If one knows the number of the various kinds of atoms in the unit cell and the volume of the unit cell, the specific gravity can be calculated. The chemical formula of the mineral gives the proportions of the different atoms but it does not provide us with a knowledge of the number of formula units per unit cell. This number, usually small, is indicated by Z. For example, in aragonite, $CaCO_3$, the ratio of the atoms is $1Ca:1C:3oxygens$, but there are four formula units

per cell or 4Ca, 4C, 12 oxygens. The molecular weight, M, of $CaCO_3$ is 100.09; the molecular weight of the contents of the unit cell ($Z = 4$) is $4 \times 100.09 = 400.36$.

The volume of the unit cell, V, in the orthogonal crystal systems is found by multiplying the cell dimensions, as $a \times b \times c = V$. In the inclined systems the angles between cell edges must also be considered in obtaining the volume. Aragonite is orthorhombic with cell dimensions: $a = 4.95$ Å, $b = 7.96$, $c = 5.73$. Therefore, $V = 225.76$ Å^3.

Converting Å^3 to cm^3, we divide by $(10^8)^3 = 10^{24}$ or $V = 225.76 \times 10^{-24}$ cm^3. Knowing the values M and V, the density, D, can be calculated using the formula:

$$D = \frac{Z \times M}{N \times V}$$

where N is Avogadro's number, 6.02338×10^{23}. Substituting values for aragonite,

$$D = \frac{4 \times 100.09}{6.02338 \times 10^{23} \times 225.76 \times 10^{-24}}$$
$$= 2.945 \text{ g/cm}^3$$

This value, 2.945, for the calculated density of aragonite is in excellent agreement with the best measured values which are 2.947 ± 0.002.

In the study of new minerals, the numerical value of Z is commonly unknown. Hence it is necessary to make successive trials of the calculations given above, using different values of Z until the best possible agreement with the measured specific gravity is secured. Z is always an integer and generally small.

LUSTER

The term luster refers to the general appearance of a mineral surface in reflected light. There are two types of luster, *metallic* and *nonmetallic,* but with no sharp division between them. Minerals with an intermediate luster are said to be *submetallic.*

A mineral having the brilliant appearance of a metal has a metallic luster. Such minerals are quite opaque to light and, as a result, give a black or very

* Although bromoform and methylene iodide are miscible, *do not mix them.* The mixture will become black.

dark streak (see page 191). Galena, pyrite, and chalcopyrite are common minerals with metallic luster.

Minerals with a nonmetallic luster are, in general, light-colored and transmit light, if not through thick portions, at least through thin edges. The streak of a nonmetallic mineral is either colorless or very light in color. The following terms are used to describe further the luster of nonmetallic minerals.

Vitreous. The luster of glass. Examples—quartz and tourmaline.

Resinous. Having the luster of resin. Examples—sphalerite and sulfur.

Pearly. An iridescent pearl-like luster. This is usually observed on mineral surfaces that are parallel to cleavage planes. Examples—basal plane of apophyllite and cleavage surface of talc.

Greasy. Appears as if covered with a thin layer of oil. This luster results from light scattered by a microscopically rough surface. Examples, nepheline and some specimens of sphalerite and massive quartz.

Silky. Silklike. It is caused by the reflection of light from a fine fibrous parallel aggregate. Examples—fibrous gypsum, malachite, and serpentine (chrysotile).

Adamantine. A hard, brilliant luster like that of a diamond. It is due to the mineral's high index of refraction (see page 196). Examples—the transparent lead minerals, such as cerussite and anglesite.

COLOR

When white light strikes the surface of a mineral, part of it is reflected and part refracted. If the light suffers no absorption, the mineral is colorless, both in reflected and transmitted light. Minerals are colored because certain wavelengths of light are absorbed, and the color results from a combination of those wavelengths that reach the eye. Some minerals show different colors when light is transmitted along different crystallographic directions. This selective absorption known as *pleochroism* is shown by transparent varieties of cordierite and spodumene. If there are only two such directions, as in tourmaline, the property is called *dichroism*.

For some minerals, color is a fundamental property directly related to one of its major constituent elements and is therefore constant and characteristic. For such minerals, called *idiochromatic*, color serves as an important means of identification. Thus malachite is always green, azurite is always blue, and rhodonite and rhodochrosite are always red or pink.

For most metallic minerals, color is constant, as: the brass-yellow of chalcopyrite, the brownish bronze of bornite, and the copper-red of niccolite. Because alteration may produce a colored tarnish on these minerals different from the true color, it is important that a fresh surface be examined. This is particularly true of copper minerals, especially bornite. This mineral is called "peacock copper ore" because on exposure its bronze color is covered by a blue-violet film.

Most minerals are composed of elements that produce no characteristic color and are colorless. Yet colored varieties of many of them are common. In some cases the color is attributable to appreciable amounts of an element, such as iron, that has a strong pigmenting power. In several solid solution series, color is directly related to the amount of iron present. For example, the amphibole tremolite, with composition $Ca_2Mg_5(Si_8O_{22})(OH)_2$ is white; but when some Fe^{2+} substitutes for Mg^{2+}, the mineral is green. With increasing amounts of iron, the color changes from pale to dark green to nearly black. Also in sphalerite, ZnS, the progressive substitution of iron for zinc changes the color from white through yellow and brown to black.

Minerals such as those mentioned that show a variation in color are called *allochromatic*. However, the presence of a major chemical constituent is only one of the causes of color in minerals. Other factors are small amounts of chemical impurities, defects in crystal structure, and finely divided inclusions of other minerals.

The ions of certain elements are strongly light-absorbing and their presence in small, even trace, amounts may cause the mineral to be deeply colored. Chief among these elements known as *chromophores* are: Fe, Mn, Cu, Cr, Co, Ni, and V. Thus the deep green of emerald results from small amounts of chromium or vanadium in beryl and the purple of amethyst is attributed to trace amounts of iron in quartz. Zoisite, normally white or gray, was found in 1967 in gem quality crystals of a rich sapphire blue; the color results from the presence of vanadium.

Structure defects in minerals having no chromophoric ions may cause white light to be selectively absorbed. The imperfections may be due to structural voids or to the presence of foreign ions. Color in minerals with structural imperfections may be changed or introduced by heat treatment or exposure to high-energy radiation. For example, X-radiation turns some colorless quartz smoky, but on subsequent heating to 400°C the quartz returns to its colorless state. Not all colorless quartz responds this way. X-radiation has no effect on some crystals and others are selectively pigmented. It is believed that the coloration is associated with the presence of foreign ions. Colorless diamonds by exposure to the proper radiation can be colored either green or blue, and the green diamond will turn yellow with heat treatment. The color of many other gem stones can be enhanced by heating.

The color of certain silicate minerals results from the presence of anions or anionic groups such as Cl^{1-}, CO_3^{2-}, SO_4^{3-} in the structure. All these ions may be present in lazurite (blue), Cl^{1-} in sodalite (blue and green), CO_3^{2-} in cancrinite (orange-yellow), SO_4^{3-} in hauynite (blue). Heating causes a change in color in all these minerals.

The mechanical admixture of impurities can give a variety of colors to otherwise colorless minerals. Quartz may be green because of the presence of chlorite; calcite may be black, colored by manganese oxide or carbon. Hematite, as the most common pigmenting impurity, imparts its red color to many minerals including some feldspar and calcite and the fine-grained variety of quartz, jasper.

Streak

The color of a finely powdered mineral is known as its *streak*. Although the color of a mineral may vary, the streak is usually constant and is thus useful in mineral identification. The streak is determined by rubbing the mineral on a piece of unglazed porcelain, a *streak plate*. The streak plate has a hardness

of about 7, and thus it cannot be used with minerals of greater hardness.

Play of Colors

Interference of light either at the surface or in the interior of a mineral may produce a series of colors as the angle of incident light changes. The striking play of colors seen in precious opal results from the interference of light reflected from submicroscopic layers of nearly spherical particles arranged in a regular pattern. Common opal lacks this layering and the scattered light produces a pearly or milky *opalescence*.

An internal iridescence is caused by light defracted and reflected from closely spaced fractures, cleavage planes, twin lamellae, exsolution lamellae, or minute foreign inclusions in parallel orientation. Some specimens of labradorite show colors ranging from blue to green or yellow with changing angle of incident light. This *iridescence* is due to extremely fine, less than $\frac{1}{10}$ of a micron or thinner in width, exsolution-type lamellae in the range of An_{47} to An_{58}. The delicate colors of some albitic feldspars (peristerite), ranging in composition from An_2 to An_{16}, are caused by exsolution-type lamellae of An_0 and An_{25} composition.

A surface iridescence similar to that produced by soap bubbles or thin films of oil on water is caused by interference of light as it is reflected from thin surface films. It is most commonly seen on metallic minerals, particularly hematite, limonite, and sphalerite.

Chatoyancy and Asterism

In reflected light some minerals have a silky appearance which results from closely packed parallel fibers or from a parallel arrangement of inclusions or cavities. When a cabochon gemstone is cut from such a mineral, it shows a band of light at right angles to the length of the fibers or direction of the inclusions. This property known as *chatoyancy* is shown particularly well by "satin spar" gypsum, *cat's eye*, a gem variety of chrysoberyl and *tiger's eye*, fibrous crocidolite replaced by quartz.

In some crystals, particularly those of the hex-

FIG. 5.10. Asterism in a sphere of rose quartz. Sphere diameter is 5.5 cm. The 6-rayed star is caused by microscopic needlelike inclusions of rutile (TiO_2) which are oriented in three directions (at 120° to each other) by the quartz structure. These inclusions reflect a spotlight source yielding the 6-rayed star.

agonal system, inclusions may be arranged in three crystallographic directions at 120° to each other. A cabochon stone cut from such a crystal shows what might be called a triple chatoyancy, that is, one beam of light at right angles to each direction of inclusions producing a six-pointed star. The phenomenon, seen in star rubies and sapphires, is termed *asterism* (see Fig. 5.10). Some phlogopite mica containing rutile needles oriented in a pseudohexagonal pattern, shows a striking asterism in transmitted light.

LUMINESCENCE

Any emission of light by a mineral that is not the direct result of incandescence is *luminescence*. This phenomenon may be brought about in several ways and is usually observed in minerals containing foreign ions called *activators*. Most luminescence is faint and can be seen only in the dark.

Fluorescence and Phosphorescence

Minerals that luminesce during exposure to ultraviolet light, X-rays, or cathode rays are *fluorescent*. If the luminescence continues after the exciting rays are cut off, the mineral is said to be *phosphorescent*. There is no sharp distinction between fluorescence and phosphorescence, for some minerals that appear only to fluoresce can be shown by refined methods to continue to glow for a small fraction of a second after the removal of the exciting radiation.

Fluorescence is produced when the energy of shortwave radiation is absorbed by impurity ions and released as longer wave radiation—visible light. Minerals vary in their ability to absorb ultraviolet light at a given wavelength. Thus some fluoresce only in shortwave uv, whereas others may fluoresce only in longwave uv, and still others will fluoresce under either wave length of uv. The color of the emitted light varies considerably with the wavelengths or source of ultraviolet light.

Fluorescence is an unpredictable property, for

some specimens of a mineral show it, whereas other apparently similar specimens even from the same locality, do not. Thus only some fluorite, the mineral from which the property receives its name, will fluoresce. Its usual blue fluorescence may result from the presence of organic material or rare earth ions. Other minerals that frequently, but by no means invariably, fluoresce are scheelite, willemite, calcite, eucryptite, scapolite, diamond, hyalite, and autunite. The pale blue fluorescence of most scheelite is ascribed to molybdenum substituting for tungsten. And the brilliant fluorescence of willemite and calcite from Franklin, New Jersey, is attributed to the presence of manganese.

With the development of synthetic phosphors, fluorescence has become a commonly observed phenomenon in fluorescent lamps, paints, cloth, and tapes. The fluorescent property of minerals also has practical applications in prospecting and ore dressing. With a portable ultraviolet light one can at night detect scheelite in an outcrop and underground the miner can quickly estimate the amount of scheelite on a freshly blasted surface. At Franklin, New Jersey, ultraviolet light has long been used to determine the amount of willemite that goes into the tailings. Eucryptite is an ore of lithium in the great pegmatite at Bikita, Rhodesia. In white light it is indistinguishable from quartz, but under ultraviolet light it fluoresces a salmon pink and can be easily separated.

Thermoluminescence

This is the property possessed by some minerals of emitting visible light when heated to a temperature below that of red heat. It is best shown by nonmetallic minerals that contain foreign ions as activators. When a thermoluminescent mineral is heated, the initial visible light, usually faint, is given off at a temperature between 50° and 100°C, and light usually ceases to be emitted at temperatures higher than 475°C. For a long time, fluorite has been known to possess this property; the variety *chlorophane* was named because of the green light emitted. Other minerals that are commonly thermoluminescent are calcite, apatite, scapolite, lepidolite, and feldspar.

Triboluminescence

This is the property possessed by some minerals of becoming luminous on being crushed, scratched, or rubbed. Most minerals showing this property are nonmetallic and possess a good cleavage. Fluorite, sphalerite, and lepidolite may be triboluminescent and, less commonly, pectolite, amblygonite, feldspar, and calcite.

ELECTRICAL AND MAGNETIC PROPERTIES

The conduction of electricity in crystals is related to the type of bonding. Minerals with pure metallic bonding, such as the native metals, are excellent electrical *conductors,* whereas those in which the bonding is partially metallic, as in some sulfide minerals, are *semiconductors.* Ionic or covalent bonded minerals are usually nonconductors. For nonisometric minerals, electrical conductivity is a vectorial property varying with crystallographic direction. For example, the hexagonal mineral, graphite, is a far better conductor at right angles to the c axis than parallel to it.

Piezoelectricity

Polar axes are present only in crystals that lack a center of symmetry. Of the 32 crystal classes, 21 have no center of symmetry and of these all but one, the gyroidal class, has at least one polar axis with different crystal forms at opposite ends. If pressure is exerted at the ends of a polar axis, a flow of electrons toward one end produces a negative electrical charge, while a positive charge is induced at the opposite end. This is *piezoelectricity* and any mineral crystallizing in one of the 20 classes with polar axes should show it. However, in some minerals the charge developed is too weak to be detected.

The property of piezoelectricity was first detected in quartz in 1881 by Pierre and Jacques Curie but nearly 40 years passed before it was used in a practical way. Toward the end of World War I it was found that sound waves produced by a submarine could be detected by the piezoelectric current generated when they impinged on a submerged quartz plate. The device was developed too late to have great value during the war, but it pointed the way to other applications. In 1921 the piezoelectric property of quartz was first used to control radio frequencies, and since then millions of quartz plates have been used for this purpose. When subjected to an alternating current, a properly cut slice of quartz is mechanically deformed and vibrates by being flexed first one way and then the other; the thinner the slice, the greater the frequency of vibration. By placing a quartz plate in the electric field generated by a radio circuit, the frequency of transmission or reception is controlled when the frequency of the quartz coincides with the oscillations of the circuit. The tiny quartz plate used in digital quartz watches serves the same function as quartz oscillators used to control radio frequencies. That is, it mechanically vibrates at a constant, predetermined frequency; this controls accurately the radio frequency of the electronic circuit in the watch. This circuit counts the crystal frequency and provides the digital time display of the watch.

The piezoelectric property of tourmaline has been known almost as long as that of quartz, but compared to quartz tourmaline is a less effective radio oscillator and is rare in occurrence. Nevertheless, small amounts of it are used today in piezoelectric pressure gauges. In tourmaline, which is hexagonal, c is the polar axis. Plates cut normal to this direction will generate an electric current when subjected to a transient pressure. The current generated is proportional to the area of the plate and to the pressure. Tourmaline gauges were developed to record the blast pressure of the first atomic bomb in 1945 and since then have been used by the United States with each atomic explosion. However, lesser pressures also can be recorded by them, such as those generated by firing a rifle or by surf beating on a sea wall.

Pyroelectricity

Temperature changes in a crystal may cause the simultaneous development of positive and negative charges at opposite ends of a polar axis. This property of *pyroelectricity* is observed, as is piezoelectricity, only on crystals with polar axes. Crystals that belong to the 10 crystal classes having a unique polar axis are considered to show "true" or *primary* pyroelectricity. For example, tourmaline has a single polar axis, c, and falls within this group, whereas quartz with its three polar a axes does not. However, a temperature gradient in all crystals having polar axes such as quartz will produce a pyroelectric effect. In such crystals the polarization is the result of the deformation resulting from unequal thermal expansion that produces piezoelectric effects. If quartz is heated to about 100°C, on cooling it will develop positive charges at three alternate prismatic edges and negative charges at the three remaining edges. These charges have been called *secondary* pyroelectric polarization.

By means of single crystal X-ray photographs some minerals can be assigned to specific point groups, but for others X-ray data are ambiguous. For example, on the basis of X-ray photographs, a mineral might belong to either of the classes $2/m\ 2/m\ 2/m$ or $mm2$. If it could be shown to be piezoelectric or pyroelectric it would definitely belong to $mm2$.

Magnetism

Magnetite, Fe_3O_4, and pyrrhotite, $Fe_{1-x}S$ are the only common minerals attracted to a small hand magnet. They are *ferromagnetic*. Lodestone, a variety of magnetite, is a natural magnet with the attracting power and polarity of a true magnet. In the field of a powerful electromagnet many other minerals, especially those containing iron, are attracted. These are called *paramagnetic*, whereas those that are repelled are called *diamagnetic*. Because of their different magnetic susceptibilities, minerals can be separated from each other by an electromagnet and magnetic separation of minerals is a standard procedure on both laboratory and commercial scale. Thus the magnetic separator is standard equipment in mineralogical laboratories and is used on a commercial scale to separate ore minerals from gangue.

6
OPTICAL PROPERTIES OF MINERALS

The optical properties are less easily determined than the other physical properties of minerals but are important in mineral characterization and identification. Because they are usually observed microscopically, only a very small amount of material is necessary. The following discussion deals with the optical properties that can be determined in transmitted light and thus applies only to nonopaque minerals.

NATURE OF LIGHT

To account for all the properties of light it is necessary to resort to two theories: the *wave theory* and the *corpuscular theory*. However, it is the wave theory that we shall consider in explaining the optical behavior of crystals. This theory assumes that visible light, as part of the electromagnetic spectrum, travels in straight lines with a transverse wave motion; that is, it vibrates at right angles to the direction of propagation. The wave motion is similar to

that generated by dropping a pebble into still water with waves moving out from the central point. The water merely rises and falls, it is only the wave front that moves forward. The *wavelength* (λ) of such a wave motion is the distance between successive crests (or troughs); the *amplitude* is the displacement on either side of the position of equilibrium; the *frequency* is the number of waves per second passing a fixed point; and the *velocity* is the frequency multiplied by the wavelength. Similarly light waves (Fig. 6.1) have length, amplitude, frequency, and velocity but their transverse vibrations, perpendicular to the direction of propagation, take place in all possible directions.

Visible light occupies a very small portion of the electromagnetic spectrum (see Fig. 3.1). The wavelength determines the color and varies from slightly more than 7000 Å at the red end to about 4000 Å at the violet end. White light is composed of all wavelengths between these limits, whereas light of a single wavelength is called monochromatic.

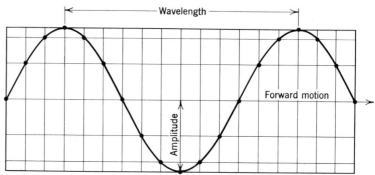

FIG. 6.1. Wave motion.

Reflection and Refraction

When a light ray passes from a rare medium, such as air, into a denser medium, such as glass, part of it is reflected from the surface back into the air and part enters the glass (Fig. 6.2). The reflected ray obeys the laws of reflection which state: (a) that the angle of incidence (*i*) equals the angle of reflection (*r′*), when both angles are measured from the surface normal, and (b) that the incident and reflected rays lie in the same plane. The light that passes into the glass travels with a lesser velocity than in air and no longer follows the path of the incident ray but is bent or *refracted*. The amount of bending depends on the obliquity of the incident ray and the relative velocity of light in the two media; the greater the angle of incidence and the greater the velocity difference, the greater the refraction.

Index of Refraction

The index of refraction, *n*, of a material can be expressed as the ratio between the velocity of light in the air (*V*) and its velocity in the denser material (*v*), that is, $n = V/v$. As a basis for comparison the velocity of light in air is considered equal to 1,

$n = 1/v$, or the index of refraction is equal to the reciprocal of the velocity.*

The precise relationship of the angle of incidence (*i*) to the angle of refraction (*r*) is given by Snell's law which states that for the same two media the ratio of sin *i* : sin *r* is a constant. This is usually expressed as sin *i*/sin *r* = *n*, where the constant, *n*, is the *index of refraction*.

The velocity of light in glass is equal to frequency multiplied by wavelength; therefore, with fixed frequency, the longer the wavelength the greater the velocity. Red light with its longer wavelength has a greater velocity than violet light and because of the reciprocal relation between velocity and refractive index, *n* for red light is less than *n* for violet light (Fig. 6.3). A crystal thus has different refractive indices for different wavelengths of light. This phenomenon is known as *dispersion of the indices,* and because of it, monochromatic light is used for accurate determination of refractive index.

FIG. 6.3. Different refraction for different wavelengths of light.

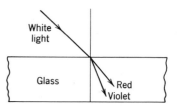

FIG. 6.2. Reflected and refracted light.

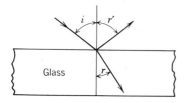

* As a standard for comparison the velocity of light in vacuum is taken as unity. Other light velocities are thus expressed in terms of $V = 1$. Because for air, $v = 0.9997$ (almost as great as in vacuum) it may also be considered unity.

TOTAL REFLECTION AND THE CRITICAL ANGLE

We have seen (Fig. 6.2) that light is reflected toward the normal when it passes from a lower to a higher refractive index medium. When the conditions are reversed, as in Fig. 6.4, and the light moves from the higher to the lower index medium, it is refracted away from the normal. In Fig. 6.4, assume that lines A, B, C, and so forth represent light rays moving through glass, and into air at point O. The greater the obliquity of the incident ray, the greater the angle of refraction. Finally, an angle of incidence is reached, as at ray D, for which the angle of refraction is 90°, and the ray then grazes the surface. The angle of incidence at which this takes place is known as the *critical angle*. Rays such as E and F, striking the interface at a greater angle, are totally reflected back into the higher index medium.

The measurement of the critical angle is a quick and easy method of determining the refractive index of both liquids and solids. The instrument used is a refractometer, of which there are many types. A description of one of these, the Pulfrich refractometer, will suffice to illustrate the underlying principles. This instrument employs a polished hemisphere of high refractive index glass (Fig. 6.5). A crystal face or polished surface of the mineral is placed on the equatorial plane of the hemisphere but separated from it by a film of liquid. The liquid is necessary to exclude the air and must have a refractive index higher than that of the hemisphere. Slightly converging light is directed upward through the hemisphere and, depending on the angle of incidence, is either partly refracted through the unknown or totally reflected back through the hemisphere. If a telescope is placed in a position to

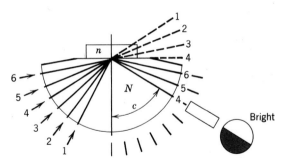

FIG. 6.5. Pulfrich refractometer and measurement of critical angle c.

receive the reflected rays, we can observe a sharp boundary between the portion of the field intensely illuminated by the totally reflected light and the remainder of the field. When the telescope is moved so that its cross hairs are precisely on the contact, the critical angle c is read on a scale. Knowing this angle and the index of refraction of the hemisphere, N, we can calculate the index of refraction of the mineral: n mineral = sin critical angle × N hemisphere.

ISOTROPIC AND ANISOTROPIC CRYSTALS

For optical considerations all transparent substances can be divided into two groups: *isotropic* and *anisotropic*. The isotropic group includes such noncrystalline substances as gases, liquids, and glass, but it also includes crystals that belong to the isometric crystal system. In them light moves in all directions with equal velocity and hence each isotropic substance has a single refractive index. In anisotropic substances, which include all crystals except those of the isometric system, the velocity of light varies with crystallographic direction and thus there is a range of refractive index.

In general, light passing through an anisotropic crystal is broken into two polarized rays vibrating in mutually perpendicular planes. Thus for a given orientation, a crystal has two indices of refraction, one associated with each polarized ray.

FIG. 6.4. Critical angle.

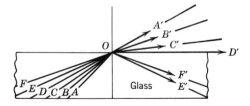

POLARIZED LIGHT

We have seen that light can be considered a wave motion with vibrations taking place in all directions at right angles to the direction of propagation. When the wave motion is confined to vibrations in a single plane, the light is said to be *plane polarized*. The three principal ways of polarizing light are by double refraction, absorption, and reflection.

Polarized Light by Double Refraction

It has been pointed out that when light passes through an anisotropic crystal it is divided into two polarized rays. The principle on which the first efficient polarizer was based was the elimination of one of these rays. The crystalline material used was the optically clear variety of calcite, Iceland spar, and the polarizer was called the *Nicol prism,* after the inventor William Nicol. Calcite has such a strong double refraction that each ray produces a separate image when an object is viewed through a

cleavage fragment (Fig. 6.6a). In making the Nicol prism (Fig. 6.7), an elongated cleavage rhombohedron of calcite is sawed at a specified angle and the two halves rejoined by cementing with Canada balsam. Faces are then ground at the ends of the prism to make angles of 90° with the cemented surface. On entering the prism light is resolved into two rays *O* and *E*. Because of the greater refraction of the *O* ray it is totally reflected at the Canada balsam surface. The *E* ray with refractive index close to that of the balsam proceeds essentially undeviated through the prism and emerges as plane polarized light.

Polarized Light by Absorption

The polarized rays into which light is divided in anisotropic crystals may be differentially absorbed. If one ray suffers nearly complete absorption and the other very little, the emerging light will be plane polarized. This phenomenon is well illustrated by some tourmaline crystals (Fig. 6.8). Light passed through the crystal at right angles to [0001] emerges

FIG. 6.6(*a*). Calcite, viewed normal to the rhombohedron face showing double refraction. The double repetition of "calcite" at the top of the photograph as seen through a face cut on the specimen parallel to the base. (*b*) Calcite showing no double refraction. Viewed parallel to the *c* axis.

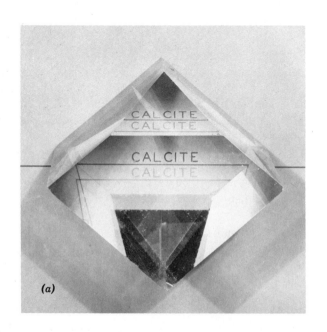

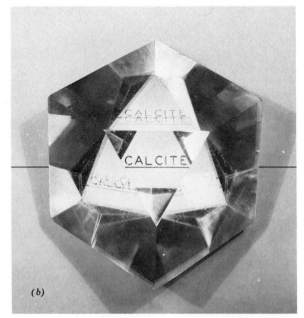

(a)

(b)

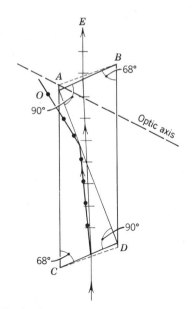

FIG. 6.7. Nicol prism.

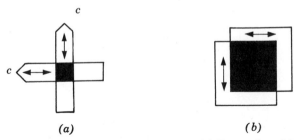

FIG. 6.8. Polarized light by absorption. (a) Tourmaline. (b) *Polaroid.* Arrows indicate directions of maximum transmission; directions of maximum absorption are at right angles.

essentially plane polarized, with vibrations parallel to the c axis. The other ray, vibrating perpendicularly to it, is almost completely absorbed. When two crystals are placed at right angles one above the other, the polarized ray emerging from one is absorbed by the other. Polarizing sheets, such as *Polaroid,* are made by aligning crystals on an acetate base. These crystals absorb very little light in one vibration direction, but are highly absorptive in the other. The light transmitted by the sheet is thus plane polarized. Because they are thin and can be made in large sheets, manufactured polarizing plates are extensively used in optical equipment, including many polarizing microscopes.

Polarized Light by Reflection

Light reflected from a smooth, nonmetallic surface is partially polarized with the vibration directions parallel to the reflecting surface. The extent of polarization depends on the angle of incidence (Fig. 6.9) and the index of refraction of the reflecting surface. It is most nearly polarized when the angle between the reflected and refracted ray is 90° (Brewster's law). The fact that reflected light is polarized can be easily demonstrated by viewing it through a polarizing filter. When the vibration direction of the filter is parallel to the reflecting surface the light passes through the filter with only slight reduction in intensity (Fig. 6.9a); when the filter is turned 90°, only a small percentage of the light reaches the eye, Fig. 6.9b.

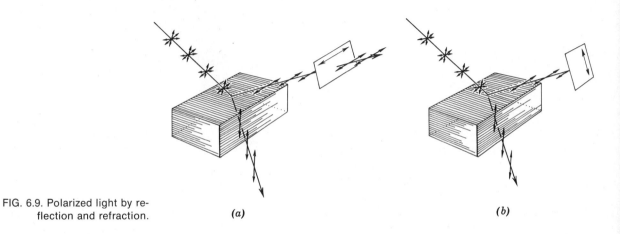

FIG. 6.9. Polarized light by reflection and refraction.

(a)

(b)

The Polarizing Microscope

The polarizing microscope is the most important instrument for determining the optical properties of crystals; with it more information can be obtained easily and quickly than with more specialized devices. Several manufacturers each make a number of models of polarizing microscopes that vary in complexity of design and sophistication and hence in price. A student model made by E. Leitz (SM-Pol) is illustrated in Fig. 6.10 with the essential parts named.

Although a polarizing microscope differs in detail from an ordinary compound microscope, its primary function is the same: to yield an enlarged image of an object placed on the stage. The magnification is produced by a combination of two sets of lenses: the objective and the ocular. The function of the objective lens, at the lower end of the microscope tube, is to produce an image that is sharp and clear. The ocular merely enlarges this image including any imperfection resulting from a poor quality objective. For mineralogical work it is desirable to have three objectives: low, medium, and high-power. In Fig. 6.10 these are shown mounted on a revolving nose piece and can be successively rotated into position. The magnification produced by an objective is usually indicated on its housing, such as 2x (low), 10x (medium), and 50x (high). Oculars also have different magnifications such as

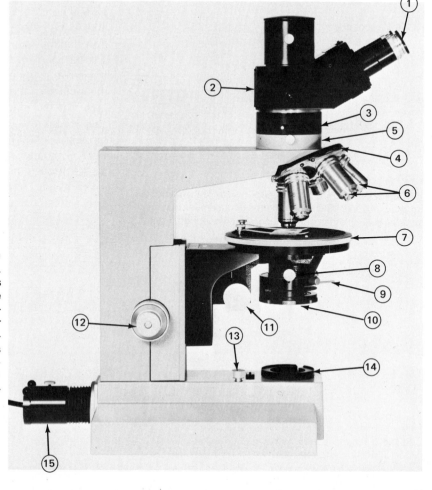

FIG. 6.10. Polarizing microscope (SM-POL manufactured by E. Leitz, Inc.). 1. Oculars. 2. Bertrand-Amici lens and pinhole stop (not in view, on opposite side). 3. Analyzer (not in view, on opposite side). 4. Revolving nosepiece for objectives. 5. Tube slot for compensators (out of sight, on opposite side). 6. Objectives. 7. Rotating stage. 8. Knob for swinging in and out condenser lens. 9. Substage iris diaphragm. 10. Rotatable polarizer. 11. Knob for vertical adjustment of condenser assembly. 12. Vertical adjustment of stage, for focusing (knobs on both sides). 13. Field diaphragm. 14. Dustglass. 15. Substage illuminator.

5x, 7x, 10x. The total magnification of the image can be determined by multiplying the magnification of the objective by that of the ocular as: 50x · 10x = 500x. Although in routine work the three objectives are frequently interchanged, a single ocular usually suffices. The ocular assembly, which slips into the upper end of the microscope tube, carries cross hairs—one N–S (front-back) the other E–W. These enable one to locate under high power a particular mineral grain that has been brought to the center of the field under low power. They are also essential in aligning cleavage fragments for making angular measurements. A condenser is located below the stage. The upper lens of the condenser, used with high-power objectives, makes the light strongly converging and can be rotated easily into or out of the optical system. The iris diaphragm, also located below the stage, can be opened or closed to control the depth of focus and to regulate the intensity of light striking the object.

In addition to the lenses, condenser, and diaphragm mentioned above which are common to all compound microscopes, the polarizing microscope has several other features. The *polarizer* below the stage is a polarizing plate, or Nicol prism, that transmits plane polarized light vibrating in a N–S (front-back) direction. The *analyzer,* fitted in the tube above the stage, is a similar plate or prism that transmits light vibrating only in an E–W direction. The polarizer and analyzer are collectively called *polars.** When both polars are in position they are said to be crossed and, if no anisotropic crystal is between them, no light reaches the eye. The polarizer remains fixed but the analyzer can be removed from the optical path at will. The Bertrand-Amici lens is an accessory that is used to observe interference figures (see page 207). In working with crystals it is frequently necessary to change their orientation. This is accomplished by means of a rotating stage, whose axis of rotation is the same as the microscope axis.

---

* Before using a microscope the vibration directions of the polars should be checked; for, although the orientation is usually as given, in some microscopes the vibration directions of transmitted light are reversed, i.e., the polarizer is E–W, the analyzer N–S.

Microscopic Examination of Minerals and Rocks

The polarizing microscope is also called the petrographic microscope because it is used in the study of rocks. In examining thin sections of rocks, the textural relationships are brought out and certain optical properties can be determined. It is equally effective in working with powdered mineral fragments. On such loose grains all the optical properties can be determined and in most cases they characterize a mineral sufficiently to permit its identification.

The optimum size of mineral grains for examination with a polarizing microscope is minus 50 mesh–plus 100 mesh, but larger or smaller sizes may be used. To prepare a mount for examination (1) put a few mineral grains on an object glass, a slide 46 mm × 27 mm, (2) immerse the grains in a drop of liquid of known refractive index, and (3) place a cover glass on top of the liquid. Using this type of mount the refractive indices of mineral grains are determined by the *immersion method.*

In using this method there should be available a series of calibrated liquids ranging in refractive index from 1.41 to 1.77, with a difference of 0.01 or less between adjacent liquids. These liquids cover the refractive index range of most of the common minerals. The immersion method is one of trial and error and involves comparing the refractive index of the unknown with that of a known liquid.

Isotropic Crystals and the Becke Line

Because light moves in all directions through an isotropic substance with equal velocities, there is no double refraction and only a single index of refraction. With a polarizing microscope, objects are always viewed in polarized light, which conventionally is vibrating N–S. If the object is an isotropic mineral, the light passes through it and continues to vibrate in the same plane. If the analyzer is inserted, darkness results for this polar permits light to pass only if it is vibrating in an E–W direction. Darkness remains as the position of the crystal is changed by rotating the microscope stage. This is a characteristic that distinguishes isotropic from anisotropic crystals.

Let us first consider how the single refractive index of isotropic substances is determined. The first mount may be made by using any liquid, but if the mineral is a complete unknown, it is well to select a liquid near the middle of the range. When the grains are brought into sharp focus using a medium-power objective and plane polarized light, in all likelihood they will stand in relief; that is, they will be clearly discernible from the surrounding liquid. This is because the light is refracted as it passes from one medium to another of different refractive index. The farther apart the indices of refraction of mineral and liquid, the greater the relief. But when the indices of the two are the same, there is no refraction at the interface, and the grains are essentially invisible. Relief shows that the index of refraction of the mineral is different from that of the liquid; but is it higher or lower? The answer to this important question can be found by means of the *Becke line* (Fig. 6.11). If the mineral grain is thrown slightly out of focus by raising the microscope tube (in most modern microscopes this is accomplished by lowering the stage), a narrow line of light will form at its edge and move toward the medium of higher refractive index. Thus if the Becke line moves into the mineral grain, a new mount must be made using a higher refractive index liquid. After several tries it may be found that there is no Becke line and the mineral grains are invisible in a given liquid. The index of refraction of the mineral is then the same as that of the calibrated liquid. More frequently, however, the refractive index of the mineral is found to be greater than that one liquid but less than that of its next higher neighbor. In such cases it is necessary to interpolate. If the Becke line moving into the mineral in the lower index liquid is more intense than the line moving out of the mineral into the next higher liquid, it can be assumed that the refractive index of the mineral is closer to that of the higher liquid. In this way it is usually possible to report a refractive index to ± 0.003.

Frequently the Becke line can be sharpened by restricting the light by means of the substage diaphragm. Most liquids have a greater dispersion than minerals. Thus, if the refractive index of liquid and mineral are matched for a wavelength near the center of the spectrum, the mineral has a higher index than the liquid for red light but a lower index for violet light. This is evidenced, when observed in white light, by a reddish line moving into the grain while a bluish line moves out.

Aside from color, the single index of refraction is the only significant optical characteristic of isotropic minerals. It is, therefore, important in mineral identification to consider other properties such as cleavage, fracture, color, hardness, and specific gravity.

UNIAXIAL CRYSTALS

We have seen that light moves in all directions through an isotropic substance with equal velocity and vibrates in all directions at right angles to the directions of propagation. In hexagonal and tetragonal crystals, there is one and only one direction in which light moves in this way. This is parallel to the c axis, with vibrations in all directions in the

FIG. 6.11. The Becke Line. When thrown out of focus by raising microscope tube (or lowering stage), white line moves into medium of higher refractive index. (a) In focus. (b) n of grain > n of liquid. (c) n of grain < n of liquid.

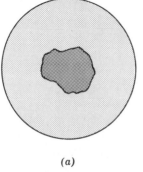

(a)

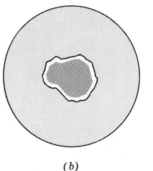

(b)

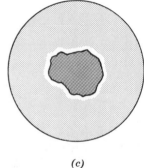

(c)

basal plane. For this reason the c axis is called the *optic axis*, and hexagonal and tetragonal crystals are called optically *uniaxial*. This distinguishes them from orthorhombic, monoclinic, and triclinic crystals which have two optic axes and are called *biaxial*.

When light moves in uniaxial crystals in any direction other than parallel to the c axis, it is broken into two rays traveling with different velocities. One, the *ordinary ray,* vibrates in the basal plane; the other, the *extraordinary ray,* vibrates at right angles to it and thus in a plane that includes the c axis. Such a plane, of which there is an infinite number, is referred to as the *principal section*. The nature of these two rays can be brought out in the following way. Assume that the direction of the incident beam is varied to make all possible angles with the crystal axes, and that the distance traveled by the resulting rays in any instant can be measured. We would find that:

1. One ray, with waves always vibrating in the basal plane, traveled the same distance in the same time. Its surface can be represented by a sphere, and since it acts much as ordinary light it is the *ordinary ray (O ray)*.
2. The other ray, with waves vibrating in the plane that includes the c axis, traveled in the same time different distances depending on the orientation of the incident beam. If the varying distances of this, the *extraordinary ray (E ray)* were plotted, they would outline an ellipsoid of revolution, with the optic axis the axis of revolution.

Uniaxial crystals are divided into two optical groups: positive and negative. They are *positive* if the O ray has the greater velocity, and *negative* if the E ray has the greater velocity. Cross-sections of the ray velocity surfaces are shown in Fig. 6.12. Note that in both positive and negative crystals the O and E rays have the same velocity when traveling along the optic (c) axis. But the difference in their velocities becomes progressively greater as the direction of light propagation moves away from the optic axis, reaching a maximum at 90°.

Because the two rays have different velocities, there are two indices of refraction in uniaxial crystals. Each index is associated with a vibration

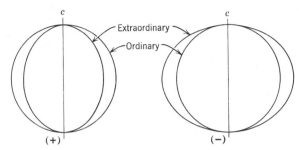

FIG. 6.12. Ray velocity surfaces of uniaxial crystals.

direction. The index related to vibration along the ordinary ray is designated ω (omega); whereas that associated with the extraordinary ray is ϵ (epsilon) or ϵ'. In positive crystals the O ray has a greater velocity than the E ray, and ω is less than ϵ. But in negative crystals with the E ray having the greater velocity, ω is greater than ϵ. The two principal indices of refraction of a uniaxial crystal are ω and ϵ and the difference between them is the *birefringence*.

The *uniaxial indicatrix* is a geometrical figure that is helpful in visualizing the relation of the refractive indices and their vibration directions that are perpendicular to the direction of propagation of light through a crystal. For positive crystals the indicatrix is a prolate spheroid of revolution; for negative crystals it is an oblate spheroid of revolution (Fig. 6.13). In their construction the direction of radial lines is proportional to the refractive indices. First consider light moving parallel to the optic axis. It is not doubly refracted but moves through the crystal as the ordinary ray with waves vibrating in all directions in the basal plane. This is why light

FIG. 6.13. Optical indicatrix, uniaxial crystals. (a) Positive. (b) Negative.

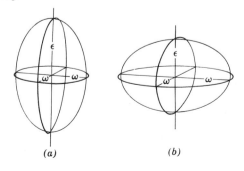

(a)   *(b)*

moving parallel to the c axis of calcite (Fig. 6.6b) produces a single image. There is a single refractive index for all these vibrations, proportional to the radius of the equatorial circle of the indicatrix. Now consider light traveling perpendicular to the optic axis. It is doubly refracted. The waves of the ordinary ray vibrate, as always, in the basal plane and the associated refractive index, ω, is again an equatorial radius of the indicatrix. The vibration direction of waves of the extraordinary ray must be at right angles to both the vibration direction of the ordinary waves and the direction of propagation. Thus, in this special case, it is parallel to the optic axis. The axis of revolution of the indicatrix is then proportional to ϵ, the greatest index in (+) crystals and the least in (−) crystals. It can be seen that light moving through a crystal in a random direction gives rise to two rays: (1) the O ray with waves vibrating in the basal section, the associated index, ω, and (2) the E ray with waves vibrating in the principal section in a direction at right angles to propagation. The length of the radial line along this vibration direction is an ϵ', a refractive index lying between ω and ϵ.

A study of the indicatrix shows that (1) ω, can be determined on *any* crystal grain; and only ω can be measured when light moves parallel to the optic axis, (2) ϵ can be measured only when light moves normal to the c axis, and (3) a randomly oriented grain yields, in addition to ω, an index intermediate to ω and ϵ, called ϵ'. The less the angle between the direction of light propagation and the normal to the optic axis, the closer is the value ϵ' to true ϵ.

Uniaxial Crystals Between Crossed Polars

Extinction

We have seen that because isotropic crystals remain dark in all positions between crossed polars, they can be distinguished from anisotropic crystals. However, there are special conditions under which uniaxial crystals present a dark field when viewed between crossed polars. One of these conditions is when light moves parallel to the optic axis. Moving in this direction, light from the polarizer passes through the crystal as through an isotropic substance and is completely cut out by the analyzer. The other special condition is when the vibration direction of light from the polarizer coincides exactly with one of the vibration directions of the crystal. In this situation, light passes through the crystal as either the O ray or the E ray to be completely eliminated by the analyzer, and the crystal is said to be at *extinction*. As the crystal is rotated from this extinction position it becomes progressively lighter, reaching a maximum brightness at 45°. There are four extinction positions in a 360° rotation, one every 90°.

Interference

Let us consider how the crystal affects the behavior of polarized light as it is rotated from one extinction position to another. Figure 6.14 represents five positions of a tiny quartz crystal elongated on the c axis and lying on a prism face. In the diagrams it is assumed that light from the polarizer is moving upward, normal to the page, and vibrating in direction P–P. The vibration direction of the analyzer is A–A. The crystal in (a) is at an extinction position and light moves through it as the E ray vibrating parallel to the c axis. At 90°, position (b), the crystal is also at extinction with light moving through it as the O ray. When the crystal is turned as in (c), (d), and (e), polarized light entering it is resolved into two components. One moves through it as the O ray, vibrating in the basal plane; the other as the E ray vibrating in the principal section. In (c) most light is transmitted as the E ray but in (e) it is transmitted mostly as the O ray. In (d), the 45° position, the amounts of light transmitted by the two rays are equal.

When these rays from the crystal enter the analyzer each is broken up into an O and E ray conforming in vibration directions to those of the analyzer. Only the components of the rays vibrating in an E–W direction, are permitted to pass. During their passage through the crystal the two rays travel with different velocities and thus on emerging there is a phase difference because one is ahead of the other. The amount it is ahead depends both on the difference in velocities and on the thickness of

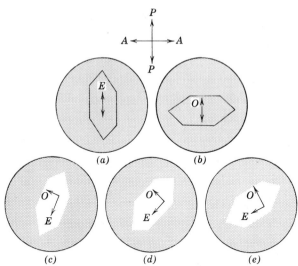

FIG. 6.14. Quartz crystal between crossed polars.

crystal traversed. Because both rays vibrate in the same plane of the analyzer, they interfere. For monochromatic light, if one ray is an integral number of wavelengths ($n\lambda$) behind the other, the interference results in darkness. On the other hand, if the path difference is $\lambda/2$, $3\lambda/2$, or in general $(2n - 1)\lambda/2$, the waves reinforce one another to produce maximum brightness.

Each wavelength has its own sets of critical conditions when interference produces darkness. Consequently, when white light is used, "darkness" for one wavelength means its elimination from the spectrum and its complimentary color appears. The colors thus produced are called *interference colors*. There are different *orders* of interference depending on whether the color results from a path difference of 1λ, 2λ, 3λ, . . . , $n\lambda$. These are called first order, second order, third order, and so forth interference colors as indicated in Fig. 6.15.

Interference colors depend on three factors: orientation, thickness, and birefringence. With a continuous change in direction of light, from parallel to perpendicular to the optic axis, there is a continuous increase in interference colors. For a given orientation, the thicker the crystal and the greater its birefringence the higher the order of interference color. If a crystal plate is of uniform thickness, as a cleavage flake may be or a grain in a rock thin sec-

tion, it will show a single interference color. As seen in immersion liquids grains commonly vary in thickness and a variation in interference colors reflects this irregularity.

Accessory Plates

The gypsum plate, mica plate, and quartz wedge are accessory plates used with the polarizing microscope; their function is to produce interference of known amounts and thus predetermined colors. They are all constructed so that the fast ray (the vibration direction of the lesser refractive index) is parallel to the long dimension. The *gypsum plate* is made by cleaving a gypsum crystal to such a thickness that in white light it produces a uniform red interference color: *red of the first order*. The *mica plate* is made with a thin mica flake, cleaved to a thickness that for yellow light it yields a path difference of a quarter of a wavelength. It is thus also called the *quarter wave plate*. The *quartz wedge* is an elongated wedge-shaped piece of quartz (see Fig. 6.16a) with the vibration direction of the fast ray (ω) parallel to its length and the slow ray (ϵ) across its length. As thicker portions of the wedge are placed in the optical path, the path difference of the rays passing through it also increases producing

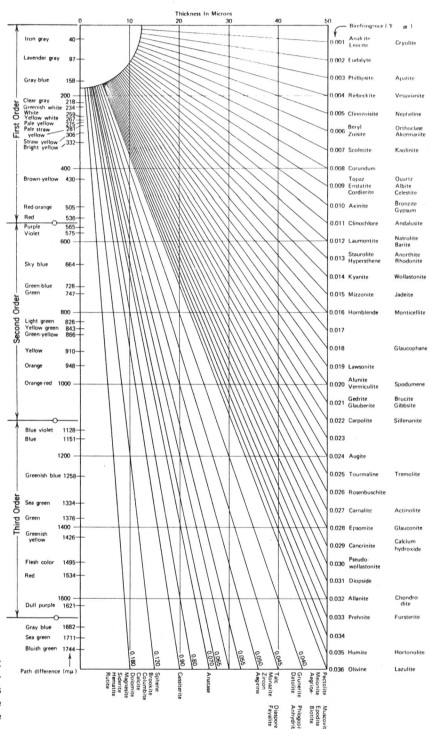

FIG. 6.15. Michel-Lévy chart showing the relation of interference colors to thickness and birefringence and the birefringence of some common minerals.

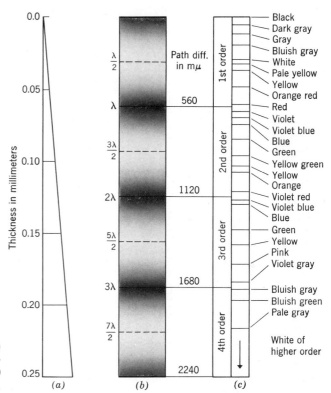

FIG. 6.16. Quartz wedge. (a) Cross section; (b) In monochromatic light, $\lambda = 560$ mμ; (c) Colors in white light.

a succession of interference colors. The number of orders depends on the wedge angle; the greater the angle the more orders per unit of length.

When the quartz wedge is viewed between crossed polars in monochromatic light, it is crossed by alternating dark and light bands; dark where the path difference is $n\lambda$ and brightest where the path difference is $(2n - 1)\lambda/2$, Fig. 6.16b. In white light a succession of interference colors is observed that resemble colors seen in thin oil films on water, Fig. 6.16c. The colors result from interference phenomena described by Sir Isaac Newton and are called *Newton's colors*.

Uniaxial Crystals in Convergent Polarized Light

What are known as *interference figures* are seen when properly oriented crystal sections are examined in convergent polarized light. To see them, the polarizing microscope (usually used as an orthoscope) is converted to a conoscope by swinging in the upper substage condensing lens, so that the section can be observed in strongly converging light, using a high-power objective. The interference figure then appears as an image just above the upper lens of the objective and can be seen between crossed polars by removing the ocular and looking down the microscope tube. If the Bertrand lens, an accessory lens located above the analyzer, is inserted, an enlarged image of the figure can be seen through the ocular.

The principal interference figure of a uniaxial crystal, the optic axis figure, Fig. 6.17, is seen when one views the crystal parallel to the c axis. Only for the central rays from the converging lens is there no double refraction, the others, traversing the crystal in directions not parallel to the c axis, are resolved into O and E rays having increasing path difference as the obliquity to the c axis increases. The interference of these rays produces concentric circles of

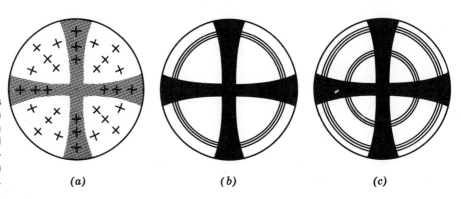

FIG. 6.17. Uniaxial optic axis interference figures. (a) Radial lines indicate vibration directions of E ray; tangential lines indicate vibration directions of O ray. (b) and (c) show isochromatic curves.

(a)　　*(b)*　　*(c)*

interference colors. The center is black with no interference but moving outward there is a progression from first-order to second-order to third-order, and so forth, interference colors. If the crystal section is of uniform thickness, no change will be noted as it is moved horizontally. If, however, the thickness varies, as in a wedge-shaped fragment, the positions of the colors change with horizontal movement. At the thin edge there may be only gray of the first-order, but as the crystal is moved so the light path through it becomes greater, all the first-order colors may appear. And with increasing crystal thickness, the path difference of the two rays may be great enough to yield second-, third-, and higher-order interference colors.

The reason for the black cross superimposed on the rings of interference colors is brought out in Fig. 6.17a. In this drawing the radial dashes indicate the vibration directions of the E ray and those at right angles the vibration directions of the O ray. It will be seen that where these vibration directions are parallel or nearly parallel to the vibration directions of the polarizer and analyzer no light passes and thus the formation of the dark cross.

Figure 6.17 illustrates a centered optic axis figure as obtained on a crystal plate whose c axis coin-

cides with the axis of the microscope; as the stage is rotated, no movement of the figure is seen. If the optic axis of the crystal makes an angle with the axis of the microscope, the black cross is no longer symmetrically located in the field of view (Fig. 6.18). When the stage is rotated, the center of the cross moves in a circular path, but the bars of the cross remain parallel to the vibration directions of the polarizer and analyzer. Even if the inclination of the optic axis is so great that the center of the cross does not appear, on rotation of the crystal the bars move across the field maintaining their parallelism to the vibration directions of the polars.

The *flash figure* is an interference figure produced by a uniaxial crystal when its optic axis is normal to the axis of the microscope. That is, a hexagonal or tetragonal crystal lying on a face in the prism zone. When the crystal is at an extinction position the figure is an ill-defined cross occupying much of the field. On rotation of the stage, the cross breaks into two hyperbolas that rapidly leave the field in those quadrants containing the optic axis. The cross forms because the converging light is broken into O and E rays with vibration directions mostly parallel or nearly parallel to the vibration directions of the polarizer and analyzer. A centered

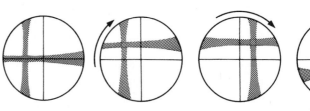

FIG. 6.18. Positions of off-centered uniaxial optic axis figure on clockwise rotation of microscope stage.

flash figure not only indicates the vibration direction of the *E* ray, but assures one that in this direction a true value of ε can be obtained in plane polarized light.

Determination of Optic Sign

The mica plate, the gypsum plate, and the quartz wedge may be used with a uniaxial optic axis figure to determine the optic sign; that is, whether the crystal is positive or negative. They are inserted below the analyzer in a slot in the microscope tube so positioned that when the plates are in place, their vibration directions make angles of 45° with the vibration directions of the polars.

From the previous discussion we have learned that in the optic axis interference figure, the *E* ray vibrates radially and the *O* ray tangentially. By use of an accessory plate in which vibration directions of the slow and fast rays are known, one can tell whether the *E* ray of the crystal is slower (positive crystals) or faster (negative crystals) than the *O* ray and thus determine the optic sign. For most American-made equipment the vibration of the slow ray is at right angles to the length of the plate and is so marked on the metal carrier. However, before using an accessory the vibration directions should be checked. The principle in the use of all the plates is the same: to add or to subtract from the path difference of the *O* and *E* rays of the crystal.

If the *mica plate* is superimposed on a uniaxial optic axis figure in which the ordinary ray is slow, (negative crystal) the interference of the plate rein-

forces the interference colors in the SE and NW quadrants, causing them to shift slightly toward the center. At the same time subtraction causes the colors in the NE and SW quadrants to shift slightly away from the center. The most marked effect produced by the mica plate is the formation of two black spots near the center of the black cross in the quadrants where subtraction occurs (Fig. 6.19).

The *gypsum plate* is usually used to determine the optic sign when low-order interference colors or no colors at all are seen in the optic axis figure. It has the effect of superimposing red of the first-order on the interference figure. If the figure shows several orders of interference colors, one should consider the color effect on the grays of the first-order near the center. In the quadrants where there is addition, the red plus the gray gives blue; in the alternate quadrants the red minus the gray gives yellow. The arrangement of colors in positive crystals is: yellow SE–NW; blue NE–SW; and in negative crystals, yellow NE–SW, blue SE–NW. It is suggested that the student insert the colors in Fig. 6.20.

The *quartz wedge* is most effective in determining optic sign when high-order interference colors are present in the optic axis figure. The wedge is usually inserted with the thin edge first. If its retardation is added to that of the crystal, the interference colors in two opposite quadrants will increase progressively as the wedge moves through the microscope tube. If the retardation is subtracted from that produced by the crystal, the order of colors will decrease. Thus as the quartz wedge is slowly inserted over an optic axis figure of a negative crystal, the color bands in the SE–NW quad-

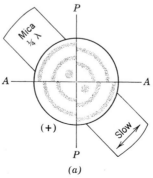

FIG. 6.19. Determination of optic sign with mica plate.

(a)

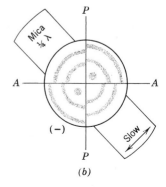

(b)

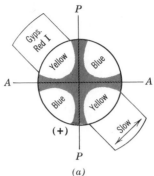

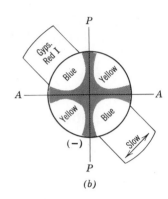

FIG. 6.20. Determination of optic sign with gypsum plate.

(a)

(b)

rants move toward the center and disappear. At the same time in the NE–SW quadrants the colors move outward to the edge of the field. In a positive crystal similar phenomena are observed but the colors move in the opposite directions, that is, away from the center in the SE–NW quadrants and toward the center in the NE–SW quadrants.

Sign of Elongation

Hexagonal and tetragonal crystals are frequently elongated on the c axis or have prismatic cleavage that permits them to break into splintery fragments also elongated parallel to c. If such an orientation is known, one can determine the optic sign by turning the elongated grain to the 45° position and inserting the gypsum plate. If the interference colors rise (gray of mineral plus first-order red equals blue), the slow ray of the gypsum has been superimposed on the slow ray of the mineral. If this is also the direction of

elongation, it means the ϵ ray is slow (ϵ is the higher refractive index) and the mineral has positive elongation and is optically positive. When the slow ray of the gypsum plate is parallel to the elongation of the mineral grain and the interference colors fall (gray of mineral minus first-order red equals yellow), the mineral has negative elongation and is optically negative (Fig. 6.21).

Very commonly the interference colors of small grains are grays of the first order. Thus, on superimposing red of the first order, addition gives a blue color and subtraction a yellow color.

Absorption and Dichroism

In the discussion of polarized light (page 198) it was pointed out that in some tourmaline the absorption of one ray is nearly complete but for the other ray it is negligible. Although less striking, many crystals show a similar phenomenon: more light is

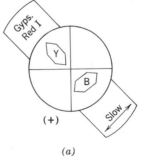

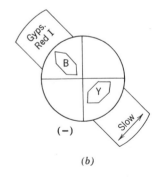

FIG. 6.21. Determination of sign of elongation with gypsum plate. In diagrams Y = yellow, B = blue. (a) Positive elongation, (b) negative elongation.

(a)

(b)

absorbed in one vibration direction than in the other. In tourmaline where absorption of the O ray is greatest, it is expressed as, absorption: $O > E$ or $\omega > \epsilon$. In other crystals certain wavelengths may be absorbed in one direction and different wavelengths in the other direction and the complimentary colors are transmitted. Thus the crystal has different colors in different vibration directions and is said to be *dichroic*. Dichroism is expressed by giving the colors, for example, O or ω = yellow, E or ϵ = pink. Absorption is independent of other properties and is considered, as are refractive indices, a fundamental optical property of crystals.

BIAXIAL CRYSTALS

Orthorhombic, monoclinic, and triclinic crystals are called optically biaxial for they have two directions in which light travels with zero birefringence. In uniaxial crystals there is only one such direction.

Light moving through a biaxial crystal, except along an optic axis, travels as two rays with mutually perpendicular vibrations. The velocities of the rays differ from each other and change with changing crystallographic direction. The vibration directions of the fastest ray, X, and the slowest ray, Z, are at right angles to each other. The direction perpendicular to the plane defined by X and Z is designated as Y. For biaxial crystals there are thus three indices of refraction resulting from rays vibrating in each of these principal optical directions. The numerical difference between the greatest and least refractive indices is the *birefringence*. Various letters and symbols have been used to designate the refractive indices, but the most generally accepted are the Greek letters as follows:

Index[a]		Direction	Ray Velocity
(alpha)	α lowest	X	highest
(beta)	β middle	Y	intermediate
(gamma)	γ highest	Z	lowest

[a] Other equivalent designations are: $\alpha = nX, n_x, N_x, N_p$; $\beta = nY, n_y, N_y, N_m$; $\gamma = nZ, n_z, N_z, N_g$.

The Biaxial Indicatrix

The biaxial indicatrix is a triaxial ellipsoid with its three axes the mutually perpendicular optical directions X, Y, and Z. The lengths of the semiaxes are proportional to the refractive indices: α along X, β along Y, and γ along Z. Figure 6.22 shows the three principal sections through the indicatrix; these are the planes XY, YZ, and XZ. They all are ellipses and each has the length of its semimajor and semiminor axes proportional to refractive indices as shown. Of most interest is the XZ section. With its semimajor axis proportional to γ and its semiminor axis proportional to α, there must be points on the ellipse between these extremes where the radius is proportional to the intermediate index, β. In Fig. 6.22 this radius is marked S. With two exceptions every sec-

FIG. 6.22. Principal sections through the biaxial indicatrix of a positive crystal.

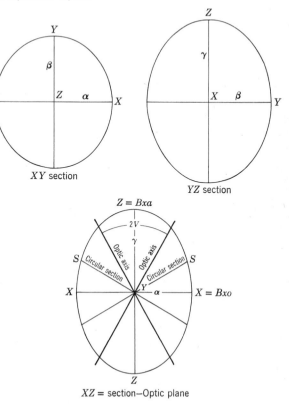

XY section

YZ section

XZ = section—Optic plane

tion passing through the center of a triaxial ellipsoid is an ellipse. The exceptions are *circular sections* of which S is the radius. The two directions normal to these sections are the *optic axes;* and the XZ plane in which they lie is called the *optic plane.* The Y direction perpendicular to this plane is the *optic normal.* Light moving along the optic axes and vibrating in the circular sections shows no birefringence and gives the constant refractive index, β. The optic axis of a uniaxial crystal is analogous to these directions, for light moving parallel to it also vibrates in a circular section with constant refractive index.

With variation in refractive indices there is a corresponding variation in the axial lengths of the biaxial indicatrix. Some crystals are nearly uniaxial and in these the intermediate index, β, is very close to either α or γ. If β is close to α, the circular sections make only a small angle with the XY plane and the optic axes make the same angle with the Z direction. This is angle V and the angle between the two optic axes, known as the *optic angle,* is $2V$. The optic angle is always acute and because, in this case, it is bisected by Z, Z is called the *acute bisectrix* (Bxa); X is the *obtuse bisectrix* (Bxo) because it bisects the obtuse angle between the optic axes. When Z is the Bxa, the crystal is optically positive.

If β is closer to γ than to α, the acute angle between the optic axes is bisected by X and the obtuse angle bisected by Z. In this case, with X the Bxa, the crystal is negative. When β lies exactly halfway between α and γ, the optic angle is $90°$.

The relation between the optic angle and the indices of refraction is expressed by formula (a) below. A close approximation to the optic angle can be made using formula (b).

(a) $\cos^2 V_x = \dfrac{\gamma^2(\beta^2 - \alpha^2)}{\beta^2(\gamma^2 - \alpha^2)}$

(b) $\cos^2 V_x = \dfrac{\beta - \alpha}{\gamma - \alpha}$

The error using the simplified formula increases with increase of both birefringence and V and always yields values for V' less than true V. It should be noted that in using either formula *half* the optic angle is calculated, and that it is determined with X the bisectrix. Thus, when $V < 45°$ the crystal is negative but when $V > 45°$ the crystal is positive.

Biaxial Crystals in Convergent Polarized Light

Biaxial interference figures are obtained and observed in the same manner as uniaxial figures, that is, with converging light, high-power objective, and Bertrand lens. Although interference figures can be observed on random sections of biaxial crystals, the most symmetrical and informative are obtained on sections normal to the optical directions X, Y, and Z and to an optic axis.

The *acute bisectrix figure* is observed on a crystal plate cut normal to the acute bisectrix. If $2V$ is very small there are four positions during a $360°$ rotation at which the figure resembles the uniaxial optic axis figure. That is, a black cross is surrounded by circular bands of interference colors. However, as the stage is turned, the black cross breaks into two hyperbolas that have a slight but maximum separation at a $45°$ rotation; and the color bands, known as *isochromatic curves,* assume an oval shape. The hyperbolas are called *isogyres* and the dark spots, called *melatopes,* at their vertices in the $45°$ position result from light rays that traveled along the optic axes in the crystal. Thus with increasing optic angle the separation of the isogyres increases, and the isochromatic curves are arranged symmetrically about the melatopes as shown in Fig. 6.23. For most crystals when $2V$ exceeds $60°$ the isogyres leave the field at the $45°$ position; the larger the optic angle, the faster they leave.

The portion of the interference figure occupied by the isogyres is dark, for here, light as it emerges from the section has vibration directions parallel to those of the polarizer and analyzer. The dark cross is thus present when the obtuse bisectrix and optic normal coincide with the vibration directions of the polars. The bar of the cross parallel to the optic plane is narrower and better defined than the other bar (Fig. 6.23). Because light travels along the optic axes with no birefringence, their points of emer-

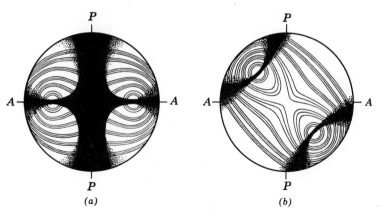

FIG. 6.23. Acute bisectrix interference figure. (a) Parallel position. (b) 45° position.

(a)

(b)

gence are, of course, dark in all positions of the figure.

The Apparent Optic Angle

The distance between the points of emergence of the optic axes is dependent not only on 2V but on β as well. The refractive index of the crystals is β for light rays moving along the optic axes. These rays are refracted on leaving the crystal, giving an apparent optic angle, 2E, greater than the real angle, 2V (Fig. 6.24). The higher the β refractive index, the greater the refraction. Thus if two crystals have the same 2V, the one with the higher β index has the larger apparent angle and the farther apart the optic axes emerge in the interference figure.

The *optic axis figure* is observed on mineral grains cut normal to an optic axis. Such grains are easy to select for they remain essentially dark between crossed polars on complete rotation. The figure consists of a single isogyre at the center of which is the emergence of the optic axis. When the

optic plane is parallel to the vibration direction of either polar, the isogyre crosses the center of the field as a straight bar. On rotation of the stage it swings across the field forming a hyperbola in the 45° position. In this position the figure can be pictured as half an acute bisectrix figure with the convex side of the isogyre pointing toward the acute bisectrix. As 2V increases, the curvature of the isogyre decreases and when 2V = 90°, the isogyre is straight (Fig. 6.25).

The *obtuse bisectrix figure* is obtained on a crystal section cut normal to the obtuse bisectrix. When the plane of the optic axes is parallel to the vibration direction of either polar, there is a black cross. On rotation of the stage the cross breaks into two isogyres that move rapidly out of the field in the direction of the acute bisectrix. Although not as informative as an acute bisectrix figure, a centered obtuse bisectrix figure indicates that an accurate determination of β and of either α or γ can be made on the mineral section producing it.

FIG. 6.24. Relation of 2V to 2E.

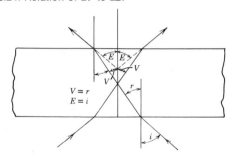

$$V = r$$
$$E = i$$

FIG. 6.25. Curvature of isogyre in optic axis figure from 0° to 90° 2V.

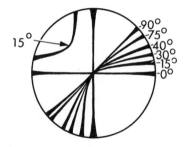

The *optic normal figure* is obtained on sections cut parallel to the plane of the optic axes and resembles the flash figure of a uniaxial crystal. When the X and Z optical directions are parallel to the vibration directions of the polars, the figure is a poorly defined cross. On slight rotation of the stage it splits into hyperbolas that move rapidly out of the field in the quadrants containing the acute bisectrix. An optic normal figure is obtained on sections with maximum birefringence and indicates that α and γ can be determined on this section.

Determination of Optic Sign of a Biaxial Crystal

The optic sign of biaxial crystals can best be determined on acute bisectrix or optic axis figures with the aid of accessory plates. Let us assume that a in Fig. 6.26 represents an acute bisectrix figure of a negative crystal in the 45° position. By definition, X is the acute bisectrix and Z the obtuse bisectrix. OP is the trace of the optic plane and Y the vibration direction of β, at right angles. The velocity is constant for all rays moving along the optic axes, and for them the refractive index of the crystal is β, including those vibrating in the optic plane. Consider the velocities of other rays vibrating in the optic plane. In a negative crystal, those emerging between the isogyres of an acute bisectrix figure have a lesser velocity, those emerging outside the isogyres have a greater velocity. If the gypsum plate is superimposed over such a figure, the slow ray of the plate combines with the fast ray of the crystal and subtraction of interference colors produces a yellow color on the convex sides of the isogyres. On the concave sides of the isogyres a blue color is produced by addition, the slow ray of the plate over the slow ray of the crystal. In a positive crystal the reverse color effect is seen, for here Z is the acute bisectrix.

An optic axis figure in the 45° position can be used in a manner similar to the acute bisectrix figure in determining optic sign. Insertion of the gypsum plate yields for (−) crystal: convex side yellow, concave side blue (Fig. 6.26b), for (+) crystal: convex side blue, concave side yellow.

The quartz wedge may be used to determine optic sign if several isochromatic bands are present. As the wedge is inserted the colors move out in those portions of the figure where there is subtraction and move in where there is addition. In other words, in the areas that a gypsum plate would render blue, the colors move in; in those that it would render yellow, they move out.

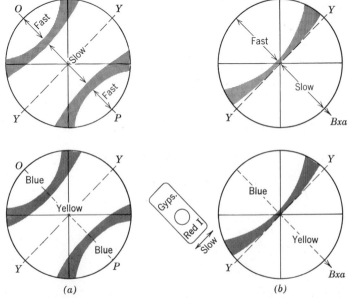

FIG. 6.26. Optic sign determination of negative crystal with gypsum plate. (*a*) Acute bisectrix figure. (*b*) Optic axis figure.

There is a tendency for students learning optical techniques to search for interference figures in the initial immersion when, in all probability, the indices of the mineral are far removed from the refractive index of the liquid. It is time saving, particularly in working with biaxial minerals, to first match as closely as possible the refractive indices of the unknown with n of the liquid. An interference figure, then, in addition to the optic sign, will yield other useful information. Using a grain giving an optic axis figure, β can be compared with n of the liquid. On a grain yielding an acute bisectrix figure, β and either α or γ can be compared with the refractive index of the liquid. The indices of a grain showing highest interference colors are most likely to be close to α and γ. Therefore, in the same mount, one should check such a grain to estimate how far α and γ are from the refractive index of the liquid.

Optical Orientation in Biaxial Crystals

The orientation of the optical indicatrix is one of the fundamental optical properties. It is given by expressing the relationship of X, Y, and Z optical directions to the crystallographic axes a, b, and c.

In *orthorhombic crystals* each of the crystallographic axes is coincident with one of the principal optical directions. For example, the optical orientation of anhydrite is: $X = c$, $Y = b$, $Z = a$. Usually the optical directions coinciding with only two axes are given, for this completely fixes the position of the indicatrix.

It is difficult or impossible to determine the axial directions on microscopic grains of some min-

erals. To do it one must use a fragment oriented by X-ray study or broken from a faced crystal. However, even in small particles, orientation can be expressed relative to cleavages. Powdered fragments tend to lie on cleavages, which in orthorhombic crystals are commonly pinacoidal or prismatic. For example, barite has {001} and {210} cleavage and most grains lie on faces of these forms. Those lying on {001} will be diamond shaped (Fig. 6.27a) and have *symmetrical extinction*. That is, the extinction position makes equal angles with the bounding cleavage faces. Fragments lying on {210} will have *parallel extinction* (Fig. 6.27b). Symmetrical and parallel extinction are characteristic of orthorhombic crystals.

Barite is (+), $X = c$, $Y = b$, $Z = a$. Thus, grains lying on {001} yield a centered Bxo figure and one can determine β in the b direction and γ in the a direction. Grains showing parallel extinction do not give a centered interference figure but α can be measured parallel to c.

In *monoclinic crystals* one of the principal optical directions (X, Y, or Z) of the indicatrix coincides with the b axis; the other two lie in the a–c plane of the crystal. The orientation is given by stating which optical direction equals b and indicating the *extinction angle*, the angle between the c axis and one of the other optical directions. If the extinction lies between the $+$ ends of the a and c axes, the angle is positive; between $+c$ and $-a$, the angle is negative. In gypsum $Y = b$ and $Z \wedge c = 53°$. Thus a fragment lying on the {010} cleavage would yield an optic normal interference figure and α and γ could be determined; γ at the extinction position $+53°$; α at extinction position $-37°$ (see Fig. 6.27c). A grain lying on the {100} cleavage would show parallel extinc-

FIG. 6.27. Optical orientation. (a) Barite on {001} showing symmetrical extinction. (b) Barite on {210} showing parallel extinction. (c) Gypsum on {010} showing extinction angle.

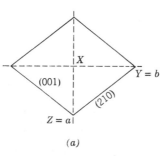

(a)

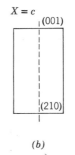

(b)

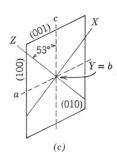

(c)

tion and β could be measured at right angles to the trace of {010}. Crystal fragments lying on any other face in the [001] zone, as on {110} will show an extinction angle but the angle to be recorded, usually that of *maximum extinction* is observed on {010}. Parallel extinction indicates the grain is lying on a face in the [010] zone and the index of the ray vibrating parallel to b can be measured.

In *triclinic crystals* the optic indicatrix can occupy any position relative to the crystallographic axes. Thus a complete optical orientation necessitates giving ϕ and ρ angles of the principal optical directions. But in most cases it suffices to give the extinction angles observed on grains lying on known cleavage faces.

Dispersion

We have seen that the refractive indices of a mineral vary with the wavelength of light. This dispersion of the indices means that, with a variation in the color of light, there is a variation in the indicatrix. The resulting change in the positions of the optic axes, and the accompanying change in $2V$, is known as dispersion of the optic axes, but is usually called just *dispersion*. Instead of giving different values for $2V$ for different wavelengths of light, dispersion is usually expressed by stating whether $2V$ is greater or lesser for red light than for violet light.

Using white light, dispersion can be observed in acute bisectrix figures and optic axis figures and is evidenced as a red fringe on one side of an isogyre and a blue fringe on the other side. Let us assume that $2V$ is greater for red then for violet light. Red light moving along the "red" optic axis has zero path difference. Thus where this axis emerges, red has been removed from the white light and a blue color appears. Similarly, violet light has been removed at the point of emergence of the "violet" axis and a red color appears. In this case the red fringe would appear on the convex side of the isogyre, the blue fringe on the concave side and the dispersion is expressed as: $r > v$. If the positions of the color fringes were reversed, the dispersion would be:

$r < v$. In most interference figures the color fringes are subtle and the isogyre is essentially black.

The foregoing explanation of dispersion is strictly true only for orthorhombic crystals where the plane of the optic axis is an axial plane of the crystal and the acute bisectrix a crystal axis. In monoclinic crystals there is dispersion of the bisectrices as well as of the optic axes giving rise to three types of dispersion. The dispersion is called: *crossed* when Bxa = b; *horizontal* when Bxo = b; and *inclined* when the optic normal = b. The distribution of the color fringes indicating these special types of dispersion are rarely seen, and monoclinic dispersion is usually expressed as $r >$ or $< v$.

Absorption and Pleochroism

The absorption of light in biaxial crystals may differ in the X, Y, and Z optical directions. If the difference is only in intensity and X has the greatest absorption and Z the least, it is expressed as $X > Y > Z$. If different wave lengths are absorbed in different directions the mineral is said to be pleochroic and the color of the transmitted light is given. For example, in hypersthene the *pleochroism* is: X = brownish red, Y = reddish yellow, Z = green. The term pleochroism is commonly used to denote all differential absorption in both uniaxial and biaxial crystals.

References and Suggested Reading

Bloss, F. D., 1961, *An Introduction to the Methods of Optical Crystallography.* Holt, Rinehart and Winston, New York, 294 pp.

Heinrich, E. Wm., 1965, *Microscopic Identification of Minerals.* McGraw-Hill Book Co., New York, 414 pp.

Larsen, E. S. and H. Berman, 1934, The Microscopic Determination of the Nonopaque Minerals, 2nd ed., U.S. Geological Survey Bulletin 848, Government Printing Office, Washington, D.C., 266 pp.

Stoiber, R. E. and S. A. Morse, 1972, *Microscopic Identification of Crystals.* The Ronald Press Co., New York, 278 pp.

Wahlstrom, E. E., 1969, *Optical Crystallography.* 4th ed., John Wiley & Sons, New York, 489 pp.

7
SYSTEMATIC MINERALOGY

PART I: NATIVE ELEMENTS, SULFIDES AND SULFOSALTS

This chapter and three subsequent chapters include descriptions of all the common minerals and those rarer ones of most economic importance. The list, numbering about 200, is relatively small, because approximately 2000 minerals are recognized as valid species. The names and chemical compositions of some other minerals are also given.

The treatment of all minerals* follows a common scheme of presentation. The headings used and the data given under each are as follows:

Crystallography. Under this heading is given the following crystallographic information:

The crystal system and symbol of the crystal class.
Crystallographic features usually observed by inspection, such as habit, twinning, and crystal forms.

"Angles," interfacial angles for those minerals frequently found in well-developed crystals.

Other pertinent data are listed without subheadings. For example, consider the following data for cerussite:

Pmcn (space group), $a = 5.15$, $b = 8.47$, $c = 6.11$ Å (unit cell dimensions), $a:b:c = 0.608:1:0.721$ (axial ratios),* $Z = 4$ (formula units per unit cell).
d's: 3.59(10), 3.50(4), 3.07(2), 2.49(3), 2.08(3) (strongest X-ray lines in Ångstrom units with relative intensities in parentheses).

Physical Properties. The physical properties: *Cleavage,* **H** (hardness), **G** (specific gravity), *Luster,*

---

* Mineral names are given in boldface capitals (e.g., **GOLD**) and smaller letters (e.g., **Arsenic**). Those in boldface capitals are considered most common or important.

* The axial ratios are derived from the cell dimensions and thus may be slightly inconsistent with the interfacial angles obtained from morphological measurements (see p. 49).

Color, and *Optics* (brief summary of optical data) are listed.

Composition and Structure. Chemical composition; frequently with the percentage of elements or oxides and the elements that may substitute for those given in the chemical formula. Brief description of the most important aspects of the crystal structure.

Diagnostic Features. The outstanding properties and tests that aid one in recognizing the mineral and distinguishing it from others.

Occurrence. A brief statement of the mode of occurrence and characteristic mineral associations are given.* The localities where a mineral is or has been found in notable amount or quality are mentioned. For minerals found abundantly the world over, emphasis is on American localities.

Use. For a mineral of economic value, there is a brief statement of its uses.

Similar Species. The similarity of the species listed to the mineral whose description precedes may be either on the basis of chemical composition or crystal structure.

Mineral Classification

Chemical composition has been the basis for the classification of minerals since the middle of the nineteenth century. According to this scheme, minerals are divided into classes depending on the dominant anion or anionic group (e.g., oxides, halides, sulfides, silicates, etc.). There are a number of reasons why this criterion is a valid basis for the broad framework of mineral classification. First, minerals having the same anion or anionic group dominant in their composition have unmistakable family resemblances, in general stronger and more clearly marked than those shared by minerals containing the same dominant cation. Thus the carbonates resemble each other more closely than do the minerals of copper. Second, minerals related by dominance of the same anion tend to occur together or in the same or similar geologic environment. Thus the sulfides occur in close mutual association

---

* See chapter 11 for a discussion of rock types, mineral associations and some terms relating to ore deposits.

---

in deposits of vein or replacement type, whereas the silicates make up the great bulk of the rocks of the earth's crust. Third, such a scheme of mineral classification agrees well with the current chemical practice in the naming and classification of inorganic compounds.

However, it was early recognized that chemistry alone does not adequately characterize a mineral. A full appreciation of the nature of minerals was to wait until X-rays were used to determine internal structures. It is now clear that mineral classification must be based on chemical composition *and* internal structure, because these together represent the essence of a mineral and determine its physical properties. Crystallochemical principles were first used by W. L. Bragg and V. M. Goldschmidt for silicate minerals. This large mineral group was divided into subclasses partially on the basis of chemical composition but principally in terms of internal structure. Within the silicate class, therefore, framework, chain and sheet silicate (etc.) subclasses exist on the basis of the structural arrangement of SiO_4 tetrahedra. Such structural principles in combination with chemical composition provide a logical classification. It is this scheme of classification that is used in the subsequent sections on Systematic Mineralogy.

The broadest divisions of the classification used in this book (as based on C. Palache, H. Berman, and C. Frondel *Dana's System of Mineralogy,* 7th ed., and H. Strunz, *Mineralogische Tabellen,* 5th ed., see references) are:

1. Native elements	⎫	
2. Sulfides	⎬	Chapter 7
3. Sulfosalts	⎭	
4. Oxides	⎫	
(a) Simple and multiple	⎬	Chapter 8
(b) Hydroxides		
5. Halides	⎭	
6. Carbonates	⎫	
7. Nitrates		
8. Borates		
9. Phosphates	⎬	Chapter 9
10. Sulfates		
11. Tungstates	⎭	
12. Silicates		Chapter 10

The above classes are subdivided into *families* on the basis of chemical types, and the family in turn may be divided into *groups* on the basis of structural similarity. A group is made up of *species,* which may form *series* with each other, and finally a species may have several *varieties.* In each of the classes, the mineral with the highest ratio of metal to nonmetal is given first, followed by those containing progressively less metal. Because a relatively small number of minerals is described in this book, often only one member of a group or family is represented, and thus a rigorous adherence to division and subdivision is impractical.

The discussion of each of the chemical classes, families or groups, is preceded by an introduction regarding the most important aspects of the crystal chemistry and crystal structure.

NATIVE ELEMENTS

With the exception of the free gases of the atmosphere, only about 20 elements are found in the native state. These elements can be divided into (1) metals; (2) semimetals; (3) nonmetals. The more common *native metals,* which display very simple structures, constitute three groups: the *Gold Group* (space group: $Fm3m$), gold, silver, copper, and lead, all of which are isostructural; the *Platinum Group* (space group: $Fm3m$), platinum, palladium, iridium, and osmium, all of which are isostructural; and the *Iron Group,* iron and nickel-iron, of which pure Fe as well as kamacite have space group $Im3m$ and the more Ni-rich variety of nickel-iron (taenite) $Fm3m$.

NATIVE ELEMENTS

Metals		Semimetals	
Gold Group		Arsenic Group	
Gold	Au	Arsenic	As
Silver	Ag	Bismuth	Bi
Copper	Cu	Nonmetals	
Platinum Group		Sulfur	S
Platinum	Pt	Diamond	C
Iron Group		Graphite	C
Iron	Fe		
(Kamacite	Fe, Ni)		
(Taenite	Fe, Ni)		

In addition, mercury, tantalum, tin, and zinc have been found. The *native semimetals* form two isostructural groups: arsenic, antimony, and bismuth (space group $R3m$), and the less common selenium and tellurium (space group $P3_121$). The important *nonmetals* are sulfur and carbon, in the form of diamond and graphite.

Native Metals

It is fitting that *systematic mineralogy* begin with a discussion of the gold group, for our knowledge of the properties and usefulness of metals arose from the chance discovery of nuggets and masses of these minerals. Many relatively advanced early cultures were restricted in their use of metal to that found in the native state.

The elements of the *Gold Group* belong to the same group in the periodic table of elements and, hence, their atoms have somewhat similar chemical properties; all are sufficiently inert to occur in an elemental state in nature. When uncombined with other elements, the atoms of these metals are united into crystal structures by the rather weak metallic bond. The minerals are isostructural and are built on the face-centered cubic lattice with atoms in 12 coordination (see Fig. 7.1a). Complete solid solution takes place between gold and silver, as these two elements have the same atomic radii (1.44 Å). Copper, because of its smaller ionic radius (1.28 Å), exhibits only limited solid solution in gold and silver. Conversely, native copper carries only traces of gold and silver in solid solution.

The similar properties of the members of this group arise from their common structure. All are rather soft, malleable, ductile, and sectile. All are excellent conductors of heat and electricity, display metallic luster and hackly fracture, and have rather low melting points. These are properties conferred by the metallic bond. All are isometric hexoctahedral and have high densities resulting from the cubic closest packing of their structures.

Those properties with respect to which the minerals of this group differ arise from the properties of the atoms of the individual elements. Thus, the yellow of gold, the red of copper, and the white of

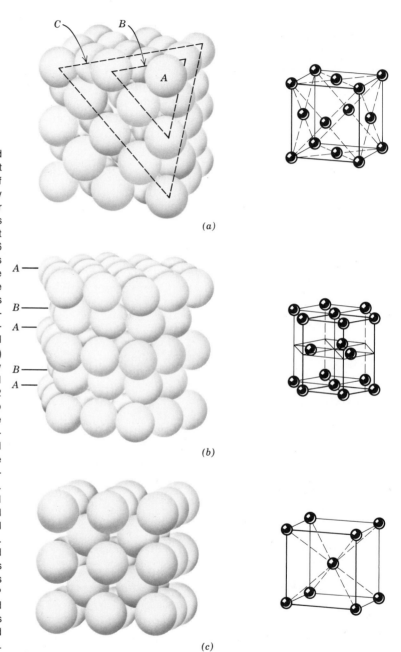

FIG. 7.1 (a) Close-packed model of cubic closest packing (*ABCABC* . . .) of equal spheres as shown by Cu, Au, Pt, and many other metals. Each metal atom is surrounded by 12 closest neighbors (see also Figs. 4.6 and 4.7). Close-packed layers are parallel to {111}. The face-centered cubic lattice (*F*) compatible with this packing sequence is illustrated on the right. (b) Close-packed model of hexagonal closest packing (*ABAB* . . .) of equal spheres as shown by Mg, Zn, and Cd. Each metal atom is surrounded by 12 closest neighbors (see also Figs. 4.6 and 4.7). This type of stacking leads to a hexagonal (*H*) lattice, as illustrated on the right, which can be reinterpreted as a rhombohedral (*R*) lattice (see Fig. 2.17c). (c) Close-packed model of body-centered cubic packing of equal spheres, as shown by Fe. Each sphere is surrounded by 8 closest neighbors. This packing is not as close as exhibited by *CCP* and *HCP* (above). The body-centered (*I*) lattice compatible with this packing model is illustrated on the right.

silver are atomic properties. The specific gravities likewise depend on atomic properties and show a rough proportionality to the atomic weights.

Although only platinum is discussed here, the *Platinum Group* also includes the rarer minerals palladium, platiniridium, and iridosmine. The last two are respectively alloys of iridium and platinum and iridium and osmium, with hexagonally close-packed structures and space group $P6_3/mmc$. Platinum and iridium, however, have cubic closest packed structures, similar to metals of the Gold Group, with space group $Fm3m$. The platinum metals are harder and have higher melting points than metals of the Gold Group.

Members of the *Iron Group* metals are isometric and include pure iron (Fe), which occurs only rarely on the earth's surface, and two species of nickel-iron (*kamacite* and *taenite*), which are common in meteorites. Iron and nickel both have atomic radii of 1.24 Å, and thus nickel can and usually does substitute for some of the iron. Pure iron and kamacite, which contains about 5.5 weight percent Ni, show cubic close packing with space group $Im3m$. Taenite, which shows a range in Ni content from 27 to 65 weight percent, is cubic closest packed with space group $Fm3m$ (see Figs. 7.1a and 7.1c). These two minerals are characteristic of iron meteorites and it is believed that Fe-Ni alloys of this type constitute a large part of the earth's core.

GOLD—Au

Crystallography. Isometric; $4/m\bar{3}2/m$. Crystals are commonly octahedral (Fig. 7.2a), rarely showing the faces of the dodecahedron, cube, and trapezohedron {113}. Often in arborescent crystal groups with crystals elongated in the direction of a 3-fold symmetry axis, or flattened parallel to an octahedron face. Crystals are irregularly formed; passing into filiform, reticulated, and dendritic shapes (Fig. 7.2b). Seldom shows crystal forms; usually in irregular plates, scales, or masses.

$Fm3m$; $a = 4.079$ Å; $Z = 4$. $d's$: 2.36(10), 2.04(7), 1.443(6), 1.229(8), 0.785(5).

Physical Properties. H $2\frac{1}{2}$–3, **G** 19.3 when pure. The presence of other metals decreases the specific gravity, which may be as low as 15. *Fracture* hackly. Very malleable and ductile. Opaque. *Color* various shades of yellow depending on the purity, becoming paler with increase of silver.

Composition and Structure. A complete solid solution series exists between Au and Ag, and most gold contains some Ag. California gold carries 10 to 15% Ag. When Ag is present in amounts of 20% or greater, the alloy is known as *electrum*. Small amounts of Cu and Fe may be present as well as traces of Bi, Pb, Sn, Zn, and the platinum metals. The purity or *fineness* of gold is expressed in parts

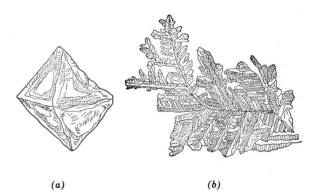

FIG. 7.2 (a) Malformed gold octahedron. (b) Dendritic gold.

per 1000. Most gold contains about 10% of other metals and thus has a fineness of 900. The structure of gold is based on cubic closest packing of Au atoms (See Fig. 7.1a).

Diagnostic Features. Gold is distinguished from the yellow sulfides pyrite and chalcopyrite and from yellow flakes of altered micas by its sectility and its high specific gravity. It fuses at 3 (1063°C) and is soluble only in aqua regia.

Occurrence. Although gold is a rare element, it occurs in nature widely distributed in small amounts. It is found most commonly in veins that bear a genetic relation to silicic types of igneous rocks. In places it has been found intimately associated with igneous rocks as in the Bushveld igneous complex in the Transvaal, South Africa. Most gold occurs as the native metal; tellurium and possibly selenium are the only elements that are combined with it in nature.

The chief source of gold are hydrothermal gold-quartz veins where, together with pyrite and other sulfides, gold was deposited from ascending mineral-bearing solutions. The gold is merely mechanically mixed with the sulfides and is not in chemical substitution. At and near the surface of the earth oxidation of the gold-bearing sulfides sets the gold free making its extraction easy. Such ores are called "free-milling" because their gold content can be recovered by amalgamation. Finely crushed ore is washed over copper plates coated with mercury. When sulfides are present in any quantity, not all the gold can be recovered by amalgamation, and either the cyanide or chlorination process must be used. In the cyanide process finely crushed ore is treated with a solution of potassium or sodium cyanide, forming a soluble cyanide. The gold is then recovered by precipitation with zinc or by electrolysis. The chlorination process renders the gold in a soluble form by treating the crushed and roasted ore with chlorine. In the majority of veins the gold is so finely divided and uniformly distributed that its presence in the ore cannot be detected with the eye.

When gold-bearing veins are weathered, the gold liberated either remains in the soil mantle as an eluvial deposit, or is washed into the neighboring streams to form a placer deposit. Because of its high specific gravity, gold works its way through the lighter sands and gravels to lodge behind irregularities or to be caught in crevices in the bedrock. The rounded or flattened nuggets of placer gold can be removed by panning, a process of washing away all but the heavy concentrate from which the gold can be easily separated. On a larger scale, gold-bearing sand is washed through sluices where the gold collects behind cross-bars or riffles and amalgamates with mercury placed behind the riffles. Most placer mining is today carried on with dredges, some of which are gigantic and can extract the gold from thousands of cubic yards of gravel a day. Some dredges can operate at a profit handling gravels that average no more than 10 cents in gold per cubic yard.

Of the estimated world production of gold in 1975 of 38,710,000 ounces, approximately two-thirds comes from the Republic of South Africa. The principal sources of South African gold are the Precambrian Witwatersrand conglomerate "the Rand," in the Transvaal and similar conglomerates in the Orange Free State. The reefs in these conglomerates in which the gold is concentrated are thought to be fossil placers. Although no accurate figures are available, the U.S.S.R. is believed to be the second-ranking country in gold production. Gold is mined in the Ural Mountains but it is estimated that two-thirds of the production is from placers, mostly in Siberia. Canada ranks third in gold production followed by the United States, Ghana, Australia, and the Philippines.

The most important gold-producing states of the United States, in order of the amounts of gold produced at present, are: South Dakota, Nevada, Utah, Arizona, Colorado, and Montana. The Homestake mine, South Dakota, and the Carlin mine, Nevada account for over 60% of the country's production. The gold from Arizona and Utah is essentially a by-product of copper mining. For 100 years following the discovery of gold in the streams of California in 1848, that state was the leading producer. The most important gold-producing districts were those of the Mother Lode, gold-quartz veins along the western slope of the Sierra Nevada.

Use. The principal use of gold is as a monetary standard. Other uses are in jewelry, scientific instruments, electroplating, gold leaf, dental appliances, and as small gold bars for investment purposes.

SILVER—Ag

Crystallography. Isometric; $4/m\bar{3}2/m$. Crystals are commonly malformed and in branching, arborescent, or reticulated groups. Found usually in irregular masses, plates, and scales; in places as coarse or fine wire (Fig. 7.3).

FIG. 7.3. Silver, Chañarcillo,

$Fm3m$; a = 4.09 Å; Z = 4. $d's$: 2.34(10), 1.439(6), 1.228(8), 0.936(7), 0.934(8).

Physical Properties. **H** $2\frac{1}{2}$–3. **G** 10.5 when pure, 10–12 when impure. *Fracture* hackly. Malleable and ductile. *Luster* metallic. *Color* and *streak* silver-white, often tarnished to brown or gray-black.

Composition and Structure. Native silver frequently contains alloyed Au, Hg, and Cu, more rarely traces of Pt, Sb, and Bi. *Amalgam* is a solid solution of Ag and Hg. The structure of silver is based on cubic closest packing of Ag atoms (see Fig. 7.1a).

Diagnostic Features. Silver can be distinguished by its malleability, color on a fresh surface, and high specific gravity. Fusible at 2 (960°C) to a bright globule (see Table 4.17).

Occurrence. Native silver is widely distributed in small amounts in the oxidized zone of ore deposits. However, native silver in larger deposits is the result of deposition from primary hydrothermal solutions. There are three types of primary deposits: (1) Associated with sulfides, zeolites, calcite, barite, fluorite, and quartz as typified by the occurrence in Kongsberg, Norway. Mines at this locality were worked for several hundred years and produced magnificient specimens of crystallized wire silver. (2) With arsenides and sulfides of cobalt, nickel, and silver and with native bismuth. Such was the type at the old silver mines at Freiberg and Schneeberg in Saxony and at Cobalt, Ontario. (3) With uraninite

and cobalt-nickel minerals. Deposits at the old but still active mines of Joachimsthal, Bohemia, and at Great Bear Lake, Northwest Territories, are of this type.

In the United States small amounts of primary native silver are associated with native copper on the Keweenaw Peninsula, Michigan. Elsewhere in the United States native silver is largely secondary.

Use. An ore of silver, although most of the world's supply comes from other minerals such as *acanthite*, Ag_2S, *proustite*, Ag_3AsS_3 and *pyrargyrite*, Ag_3SbS_3. Silver has many uses chief of which are in photographic film emulsions, plating, brazing alloys, tableware, and electronic equipment. Because of declining production and increased price, silver in coinage has been largely replaced by other metals such as nickel and copper.

COPPER—Cu

Crystallography. Isometric; $4/m\overline{3}2/m$. Tetrahexahedron faces are common, as well as the cube, dodecahedron, and octahedron. Crystals are usually malformed and in branching and arborescent groups (Figs. 7.4a and b). Usually occurs in irregular masses, plates, and scales, and in twisted and wirelike forms.

$Fm3m$; a = 3.615 Å, Z = 4. $d's$: 2.09(10), 1.81(10), 1.28(10), 1.09(10), 1.05(8).

FIG. 7.4 (a) Dendritic copper. (b) Native copper, Keweenaw Peninsula, Michigan.

(a)

(b)

Physical Properties. H $2\frac{1}{2}$–3. G 8.9. Highly ductile and malleable. *Fracture* hackly. *Luster* metallic. *Color* copper-red on fresh surface, usually dark with dull luster because of tarnish.

Composition and Structure. Native copper often contains small amounts of Ag, Bi, Hg, As, and Sb. The structure of copper is based on cubic closest packing of Cu atoms (see Fig. 7.1a).

Diagnostic Features. Native copper can be recognized by its red color on fresh surfaces, hackly fracture, high specific gravity, and malleability. Fuses at 3 (1083°C) to a globule. It dissolves readily in nitric acid, and the solution is colored a deep blue on addition of an excess of ammonium hydroxide.

Occurrence. Small amounts of native copper have been found at many localities in the oxidized zones of copper deposits associated with cuprite, malachite, and azurite.

Most primary deposits of native copper are associated with basaltic lavas, where deposition of copper resulted from the reaction of hydrothermal solutions with iron oxide minerals. The only major deposit of this type is on the Keweenaw Peninsula, Michigan, on the southern shore of Lake Superior. The rocks of the area are composed of Precambrian basic lava flows interbedded with conglomerates. The series, exposed for 100 miles along the length of the peninsula, dips to the northwest beneath Lake Superior to emerge on Isle Royale 50 miles away. Some copper is found in veins cutting these rocks, but it occurs principally either in the lava flows filling cavities or in the conglomerates filling interstices or replacing pebbles. Associated minerals are prehnite, datolite, epidote, calcite, and zeolites. Small amounts of native silver are present. Michigan copper had long been used by the American Indians but it was not until 1840 that its source was located. Exploitation began shortly thereafter and during the next 75 years there was active mining through the length of the Keweenaw peninsula. Most of the copper in the district was in small irregular grains, but notable large masses were found; one found in 1857 weighed 420 tons.

Sporadic occurrences of copper similar to the Lake Superior district have been found in the sandstone areas of the eastern United States, notably in New Jersey, and in the glacial drift overlying a similar area in Connecticut. In Bolivia at Corocoro, southwest of La Paz, there is a noted occurrence in sandstone. Native copper occurs in small amounts associated with the oxidized copper ores of Arizona, New Mexico, and northern Mexico.

Use. A minor ore of copper; copper sulfides are today the principal ores of the metal. The greatest use of copper is for electrical purposes, mostly as wire. It is also extensively used in alloys, such as brass (copper and zinc), bronze (copper and tin with some zinc), and German silver (copper, zinc, and nickel). These and many other minor uses make copper second only to iron as a metal essential to modern civilization.

PLATINUM—Pt

Crystallography. Isometric; $4/m\overline{3}2/m$. Cubic crystals are rare and commonly malformed. Usually found in small grains and scales. In some places it occurs in irregular masses and nuggets of larger size.

$Fm3m$; $a = 3.923$ Å, $Z = 4$. $d's$: 2.27(9), 1.180(10), 1.956(8), 1.384(8).

Physical Properties. H 4–$4\frac{1}{2}$. (Unusually high for a metal.) G 21.45 when pure; 14–19 when native. Malleable and ductile. *Color* steel-gray, with bright luster. Magnetic when rich in iron.

Composition and Structure. Platinum is usually alloyed with several percent Fe and with smaller amounts of Ir, Os, Rh, Pd; also Cu, Au, Ni. The structure of platinum is based on cubic closest packing at Pt atoms (see Fig. 7.1a).

Diagnostic Features. Determined by its high specific gravity, malleability, infusibility in the blowpipe flame, and insolubility except in aqua regia.

Occurrence. Most platinum occurs as the native metal in ultrabasic rocks, especially dunites, associated with olivine, chromite, pyroxene, and magnetite. It has been mined extensively in placers which are usually close to the platinum-bearing igneous parent rock.

Platinum was first discovered in the United States of Colombia, South America. It was taken to

Europe in 1735 where it received the name *platina* from the word *plata* (Spanish for silver) because of its resemblance to silver. A small amount of platinum is still produced in Colombia from placers in two districts near the Pacific Coast. In 1822 platinum was discovered in placers on the Upper Tura River on the eastern slope of the Ural Mountains, U.S.S.R. From that time until 1934 most of the world's supply of platinum came from placers of that district, which centered around the town of Nizhne Tagil. In 1934, because of the large amount of platinum recovered from the copper-nickel ore of Sudbury, Ontario, Canada became the leading producer. At Sudbury platinum occurs in *sperrylite*, $PtAs_2$. In 1954 the Republic of South Africa moved into first place. Part of the South African production is a by-product from gold mining on the Rand, but the chief source is the Merensky Reef in the ultrabasic rocks of the Bushveld igneous complex. The Merensky Reef is a horizon of this layered intrusive about 12 in. thick extending for many miles with a uniform platinum content of about one-half ounce per ton of ore. The chief mining operations are near Rustenburg in the Transvaal. Although Canadian and South African production have steadily increased, Russia once again took the lead in 1963. It is now estimated that over 50% of the world's platinum metals comes from this country. In smaller amounts platinum has come from Borneo, New South Wales, New Zealand, Brazil, Peru, and Madagascar.

In the United States small amounts of platinum have been recovered from the gold-bearing sands of North Carolina and from black-sand placers in California, Oregon, and Alaska.

Use. The uses of platinum depend chiefly on its high melting point (1755°C), resistance to chemical attack, and superior hardness. It is used as a catalyst in the chemical and petroleum industry, chemical apparatus, electrical equipment, and jewelry. It is also used in dentistry, surgical instruments, pyrometry, and photography.

Similar species. *Iridium; iridosmine,* an alloy of iridium and osmium; and *palladium* are rare minerals of the platinum group associated with platinum.

Iron—Fe

Crystallography. Isometric; $4/m\bar{3}2/m$. Crystals are rare. Terrestrial: in blebs and large masses; meteoritic (*kamacite*): in plates and lamellar masses, and in regular intergrowth with nickel-iron (*taenite*).

Ordinary iron (α-Fe and kamacite) body-centered cubic, $Im3m$, $a = 2.86$ Å, $Z = 2$. $d's$: 2.02(9), 1.168(10), 1.430(7), 1.012(7). Nickel-iron (taenite), face-centered cubic, $Fm3m$, $a = 3.56$ Å, $Z = 4$. $d's$: 2.06(10), 1.073(4), 1.78(3), 1.27(2).

Physical Properties. (α-Fe). *Cleavage* {001} poor. **H** $4\frac{1}{2}$. **G** 7.3–7.9. *Fracture* hackly. Malleable. Opaque. *Luster* metallic. *Color* steel-gray to black. Strongly magnetic.

Composition and Structure. α-Fe always contains some Ni and frequently small amounts of Co, Cu, Mn, S, C. *Kamacite* contains approximately 5.5 weight percent Ni. *Taenite* is highly variable in Ni content and ranges from about 27 to 65 weight percent Ni. The structures of α-Fe and kamacite are based on body-centered cubic packing of atoms (see Fig. 7.1c) whereas the structure of taenite is based on face-centered cubic packing (see Fig. 7.1a).

Diagnostic Features. Iron can be recognized by its strong magnetism, its malleability, and the oxide coating usually on its surface. It is infusible but soluble in HCl.

Occurrence. Occurs sparingly as terrestrial iron but commonly in meteorites. Iron in the elemental (native) state is highly unstable in the oxidizing conditions in rocks of the upper crust and in the earth's atmosphere. Iron is normally present as Fe^{2+} or Fe^{3+} in oxides such as magnetite, Fe_3O_4, or hematite, Fe_2O_3, or as the hydroxide, goethite, FeO.OH. Terrestrial iron is regarded as a primary magmatic constituent or a secondary product formed by the reduction of iron compounds by assimilated carbonaceous material. The most important locality is Disko Island, Greenland, where fragments ranging from small grains to masses of many tons are included in basalts. Masses of nickel-iron have been found in Josephine and Jackson counties, Oregon.

Iron meteorites are composed largely of a regular intergrowth of *kamacite* and *taenite*. Evidence of

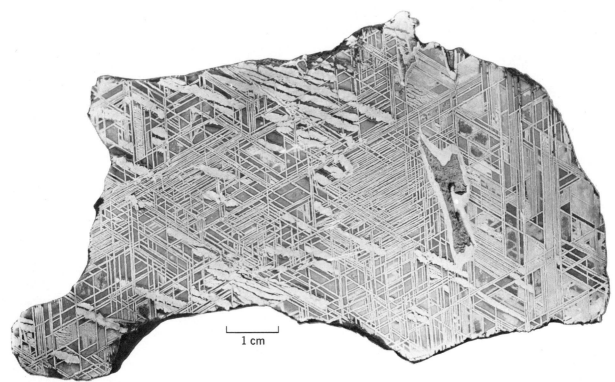

1 cm

FIG. 7.5 Widmanstätten pattern in the Edmonton (Kentucky) iron meterorite. The octahedral pattern is outlined by coarse kamacite and very thin taenite lamellae. The irregular inclusion at the right center consists of troilite (FeS). (*Smithsonian Astrophysical Observatory*, courtesy J. A. Wood.)

the intergrowth, the Widmanstätten pattern (Fig. 7.5), is seen on polished and etched surfaces of such meteorites. Stony-iron meteorites consist of mixtures of silicates, kamacite, taenite and troilite (FeS).

Native Semimetals

The semimetals arsenic, antimony, and bismuth belong to an isostructural group with space group $R\bar{3}m$. Their structures, unlike those of the metals, cannot be represented as a simple packing of spheres, because each atom is somewhat closer to three of its neighbors than to the remainder of the surrounding atoms. In Fig. 7.6a, which represents the structure of arsenic (or antimony), the spheres intersect each other along the flat circular areas. The bonding between the four closest atoms, forming pyramidal groups (see Fig. 7.6b) is due to

the covalent nature of these bonds. This covalency is related to the position of these semimetals in group Va of the periodic table toward the electronegative end of a row. The relatively strong bonding to four closest neighbors results in a layered structure as shown in Fig. 7.6b. These layers are parallel to {0001} and the weak bonding between them gives rise to a good cleavage. Members of this group have similar physical properties. They are rather brittle and much poorer conductors of heat and electricity than the native metals. These properties reflect a bond type intermediate between metallic and covalent; hence it is stronger and more directional than pure metallic in its properties, leading to lower symmetry.

Of the three native elements in this group arsenic and bismuth are more common in nature than antimony.

Arsenic—As

Crystallography. Hexagonal-R; $\bar{3}2/m$. Pseudocubic crystals rare. Usually granular massive, reniform, and stalactitic.

Physical Properties. *Cleavage* {0001} perfect. **H** $3\frac{1}{2}$. **G** 5.7. *Luster* nearly metallic on fresh surface. *Color* tin-white on fresh fracture, tarnishes to dark gray on exposure. *Streak* gray. Brittle.

$R\bar{3}m$; $a = 3.76$, $c = 10.55$ Å; $a:c = 1:2.805$, $Z = 6$. $d's$: 3.52(4), 2.77(10), 2.05(6), 1.88(7), 1.768(4).

Composition and Structure. Native arsenic often shows a limited substitution by Sb (atomic radii of As and Sb are 1.25 and 1.45 Å, respectively) and traces of Fe, Ag, Au, and Bi. The structure is shown in Fig. 7.6.

Diagnostic Features. Diagnostic blowpipe and chemical tests are given in Table 4.18 and on page 178.

Occurrence. Arsenic is a comparatively rare mineral found in veins in crystalline rocks associated with silver, cobalt, or nickel ores. Found in the silver mines of Freiberg, in Saxony; at Andreasberg in the Harz Mountains; in Bohemia; Rumania; and Alsace. Very little found in the United States.

Use. Very minor ore of arsenic; most arsenic of commerce is obtained from other arsenic-bearing minerals, such as arsenopyrite, FeAsS, and enargite, Cu_3AsS_4.

Name. The name arsenic is derived from a Greek word meaning *masculine* from the belief that metals were of different sexes. The term was first applied to the arsenic sulfide because of its potent properties.

Similar Species. *Allemontite*, AsSb, an intermediate compound of arsenic and antimony.

Bismuth—Bi

Crystallography. Hexagonal-R; $\bar{3}2/m$. Distinct crystals are rare. Usually laminated and granular; may be reticulated or arborescent. Artificial crystals pseudocubic {$01\bar{1}2$}.

$R\bar{3}m$; $a = 4.54$, $c = 11.85$ Å; $a:c = 1:2.609$, $Z = 6$. $d's$: 3.28(10), 2.35(5), 2.27(5), 1.86(3), 1.44(3).

Physical Properties. *Cleavage* {0001} perfect. **H** $2-2\frac{1}{2}$. **G** 9.8. Sectile. Brittle. *Luster* metallic. *Color* silver-white with a decided reddish tone. *Streak* silver-white, shining.

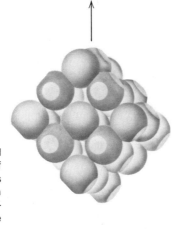

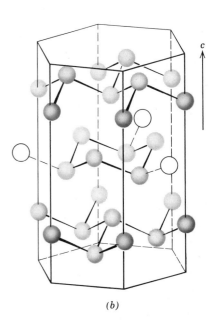

FIG. 7.6 (a) Close-packed model of the structure of As and Sb. Flat areas represent overlap between adjoining atoms. (b) Expanded view of the structure of As or Sb, showing corrugated layers of covalently bonded groups, parallel to {0001}.

(a)

(b)

Composition and Structure. Small amounts of As, S, Te, As may be present. Structure is very similar to that of arsenic or antimony (see Fig. 7.6).

Diagnostic Features. Bismuth is recognized chiefly by its laminated nature, its reddish-silver color, its perfect cleavage, and its sectility. Fusible at 1 (271°C). It gives a diagnostic blowpipe test, see Table 4.18 and page 178.

Occurrence. Bismuth is a comparatively rare mineral occurring with ores of silver, cobalt, nickel, lead, and tin. Found in the silver veins of Saxony; in Norway and Sweden; and in Cornwall, England. Important deposits are found in Australia, but the most productive deposits are in Bolivia. It is found in small veins associated with silver and cobalt minerals at Cobalt, Ontario, Canada.

Use. The chief ore of bismuth. Bismuth forms low-melting alloys with lead, tin, and cadmium, which are used for electric fuses and safety plugs in water sprinkling systems. About 75% of the bismuth produced is for medicine and cosmetics. Bismuth nitrate is relatively opaque to X-rays and is taken internally when the digestive organs are to be photographed.

Name. Etymology in dispute; possibly from the Greek meaning *lead white*.

Native Nonmetals

The structure of the nonmetals sulfur, diamond, and graphite are very different from those of the metals and semimetals. *Sulfur* ordinarily occurs in orthorhombic crystal form (with space group *Fddd*) in nature; two monoclinic polymorphs are very rare in nature but are commonly produced synthetically. The orthorhombic structure of S is stable at atmospheric pressure below 95.5°C, and a monoclinic polymorph is stable from 95.5 to 119°C. This monoclinic polymorph melts above 119°C. The unit cell of the orthorhombic form of sulfur contains a very large number of S atoms, namely 128. The sulfur atoms are arranged in puckered rings of eight atoms which form closely bonded S_8 molecules (see Fig. 7.7a). These rings are bonded to each other with a relatively large spacing between the rings (see Fig. 7.7b), and 16 of such rings are present in the unit cell. Small amounts of selenium (atomic radius 1.16 Å) may substitute for S (atomic radius 1.04 Å) in the structure.

The structures of the two carbon polymorphs, *diamond* and *graphite*, are shown in Figs. 7.8a and b. Diamond has an exceptionally close-knit and

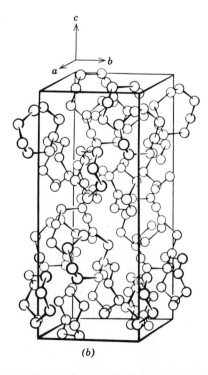

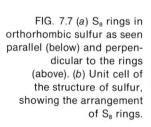
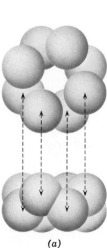

FIG. 7.7 (a) S_8 rings in orthorhombic sulfur as seen parallel (below) and perpendicular to the rings (above). (b) Unit cell of the structure of sulfur, showing the arrangement of S_8 rings.

(a)

(b)

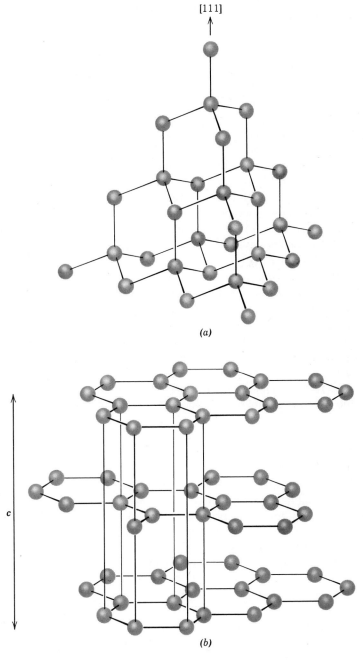

(a)

(b)

FIG. 7.8 (*a*) Partial representation of the structure of diamond. The horizontal plane
is (111). (*b*) The structure of
graphite with sheets // {0001}.
Vertical lines link atoms in
successive sheets.

strongly bonded structure in which each carbon
atom is bound by powerful and highly directive covalent bonds to four carbon neighbors at the apices
of a regular tetrahedron. The resulting structure,
although strongly bonded, is not close packed, and
only 34% of the available space if filled. The presence of rather widely spaced sheets of carbon atoms
parallel to {111} (see Fig. 7.8*a*) accounts for prominent octahedral cleavage in diamond. The {111}
sheets are planes of maximum atomic population.

The structure of graphite, illustrated in Fig. 7.8*b*, consists of six-membered rings in which each carbon atom has three near neighbors arranged at the apices of an equilateral triangle. Three of the four valence electrons in each carbon atom may be considered to be locked up in tight covalent bonds with its three close neighbors in the plane of the sheet. The fourth is free to wander over the surface of the sheet, creating a dispersed electrical charge which bestows on graphite its relatively high electrical conductivity. In contrast, diamond, in which all four valence electrons are locked up in covalent bonds, is among the best electrical insulators.

The sheets composing the graphite crystal are stacked in such a way that alternate sheets are in identical position, with the intervening sheet translated a distance of one-half the identity period in the plane of the sheets (see Fig. 7.8*b*). The distance between sheets is much greater than one atomic diameter, and van der Waals' bonding forces perpendicular to the sheets are very weak. This wide separation and weak binding give rise to the perfect basal cleavage and easy gliding parallel to the sheets. Because of this open structure only about 21% of the available space in graphite is filled, and the specific gravity is proportionately less than that of diamond.

SULFUR—S

Crystallography. Orthorhombic; $2/m2/m2/m$. Pyramidal habit common (Fig. 7.9*a*), often with two dipyramids, first-order prism and base in combination (Fig. 7.9*b*). Commonly found in irregular masses imperfectly crystallized. Also massive, reniform, stalactitic, as incrustations, earthy.

Angles: $c(001) \wedge n(011) = 62°18'$, $c(001) \wedge p(111) = 70°40'$, $c(001) \wedge s(113) = 45°10'$, $p(111) \wedge p(1\bar{1}1) = 73°34'$.

Fddd; $a = 10.47$, $b = 12.87$, $c = 24.49$ Å; $a:b:c = 0.813:1:1.903$; $Z = 128$. d's: 7.76(4), 5.75(5), 3.90(10), 3.48(4), 3.24(6).

Physical Properties. *Fracture* conchoidal to uneven. Brittle. **H** $1\frac{1}{2}$–$2\frac{1}{2}$. **G** 2.05–2.09. *Luster* resinous. *Color* sulfur-yellow, varying with impurities to yellow shades of green, gray, and red. Transparent

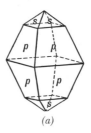 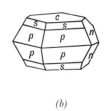

FIG. 7.9. Sulfur crystals.

to translucent. *Optics:* (+); $\alpha = 1.957$, $\beta = 2.037$, $\gamma = 2.245$, $2V = 69°$.

Sulfur is a poor conductor of heat. When a crystal is held in the hand close to the ear it will be heard to crack. This is due to the expansion of the surface layers because of the heat from the hand, while the interior, because of the slow heat conductivity, is unaffected. Crystals of sulfur should, therefore be handled with care.

Composition and Structure. Native sulfur may contain small amounts of selenium in substitution for S. The structure of the orthorhombic polymorph is given in Fig. 7.7. The monoclinic polymorphs (α and β sulfur) are very rare in nature.

Diagnostic features. Sulfur can be told by its yellow color and the ease with which it burns. The absence of a good cleavage distinguishes it from orpiment. Fusible at 1 (112.8°C) and burns with a blue flame to sulfur dioxide. Sublimates in closed tube are diagnostic (see Table 4.19).

Occurrence. Sulfur often occurs at or near the crater rims of active or extinct volcanoes where it has been derived from the gases given off in fumaroles. These may furnish sulfur as a direct sublimation product or by the incomplete oxidation of hydrogen sulfide gas. It is also formed from sulfates, by the action of sulfur-forming bacteria. Sulfur may be found in veins associated with metallic sulfides and formed by the oxidation of the sulfides. It is most commonly found in the Tertiary sedimentary rocks and most frequently associated with anhydrite, gypsum, and limestone; often in clay rocks; frequently with bituminous deposits. The large deposits near Girgenti, Sicily, are noteworthy for the fine crystals associated with celestite, gypsum, calcite, aragonite. Sulfur is also found associated with the volcanoes of Mexico, Hawaii, Japan, Argentina, and at Ollague, Chile, where it is minded at an elevation of 19,000 feet.

In the United States the most productive deposits are in Texas and Louisana where sulfur is associated with anhydrite, gypsum, and calcite in the cap rock of salt domes. There are 10 to 12 producing localities, but the largest are Boling Dome in Texas and Grand Ecaille, Plaquemines Parish, Louisiana. Sulfur is also recovered from salt domes in Mexico and offshore in the shallow waters of the Gulf of Mexico. Sulfur is obtained from these deposits by the Frasch method. Superheated water is pumped down to the sulfur horizon, where it melts the sulfur; compressed air then forces the sulfur to the surface.

About half of the world's sulfur is produced as the native element; the remainder is recovered as a by-product in smelting sulfide ores, from sour natural gas, and from pyrite.

Use. Sulfur is used in the chemical industry chiefly in the manufacture of sulfuric acid. It is also used in fertilizers, insecticides, explosives, coal-tar products, rubber, and in the preparation of wood pulp for paper manufacture.

DIAMOND—C

Crystallography. Isometric; $4/m\overline{3}2/m$. Crystals usually octahedral but may be cubic or dodecahedral. Curved faces, especially of the hexoctahedron are frequently observed (Figs. 7.10a and b). Crystals may be flattened on $\{111\}$. Twins on $\{111\}$ (spinel law) are common, usually flattened parallel to the twin plane (Fig. 7.10c). *Bort,* a variety of diamond,

has rounded forms and rough exterior resulting from a radial or cryptocrystalline aggregate. The term is also applied to badly colored or flawed diamonds without gem value.

$Fd3m$; $a = 3.567$ Å, $Z = 8$. $d's$: 2.06(10) 1.26(8), 1.072(7), 0.813(6), 0.721(9).

Physical Properties. *Cleavage* $\{111\}$ perfect. **H** 10 (hardest known mineral). **G** 3.51. *Luster* adamantine; uncut crystals have a characteristic greasy appearance. The very high refractive index, 2.42, and the strong dispersion of light account for the brilliancy and "fire" of the cut diamond. *Color* usually pale yellow or colorless; also pale shades of red, orange, green, blue, and brown. Deeper shades are rare. *Carbonado* or *carbon* is black or grayish black bort. It is noncleavable, opaque, and less brittle than crystals.

Composition, Structure and Synthesis. Pure carbon. The structure of diamond is illustrated in Fig. 7.8a.

The synthesis of diamond in 1955 was the realization of an age-old dream and established firmly the stability relations between the polymorphs of carbon over a wide range of pressure and temperature (Fig. 7.11). Diamond, as is to be expected from its high specific gravity and fairly close packing, is the high pressure polymorph. At low pressures or temperatures it is unstable with respect to graphite and may be converted to graphite at moderate temperatures. The reason that diamond and graphite can coexist at room temperatures and pressures is because the reconstructive polymorphic reaction between the two minerals is very sluggish. In order to permit the change from graphite to diamond, extremely high temperatures are needed to cause the carbon atoms in the graphite structure to break

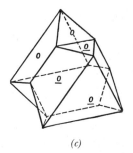

FIG. 7.10. Diamond crystals. *(a)* *(b)* *(c)*

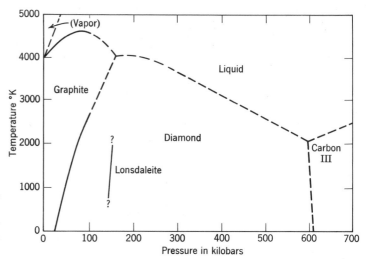

FIG. 7.11. Pressure-temperature phase diagram for carbon, based on experimental data from various sources.

loose by thermal agitation and to make them available for building the diamond structure. Such temperatures also increase the pressure required to bring about the polymorphic reaction. Therefore, diamonds were not synthesized until apparatus could be built that would simultaneously exert a very high pressure and withstand the high temperature. Such a press was built by General Electric engineers, and the first diamonds were synthesized in 1955 using pressures of 600,000 to 1,500,000 pounds per square inch and temperatures of 750 to 2750°C!

The success of the diamond syntheses encouraged experimentation with boron nitride, BN, whose structure is similar to graphite. In 1956 a high-pressure, high-temperature polymorph of boron nitride having the structure and hardness of diamond was prepared. This compound, known by the trade name of *Borazon*, is not a mineral, but its synthesis illustrates the application of crystal chemical concepts first evolved and tested in minerals.

After the initial synthesis of diamond, scientists at the General Electric Company continued experimentation. In 1967 they produced a hexagonal transparent polymorph of carbon with specific gravity and refractive indices close to those of diamond. Almost simultaneously a hexagonal diamond was found in the Canyon Diablo, Arizona, iron meteorite. The name *lonsdaleite* is given to this naturally occurring hexagonal polymorph of diamond.

Diagnostic Features. Diamond is distinguished from minerals that resemble it by its great hardness, adamantine luster, and cleavage. Insoluble in acids and alkalis. At a high temperature in oxygen will burn to CO_2 gas, leaving no ash.

Occurrence. Diamonds have been discovered in many different localities but in only a few in notable amount. Most commonly diamond is found in alluvial deposits, where it accumulates because of its inert chemical nature, its great hardness, and its fairly high specific gravity. In several countries in Africa and more recently in Siberia diamonds have been found *in situ*. The rock in which they occur is an altered peridotite called *kimberlite*. Although these intrusive bodies vary in size and shape, many are roughly circular with a pipelike shape and are referred to as "diamond pipes." In the deepest mine, in Kimberley, South Africa, the diameter of the pipe decreased with depth and mining stopped at a depth of 3500 feet although the pipe continued. At the surface the kimberlite is weathered to a soft yellow rock, "yellow ground," that in depth gives way to a harder "blue ground." The ratio of diamonds to barren rock varies from one pipe to another. In the Kimberley mine it was 1 : 8,000,000 but in some it may be as high as 1 : 30,000,000.

Diamonds were first found in India which remained virtually their only source until they were discovered in Brazil in 1725. The early diamonds came from stream gravels in southern and central India, and it is estimated that 12 million carats* were produced from this area. Today the Indian production is only a few hundred carats a year.

Following their discovery in stream gravels in the state of Minas Gerais, Brazil, diamonds were also found in the states of Bahia, Goyaz, and Mato Grasso. Brazil today produces about 160,000 carats annually, chiefly from the stream gravels near the city of Diamantina, Minas Gerais. Extensive upland deposits of diamond-bearing gravels and clays are also worked. The black carbonado comes only from Bahia.

In 1866 diamonds were discovered in the gravels of the Vaal River, South Africa, and in 1871 in the "yellow ground" of several pipes located near the present city of Kimberley. Although some diamonds are still recovered from gravels, the principal South African production is from kimberlite pipes. The deposits were originally worked as open pits, but as they deepened, underground methods were adopted. The world's largest and most productive diamond mine is the Premier, 24 miles east of Pretoria, South Africa. Since mining began there in 1903, nearly 30 million carats or six tons of diamonds have been produced. It was at the Premier mine in 1905 that the world's largest diamond, the Cullinan, weighing 3024 carats, was found. Prospecting for diamonds has located in South Africa over 700 kimberlite pipes, dikes, and sills most of which are barren. As recently as 1966 the Finsch Mine, 80 miles west of Kimberley, came into production. Even more recently diamond has been found in kimberlite in Botswana. The most notable pipe outside of South Africa is the Williamson Diamond Mine in Tanzania, discovered in 1940; because of it, this country is a major diamond producer.

The early method of treatment was to crush the "blue ground" into coarse fragments and spread it out on platforms to disintegrate gradually under

atmospheric influences. The present method is to crush the rock fine enough to permit immediate concentration. The diamonds are finally separated on tables coated with grease, to which the diamonds adhere, whereas the rest of the material is washed away.

In Cape Province, South Africa, on the desert coast just south of the mouth of the Orange River, terrace deposits containing high-quality stones were discovered in 1927. Later similar deposits were found along the coast north of the Orange River in South-West Africa extending 50 to 60 miles up the coast. Elsewhere in Africa alluvial diamonds have been found in Zaire, Angola, Ghana, and Sierra Leone. Zaire is by far the largest producer and furnishes from placer deposits about 30% of the world's supply. These Zaire diamonds are mostly of industrial grade, with only 10% of gem quality. About 75% of the world's output of diamonds comes from the African continent. The U.S.S.R. is a major producer of diamonds from both pipes and placers. Although figures are not available, it is estimated that the Soviet Union accounts for nearly 25% of the world production. British Guiana, Venezuela, and Australia are small producers.

Diamonds have been found sparingly in various parts of the United States. Small stones have occasionally been discovered in the stream sands along the eastern slope of the Appalachian Mountains from Virginia south to Georgia. Diamonds have also been reported from the gold sands of northern California and southern Oregon and sporadic occurrences have been noted in the glacial drift in Wisconsin, Michigan, and Ohio. In 1906 diamonds were found in a kimberlite pipe near Murfreesboro, Pike County, Arkansas. This locality, resembling the diamond pipes of South Africa, has yielded about 40,000 stones but is at present unproductive. In 1951 the old mine workings were opened to tourists who, for a fee, were permitted to look for diamonds.

Use. *In Industry.* Fragments of diamond crystals are used to cut glass. The fine powder is employed in grinding and polishing diamonds and other gem stones. Wheels are impregnated with diamond powder for cutting rocks and other hard materials.

---

* 1 carat = 200 milligrams; a unit of weight in gemstones.

Steel bits are set with diamonds, especially the cryptocrystalline variety, *carbonado,* for diamond drilling in exploratory mining work. Diamond is also used in wire drawing and in tools for the truing of grinding wheels.

In Gems. The diamond is the most important of the gem stones. Its value depends on its hardness, its brilliancy which results from its high index of refraction, and its "fire," resulting from its strong dispersion. In general, the most valuable are those flawless stones that are colorless or possess a "blue-white" color. A faint straw-yellow color, which diamond often shows, detracts from its value. Diamonds colored deep shades of yellow, red, green, or blue, known as fancy stones, are greatly prized and bring very high prices. Diamonds can be colored deep shades of green by irradiation with high-energy nuclear particles, neutrons, deuterons, and alpha particles, and blue by exposing them to fast-moving electrons. A stone colored green by irradiation can be made a deep yellow by proper heat treatment. These artificially colored stones are difficult to distinguish from those of natural color.

The value of a cut diamond depends on its color and purity, on the skill with which it has been cut, and on its size. A stone weighing 1 carat (0.2 g) cut in the form of a brilliant would be 6.25 mm in diameter and 4 mm in depth. A 2-carat stone of the same quality would have a value three or four times as great.

Artificial. So far synthetic diamonds are small and not suitable for cutting into gems, but several million carats are produced each year for industrial purposes.

Name. The name diamond is a corruption of the Greek word *adamas,* meaning *invincible.*

GRAPHITE—C

Crystallography. Hexagonal; $6/m2/m2/m$. In tabular crystals of hexagonal outline with prominent basal plane. Distinct faces of other forms very rare. Triangular markings on the base are the result of gliding along an undetermined second-order pyramid. Usually in foliated or scaly masses, but may be radiated or granular.

$C6_3/mmc$. $a = 2.46$, $c = 6.74$ Å, $Z = 4$. $a:c = 1:2.740$. $d's$: 3.36(10), 2.03(5), 1.675(8), 1.232(3), 1.158(5).

Physical Properties. *Cleavage.* {0001} perfect. **H** 1–2 (readily marks paper and soils the fingers). **G** 2.23. *Luster* metallic, sometimes dull earthy. *Color* and streak, black. Greasy feel. Folia flexible but not elastic.

Composition and Structure. Carbon. Some graphite impure with iron oxide, clay, or other minerals. The structure is illustrated in Fig. 7.8b.

Diagnostic Features. Graphite is recognized by its color, foliated nature, and greasy feel. Distinguished from molybdenite by its black color (molybdenite has a blue tone), and black streak on glazed porcelain. Infusible, but may burn to CO_2 at a high temperature. Unattacked by acids.

Occurrence. Graphite most commonly occurs in metamorphic rocks such as crystalline limestones, schists, and gneisses. It may be found as large, crystalline plates or disseminated in small flakes in sufficient amount to form a considerable proportion of the rock. In these cases, it has probably been derived from carbonaceous material of organic origin that has been converted into graphite during metamorphism. Metamorphosed coal beds may be partially converted into graphite, as the graphite coals of Rhode Island and in the coal fields of Sonora, Mexico. Graphite also occurs in hydrothermal veins associated with quartz, biotite, orthoclase, tourmaline, apatite, pyrite, and sphene, as in the deposits at Ticonderoga, New York. The graphite in these veins may have been formed from hydrocarbons introduced into them during the metamorphism of the region and derived from the surrounding carbon-bearing rocks. Graphite occurs occasionally as an original constituent of igneous rocks as in the basalts of Ovifak, Greenland, in a nepheline syenite in India, in a graphite pegmatite in Maine. It is also found in some iron meteorites as graphite nodules.

The principal countries producing natural graphite are: North and South Korea, U.S.S.R., Mexico, Austria, Malagasy Republic, and Sri Lanka. Deposits in the United States are in the Adirondack region of New York, in Essex, Warren, and Washington counties, particularly at Ticonderoga, and in Clay County, Alabama, and in Texas.

Artificial. Graphite is manufactured on a large scale in electrical furnaces using anthracite coal or petroleum coke as the raw materials. The use of artificial graphite in the United States is considerably in excess of that of the natural mineral.

Use. Use in the manufacture of refractory crucibles for the steel, brass, and bronze industries. Flake graphite for crucibles comes mostly from Sri Lanka and the Malagasy Republic. Mixed with oil, graphite is used as a lubricant, and mixed with fine clay, it forms the "lead" of pencils. It is employed in the manufacture of protective paint for structural steel and is used in foundry facings, batteries, electrodes, generator brushes, and in electrotyping.

Name. Derived from the Greek word meaning *to write,* in allusion to its use in pencils.

SULFIDES

The sulfides form an important class of minerals that includes the majority of the ore minerals. With them are classed the similar but rarer sulfarsenides, arsenides, and tellurides.

Most of the sulfide minerals are opaque with distinctive colors and characteristically colored streaks. Those that are nonopaque, such as cinnabar, realgar, and orpiment, have high refractive indices and transmit light only on thin edges.

The general formula for the sulfides is given as $X_m Z_n$ in which X represents the metallic elements and Z the nonmetallic element. The general order of listing of the various minerals is in a decreasing ratio of $X : Z$.

The sulfides can be divided into small groups of similar structures but it is difficult to make broad generalizations about their structure. Regular octahedral or tetrahedral coordination about sulfur is found in many simple sulfides such as in galena, PbS, which has an NaCl type structure, and in sphalerite (see Fig. 7.12a). In more complex sulfides, as well as sulfosalts, distorted coordination polyhedra may be found (see tetrahedrite, Fig. 7.12c). Many of the sulfides have ionic and covalent bonding,

SULFIDES, SULFARSENIDES, ARSENIDES, AND TELLURIDES

Acanthite (Argentite)	Ag_2S	Cinnabar	HgS
Chalcocite	Cu_2S	Realgar	AsS
Bornite	Cu_5FeS_4	Orpiment	As_2S_3
Galena	PbS	Stibnite	Sb_2S_3
Sphalerite	ZnS	Pyrite	FeS_2
Chalcopyrite	$CuFeS_2$	Marcasite	FeS_2
Pyrrhotite	$Fe_{1-x}S$	Molybdenite	MoS_2
Niccolite	$NiAs$	Cobaltite	$(Co, Fe)AsS$
Millerite	NiS	Arsenopyrite	$FeAsS$
Pentlandite	$(Fe, Ni)_9S_8$	Skutterudite	$(Co, Ni)As_3$
Covellite	CuS	Calaverite	$AuTe_2$
		Sylvanite	$(Au, Ag)Te_2$

whereas others, displaying most of the properties of metals, have metallic bonding characteristics. The structures and some aspects of the crystal chemistry of a few of the most common sulfides (e.g., sphalerite, chalcopyrite, pyrite, marcasite, and covellite) will be discussed.

ZnS occurs in two polymorphic structure types: the sphalerite structure (Fig. 7.12a) and the wurtzite structure (Fig. 7.20). In both the sphalerite and wurtzite structures Zn is surrounded by four sulfurs in tetrahedral coordination, but in sphalerite the Zn atoms are arranged in a face-centered cubic lattice, whereas in wurtzite they are approximately in positions of hexagonal closest packing. The atomic arrangement in sphalerite is like that in diamond (see Fig. 7.8a) in which the carbon has been replaced by equal amounts of Zn and S. Greenockite, CdS, a relatively rare sulfide, is isostructural with ZnS and occurs in both the sphalerite and wurtzite type structures. Natural sphalerite shows extensive Fe^{2+} and very limited Cd^{2+} substitution for Zn (see Table 4.12).

Chalcopyrite, $CuFeS_2$, has a structure (Fig. 7.12b) that can be derived from the sphalerite structure by regularly substituting Cu and Fe ions for Zn in sphalerite; this leads to a doubling of the unit cell. Stannite, Cu_2FeSnS_4, has a structure based on sphalerite in which layers of ordered Fe and Sn alternate with layers of Cu (chalcopyrite can be rewritten as $Cu_2Fe_2S_4$; this illustrates chemically the close similiarity to Cu_2FeSnS_4).

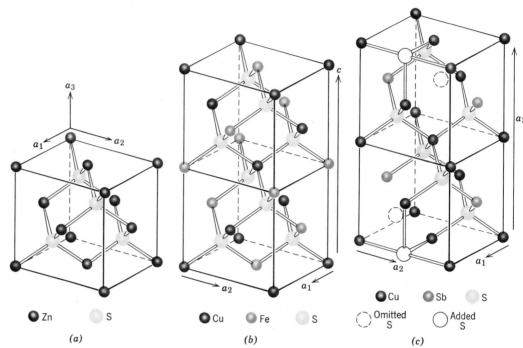

FIG. 7.12. The sphalerite structure and derivatives. (a) Sphalerite, ZnS. (b) Chalcopyrite, $CuFeS_2$. (c) Tetrahedrite, $Cu_{12}Sb_4S_{13}$ (after B. J. Wuensch, 1974, in *Sulfide Mineralogy*).

Covellite, CuS, is an example of a chemically very simple substance with a rather complex structure (see Fig. 7.13) in which part of the Cu is tetrahedrally coordinated by four S, and part coordinated by three S in the form of a triangle. The structure can, therefore, be viewed as made of sheets of CuS_3 triangles, between double layers of CuS_4 tetrahedra; covalent sulfur-sulfur bonds link the layers.

Pyrite, FeS_2, has a cubic structure as shown in Fig. 7.14a. The structure contains covalently bonded S_2 pairs, which occupy the position of Cl in the NaCl structure type. The pyrite structure may be considered as derived from the NaCl structure in which Fe is found in the Na positions of NaCl. Another polymorphic form of FeS_2 is marcasite (see Fig. 7.14b) with an orthorhombic structure. This structure, as pyrite, contains closely spaced S_2 pairs. It is still unclear what the stability fields of the two polymorphs, pyrite and marcasite, are. From geological occurrences one would conclude that marcasite occurs over a range of low to medium temperature

in sedimentary rocks and metalliferous veins. Pyrite, however, occurs also as a magmatic (high T) constituent.

The structure of arsenopyrite, FeAsS, can be derived from the marcasite (FeS_2 or FeSS) structure (see Fig. 7.14c) in which one S per unit formula is replaced by As. The general coordination of the atoms in marcasite and FeAsS are approximately the same in both structures.

Acanthite—Ag₂S

Crystallography. Monoclinic, $2/m$ below 173°C; isometric, $4/m\bar{3}2/m$ above 173°C. Crystals, twinned polymorphs of the high-temperature form, commonly show the cube, octahedron, and dodecahedron but frequently are arranged in branching or reticulated groups. Most commonly massive or as a coating. Historically acanthite has been referred to as argentite.

$P2_1/n$; $a = 4.23$, $b = 6.93$, $c = 7.86$ Å, $\beta =$

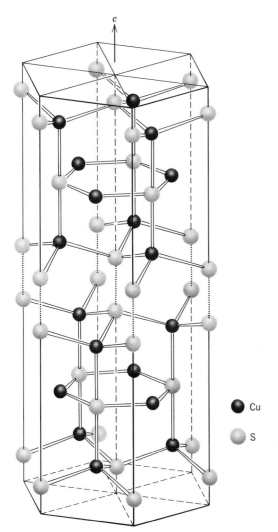

FIG. 7.13. Structure of covellite, CuS. The dotted lines indicate covalent S-S bonds (after B. J. Wuensch, 1974, in *Sulfide Mineralogy*).

Composition and Structure. Ag 87.1, S 12.9%. Ag_2S has space group $P\ 2_1/n$ below 173°C, and space group $Im3m$ above 173°C. On cooling Ag_2S from above 173°C the structure twins pervasively to produce apparently cubic crystals of twinned acanthite, known to mineralogists as *argentite*. However, it seems that acanthite is the only stable form of Ag_2S at ordinary temperatures.

Diagnostic Features. Acanthite can be distinguished by its color, sectility, and high specific gravity. Fusible at $1\frac{1}{2}$ with intumescence. When fused on charcoal in the oxidizing flame it emits the odor of sulfur dioxide and yields a globule of silver.

Occurrence. Acanthite is an important primary silver mineral found in veins associated with native silver, the ruby silvers, polybasite, stephanite, galena, and sphalerite. It may also be of secondary origin. It is found in microscopic inclusions in argentiferous galena. Acanthite is an important ore in the silver mines of Guanajuato and elsewhere in Mexico; in Peru, Chile, and Bolivia. Important European localities are Freiberg, Saxony; Joachimsthal, Bohemia; Schemnitz and Kremnitz, Czechoslovakia, and Kongsberg, Norway. In the United States it has been an important ore mineral in Nevada, notably at the Comstock Lode and at Tonopah. It is also found in the silver districts of Colorado, and in Montana at Butte associated with copper ores.

Use. An important ore of silver.

Name. The name acanthite comes from the Greek word meaning *thorn,* in allusion to the shape of the crystals. The name argentite comes from the Latin *argentum,* meaning silver.

99°35′, $a:b:c\ =\ 0.610:1:1.134.\ Z\ =\ 4.\ d's$: 2.60(10), 2.45(8), 2.38(5), 2.22(3), 2.09(4).

Physical Properties. H $2-2\frac{1}{2}$. **G** 7.3. Very sectile; can be cut with a knife like lead. *Luster* metallic. *Color* black. *Streak* black, shining. Opaque. Bright on fresh surface but on exposure becomes dull black, owing to the formation of an earthy sulfide.

CHALCOCITE—Cu_2S

Crystallography. Orthorhombic; $2/m2/m2/m$ (below 105°C); above 105°C, hexagonal. Crystals are very rare, usually small and tabular with hexagonal outline; striated parallel to the a axis (Fig. 7.15). Commonly fine-grained and massive.

$Abm2$; $a\ =\ 11.92,\ b\ =\ 27.33,\ c\ =\ 13.44$Å. $a:b:c\ =\ 0.436:1:0.492,\ Z\ =\ 96.\ d's$: 3.39(3), 2.40(7), 1.969(8), 1.870(10), 1.695(4).

Physical Properties. *Cleavage* {110} poor. *Fracture* conchoidal. **H** $2\frac{1}{2}-3$. **G** 5.5–5.8. *Luster* me-

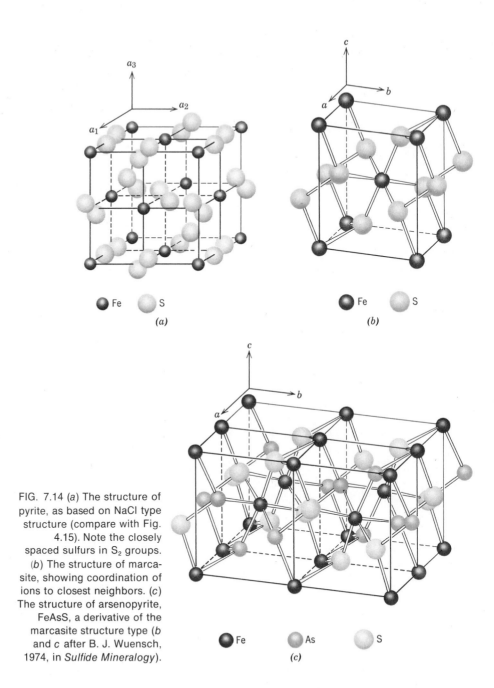

FIG. 7.14 (a) The structure of pyrite, as based on NaCl type structure (compare with Fig. 4.15). Note the closely spaced sulfurs in S_2 groups. (b) The structure of marcasite, showing coordination of ions to closest neighbors. (c) The structure of arsenopyrite, FeAsS, a derivative of the marcasite structure type (b and c after B. J. Wuensch, 1974, in *Sulfide Mineralogy*).

tallic. Imperfectly sectile. *Color* shining lead-gray, tarnishing to dull black on exposure. *Streak* grayish black. Some chalcocite is soft and sooty.

Composition and Structure. Cu 79.8, S 20.2%. May contain small amounts of Ag and Fe. Below 105°C the structure is based on hexagonal closest packing of sulfur atoms with an orthorhombic space group. Above 105° it inverts to high chalcocite with space group $P6_3/mmc$.

Diagnostic Features. Chalcocite is distinguished by its lead-gray color and sectility. Fusible at 2–2½. When heated on charcoal it gives odor of

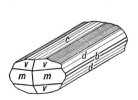

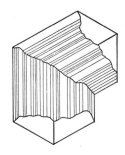

FIG. 7.15. Chalcocite crystals.

sulfur dioxide and is easily reduced to metallic copper. Diagnostic copper flame test, see Table 4.20.

Occurrence. Chalcocite is one of the most important copper-ore minerals. Fine crystals are rare but have been found in Cornwall, England, and Bristol, Connecticut. Chalcocite may occur as a primary mineral in veins with bornite, chalcopyrite, enargite, and pyrite. But its principal occurrence is as a supergene mineral in enriched zones of sulfide deposits. Under surface conditions the primary copper sulfides are oxidized; the soluble sulfates formed move downward reacting with the primary minerals to form chalcocite and thus enriching the ore in copper. The water table is the lower limit of the zone of oxidation and here a "chalcocite blanket" may form (see Fig. 11.19). Many famous copper mines owe their greatness to this process of secondary enrichment as: Rio Tinto, Spain; Ely, Nevada; Morenci, Miami, and Bisbee, Arizona; and Butte, Montana.

Much of the world's copper is today produced from what is called "porphyry copper" ore. In these deposits primary copper minerals disseminated through the rock, usually a prophyry, have been altered, at least in part, to chalcocite and thus enriched to form a workable ore body. The amount of copper in such deposits is small, rarely greater than 1 or 2% and may be as low as 0.40%. The largest copper producer in the United States is a porphyry copper deposit at Bingham, Utah.

Use. An important copper ore.

Similar Species. *Digenite,* Cu_9S_5, is blue to black, associated with chalcocite. *Stromeyerite,* $(Ag, Cu)_2S$, is a steel-gray mineral found in copper-silver veins.

BORNITE—Cu₅FeS₄

Crystallography. Tetragonal, $\overline{4}2m$, below 228°C; isometric, $4/m\overline{3}2/m$, above 228°C. The most common form in ores is tetragonal. Rarely in rough pseudocubic and less commonly dodecahedral and octahedral crystals. Usually massive.

$P\overline{4}2_1c$. $a = 10.94$, $c = 21.88$ Å; $a:c = 1:0.500$; $Z = 16$. $d's$: 3.31(4), 3.18(6), 2.74(5), 2.50(4), 1.94(10).

$Fm3m$, $a = 5.50$ Å; $Z = 1$. $d's$: 3.17(5), 2.75(5), 1.94(10), 1.66(1).

Physical Properties. H 3. **G** 5.06–5.08. *Luster* metallic. *Color* brownish-bronze on fresh fracture but quickly tarnishing to variegated purple and blue (hence called *peacock ore*) and finally to almost black on exposure. *Streak* grayish-black.

Composition and Structure. Cu 63.3, Fe 11.2, S 25.5% for stoichiometric Cu_5FeS_4. Shows extensive solid solution within the Cu-Fe-S system. Its stoichiometric composition in terms of Cu-Fe-S is shown in Fig. 7.16, with the compositions of other common sulfides in this system. The structure of the high T polymorph of bornite is relatively complex with sulfur atoms in a face-centered cubic lattice and the Cu and Fe atoms in tetrahedral coordination to S. The structure of the low temperature form is derived from the high T form with the addition of defects. This defect structure gives rise to large variations in Cu, Fe, and S contents.

Diagnostic Features. Bornite is distinguished by its characteristic bronze color on the fresh fracture and by the purple tarnish. Fusible at $2\frac{1}{2}$. When heated on charcoal it gives off the odor of sulfur dioxide and becomes magnetic. For diagnostic copper flame test see Table 4.20.

Alteration. Bornite alters readily to chalcocite and covellite.

Occurrence. Bornite is a widely occurring copper ore usually found associated with other sulfides (chalcocite, chalcopyrite, covellite, pyrrhotite and pyrite; see Fig. 7.16) in hypogene deposits. It is less frequently found as a supergene mineral, in the upper, enriched zone of copper veins. It occurs disseminated in basic rocks, in contact metamorphic deposits, in replacement deposits, and in pegmatites. It is not as important an ore of copper as chalcocite and chalcopyrite.

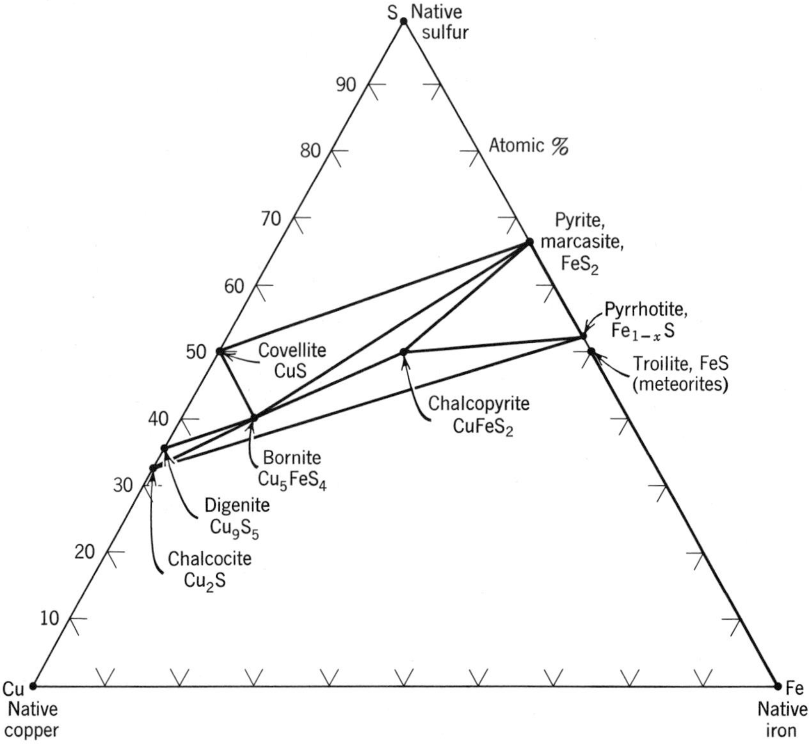

FIG. 7.16. Some of the most common sulfides represented in the Cu-Fe-S system. Many of these sulfides (e.g., bornite and chalcopyrite) show some solid solution of especially Cu and Fe; this is not shown in the diagram. Tielines connect commonly occurring pairs of minerals. Triangles indicate coexistences of three sulfides. The Fe-FeS coexistence is common in iron meteorites.

Good crystals of bornite have been found associated with crystals of chalcocite at Bristol, Connecticut, and in Cornwall, England. Found in large masses in Chile, Peru, Bolivia, and Mexico. In the United States it has been an ore mineral at Magma mine, Pioneer, Arizona; Butte, Montana; Engels mine, Plumas County, California; Halifax County, Virginia; and Superior, Arizona.

Use. An ore of copper.

Name. Bornite was named after the German mineralogist von Born (1742–1791).

GALENA—PbS

Crystallography. Isometric; $4/m\bar{3}2/m$. The most common form is the cube, sometimes truncated by the octahedron (Figs. 7.17 and 7.18). Dodecahedron and trisoctahedron rare.

$Fm3m$; $a = 5.936\text{Å}$, $Z = 4$. d's: 3.44(9), 2.97(10), 2.10(10), 1.780(9), 1.324 (10).

Physical Properties. *Cleavage* perfect {001}. **H** $2\frac{1}{2}$. **G** 7.4–7.6. *Luster* bright metallic. *Color* and *streak* lead-gray.

Composition and Structure. Pb 86.6, S 13.4%. Silver is usually present as admixtures of silver minerals such as acanthite or tetrahedrite but also in solid solution. Inclusions probably also account for the small amounts of Zn, Cd, Sb, As, and Bi that may be present. Selenium may substitute for sulfur and a complete series of PbS-PbSe has been reported. Galena has a NaCl type structure with Pb in place of Na and S in place of Cl.

FIG. 7.17. Galena crystals.

FIG. 7.18. (a) Galena crystals, Joplin, Missouri. (b) Galena on sphalerite and dolomite crystals.

Diagnostic Features. Galena can be easily recognized by its good cleavage, high specific gravity, softness, and lead-gray streak. Fusible at 2. For charcoal test see Table 4.18.

Alteration. By oxidation galena is converted into anglesite, $PbSO_4$, and cerussite, $PbCO_3$.

Occurrence. Galena is a very common metallic sulfide, found in veins associated with sphalerite, pyrite, marcasite, chalcopyrite, cerussite, anglesite, dolomite, calcite, quartz, barite, and fluorite. When found in hydrothermal veins galena is frequently associated with silver minerals; it often contains silver itself and so becomes an important silver ore. A large part of the supply of lead comes as a secondary product from ores mined chiefly for their silver. In a second type of deposit typified by the lead-zinc ores of the Mississippi Valley, galena, associated with sphalerite, is found in veins, open space filling, or replacement bodies in limestones. These are low-temperature deposits, located at shallow depths, and usually contain little silver. Galena is also found in contact metamorphic deposits, in pegmatites and as disseminations in sedimentary rocks.

Famous world localities are Freiberg, Saxony; the Harz Mountains; Westphalia and Nassau; Pribram, Bohemia; Cornwall, Derbyshire, and Cumberland, England; Sullivan mine, British Columbia; and Broken Hill, Australia.

In the United States there are many lead-producing districts; only the most important are mentioned here. In the Tri-State district of Missouri, Kansas, and Oklahoma centering around Joplin, Missouri, galena is associated with zinc ores and is found in irregular veins and pockets in limestone and chert. It is found in a similar manner but in smaller amount in Illinois, Iowa, and Wisconsin. The deposits of southeast Missouri where galena is disseminated through limestone are particularly productive. In the Coeur d'Alene district, Idaho, galena is the chief ore mineral in the lead-silver vein deposits. Lead is produced in Utah at Bingham and from the silver deposits of the Tintic and Park City districts; and in Colorado, chiefly from the lead-silver ores of the Leadville district.

Use. Practically the only source of lead and an important ore of silver. The largest use of lead is in storage batteries, but nearly as much is consumed in making metal products such as pipe, sheets, and shot. Lead is converted into the oxides (*litharge,* PbO, and *minimum,* Pb_3O_4) used in making glass and in giving a glaze to earthenware, and into white lead (the basic carbonate), the principal ingredient of many paints. However, this latter use is diminishing because of the poisonous nature of lead-based paints. Diminishing also is its use in gasoline antiknock additives because of environmental restrictions. Lead is a principal metal of several

alloys as solder (lead and tin), type metal (lead and antimony), and low-melting alloys (lead, bismuth, and tin). Lead is used as shielding around radioactive materials.

Name. The name galena is derived from the Latin *galena,* a name originally given to lead ore.

Similar Species. *Altaite,* $PbTe$, and *alabandite,* MnS, like galena, have a NaCl type of structure.

SPHALERITE (*Zinc Blende*)—ZnS

Crystallography. Isometric; $\overline{4}3m$. Tetrahedron, dodecahedron, and cube common forms (Fig. 7.19), but the crystals, frequently highly complex and usually malformed or in rounded aggregates, often show polysynthetic twinning on {111}. Usually found in cleavable masses, coarse to fine granular. Compact, botryoidal, cryptocrystalline.

$F\overline{4}3m$; $a = 5.41$ Å; $Z = 4$. $d's$: 3.12(10), 1.910(8), 1.631(7), 1.240(4), 1.106(5).

Physical Properties. *Cleavage,* {011} perfect but some sphalerite is too fine grained to show cleavage. H $3\frac{1}{2}$–4. G 3.9–4.1. *Luster* nonmetallic and resinous to submetallic; also adamantine. *Color* colorless when pure, and green when nearly so. Commonly yellow, brown to black, darkening with increase in iron. Also red (ruby zinc). Transparent to translucent. *Streak* white to yellow and brown.

Composition and Structure. Zn 67, S 33% when pure. Nearly always contains some Fe with the amount of Fe dependent on the temperature and chemistry of the environment. If Fe is in excess as indicated by association with pyrrhotite the amount of FeS in sphalerite can reach 50 mole percent (see

FIG. 7.19. Sphalerite crystals.

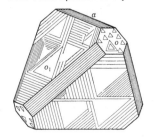
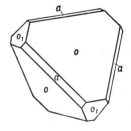

also Table 4.12). If sphalerite and pyrrhotite crystallize together, the amount of iron is an indication of the temperature of formation, and sphalerite can be used as a geologic thermometer. Mn and Cd are usually present in small amounts in solid solution (see Table 4.12).

The structure of sphalerite is similar to that of diamond with one half of the carbon atoms replaced by Zn and the other half by S (see Figs. 7.8a and 7.12a). Sphalerite is considered the low-temperature cubic polymorph of ZnS, and wurtzite the high-temperature polymorph stable above 1020°C at 1 atmosphere pressure. The *wurtzite* polymorph (space group $P6_3mc$), with Zn atoms in hexagonal closest packing (see Fig. 7.20), shows a very large number of stacking sequences, referred to as wurtzite *polytypes,* which differ in the length of their *c* axes. For example the *4H* polytype has a *c* axis of 12.46 Å and the *8H* polytype a *c* axis of 24.96 Å (*H* = hexagonal).

Diagnostic Features. Sphalerite can be recognized by its striking resinous luster and perfect cleavage. The dark varieties (black jack) can be told by the reddish-brown streak, always lighter than the massive mineral. Pure ZnS is infusible; it becomes fusible, with difficulty, with an increase in the amount of Fe. Diagnostic sulfur tests are given in Tables 4.18 and 4.19. A Zn test is given on page 179.

Occurrence. Sphalerite, the most important ore mineral of zinc, is extremely common. Its occurrence and mode of origin are similar to those of galena, with which it is commonly found. In the shallow seated lead-zinc deposits of the Tri-State district of Missouri, Kansas, and Oklahoma these

minerals are associated with marcasite, chalcopyrite, calcite, and dolomite. Sphalerite with only minor galena occurs in hydrothermal veins and replacement deposits associated with pyrrhotite, pyrite, and magnetite. Sphalerite is also found in veins in igneous rocks and in contact metamorphic deposits.

Zinc is mined in significant amounts in over 40 countries. Although in a few places the ore minerals are hemimorphite and smithsonite and at Franklin, New Jersey, willemite, zincite, and franklinite, most of the world's zinc comes from sphalerite. The principal producing countries are: Canada, U.S.S.R., United States, Australia, Peru, Mexico, and Japan. In the United States nearly 60% of the zinc is produced east of the Mississippi River with Tennessee, New York, Pennsylvania, and New Jersey the principal producing states. In the western United States, Idaho, Colorado, and Utah are the chief producers. The mines in the Tri-State district, formerly major zinc producers, are now largely exhausted.

Use. The most important ore of zinc. The chief uses for metallic zinc, or *spelter,* are in galvanizing iron; making brass, an alloy of copper and zinc; in electric batteries; and as sheet zinc. Zinc oxide, or zinc white, is used extensively for making paint. Zinc chloride is used as a preservative for wood. Zinc sulfate is used in dyeing and in medicine. Sphalerite also serves as the most important source of cadmium, indium, gallium, and germanium.

Name. Sphalerite comes from the Greek meaning *treacherous.* Blende because, although often resembling galena, it yielded no lead; from the German word meaning *blind* or *deceiving.*

Similar Species. *Greenockite,* CdS, a rare mineral mined as a source of cadmium, is isostructural in two polymorphic forms with sphalerite and wurtzite.

FIG. 7.20. The wurtzite type polymorph of ZnS. See Fig. 7.12a for the sphalerite polymorph.

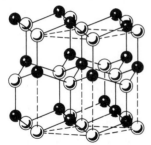

CHALCOPYRITE—CuFeS₂

Crystallography. Tetragonal; $\bar{4}2m$. Commonly tetrahedral in aspect with the disphenoid p {112} dominant. Other forms shown in Fig. 7.21 are rare. Usually massive.

Angles: $p(112)$ $\wedge$ $p'(1\bar{1}2)$ = 108°40′,

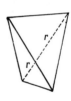
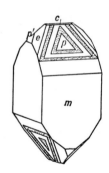

FIG. 7.21. Chalcopyrite crystals.

$c(001) \wedge z(011) = 63°06'$, $c(001) \wedge p(112) = 54°20'$, $c(001) \wedge e(102) = 44°34'$.

$I42d$; $a = 5.25$, $c = 10.32$ Å, $a:c = 1:1.966$; $Z = 4$. $d's$: 3.03(10), 1.855(10), 1.586(10), 1.205(8), 1.074(8).

Physical Properties. H $3\frac{1}{2}$–4. **G** 4.1–4.3. *Luster* metallic. *Color* brass-yellow; often tarnished to bronze or iridescent. *Streak* greenish black. Brittle.

Composition and Structure. Cu 34.6, Fe 30.4, S 35.0%. It deviates very little from ideal $CuFeS_2$. See Fig. 7.16 for calcopyrite composition in the Cu-Fe-S system. Its structure can be regarded as a derivative of the sphalerite structure in which half the Zn is replaced by Cu and the other half by Fe. This leads to a doubling of the unit cell (see Fig. 7.12b).

Diagnostic Features. Recognized by its brass-yellow color and greenish-black streak. Distinguished from pyrite by being softer than steel and from gold by being brittle. Known as "fool's gold," a term also applied to pyrite. On charcoal it fuses at 2 to a magnetic globule, and emits the odor of sulfur dioxide. Diagnostic sulfur tests are given in Tables 4.18 and 4.19 and a Cu flame test in Table 4.20.

Occurrence. Chalcopyrite is the most widely occurring copper mineral and one of the most important sources of that metal. Most sulfide ores contain some chalcopyrite but the most important economically are the hydrothermal vein and replacement deposits. In the low-temperature deposits as in the Tri-State district, it occurs as small crystals associated with galena, sphalerite, and dolomite. Associated with pyrrhotite and pentlandite, it is the chief copper mineral in the ores of Sudbury,

Ontario, and similar high-temperature deposits. Chalcopyrite is the principal primary copper mineral in the "porphyry-copper" deposits. Also occurs as an original constituent of igneous rocks; in pegmatite dikes; in contact metamorphic deposits; and disseminated in schistose rocks. It may carry gold or silver and become an ore of those metals. Often in subordinate amount with large bodies of pyrite, making them serve as low-grade copper ores.

A few of the localities at which chalcopyrite is the chief ore of copper are: Cornwall, England; Falun, Sweden; Schemnitz, Czechoslovakia; Schlaggenwald, Bohemia; Freiberg, Saxony; Rio Tinto, Spain; South Africa; Zambia, and Chile. Found widely in the United States but usually with other copper minerals in equal or greater amount; found at Butte, Montana; Bingham, Utah; Jerome, Arizona; Ducktown, Tennessee; and various districts in California, Colorado, and New Mexico. In Canada the most important occurrences of chalcopyrite are at Sudbury, Ontario, and in the Rouyn district, Quebec.

Alteration. Chalcopyrite is the principal source of copper for the secondary minerals malachite, azurite, covellite, chalcocite, and cuprite. Concentrations of copper in the zone of supergene enrichment are often the result of such alteration and removal of copper in solution with its subsequent deposition.

Use. Important ore of copper.

Name. Derived from Greek word meaning *brass* and from *pyrites*.

Similar Species. *Stannite*, Cu_2FeSnS_4, tetragonal, is a rare mineral and a minor ore of tin. The crystal structure of stannite can be derived from that of chalcopyrite by substitution of Fe and Sn in place of part of the Cu in chalcopyrite, see page 235.

PYRRHOTITE (*Magnetic Pyrites*)—$Fe_{1-x}S$

Crystallography. Monoclinic; $2/m$, for low-temperature form stable below about 250°C; hexagonal, $6/m3/m2/m$, for high-temperature form, stable above about 300°C. Hexagonal crystals, usually tabular but in some cases pyramidal (Fig. 7.22), indicate formation as the high-temperature polymorph.

Angles: $c(0001) \wedge s(10\bar{1}2) = 45°08'$, $c(0001) \wedge u(20\bar{2}1) = 76°0'$.

$A2/a$; $a = 12.78$, $b = 6.86$, $c = 11.90$, $\beta = 117°17'$, $a:b:c = 1.863:1:1.734$; $Z = 4$. No specific d's are given here because there are several monoclinic forms.

$C6/mmc$; $a = 3.44$, $c = 5.73$ Å, $a:c = 1:1.666$; $Z = 2$. d's: 2.97(6), 2.63(8), 1.06(10), 1.718(6), 1.045(8).

Physical Properties. H 4. G 4.58–4.65. *Luster* metallic. *Color* brownish bronze. *Streak* black. Magnetic, but varying in intensity; the greater the amount of iron, the lesser the magnetism. Opaque.

Composition and Structure. Most pyrrhotites have a deficiency of iron with respect to sulfur, as indicated by the formula $Fe_{1-x}S$, with x between 0 and 0.2. This is referred to as omission solid solution (see page 162). The structure of pyrrhotite is a complex derivative of the NiAs structure type. The sulfur atoms are arranged in approximate hexagonal closest packing. A large number of ordered structures exist for this nonstoichiometric sulfide. Several monoclinic forms are stable up to about 250°C whereas a hexagonal form is stable from about 300°C to the melting temperature of 1190°C. The structural relations within these various poly-

morphs is complex. The mineral *troilite* is essentially FeS in composition.

Diagnostic Features. Recognized usually by its massive nature, bronze color, and magnetism. Fusible at 3. When heated on charcoal, it gives off the odor of sulfur dioxide and becomes strongly magnetic.

Occurrence. Pyrrhotite is commonly associated with basic igneous rocks, particularly norites. It occurs in them as disseminated grains or, as at Sudbury, Ontario, as large masses associated with pentlandite, chalcopyrite, or other sulfides. At Sudbury vast tonnages of pyrrhotite are mined principally for the copper, nickel, and platinum that are extracted from associated minerals. Pyrrhotite is also found in contact metamorphic deposits, in vein deposits, and in pegmatites.

Large quantities are known in Finland, Norway, and Sweden; U.S.S.R.; in Germany from Andreasberg in the Harz Mountains; at Schneeberg, Saxony; and Bodenmais, Bavaria. In the United States in crystals from Standish, Maine; at the Gap mine, Lancaster County, Pennsylvania; and in considerable amount at Ducktown, Tennessee.

Use. It is mined for its associated nickel, copper, and platinum. At Sudbury, Ontario, it is also a source of sulfur and an ore of iron.

Name. The name pyrrhotite comes from the Greek meaning *reddish*.

Similar Species. *Troilite*, FeS, common in iron meteorites.

Niccolite—NiAs

Crystallography. Hexagonal; $6/m2/m2/m$. Rarely in tabular crystals. Usually massive, reniform with columnar structure.

$C6/mmc$; $a = 3.61$, $c = 5.02$ Å; $a:c = 1:1.391$; $Z = 2$. d's: 2.66(10), 1.961(9), 1.811(8), 1.328(3), 1.071(4).

Physical Properties. H 5–5½. G 7.78. *Luster* metallic. *Color* pale copper-red (hence called *copper nickel*), with gray to blackish tarnish. *Streak* brownish-black. Opaque.

Composition and Structure. Ni 43.9, As 56.1%. Usually a little Fe, Co, and S. As frequently

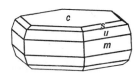

replaced in part by Sb. The structure of NiAs is based on hexagonal closest packing of arsenic atoms giving a sequence ABABAB . . . . The nickel atoms are coordinated to six closest As atoms.

Diagnostic Features. Characterized by its copper-red color. Fusible at 2. On charcoal gives diagnostic As test (Table 4.18). Gives nickel test with dimethylglyoxime.

Alteration. Quickly alters to *annabergite* (green nickel bloom, $Ni_3(AsO_4)_2 \cdot 8H_2O$) in moist atmosphere.

Occurrence. Niccolite, with other nickel arsenides and sulfides, pyrrhotite, and chalcopyrite, frequently occurs in, or is associated with, norites. Also found in vein deposits with cobalt and silver minerals.

Found in Germany in the silver mines of Saxony, the Harz Mountains, in Hessen-Nassau; and at Cobalt, Ontario.

Use. A minor ore of nickel.

Name. The first name of this mineral, *kupfernickel,* gave the name *nickel* to the metal. Niccolite is from the Latin for nickel.

Similar Species. *Breithauptite,* NiSb, is isostructural with niccolite with similar occurrence and association.

Millerite—NiS

Crystallography. Hexagonal-R; $\overline{3}2/m$ (low-temperature polymorph stable below 379°C). Usually in hairlike tufts and radiating groups of slender to capillary crystals (*capillary pyrites*). In velvety incrustations. Rarely in coarse, cleavable masses.

$R3m$; $a = 9.62$, $c = 3.16$ Å; $a:c = 1:0.328$; $Z = 9$. $d's$: 4.77(8), 2.75(10), 2.50(6), 2.22(6), 1.859(10).

Physical Properties. *Cleavage* {$10\overline{1}1$}, {$01\overline{1}2$} good. **H** 3–3½. **G** 5.5 ± 0.2. *Luster* metallic. *Color* pale brass-yellow; with a greenish tinge when in fine hairlike masses. *Streak* black, somewhat greenish.

Composition and Structure. Ni 64.7, S 35.3%. The low-temperature form, stable below 379°C shows little if any solid solution. Ni and S are both in 5 coordination in the low-temperature form. The high-temperature form (stable above 379°C) has an NiAs type structure and exhibits considerable metal deficiency, $Ni_{1-x}S$, similar to pyrrhotite.

Diagnostic Features. Characterized by its capillary crystals and distinguished from minerals of similar color by nickel tests. Fusible at 1½–2 to a magnetic globule. Gives odor of sulfur dioxide when heated on charcoal.

Occurrence. Millerite forms as a low-temperature mineral often in cavities and as an alteration of other nickel minerals, or as crystal inclusions in other minerals.

Occurs in various localities in Saxony, Westphalia, and Hessen-Nassau, and in Bohemia. In the United States, it is found with hematite and ankerite at Antwerp, New York; with pyrrhotite at the Gap mine, Lancaster County, Pennsylvania; in geodes in limestone in St. Louis, Missouri; Keokuk, Iowa; and Milwaukee, Wisconsin. In coarse, cleavable masses it is a major ore mineral at the Marbridge Mine, Lamotte Township, Quebec.

Use. A subordinate ore of nickel.

Name. In honor of the mineralogist, W. H. Miller (1801–1880), who first studied the crystals.

Pentlandite—(Fe, Ni)$_9$S$_8$

Crystallography. Isometric; $4/m\overline{3}2/m$. Massive, usually in granular aggregates with octahedral parting.

$Fm3m$; $a = 10.07$ Å; $Z = 4$. $d's$: 5.84(2), 3.04(6), 2.92(2), 2.31(3), 1.781(10).

Physical Properties. Parting on {111}. **H** 3½–4. **G** 4.6–5.0. Brittle. *Luster* metallic. *Color* yellowish-bronze. *Streak* light bronze-brown. Opaque. Nonmagnetic.

Composition and Structure. (Fe, Ni)$_9$S$_8$. Usually the ratio of Fe:Ni is close to 1:1. Commonly contains small amounts of Co. A rather complicated face-centered cubic structure with the metal atoms in octahedral as well as tetrahedral coordination with sulfur. Pure (Fe, Ni)$_9$S$_8$, without Co, is stable up to 610°C in the Fe-Ni-S system. Pentlandite with as much as 40.8 weight percent Co is stable up to 746°C. Pentlandite commonly occurs as exsolution lamellae within pyrrhotite.

Diagnostic Features. Pentlandite closely resembles pyrrhotite in appearance but can be distinguished from it by octahedral parting and lack of magnetism. Fusible at $1\frac{1}{2}$–2. On heating gives odor of sulfur dioxide and becomes magnetic. Gives nickel test with dimethylglyoxime.

Occurrence. Pentlandite usually occurs in basic igneous rocks where it is commonly associated with other nickel minerals and pyrrhotite, and chalcopyrite, and has probably accumulated by magmatic segregation.

Found at widely separated localities in small amounts but its chief occurrences are in Canada where, associated with pyrrhotite, it is the principal source of nickel at Sudbury, Ontario, and the Lynn Lake area, Manitoba. It is also an important ore mineral in similar deposits in the Petsamo district of U.S.S.R.

Use. The principal ore of nickel. The chief use of nickel is in steel. Nickel steel contains $2\frac{1}{2}$–$3\frac{1}{2}$% nickel, which greatly increases the strength and toughness of the alloy, so that lighter machines can be made without loss of strength. Nickel is also an essential constituent of stainless steel. The manufacture of Monel metal (68% Ni, 32% Cu) and Nichrome (38–85% Ni) consumes a large amount of the nickel produced. Other alloys are German silver (Ni, Zn, and Cu); metal for coinage—the 5-cent coin of the United States is 25% Ni and 75% Cu, low-expansion metals for watch springs and other instruments. Nickel is used in plating; although chromium now largely replaces it for the surface layer, nickel is used for a thicker underlayer.

Name. After J. B. Pentland, who first noted the mineral.

Covellite—CuS

Crystallography. Hexagonal; $6/m2/m2/m$. Rarely in tabular hexagonal crystals. Usually massive as coatings or disseminations through other copper minerals.

$P6_3/mmc$; $a = 3.80$, $c = 16.36$ Å; $a:c = 1:4.305$; $Z = 6$. d's: 3.06(4), 2.83(6), 2.73(10), 1.899(8), 1.740(5).

Physical Properties. *Cleavage* {0001} perfect giving flexible plates. **H** $1\frac{1}{2}$–2. **G** 4.6–4.76. *Luster* metallic. *Color* indigo-blue or darker. *Streak* lead-gray to black. Often iridescent. Opaque.

Composition and Structure. Cu 66.4, S 33.6%. A small amount of Fe may be present. Covellite has a rather complex structure (see Fig. 7.13). One type of Cu atom is in tetrahedral coordination with S, with the tetrahedra sharing corners to form layers. A second type of Cu is in trigonal coordination with S to build planar layers. The excellent {0001} cleavage is parallel to this layer structure. Covellite is stable up to 507°C, its temperature of decomposition.

Diagnostic Features. Characterized by the indigo-blue color, micaceous cleavage yielding flexible plates, and association with other copper sulfides. Fusible at $2\frac{1}{2}$. For sulfur tests see Tables 4.18 and 4.19; green copper flame test with HCl.

Occurrence. Covellite is not an abundant mineral but is found in most copper deposits as a supergene mineral, usually as a coating, in the zone of sulfide enrichment. It is associated with other copper minerals, principally chalcocite, chalcopyrite, bornite, and enargite, and is derived from them by alteration. Primary covellite is known but uncommon.

Found at Bor, Serbia, Yugoslavia; and Leogang, Austria. In large iridescent crystals from the Calabona mine, Alghero, Sardinia. In the United States covellite is found in appreciable amounts at Butte, Montana; Summitville, Colorado; and La Sal district, Utah. Formerly found at Kennecott, Alaska.

Use. A minor ore of copper.

Name. In honor of N. Covelli (1790–1829), the discoverer of the Vesuvian covellite.

CINNABAR—HgS

Crystallography. Hexagonal-R; 32 (low-temperature polymorph stable below approximately 344°C). High-temperature form, known as *metacinnabar*, isometric; $\bar{4}3m$. Crystals of cinnabar usually rhombohedral, often in penetration twins. Trapezohedral faces rare. Usually fine granular, massive; also earthy, as incrustations and disseminations through the rock. Metacinnabar crystals are tetrahedral, with rough faces; usually massive.

$P3_121$ or $P3_221$; $a = 4.146$, $c = 9.497$ Å;

$a:c = 1:2.291$; $Z = 3$. $d's$: 3.37(10), 3.16(8), 2.87(10), 2.07(8), 1.980(8).

$F\bar{4}3m$; $a = 5.852$; $Z = 4$. $d's$: 3.38(10), 2.93(3), 2.07(5), 1.764(4).

Physical Properties. *Cleavage* $\{10\bar{1}0\}$ perfect. **H** $2\frac{1}{2}$. **G** 8.10. *Luster* adamantine when pure to dull earthy when impure. *Color* vermilion-red when pure to brownish-red when impure. *Streak* scarlet. Transparent to translucent. *Metacinnabar* has a metallic luster and grayish black color. *Hepatic cinnabar* is an inflammable liver-brown variety of cinnabar with bituminous impurities; usually granular or compact, sometimes with a brownish streak.

Composition and Structure. Hg 86.2, S 13.8%, with small variations in Hg content. Traces of Se and Te may replace S. Frequently impure from admixture of clay, iron oxide, bitumen. The structure of cinnabar, which is different from that of any other sulfide, is based on infinite spiral Hg-S-Hg chains which extend along the c axis; it can be represented by one of two enantiomorphic space groups (see above). *Metacinnabar,* isometric, is stable above about 344°C. Some Hg deficiency is found in this polymorph and its composition may be represented as $Hg_{1-x}S$, with x ranging from 0 to 0.08.

Diagnostic Features. Recognized by its red color and scarlet streak, high specific gravity, and cleavage. Wholly volatile before the blowpipe. For closed tube tests see Table 4.19.

Occurrence. Cinnabar is the most important ore of mercury but is found in quantity at comparatively few localities. Occurs as impregnations and as vein fillings near recent volcanic rocks and hot springs and evidently deposited near the surface from solutions which were probably alkaline. Associated with pyrite, marcasite, stibnite, and sulfides of copper in a gangue of opal, chalcedony, quartz, barite, calcite, and fluorite.

The important localities for the occurrence of cinnabar are at Almaden, Spain; Idria, Yugoslavia; Huancavelica in southern Peru; and the provinces of Kweichow and Hunan, China. In the United States the important deposits are in California at New Idria in San Benito County, in Napa County, and at New Almaden in Santa Clara County. It also occurs in Nevada, Utah, Oregon, Arkansas, Idaho, and Texas.

Use. The only important source of mercury. The principal uses of mercury are in electrical apparatus, industrial control instruments, electrolytic preparation of chlorine and caustic soda, and in mildew proofing of paint. Other lesser but important uses are in dental preparations, scientific instruments, drugs, catalysts, and in agriculture. Formerly a major use of mercury was in the amalgamation process for recovering gold and silver from their ores, but amalgamation has been essentially abandoned in favor of other methods of extraction.

Name. The name cinnabar is supposed to have come from India, where it is applied to a red resin.

REALGAR—AsS

Crystallography. Monoclinic; $2/m$. Found in short, vertically striated, prismatic crystals (Fig. 7.23). Frequently coarse to fine granular and often earthy and as an incrustation.

Angles: $b(010) \wedge m(110) = 56°36'$, $b(010) \wedge l(120) = 37°10'$, $c(001) \wedge n(011) = 24°56'$, $c(001) \wedge z(101) = 29°25'$.

$P2_1/n$; $a = 9.29$, $b = 13.53$, $c = 6.57$ Å; $a:b:c = 0.687:1:0.486$, $\beta = 106°33'$; $Z = 16$. $d's$: 5.40(10), 3.19(9), 2.94(8), 2.73(8), 2.49(5).

Physical Properties. *Cleavage* $\{010\}$ good. **H** $1\frac{1}{2}$–2. **G** 3.48. Sectile. *Luster* resinous. *Color* and *streak* red to orange. Translucent to transparent.

FIG. 7.23. Realgar.

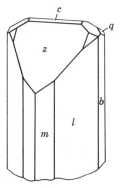

Composition and Structure. As 70.1; S 29.9%. The realgar structure contains ring-like groups of As_4S_4, somewhat similar to the rings of S_8 in native sulfur. Each As is bonded covalently to another arsenic as well as to two sulfur atoms.

Diagnostic Features. Realgar is distinguished by its red color, resinous luster, orange-red streak, and almost invariable association with orpiment. Fusible at 1. See Table 4.18 for reaction on charcoal.

Alteration. On long exposure to light disintegrates to a reddish-yellow powder.

Occurrence. Realgar is found in veins of lead, silver, and gold ores associated with orpiment, other arsenic minerals, and stibnite. It also occurs as a volcanic sublimation product and as a deposit from hot springs.

Realgar is found associated with silver and lead ores in Hungary, Bohemia, and Saxony. Found in good crystals at Nagyág, Transylvania; Binnenthal, Switzerland; and Allchar, Macedonia. In the United States realgar is found at Mercur, Utah; at Manhattan, Nevada; and deposited from the geyser waters in the Norris Geyser Basin, Yellowstone National Park.

Use. Realgar was used in fireworks to give a brilliant white light when mixed with saltpeter and ignited. Today, artificial arsenic sulfide is used for this purpose. It was formerly used as a pigment.

Name. The name is derived from the Arabic, Rahj al ghar, *powder of the mine.*

ORPIMENT—As₂S₃

Crystallography. Monoclinic; $2/m$. Crystals small tabular or short prismatic, and rarely distinct; many pseudo-orthorhombic. Usually in foliated or columnar masses (Fig. 7.24).

Angles: $b(010) \wedge m(110) = 39°59'$, $b(010) \wedge u(210) = 59°12'$, $b(010) \wedge x(\overline{3}11) = 73°27'$.

$P2_1/n$; $a = 11.49$, $b = 9.59$, $c = 4.25$ Å; $a:b:c = 1.198:1:0.443$, $\beta = 90°27'$; $Z = 4$. $d's$: 4.78(10), 2.785(4), 2.707(6), 2.446(6), 2.085(4).

Physical Properties. *Cleavage* {010} perfect; cleavage laminae flexible but not elastic. Sectile. **H** $1\frac{1}{2}$–2. **G** 3.49. *Luster* resinous, pearly on cleavage

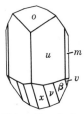

FIG. 7.24. Orpiment.

Allchar, Macedonia

face. *Color* lemon-yellow. *Streak* pale yellow. Translucent.

Composition and Structure. As 61, S 39%. Contains up to 2.7% Sb. In the structure of orpiment trigonal pyramids of AsS_3 can be recognized that share corners to form six-membered rings. These rings are linked to produce a corrugated As_2S_3 layer structure. The bonds within the layers are essentially covalent while those between the layers are residual in nature. The perfect {010} cleavage is parallel to these layers.

Diagnostic Features. Characterized by its yellow color and foliated structure. Distinguished from sulfur by its perfect cleavage. Gives the same tests as realgar.

Occurrence. Orpiment is a rare mineral, associated usually with realgar and formed under similar conditions. Found in various places in Rumania, Kurdistan, Peru, and Japan. In the United States it occurs at Mercur, Utah, and at Manhattan, Nevada. Deposited with realgar from geyser waters in the Norris Geyser Basin, Yellowstone National Park.

Use. Used in dyeing and in a preparation for the removal of hair from skins. Artificial arsenic sulfide is largely used in place of the mineral. Both realgar and orpiment were formerly used as pigments but this use has been discontinued because of their poisonous nature.

Name. Derived from the Latin, *auripigmentum*, "golden paint," in allusion to its color and because the substance was supposed to contain gold.

STIBNITE—Sb₂S₃

Crystallography. Orthorhombic; $2/m2/m2/m$. Slender prismatic habit, prism zone vertically

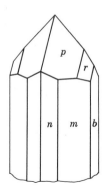

FIG. 7.25. Stibnite.

striated. Crystals often steeply terminated (Fig. 7.25) and sometimes curved or bent (Fig. 7.26). Often in radiating crystal groups or in bladed forms with prominent cleavage. Massive, coarse to fine granular.

Angles: $b(010) \wedge m(110) = 45°12'$; $b(010) \wedge n(210) = 63°36'$; $s(111) \wedge m(110) = 64°17'$; $p(331) \wedge m(110) = 34°42'$.

Pbnm; $a = 11.22$, $b = 11.30$, $c = 3.84$ Å; $a:b:c = 0.993:1:0.340$; $Z = 4$. *d's*: 5.07(4), 3.58(10), 2.76(3), 2.52(4), 1.933(5).

Physical Properties. *Cleavage* {010} perfect, showing striations parallel to [100]. **H** 2.0. **G** 4.52–4.62. *Luster* metallic, splendent on cleavage surfaces. *Color* and *streak* lead-gray to black. Opaque.

Composition and Structure. Sb 71.4, S 28.6%. May carry small amounts of Au, Ag, Fe, Pb, and Cu. The structure of stibnite is composed of bands of closely bonded Sb and S atoms which are parallel to the c axis. Sb-S distances in the bands are about 2.5Å; these represent covalent linking. Dis-tances between adjoining bands are larger, 3.2 Å. The long, striated prisms (//c) of stibnite are parallel to these structural bands. The excellent {010} cleavage occurs between Sb-S bands.

Diagnostic Features. Characterized by its easy fusibility, bladed habit, perfect cleavage in one direction, lead-gray color, and soft black streak. Fusible at 1. For charcoal and open tube tests for Sb and S see Table 4.18.

Occurrence. Stibnite is found in low-temperature hydrothermal veins or replacement deposits and in hot spring deposits. It is associated with other antimony minerals that have formed as the product of its decomposition, and with galena, cinnabar, sphalerite, barite, realgar, orpiment, and gold.

The finest crystals of stibnite have come from the province of Iyo, Island of Shikoku, Japan. The world's most important producing district is in the province of Hunan, China. It occurs also in Algeria, Borneo, Bolivia, Peru, and Mexico. Found in quantity at only a few localities in the United States, the chief deposits being in California, Nevada, and Idaho.

Use. The chief ore of antimony. The metal is used in various alloys, as antimonial lead for storage batteries, type metal, pewter, babbitt, britannia metals, and antifriction metal. The sulfide is employed in fireworks, matches, percussion caps, vulcanizing rubber, and in medicines. Antimony trioxide is used as a pigment and for making glass.

Name. The name stibnite comes from an old Greek word that was applied to the mineral.

Similar Species. *Bismuthinite*, Bi_2S_3, is a rare mineral isostructural with stibnite and with similar physical properties.

FIG. 7.26. Curved stibnite crystal, Ischinokowa, Japan.

PYRITE—FeS$_2$

Crystallography. Isometric; $2/m\bar{3}$. Frequently in crystals (Fig. 7.27). The most common forms are the cube, the faces of which are usually striated, the pyritohedron, and the octahedron. Figure 7.27f shows a penetration twin, known as the *iron cross* with [011] the twin axis. Also massive, granular, reniform, globular, and stalactitic.

$Pa3$; a = 5.42 Å, Z = 4. d's: 2.70(7), 2.42(6), 2.21(5), 1.917(4), 1.632(10).

Physical Properties. *Fracture* conchoidal. Brittle. **H** 6–6$\frac{1}{2}$ (unusually hard for a sulfide). **G** 5.02. *Luster* metallic, splendent. *Color* pale brass-yellow; may be darker because of tarnish. *Streak* greenish or brownish black. Opaque. Paramagnetic.

Composition and Structure. Fe 46.6, S 53.4%. May contain small amounts of Ni and Co. Some analyses show considerable Ni, and a complete solid solution series exists between pyrite and *bravoite*, (Fe, Ni)S$_2$. Frequently carries minute quantities of Au and Cu as microscopic impurities. The structure of pyrite can be considered as a modified NaCl type structure with Fe in the Na and S$_2$ in the Cl positions (Fig. 7.14a). FeS$_2$ occurs in two polymorphs, pyrite and *marcasite* (see p. 252).

Diagnostic Features. Distinguished from chalcopyrite by its paler color and greater hardness, from gold by its brittleness and hardness, and from marcasite by its deeper color and crystal form. Fusible at 2$\frac{1}{2}$–3 to a magnetic globule. Yields much sulfur in the closed tube and sulfur dioxide when heated in the open tube.

Alteration. Pyrite is easily altered to oxides of iron, usually limonite. In general, however, it is much more stable than marcasite. Pseudomorphic crystals of limonite after pyrite are common. Pyrite veins are usually capped by a cellular deposit of limonite, termed *gossan*. Rocks that contain pyrite are unsuitable for structural purposes because the ready oxidation of pyrite would serve both to disintegrate the rock and to stain it with iron oxide.

Occurrence. Pyrite is the most common and widespread of the sulfide minerals. It has formed at both high and low temperatures, but the largest masses probably at high temperature. It occurs as a magmatic segregation, as an accessory mineral in igneous rock, and in contact metamorphic deposits and hydrothermal veins. Pyrite is a common mineral in sedimentary rocks, being both primary and secondary. It is associated with many minerals but

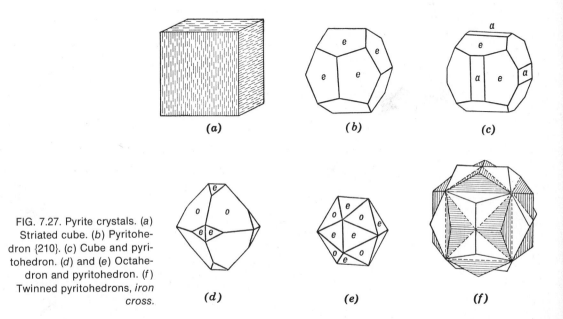

FIG. 7.27. Pyrite crystals. (a) Striated cube. (b) Pyritohedron {210}. (c) Cube and pyritohedron. (d) and (e) Octahedron and pyritohedron. (f) Twinned pyritohedrons, *iron cross*.

found most frequently with chalcopyrite, sphalerite, galena.

Large and extensively developed deposits occur at Rio Tinto and elsewhere in Spain and also in Portugal. Important deposits of pyrite in the United States are in Prince William, Louisa, and Pulaski counties, Virginia, where it occurs in large lenticular masses that conform in position to the foliation of the enclosing schists; in St. Lawrence County, New York, at the Davis Mine, near Charlemont, Massachusetts; and in various places in California, Colorado, and Arizona.

Use. Pyrite is often mined for the gold or copper associated with it. Because of the large amount of sulfur present in the mineral it is used as an iron ore only in those countries where oxide ores are not available. It's chief use is a source of sulfur for sulfuric acid and *copperas* (ferrous sulfate). Copperas is used in dyeing, in the manufacture of inks, as a preservative of wood, and as a disinfectant.

Name. The name *pyrite* is from a Greek word meaning *fire,* in allusion to the brilliant sparks emitted when struck by steel.

MARCASITE—FeS$_2$

Crystallography. Orthorhombic; $2/m2/m2/m$. Crystals commonly tabular {010}; less commonly prismatic [001] (Fig. 7.28). Often twinned, giving cockscomb and spear-shaped groups (Fig. 7.29). Usually in radiating forms. Often stalactitic, having an inner core with radiating structure and covered with irregular crystal groups. Also globular and reniform.

FIG. 7.28. Marcasite crystals.

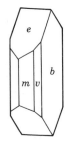 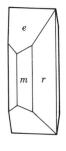

Angles: $b(010) \wedge m(110) = 50°40'$, $b(010) \wedge r(140) = 16°15'$, $e(101) \wedge e'(\overline{1}01) = 105°21'$.

Pmnn; $a = 4.45$, $b = 5.42$, $c = 3.39$ Å; $a:b:c = 0.821:1:0.625$. $Z = 2$. $d's$: 2.70(10), 2.41(6), 2.32(6), 1.911(5). 1.755(9).

Physical Properties. H 6–6½. G 4.89. *Luster* metallic. *Color* pale bronze-yellow to almost white on fresh fracture, hence called *white iron pyrites.* Yellow to brown tarnish. *Streak* grayish-black. Opaque.

Composition and Structure. Of constant composition, FeS$_2$, dimorphous with pyrite. The marcasite structure is shown in Fig. 7.14*b*. The configuration of nearest neighbor atoms is the same as in pyrite. The stability relationships of pyrite and marcasite are still unclear. Experimental evidence indicates that marcasite is metastable relative to pyrite and pyrrhotite above about 157°C. Geological occurrences of marcasite indicate a lower temperature of stability range than for pyrite which may occur in magmatic segregations.

Diagnostic Features. Usually recognized and distinguished from pyrite by its pale yellow color, its crystals or its fibrous habit, and the following chemical test. When fine powder is treated with cold nitric acid, and the solution is allowed to stand until vigorous action ceases and is then boiled, the mineral is decomposed with separation of sulfur. Pyrite treated in the same manner would have been completely dissolved. Blowpipe tests same as for pyrite.

Alteration. Marcasite usually disintegrates more easily than pyrite with the formation of ferrous sulfate and sulfuric acid. The white powder that forms on marcasite is *melanterite,* FeSO$_4$·7H$_2$O.

Occurrence. Marcasite is found in metalliferous veins, frequently with lead and zinc ores. It is less stable than pyrite, being easily decomposed, and is much less common. It is deposited at low temperatures from acid solutions and commonly formed under surface conditions as a supergene mineral. Marcasite most frequently occurs as replacement deposits in limestone, and often in concretions embedded in clays, marls, and shales.

Found abundantly in clay near Carlsbad and elsewhere in Bohemia; in various places in Saxony; and in the chalk marl of Folkestone and Dover, Eng-

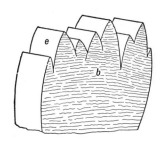

FIG. 7.29. "Cockscomb" marcasite.

land. In the United States marcasite is found with zinc and lead deposits of the Joplin, Missouri district; at Mineral Point, Wisconsin; and at Galena, Illinois.

Use. Marcasite is used to a slight extent as a source of sulfur.

Name. Derived from an Arabic word, at one time applied generally to pyrite.

MOLYBDENITE—MoS$_2$

Crystallography. Hexagonal; $6/m2/m2/m$. Crystals in hexagonal plates or short, slightly tapering prisms. Commonly foliated, massive, or in scales.

$P6_3/mmc$; $a = 3.16$, $c = 12.32$ Å; $a:c = 1:3.889$; $Z = 2$. $d's$: 6.28(10), 2.28(9), 1.824(6), 1.578(4), 1.530(4).

Physical Properties. *Cleavage* {0001} perfect, laminae flexible but not elastic. Sectile. **H** $1-1\frac{1}{2}$. **G** 4.62–4.73. Greasy feel. *Luster* metallic. *Color* lead-gray. *Streak* grayish-black. Opaque.

Composition and Structure. Mo 59.9, S 40.1%, of essentially constant composition. In the structure of molybdenite a sheet of Mo atoms is sandwiched between two sheets of S atoms, the three sheets together forming a layered structure. The bond strengths within the layer are much stronger than between the layers, giving rise to the excellent {0001} cleavage. MoS$_2$ occurs as two polytypes of which one is hexagonal (2*H*) and the other rhombohedral (3*R*).

Diagnostic Features. Resembles graphite but is distinguished from it by higher specific gravity; by a blue tone to its color, whereas graphite has a brown tinge. On glazed porcelain, it gives a greenish streak, graphite a black streak. Infusible. See Table 4.18 for a diagnostic reaction on plaster.

Alteration. Molybdenite alters to yellow *ferrimolybdite*, Fe$_2$(MoO$_4$)$_3$·8H$_2$O.

Occurrence. Molybdenite forms as an accessory mineral in certain granites; in pegmatites and aplites. Commonly in high-temperature vein deposits associated with cassiterite, scheelite, wolframite, and fluorite. Also in contact metamorphic deposits with lime silicates, scheelite, and chalcopyrite.

Occurs with the tin ores of Bohemia, from various places in Norway and Sweden, from England, China, Mexico, and New South Wales, Australia. In the United States molybdenite is found in many localities; at Blue Hill, Maine; Westmoreland, New Hampshire; in Okanogan County, Washington; Questa, New Mexico. From various pegmatites in Ontario, Canada. The bulk of the world's supply comes from Climax, Colorado, where molybdenite occurs in quartz veinlets in silicified granite with fluorite and topaz. Much molybdenum is produced at Bingham Canyon, Utah as a by-product of the copper mining.

Use. The principal ore of molybdenum.

Name. The name molybdenite comes from the Greek word meaning *lead*.

COBALTITE—(Co, Fe) AsS

Crystallography. Orthorhombic; *mm*2. Pseudoisometric with forms that appear isometric, that is, cubes and pyritohedrons with the faces striated as in pyrite. Also granular.

$Pca2_1$; $a = 5.58$, $b = 5.58$, $c = 5.58$ Å;

$a:b:c = 1:1:1$; $Z = 4$. $d's$: 2.77(8), 2.49(10), 2.27(9), 1.680(10), 1.490(8).

Physical Properties. *Cleavage* pseudocubic, perfect. Brittle. **H** $5\frac{1}{2}$. **G** 6.33. *Luster* metallic. *Color* silver-white, inclined to red. *Streak* grayish black.

Composition and Structure. Usually contains considerable Fe (maximum about 10%) and lesser amounts of Ni. *Gersdorffite,* NiAsS, and cobaltite form a complete solid solution series, but intermediate compositions are rare. The structure of cobaltite is closely related to that of pyrite (Fig. 7.14a) in which half the S_2 pairs are replaced by As. Such a substitution causes a lowering of the symmetry of the structure of cobaltite as compared to that of pyrite. Natural cobaltites have a somewhat disordered distribution as As and S in the structure.

Diagnostic Features. Although in crystal form cobaltite resembles pyrite, it can be distinguished by its silver color and cleavage. Fusible at 2–3. For As and S tests see Table 4.18; for a diagnostic Co bead test see Table 4.21.

Occurrence. Cobaltite is usually found in high-temperature deposits, as disseminations in metamorphosed rocks, or in vein deposits with other cobalt and nickel minerals. Notable occurrences of cobaltite are at Tunaberg, Sweden, and Cobalt, Ontario. The largest producer of cobalt today is Zaire, where oxidized cobalt and copper ores are associated.

Use. An ore of cobalt.

Name. In allusion to its chemistry.

ARSENOPYRITE—FeAsS

Crystallography. Monoclinic; $2/m$, pseudo-orthorhombic. Crystals are commonly prismatic elongated on c and less commonly on b (Fig. 7.30). Twinning on {100} and {001} produces pseudo-orthorhombic crystals; on {110} generates contact or penetration twins; may be repeated, as in marcasite.

Angles: $q(210) \wedge q'(2\bar{1}0) = 80°10'$, $u(120) \wedge u'(1\bar{2}0) = 146°55'$, $n(101) \wedge n'(\bar{1}01) = 68°13'$.

$P2_1c$; $a = 5.74$, $b = 5.67$, $c = 5.78$ Å, $\beta = 112°17'$; $a:b:c = 1.012:1:1.019$; $Z = 4$. $d's$: 2.68(10), 2.44(9), 2.418(9), 2.412(9), 1.814(9).

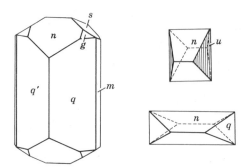

FIG. 7.30. Arsenopyrite crystals.

Physical Properties. *Cleavage* {101} poor. **H** $5\frac{1}{2}$–6. **G** 6.07. *Luster* metallic. *Color* silver-white. *Streak* black. Opaque.

Composition and Structure. Close to FeAsS with some variation in As and S contents ranging from $FeAs_{0.9}S_{1.1}$ to $FeAs_{1.1}S_{0.9}$. Cobalt may replace part of the Fe and a series extends to *glaucodot,* (Co, Fe)AsS. The structure of arsenopyrite is a derivative from the marcasite structure type (See Fig. 7.14c) in which one half of the S is replaced by As.

Diagnostic Features. Distinguished from marcasite by its silver-white color. Its crystal form and lack of Co test distinguish it from skutterudite. Fusible at 2 to magnetic globule. See Table 4.18 for As and S tests.

Occurrence. Arsenopyrite is the most common mineral containing arsenic. It occurs with tin and tungsten ores in high-temperature hydrothermal deposits, associated with silver and copper ores, galena, sphalerite, pyrite, chalcopyrite. Frequently associated with gold. Often found sparingly in pegmatites, in contact metamorphic deposits, disseminated in crystalline limestones.

Arsenopyrite is a widespread mineral and is found in considerable abundance in many localities, as at Freiberg and Munzig, Saxony; with tin ores in Cornwall, England; from Tavistock, Devonshire; in various places in Bolivia. In the United States it is found in fine crystals at Franconia, New Hampshire, Roxbury, Connecticut; and Franklin, New Jersey. It is associated with gold at Lead, South Dakota. Large quantities occur at Deloro, Ontario.

Use. The principal source of arsenic. Most of the arsenic produced is recovered in the form of the oxide, as a by-product in the smelting of arsenical ores for copper, gold, lead, and silver. Metallic arsenic is used in some alloys, particularly with lead in shot metal. Arsenic is used chiefly, however, in the form of white arsenic or arsenious oxide in medicine, insecticides, preservatives, pigments, and glass. Arsenic sulfides are used in paints and fireworks.

Name. Arsenopyrite is a contraction of the older term *arsenical pyrites*.

SKUTTERUDITE—(Co, Ni)As$_3$

Crystallography. Isometric; $2/m\bar{3}$. Common crystal forms are cube and octahedron, more rarely dodecahedron and pyritohedron. Usually massive, dense to granular.

*Im*3; $a = 8.21$–8.29 Å, $Z = 8$; $d's$: 2.61(10), 2.20(8), 1.841(9), 1.681(7), 1.616(9).

Physical Properties. H $5\frac{1}{2}$–6. **G** 6.5 ± 0.4. Brittle. *Luster* metallic. *Color* tin-white to silver gray. *Streak* black. *Opaque.*

Composition and Structure. Essentially (Co, Ni)As$_3$ but Fe usually substitutes for some Ni or Co. The high nickel varieties are called *nickel skutterudite*, (Ni, Co)As$_3$. A remarkable feature of the structure of skutterudite is the square grouping of As as As$_4$. The unit cell contains 8(Co + Ni) atoms and six As$_4$ groups which leads to the formula (Co, Ni)As$_3$. *Smaltite*, (Co, Ni)As$_{3-x}$ and *chloanthite*, (Ni, Co)As$_{3-x}$ (with $x = 0.5$–1.0) have skutterudite-like structures but a deficiency of As.

Diagnostic Features. Fusible at 2–$2\frac{1}{2}$. See Table 4.18 for As test on charcoal and Table 4.21 for Co bead test. Presence of Co distinguishes skutterudite from massive arsenopyrite.

Occurrence. Skutterudite is usually found with cobaltite and niccolite in veins formed at moderate temperature. Native silver, bismuth, arsenopyrite, and calcite are also commonly associated with it.

Notable localities are Skutterude, Norway; Annaberg, Schneeberg, and Freiberg, in Saxony; and Cobalt, Ontario, where skutterudite is associated with silver ores.

Use. An ore of cobalt and nickel. Cobalt is chiefly used in alloys for making permanent magnets and high-speed tool steel. Cobalt oxide is used as blue pigment in pottery and glassware.

Name. Skutterudite from the locality, Skutterude, Norway.

Similar Species. *Linnaeite*, Co$_3$S$_4$, associated with cobalt and nickel minerals.

Calaverite—AuTe$_2$

Crystallography. Monoclinic; $2/m$. Rarely in distinct crystals elongated parallel to b; faces of this zone are deeply striated. Terminated at the ends of the b axis with a large number of faces. Twinning on {101}, {310}, {111} frequent. Usually granular.

$C2/m$; $a = 7.19$, $b = 4.40$, $c = 5.07$ Å; $\beta = 90°13'$; $a:b:c = 1.634:1:1.152$; $Z = 2$. $d's$: 2.99(10), 2.91(7), 2.20(5), 2.09(9), 2.06(6).

Physical Properties. H $2\frac{1}{2}$. **G** 9.35. *Luster* metallic. *Color* brass-yellow to silver-white. *Streak* yellowish to greenish gray. Opaque. Very brittle.

Composition. Au 44.03, Te 55.97%. Ag usually replaces Au to a small extent.

Diagnostic Features. Distinguished from sylvanite by the presence of only a small amount of Ag and by the lack of cleavage. Fusible at 1. On charcoal it fuses with a bluish-green flame, yielding globules of metallic gold. When decomposed in hot concentrated sulfuric acid the solution assumes a deep red color (Te), and a spongy mass of gold separates.

Occurrence. Calaverite is usually found in low-temperature hydrothermal veins, but also in some higher temperature deposits. It is associated with sylvanite and other tellurides at Cripple Creek, Colorado, at Kalgoorlie, Western Australia and Nagyág, Transylvania.

Use. An ore of gold.

Name. Named from Calaveras County, California, where it was originally found in the Stanislaus mine.

Similar Species. Other rare tellurides are *krennerite*, AuTe$_2$; *altaite*, PbTe; *hessite*, Ag$_2$Te; *petzite*, (Ag, Au)$_2$Te; and *nagyagite*, a sulfotelluride of Pb and Au.

Sylvanite—(Au, Ag)Te$_2$

Crystallography. Monoclinic; $2/m$. Distinct crystals rare. Usually bladed or granular. Often in skeleton forms on rock surfaces resembling writing in appearance.

$P2/c$; $a = 8.94$, $b = 4.48$, $c = 14.59$ Å, $\beta = 145°26'$; $a:b:c = 1.995:1:3.256$; $Z = 2$. d's: 3.05(10), 2.25(3), 2.14(5), 1.988(3), 1.796(2).

Physical Properties. *Cleavage* {010} perfect. **H** $1\frac{1}{2}$–2. **G** 8–8.2. *Luster* brilliant metallic. *Color* silver-white. *Streak* gray. Opaque.

Composition. The ratio of gold to silver varies somewhat; when Au:Ag = 1:1, Te 62.1, Au 24.5, Ag 13.4%.

Diagnostic Features. Fusible at 1. When heated in concentrated sulfuric acid, the solution assumes a deep red color (Te). When heated on charcoal, it gives a gold-silver globule and blue-green flame of Te. Distinguished from calaverite by its good cleavage.

Occurrence. Sylvanite is a rare mineral associated with calaverite and other tellurides, pyrite and other sulfides in small amounts, gold, quartz, chalcedony, fluorite, and carbonates. Usually in veins formed at low temperatures but may be in higher temperature veins.

It is found at Offenbanya and Nagyág in Transylvania; at Kalgoorlie and Mulgabbie, Western Australia. In the United States it is found sparingly at several localities in California and Colorado, but the most notable occurrence is at Cripple Creek, Colorado.

Use. An ore of gold and silver.

Name. Derived from Transylvania, where it was first found, and in allusion to *sylvanium,* one of the names first proposed for the element tellurium.

SULFOSALTS

The sulfosalts comprise a very diverse and relatively large group of minerals with about 100 species; only a few of the important ones will be considered here. The sulfosalts differ from the sulfides, sulfarsenides, and arsenides in that As and Sb play a role more or less like that of metals in the structure; in sulfarsenides and arsenides the semimetals take the place of sulfur in the structure. For example, in arsenopyrite, FeAsS, (Fig. 7.14c) the As is substituted for S in a marcasite type structure. In enargite, Cu_3AsS_4, on the other hand, As enters the metal po-

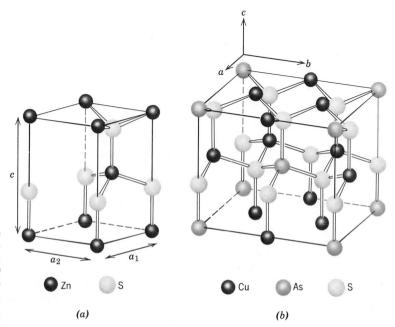

FIG. 7.31 (a) Wurtzite structure type of ZnS. (b) Enargite, Cu_3AsS_4, an orthorhombic derivative of the wurtzite structure (after B. J. Wuensch, 1974, in *Sulfide Mineralogy*).

Zn   S

Cu   As   S

(a)   (b)

SULFOSALTS

Pyrargyrite	Ag_3SbS_3	Enargite	Cu_3AsS_4
Proustite	Ag_3AsS_3	Bournonite	$PbCuSbS_3$
Tetrahedrite	$Cu_{12}Sb_4S_{13}$	Jamesonite	$Pb_4FeSb_6S_{14}$
Tennantite	$Cu_{12}As_4S_{13}$		

sition of the wurtzite type structure and is coordinated to four neighboring S ions (see Fig. 7.31). Sulfosalts may be considered double sulfides. As such enargite, Cu_3AsS_4 may be considered $3Cu_2S \cdot As_2S_5$.

The term sulfosalt was originally proposed to indicate that a compound was a salt of one of a series of acids in which sulfur had replaced the oxygen of an ordinary acid. Because such acids may be purely hypothetical, it is somewhat misleading to endeavor to thus explain this class of minerals. Nevertheless, the term sulfosalt is retained for indicating a certain type of unoxidized sulfur mineral that is structurally distinct from a sulfide.

Pyrargyrite—Ag_3SbS_3 (*Dark Ruby Silver*)
Proustite—Ag_3AsS_3 (*Light Ruby Silver*)
These minerals, the *ruby silvers,* are isostructural with similar crystal forms, physical properties, and occurrence; thus the general description that follows applies to both.

Crystallography. Hexagonal-*R*; $3m$. Crystals commonly prismatic with hemimorphic (polar) development of ditrigonal pyramid $\{21\overline{3}1\}$ and trigonal pyramid $\{10\overline{1}1\}$. Usually distorted with complex development. Commonly massive, compact, in disseminated grains.

R3c. Pyrargyrite: $a = 11.06$, $c = 8.74$ Å; $a:c = 1:0.790$; $Z = 6$. $d's$: 3.35(5), 3.21(8), 2.79(10), 2.58(5), 2.54(5). Proustite: $a = 10.79$, $c = 8.69$ Å; $a:c = 1:0.805$; $Z = 6$. $d's$: 3.28(8), 3.18(8), 2.76(10), 2.56(8), 2.48(8).

Physical Properties. *Cleavage* $\{10\overline{1}1\}$ distinct. *Luster* adamantine. Translucent. **H** $2-2\frac{1}{2}$. **G** 5.85 (pyrargyrite), 5.57 (proustite). *Color* and *streak* red (pyrargyrite), scarlet-vermilion (proustite).

Composition and Structure. Pyrargyrite: Ag 59.7, Sb 22.5, S 17.8%. Proustite: Ag 65.4, As 15.2, S 19.4%. There is very little solid solution between the two minerals. The structures of pyrargyrite and proustite are based on a rhombohedral lattice in which AsS_3 (or SbS_3) groups are located at each corner of the rhombohedral cell. The AsS_3 (or SbS_3) groups are low pyramids linked by Ag atoms; each S atom has 2 Ag atoms as nearest neighbors (see Fig. 7.32).

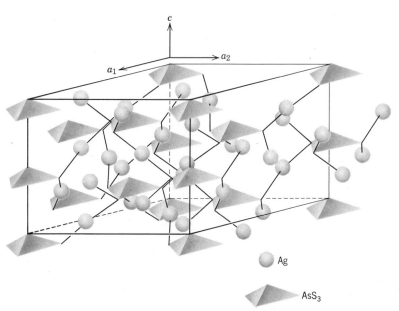

FIG. 7.32. Structure of proustite, Ag_3AsS_3. Pyramids represent AsS_3 groups and joins represent covalent bonds. This drawing is based on a hexagonal unit cell containing six formula units of Ag_3AsS_3.

Ag

AsS_3

Diagnostic Features. Fusible at 1. Characterized by brilliant luster and red color. Pyrargyrite is distinguished from proustite by darker red color and tests for antimony on charcoal (dense white coating) and in open tube. Proustite gives diagnostic As test on charcoal (see Table 4.18).

Occurrence. Pyrargyrite is the more common of these two minerals, both of which are found in low-temperature silver veins as minerals that crystallize late in the sequence of primary deposition. They are associated with other silver sulfosalts, acanthite, tetrahedrite, and native silver.

Notable localities are: Andreasberg, Harz Mountains; Freiberg, Saxony; Pribram, Bohemia; Guanajuato, Mexico; Chañarcillo, Chile; and in Bolivia. In the United States they are found in various silver veins in Colorado, Nevada, New Mexico, and Idaho. In Canada they occur in the silver veins at Cobalt, Ontario.

Use. Ores of silver.

Name. Pyrargyrite derived from two Greek words meaning *fire* and *silver,* in allusion to its color and composition. Proustite named in honor of the French chemist, J. L. Proust (1755–1826).

Similar Species. *Polybasite,* $Ag_{16}Sb_2S_{11}$, and *stephanite,* Ag_5SbS_4, are rare minerals of primary origin deposited at low to moderate temperatures. They are associated with other silver sulfosalts, acanthite, tetrahedrite, and the commoner sulfides.

Tetrahedrite—$Cu_{12}Sb_4S_{13}$
Tennantite—$Cu_{12}As_4S_{13}$

These two isostructural minerals form a complete solid solution series. They are so similar in crystallographic and physical properties that it is impossible to distinguish them by inspection.

Crystallography. Isometric; $\bar{4}3m$. Habit tetrahedral (Fig. 7.33), may occur in groups of parallel crystals. Tetrahedron, tristetrahedron, dodecahedron, and cube are the common forms. Frequently in crystals. Contact and penetration twins on [111]. Also massive, coarse, or fine, granular.

$I\bar{4}3m$. Tetrahedrite: $a = 10.37$ Å; $Z = 2$; $d's$: 3.00(10), 2.60(6), 2.45(4), 1.839(10), 1.568(8). With increase in silver a increases. Tennantite: $a =$

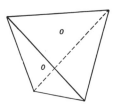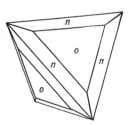

FIG. 7.33. Tetrahedrite crystals.

10.21 Å; $Z = 2$; $d's$: 2.94(10), 2.55(6), 2.40(4), 1.803(10), 1.537(8).

Physical Properties. H $3–4\frac{1}{2}$, **G** 4.6–5.1 (tennantite harder but of lower specific gravity than tetrahedrite). *Luster* metallic to submetallic. *Color* grayish-black to black. *Streak* black to brown. Opaque.

Composition and Structure. Although Cu is the predominant metal, Fe is always present (1 to 13%) and Zn usually (0 to 8%). Less commonly Ag, Pb, and Hg substitute for Cu. The argentiferous variety, *freibergite,* may contain as much as 18% Ag. The structure of tetrahedrite (and tennantite) can be regarded as a derivative of the sphalerite structure (see Fig. 7.12c). One quarter of the metal atoms are replaced by As or Sb, which form only three bonds with S. This causes the omission of one quarter of the S atoms per unit cell. Cu is surrounded by three S atoms in triangular coordination.

Diagnostic Features. Recognized by their tetrahedral crystals and when massive by their brittleness, luster, and gray color. Fusible at $1\frac{1}{2}$. On charcoal or in open tube they usually give tests for As and Sb. Therefore, a quantitative analysis for these elements may be necessary to determine at which end of the series a specimen belongs. Both give diagnostic copper flame test (Table 4.20).

Occurrence. Tetrahedrite, the most common member of the sulfosalt group, is widespread in occurrence and varied in association. Tennantite is less widely distributed. Commonly found in hydrothermal veins of copper, silver, lead, and zinc minerals formed at low to moderate temperatures. Rarely in higher-temperature veins or in contact metamorphic deposits. Usually associated with chalcopyrite, pyrite, sphalerite, galena, and various other silver, lead, and copper minerals. May carry sufficient silver to become an important ore of that metal.

Notable localities: Cornwall, England; the Harz Mountains, Germany; Freiberg, Saxony; Pribram, Bohemia; various places in Rumania; and the silver mines in Colorado, Montana, Nevada, Arizona, and Utah.

Use. An ore of silver and copper.

Name. Tetrahedrite in allusion to the tetrahedral form of the crystals. Tennantite after the English chemist, Smithson Tennant (1761–1815).

ENARGITE—Cu₃AsS₄

Crystallography. Orthorhombic; $mm2$. Crystals elongated parallel to c and vertically striated, also tabular parallel to {001}. Columnar, bladed, massive.

$Pnm2_1$; $a = 6.41$, $b = 7.42$, $c = 6.15$ Å; $a:b:c = 0.864:1:0.829$; $Z = 2$. $d's$: 3.22(10), 2.87(8), 1.859(9), 1.731(6), 1.590(5).

Physical Properties. *Cleavage* {110} perfect, {100} and {010} distinct. **H** 3. **G** 4.45. *Luster* metallic. *Color* and *streak* grayish black to iron-black. Opaque.

Composition and Structure. In Cu₃AsS₄: Cu 48.3, As 19.1, S 32.6%. Sb substitutes for As up to 6% by weight, and some Fe and Zn are usually present. The structure of enargite can be regarded as a derivative of the wurtzite structure (Fig. 7.31), in which $\frac{1}{4}$ of Zn is replaced by Cu and $\frac{3}{4}$ of Zn by As. In chemical terms Zn₄S₄ becomes Cu₃AsS₄. The low-temperature polymorph of Cu₃AsS₄ (stable below 320°C) is *luzonite* with a tetragonal structure ($I\bar{4}2m$). *Famatinite,* Cu₃SbS₄, the antimony analogue of enargite is isostructural with luzonite. Extensive solid solution exists in famatinite toward luzonite, and famatinite toward enargite.

Diagnostic Features. Characterized by its color and cleavage. Distinguished from stibnite by a test for Cu. Fusible at 1. Diagnostic tests for As and Cu are given in Tables 4.18 and 4.20 respectively.

Occurrence. Enargite is a comparatively rare mineral, found in vein and replacement deposits formed at moderate temperatures associated with pyrite, sphalerite, bornite, galena, tetrahedrite, covellite, and chalcocite.

Notable localities: Bor, near Zajecar, Yugoslavia. Found abundantly at Morococha and Cerro de Pasco, Peru; also from Chile and Argentina; Island of Luzon, Philippines. In the United States is an important ore mineral at Butte, Montana, and to a lesser extent at Bingham Canyon, Utah. Occurs in the silver mines of the San Juan Mountains, Colorado.

Use. An ore of copper. Arsenic oxide also obtained from it at Butte, Montana.

Name. From the Greek meaning *distinct,* in allusion to the cleavage.

Bournonite—PbCuSbS₃

Crystallography. Orthorhombic; $2/m2/m2/m$. Crystals usually short prismatic to tabular. May be complex with many vertical prism and pyramid faces. Frequently twinned on {110}, giving tabular crystals with recurring reentrant angles in the [001] zone (Fig. 7.34), whence the common name of *cogwheel ore.* Also massive; granular to compact.

Angles: $b(010) \wedge m(110) = 46°50'$, $(001) \wedge (101) = 43°45'$.

$Pn2_1m$; $a = 8.16$, $b = 8.71$, $c = 7.81$ Å; $a:b:c = 0.937:1:0.897$; $Z = 4$. $d's$: 4.37(4), 3.90(8), 2.99(4), 2.74(10), 2.59(5), 1.765(6).

Physical Properties. *Cleavage* {010} imperfect. **H** 2½–3. **G** 5.8–5.9. *Luster* metallic. *Color* and *streak* steel-gray to black. Opaque.

Composition and Structure. The percentages of the elements in PbCuSbS₃: Pb 42.4, Cu 13.0, Sb 24.9, S 19.7. As may substitute for Sb to about Sb:As = 4:1. In the bournonite structure CuS₄ tet-

FIG. 7.34. Twinned bournonite.

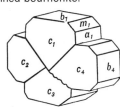

rahedral groups are connected at the vertices to AsS_3 pyramidal groups and along the edges with Pb polyhedra. The Pb polyhedra have 7 and 8 coordination.

Diagnostic Features. Recognized by its characteristic crystals and high specific gravity. Fusible at 1. For Pb and Sb tests see Table 4.18.

Occurrence. Bournonite, one of the commonest of the sulfosalts, occurs typically in hydrothermal veins formed at moderate temperature. It is associated with galena, tetrahedrite, chalcopyrite, sphalerite, and pyrite. Frequently noted as microscopic inclusions in galena.

Notable localities are: the Harz Mountains; Kapnik and elsewhere in Rumania; Liskeard, Cornwall; also found in Australia, Mexico and Bolivia. In the United States, it has been found in various places in Arizona, Utah, Nevada, Colorado, and California, but not in notable amount or quality.

Use. An ore of copper, lead, and antimony.

Name. After Count J. L. de Bournon (1751–1825), French crystallographer and mineralogist.

Jamesonite—$Pb_4FeSb_6S_{14}$

Crystallography. Monoclinic; $2/m$. Usually in acicular crystals or in capillary forms with a feather-like appearance; hence called *feather ore*. Also fibrous to compact massive. Several different compounds are often designated "feather ore."

$P2_1/a$; $a = 15.71$, $b = 19.05$, $c = 4.04$ Å; $\beta = 91°48'$; $a:b:c = 0.825:1:0.212$; $Z = 2$. d's: 3.44(10), 3.18(5), 3.09(5), 2.84(9), 2.75(8).

Physical Properties. *Cleavage* {001} good. Brittle. **H** 2-3. **G** 5.63. *Luster* metallic. *Color* and *streak* steel-gray to grayish-black. Opaque.

Composition and Structure. In $Pb_4FeSb_6S_{14}$: Pb 40.16, Fe 2.71, Sb 35.39, S 21.74%. Cu and Zn may be present in small amounts. Jamesonite has a complex structure with irregular coordination. SbS_3 groups are linked to form complex chains of Sb_3S_7 parallel to the c axis. Fe is in 6 coordination with S and Pb shows 7 and 8 coordination with S.

Diagnostic Features. Recognized by its characteristic fibrous (feathery) appearance, and distinguished from stibnite by lack of good lengthwise cleavage. Difficult to distinguish from similar species (see below).

Occurrence. Jamesonite is found in ore veins formed at low to moderate temperatures, associated with other lead sulfosalts, with galena, stibnite, tetrahedrite, sphalerite.

Found in Cornwall, England, and in various localities in Czechoslovakia, Rumania, and Saxony; also found in Tasmania and Bolivia. In the United States from Sevier County, Arkansas, and at Silver City, South Dakota.

Use. A minor ore of lead.

Name. After the mineralogist Robert Jameson (1774–1854), of Edinburgh.

Similar Species. A number of minerals similar to jamesonite in composition and general physical characteristics are included under the term *feather ore*. These include such minerals as *zinkenite*, $Pb_6Sb_{14}S_{27}$; *boulangerite*, $Pb_5Sb_4S_{11}$; and *meneghinite*, $CuPb_{13}Sb_7S_{24}$. Other minerals of similar composition but different habit are *plagionite*, $Pb_5Sb_8S_{17}$; *semseyite*, $Pb_9Sb_8S_{21}$; and *geocronite*, $Pb_5(Sb, As)_2S_8$.

References and Suggested Reading

Bragg, L. and G. F. Claringbull, 1965, *Crystal Structures of Minerals*. G. Bell and Sons, Ltd., London, 409 pp.

Deer, W. A., Howie, R. A., and J. Zussman, 1962, *Rock-Forming Minerals,* vol. 5, Non-Silicates. John Wiley & Sons, Inc., New York, 371 pp.

Mineralogical Society of America, 1974, *Sulfide Mineralogy* (Short Course Notes). P. H. Ribbe, editor, Southern Printing Co., Blacksburg, Virginia.

Palache, C., Berman, H., and C. Frondel, 1944, *The System of Mineralogy,* 7th ed., vol. 1, Elements, Sulfides, Sulfosalts, Oxides. John Wiley & Sons, Inc., New York, 843 pp.

Strunz, H., 1970, *Mineralogische Tabellen,* 5th ed. Akademische Verlagsgesellschaft, Geest und Portig K. G., Leipzig, 621 pp.

8
SYSTEMATIC MINERALOGY

PART II: OXIDES, HYDROXIDES, AND HALIDES

The *oxides* are a group of minerals that are relatively hard, dense, and refractory and generally occur as accessory minerals in igneous and metamorphic rocks and as resistant detrital grains in sediments. The *hydroxides,* on the other hand, tend to be lower in hardness and density and are found mainly as secondary alteration or weathering products. The *halides* comprise about 80 chemically related minerals with diverse structure types and of diverse geologic origins. Here we will discuss only six of the most common halides.

OXIDES

The oxide minerals include those natural compounds in which oxygen is combined with one or more metals. They are here grouped as simple oxides and multiple oxides. The simple oxides, compounds of one metal and oxygen, are of several types with different $X:O$ ratios (the ratio of metal to oxygen) as X_2O, XO, X_2O_3. Although not described

on the following pages, ice, H_2O (see Fig. 4.30), is a simple oxide of the X_2O type in which hydrogen is the cation. The most common of all oxides, SiO_2, quartz and its polymorphs, is not considered in this chapter, but instead is treated in Chapter 10 with the silicates because the structure of quartz (and its polymorphs) is most closely related to that of other Si-O compounds. The multiple oxides, XY_2O_4, have two nonequivalent metal atom sites (A and B). Within the oxide class are several minerals of great economic importance. These include the chief ores of iron (hematite and magnetite), chromium (chromite), manganese (pyrolusite, as well as the hydroxides, manganite, and psilomelane), tin (cassiterite), and uranium (uraninite).

The bond type in oxide structures is generally strongly ionic in contrast to the sulfide structures with ionic, covalent, as well as metallic bonding. One of the first structure determinations reported by Sir Lawrence Bragg in 1916 was that of cuprite, Cu_2O, in which the oxygen atoms are arranged at the corners and center of tetrahedral groups. The Cu

OXIDES

X_2O and XO Types		XY_2O_4 Type	
Cuprite	Cu_2O		
Zincite	ZnO		
X_2O_3 Type		*Spinel Group*	
Hematite Group		Spinel	$MgAl_2O_4$
Corundum	Al_2O_3	Gahnite	$ZnAl_2O_4$
Hematite	Fe_2O_3	Magnetite	Fe_3O_4
Ilmenite	$FeTiO_3$	Franklinite	(Zn, Fe, Mn)
XO_2 Type			$(Fe, Mn)_2O_4$
(excluding SiO_2)		Chromite	$FeCr_2O_4$
Rutile Group		Chrysoberyl	$BeAl_2O_4$
Rutile	TiO_2	Columbite	(Fe, Mn)
Pyrolusite	MnO_2		$(Nb, Ta)_2O_6$
Cassiterite	SnO_2		
Uraninite	UO_2		

atoms lie halfway between the oxygens (see Fig. 8.1) within the tetrahedral groups.

The structures of the *Hematite Group* are based on hexagonal closest packing of oxygens with the cations in octahedral coordination between them (see Fig. 8.2a). As can be seen in a basal projection of this structure only $\frac{2}{3}$ of the octahedral spaces are occupied by Fe^{3+} or Al^{3+}. The presence of $\frac{1}{3}$ oxygen octahedra without central Al^{3+} or Fe^{3+} ions is related to the electrostatic valency (e.v.) or bond strength of the Al^{3+}-O^{2-} and Fe^{3+}-O^{2-} bonds. As Al^{3+} is surrounded by six oxygens, the e.v. of each of the six Al-O bonds $= \frac{1}{2}$. Each oxygen is shared between four octahedra which means that four bonds of

FIG. 8.1. Structure of cuprite. (a) Oxygen atoms at corners and center of tetrahedral group; Cu halfway between two oxygens. (b) Oxygens in a cubic lattice as shown in a polyhedral representation of the cuprite structure.

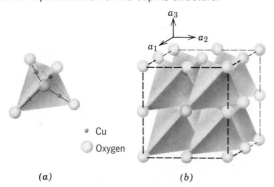

a_3
a_2
a_1

• Cu
◯ Oxygen

(a)　　　　(b)

e.v. $= \frac{1}{2}$ can radiate from an oxygen position. In the (0001) plane, as in Fig. 8.2a this allows for only *two* Al-O bonds from each oxygen, as shown by the geometry of two octahedra sharing one oxygen corner. A vertical section through the corundum (or hematite) structure (Fig. 8.2b) shows the arrangement of the Al^{3+} (or Fe^{3+}) ions and of the omitted cations as well. In the vertical stacking of octahedra each octahedron shares a face between two adjoining layers. The cations, Al^{3+} or Fe^{3+}, within the octahedra that share faces will tend to move away from the shared face on account of the repulsive forces between them (see Fig. 8.2b). The corundum and hematite structures have space group $R\overline{3}m$. The ilmenite structure, in which Fe and Ti are arranged in alternating Fe-O and Ti-O layers (Fig. 8.2c) has a lower symmetry (space group $R\overline{3}$) because of the ordered Fe and Ti substitution for Al^{3+} (as in corundum) or Fe^{3+} (as in hematite).

The structure of periclase, MgO, is identical to that of NaCl with cubic space group $Fm3m$. In a section parallel to {111} this structure appears very similar (see Fig. 8.3) to that of members of the *Hematite Group*. In such a section a "layer" of octahedra is seen without any cation omissions in the octahedral sites. With Mg in 6 coordination, the e.v. of each of the Mg-O bonds is $\frac{1}{3}$. This in turn allows for six bonds, each with e.v. $= \frac{1}{3}$, to radiate from every oxygen ion in the structure. In other words, each oxygen is shared between six Mg-O octahedra, and not between just four Al-O (or Fe-O) octahedra as in the structures of the *Hematite Group*. The MgO type structure, therefore, shows no cation vacancies.

The structures of XO_2 type oxides that we will consider fall into two structure types. One is the structure typified by *rutile* in which the cation is in 6 coordination with oxygen. The radius ratio, $R_X:R_O$, in this structure lies between the limits of 0.732 and 0.414 (see Table 8.1). The other has the *fluorite* structure (see Fig. 8.25) in which each oxygen has four cation neighbors arranged about it at the apices of a more or less regular tetrahedron, whereas each cation has eight oxygens surrounding it at the corners of a cube. The oxides of 4-valent uranium, thorium, and cerium, of great interest because of

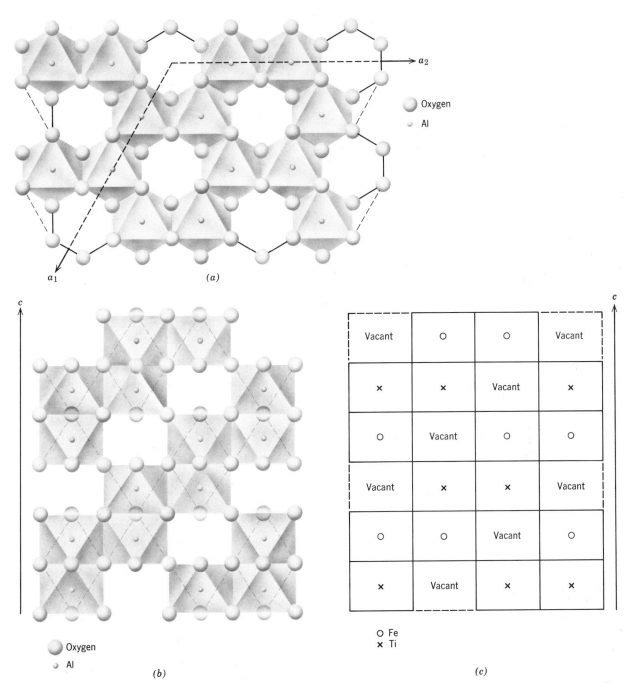

FIG. 8.2. Structures of the hematite group. (*a*) Basal sheet of octahedra in corundum, Al_2O_3, or hematite, Fe_2O_3, with one octahedron vacant for every two octahedra with Al or Fe^{3+} in center. (*b*) Vertical section through the corundum structure showing the locations of filled and empty octahedra. Note locations of Al^{3+} within octahedra. (*c*) Schematic vertical cross-section through the ilmenite structure. This representation is the same as in Fig. 8.2*b* with oxygen positions and octahedral outlines eliminated; only cation locations are shown.

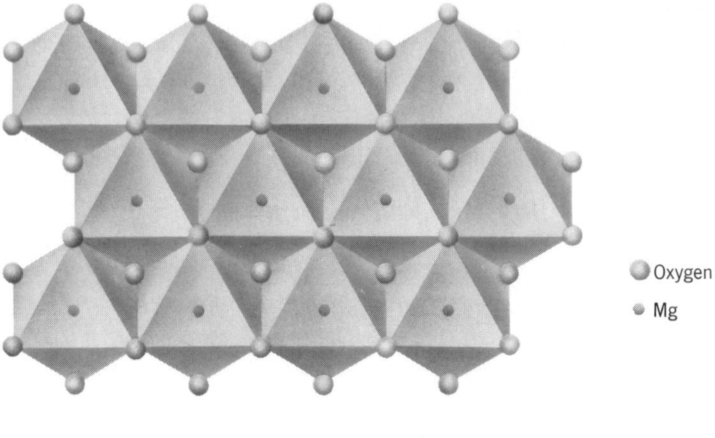

FIG. 8.3. Structure of periclase, MgO, as seen parallel to (111). Compare this with Fig. 8.2a as well as with Fig. 8.19, of brucite.

their connection with nuclear chemistry, have this structure. In general, all dioxides in which the radius ratio of cation to oxygen lies within and close to the limits for 8 coordination (0.732-1) may be expected to have this structure and to be isometric hexoctahedral (see Table 8.1).

The members of the *Rutile Group* are isostructural with space group $P4_2/mnm$. (Rutile itself has two additional polymorphs, anatase and brookite, with slightly different structures). In the rutile structure (see Fig. 8.4) Ti^{4+} is located at the center of oxygen octahedra the edges of which are shared forming chains parallel to the c axis. These chains are cross-linked by the sharing of octahedral corners (see Fig. 8.4b). Each oxygen is linked to three titanium ions and the e.v. of each of the Ti-O bonds $= \frac{2}{3}$. The prismatic habit of minerals of the rutile group is a reflection of the chainlike arrangement of octahedra.

Table 8.1
RADIUS RATIOS IN *XO*$_2$ TYPE OXIDES (R_O = 1.4Å)

R_X	R_X/R_O	Ion	Mineral	Structure Type
0.60	0.43	Mn^{4+}	Pyrolusite	Rutile
0.68	0.49	Ti^{4+}	Rutile	Rutile
0.71	0.51	Sn^{4+}	Cassiterite	Rutile
0.94	0.67	Ce^{4+}	Cerianite	Fluorite
0.97	0.69	U^{4+}	Uraninite	Fluorite
1.02	0.73	Th^{4+}	Thorianite	Fluorite

The minerals of the *Spinel Group* with the general formula $X^{2+}Y_2^{3+}O_4$ are based on the arrangement of oxygens in approximate cubic closest packing. The interstitial divalent cations (X^{2+}) may be Mg, Fe, Zn, or Mn, and the trivalent cations (Y^{3+}) Al, Fe, Mn, or Cr (see Table 8.2). Solid solution is extensive among the divalent cations, but incomplete among the trivalent cations. Because of the extensive solid solution the minerals of the spinel group show a wide range in physical properties.

In a unit cell of spinel, with edge length approximately 8 Å, there are 64 possible tetrahedral *A* sites and 32 possible octahedral *B* sites. The spinel structure in which only 8 of the *A* sites and 16 *B* sites per unit cell are occupied by cations, is difficult to represent in graphical form. For this reason a photograph of a polyhedral model of the structure is shown in Fig. 8.5. Occupied octahedra are joined along common edges to form rows and planes parallel to {111} of the structure. Occupied tetrahedra with their apices along $\overline{3}$ axes provide cross links between layers of octahedra. In spinel ($MgAl_2O_4$) each oxygen has three Al neighbors and one Mg. The sum of the bond strengths reaching the oxygen is $3 \times (\frac{3}{6}) + \frac{2}{4} = 2$ which neutralizes the oxygen.

The coordination polyhedra about the various cations in spinel are not what one might predict on the basis of the ionic sizes of the cations. Because Mg^{2+} is larger than Al^{3+}, one would expect Mg to occur in octahedral and Al in tetrahedral coordination with oxygen. However, in the *normal spinel structure* the general concepts of radius-ratio do not

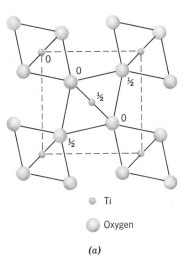

FIG. 8.4. Structure of rutile. (a) Projection on (001) showing location of Ti and O atoms and the outline of the unit cell (dashed lines). (b) Projection on (110) showing chains of octahedra parallel to c.

● Ti

◯ Oxygen

(a)

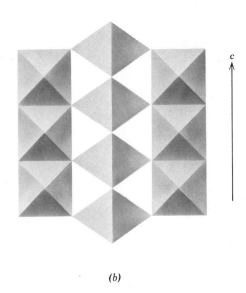

(b)

apply; indeed the larger cation is in the smaller polyhedron, and vice versa. When crystal field stabilization energies are considered instead of the geometric aspects of the ions, it becomes clear that the larger cation may occupy tetrahedral sites.

FIG. 8.5. Polyhedral model of the structure of spinel, $X^{2+}Y_2^{3+}O_4$. In the *normal spinel* structure diavalent cations (Mg, Fe^{2+}, etc.) occur in tetrahedral coordination (A site) and trivalent cations (Al, Fe^{3+}, etc.) in octahedral coordination (B site).

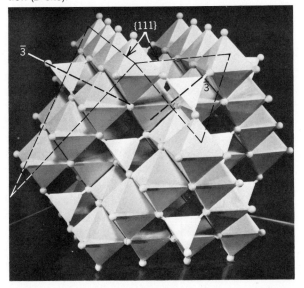

When trivalent ions (Y^{3+}) in the spinel structure occupy the eight tetrahedral sites (A) and the remaining eight trivalent (Y^{3+}) and eight divalent (X^{2+}) cations occupy the octahedral sites (B), the structure is referred to as an *inverse spinel structure*. For example, magnesioferrite (see Table 8.2) $MgFe_2O_4$ can be rewritten as $Fe^{3+}(Mg, Fe^{3+})_2O_4$ to represent more clearly its inverse spinel structure.

The spinel structure has a coordination scheme that is similar to that of the silicates of the olivine series. The composition of members of the forsterite-fayalite series with a total solid solution series from Mg_2SiO_4 to Fe_2SiO_4 can be represented as $X_2^{+2}Y^{+4}O_4$. Although this is not the same as $X^{+2}Y_2^{+3}O_4$ as in spinel, in both cases the total cation charge is identical. If one compares the structure of an Fe-Mg olivine with that of a possible Fe-Mg spinel, one finds that the spinel structure is about 12% denser than the olivine structure for the same composition. This observation has led petrologists to suggest that olivine, which is thought to be abundant in the mantle, is converted to a spinel type structure at great depth. In 1969 an olivine, *ringwoodite,* from a stony meteorite was found to have a spinel structure. This same olivine composition $(Fe_{0.7}Mg_{0.3})_2SiO_4$ had been previously synthesized as a spinel in the laboratory of A. E. Ringwood, Australian National University.

Table 8.2

MEMBERS OF THE SPINEL GROUP (XY_2O_4)	X	Y = Al Spinel Series[a]		Magnetite Series[b]	Y = Fe³⁺	Chromite Series[a]	Y = Cr³⁺
	Mg	Spinel	$MgAl_2O_4$	Magnesioferrite	$Fe^{3+}(Mg, Fe^{3+})_2O_4$	Magnesiochromite	$MgCr_2O_4$
	Fe²⁺	Hercynite	$FeAl_2O_4$	Magnetite	$Fe^{3+}(Fe^{2+}, Fe^{3+})O_4$	Chromite	$Fe^{2+}Cr_2O_4$
	Zn	Gahnite	$ZnAl_2O_4$	Franklinite	$ZnFe_2^{3+}O_4$		
	Mn	Galaxite	$MnAl_2O_4$	Jacobsite	$Fe^{3+}(Mn, Fe^{3+})_2O_4$		

[a] All are normal spinels.
[b] All are inverse spinels except franklinite.

CUPRITE—Cu_2O

Crystallography. Isometric; $4/m\bar{3}2/m$. Commonly occurs in crystals showing the cube, octahedron, and dodecahedron, frequently in combination (Fig. 8.6). Sometimes in elongated capillary crystals, known as "plush copper" or *chalcotrichite*.

$Pn3m$; $a = 4.27$ Å; $Z = 2$. d's: 2.46(10), 2.13(6), 1.506(5), 1.284(4), 0.978(3).

Physical Properties. H $3\frac{1}{2}$–4. G 6.1. *Luster* metallic-adamantine in clear crystallized varieties. *Color* red of various shades; ruby-red in transparent crystals, called "ruby copper." *Streak* brownish-red.

Composition and Structure. Cu 88.8, O 11.2%. Usually pure, but FeO may be present as an impurity. The structure, as based on Cu atoms at the corners and centers of tetrahedral groups is shown in Fig. 8.1.

Diagnostic Features. Usually distinguished from other red minerals by its crystal form, high luster, streak, and association with limonite. Fusible at 3. Gives diagnostic Cu flame test (Table 4.20) and copper globule on charcoal (Table 4.17). When dissolved in concentrated HCl and solution diluted with cold water, it gives a white precipitate of cuprous chloride.

Occurrence. Cuprite is a supergene copper mineral and in places an important copper ore. It is found in the upper oxidized portions of copper veins, associated with limonite and secondary copper minerals such as native copper, malachite, azurite, and chrysocolla (see Figs. 11.18 and 11.19).

Noteworthy foreign countries where cuprite is an ore are Chile, Bolivia, Australia, and Zaire. Fine crystals have been found at Cornwall, England; Chessy, France; and the Ural Mountains. Found in the United States in excellent crystals in the copper deposits at Bisbee, Arizona. Also found at Clifton and Morenci, Arizona.

Use. A minor ore of copper.

Name. Derived from the Latin *cuprum,* copper.

Similar Species. *Tenorite* (also known as *melaconite*), CuO, is a black supergene mineral.

Zincite—ZnO

Crystallography. Hexagonal; *6mm*. Crystals are rare, terminated at one end by faces of a steep pyramid and the other by a pedion (see Fig. 2.109). Usually massive with platy or granular appearance.

$P6_3mc$; $a = 3.25$, $c = 5.19$ Å; $a:c = 1 : 1.625$; $Z = 2$. d's: 2.83(7), 2.62(5), 2.49(10), 1.634(6), 1.486(6).

Physical Properties. *Cleavage* $\{10\bar{1}0\}$ perfect; $\{0001\}$ parting. H 4. G 5.68. *Luster* subadamantine. *Color* deep red to orange-yellow. *Streak* orange-yellow. Translucent. *Optics:* (+); $\omega = 2.013$, $\epsilon = 2.029$.

Composition and Structure. Zn 80.3, O 19.7%. Mn²⁺ often present and probably colors the mineral; pure ZnO is white. The structure of zincite is similar to that of wurtzite (see Fig. 7.20). Zn is in hexagonal closest packing and each oxygen lies within a tetrahedral group of four Zn ions.

FIG. 8.6. Cuprite crystals.

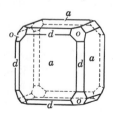 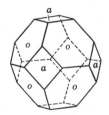

Diagnostic Features. Distinguished chiefly by its red color, orange-yellow streak, and the association with franklinite and willemite. Infusible. Soluble in HCl. When mixed with reducing mixture and intensely heated on charcoal it gives a nonvolatile coating of zinc oxide, yellow when hot, white when cold. Usually gives a blue color to the sodium carbonate bead due to Mn^{2+} in solid solution.

Occurrence. Zincite is confined almost exclusively to the zinc deposits at Franklin, New Jersey, where it is associated with franklinite and willemite in calcite, often in an intimate mixture. Reported only in small amounts from other localities.

Use. An ore of zinc, particularly used for the production of zinc white (zinc oxide).

Similar Species. *Periclase,* MgO, is another oxide of *XO* composition. The structure of MgO is of the NaCl type with space group *Fm3m*. Continuous octahedral sheets can be identified in the structure parallel to {111}, see Fig. 8.3. It is found in contact metamorphosed Mg-rich limestones by the reaction: $CaMg(CO_3)_2 \rightarrow CaCO_3 + MgO + CO_2$.

Hematite Group X_2O_3*

✳ **CORUNDUM—Al₂O₃**
Crystallography. Hexagonal-R; $\bar{3}2/m$. Crystals commonly tabular on {0001} or prismatic {11$\bar{2}$0}. Often tapering hexagonal pyramids (Fig. 8.7), rounded into barrel shapes with deep horizontal

* The long-established morphological unit with *c* one-half the length of *c* of the unit cell is retained here for members of this group; the indices of forms of these minerals are expressed accordingly.

striations. May show rhombohedral faces. Usually rudely crystallized or massive; coarse or fine granular. Polysynthetic twinning is common on {10$\bar{1}$1} and {0001}.

Angles: $c(0001) \wedge r(10\bar{1}1) = 57°35'$, $c(0001) \wedge n(22\bar{4}3) = 61°12'$, $r(10\bar{1}1) \wedge r'(\bar{1}101) = 93°56'$, $n(22\bar{4}3) \wedge n'(\bar{2}4\bar{2}3) = 51°58'$.

$R\bar{3}c$; $a = 4.76$, $c = 12.98$ Å; $a:c/2 = 1:1.363$, $Z = 6$. $d's$: 2.54(6), 2.08(9), 1.738(5), 1.599(10), 1.374(7).

Physical Properties. Parting {0001} and {10$\bar{1}$1}, the latter giving nearly cubic angles; more rarely prismatic parting. **H** 9. Corundum may alter to mica, and care should be exercised in obtaining a fresh surface for hardness test. **G** 4.02. *Luster* adamantine to vitreous. Transparent to translucent. *Color* various; usually some shade of brown, pink, or blue. May be white, gray, green, red (*ruby*), or blue (*sapphire*). Rubies and sapphires having a stellate opalescence when viewed in the direction of the *c* crystal axis are termed *star ruby* or *star sapphire*. *Emery* is a black granular corundum intimately mixed with magnetite, hematite, or hercynite. *Optics:* (−), $\omega = 1.769$, $\epsilon = 1.760$.

Composition and Structure. Al 52.9, O 47.1%. Rubies contain trace amounts of Cr (ppm) as a coloring agent and sapphires are blue probably due to trace amounts of Fe or Ti. The structure of corundum is illustrated in Fig. 8.2 and consists of oxygen in hexagonal closest packing and Al in octahedral coordination. Two-thirds of the octahedra are occupied by Al and $\frac{1}{3}$ are vacant (see also discussion on page 262).

Diagnostic Features. Characterized chiefly by its great hardness, high luster, specific gravity,

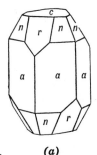
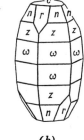
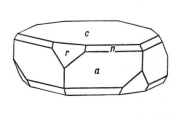

FIG. 8.7. Corundum crystals.　　*(a)*　　　*(b)*　　　*(c)*

and parting. Infusible. Insoluble. Finely pulverized material moistened with cobalt nitrate and intensely ignited becomes blue (Al).

Occurrence. Corundum is common as an accessory mineral in some metamorphic rocks, such as crystalline limestone, mica-schist, gneiss. Found also as an original constituent of silica deficient (undersaturated) igneous rocks as syenites and nepheline syenites. May be found in large masses in the zone separating peridotites from adjacent country rocks. It is disseminated in small crystals through certain lamprophyric dikes and is found in large crystals in pegmatites. Found frequently in crystals and rolled pebbles in detrital soil and stream sands, where it has been preserved through its hardness and chemical inertness. Associated minerals are commonly chlorite, micas, olivine, serpentine, magnetite, spinel, kyanite, and diaspore.

The finest rubies have come from Burma; the most important locality is near Mogok, 90 miles north of Mandalay. Some stones are found here in the metamorphosed limestone that underlies the area, but most of them have been recovered from the overlying soil and associated stream gravels. Darker, poorer quality rubies have been found in alluvial deposits near Bangkok, Thailand, and at Battambang, Cambodia. In Sri Lanka rubies of relatively inferior grade, associated with more abundant sapphires and other gem stones, are recovered from stream gravels. In the United States a few rubies have been found associated with the large corundum deposits of North Carolina.

Sapphires are found in the alluvial deposits of Thailand, Sri Lanka, and Cambodia associated with rubies. The stones from Cambodia of a cornflower-blue color are most highly prized. Sapphires occur in Kashmir, India, and are found over an extensive area in central Queensland, Australia. In the United States small sapphires of fine color are found in various localities in Montana. They were first discovered in the river sands east of Helena during placer operations for gold, and later found embedded in the rock of a lamprophyre dike at Yogo Gulch. The rock was formerly mined for sapphires but mining operations ceased in 1929.

Common abrasive corundum is found in large crystals in the Transvaal, Republic of South Africa. These crystals found loose in the soil or in pegmatites constitute a major source of abrasive corundum. Common corundum is found in the United States in various localities along the eastern edge of the Appalachian Mountains in North Carolina and Georgia. At one time it was extensively mined in southwestern North Carolina. It occurs here in large masses lying at the edges of intruded masses of an olivine rock (dunite) and is thought to represent a segregation from the original magma. Found as an original constituent of a nepheline syenite in Ontario, Canada, where in places it forms more than 10% of the rock mass.

Emery is found in large quantities on Cape Emeri on the island of Naxos and in various localities in Asia Minor, where it has been extensively mined for centuries. In the United States emery has been mined at Chester, Massachusetts, and Peekskill, New York.

Artificial. Artificial corundum is manufactured from bauxite on a large scale. This synthetic material, together with other manufactured abrasives, notably silicon carbide, has largely taken the place of natural corundum as an abrasive.

Synthetic rubies and sapphires, colored with small amounts of Cr and Ti are synthesized by fusing alumina powder in an oxy-hydrogen flame which on cooling forms single crystal "boules." This, the Verneuil process, has been in use since 1902. In 1947 the Linde Air Products of the United States succeeded in synthesizing star rubies and star sapphires. This was accomplished by introducing titanium which, during proper heat treatment, exsolves as oriented rutile needles to produce the star. The artificial rival the natural stones in beauty, and it is difficult for the untrained person to distinguish them.

Use. As a gem stone and abrasive. The deep red ruby is one of the most valuable of gems, second only to emerald. The blue sapphire is also valuable, and tones of other colors may command good prices. Stones of gem quality are used as watch jewels and as bearings in scientific instruments.

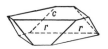
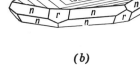

FIG. 8.8. Hematite crystals. **(a)** **(b)** **(c)**

Corundum is used as an abrasive, either ground from the pure massive mineral or in its impure form as emery.

Name. Probably from Kauruntaka, the Indian name for the mineral.

✷ **HEMATITE—Fe$_2$O$_3$**

Crystallography. Hexagonal-R; $\bar{3}2/m$. Crystals are usually thick to thin tabular on {0001}, basal planes often show triangular markings, and the edges of the plates may be beveled with rhombohedral forms (Fig. 8.8b). Thin plates may be grouped in rosettes (*iron roses*) (Fig. 8.9). More rarely crystals are distinctly rhombohedral, often with nearly cubic angles and may be polysynthetically twinned on {0001} and {10$\bar{1}$1}. Also in botryoidal to reinform shapes with radiating structure, *kidney ore* (see Fig. 5.1). May also be micaceous and foliated, *specular*. Usually earthy; called *martite* when in octahedral pseudomorphs after magnetite.

Angles: $c(0001) \wedge r(10\bar{1}1) = 57°37'$, $c(0001) \wedge n(22\bar{4}3) = 61°13'$, $r(10\bar{1}1) \wedge r'(\bar{1}101) = 94°$, $n(22\bar{4}3) \wedge n'(\bar{2}4\bar{2}3) = 51°59'$.

$R\bar{3}c$; $a = 5.04$, $c = 13.76$ Å; $a:c/2 = 1:1.365$; $Z = 6$. $d's$: 2.69(10), 2.52(8), 2.21(4), 1.843(6), 1.697(7).

Physical Properties. *Parting* on {10$\bar{1}$1} with nearly cubic angles, also on {0001}. **H** $5\frac{1}{2}$–$6\frac{1}{2}$, **G** 5.26 for crystals. *Luster* metallic in crystals and dull in earthy varieties. *Color* reddish-brown to black. Red earthy variety is known as *red ocher,* platy and metallic variety as *specularite.* *Streak* light to dark red which becomes black on heating. Translucent.

Composition and Structure. Fe 70, O 30%; essentially a pure substance at ordinary temperatures, although small amounts of Mn and Ti have been reported. Forms a complete solid solution series with ilmenite above 950°C (see Fig. 8.10). The structure is similar to that of corundum (see Fig. 8.2 and page 262).

Diagnostic Features. Distinguished chiefly by

FIG. 8.9. Hematite, iron rose.

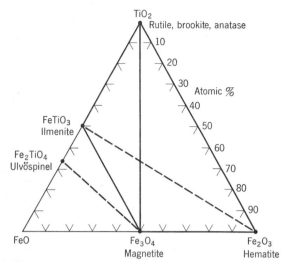

FIG. 8.10. Compositions of naturally occurring oxide minerals in the system FeO-Fe$_2$O$_3$-TiO$_2$. Solid lines (tielines) indicate common geological coexistences at relatively low temperatures. Dashed lines indicate complete solid solution between end members. Hematite-ilmenite is a complete series above 950°C. Magnetite-ulvöspinel is a complete series above about 600°C. Magnetite commonly contains ilmenite lamellae. It is possible that such intergrowths result from the oxidation of members of the magnetite-ulvöspinel series.

its characteristic red streak. Infusible. Becomes strongly magnetic on heating in the reducing flame. Slowly soluble in HCl; solution with potassium ferrocyanide gives dark blue precipitate (test for Fe^{3+}).

Occurrence. Hematite is widely distributed in rocks of all ages and forms the most abundant and important ore of iron. It may occur as a sublimation product in connection with volcanic activities. Occurs in contact metamorphic deposits and as an accessory mineral in feldspathic igneous rocks such as granite. Found from microscopic scales to enormous masses in regionally metamorphosed rocks where it may have originated by the oxidation of limonite, siderite, or magnetite. It is found in red sandstones as the cementing material that binds the quartz grains together. The quantity of hematite in a deposit that can be mined economically must be measured in tens of millions of tons. Such major accumulations are largely sedimentary in origin and many of them, through leaching of associated silica

by meteoric waters, have been enriched to high-grade (direct shipping) ores (over 50% Fe). Like limonite, it may be formed in irregular masses and beds as the result of the weathering and oxidation of iron-bearing rocks. The oölitic ores are of sedimentary origin and may occur in beds of considerable size.

Noteworthy localities for hematite crystals are the island of Elba; St. Gothard, Switzerland, in "iron roses"; in the lavas of Vesuvius; at Cleator Moor, Cumberland, England; Minas Gerais, Brazil.

In the United States the columnar and earthy varieties are found in enormous bedded, Precambrian deposits that have furnished a large proportion of the iron ore of the world. The chief iron ore districts of the United States are grouped around the southern and northwestern shores of Lake Superior in Michigan, Wisconsin, and Minnesota. The chief districts, which are spoken of as iron ranges, are, from east to west, the Marquette in northern Michigan; the Menominee in Michigan to the southwest of the Marquette; the Penokee-Gogebic in northern Wisconsin. In Minnesota the Mesabi, northwest of Duluth; the Vermilion, near the Canadian boundary; and the Cuyuna, southwest of the Mesabi. The iron ore of these different ranges varies from the hard specular variety to the soft, red earthy type. Of the several ranges the Mesabi is the largest, and since 1892 has yielded over 2.5 billion tons of high-grade ore, over twice the total production of all the other ranges.

Oölitic hematite is found in the United States in the rocks of the Clinton formation, which extends from central New York south along the line of the Appalachian Mountains to central Alabama. The most important deposits lie in eastern Tennessee and northern Alabama, near Birmingham. Hematite has been found at Iron Mountain and Pilot Knob in southeastern Missouri. Deposits of considerable importance are located in Wyoming in Laramie and Carbon counties.

Although the production of iron ore within the United States remains large, the rich deposits have been largely worked out. In the future much of the iron must come from low-grade deposits or must be imported. The low-grade, silica-rich iron formation from which the high-grade deposits have been derived is known as *taconite* which contains about 25% iron. The iron reserves in taconite are far greater than were the original reserves of high-grade ore, and at present many millions of tons of iron are being concentrated from it each year.

Exploration outside the United States has in recent years been successful in locating several ore bodies with many hundreds of millions of tons of high-grade ore. These are notably in Venezuela, Brazil, Canada, and Australia. Brazil's iron mountain, Itabira, is estimated to have 15 billion tons of very pure hematite. In 1947 Cerro Bolivar in Venezuela was discovered as an extremely rich deposit of hematite and by 1954 ore was being shipped from there to the United States. In Canada major Precambrian iron ore deposits have been located in the Labrador Trough close to the boundary between Quebec and Labrador. A 350-mile railroad built into this previously inaccessible area began in 1954 to deliver iron ore to Seven Islands, a major port on the St. Lawrence River. Large Australian deposits are located in the Precambrian rocks of the Hamersley Range, Western Australia. The main producers of iron ore are Australia, the U.S.S.R., the United States, Brazil, Canada, China, and Sweden.

Use. Most important ore of iron for steel manufacture. Also used in pigments, red ocher, and as polishing powder.

Name. Derived from a Greek word meaning *blood,* in allusion to the color of the powdered mineral.

✳ ILMENITE—FeTiO$_3$

Crystallography. Hexagonal-*R*; $\bar{3}$. Crystals are usually thick tabular with prominent basal planes and small rhombohedral truncations. Crystal constants close to those for hematite. Often in thin plates. Usually massive, compact; also in grains or as sand.

Angles: (0001) $\wedge$ (10$\bar{1}$1) = 57°59′, (0001) $\wedge$ (22$\bar{4}$3) = 61°33′, (10$\bar{1}$1) $\wedge$ ($\bar{1}$101) = 94°29′.

$R\bar{3}$; $a = 5.09$, $c = 14.06$ Å; $a:c/2 = 1:1.381$. $Z = 6$. $d's$: 2.75(10), 2.54(7), 1.867(5), 1.726(8), 1.507(4).

Physical Properties. H 5½–6. G 4.7. *Luster* metallic to submetallic. *Color* iron-black. *Streak* black to brownish-red. May be magnetic without heating. Opaque.

Composition and Structure. Fe 36.8, Ti 31.6, O 31.6% for stoichiometric FeTiO$_3$. The formula can be more realistically expressed as (Fe, Mg, Mn)TiO$_3$ with limited Mg and Mn substitution. Ilmenite may contain limited amounts of Fe$_2$O$_3$ (less than 6 weight percent) at ordinary temperatures. However, a complete solid solution exists between Fe$_2$O$_3$ and FeTiO$_3$ above 950° (see Fig. 8.10). Related species are *geikielite,* MgTiO$_3$ and *pyrophanite,* MnTiO$_3$. The structure of ilmenite is very similar to that of corundum (Fig. 8.2) with Fe and Ti ordered in alternate octahedrally coordinated layers perpendicular to the *c* axis (see Fig. 8.2*c*).

Diagnostic Features. Ilmenite can be distinguished from hematite by its streak and from magnetite by its lack of strong magnetism. Infusible. Magnetic after heating. After fusion with sodium carbonate it can be dissolved in sulfuric acid and on the addition of hydrogen peroxide the solution turns yellow.

Occurrence. Ilmenite is a common accessory mineral in igneous rocks. It may be present in large masses in gabbros, diorites, and anorthosites as a product of magmatic segregation intimately associated with magnetite. It is also found in some pegmatites and vein deposits. As a constituent of black sands it is associated with magnetite, rutile, zircon, and monazite.

Found in large quantities at Krägerö and other localities in Norway; in Finland; and in crystals at Miask in the Ilmen Mountains, U.S.S.R. It is mined in considerable quantities from beach sands, notably in Australia, India, and Brazil. In the United States found at Washington, Connecticut; in Orange County, New York; and in many of the magnetite deposits of the Adirondack region, notably at Tahawus, Essex County, where it is actively mined. A large ilmenite-hematite deposit is mined at Allard Lake, Quebec.

Use. The major source of titanium. World production of ilmenite is approximately 3 million tons. It is used principally in the manufacture of titanium dioxide which is being used in increasingly large amounts as a paint pigment replacing older pigments, notably lead compounds. Also the use of metallic titanium as a structural material is increasing. Because of its high strength-to-weight ratio, titanium is proving to be a desirable metal for aircraft and space vehicle construction in both frames and engines. Ilmenite cannot be used as an iron ore because of difficulties in smelting it, but mixtures of ilmenite-magnetite and ilmenite-hematite are separated so that both titanium and iron can be recovered.

Name. From the Ilmen Mountains, U.S.S.R.

Similar Species. *Perovskite,* $CaTiO_3$, a pseudocubic titanium mineral found usually in nepheline syenites and carbonatites.

Rutile Group, XO_2

RUTILE—TiO_2

Crystallography. Tetragonal; $4/m2/m2/m$. Usually in prismatic crystals with dipyramid terminations common and vertically striated prism faces. Frequently in elbow twins, often repeated (Fig. 8.11) with twin plane $\{011\}$. Crystals frequently slender acicular. Also compact, massive.

Angles: $e(101) \wedge e'(011) = 45°2'$, $e(101) \wedge e''(\bar{1}01) = 65°35'$, $s(111) \wedge s'(\bar{1}11) = 56°26'$, $s(111) \wedge s''(\bar{1}\bar{1}1) = 84°40'$.

$P4_2/mnm$; $a = 4.59$, $c = 2.96$ Å; $a:c = 1:0.645$. $Z = 2$. d's: 3.24(10), 2.49(5), 2.18(3), 1.687(7), 1.354(4).

Physical Properties. *Cleavage* $\{110\}$ distinct. **H** 6–$6\frac{1}{2}$. **G** 4.18–4.25. *Luster* adamantine to submetallic. *Color* red, reddish-brown to black. *Streak* pale brown. Usually subtranslucent, may be transparent. *Optics:* (+); $\omega = 2.612$, $\epsilon = 2.899$.

Composition and Structure. Ti 60, O 40%. Although rutile is essentially TiO_2 some analyses report considerable amounts of Fe^{2+}, Fe^{3+}, Nb, and Ta. When Fe^{2+} substitutes for Ti^{4+}, electrical neutrality is maintained by twice as much Nb^{5+} and Ta^{5+} entering the structure. This leads to the formulation: $Fe_x (Nb, Ta)_{2x} Ti_{1-3x} O_2$. The structure of rutile is given in Fig. 8.4 and described on page 264. TiO_2 occurs in two additional, relatively rare polymorphs, tetragonal *anatase* (space group $I4_1/amd$) and orthorhombic *brookite* (space group $Pbca$). In anatase TiO_6 octahedra share four edges, in brookite the octahedra share three edges whereas in rutile they share only two edges. The stability fields of the three polymorphs are still not well defined. Rutile is generally considered to be a high T polymorph of TiO_2 and occurs in rocks which formed over a wide range of P and T. Its large stability range may suggest that the polymorph in which only two edges are shared between adjoining octahedra is indeed the most stable structure.

Diagnostic Features. Characterized by its peculiar adamantine luster and red color. Lower specific gravity distinguishes it from cassiterite. Infusible. Insoluble. After fusion with sodium carbonate it can be dissolved in sulfuric acid; the solution turns yellow on addition of hydrogen peroxide (Ti).

Occurrence. Rutile is found in granite, granite pegmatites, gneiss, mica schist, metamorphic limestone, and dolomite. It may be present as an accessory mineral in the rock, or in quartz veins traversing it. Often occurs as slender crystals in quartz

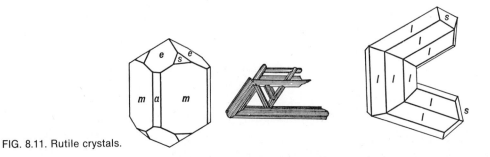

FIG. 8.11. Rutile crystals.

and micas. Is found in considerable quantities in black sands associated with ilmenite, magnetite, zircon, and monazite.

Notable European localities are: Krägerö, Norway; Yrieix, near Limoges, France; in Switzerland; and the Tyrol. Rutile from beach sands of northern New South Wales and southern Queensland makes Australia the largest producer of rutile. In the United States remarkable crystals have come from Graves Mountain, Lincoln County, Georgia. Rutile is found in Alexander County, North Carolina, and at Magnet Cove, Arkansas. It has been mined in Amherst and Nelson counties, Virginia, and derived in commercial quantities from the black sands of northeastern Florida.

Artificial. Single crystals of rutile have been manufactured by the Verneuil process. With the proper heat treatment they can be made transparent and nearly colorless, quite different from the natural mineral. Because of its high refractive index and dispersion, this synthetic material makes a beautiful cut gem stone with only a slight yellow tinge. It is sold under a variety of names, some of the better known are *titania, kenya gem,* and *miridis.*

Use. Most of the rutile produced is used as a coating of welding rods. Some titanium derived from rutile is used in alloys; for electrodes in arc lights; to give a yellow color to porcelain and false teeth. Manufactured oxide is used as a paint pigment. (See ilmenite.)

Name. From the Latin *rutilus,* red, an allusion to the color.

Similar Species. *Stishovite,* a very high-pressure polymorph of SiO_2, isostructural with rutile, is found rarely in meteorite impact craters. This dense form of SiO_2 is the only example of Si in 6 coordination rather than its common tetrahedral coordination. *Leucoxene,* a fine-grained, yellow to brown alteration product of minerals high in Ti (ilmenite, perovskite, sphene) consists mainly of rutile, less commonly anatase.

✻PYROLUSITE—MnO₂

Crystallography. Tetragonal; $4/m\,2/m\,2/m$. Rarely in well-developed crystals, *polianite.* Usually in radiating fibers or columns. Also granular massive; often in reniform coats and dendritic shapes (Fig. 8.12). Frequently pseudomorphous after manganite.

$P\,4_2/mnm$;   a = 4.39.   c = 2.86 Å;   a : c =

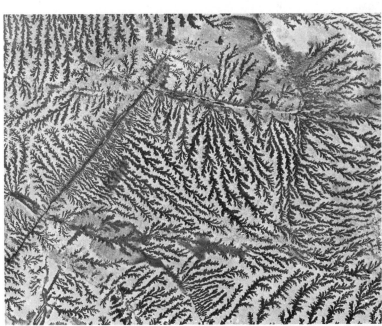

FIG. 8.12. Dendrites of pyrolusite on limestone, Sardinia.

$1:0.651$; $Z = 2$. $d's$: 3.11(10), 2.40(5), 2.11(4), 1.623(7), 1.303(3).

Physical Properties. *Cleavage* {110} perfect. **H** 1–2 (often soiling the fingers). For coarsely crystalline *polianite* the hardness is 6–6½. **G** 4.75. *Luster* metallic. *Color* and streak iron-black. Fracture splintery. Opaque.

Composition and Structure. Mn 63.2, O 36.8%. Commonly contains some H_2O. The structure is the same as that of rutile (see Fig. 8.4) with Mn in 6 coordination with oxygen.

Diagnostic Features. Characterized by and distinguished from other manganese minerals by its black streak, low hardness, and small amount of water. Infusible. Gives a bluish-green opaque bead with sodium carbonate. When heated in the closed tube it gives off oxygen which causes a splinter of charcoal to ignite when placed in tube above the mineral.

Occurrence. Manganese is present in small amounts in most crystalline rocks. When dissolved from these rocks, it may be redeposited as various minerals, but chiefly as pyrolusite. Nodular deposits of pyrolusite are found in bogs, on lake bottoms, and on the floors of seas and oceans. Nests and beds of manganese ores are found enclosed in residual clays, derived from the decay of manganiferous limestones. Also found in veins with quartz and various metallic minerals.

Pyrolusite is the most common manganese ore and is widespread in its occurrence. The chief manganese-producing countries are the U.S.S.R., Gabon, Brazil, Republic of South Africa, Australia, India, and Cuba. In the United States, manganese ores in small amount are found in Virginia, West Virginia, Georgia, Arkansas, Tennessee, with the iron ores of the Lake Superior districts, and in California.

Use. Most important manganese ore. Manganese is used with iron in the manufacture of *spiegeleisen* and *ferromanganese*, employed in making steel. This is its principal use for about 13.2 pounds of manganese are consumed in the production of one ton of steel. Also used in various alloys with copper, zinc, aluminum, tin, and lead. Pyrolusite is used as an oxidizer in the manufacture of chlorine,

bromine, and oxygen; as a disinfectant in potassium permanganate; as a drier in paints; as a decolorizer of glass; and in electric dry-cells and batteries. Manganese is also used as a coloring material in bricks, pottery, and glass.

Name. *Pyrolusite* is derived from two Greek words meaning *fire* and *to wash,* because it is used to free glass through its oxidizing effect of the colors due to iron.

Similar Species. *Alabandite*, MnS, is comparatively rare, associated with other sulfides in veins. *Wad* is the name given to manganese ore composed of an impure mixture of hydrous manganese oxides.

CASSITERITE—SnO$_2$

Crystallography. Tetragonal; $4/m\,2/m\,2/m$. The common forms are the prisms {110} and {010} and the dipyramids {111} and {011} (Fig. 8.13a). Frequently in elbow-shaped twins with a characteristic notch, giving rise to the miner's term *visor tin* (Fig. 8.13b); the twin plane is {011}. Usually massive granular; often in reniform shapes with radiating fibrous appearance, *wood tin*.

Angles: $m(110) \wedge s(111) = 46°27'$, $a(100) \wedge e(101) = 56°05'$, $s(111) \wedge s'(\bar{1}\bar{1}1) = 87°7'$, $e(101) \wedge e'(\bar{1}01) = 67°50'$.

$P4_2/mnm$; $a = 4.73$, $c = 3.18$ Å; $a:c = 1:0.672$, $Z = 2$. $d's$: 2.36(8), 2.64(7), 1.762(10), 1.672(4), 1.212(5).

Physical Properites. *Cleavage* {010} imperfect. **H** 6–7. **G** 6.8–7.1 (unusually high for a nonmetallic mineral). *Luster* adamantine to submetallic and dull. *Color* usually brown or black; rarely

FIG. 8.13. Cassiterite crystals.

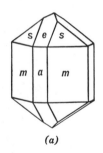

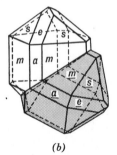

(a)   *(b)*

yellow or white. *Streak* white. Translucent, rarely transparent. *Optics:* (+); $\omega = 1.997$, $\epsilon = 2.093$.

Composition and Structure. Close to SnO_2 with Sn 78.6, O 21.4%. Small amounts of Fe^{3+} may be present and lesser amounts of Nb and Ta substitute for Sn. The structure is that of rutile (see Fig. 8.4 and page 264).

Diagnostic Features. Recognized by its high specific gravity, adamantine luster, and light streak. Infusible. Insoluble. Gives globule of tin with coating of white tin oxide when finely powdered mineral is fused on charcoal with reducing mixture. When cassiterite is placed in dilute HCl with metallic Zn the surface of the cassiterite is reduced and the specimen becomes coated with a dull gray deposit of metallic tin that becomes bright on rubbing.

Occurrence. Cassiterite is widely distributed in small amounts but is produced on a commercial scale in only a few localities. It has been noted as an original constituent of igneous rocks and pegmatites, but it is more commonly found in high-temperature hydrothermal veins in or near granitic rocks. Tin veins usually have minerals that contain fluorine or boron, such as tourmaline, topaz, fluorite, and apatite, and the minerals of the wall rocks are commonly much altered. Frequently associated with wolframite, molybdenite and arsenopyrite. Cassiterite is commonly found in the form of rolled pebbles in placer deposits, *stream tin*.

Most of the world's supply of tin comes from alluvial deposits chiefly in Malaysia, Bolivia, Indonesia, U.S.S.R., Thailand, and China. Today Bolivia is the only country where a significant production comes from vein deposits; but in the past the mines of Cornwall, England, were major producers. In the United States cassiterite is not found in sufficient quantities to warrant mining but is present in small amounts in numerous pegmatites.

Use. Principal ore of tin. The chief use of tin was in the manufacture of *tin plate* and *tern plate* for food containers. Tern plate is made by applying a coating of tin and lead instead of pure tin. At present aluminum, glass, paper, plastic, and tin-free steel are replacing tin plated containers. Tin is also used with lead in solders, in babbitt metal with antimony

and copper, and in bronze and bell-metal with copper. "Phosphor bronze" contains 89% Cu, 10% Sn, and 1% P. Artificial tin oxide is a polishing powder.

Name. From the Greek word meaning *tin*.

Similar Species. *Stannite,* Cu_2FeSnS_4, is structurally similar to chalcopyrite and sphalerite. It is a minor ore of tin.

URANINITE—UO$_2$

Crystallography. Isometric; $4/m\bar{3}2/m$. The rare crystals are usually octahedral with subordinate cube and dodecahedron faces. More commonly as the variety *pitchblende*: massive or botryoidal with a banded structure. *Kidney shaped*

$Fm3m$; $a = 5.46$ Å; $Z = 4$. *d's:* 3.15(7), 1.926(6), 1.647(10), 1.255(5), 1.114(5).

Physical Properties. H $5\frac{1}{2}$. G 7.5–9.7 for crystals; 6.5–9 for pitchblende. The specific gravity decreases with oxidation of U^{4+} to U^{6+}. *Luster* submetallic to pitchlike, dull. *Color* black. *Streak* brownish-black.

Composition and Structure. Uraninite is always partially oxidized, and thus the actual composition lies between UO_2 and U_3O_8 ($= U^{+4}O_2 + 2U^{+6}O_3$). Th can substitute for U and a complete series between uraninite and *thorianite,* ThO_2, has been prepared synthetically. In addition to Th, analyses usually show the presence of small amounts of Pb, Ra, Ce, Y, N, He, and A. The lead occurs as one of two stable isotopes (Pb^{206} and Pb^{207}) which result from the radioactive decay of uranium. For example: $U^{238} \rightarrow Pb^{206} + 8He^4$ and $U^{235} \rightarrow Pb^{207} + 7 He^4$. Thorium decays as follows: $Th^{232} \rightarrow Pb^{208} + 6He^4$. In addition to ionized helium atoms (α particles), electrons (β particles) are emitted during the decay process. Because the radioactive disintegration proceeds at a uniform and known rate, the accumulation of both helium and lead can be used as a measure of the time elapsed since the mineral crystallized. For example, for the deacy of U^{238}, the *half-life, T,* which is the time needed to reduce the number of U^{238} atoms by one-half, is 4.53×10^9 years. Both lead-uranium and helium-uranium ratios have been used by geologists to determine the age of rocks.

It was in uraninite that helium was first discovered on earth, having been previously noted in the sun's spectrum. Also radium was discovered in uraninite. The structures of uraninite and thorianite are like that of fluorite (Fig. 8.25) in which the metal atom is at the center of eight oxygens at the corner of a cube, with each oxygen at the center of a tetrahedral group of metal atoms.

Diagnostic Features. Characterized chiefly by its pitchy luster, high specific gravity, color, and streak. Infusible. See Table 4.21 for diagnostic U bead test. Because of its radioactivity, uraninite, as well as other uranium compounds, can be detected in small amounts by Geiger-Müller counters, ionization chambers, and similar instruments.

Occurrence. Uraninite occurs as a primary constituent of granitic rocks and pegmatites. It also is found in high-temperature hydrothermal veins associated with cassiterite, chalcopyrite, pyrite, and arsenopyrite as at Cornwall, England; or in medium-temperature veins with native silver and Co-Ni-As minerals as at Joachimsthal, Czechoslovakia, and Great Bear Lake, Canada. The world's most productive uranium mine during and immediately following World War II was the Shinkolobwe Mine in Zaire. Here uraninite and secondary uranium minerals are in vein deposits associated with cobalt and copper minerals. The ores also carry significant amounts of Mo, W, Au, and Pt.

Wilberforce, Ontario, has been a famous Canadian locality, but the present major Canadian sources are the mines at Great Bear Lake, the Beaverlodge region, Saskatchewan; and the Blind River area, Ontario. At Blind River uraninite occurs as detrital grains in a Precambrian quartz conglomerate. In the same type of deposit it is found in the gold-bearing Witwatersrand conglomerates, Republic of South Africa.

In the United States uraninite was found long ago in isolated crystals in pegmatites at Middletown, Glastonbury, and Branchville, Connecticut, and the mica mines of Mitchell County, North Carolina. A narrow vein of it was mined near Central City, Gilpin County, Colorado. Recent exploration has located many workable deposits of uraninite and associated uranium minerals on the Colorado Plateau in Arizona, Colorado, New Mexico, and Utah.

Use. Uraninite is the chief ore of uranium although other minerals are important sources of the element such as *carnotite* (page 334), *tyuyamunite* (page 335), *torbernite* (page 334), and *autunite* (page 334).

Uranium has assumed an important place among the elements because of its susceptibility to nuclear fission, a process by which the nuclei of uranium atoms are split apart with the generation of tremendous amounts of energy. This energy, first demonstrated in the atomic bomb, is now produced by nuclear-power reactors for generating electricity.

Uraninite is also the source of radium but contains it in very small amounts. Roughly 750 tons of ore must be mined in order to furnish 12 tons of concentrates; chemical treatment of these concentrates yields about 1 gram of a radium salt. In the form of various compounds uranium has a limited use in coloring glass and porcelain, in photography, and as a chemical reagent.

Name. Uraninite in allusion to the composition.

Similar Species. *Thorianite*, ThO_2, is dark gray to black with submetallic luster. Found chiefly in pegmatites and as water-worn crystals in stream gravels. *Cerianite*, $(Ce, Th)O_2$, an extremely rare mineral, has, as does thorianite, the fluorite structure.

Spinel Group, XY_2O_4

✳ **SPINEL—$MgAl_2O_4$**

Crystallography. Isometric; $4/m\bar{3}2/m$. Usually in octahedral crystals or in twinned octahedrons (spinel twins) (Figs. 8.14a and b). Dodecahedron may be present as small truncations (Fig. 8.14c), but other forms rare. Also massive and as irregular grains.

$Fd3m$; $a = 8.10$ Å; $Z = 8$; $d's$: 2.44(9), 2.02(9), 1.552(9), 1.427(10), 1.053(10).

Physical Properties. **H** 8. **G** 3.5–4.1. **G** 3.55 for the composition as given. Nonmetallic. *Luster* vitreous. *Color* various: white, red, lavender, blue,

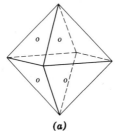 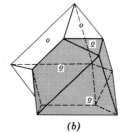 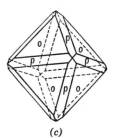

FIG. 8.14. Spinel crystals.　　**(a)**　　　　　　**(b)**　　　　　　**(c)**

green, brown, black. *Streak* white. Usually translucent, may be clear and transparent. *Optics:* $n = 1.718$.

Composition and Structure. MgO 28.2, Al_2O_3 71.8%. Fe^{2+}, Zn, and less commonly Mn^{2+} substitute for Mg in all proportions. Fe^{3+} and Cr may substitute in part for Al. The clear red, nearly pure magnesium spinel is known as *ruby spinel*. *Pleonaste,* an Fe-rich spinel is dark green to black, and *picotite* a Cr-rich spinel, yellowish to greenish-brown. The structure of spinel is illustrated in Fig. 8.5 and discussed on page 264.

Diagnostic Features. Recognized by its hardness (8), its octahedral crystals, and its vitreous luster. The iron spinel can be distinguished from magnetite by its nonmagnetic character and white streak. Infusible. The finely powdered mineral dissolves completely in the salt of phosphorus bead (proving the absence of silica).

Occurrence. Spinel is a common high-temperature mineral occurring in contact metamorphosed limestones and metamorphic argillaceous rocks poor in SiO_2. Occurs also as an accessory mineral in many dark igneous rocks. In contact metamorphic rocks it is associated with phlogopite, pyrrhotite, chondrodite, and graphite. Found frequently as rolled pebbles in stream sands, where it has been preserved because of its resistant physical and chemical properties. The ruby spinels are found in this way, often associated with the gem corundum, in the sands of Sri Lanka, Thailand, Upper Burma, and Malagasy Republic. Ordinary spinel is found in various localities in New York and New Jersey.

Use. When transparent and finely colored, spinel is used as a gem. Usually red and known as

ruby spinel or *balas ruby*. Some stones are blue. The largest cut stone known weighs about 80 carats. The stones are usually comparatively inexpensive.

Artificial. Synthetic spinel has been made by the Verneuil process (see corundum) in various colors rivaling the natural stones in beauty. Synthetic spinel is also used as a refractory.

Similar Species. *Hercynite,* $FeAl_2O_4$, is associated with corundum in some emery; also found with andalusite, sillimanite, and garnet. *Ferroan galaxite,* $(Mn, Fe)Al_2O_4$, an Fe^{2+} rich variety of galaxite, $MnAl_2O_4$, occurs sporadically in nature.

Gahnite—$ZnAl_2O_4$

Crystallography. Isometric; $4/m\overline{3}2/m$. Commonly octahedral with faces striated parallel to the edge between the dodecahedron and octahedron. Less frequently showing well-developed dodecahedrons and cubes.

$Fd3m$; $a = 8.12$ Å; $Z = 8$. $d's$: 2.85(7), 2.44(10), 1.65(4), 1.48(6), 1.232(8).

Physical Properties. H $7\frac{1}{2}$–8. G 4.55. *Luster* vitreous. *Color* dark green. *Streak* grayish. Translucent. *Optics:* $n = 1.80$.

Composition and Structure. Fe^{2+} and Mn^{2+} may substitute for Zn; and Fe^{3+} for Al. Structure is that of spinel (see Fig. 8.5 and page 264).

Diagnostic Features. Characterized by crystal form (striated octahedrons) and hardness. Infusible. The fine powder fused with sodium carbonate on charcoal gives a white nonvolatile coating of zinc oxide.

Occurrence. Gahnite is a rare mineral. It occurs in granitic pegmatites, in zinc deposits, and also as a contact metamorphic mineral in crystalline limestones. Found in large crystals at Bodenmais, Bavaria; in a talcose schist near Falun, Sweden. In the United States found at Charlemont, Massachusetts, and Franklin, New Jersey.

Name. After the Swedish chemist J. G. Gahn, the discoverer of manganese.

✹ MAGNETITE—Fe_3O_4

Crystallography. Isometric; $4/m\bar{3}2/m$. Magnetite is frequently in octahedral crystals (Fig. 8.15a), occasionally twinned on {111}. More rarely in dodecahedrons (Fig. 8.15b). Dodecahedrons may be striated parallel to the intersection with the octahedron (Fig. 8.15c). Other forms rare. Usually granular massive, coarse, or fine grained.

$Fd3m$; $a = 8.40$ Å; $Z = 8$. $d's$: 2.96(6), 2.53(10), 1.611(8), 1.481(9). 1.094(8).

Physical Properties. Octahedral parting on some specimens. **H** 6. **G** 5.18. *Luster* metallic. *Color* iron-black. *Streak* black. Strongly magnetic; may act as a natural magnet, known as *lodestone*. Opaque.

Composition and Structure. Fe 72.4, O 27.6%. The composition of most magnetite corresponds closely to Fe_3O_4. However, some analyses show a few percent of Mg and Mn^{2+} substituting for Fe^{2+} and Al, Cr, Mn^{3+}, and Ti^{4+} substituting for Fe^{3+}. Above 600°C a complete solid solution series is possible (see Fig. 8.10) between magnetite and *ulvöspinel*, Fe_2TiO_4, which can be rewritten as $Fe^{3+}(Ti, Fe^{2+})_2O_4$. The structure of magnetite is that of an *inverse spinel* (see page 265), which can be expressed by rewriting the formula as $Fe^{3+}(Fe^{2+}, Fe^{3+})_2O_4$.

Diagnostic Features. Characterized chiefly by its strong magnetism, its black color, and its hardness (6). Can be distinguished from magnetic franklinite by streak. Infusible. Slowly soluble in HCl, and solution reacts for both ferrous and ferric iron.

Occurrence. Magnetite is a common mineral found disseminated as an accessory through most igneous rocks. In certain types of rocks, through magmatic segragation, magnetite may become one of the chief constituents of the rock and may thus form large ore bodies. Such bodies are often highly titaniferous. Commonly associated with crystalline metamorphic rocks and may occur as large beds and lenses. A common constituent of sedimentary and metamorphic banded Precambrian iron-formations. In such occurrences it is probably of chemical, sedimentary origin. Found in the black sands of the seashore. Occurs as thin plates and dendritic growths between plates of mica. Often intimately associated with corundum, forming emery.

The largest magnetite deposits in the world are in northern Sweden at Kiruna and Gellivare associated with apatite, and are believed to have formed by magmatic segregation. Other important deposits are in Norway, Rumania, and U.S.S.R. The most powerful natural magnets are found in Siberia, in the Harz Mountains, on the Island of Elba, and in the Bushveld igneous complex, Republic of South Africa. The extensive Precambrian iron deposits in the Lake Superior region and in the Labrador Trough, Canada, carry considerable magnetite as an ore mineral.

The Precambrian iron-formation of the Lake Superior region, containing about 25% iron mostly in the form of magnetite, is today a major source of iron in the United States. This low-grade ore, *taconite*, is actively mined and the magnetite sepa-

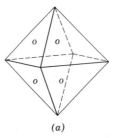 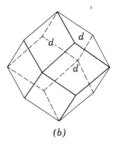 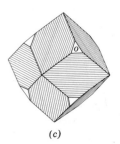

FIG. 8.15. Magnetite crystals. (a) (b) (c)

rated magnetically from the waste material. Commercial quantities also occur in several localities in the Adirondack region in New York, Utah, California, New Jersey, and Pennsylvania. Found as lodestone and in crystals at Magnet Cove, Arkansas.

Use. An important iron ore.

Name. Probably derived from the locality Magnesia, bordering on Macedonia. A fable, told by Pliny, ascribes its name to a shepherd named Magnes, who first discovered the mineral on Mount Ida by noting that the nails of his shoes and the iron ferrule of his staff adhered to the ground.

Similar Species. *Magnesioferrite,* $Fe^{3+}(Mg, Fe^{3+})_2O_4$ (or $MgFe_2O_4$) with an inverse spinel structure, is a rare mineral found chiefly in fumaroles. *Jacobsite,* $Fe^{3+}(Mn, Fe^{3+})_2O_4$ (or $MnFe_2O_4$), an inverse spinel and a rare mineral found at Langban, Sweden. *Ulvöspinel,* $Fe^{3+}(Ti, Fe^{2+})_2O_4$ (or Fe_2TiO_4), also with an inverse spinel structure, is not uncommonly present as exsolution blebs and lamellae within magnetite.

Franklinite—(Zn, Fe, Mn) (Fe, Mn)$_2$O$_4$

Crystallography. Isometric; $4/m\overline{3}2/m$. Crystals octahedral with dodecahedral truncations; often rounded. Also massive, coarse, or fine granular, in rounded grains.

$Fd3m$; $a = 8.42$ Å; $Z = 8$. $d's$: 2.51(10), 1.610(10), 1.480(9), 1.278(6), 1.091(6).

Physical Properties. **H** 6. **G** 5.15. *Luster* metallic. *Color* iron-black. *Streak* reddish-brown to dark brown. Slightly magnetic.

Composition and Structure. Dominantly $ZnFe_2O_4$ but always with some substitution of Fe^{2+} and Mn^{2+} for Zn, and Mn^{3+} for Fe^{3+}. Analyses show a wide range in the proportions of the various elements. The structure of franklinite is that of a normal spinel (see page 264 and Fig. 8.5).

Diagnostic Features. Resembles magnetite but is only slightly magnetic and has a dark brown streak. Usually identified by its characteristic association with willemite and zincite. Infusible. Becomes strongly magnetic on heating in the reducing flame. Gives an Mn bead test (Table 4.21).

Occurrence. Franklinite, with only minor exceptions, is confined to the zinc deposits at Franklin, New Jersey, where, enclosed in a granular limestone, it is associated with zincite and willemite.

Use. As an ore of zinc and manganese. The zinc is converted into zinc white, ZnO, and the residue is smelted to form an alloy of iron and manganese, *spiegeleisen,* used in the manufacture of steel.

Name. From Franklin, New Jersey.

CHROMITE—$\overset{(Mg)}{Fe}Cr_2O_4$

Crystallography. Isometric; $4/m\overline{3}2/m$. Habit octahedral but crystals small and rare. Commonly massive, granular to compact.

$Fd3m$; $a = 8.36$ Å; $Z = 8$. $d's$: 4.83(4), 2.51(10), 2.08(5), 1.602(6), 1.473(8).

Physical Properties. **H** $5\frac{1}{2}$. **G** 4.6. *Luster* metallic to submetallic; frequently pitchy. *Color* iron-black to brownish black. *Streak* dark brown. Subtranslucent. *Optics:* $n = 2.16$.

Composition and Structure. For $FeCr_2O_4$, FeO 32.0, Cr_2O_3 68.0%. Some Mg is always present substituting for Fe^{2+} and some Al and Fe^{3+} may substitute for chromium. A complete series probably extends to $MgCr_2O_4$. The structure of chromite is that of a normal spinel (see page 264 and Fig. 8.5).

Diagnostic Features. The submetallic luster usually distinguishes chromite. A green borax bead is diagnostic of Cr (see Table 4.21). When fused on charcoal with sodium carbonate it gives a magnetic residue.

Occurrence. Chromite is a common constituent of peridotites and other ultrabasic rocks and of serpentines derived from them. One of the first minerals to separate from a cooling magma; large chromite ore deposits are thought to have been derived by such magmatic differentiation. For example, the Bushveld igneous complex of South Africa and the Greak Dike of Rhodesia contain many seams of chromite enclosed in pyroxenites. Associated with olivine, serpentine, and corundum.

The important countries for its production are U.S.S.R., Republic of South Africa, Turkey, the Philippines, and Rhodesia. Found only sparingly in the United States. Pennsylvania, Maryland, North Carolina, and Wyoming have produced it in the past. California, Alaska, and Oregon are small producers. During World War II, bands of chromite were mined in the Stillwater igneous complex, Montana.

Use. The only ore of chromium. Chromite ores are grouped into three categories—metallurgical, refractory, and chemical—on the basis of their chrome content and their Cr/Fe ratio. As a metal, chromium is used as a ferroalloy to give steel the combined properties of high hardness, great toughness, and resistance to chemical attack. Chromium is a major constituent in stainless steel. *Nichrome,* an alloy of Ni and Cr, is used for resistance in electrical heating equipment. Chromium is widely used in plating plumbing fixtures, automobile accessories, and so forth.

Because of its refractory character, chromite is made into bricks for the linings of metallurgical furnaces. The bricks are usually made of crude chromite and coal tar but sometimes of chromite with kaolin, bauxite, or other materials. Chromium is a constituent of certain green, yellow, orange, and red pigments and in $K_2Cr_2O_7$ and $Na_2Cr_2O_7$, which are used as mordants to fix dyes.

Similar Species. *Magnesiochromite,* $MgCr_2O_4$, is in both occurrence and appearance similar to chromite.

Name. Named in allusion to the composition.

Chrysoberyl—$BeAl_2O_4$

Crystallography. Orthorhombic; $2/m2/m2/m$. Usually in crystals tabular on {001}, the faces of which are striated parallel to [100]. Commonly twinned on {130} giving pseudohexagonal appearance (Fig. 8.16).

FIG. 8.16. Twinned chrysoberyl.

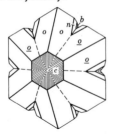

Angles: $c(001) \wedge o(111) = 43°5'$, $b(010) \wedge m(110) = 59°48'$, $c(001) \wedge n(121) = 51°8'$.

Pmnb: $a = 5.47$, $b = 9.39$, $c = 4.42$ Å; $a:b:c = 0.583:1:0.471$; $Z = 4$. *d's:* 3.24(8), 2.33(8), 2.57(8), 2.08(10), 1.61(10).

Physical Properties. *Cleavage* {110}. **H** $8\frac{1}{2}$. **G** 3.65–3.8. *Luster* vitreous. *Color* various shades of green, brown, yellow; may be red by transmitted light. *Optics:* (+); $\alpha = 1.746$, $\beta = 1.748$, $\gamma = 1.756$; $2V = 45°$; $X = c$, $Y = b$.

Alexandrite is a gem variety, emerald-green in daylight, but red by transmitted light and usually red by artificial light. *Cat's eye,* or *cymophane,* is a chatoyant variety which, when cut as an oval or round *cabochon* gem, shows a narrow band of light on its surface. The effect results from minute tubelike cavities or needlelike inclusions with parallel orientation.

Composition and Structure. BeO 19.8, Al_2O_3 80.2%. Be 7.1%. From its formula, $BeAl_2O_4$, it would appear that chrysoberyl is a member of the spinel group. However, because of the small size of Be^{2+}, chrysoberyl has a structure of lower symmetry. It consists of oxygen atoms in hexagonal closest packing with Be in 4 coordination with O and Al in 6 coordination with O. Morphologically and structurally chrysoberyl is very similar to olivine $(Mg, Fe)_2SiO_4$. Be is in the same coordination as Si, and Al in the same coordination as Fe^{2+} or Mg. The hexagonal close-packed arrangement of O in chrysoberyl leads to pseudohexagonal lattice, angles, and twinning.

Diagnostic Features. Characterized by its high hardness, its yellowish to emerald-green color, and its twin crystals. Infusible. Insoluble. Wholly soluble in the salt of phosphorus bead (absence of silica).

Occurrence. Chrysoberyl is a rare mineral. It occurs in granitic rocks, pegmatites, and mica schists. Frequently in river sands and gravels. The outstanding alluvial gem deposits are found in Brazil and Sri Lanka; the *alexandrite* variety comes from the Ural Mountains. In the United States chrysoberyl of gem quality is rarely found. It has been found in Oxford County and elsewhere in Maine; Haddam, Connecticut; and Greenfield, New York. Recently found in Colorado.

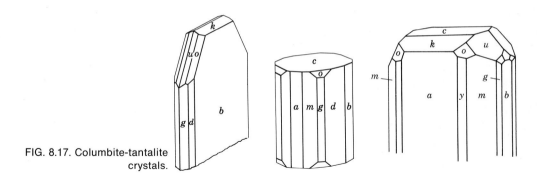

FIG. 8.17. Columbite-tantalite crystals.

Use. As a gemstone. The ordinary yellowish-green stones are inexpensive; the varieties alexandrite and cat's eye are highly prized gems.

Name. *Chrysoberyl* means *golden beryl.* *Cymophane* is derived from two Greek words meaning *wave* and *to appear,* in allusion to the chatoyant effect. *Alexandrite* was named in honor of Alexander II of Russia.

Columbite-Tantalite— (Fe, Mn)Nb$_2$O$_6$-(Fe, Mn)Ta$_2$O$_6$

Crystallography. Orthorhombic; $2/m\,2/m\,2/m$. Commonly in crystals. The habit is short prismatic or thin tabular on {010}; often in square prisms because of prominent development of {100} and {010} (Fig. 8.17). Also in heart-shaped twins, twinned on {201}.

Angles: $b(010) \wedge m(110) = 68°05'$, $b(010) \wedge g(130) = 39°39'$, $c(001) \wedge u(111) = 43°49'$.

Pcan; $a = 5.74$, $b = 14.27$, $c = 5.09$ Å; $a:b:c = 0.402:1:0:0.357$ (for 17% Ta$_2$O$_5$);

$Z = 4$. $d's$ for columbite: 3.66(7), 2.97(10), 1.767(6), 1.735(7), 1.712(8).

Physical Properties. *Cleavage* {010} good. **H** 6. **G** 5.2–7.9, varying with the composition, increasing with rise in percentage of Ta$_2$O$_5$ (Fig. 8.18). *Luster* submetallic. *Color* iron-black, frequently iridescent. *Streak* dark red to black. Subtranslucent.

Composition and Structure. A complete solid solution exists from *columbite,* (Fe, Mn)Nb$_2$O$_6$, to *tantalite,* (Fe, Mn)Ta$_2$O$_6$. Often contains small amounts of Sn and W. A variety known as *manganotantalite* is essentially tantalite with Mn^{2+} substituting for most of the Fe^{2+}. In the structure octahedral chains of (Mn, Fe)O$_6$ and (Ta, Nb)O$_6$ exist in which octahedra join through sharing of edges. The chains are linked to each other by common apices.

Diagnostic Features. Recognized usually by its black color with lighter colored streak, and high specific gravity. Distinguished from wolframite by lower specific gravity and less distinct cleavage. Fuses with difficulty (5–5½). Fused with borax; the

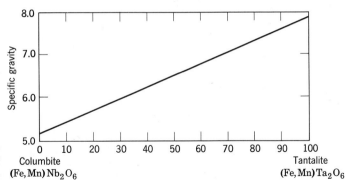

FIG. 8.18. Columbite-tantalite. Variation of specific gravity with composition.

bead dissolved in HCl; the solution boiled with Sn gives a blue color (Nb). Generally gives a bead test for Mn (Table 4.21).

Occurrence. The minerals occur in granitic rocks and pegmatites, associated with quartz, feldspar, mica, tourmaline, beryl, spodumene, cassiterite, samarskite, wolframite, microlite, and monazite.

Notable localities for their occurrence are Canada; Zaire; Nigeria; Brazil; near Moss, Norway; Bodenmais, Bavaria; Ilmen Mountains, U.S.S.R.; Western Australia (*manganotantalite*); and Malagasy Republic. In the United States it is found at Standish, Maine; Haddam, Middletown, and Branchville, Connecticut; in Amelia County, Virginia; Mitchell County, North Carolina; Black Hills, South Dakota; and near Canon City, Colorado.

Use. Source of tantalum and niobium. Because of its resistance to acid corrosion, tantalum is employed in chemical equipment, in surgery for skull plates and sutures, also in some tool steels and in electronic tubes. Niobium has its chief use in alloys in weldable high-speed steels, stainless steels, and alloys resistant to high temperatures, such as used in the gas turbine of the aircraft industry.

Name. *Columbite* from Columbia, a name for America, whence the original specimen was obtained. *Tantalite* from the mythical Tantalus in allusion to the difficulty in dissolving in acid.

Similar Species. *Microlite,* $Ca_2Ta_2O_6(O, OH,F)$, is found in pegmatites; *pyrochlore,* $(Ca, Na)_2(Nb,Ta)_2O_6(O,OH,F)$, and *fergusonite,* $YNbO_4$, an oxide of niobium, tantalum, and rare earths (represented by Y in the fergusonite formula) are found associated with alkalic rocks.

HYDROXIDES

The hydroxide minerals to be considered in this section are:

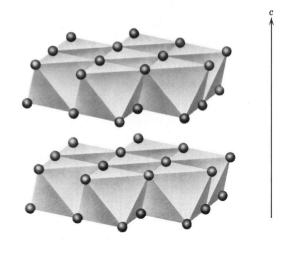

FIG. 8.19. Structure of brucite, composed of parallel layers of Mg^{2+} in octahedral coordination with $(OH)^-$. Large spacing along the c axis is caused by weak, residual bonding between adjacent layers.

All structures in this group are characterized by the presence of the hydroxyl $(OH)^-$ group, or H_2O molecules. The presence of $(OH)^-$ groups causes the bond strengths in these structures generally to be much weaker than in the oxides.

The structure of *brucite* (see Fig. 8.19) consists of Mg^{2+} octahedrally coordinated to $(OH)^-$, with the octahedra sharing edges to form a layer. Because each $(OH)^-$ group is shared between three adjoining octahedra, the Mg^{2+} to $(OH)^-$ bond strengths have an e.v. $= \frac{1}{3}$. With three such bonds ($3 \times \frac{1}{3} = 1$) the $(OH)^-$ group is neutralized. For this reason, the layers in the brucite structure are held together by only weak residual bonds (compare Fig. 8.19 with Fig. 8.3). The structure of *gibbsite*, $Al(OH)_3$, is in principle identical to that of brucite except that, because of charge requirements, $\frac{1}{3}$ of the octahedrally coordinated cation positions are vacant (see Fig.

HYDROXIDES

		GOETHITE GROUP	
Brucite	$Mg(OH)_2$	Diaspore	$\alpha AlO.OH$
Manganite	$MnO(OH)$	Goethite	$\alpha FeO.OH$
Psilomelane	$(Ba, Mn)_3(O, OH)_6Mn_8O_{16}$	Bauxite—mixture of diaspore, gibbsite, and boehmite	

8.2a). A basal structural layer in corundum (as in Fig. 8.2) is structurally equivalent to the Al-OH sheets in gibbsite. The brucite structure type is referred to as *trioctahedral* (each (OH)⁻ group is surrounded by three occupied octahedral positions) and the gibbsite structure type as *dioctahedral* (only two out of three octahedrally coordinated cation sites are filled). Such trioctahedral and dioctahedral layers are essential building units of the phyllosilicates (see page 392).

The structure of *diaspore* (αAlO·OH; space group *Pbnm*) is shown in Fig. 8.20. Oxygen and (OH)⁻ groups are arranged in hexagonal closest packing with Al^{3+} in octahedral coordination between them. A chainlike pattern is produced by $Al(O,OH)_6$ octahedra extending along the *c* axis. The octahedra in each chain share edges and the chains are joined to each other by adjoining apical oxygens. *Goethite* (αFeO·OH) is isostructural with diaspore. Both compositions, AlO·OH and FeO·OH, occur in nature in two crystal forms. γAlO·OH, *boehmite*, and γFeO·OH, *lepidochrocite*, both with space group *Amam*, show a somewhat different linkage of the octahedrally coordinated cations from that found in diaspore or goethite as seen in Fig. 8.21. The octahedra are linked by their apices to form chains; the chains, in turn, are joined by the sharing of octahedral edges which results in corrugated sheets parallel to {010}. The sheets are weakly held together by hydrogen bonds between pairs of oxygens. Such bonding is represented as (OH)⁻ groups in Fig. 8.21.

✷Brucite—Mg(OH)₂

Crystallography. Hexagonal-*R*; $\bar{3}2/m$. Crystals usually tabular on {0001} and may show small rhombohedral truncations. Commonly foliated, massive.

$C\bar{3}m$; $a = 3.13$, $c = 4.74$ Å; $a:c = 1:1.514$; $Z = 1$. $d's$: 4.74(8), 2.37(10), 1.793(10), 1.372(7), 1.189(9).

Physical Properties. *Cleavage* {0001} perfect. Folia flexible but not elastic. Sectile. **H** 2½. **G** 2.39. *Luster* on base pearly, elsewhere vitreous to waxy. *Color* white, gray, light green. Transparent to translucent. *Optics.* (+), $\omega = 1.566$, $\epsilon = 1.581$.

Composition and Structure. For Mg(OH)₂: MgO 69.0, H₂O 31.0%. Fe^{2+} and Mn^{2+} may substitute for Mg. The structure is illustrated in Fig. 8.19. The perfect {0001} cleavage is parallel to the octahedral sheets. Upon heating, brucite transforms into periclase (MgO).

Diagnostic Features. Recognized by its foliated nature, light color, and pearly luster on cleavage face. Distinguished from talc by its greater hardness and lack of greasy feel, and from mica by being inelastic. Infusible. Glows before the blowpipe. Gives water in the closed tube. Easily soluble in HCl and gives test for Mg.

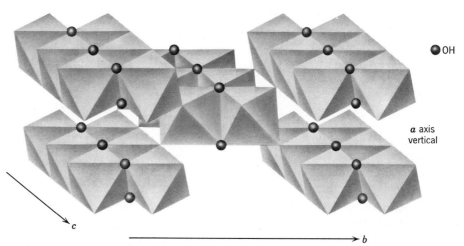

FIG. 8.20. The structure of diaspore (αAlO·OH) and goethite (αFeO·OH). The double chains of $AlO_3(OH)_3$ or $FeO_3(OH)_3$ octahedra run parallel to the *c* axis. Only (OH) groups are indicated; all unmarked apices of octahedra represent oxygen.

● OH

a axis vertical

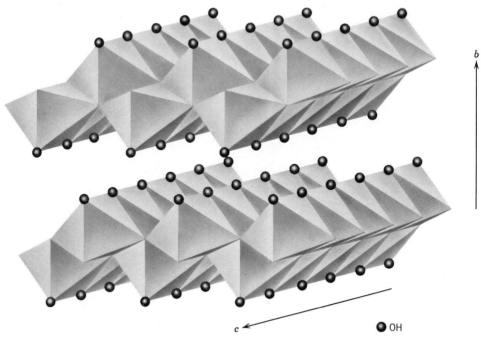

FIG. 8.21. The structure of boehmite (γAlO·OH) and lepidochrocite (γFeO·OH) showing the arrangement of $AlO_4(OH)_2$ or $FeO_4(OH)_2$ octahedra. The corrugated sheets are parallel to (010). Only (OH) groups are indicated; all unmarked apices of octahedra represent oxygen.

Occurrence. Brucite is found associated with serpentine, dolomite, magnesite, and chromite; as an alteration product of periclase and magnesium silicates, especially serpentine. It is also found in crystalline limestone.

Notable foreign localities for its occurrence are at Unst, one of the Shetland Islands, and Aosta, Italy. In the United States found at Tilly Foster Iron Mine, Brewster, New York; at Wood's Mine, Texas, Pennsylvania; and Gabbs, Nevada.

Use. Brucite is used as a raw material for magnesia refractories and is a minor source of metallic magnesium.

Name. In honor of the early American mineralogist Archibald Bruce.

Similar Species. *Gibbsite*, $Al(OH)_3$, one of the three hydroxides of Al that are the main constituents of bauxite. The structure of gibbsite is like that of brucite, with one-third of the octahedral cation positions vacant (compare with Fig. 8.2a).

MANGANITE—MnO(OH)

Crystallography. Monoclinic; $2/m$ (pseudoorthorhombic). Crystals usually prismatic parallel to c and vertically striated (Fig. 8.22). Often columnar to coarse fibrous. Twinned on {011} as both contact and penetration twins.

Angles: $m(210) \wedge m(2\bar{1}0) = 99°40'$, $u(101) \wedge u'(\bar{1}01) = 65°41'$.

$B2_1/d$; $a = 8.84$, $b = 5.23$, $c = 5.74$ Å, $\beta = 90°$, $a:b:c = 1.690:1:1.098$; $Z = 8$. $d's$: 3.38(10), 2.62(9), 2.41(6), 2.26(7), 1.661(9).

Physical Properties. Cleavage {010} perfect. {110} and {001} good. **H** 4. **G** 4.3. *Luster* metallic. *Color* steel-gray to iron-black. *Streak* dark brown. Opaque.

Composition and Structure. Mn 62.4, O 27.3, H_2O 10.3%. The structure consists of hexagonal closest packing of oxygen and $(OH)^-$ groups with Mn^{3+} in octahedral coordination with O^{2-} and $(OH)^-$. In this respect the structure is similar to that

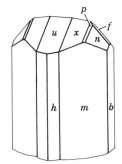

FIG. 8.22. Manganite.

of diaspore; however the distribution of the cations is quite different.

Diagnostic Features. Recognized chiefly by its black color and prismatic crystals. Hardness (4) and brown streak distinguish it from pyrolusite. Infusible. Powdered mineral gives a bluish-green bead with sodium carbonate. Much water when heated in the closed tube.

Occurrence. Manganite is found associated with other manganese oxides in deposits formed by meteoric waters. Found often in low-temperature hydrothermal veins associated with barite, siderite, and calcite. It frequently alters to pyrolusite.

Occurs at Ilfeld, Harz Mountains, in fine crystals; also at Ilmenau, Thuringia, and Cornwall, England. In the United States at Negaunee, Michigan. In Nova Scotia.

Use. A minor ore of manganese.

Name. Named in allusion to the composition.

PSILOMELANE—$(Ba^{2+},Mn^{2+})_3(O,OH)_6Mn_8^{4+}O_{16}$
Crystallography. Orthorhombic, 222. Massive, botryoidal, stalactitic. Appears amorphous.

$P222$; $a = 9.45$, $b = 13.90$, $c = 5.72$; $a:b:c = 0.680:1:0.411$; $Z = 2$. d's: 3.46(7), 2.87(7), 2.41(7), 2.190(10), 1.820(7).

Physical Properties. H 5–6. **G** 3.7–4.7. *Luster* submetallic. *Color* black. *Streak* brownish-black. Opaque.

Composition and Structure. Small amounts of Mg, Ca, Ni, Co, Cu, and Si may be present. The structure is somewhat similar to that of rutile (Fig. 8.4) with complex chains of MnO,OH octahedra and large channels between adjoining chains. Both Ba and adsorbed H_2O are located in these channels. Upon heating to about 600°C, psilomelane transforms into *hollandite,* $Ba_2Mn_8O_{16}$.

Diagnostic Features. Distinguished from the other manganese oxides by its greater hardness and botryoidal form, and from limonite by its black streak. Infusible. With sodium carbonate gives an opaque bluish-green bead.

Occurrence. Psilomelane is a secondary mineral; it occurs usually with pyrolusite, and its origin and associations are similar to those of that mineral.

Use. An ore of manganese. (See pyrolusite.)

Name. Derived from two Greek words meaning *smooth* and *black,* in allusion to its appearance.

Diaspore—$\alpha AlO·OH$
Crystallography. Orthorhombic; $2/m2/m2/m$. Usually in thin crystals, tabular parallel to {010}; sometimes elongated on [001]. Bladed, foliated massive, disseminated.

Angles: (010) $\wedge$ (110) = 64°53'.

Pbnm; $a = 4.41$, $b = 9.40$, $c = 2.84$ Å; $a:b:c = 0.469:1:0.302$; $Z = 4$. d's: 3.98(10), 2.31(8), 2.12(7), 2.07(7), 1.629(8).

Physical Properties. *Cleavage* {010} perfect. **H** $6\frac{1}{2}$–7. **G** 3.35–3.45. *Luster* vitreous except on cleavage face, where it is pearly. *Color* white, gray, yellowish, greenish. Transparent to translucent. *Optics:* (+), $\alpha = 1.702$, $\beta = 1.722$, $\gamma = 1.750$; $2V = 85°$; $X = c, Y = b; r < v$.

Composition and Structure. Al_2O_3 85, H_2O 15%. The structure is illustrated in Fig. 8.20 and consists of Al in 6 coordination with oxygen and OH^- forming $AlO_3(OH)_3$ octahedra. The structure of *boehmite,* $\gamma AlO·OH$ is shown in Fig. 8.21.

Diagnostic Features. Characterized by its good cleavage, its bladed habit, and its high hardness. Infusible. Insoluble. Decrepitates and gives water when heated in the closed tube. Ignited with cobalt nitrate turns blue (Al).

Occurrence. Diaspore is commonly associated with corundum in emery rock, in dolomite, and in chlorite schist. In a fine-grained massive form it is a major constituent of much bauxite.

Notable localities are the Ural Mountains; Schemnitz, Czechoslovakia; Campolungo in Switzerland; and the island of Naxos, Greece. In the United States it is found in Chester County, Pennsylvania; at Chester, Massachusetts; with alunite, forming rock masses at Mt. Robinson, Rosite Hills, Colorado. It is found abundantly in the bauxite and aluminous clays of Arkansas, Missouri, and elsewhere in the United States.

Use. As a refractory.

Name. Derived from a Greek word meaning *to scatter,* in allusion to its decrepitation when heated.

GOETHITE—αFeO·OH

Crystallography. Orthorhombic; $2/m\,2/m\,2/m$. Rarely in distinct prismatic, vertically striated

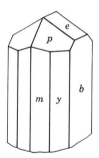

FIG. 8.23. Goethite.

crystals (Fig. 8.23). Often flattened parallel to {010}. In acicular crystals. Also massive, reniform, stalactitic in radiating fibrous aggregates (Fig. 8.24). Foliated. The so-called bog ore is generally loose and porous.

Angles: $b(010) \wedge m(110) = 65°20'$, $b(010) \wedge y(120) = 47°26'$, $e(021) \wedge e'(0\overline{2}1) = 62°30'$.

FIG. 8.24. Goethite from an iron mine near Negaunee, Michigan.

Pbnm; $a = 4.65$, $b = 10.02$, $c = 3.04$ Å; $a:b:c = 0.464:1:0.303$; $Z = 4$. $d's$: 4.21(10), 2.69(8), 2.44(7), 2.18(4), 1.719(5).

Physical Properties. *Cleavage* {010} perfect. **H** 5–5½. **G** 4.37; may be as low as 3.3 for impure material. *Luster* adamantine to dull, silky in certain fine, scaly or fibrous varieties. *Color* yellowish-brown to dark brown. *Streak* yellowish-brown. Sub-translucent.

Composition and Structure. Fe 62.9, O 27.0, H_2O 10.1%. Mn is often present in amounts up to 5%. The massive varieties often contain adsorbed or capillary H_2O. Goethite is isostructural with dia-spore (see Fig. 8.20). *Lepidochrosite*, $\gamma FeO \cdot OH$, a polymorph of goethite is a platy mineral and is often associated with goethite; it is isostructural with boehmite (see Fig. 8.21).

Diagnostic Features. Distinguished from hematite by its streak. Fusible with difficulty (5–5½). Becomes magnetic in the reducing flame and gives water in the closed tube.

Occurrence. Goethite is one of the com-monest minerals and is typically formed under oxi-dizing conditions as a weathering product of iron-bearing minerals. It also forms as a direct inorganic or biogenic precipitate from water and is wide-spread as a deposit in bogs and springs. Goethite forms the gossan or "iron hat" over metalliferous veins. Large quantities of goethite have been found as residual lateritic mantles resulting from the weathering of serpentine.

Goethite in some localities constitutes an im-portant ore of iron. It is the principal consistuent of the valuable minette ores of Alsace-Lorraine. Other notable European localities are: Eiserfeld in West-phalia; Pribram, Bohemia; and Cornwall, England. Large deposits of iron-rich laterites composed essen-tially of goethite are found in the Mayari and Moa districts of Cuba.

In the United States goethite is common in the Lake Superior hematite deposits and has been ob-tained in fine specimens at Negaunee, near Mar-quette, Michigan. Goethite is found in iron-bearing limestones along the Appalachian Mountains, from western Massachusetts as far south as Alabama. Such deposits are particularly important in Ala-bama, Georgia, Virginia, and Tennessee. Finely crystallized material occurs with smoky quartz and microcline in Colorado at Fluorissant and in the Pikes Peak region.

Use. An ore of iron.

Name. In honor of Goethe, the German poet.

Similar Species. *Limonite*, $FeO \cdot OH \cdot nH_2O$, is used mainly as a field term to refer to natural hydrous iron oxides of uncertain identity.

✳BAUXITE*—a mixture of diaspore, gibbsite, and boehmite

Crystallography. A mixture. Pisolitic, in round concretionary grains; also massive, earthy, claylike.

Physical Properties. **H** 1–3. **G** 2–2.55. *Luster* dull to earthy. *Color* white, gray, yellow, red. Trans-lucent.

Composition. A mixture of hydrous aluminum oxides in varying proportions. Some bauxites closely approach the composition of *gibbsite*, $Al(OH)_3$ (see brucite, similar species), but most are a mixture and usually contain some Fe. As a result, bauxite is not a mineral and should be used only as a rock name. The principal constituents of the rock bauxite are *gibbsite; boehmite*, $\gamma AlO \cdot OH$ (see Fig. 8.21), and *diaspore*, $\alpha AlO \cdot OH$ (see Fig. 8.20), any one of which may be dominant.

Diagnostic Features. Can usually be recog-nized by its pisolitic character. Infusible. Insoluble. Assumes a blue color when moistened with cobalt nitrate and then ignited (Al). Gives water in the closed tube.

Occurrence. Bauxite is of supergene origin, commonly produced under subtropical to tropical climatic conditions by prolonged weathering and leaching of silica from aluminum-bearing rocks. Also may be derived from the weathering of clay-bearing limestones. It has apparently originated as a colloidal precipitate. It may occur in place as a direct derivative of the original rock, or it may have been transported and deposited in a sedimentary formation. In the tropics deposits known as *laterites*,

---

* Although bauxite is not a mineral species, it is described here because of its importance as the ore of aluminum.

consisting largely of hydrous aluminum and ferric oxides, are found in the residual soils. These vary widely in composition and purity but many are valuable as sources of aluminum and iron.

Bauxite occurs over a large area in the south of France, an important district being at Baux, near Arles, France. The principal world producers are Surinam, Jamaica, and Guiana. Other major producing countries are Indonesia, U.S.S.R., Australia, and Hungary. In the United States, the chief deposits are found in Arkansas, Georgia, and Alabama. In Arkansas bauxite has formed by the alteration of a nepheline syenite.

Use. The ore of aluminim. Eighty-five percent of the bauxite produced is consumed as aluminum ore. Because of its low density and great strength, aluminum has been adapted to many uses. Sheets, tubes, and castings of aluminum are used in automobiles, airplanes, and railway cars, where light weight is desirable. It is manufactured into cooking utensils, food containers, household appliances, and furniture. Aluminum is replacing copper to some extent in electrical transmission lines. Aluminum is alloyed with copper, magnesium, zinc, nickel, silicon, silver, and tin. Other uses are in paint, aluminum foil, and numerous salts.

The second largest use of bauxite is in the manufacture of Al_2O_3, which is used as an abrasive. It is also manufactured into aluminous refractories. Synthetic alumina is also used as the principal ingredient in heat-resistant porcelain such as spark plugs.

Name. From its occurrence at Baux, France.

HALIDES

The chemical class of halides is characterized by the dominance of the electronegative halogen ions, Cl^-, Br^-, F^-, and I^-. These ions are large, have a charge of only -1, and are easily polarized. When they combine with relatively large, weakly polarized cations of low valence, both cations and anions behave as almost perfectly spherical bodies. The packing of these spherical units leads to structures of the highest possible symmetry.

The structure of NaCl, shown in Figs. 8.25 and 4.15, was the first to be determined by X-ray diffrac-

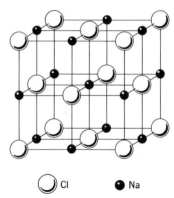

FIG. 8.25. The structure of halite.

tion techniques by W. H. and W. L. Bragg in 1913. The arrangement of the ions in the structure showed unambiguously that no molecules exist in the NaCl structure. Each cation and each anion is surrounded by 6 closest neighbors in octahedral coordination. Many halides of the XZ type crystallize with the *NaCl structure;* some mineral examples are, *sylvite,* KCl, *carobbiite,* KF, and *chlorargyrite,* AgCl. Some XZ sulfides and oxides that have the NaCl type structure are, *galena,* PbS, *alabandite,* MnS, and *periclase,* MgO. Several alkali halides such as CsCl, CsBr, and CsI, none of which is naturally occurring, do not have the NaCl type structure but crystallize with a geometric arrangement of 8 closest neighbors (cubic coordination) around the cation as well as the anion. This is known as the *CsCl structure type.* The radius ratio ($R_x : R_z$) is the primary factor in determining which of the two structure types is adopted by a given alkali halide of the XZ type.

The structures of several of the XZ_2 halides are identical to that of fluorite, CaF_2 (see Fig. 8.26), in which the Ca^{2+} ions are arranged at the corners and face centers of a cubic unit cell. F^- ions are tetrahedrally coordinated to 4 Ca^{2+}. Each Ca^{2+} is coordinated to 8 F^- which surround it at the corners of a cube. The radius ratio ($R_{Ca} : R_F = 0.979$) leads to 8 coordination for Ca. Several oxides such as *uraninite,* UO_2, and *thorianite,* ThO_2, have a *fluorite type structure.*

Because the weak electrostatic charges are spread over the entire surface of the nearly spherical ions, the halides are the most perfect examples of pure ionic bonding. The isometric halides all have

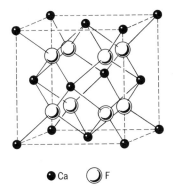

Ca ○ F

FIG. 8.26. The structure of fluorite.

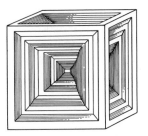

FIG. 8.27. Hopper-shaped halite crystal.

relatively low hardness and moderate to high melting points and are poor conductors of heat and electricity in the solid state. Such conduction of electricity as takes place does so by electrolysis, that is, by transport of charges by ions rather than by electrons. As the temperature increases and ions are liberated by thermal disorder, electrical conductivity increases rapidly, becoming excellent in the molten state. Advantage is taken of this conductivity of halide melts in the commercial methods for the preparation of sodium and chlorine by electrolysis of molten sodium chloride in the Downs cells, and in the Hall process for the electrolytic preparation of aluminum using molten cryolite. These properties are those conferred by the ionic bond.

When the halogen ions are combined with smaller and more strongly polarizing cations than those of the alkali metals, structures of lower symmetry result, and the bond has somewhat more covalent properties. In such structures, water or hydroxyl may enter as essential constituents, as in atacamite. Although there are over 85 species we will consider only the following in detail:

HALIDES

Halite	NaCl	Cryolite	Na$_3$AlF$_6$
Sylvite	KCl	Fluorite	CaF$_2$
Cerargyrite	AgCl	Atacamite	Cu$_2$Cl(OH)$_3$

✳ HALITE—NaCl

Crystallography. Isometric; $4/m\bar{3}2/m$. Habit cubic; other forms very rare. Some crystals hopper-

shaped (Fig. 8.27). Found in crystals or granular crystalline masses showing cubic cleavage, known as *rock salt.* Also massive, granular to compact.

$Fm3m$; $a = 5.640$ Å, $Z = 4$. $d's$: 2.82(10), 1.99(4), 1.628(2), 1.261(2), 0.892(1).

Physical Properties. *Cleavage* {001} perfect. **H** $2\frac{1}{2}$. **G** 2.16. *Luster* transparent to translucent. *Color* colorless or white, or when impure may have shades of yellow, red, blue, purple. Salty taste. Diathermanous. *Refractive index* = 1.544.

Composition and Structure. Na 39.3, Cl 60.7%. Commonly contains impurities, such as calcium and magnesium sulfates and calcium and magnesium chlorides. The structure of halite is illustrated in Figs. 8.24 and 4.19. This structure exists in a large number of *XZ* compounds with a radius ratio of between 0.41 and 0.73.

Diagnostic Features. Characterized by its cubic cleavage and taste, and distinguished from sylvite by its yellow flame color and by less bitter taste. Fusible at $1\frac{1}{2}$.

Occurrence. Halite is a common mineral, occurring often in extensive beds and irregular masses, precipitated by evaporation with gypsum, sylvite, anhdyrite, calcite, clay, and sand. Halite is dissolved in the waters of salt springs, salt lakes, and the ocean. It is a major salt in playa deposits of enclosed basins.

The deposits of salt have been formed by the gradual evaporation and ultimate drying up of enclosed bodies of salt water. The salt beds formed in this way may have subsequently been covered by other sedimentary deposits and gradually buried beneath the rock strata formed on them. Salt beds range between a few feet to over 200 feet in thickness and it is estimated that some are buried beneath as much as 35,000 feet of overlying strata.

Extensive bedded deposits of salt are widely distributed throughout the world and are mined in many countries. Important production comes from China, U.S.S.R., Great Britain, West Germany, Canada, and Italy.

The United States is the largest world producer and salt is recovered on a commercial scale in some fifteen states, either from rock-salt deposits or by evaporation of saline waters. Thick beds of rock salt are found in New York State which extend through Ontario, Canada, into Michigan. Salt is recovered from these beds at many localities. Notable deposits are also found in Ohio, Kansas, New Mexico, and in Canada in Nova Scotia and Saskatchewan. Salt is obtained by the evaporation of sea waters in California and Texas and from the waters of the Great Salt Lake in Utah.

Salt is also produced from *salt domes,* nearly vertical pipelike masses of salt that appear to have punched their way upward to the surface from an underlying salt bed. Anhydrite, gypsum, and native sulfur are commonly associated with salt domes. Geophysical prospecting for frequently associated petroleum has located several hundred salt domes along the Gulf coast of Louisiana and Texas and far out into the Gulf itself. Salt domes are also found in Germany, Rumania, Spain, and Iran. In Iran, in an area of complete aridity, salt that has punched its way to the surface is not dissolved but moves down slope as a salt glacier.

Use. Halite finds its greatest use in the chemical industry, where it is the source of sodium and chlorine for the manufacture of hydrochloric acid and a large number of sodium compounds.

Salt is used extensively in the natural state in tanning hides, in fertilizers, in stock feeds, in salting icy highways, and as a weed killer. In addition to its familiar functions in the home, salt enters into the preparation of foods of many kinds, such as the preservation of butter, cheese, fish, and meat.

Name. Halite comes from the Greek word meaning *salt.*

SYLVITE—KCl

Crystallography. Isometric; $4/m\bar{3}2/m$. Cube and octahedron frequently in combination. Usually in granular crystalline masses showing cubic cleavage; compact.

$Fm3m$; $a = 6.293$ Å. $Z = 4$. $d's$: 3.15(10), 2.22(6), 1.816(2), 1.407(2), 1.282(4).

Physical Properties. Cleavage {001} perfect. **H** 2. **G** 1.99. Transparent when pure. *Color* colorless or white; also shades of blue, yellow, or red from impurities. Readily soluble in water. Salty taste but more bitter than halite. *Optics:* $n = 1.490$.

Composition and Structure. K 52.4, Cl 47.6%. May contain admixed NaCl. Sylvite has the NaCl structure (see Fig. 8.25) but because of the difference in the ionic radii of Na$^+$ (0.97 Å) and K$^+$ (1.33 Å) there is little solid solution between KCl and NaCl.

Diagnostic Features. Fusible at $1\frac{1}{2}$. Distinguished from halite by the violet flame color of K and its more bitter taste. Readily soluble in water; a solution made acid with nitric acid gives, with silver nitrate, a heavy precipitate of silver chloride.

Occurrence. Sylvite has the same origin, mode of occurrence, and associations as halite but is much rarer. It remains in the mother liquor after precipitation of halite and is one of the last salts to be precipitated. (See evaporites, p 466).

It is found in quantity and frequently well crystallized, associated with the salt deposits at Stassfurt, Germany, from Kalusz in Galicia. In the United States it is found in large amount in the Permian salt deposits near Carlsbad, New Mexico, and in western Texas. More recently, deposits have been located in Utah. The most important world reserves are in Saskatchewan, Canada where extensive bedded deposits have been found at depths greater than 3000 feet.

Use. The chief source of potassium compounds, which are principally used as fertilizers.

Name. Potassium chloride is the *sal digestivus Sylvii* of early chemistry, whence the name for the species.

Other potassium salts. Several other potassium minerals are commonly associated with sylvite and are found in Germany and Texas in sufficient amount to make them valuable as sources of potassium salts. These are: *carnallite,* KMgCl$_3$·6H$_2$O, a usually massive to granular, generally light colored mineral; *kainite*, $KMg(Cl,SO_4) \cdot 2\frac{3}{4}H_2O$ and *polyhalite*, $K_2Ca_2Mg(SO_4)_4 \cdot 2H_2O$.

CERARGYRITE—AgCl

Crystallography. Isometric; $4/m\bar{3}2/m$. Habit cubic but crystals rare. Usually massive, resembling wax, often in plates and crusts.

$Fm3m$; $a = 5.55$ Å; $Z = 4$. d's: 3.20(5), 2.80 (10), 1.97(5), 1.67(2), 1.61(2).

Physical Properties. H 2–3. **G** 5.5±. Sectile, can be cut with a knife; hornlike appearance, hence the name *horn silver*. Transparent to translucent. *Color* pearl-gray to colorless. Rapidly darkens to violet-brown on exposure to light. *Optics:* $n = 2.07$.

Composition and Structure. Ag 75.3, Cl 24.7%. A complete solid solution series exists between AgCl and *bromyrite*, AgBr. Small amounts of F may be present in substitution for Cl or Br. Some specimens contain Hg. Cerargyrite is isostructural with NaCl (see Fig. 8.25).

Diagnostic Features. Distinguished chiefly by its waxlike appearance and its sectility. Fusible at 1. Before the blowpipe on charcoal it gives a globule of silver.

Occurrence. Cerargyrite is an important supergene ore of silver found in the upper, enriched zone of silver deposits. It is found associated with native silver, cerussite, and secondary minerals in general.

Notable amounts have been found at Broken Hill, Australia; and in Peru, Chile, Bolivia, and Mexico. In the United States cerargyrite was an important mineral in the mines at Leadville and elsewhere in Colorado, at the Comstock Lode in Nevada, and in crystals at the Poorman's Lode in Idaho.

Use. A silver ore.

Name. Cerargyrite is derived from two Greek words meaning *horn* and *silver,* in allusion to its hornlike appearance and characteristics.

Similar Species. Other closely related minerals that are less common but form under similar conditions, are *bromyrite,* AgBr, and *iodobromite,* Ag(Cl, Br, I), isostructural with cerargyrite; and *iodyrite,* AgI, which is hexagonal.

CRYOLITE—Na₃AlF₆

Crystallography. Monoclinic; $2/m$. Prominent forms are $\{001\}$ and $\{110\}$. Crystals rare, usually cubic in aspect, and in parallel groupings growing out of massive material. Usually massive.

Angles: $(110) \wedge (1\bar{1}0) = 88°02'$, $(001) \wedge (101) = 55°03'$

$P2_1/n$; $a = 5.47$, $b = 5.62$, $c = 7.82$ Å, $\beta = 90°11'$; $a:b:c = 0.973:1:1.391$; $Z = 2$. d's: 4.47(2), 3.87(2), 2.75(7), 2.33(4), 1.939(10).

Physical Properties. Parting on $\{110\}$ and $\{001\}$ produces cubical forms. **H** $2\frac{1}{2}$. **G** 2.95–3.0. *Luster* vitreous to greasy. *Color* colorless to snow-white. Transparent to translucent. *Optics:* (+); $\alpha = 1.338$, $\beta = 1.338$, $\gamma = 1.339$; $2V = 43°$; $X = b$, $Z \wedge c = -44°$, $r < v$. The low refractive index, near that of water, gives the mineral the appearance of watery snow or paraffin, and causes the powdered mineral to almost disappear when immersed in water.

Composition and Structure. Na 32.8, Al 12.8, F 54.4%. In the structure of cyrolite Al is octahedrally coordinated to six F^-. The Na^+ ions are also surrounded by six F^- ions, but in a somewhat less regular pattern. At high temperature (above 550°C) cryolite transforms to an isometric form with space group $Fm3m$.

Diagnostic Features. Characterized by pseudocubic parting, white color, and peculiar luster; and for the Greenland cryolite, the association of siderite, galena, and chalcopyrite. Fusible at $1\frac{1}{2}$ with strong yellow sodium flame.

Occurrence. The only important deposit of cryolite is at Ivigtut, on the west coast of Greenland. Here, in a large mass in granite, it is associated with siderite, galena, sphalerite, and chalcopyrite; and less commonly quartz, wolframite, fluorite, cassiterite, molybdenite, arsenopyrite, columbite. It is found at Miask, U.S.S.R., and in the United States, at the foot of Pikes Peak, Colorado.

Use. Cryolite is used for the manufacture of sodium salts, of certain kinds of glass and porcelain, and as a flux for cleansing metal surfaces. It was early used as a source of aluminum. When bauxite became the ore of aluminum, cryolite was used as a flux in the electrolytic process. Today most of the sodium aluminum fluoride used in the aluminum industry is produced artificially.

Name. Name is derived from two Greek words meaning *frost* and *stone,* in allusion to its icy appearance.

✳ FLUORITE—CaF₂

Crystallography. Isometric; $4/m\bar{3}2/m$. Usually in cubes, often as penetration twins twinned on [111] (Fig. 8.28*a*). Other forms are rare, but examples of all the forms of the hexoctahedral class have been observed; the tetrahexahedron (Fig. 8.28*b*) and hexoctahedron (Fig. 8.28*c*) are characteristic. Usually in crystals or in cleavable masses. Also massive; coarse or fine granular; columnar.

$Fm3m$; $a = 5.46$ Å; $Z = 4$. $d's$: 3.15(9), 1.931(10), 1.647(4), 1.366(1), 1.115(2).

Physical Properties. *Cleavage* {111} perfect. **H** 4. **G** 3.18. Transparent to translucent. *Luster* vitreous. *Color* varies widely; most commonly light green, yellow, bluish-green, or purple; also colorless, white, rose, blue, brown. The color in some fluorite results from the presence of a hydrocarbon. A single crystal may show bands of varying colors; the massive variety is also often banded in color. The phenomenon of fluorescence (see page 192) received its name because it was early observed in some varieties of fluorite. *Optics:* $n = 1.433$.

Composition and Structure. Ca 51.3, F 48.7%. The rare earths, particularly Y and Ce, may substitute for Ca. The fluorite structure is shown in Fig. 8.26.

Diagnostic Features. Determined usually by its cubic crystals and octahedral cleavage; also vitreous luster and usually fine coloring, and by the fact that it can be scratched with a knife. Fusible at 3 giving a reddish flame (Ca).

Occurrence. Fluorite is a common and widely distributed mineral. Usually found in hydrothermal veins in which it may be the chief mineral or as a gangue mineral with metallic ores, especially those of lead and silver. Common in dolomites and limestone and has been observed also as a minor accessory mineral in various igneous rocks and pegmatites. Associated with many different minerals, as calcite, dolomite, gypsum, celestite, barite, quartz, galena, sphalerite, cassiterite, topaz, tourmaline, and apatite.

Fluorite is found in quantity in England, chiefly from Cumberland, Derbyshire, and Durham; the first two localities are famous for their magnificent crystallized specimens (Fig. 8.29). Found commonly in the mines of Saxony. Fine specimens come from Switzerland, the Tyrol, Bohemia, and Norway. The large producers of commercial fluorite (fluorspar), aside from the United States, are Mexico, U.S.S.R., Spain, Thailand, China, Italy, and France. The most important deposits in the United States are in southern Illinois near Rosiclare and Cave-in-Rock, and in the adjacent part of Kentucky. At Rosiclare fluorite, which is without crystal form, occurs in limestone, in fissure veins that in places are 40 feet in width. Twenty miles away at Cave-in-Rock the fluorite is in coarsely crystalline aggregates lining flat, open spaces. Fluorite is also mined in Colorado, New Mexico, Montana, and Utah.

Use. Fluorite is used mainly as a flux in the making of steel, in the manufacture of opalescent glass, in enameling cooking utensils, for the preparation of hydrofluoric acid. Formerly used exten-

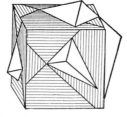

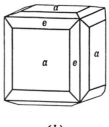

FIG. 8.28. Fluorite. (*a*) Penetration twin. (*b*) Cube and tetrahexahedron. (*c*) Cube and hexoctahedron.

(a)   **(b)**   **(c)**

FIG. 8.29. Fluorite crystals
coated with quartz,
Northumberland, England.

sively as an ornamental material and for carving vases and dishes. Small amounts of fluorite are used for lenses and prisms in various optical systems, but most of the optical material is now made synthetically.

Name. From the Latin *fluere,* meaning to flow, because it melts more easily than other minerals with which it was, in the form of cut stones, confused.

Atacamite—$Cu_2Cl(OH)_3$

Crystallography. Orthorhombic; $2/m\,2/m\,2/m$. Commonly in slender prismatic crystals with vertical striations. Also tabular parallel to $\{010\}$. Usually in confused crystalline aggregates; fibrous; granular.

Pnam; $a = 6.02$, $b = 9.15$, $c = 6.85$ Å; $a:b:c = 0.658:1:0.749$; $Z = 4$. $d's$: 5.40(10), 5.00(10), 2.82(10), 2.75(10), 2.26(10).

Physical Properties. *Cleavage* $\{010\}$ perfect. **H** 3–3½. **G** 3.75–3.77. *Luster* adamantine to vitreous. *Color* various shades of green. Transparent to

translucent. *Optics:* $(-)$, $\alpha = 1.831$, $\beta = 1.861$, $\gamma = 1.880$; $2V = 75°$; $r < v$. $X = b$, $Y = a$.

Composition and Structure. Cu 14.88, CuO 55.87, Cl 16.60, H_2O 12.65%. In the structure of atacamite part of the Cu atoms is in 6 coordination with five (OH) groups and one Cl. The remaining Cu is in 6 coordination with four (OH) groups and two Cl.

Diagnostic Features. Characterized by its green color and granular crystalline aggregates. Distinguished from malachite by its lack of effervescence in acids, and from brochantite and antlerite by its azure-blue copper chloride flame without the use of HCl. Fuses at 3–4 and on charcoal with sodium carbonate gives a copper globule. Gives acid water in the closed tube.

Occurrence. Atacamite is a comparatively rare copper mineral. Found originally as sand in the province of Atacama in Chile. Occurs in arid regions as a supergene mineral in the oxidized zone of copper deposits. It is associated with other secondary minerals in various localities in Chile, Bolivia, Mexico, and in some of the copper districts of

South Australia. In the United States occurs sparingly in the copper districts of Arizona.

Use. A minor ore of copper.

Name. From the province of Atacama, Chile.

References and Suggested Reading

Bragg, L. and G. F. Claringbull, 1965, *Crystal Structures of Minerals*. G. Bell and Sons, Ltd., London, 409 pp.

Deer, W. A., Howie, R. A., and J. Zussman, 1962, *Rock-Forming Minerals,* vol. 5, Non-silicates. John Wiley & Sons, Inc., New York, 371 pp.

Evans, R. C., 1966, *An Introduction to Crystal Chemistry,* 2nd ed. Cambridge Univ. Press, Cambridge, England, 410 pp.

Palache, C., Berman, H., and C. Frondel, 1944 and 1951, *The System of Mineralogy,* 7th ed., vols. I and II. John Wiley & Sons, New York, 834 and 1124 pp.

9
SYSTEMATIC MINERALOGY

PART III: CARBONATES, NITRATES, BORATES, SULFATES, CHROMATES, TUNGSTATES, MOLYBDATES, PHOSPHATES, ARSENATES, AND VANADATES

In this chapter are discussed representative species of the carbonates, nitrates, borates, sulfates, chromates, tungstates, molybdates, phosphates, arsenates, and vanadates. These groups together contain over 650 species but because most of them are uncommon minerals, only a small number will be considered.

These chemically diverse minerals are treated together because most of them contain *anionic complexes*, which can be recognized as strongly bonded units in their structures. Examples of such anionic complexes are: $(CO_3)^{-2}$ in carbonates, $(NO_3)^{-1}$ in nitrates, $(PO_4)^{-3}$ in phosphates, $(SO_4)^{-2}$ in sulfates, $(CrO_4)^{-1}$ in chromates, $(WO_4)^{-2}$ in tungstates, and $(AsO_4)^{-3}$ in arsenates. The bond strengths within such anionic complexes are always stronger than those between the anionic complex and other ions of the structure; these compounds are therefore referred to as *anisodesmic* (see page 145). For example, in the carbonates, radius ratio considerations predict three closest oxygen neighbors about

carbon. This arrangement is triangular with carbon at the center and oxygen at each of the corners of the triangle (see Fig. 9.1; the (NO_3) group is also triangular). The e.v. of the bonds between carbon and each of the three closest oxygens is $\frac{1}{3} \times 4 = 1\frac{1}{3}$. This means that each oxygen has a residual charge of e.v. $= \frac{2}{3}$ for bonding other ions in the carbonate structure. Figure 9.1 illustrates the anisodesmic character of structures with tetrahedral (PO_4) and (SO_4) anionic groups as well. Borates with triangular (BO_3) groups and silicates with tetrahedral (SiO_4) groups are examples of *mesodesmic* bonding (see page 146).

CARBONATES

The anionic $(CO_3)^{-2}$ complexes of carbonates are strongly bonded units and do not share oxygens with each other (as noted above, the residual e.v. of $\frac{2}{3}$ does not allow this). The triangular carbonate

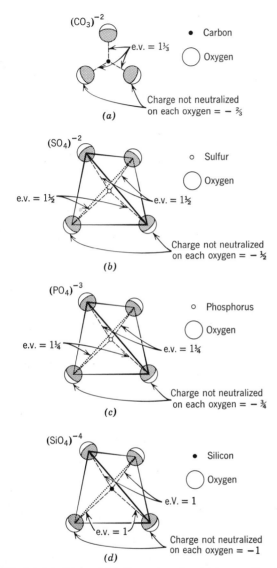

$(CO_3)^{-2}$

e.v. = $1\frac{1}{3}$

● Carbon

○ Oxygen

Charge not neutralized on each oxygen = $-\frac{2}{3}$

(a)

$(SO_4)^{-2}$

e.v. = $1\frac{1}{2}$ e.v. = $1\frac{1}{2}$

○ Sulfur

○ Oxygen

Charge not neutralized on each oxygen = $-\frac{1}{2}$

(b)

$(PO_4)^{-3}$

e.v. = $1\frac{1}{4}$ e.v. = $1\frac{1}{4}$

○ Phosphorus

○ Oxygen

Charge not neutralized on each oxygen = $-\frac{3}{4}$

(c)

$(SiO_4)^{-4}$

e.V. = 1

e.v. = 1

● Silicon

○ Oxygen

Charge not neutralized on each oxygen = -1

(d)

FIG. 9.1 (a), (b), and (c) Examples of anionic complexes, their bond strengths between the central cation and oxygen, and the residual charges on the oxygens. (d) The tetrahedral (SiO_4) group in which the e.v.'s between oxygen and the central cation are the same as the residual charge on the oxygen (=1).

groups are the basic building units of all carbonate minerals and are largely responsible for the properties peculiar to the group.

Although the bond between the central carbon and its coordinated oxygens in the (CO_3) group is strong, it is not as strong as the covalent bond in CO_2. In the presence of hydrogen ion, the carbonate group becomes unstable and breaks down to yield CO_2 and water, according to $2H^+ + CO_3 \rightarrow H_2O + CO_2$. This reaction is the cause of the familiar "fizz" test with acid which is widely used in the identification of carbonates.

The important anhydrous carbonates fall into three structurally different groups: the *Calcite Group*, the *Aragonite Group*, and the *Dolomite Group*. Aside from the minerals in these groups, the hydrous copper carbonates, azurite and malachite, are the only important carbonates.

CARBONATES

Calcite Group (Hexagonal-R; $R\bar{3}c$)		Aragonite Group (Orthorhombic; $Pmcn$)	
Calcite	$CaCO_3$	Aragonite	$CaCO_3$
Magnesite	$MgCO_3$	Witherite	$BaCO_3$
Siderite	$FeCO_3$	Strontianite	$SrCO_3$
Rhodochrosite	$MnCO_3$	Cerussite	$PbCO_3$
Smithsonite	$ZnCO_3$		

Dolomite Group (Hexagonal-R; $R\bar{3}$)	
Dolomite	$CaMg(CO_3)_2$
Ankerite	$CaFe(CO_3)_2$

Monoclinic carbonates with (OH)⁻	
Malachite	$Cu_2CO_3(OH)_2$
Azurite	$Cu_3(CO_3)_2(OH)_2$

Calcite Group

The above five members of the calcite group are isostructural with space group $R\bar{3}c$. The structure of calcite, one of the earliest to be analyzed by X-rays by W. L. Bragg in 1914 (see Fig. 9.2a), can be thought of as a derivative of the NaCl structure in which triangular (CO_3) groups replace the spherical Cl and Ca is in place of Na. The triangular shape of the (CO_3) groups causes the resulting structure to be rhombohedral instead of isometric as in NaCl. The (CO_3) groups lie in planes at right angles to the 3-fold (c) axis (Fig. 9.2) and the Ca ions, in alternate planes, are in 6 coordination with oxygens of the (CO_3) groups. Each oxygen is coordinated to two Ca ions as well as to a carbon ion at the center of the (CO_3) group.

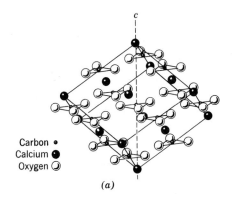

Carbon •
Calcium ●
Oxygen ◖

(a)

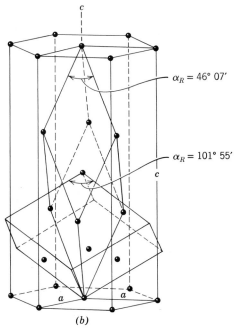

$\alpha_R = 46°\ 07'$

$\alpha_R = 101°\ 55'$

(b)

FIG. 9.2 (a) Structure of calcite, $CaCO_3$. (b) The relation of the steep, true unit cell to the cleavage rhombohedron, which is face-centered. A hexagonal cell (rhomb-based prism) is also shown.

Calcite shows perfect rhombohedral cleavage to which, traditionally, the indices of $\{10\bar{1}1\}$ have been assigned; axial ratios have been expressed accordingly. In the morphological descriptions and indexing of forms of the calcite and dolomite group minerals that follow, this convention has been preserved. However, X-ray structural determinations have shown that this rhombohedron does not corre-

spond to the correct unit cell and that the simplest unit cell is a much steeper rhombohedron (Fig. 9.2b). Therefore, the structural axial ratios differ from the morphological. It should be noted that the radius ratio of Ca:O ($=0.707$) in $CaCO_3$ is so close to the limiting value between 6 and 8 coordination (0.732) that $CaCO_3$ can occur in two structure types: *calcite* with 6 coordination of Ca to O and *aragonite*, with 9 coordination of Ca to O (see aragonite, page 303).

✳CALCITE—CaCO₃

Crystallography. Hexagonal-*R*; $\bar{3}2/m$. Crystals are extremely varied in habit and often highly complex. Over 300 different forms have been described (Fig. 9.3). Three important habits: (1) prismatic, in long or short prisms, in which the prism faces are prominent, with base or rhombohedral terminations; (2) rhombohedral, in which rhombohedral forms predominate; the unit (cleavage) form r is not common; (3) scalenohedral, in which scalenohedrons predominate, often with prism faces and rhombohedral truncations. The most common scalenohedron is $\{21\bar{3}1\}$. All possible combinations and variations of these types are found.

Twinning with the twin plane, $\{01\bar{1}2\}$, very common (Fig. 9.4); often produces twinning lamellae that may, as in crystalline limestones, be of secondary origin. This twinning may be produced artificially (see page 104). Twins with $\{0001\}$, the twin plane common. Calcite is usually in crystals or in coarse to fine grained aggregates. Also fine grained to compact, earthy, and stalactitic.

Angles: $r(10\bar{1}1) \wedge r'(\bar{1}101) = 74°55'$, $c(0001) \wedge r(10\bar{1}1) = 44°37'$, $e(01\bar{1}2) \wedge e'(\bar{1}012) = 45°3'$, $c(0001) \wedge e(01\bar{1}2) = 26°15'$, $v(21\bar{3}1) \wedge v'(\bar{2}3\bar{1}1) = 75°22'$, $c(0001) \wedge M(40\bar{4}1) = 75°47'$.

$R\bar{3}c$. Morphology $a:c = 1:0.855$. Structure (hexagonal cell) $a = 4.99$, $c = 17.06$ Å; $a:c = 1:3.419$; $Z = 6$; (rhombohedral cell) $a = 6.37$, α (rhombohedral angle) $= 46°05'$, $Z = 2$. d's: 3.04(10), 2.29(2), 2.10(2), 1.913(2), 1.875(2).

Physical Properties. *Cleavage* $\{10\bar{1}1\}$ perfect (cleavage angle = 74°55'). Parting along twin lamellae on $\{01\bar{1}2\}$. **H** 3 on cleavage, $2\frac{1}{2}$ on base. **G**

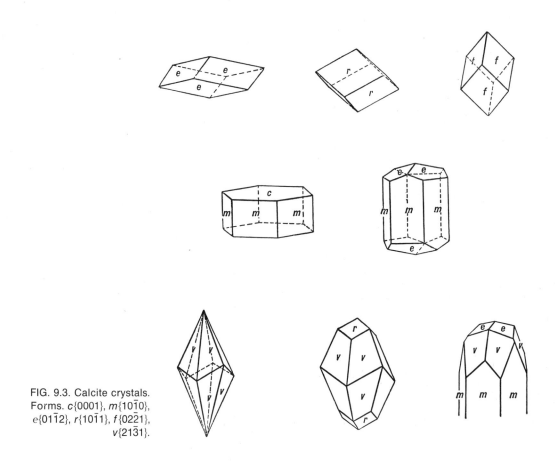

FIG. 9.3. Calcite crystals.
Forms. $c\{0001\}$, $m\{10\bar{1}0\}$,
$e\{01\bar{1}2\}$, $r\{10\bar{1}1\}$, $f\{02\bar{2}1\}$,
$v\{21\bar{3}1\}$.

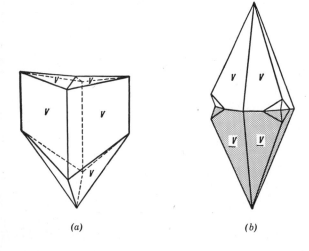

FIG. 9.4. Twinned calcite crystals. Twin planes: (a) $\{01\bar{1}2\}$.
(b) $\{0001\}$.

(a) (b)

2.71. *Luster* vitreous to earthy. Color usually white to colorless, but may be variously tinted, gray, red, green, blue, yellow; also, when impure, brown to black. Transparent to translucent. The chemically pure and optically clear, colorless variety is known as *Iceland spar* because of its occurrence in Iceland. *Optics:* $(-)$; $\omega = 1.658$, $\epsilon = 1.486$.

Composition and Structure. Most calcites tend to be relatively close to pure $CaCO_3$ with CaO 56.0 and CO_2 44.0%. Mn^{2+}, Fe^{2+}, and Mg may substitute for Ca and a complete solid solution series extends to rhodochrosite, $MnCO_3$, above 550°C; a very partial series, with up to 5 weight percent FeO in calcite, exists between calcite and siderite, $FeCO_3$. Some inorganic calcites may contain from 0 to about 2 weight percent MgO. Calcites in the hard parts of living organisms, however, may show a

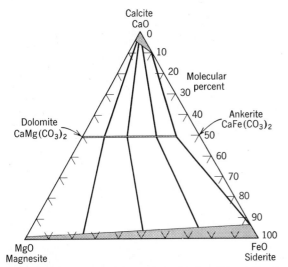

FIG. 9.5. Carbonates and the extent of their solid solution in the system CaO-MgO-FeO-(CO$_2$). Tielines connect commonly coexisting carbonate species. Calcite-dolomite associations are common in Mg-containing limestones; ankerite-siderite associations are found in iron-formations.

range of MgCO$_3$ of 2 to 16 molecular percent. See Fig. 9.5 for solid solution series in the system CaO-MgO-FeO-CO$_2$. The structure of calcite is shown in Fig. 9.2a and discussed on page 296.

Diagnostic Features. Infusible. Fragments effervesce readily in cold dilute HCl. Characterized by its hardness (3), rhombohedral cleavage, light color, vitreous luster. Distinguished from dolomite by the fact that coarse fragments of calcite effervesce freely in cold HCl and distinguished from aragonite by lower specific gravity and rhombohedral cleavage.

Occurrence. As a *rock-forming mineral:* Calcite is one of the most common and widespread minerals. It occurs in extensive sedimentary rock masses in which it is the predominant mineral; in *limestones,* it is essentially the only mineral present. Crystalline, metamorphosed limestones are *marbles. Chalk* is a fine-grained pulverulent deposit of calcium carbonate. Calcite is an important constituent of calcareous marls and calcareous sandstones. Limestone has, in great part, been formed by the deposition on a sea bottom of great thicknesses

of calcareous material in the form of shells and skeletons of sea animals. A smaller proportion of these rocks has been formed directly by precipitation of calcium carbonate.

As cave deposits, etc.: Waters carrying calcium carbonate in solution and evaporating in limestone caves often deposit calcite as stalactites, stalagmites, and incrustations. Such deposits, usually semitranslucent and of light yellow colors, are often beautiful and spectacular. Both hot and cold calcareous spring waters may form cellular deposits of calcite known as *travertine,* or *tufa,* around their mouths. The deposit at Mammoth Hot Springs, Yellowstone Park, is more spectacular than most, but of similar origin. *Onyx marble* is banded calcite and/or aragonite used for decorative purposes. Because much of this material comes from Baja California, Mexico, it is also called Mexican onyx.

Siliceous calcites: Calcite crystals may enclose considerable amounts of quartz sand (up to 60%) and form what are known as sandstone crystals. Such occurrences are found at Fontainebleau, France (Fontainebleau limestone) and in the Bad Lands, South Dakota.

Calcite occurs as a primary mineral in some igneous rocks such as carbonatites and nepheline syenites. It is a late crystallization product in the cavities in lavas. It is also a common mineral in hydrothermal veins associated with sulfide ores.

It is impossible to specify all the important districts for the occurrence of calcite in its various forms. Some of the more notable localities in which finely crystallized calcite is found are as follows: Andreasberg in the Harz Mountains; various places in Saxony; in Cumberland (Fig. 9.6), Derbyshire, Durham, Cornwall, and Lancashire, England; Iceland; and Guanajuato, Mexico. In the United States at Joplin, Missouri; Lake Superior copper district; and Rossie, New York.

Use. The most important use for calcite is for the manufacture of cements and lime for mortars. Limestone is the chief raw material, which when heated to about 900°C forms *quicklime,* CaO, by the reaction: CaCO$_3$ → CaO + CO$_2$. The CaO, when mixed with water, forms one or several CaO-hydrates (slaked lime), swells, gives off much

FIG. 9.6. Group of calcite crystals, Cumberland, England.

heat, and hardens or, as commonly termed, "sets." Quicklime when mixed with sand forms common mortar.

The greatest consumption of limestone is in the manufacture of cements. The type known as Portland cement is most widely produced. It is composed of about 75% calcium carbonate (limestone) with the remainder essentially silica and alumina. Small amounts of magnesium carbonate and iron oxide are also present. In some limestones, known as *cement rocks,* the correct proportions of silica and alumina are present as impurities. In others these oxides are contributed by clay or shale mixed with the limestone before "burning." When water is mixed with cement, hydrous calcium silicates and calcium aluminates are formed.

Limestone is a raw material for the chemical industry, and finely crushed is used as a soil conditioner, for whiting and whitewash. Great quantities are quarried each year as a flux for smelting various metallic ores, as an aggregate in concrete and as road metal. A fine grained limestone is used in lithography.

Calcite in several forms is used in the building industry. Limestone and marble as dimension stone are used both for construction purposes and deco-

rative exterior facings. Polished slabs of travertine and Mexican onyx are commonly used as ornamental stone for interiors. Indiana is the chief source of building limestone in the United States, with the most productive quarries in Lawrence and Monroe Counties. Many of the federal buildings in Washington, D.C., have been constructed from this limestone. The most important marble quarries are in Vermont, New York, Georgia, and Tennessee.

Iceland spar is valuable for various optical instruments; its best known use was in the form of the Nicol prism to produce polarized light, prior to the use of Polaroid plates.

Name. From the Latin word, *calx,* meaning burnt lime.

☀ MAGNESITE—MgCO₃

Crystallography. Hexagonal-R; $\bar{3}2/m$. Crystals rhombohedral, $\{10\bar{1}1\}$, but rare. Usually cryptocrystalline in white, compact, earthy masses, less frequently in cleavable granular masses, coarse to fine.

Angles: $(10\bar{1}1) \wedge (\bar{1}101) = 72°33'$.

$R\bar{3}c$. Morphology $a:c = 1:0.810$. Structure (hexagonal cell) $a = 4.63$, $c = 15.02$ Å; $a:c =$

$1:3.244$, $Z = 6$; (rhombohedral cell) $a = 5.62$, α (rhombohedral angle) $= 48°10'$, $Z = 2$. $d's$: 2.74(10), 2.50(2), 2.10(4), 1.939(1), 1.700(3).

Physical Properties. *Cleavage* $\{10\bar{1}1\}$ perfect. **H** $3\frac{1}{2}$–5. **G** 3.0–3.2. *Luster* vitreous. *Color* white, gray, yellow, brown. Transparent to translucent. *Optics:* (−); $\omega = 1.700$, $\epsilon = 1.509$.

Composition and Structure. MgO 47.8, CO_2 52.2%. Fe^{2+} substitutes for Mg and a complete series extends to siderite (see Fig. 9.5). Small amounts of Ca and Mn may be present. Magnesite is isostructural with calcite (see Fig. 9.2).

Diagnostic Features. Cleavable varieties are distinguished from dolomite by higher specific gravity and absence of abundant calcium. The white massive variety resembles chert and is distinguished from it by inferior hardness. Infusible. Scarcely acted upon by cold HCl, but dissolves with effervescence in hot HCl.

Occurrence. Magnesite commonly occurs in veins and irregular masses derived from the alteration of Mg-rich metamorphic and igneous rocks (serpentinites and peridotites) through the action of waters containing carbonic acid. Such magnesites are compact, cryptocrystalline and often contain opaline silica. Beds of crystalline cleavable magnesite are (1) of metamorphic origin associated with talc schists, chlorite schists, and mica schists, and (2) of sedimentary origin, formed as a primary precipitate or as a replacement of limestones by Mg-containing solutions, dolomite being formed as an intermediate product.

Notable deposits of the sedimentary type of magnesite are in Manchuria; at Satka in the Ural Mountains; and at Styria, Austria. The most famous deposit of the cryptocrystalline type is on the Island of Euboea, Greece.

In the United States the compact variety is found in irregular masses in serpentine in the Coast Range, California. The sedimentary type is mined at Chewelah in Stevens County, Washington, and in the Paradise Range, Nye County, Nevada. There are numerous minor localities in the eastern United States in which magnesite is associated with serpentine, talc, or dolomite rocks.

Use. Dead-burned magnesite, MgO, that is, magnesite that has been calcined at a high tempera-ture and contains less than 1% CO_2, is used in manufacturing bricks for furnace linings. Magnesite is the source of magnesia for industrial chemicals. It has also been used as an ore of metallic Mg, but at present the entire production of Mg comes from brines and sea water.

Name. Magnesite is named in allusion to the composition.

SIDERITE—$FeCO_3$

Crystallography. Hexagonal-*R*; $\bar{3}2/m$. Crystals usually unit rhombohedrons, frequently with curved faces. In globular concretions. Usually cleavable granular. May be botryoidal, compact, and earthy.

Angles: $(10\bar{1}1) \wedge (1\bar{1}01) = 73°00'$.

$R\bar{3}c$. Morphology $a:c = 1:0.819$. Structure (hexagonal cell) $a = 4.72$, $c = 15.45$ Å; $a:c = 1:3.275$; $Z = 6$; (rhombohedral cell) $a = 5.83$, $\alpha = 47°45'$, $Z = 2$. $d's$: 3.59(6), 2.79(10), 2.13(6), 1.963(6), 1.73(8).

Physical Properties. *Cleavage* $\{10\bar{1}1\}$ perfect. **H** $3\frac{1}{2}$–4. **G** 3.96 for pure $FeCO_3$, but decreases with presence of Mn^{2+} and Mg. *Luster* vitreous. *Color* usually light to dark brown. Transparent to translucent. *Optics:* (−); $\omega = 1.875$, $\epsilon = 1.633$.

Composition and Structure. For pure $FeCO_3$, FeO 62.1, CO_2 37.9%. Fe 48.2%. Mn^{2+} and Mg substitute for Fe^{2+} and complete series extend to rhodochrosite and magnesite (see Fig. 9.5). The substitution of Ca for Fe^{2+} is limited due to the large difference in size of the two ions. Siderite is isostructural with calcite (see Fig. 9.2 and page 296).

Diagnostic Features. Distinguished from other carbonates by its color and high specific gravity, and from sphalerite by its rhombohedral cleavage. Difficultly fusible ($4\frac{1}{2}$–5). Becomes strongly magnetic on heating. Soluble in hot HCl with effervescence.

Alteration. Pseudomorphs of limonite after siderite are common.

Occurrence. Siderite is frequently found as *clay ironstone,* impure by admixture with clay materials, in concretions with concentric layers. As *black-band ore* it is found, contaminated by carbonaceous material, in extensive stratified formations lying in shales and commonly associated with coal measures. These ores have been mined extensively in Great Britain in the past, but at present are mined only in North Staffordshire and Scotland. Clay ironstone is also abundant in the coal measures of western Pennsylvania and eastern Ohio, but it is not used to any great extent as an ore. Siderite is also formed by the replacement action of Fe-rich solutions upon limestones, and if such occurrences are extensive they may be of economic value. The most notable deposit of this type is in Styria, Austria, where siderite is mined on a large scale. Siderite, in its crystallized form, is a common vein mineral associated with various metallic ores containing silver minerals, pyrite, chalcopyrite, tetrahedrite, and galena. When siderite predominates in such veins it may be mined, as in southern Westphalia, Germany. Siderite is also a common constituent of banded Precambrian iron deposits, as in the Lake Superior region.

Use. An ore of iron. Important in Great Britain and Austria, but unimportant elsewhere.

Name. From the Greek word meaning *iron*. The name *spherosiderite* of the concretionary variety was shortened to siderite to apply to the entire species. *Chalybite,* used by some mineralogists, was derived from the Chalybes, ancient iron workers, who lived by the Black Sea.

RHODOCHROSITE—MnCO₃

Crystallography. Hexagonal-R; $\bar{3}2/m$. Only rarely in crystals of the unit rhombohedron; frequently with curved faces. Usually cleavable, massive; granular to compact.

Angles: $(10\bar{1}1) \wedge (\bar{1}101) = 73°02'$.

$R\bar{3}c$. Morphology $a:c = 1:0.818$. Structure (hexagonal cell) $a = 4.78$, $c = 15.67$ Å; $a:c = 1:3.278$; $Z = 6$; (rhombohedral cell) $a = 5.85$, $\alpha = 47°46'$, $Z = 2$. d's: 3.66(4), 2.84(10), 2.17(3), 1.770(3), 1.763(3).

Physical Properties. *Cleavage* $\{10\bar{1}1\}$ perfect. **H** 3½–4. **G** 3.5–3.7. *Luster* vitreous. *Color* usually some shade of rose-red; may be light pink to dark brown. *Streak* white. Transparent to translucent. *Optics:* (−); $\omega = 1.816$, $\epsilon = 1.597$.

Composition and Structure. For pure

MnCO₃, MnO 61.7, CO₂ 38.3%. Fe²⁺ substitutes for Mn²⁺ forming a complete solid solution series between rhodochrosite and siderite. Ca²⁺ shows some substitution for Mn²⁺. The occurrence of *kutnahorite,* CaMn(CO₃)₂, with an ordered structure of the dolomite type, suggests that only limited solid solution occurs, at ordinary temperatures, between CaCO₃ and MnCO₃. Mg may also substitute for Mn but the MnCO₃-MgCO₃ series is incomplete. Considerable amounts of Zn may substitute for Mn (see smithsonite). Rhodochrosite is isostructural with calcite (see Fig. 9.2 and page 296).

Diagnostic Features. Characterized by its pink color and rhombohedral cleavage; the hardness (4) distinguishes it from rhodonite, MnSiO₃, with hardness of 6. Infusible. Soluble in hot HCl with effervescence. Gives blue-green color to sodium carbonate bead. When heated on charcoal it turns black but it is nonmagnetic.

Occurrence. Rhodochrosite is a comparatively rare mineral, occurring in hydrothermal veins with ores of silver, lead, and copper, and with other manganese minerals. Found in the silver mines of Rumania and Saxony. Beautiful banded rhodochroiste is mined for ornamental and decorative purposes at Capillitas, Catamarca, Argentina. In the United States it is found at Branchville, Connecticut; Franklin, New Jersey; and at Butte, Montana has been mined as a manganese ore. In good crystals at Alicante, Lake County; Alma Park County; and elsewhere in Colorado.

Use. A minor ore of manganese. Small amounts used for ornamental purposes.

Name. Derived from two Greek words meaning *rose* and *color,* in allusion to its rose-pink color.

SMITHSONITE—ZnCO₃

Crystallography. Hexagonal-R; $\bar{3}2/m$. Rarely in small rhombohedral or scalenohedral crystals. Usually reniform, botryoidal, or stalactitic, and in crystalline incrustations or in honeycombed masses known as *dry-bone ore.* Also granular to earthy.

Angles: $(10\bar{1}1) \wedge (\bar{1}101) = 72°20'$.

$R\bar{3}c$. Morphology $a:c = 1:0.806$. Structure

(hexagonal cell) $a = 4.66$, $c = 15.02$ Å; $a:c = 1:3.224$, $Z = 6$; (rhombohedral cell) $a = 5.63$, $\alpha = 48°20'$, $Z = 2$. $d's$: 2.75(10), 3.55(5). 2.33(3), 1.946(3), 1.703(4).

Physical Properties. *Cleavage* $\{10\bar{1}1\}$ perfect. **H** $4-4\frac{1}{2}$. **G** 4.30–4.45. *Luster* vitreous. *Color* usually dirty brown. May be colorless, white, green, blue, or pink. The yellow variety contains Cd and is known as *turkey-fat ore*. *Streak* white. Translucent. *Optics:* (−); $\omega = 1.850$, $\epsilon = 1.623$.

Composition and Structure. For pure $ZnCO_3$, ZnO 64.8, CO_2 35.2%. Considerable Fe^{2+} may substitute for Zn, but there appears to be a gap in the $ZnCO_3$-$FeCO_3$ series. Mn^{2+} is generally present in only a few percent but the occurrence of a zincian rhodochrosite with Zn:Mn = 1:1.2 suggests there may be a complete series between $ZnCO_3$ and $MnCO_3$. Ca and Mg are present in amounts of only a few weight percent. Small amounts of Co are found in a pink, and small amounts of Cu in a blue-green variety of smithsonite. Smithsonite is isostructural with calcite (see Fig. 9.2 and page 296).

Diagnostic Features. Infusible. Soluble in cold HCl with effervescence. When heated before the blowpipe it gives bluish-green streaks in the flame, due to the burning of the volatilized Zn, and in the reducing flame on charcoal it gives a nonvolatile coating of zinc oxide, yellow when hot, white when cold. The coating moistened with cobalt nitrate and again heated, turns green. Distinguished by its effervescence in acids, its tests for zinc, its hardness, and its high specific gravity.

Occurrence. Smithsonite is a zinc ore of supergene origin, usually found with zinc deposits in limestones. Associated with sphalerite, galena, hemimorphite, cerussite, calcite, and limonite. Often found in pseudomorphs after calcite. Smithsonite is found in places in translucent green or greenish-blue material which is used for ornamental purposes. Laurium, Greece, is noted for this ornamental smithsonite; and Sardinia for yellow stalactites with concentric banding. Fine crystallized specimens have come from the Broken Hill mine, Zambia, and from Tsumeb, South-West Africa. In the United States smithsonite occurs as an ore in the zinc deposits of Leadville, Colorado; Missouri; Arkansas; Wisconsin; and Virginia. Fine greenish-blue material has been found at Kelly, New Mexico.

Use. An ore of zinc. A minor use is for ornamental purposes.

Name. Named in honor of James Smithson (1754–1829), who founded the Smithsonian Institution in Washington, D.C. English mineralogists formerly called the mineral *calamine*.

Similar Species. *Hydrozincite*, Zn_5-$(CO_3)_2(OH)_6$, occurs as a secondary mineral in zinc deposits.

Aragonite Group

When the (CO_3) group is combined with large divalent cations (ionic radii greater than 1.0 Å) the radius ratios generally do not permit stable 6 coordination and orthorhombic structures result. This is the *aragonite structure type* (Fig. 9.7) with space group *Pmcn*. $CaCO_3$ occurs in both the *calcite* and *aragonite structure types* because although Ca is somewhat large for 6 coordination (calcite), it is relatively small, at room temperature, for 9 coordination (aragonite); calcite is the stable form of $CaCO_3$ at room temperature (see Fig. 9.8). Carbonates with larger cations such as $BaCO_3$, $SrCO_3$, and $PbCO_3$, however, have the aragonite structure, stable at room temperature. In the aragonite struc-

FIG. 9.7. The structure of aragonite, $CaCO_3$, as projected on (100). Oxygens that would normally superimpose have been made visible by some displacement. Numbers represent heights of atomic positions above the plane of origin, marked with respect to *a*. Dashed rectangle outlines unit cell. Ca-O bonds are also shown.

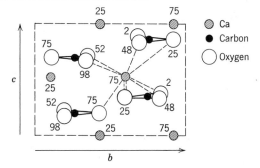

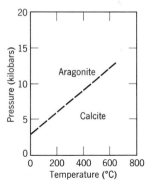

FIG. 9.8. The approximate location of the experimentally determined stability fields of calcite and aragonite.

ture the (CO_3) groups as in calcite, lie perpendicular to the c axis, but in two structural planes, with the (CO_3) triangular groups of one plane pointing in opposite directions to those of the other. In calcite all (CO_3) groups lie in a single structural plane and point in the same direction (Fig. 9.2a). Each Ca is surrounded by nine closest oxygens. The cations have an arrangement in the structure approximating hexagonal closest packing, which gives rise to marked pseudohexagonal symmetry. This is reflected in both the crystal angles and in the pseudo-hexagonal twinning which is characteristic of all members of the group.

Solid solution within the aragonite group is somewhat more limited than in the calcite group, and it is interesting to note that Ca and Ba respectively the smallest and largest ions in the group, form an ordered compound *barytocalcite*, $BaCa(CO_3)_2$, analogous to the occurrence of dolomite, $CaMg(CO_3)_2$, in the system $CaCO_3$-$MgCO_3$.

The differences in physical properties of the minerals of the aragonite group are conferred largely by the cations. Thus the specific gravity is roughly proportional to the atomic weight of the metal ions (Table 5.1).

✸ ARAGONITE—CaCO₃

Crystallography. Orthorhombic; $2/m2/m2/m$. Three habits of crystallization are common. (1) Acicular pyramidal; consisting of a vertical prism terminated by a combination of a very steep dipyramid and first-order prism (Fig. 9.9a). Usually in radiating groups of large to very small crystals. (2) Tabular; consisting of prominent {010} modified by {110} and a low prism, k\{011\} (Fig. 9.9b); often twinned on {110} (Fig. 9.9c). (3) In pseudohexagonal twins (Fig. 9.9d) showing a hexagonal-like prism terminated by a basal plane. They are formed by an intergrowth of three individuals twinned on {110} with {001} planes in common. The cyclic twins are distinguished from true hexagonal forms by noting that the basal surface is striated in three different directions and that, because the prism angle of the simple crystals is not exactly 60°, the composite prism faces for the twin will often show slight reentrant angles. Also found in reniform, columnar, and stalactitic aggregates.

Angles: $m(110) \wedge m'(1\overline{1}0) = 63°48'$, $b(010) \wedge m(110) = 58°06'$, $k(011) \wedge k'(0\overline{1}1) = 71°33'$, $b(010) \wedge k(011) = 54°13'$.

$Pmcn$; $a = 4.95$, $b = 7.96$, $c = 5.73$ Å; $a:b:c = 0.622:1:0.720$. $Z = 4$. $d's$: 3.04(9), 2.71(6), 2.36(7), 1.975(10), 1.880(8).

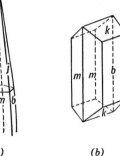
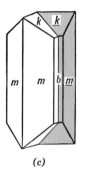
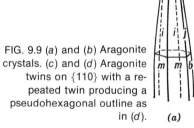
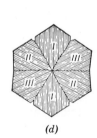

FIG. 9.9 (a) and (b) Aragonite crystals. (c) and (d) Aragonite twins on {110} with a repeated twin producing a pseudohexagonal outline as in (d).

(a)   (b)   (c)   (d)

Physical Properties. *Cleavage* {010} distinct, {110} poor. *Luster* vitreous. *Color* colorless, white, pale yellow, and variously tinted. Transparent to translucent. **H** $3\frac{1}{2}$–4. **G** 2.95 (harder and higher specific gravity than calcite). *Optics:* (−); $\alpha = 1.530$, $\beta = 1.680$, $\gamma = 1.685$; $2V = 18°$; $X = c$, $Y = a$. $r < v$ weak.

Composition, Structure and Synthesis. Most aragonite is relatively pure $CaCO_3$. Small amounts of Sr and Pb substitute for Ca. The structure of aragonite is given in Fig. 9.7 and discussed on page 303. Calcite can be transformed into the aragonite structure by extensive grinding in a mortar and pestle. This structural transformation can be determined only by X-ray diffraction techniques. The phase diagram for the polymorphs of $CaCO_3$ is given in Fig. 9.8. Aragonite, with a somewhat denser structure than calcite is stable on the high P, low T side of the curve relating the two polymorphs. If the temperature of metamorphism in a metamorphic terrane can be arrived at independently, this curve allows for the estimation of a maximum or minimum pressure of metamorphism in calcite- or aragonite-bearing rocks.

Diagnostic Features. Infusible; decrepitates on heating. Effervesces in cold HCl. Distinguished from calcite by its higher specific gravity and lack of rhombohedral cleavage. Cleavage fragments of columnar calcite are terminated by a cross cleavage that is lacking in aragonite. Distinguished from witherite and strontianite by being infusible, of lower specific gravity, and in lacking a distinctive flame color.

Alteration. Pseudomorphs of calcite after aragonite are common. $CaCO_3$ secreted by mollusks as aragonite is usually changed to calcite on the outside of the shell.

Occurrence. Aragonite is less stable than calcite under atmospheric conditions and much less common. It is precipitated in a narrow range of physicochemical conditions represented by low temperature, near-surface deposits. Experiments have shown that carbonated waters containing calcium more often deposit aragonite when they are warm and calcite when they are cold. The pearly layer of many shells and the pearl itself is aragonite.

Aragonite is deposited by hot springs; found associated with beds of gypsum and deposits of iron ore where it may occur in forms resembling coral, and is called *flos ferri*. It is found as fibrous crusts on serpentine and in amygdaloidal cavities in basalt. The occurrence of aragonite in some metamorphic rocks as in the Franciscan Formation in California and in blue schists of New Zealand is the result of recrystallization at high pressures and relatively low temperature.

Notable localities for the various crystalline types are as follows: pseudohexagonal twin crystals are found in Aragon, Spain; Bastennes, in the south of France; and at Girgenti, Sicily associated with native sulfur. Tabular type crystals are found near Bilin, Bohemia. The acicular type is found at Alston Moor and Cleator Moor, Cumberland, England. Flos ferri is found in the Styrian iron mines. Some *onyx marble* from Baja California, Mexico, is aragonite. In the United States pseudohexagonal twins are found at Lake Arthur, New Mexico. Flos ferri occurs in the Organ Mountains, New Mexico, and at Bisbee, Arizona.

Name. From Aragon, Spain where the pseudohexagonal twins were first recognized.

✳ WITHERITE—$BaCO_3$

Crystallography. Orthorhombic; $2/m\,2/m\,2/m$. Crystals always twinned on {110} forming pseudohexagonal dipyramids by the intergrowth of three individuals (Fig. 9.10). Crystals often deeply striated horizontally and by a series of reentrant angles have

FIG. 9.10. Witherite.

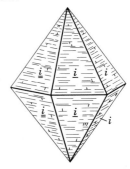

the appearance of one pyramid capping another. Also botryoidal to globular; columnar or granular.

Angles: $b(010) \wedge m(110) = 58°54'$; $i(021) \wedge i'(02\bar{1}) = 68°48'$.

Pmcn; $a = 5.26$, $b = 8.85$, $c = 6.55$ Å; $a:b:c = 0.594:1:0.740$. $Z = 4$. $d's$: 3.72(10), 2.63(6), 2.14(5), 2.03(5), 1.94(5).

Physical Properties. *Cleavage* {010} distinct, {110} poor. **H** $3\frac{1}{2}$. **G** 4.3. *Luster* vitreous. *Color* colorless, white, gray. Translucent. *Optics:* (−), $\alpha = 1.529$, $\beta = 1.676$, $\gamma = 1.677$; $2V = 16°$; $X = c$, $Y = b$; $r > v$ weak.

Composition and Structure. BaO 77.7, CO_2 22.3%. Small amounts of Sr and Ca may substitute for Ba. Witherite is isostructural with aragonite (see Fig. 9.7 and page 304).

Diagnostic Features. Fusible at $2\frac{1}{2}$–3, giving a yellowish green flame (Ba). Soluble in cold HCl with effervescence. All solutions, even the very dilute, give precipitate of barium sulfate with sulfuric acid (differentiates from Ca and Sr). Witherite is characterized by high specific gravity. It is distinguished from barite by its effervescence in acid and from strontianite by the flame test.

Occurrence. Witherite is a comparatively rare mineral, most frequently found in veins associated with galena. Found in England in fine crystals near Hexham in Northumberland and Alston Moor in Cumberland. Occurs at Leogang in Salzburg, Austria. In the United States found near Lexington, Kentucky, and in a large vein with barite at El Portal, Yosemite Park, California. Also at Thunder Bay, Lake Superior, Ontario.

Use. A minor source of barium.

Name. In honor of D. W. Withering (1741–1799), who discovered and first analyzed the mineral.

STRONTIANITE—SrCO₃

Crystallography. Orthorhombic; $2/m\,2/m\,2/m$. Crystals usually acicular, radiating like type 1 under aragonite. Twinning on {110} frequent, giving pseudohexagonal appearance. Also columnar, fibrous, and granular.

Angles: $b(010) \wedge m(110) = 58°40'$, $(001) \wedge$

$(111) = 54°18'$, $(010) \wedge (021) = 34°38'$, $(011) \wedge (0\bar{1}1) = 71°48'$.

Pmcn; $a = 5.13$, $b = 8.42$, $c = 6.09$ Å; $a:b:c = 0.609:1:0.723$; $Z = 4$. $d's$: 3.47(10), 2.42(5), 2.02(7), 1.876(5), 1.794(7).

Physical Properties. *Cleavage* {110} good. **H** $3\frac{1}{2}$–4. **G** 3.7. *Luster* vitreous. *Color* white, gray, yellow, green. Transparent to translucent. *Optics:* (−), $\alpha = 1.520$, $\beta = 1.667$, $\gamma = 1.669$; $2V = 7°$; $X = c$, $Y = b$. $r < v$ weak.

Composition and Structure. SrO 70.2, CO_2 29.8% for pure $SrCO_3$. Ca may be present in substitution for Sr to a maximum of Ca:Sr of about 1:4.5. Strontinate is isostructural with aragonite (see Fig. 9.7 and page 304).

Diagnostic Features. Infusible, but on intense ignition swells and throws out fine branches and gives a crimson flame (Sr). Effervesces in HCl and a medium dilute solution will give precipitate of strontium sulfate on addition of a few drops of sulfuric acid. Characterized by high specific gravity and effervescence in HCl. Can be distinguished from witherite and aragonite by flame test; from celestite by poorer cleavage and effervescence in acid.

Occurrence. Strontianite is a low-temperature hydrothermal mineral associated with barite, celestite, and calcite in veins in limestone or marl, and less frequently in igneous rocks and as a gangue mineral in sulfide veins. Occurs in commercial deposits in Westphalia, Germany; Spain; Mexico; and England. In the United States found in geodes and veins with calcite at Schoharie, New York; in the Strontium Hills north of Barstow, California; and near La Conner, Washington.

Use. Source of strontium. Strontium has no great commercial application; used in fireworks, red flares, military rockets, in the separation of sugar from molasses, and in various strontium compounds.

Name. From Strontian in Argyllshire, Scotland, where it was originally found.

CERUSSITE—PbCO₃

Crystallography. Orthohombic; $2/m\,2/m\,2/m$. Crystals of varied habit common and show many

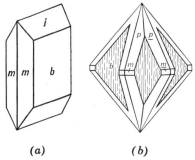

(a) **(b)**

FIG. 9.11. Cerussite.

forms. Often tabular on {010} (Fig. 9.11a). May form reticulated groups with the plates crossing each other at 60° angles; frequently in pseudohexagonal twins with deep reentrant angles in the vertical zone (Fig. 9.11b). Also in granular crystalline aggregates; fibrous; compact; earthy.

Angles: b(010) ∧ m(110) = 58°37′, b(010) ∧ i(021) = 34°40′, m(110) ∧ p(111) = 35°46′.

Pmcn; a = 5.15, b = 8.47, c = 6.11 Å; a:b:c = 0.608:1:0.721; Z = 4. d's: 3.59(10), 3.50(4), 3.07(2), 2.49(3), 2.08(3).

Physical Properties. *Cleavage* {110} good, {021} fair. **H** 3–3½. **G** 6.55. *Luster* adamantine. *Color* colorless, white, or gray. Transparent to subtranslucent. *Optics:* (−); $\alpha = 1.804$, $\beta = 2.077$, $\gamma = 2.079$; 2V = 9°; X = c, Y = b; r > v.

Composition and Structure. Most cerussite is very close in composition to $PbCO_3$, with PbO 83.5 and CO_2 16.5%. It is isostructural with aragonite (see Fig. 9.7 and page 304).

Diagnostic Features. Fusible at 1½. With sodium carbonate on charcoal gives globule of lead and yellow to white coating of lead oxide. Soluble in warm dilute nitric acid with effervescence. Recognized by its high specific gravity, white color, and adamantine luster. Crystal form and effervescence in nitric acid serve to distinguish it from anglesite.

Occurrence. Cerussite is an important and widely distributed supergene lead ore formed by the action of carbonated waters on galena. Associated with the primary minerals galena and sphalerite, and various secondary minerals such as anglesite, pyromorphite, smithsonite, and limonite.

Notable localities for its occurrence are Ems in Nassau; Mies, Bohemia; Nerchinsk, Siberia; on the island of Sardinia; in Tunis; at Tsumeb, South-West Africa; and Broken Hill, New South Wales. In the United States found at Phoenixville, Pennsylvania; Leadville, Colorado; various districts in Arizona; from the Organ Mountains, New Mexico; and in the Coeur d'Alene district in Idaho.

Use. An important ore of lead.

Name. From the Latin word meaning *white lead.*

Similar Species. *Phosgenite*, $Pb_2CO_3Cl_2$, (tetragonal) is a rare carbonate.

Dolomite Group

The dolomite group includes *dolomite,* $CaMg(CO_3)_2$, *ankerite*, $CaFe(CO_3)_2$, and *kutnahorite,* $CaMn(CO_3)_2$. These three carbonates are isostructural with space group $R\bar{3}$. The structure of dolomite is similar to that of calcite but with Ca and Mg layers alternating along the c axis. The large difference in size of the Ca^{2+} and Mg^{2+} ions (33%) causes *cation ordering* with the two cations in specific, and separate levels in the structure. With the nonequivalence of Ca and Mg layers, the 2-fold rotation axes of calcite do not exist and the symmetry is reduced to that of the rhombohedral class, $\bar{3}$. The composition of dolomite is intermediate between $CaCO_3$ and $MgCO_3$ with Ca:Mg = 1:1. The occurrence of this ordered compound, however, does not imply that solid solution exists between $CaCO_3$ and $MgCO_3$ (see Fig. 9.5). In the dolomite structure, especially at low temperatures, each of the two divalent cations occupies a structurally distinct position. At higher temperatures (above 700°C) dolomite shows small deviations from the composition with Ca:Mg = 1:1 as shown in Fig. 9.12. This diagram also shows that at elevated temperatures calcite coexisting with dolomite becomes more magnesian. Similarly dolomite in dolomite-calcite pairs becomes somewhat more calcic. At temperatures above about 1000° to 1100°C a complete solid solution exists between calcite and dolomite, but not between dolomite and magnesite. The compositions of coexisting dolomite and calcite have been used

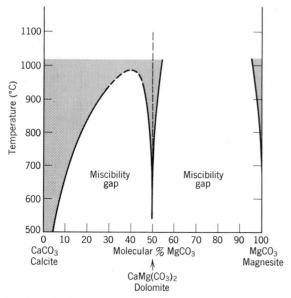

FIG. 9.12. $CaCO_3$-$MgCO_3$ system at CO_2 pressures sufficient to prevent decomposition of the carbonates. Vertical dashed line is ideal dolomite composition. Solid solution is shown by shading (after J. R. Goldsmith, 1959, *Researches in Geochemistry*, John Wiley).

for an estimate of temperatures of crystallization of rocks containing both carbonates, using the temperature scale in Fig. 9.12.

DOLOMITE—CaMg(CO$_3$)$_2$; ANKERITE—CaFe(CO$_3$)$_2$

Crystallography. For *dolomite*. Hexagonal-*R*; $\bar{3}$. Crystals are usually the unit rhombohedron (Fig. 9.13*a*), more rarely a steep rhombohedron and base (Fig. 9.13*c*). Faces are often curved, some so acutely as to form "saddle-shaped" crystals (Fig. 9.13*b*).

Other forms rare. In coarse, granular, cleavable masses to fine grained and compact. Twinning on {0001} common; lamellar twinning on {02$\bar{2}$1}. *Ankerite* generally does not occur in well-formed crystals. When it does the crystals are similar to those of dolomite.

  Angles: $r(10\bar{1}1) \wedge r'(\bar{1}101) = 73°16'$, $c(0001) \wedge M(40\bar{4}1) = 75°25'$.

  $R\bar{3}$. Morphology $a:c = 1:0.832$. Structure $a = 4.84$, $c = 15.95$ Å; $a:c = 1:3.294$; $Z = 3$. $d's$: 2.88(10), 2.19(4), 2.01(3), 1.800(1), 1.780(1).

  Physical Properties. *Cleavage* {10$\bar{1}$1} perfect. **H** $3\frac{1}{2}$–4. **G** 2.85. *Luster* vitreous; pearly in some varieties, *pearl spar*. *Color* commonly some shade of pink, flesh color; may be colorless, white, gray, green, brown, or black. Transparent to translucent. *Optics:* (−); $\omega = 1.681$, $\epsilon = 1.500$. With increasing substitution of Fe (toward ankerite composition) **G** and indices of refraction increase. Ankerite is typically yellowish white but due to oxidation of the iron, may appear yellowish brown.

  Composition and Structure. *Dolomite:* CaO 30.4, MgO 21.7, CO_2 47.9%. *Ankerite:* CaO 25.9, FeO 33.3, CO_2 40.8%. Naturally occurring dolomite deviates somewhat from Ca:Mg = 1:1 (see Fig. 9.12) with the Ca:Mg ratio ranging from 58:42 to $47\frac{1}{2}$:$52\frac{1}{2}$. A complete solid solution series extends to *ankerite* (see Fig. 9.5). Probably a complete solid solution series also exists between ankerite and *kutnahorite*, $CaMn(CO_3)_2$ (see Fig. 4.27a). Members of the dolomite group are isostructural. The structure of dolomite is discussed on page 307.

  Diagnostic Features. For *dolomite*. Infusible. In cold, dilute HCl large fragments are slowly attacked but are soluble with effervescence in hot HCl; the powdered mineral is readily soluble in

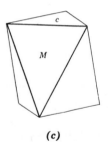

FIG. 9.13. Dolomite. *(a)*   *(b)*   *(c)*

cold acid. Crystallized variety told by its curved rhombohedral crystals and usually by its flesh-pink color. The massive rock variety is distinguished from limestone by the less vigorous reaction with HCl. *Ankerite* is similar to dolomite in its physical and chemical properties except for its color which ranges from yellow-brown to brown. Ankerite turns black on heating before the blowpipe.

Occurrence. *Dolomite* is found in many parts of the world chiefly as extensive sedimentary strata, and the crystallized equivalent, dolomitic marble. Dolomite as a rock mass is generally thought to be secondary in origin, formed from ordinary limestone by the replacement of some of the Ca by Mg. The replacement may be only partial and thus most dolomite rocks are mixtures of dolomite and calcite. The mineral occurs also as a hydrothermal vein mineral, chiefly in the lead and zinc veins that traverse limestone, associated with fluorite, calcite, barite, and siderite. *Ankerite* is a common carbonate in iron-formations

Dolomite is abundant in the Dolomite region of southern Tyrol; in crystals from the Binnenthal, Switzerland; Traversella in Piedmont; northern England; and Guanajuato, Mexico. In the United States it is found as masses of sedimentary rock in many of the middle-western states, and in crystals in the Joplin, Missouri, district.

Use of Dolomite. As a building and ornamental stone. For the manufacture of certain cements. For the manufacture of magnesia used in the preparation of refractory linings of the converters in the basic steel process. Dolomite is a potential ore of metallic Mg.

Name. *Dolomite* in honor of the French chemist, Dolomieu (1750–1801). *Ankerite* after M. J. Anker (1772–1843), Austrian mineralogist.

Similar species. *Kutnahorite*, $CaMn(CO_3)_2$, is isostructural with dolomite.

MALACHITE—$Cu_2CO_3(OH)_2$

Crystallography. Monoclinic; $2/m$. Crystals slender prismatic but seldom distinct. Crystals may be pseudomorphous after azurite. Usually in ra-

FIG. 9.14. Malachite, cut and polished.

diating fibers forming botryoidal or stalactitic masses (Fig. 9.14). Often granular or earthy.

$P2_1/a$; $a = 9.48$, $b = 12.03$, $c = 3.21$ Å, $\beta = 98°44'$; $a:b:c = 0.788:1:0.267$; $Z = 4$. $d's$: 6.00(6), 5.06(8), 3.69(9), 2.86(10), 2.52(6).

Physical Properties. *Cleavage* $\{\overline{2}01\}$ perfect but rarely seen. **H** $3\frac{1}{2}$–4. **G** 3.9–4.03. *Luster* adamantine to vitreous in crystals; often silky in fibrous varieties; dull in earthy type. *Color* bright green. *Streak* pale green. Translucent. *Optics:* (−), $\alpha = 1.655$, $\beta = 1.875$, $\gamma = 1.909$; $2V = 43°$; $Y = b$, $Z \wedge c = 24°$; pleochroism: X colorless, Y yellow green, Z deep green.

Composition and Structure. CuO 71.9, CO_2 19.9, H_2O 8.2%. Cu 57.4%. Cu^{2+} is octahedrally coordinated by O^{-2} and $(OH)^{-1}$ in $CuO_2(OH)_4$ and $CuO_4(OH)_2$ octahedra. These octahedra are linked along edges forming chains that run parallel to the c axis. The chains are crosslinked by triangular $(CO_3)^{-2}$ groups.

Diagnostic Features. Fusible at 3, giving a green flame. With fluxes on charcoal gives copper globule. Soluble in HCl with effervescence, yielding a green solution. Much water in the closed tube. Recognized by its bright green color and botryoidal forms, and distinguished from other green copper minerals by its effervescence in acid.

Occurrence. Malachite is a widely distributed supergene copper mineral found in the oxidized portions of copper veins associated with azurite, cuprite, native copper, iron oxides. It usually occurs in copper deposits associated with limestone.

Notable localities for its occurrence are at Nizhne Tagil in the Ural Mountains; at Chessy, near Lyons, France, associated with azurite; from Tsumeb, South-West Africa; Katanga, Zaire; and Broken Hill, New South Wales. In the United States, formerly an important copper ore in the southwestern copper districts; at Bisbee, Morenci, and other localities in Arizona; and in New Mexico.

Use. An ore of copper. Has been used to some extent, particularly in Russia, as an ornamental material for vases, veneer for table tops, etc. (see Fig. 9.14).

Name. Derived from the Greek word for *mallows*, in allusion to its green color.

AZURITE—$Cu_3(CO_3)_2(OH)_2$

Crystallography. Monoclinic; $2/m$. Habit varies (Fig. 9.15); crystals frequently complex and malformed. Also in radiating spherical groups.

Angles: $c(001) \wedge a(100) = 87°35'$, $m(110) \wedge a(100) = 40°33'$, $h(111) \wedge m(110) = 19°58'$, $c(001) \wedge l(013) = 30°30'$.

$P2_1c$; $a = 4.97$, $b = 5.84$, $c = 10.29$ Å, $\beta = 92°24'$. $a:b:c = 0.851:1:1.762$. $Z = 2$. $d's$: 5.15(7), 3.66(4), 3.53(10), 2.52(6), 2.34(3).

FIG. 9.15. Azurite crystals.

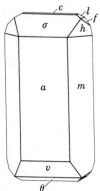 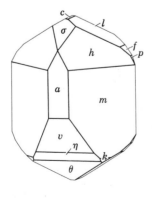

Physical Properties. *Cleavage* {011} perfect, {100} fair. **H** $3\frac{1}{2}$–4. **G** 3.77. *Luster* vitreous. *Color* intense azure-blue. Transparent to translucent. *Optics:* (+), $\alpha = 1.730$, $\beta = 1.758$, $\gamma = 1.838$; $2V = 67°$; $X = b$, $Z \wedge c = -13°$; pleochroic in blue $Z > Y > X$; $r > v$.

Composition and Structure. CuO 69.2, CO_2 25.6, H_2O 5.2%. Cu 55.3%. The structure of azurite contains Cu^{2+} ions in square, coplanar groups with 2 O^{-2} and $2(OH)^{-1}$. These square groups are linked into chains parallel to the b axis. Each $(OH)^{-1}$ group is shared by 3 Cu^{2+} and each oxygen of the triangular (CO_3) group is bonded to one copper.

Diagnostic Features. Characterized chiefly by its azure-blue color and effervescence in HCl. Gives the same tests as malachite.

Alteration. Pseudomorphs of malachite after azurite commonly observed; less common after cuprite.

Occurrence. Azurite is less common than malachite but has the same origin and associations. It is found in fine crystals at Chessy, near Lyons, France; Tsumeb, South-West Africa; in Rumania; Laurium, Greece; Siberia; and Broken Hill, New South Wales. In the United States at Copper Queen mine, Bisbee and Morenci, Arizona. Widely distributed with copper ores.

Use. A minor ore of copper.

Name. Named in allusion to its color.

Rare Hydrous Carbonates. *Aurichalcite,* $(Zn,Cu)_5(CO_3)_2(OH)_3$ pale green to blue in orthorhombic acicular crystals. *Gaylussite,* $Na_2Ca(CO_3)_2 \cdot 5H_2O$, monoclinic and *trona,* $Na_3H(CO_3)_2 \cdot 2H_2O$, monoclinic, are both found in saline-lake deposits.

NITRATES

The minerals in this group are structurally similar to the carbonates with plane, triangular $(NO_3)^{-1}$ groups, much like the $(CO_3)^{-2}$ group. Like C in the (CO_3) group, the highly charged and highly polarizing N^{5+} ion binds its three coordinated oxygens into a close-knit group in which the strength of the oxygen-nitrogen bond (e.v. = $1\frac{2}{3}$) is greater than any other possible bond in the crystal. Because of the greater strength of this N—O bond as compared with the C—O bond, nitrates are less readily decomposed by acids than carbonates.

When triangular (NO_3) groups combine in one-to-one proportions with monovalent cations whose radii permit 6 coordination, structures analogous to those of the calcite group result. Thus, *soda niter*, $NaNO_3$, and *calcite* are isostructural with the same crystallography and cleavage. Because of the lesser charge, however, soda niter is softer than calcite and melts at a lower temperature and, because of the lower atomic weight of sodium, has a lower specific gravity. *Niter*, KNO_3, is similarly a structural analogue of *aragonite*. Soda niter has an orthorhombic polymorph isostructural with niter. Of the seven known naturally occurring nitrates, we will describe only *soda niter* and *niter*.

Soda Niter—NaNO₃

Crystallography. Hexagonal-*R*; $\bar{3}2/m$. Rhombohedral crystals are rare. Usually massive, as incrustations or in beds. Soda niter is isostructural with calcite and has similar crystal constants, cleavage, and optical properties.

$R\bar{3}c$. Morphology $a:c = 1:0.828$. Structure $a = 5.07$, $c = 16.81$ Å; $a:c = 1:3.316$; $Z = 6$. $d's$: 3.03(10), 2.13(6), 1.89(6), 1.65(3), 1.46(3).

Physical Properties. *Cleavage* perfect $\{10\bar{1}1\}$. **H** 1–2. **G** 2.29. *Luster* vitreous. Colorless or white, also reddish brown, gray, yellow. Transparent to translucent. Taste cooling. Deliquescent. *Optics*: (−); $\omega = 1.587$, $\epsilon = 1.336$.

Composition and Structure. Na_2O 36.5, N_2O_5 63.5%. Soda niter is isostructural with calcite (see Fig. 9.2), the $(NO_3)^{-1}$ groups playing the same role as $(CO_3)^{-2}$ groups in the carbonates.

Diagnostic Features. Fusible at 1, giving a strong yellow flame (Na). Easily and completely soluble in water. Distinguished by its cooling taste and its deliquescence.

Occurrence. Because of its solubility in water, soda niter is found only in arid and desert regions. Found in large quantities in the provinces of Tara-

paca and Antofagasta, northern Chile, and the neighboring parts of Bolivia. Occurs over immense areas as a salt bed (*caliche*) interstratified with sand, beds of common salt, gypsum, etc. In the United States it has been noted in Humboldt County, Nevada, and in San Bernardino County, California.

Use. In Chile it is quarried, purified, and used as a source of nitrates. Soda niter now competes with nitrogen "fixed" from the air. Nitrates are used in explosives and fertilizer.

Name. From the composition, the sodium analogue of *niter*, KNO_3. See further under niter.

Niter (*saltpeter*)—KNO₃

Crystallography. Orthorhombic; $2/m2/m2/m$. Usually as thin encrustations or silky acicular crystals. Twinning on $\{110\}$ is common, giving pseudohexagonal groupings analogous to aragonite.

Pcmn; $a = 5.43$, $b = 9.19$, $c = 6.46$, $a:b:c = 0.591:1:0.703$. $Z = 4$. $d's$: 3.77(10), 3.03(6), 2.66(5), 2.19(5), 1.96(3).

Physical Properties. *Cleavage* perfect $\{011\}$. **H** 2. **G** 2.09–2.14. *Luster* vitreous. *Color* white. Translucent. *Optics*: (−); $\alpha = 1.332$, $\beta = 1.504$, $\gamma = 1.504$; $2V = 7°$; $X = c$, $Y = a$; $r < v$.

Composition and Structure. K_2O 46.5, N_2O_5 53.5%. Niter is isostructural with aragonite (see Fig. 9.7).

Diagnostic Features. Fusible at 1, giving violet flame (K). Easily soluble in water. Characterized by its cooling taste; distinguished from soda niter by the potassium test and by being nondeliquescent.

Occurrence. Niter is found as delicate crusts, as an efflorescence, on surfaces of earth, walls, rocks, and so forth. Found as a constituent of certain soils. Also in the loose soil of limestone caves. Not as common as soda niter, but produced from soils in Spain, Italy, Egypt, Arabia, Persia, and India.

Use. Used as a source of nitrogen compounds.

Name. The alkaline salt extracted by water from the ashes of vegetable matter was known to the ancients as *neter, nether* (Hebraic). The Greeks subsequently applied the name *nitron* to saline materials from trona deposits.

BORATES

Within the borate group of minerals BO_3 units are capable of polymerization (similar to the polymerization of SiO_4 tetrahedral groups in the silicates) in the form of chains, sheets, and isolated multiple groups (Fig. 9.16). This is possible because the small B^{3+} ion, which generally coordinates three oxygens in a triangular group, has bond strengths to each O with an e.v. = 1; this is exactly half the bonding energy of the oxygen ion. This permits a single oxygen to be shared between two boron ions linking the BO_3 triangles into expanded structural units (double triangles, triple rings, sheets, and chains). Because the triangular coordination of BO_3 is close to the upper stability limit of 3 coordination, boron is found also in 4 coordination in tetrahedral groups. In addition to BO_3 and BO_4 groups natural borates may contain complex ionic groups such as $[B_3O_3(OH)_5]^{-2}$ (Fig. 9.16e) that consist of one triangle and two tetrahedra. In the structure of *colemanite*, $CaB_3O_4(OH)_3 \cdot H_2O$, complex infinite chains of tetrahedra and triangles occur, and in *borax*, $Na_2B_4O_5(OH)_4 \cdot 8H_2O$, a complex ion, $[B_4O_5(OH)_4]^{-2}$ consisting of two tetrahedra and two triangles is found (Fig. 9.16f). Borates can be classified on the basis of the structural anionic linking (or lack thereof) as insular (independent single or double BO_3 or BO_4 groups), chain, sheet, and framework structures.

Although it is possible to prepare a three-dimensional boxwork made up of BO_3 triangles

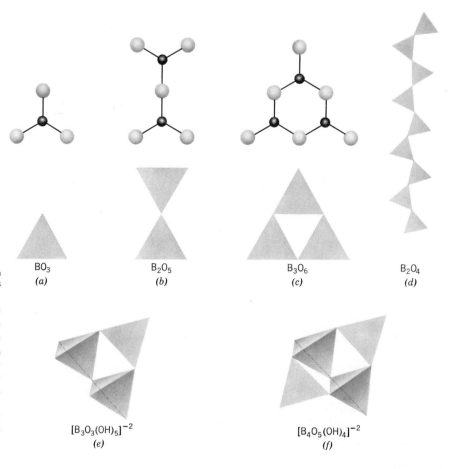

FIG. 9.16. Independent BO_3 triangles (a), multiple groups (b) and (c), and chains (d) in borates. Also complex triple and quadruple groups (e) and (f). The triple group in (e) consists of two $BO_2(OH)_2$ tetrahedra and a BO_2OH triangle. The group in (f) contains two BO_3OH tetrahedra and two BO_2OH triangles; this type of group is present in the structure of borax (see Fig. 9.18).

BO_3
(a)

B_2O_5
(b)

B_3O_6
(c)

B_2O_4
(d)

$[B_3O_3(OH)_5]^{-2}$
(e)

$[B_4O_5(OH)_4]^{-2}$
(f)

only and having the composition B_2O_3, such a configuration has a very low stability and disorders readily, yielding a glass. Because of the tendency to form somewhat disordered networks of BO_3 triangles, boron is regarded as a "network-former" in glass manufacture and is used in the preparation of special glasses of light weight and high transparency to energetic radiation.

Over 100 borate minerals are known but only the four most common are considered here:

Kernite $Na_2B_4O_6(OH)_2 \cdot 3H_2O$
Borax $Na_2B_4O_5(OH)_4 \cdot 8H_2O$
Ulexite $NaCaB_5O_6(OH)_6 \cdot 5H_2O$
Colemanite $CaB_3O_4(OH)_3 \cdot H_2O$

KERNITE—$Na_2B_4O_6(OH)_2 \cdot 3H_2O$

Crystallography. Monoclinic; $2/m$. Rarely in crystals; usually in coarse, cleavable aggregates.

$P2/a$; $a = 15.68$, $b = 9.09$, $c = 7.02$ Å; $\beta = 108°52'$; $a:b:c = 1.725:1:0.772$; $Z = 4$. $d's$: 7.41(10), 6.63(8), 3.70(3), 3.25(2), 2.88(2).

Physical Properties. *Cleavage* {001} and {100} perfect. Cleavage fragments are thus elongated parallel to the b crystallographic axis. The angle between the cleavages: (001) $\wedge$ (100) = 71°8'. **H** 3. **G** 1.95. *Luster* vitreous to pearly. *Color* colorless to white; colorless specimens become chalky white on long exposure to the air owing to a surface film of *tincalconite*, $Na_2B_4O_5(OH)_4 \cdot 3H_2O$. *Optics:* (−); $\alpha = 1.454$, $\beta = 108°52'$; $a:b:c = 1.725:1:0.772$; $Z = 4$. $X \wedge c = 39°$.

Composition and Structure. Na_2O 22.7, B_2O_3 51.0, H_2O 26.3%. The structure of kernite contains complex chains, parallel to the b axis, of composition $B_4O_6(OH)_4$. These chains consist of BO_4 tetrahedra, joined at the vertices (as in the pyroxenes of the silicates) with remaining vertices connected to BO_3 triangles. The chains are linked to each other by Na^+ (in 5 coordination) and hydroxyl-oxygen bonds.

Diagnostic Features. Characterized by long, splintery cleavage fragments and low specific gravity. Before the blowpipe it swells and then fuses (at $1\frac{1}{2}$) to a clear glass. Slowly soluble in cold water.

Occurrence. The original locality and principal occurrence of kernite is on the Mohave Desert at Boron, California. Here, associated with borax, colemanite, and ulexite in a bedded series of Tertiary clays, it is present by the million tons. This deposit of sodium borates is 4 miles long, 1 mile wide, and as much as 250 feet thick, and lies from 150 to 1000 feet beneath the surface. The kernite occurs near the bottom of the deposit and is believed to have formed from borax by recrystallization caused by increased temperature and pressure. Kernite is also found associated with borax at Tincalayu, Argentina.

Use. A source of borax and boron compounds.

Name. From Kern County, California where the mineral is found.

BORAX—$Na_2B_4O_5(OH)_4 \cdot 8H_2O$

Crystallography. Monoclinic; $2/m$. Commonly in prismatic crystals (Fig. 9.17). Also as massive, cellular material or encrustations.

Angles: $c(001) \wedge a(100) = 73°25'$, $b(010) \wedge m(110) = 43°30'$, $c(001) \wedge o(\bar{1}12) = 40°31'$, $c(001) \wedge z(\bar{1}11) = 64°8'$.

$C2/c$; $a = 11.86$, $b = 10.67$, $c = 12.20$ Å; $\beta = 106°$; $a:b:c = 1.111:1:1.143$; $Z = 4$. $d's$: 5.7(2), 4.86(5), 3.96(4), 2.84(5), 2.57(10).

Physical Properties. *Cleavage* {100} perfect. **H** 2–$2\frac{1}{2}$. **G** 1.7±. *Luster* vitreous. *Color* colorless or white. Translucent. Sweetish-alkaline taste. Clear crystals effloresce and turn white with the formation of *tincalconite*, $Na_2B_4O_5(OH)_4 \cdot 3H_2O$. *Optics:* (−); $\alpha = 1.447$, $\beta = 1.469$, $\gamma = 1.472$; $2V = 40°$. $X = b$, $Z \wedge c = -56°$. $r > v$.

FIG. 9.17. Borax.

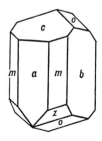

Composition and Structure. Na_2O 16.2, B_2O_3 36.6, H_2O 47.2%. The structure of borax is shown in Fig. 9.18. It contains $B_4O_5(OH)_4$ groups consisting of tetrahedra and triangles which are joined parallel to the c axis. The Na is in 6 coordination with H_2O molecules. These $Na(H_2O)_6$ octahedra are bonded along common edges making ribbons parallel to c. The boron groups are linked to the Na-H_2O ribbons by weak residual bonds.

Diagnostic Features. Characterized by its crystals and tests for boron. Fusible at $1-1\frac{1}{2}$ with much swelling and gives strong yellow flame (Na). Moistened with sulfuric acid gives the bright green flame of boron. Readily soluble in water. Much water in the closed tube.

Occurrence. Borax is the most widespread of the borate minerals. It is formed on the evaporation of enclosed lakes and as an efflorescence on the surface in arid regions. Deposits in Tibet furnished the first borax, under the name of *tincal,* to reach western civilization. In the United States it was first found in Lake County, California, and later in the desert region of southeastern California, in Death Valley, Inyo County, and in San Bernardino County. Associated with kernite, it is the principal mineral mined from the bedded deposits at Boron, California. It is also obtained commercially from the brines of Searles Lake at Trona, California.

The world's most extensive deposits of borax and other borate minerals are in Turkey, notably in

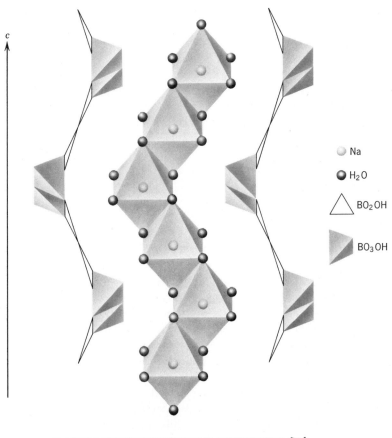

FIG. 9.18. The structure of borax as seen on (100).

the Kirka area. Associated with other borates, borax is mined in the Andes Mountains in contiguous parts of Argentina, Bolivia, and Chile. Minerals commonly associated with borax are halite, gypsum, ulexite, colemanite, and various rare borates.

Use. Although boron is obtained from several minerals, it is usually converted to borax, the principal commercial product. There are many uses of borax. It is used for washing and cleansing; as an antiseptic and preservative; in medicine; as a solvent for metallic oxides in soldering and welding; and as a flux in various smelting and laboratory operations. Elemental boron is used as a deoxidizer and alloy in nonferrous metals; in rectifiers and control tubes; and a neutron absorber in shields for atomic reactors. Boron is used in rocket fuels and as an additive in motor fuel. Boron carbide, harder than corundum, is used as an abrasive.

Name. Borax comes from an Arabic name for the substance.

Ulexite—$NaCaB_5O_6(OH)_6 \cdot 5H_2O$

Crystallography. Triclinic; $\bar{1}$. Usually in rounded masses of loose texture, consisting of fine fibers which are acicular or capillary crystals, "cotton-balls." Rarely in closely packed parallel fibers showing fiber-optical properties and called "television rock."

$P\bar{1}$; $a = 8.73$, $b = 12.75$, $c = 6.70$ Å; $\alpha = 90°16'$, $\beta = 109°8'$, $\gamma = 105°7'$; $a:b:c = 1.104:1:0.526$; $Z = 2$. $d's$: 12.3(10), 7.89(7), 6.60(7), 4.19(7), 2.67(7).

Physical Properties. *Cleavage* {010} perfect. **H** $2\frac{1}{2}$; the aggregate has an apparent hardness of 1. **G** 1.96. *Luster* silky. *Color* white. Tasteless. *Optics:* (+); $\alpha = 1.491$, $\beta = 1.504$, $\gamma = 1.520$; 2V = 73°.

Composition and Structure. Na_2O 7.7, CaO 13.8, B_2O_3 43.0, H_2O 35.5%. The structure is not well known but probably contains $B_3B_2O_7(OH)_4$ chains parallel to the c axis. Between these chains the Na and Ca are located, in 6 and 9 coordination, respectively, with OH and H_2O. Hydroxyl-hydrogen bonds link the chains and cation-OH strips.

Diagnostic Features. The soft "cotton-balls" with silky luster are characteristic of ulexite. Fusible at 1, with intumescence to a clear blebby glass, coloring the flame deep yellow. When moistened with sulfuric acid it gives momentarily the green flame of boron. Gives much water in the closed tube.

Occurrence. Ulexite crystallizes in arid regions from brines that have concentrated in enclosed basins, as in playa lakes. Usually associated with borax. It occurs abundantly in the dry basins of northern Chile and in Argentina. In the United States it has been found abundantly in certain of the enclosed basins of Nevada and California and with colemanite in bedded Tertiary deposits.

Use. A source of borax.

Name. In honor of the German chemist, G. L. Ulex (1811–1883), who discovered the mineral.

Colemanite—$CaB_3O_4(OH)_3 \cdot H_2O$

Crystallography. Monoclinic; $2/m$. Commonly in short prismatic crystals, highly modified (see Fig. 9.19). Cleavable massive to granular and compact.

Angles: $b(010) \wedge m(110) = 53°54'$, $c(001) \wedge a(100) = 69°53'$, $c(001) \wedge h(\bar{2}01) = 68°25'$, $b(010) \wedge d(\bar{1}21) = 57°56'$.

$P2_1/a$; $a = 8.74$, $b = 11.26$, $c = 6.10$ Å, $\beta = 110°7'$; $a:b:c = 0.776:1:0.542$. $Z = 4$. $d's$: 5.64(5), 4.00(4), 3.85(5), 3.13(10), 2.55(5).

Physical Properties. *Cleavage* {010} perfect. **H** $4–4\frac{1}{2}$. **G** 2.42. *Luster* vitreous. *Color* colorless to white. Transparent to translucent. *Optics:* (+); $\alpha = 1.586$, $\beta = 1.592$, $\gamma = 1.614$; 2V = 55°; $X = b$, $Z \wedge c = 84°$. $r > v$.

FIG. 9.19. Colemanite.

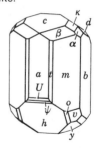

Composition and Structure. CaO 27.2, B_2O_3 50.9, H_2O 21.9%. The structure contains infinite chains with composition $[B_3O_4(OH)_2]^{-2}$ that are parallel to the *a* axis. The chains contain BO_3 triangles and $BO_2(OH)_2$ and $BO_3(OH)$ tetrahedra. Interspersed between the chains are Ca ions and H_2O molecules.

Diagnostic Features. Fusible at $1\frac{1}{2}$. Before the blowpipe it exfoliates, crumbles, and gives a green flame (B). Water in the closed tube. Characterized by one direction of highly perfect cleavage and exfoliation on heating.

Occurrence. Colemanite deposits are interstratified with lake bed deposits of Tertiary age. Ulexite and borax are usually associated, and colemanite is believed to have originated by their alteration. Found in California in Los Angeles, Ventura, San Bernardino, and Inyo counties, and in Nevada in the Muddy Mountains and White Basin, Clark County. Also occurs extensively in Argentina and Turkey.

Use. A source of borax that, at the time of the discovery of kernite, yielded over half of the world's supply.

Name. In honor of William T. Coleman, merchant of San Francisco, who marketed the product of the colemanite mines.

Similar Species. Other borates, locally abundant are: *boracite*, $Mg_3ClB_7O_{13}$; *hydroboracite*, $CaMgB_6O_8(OH)_6 \cdot 3H_2O$ and *inyoite*, $CaB_3O_3(OH)_5 \cdot 4H_2O$.

SULFATES AND CHROMATES

In the discussion of the structures of sulfide minerals we have seen that sulfur occurs as the large, divalent sulfide anion. This ion results from the filling by captured electrons of the two vacancies in the outer, or valence, electron shell. The six electrons normally present in this shell may be lost, giving rise to a small, highly charged and highly polarizing positive ion (radius = 0.30 Å). The ratio of the radius of this S^{6+} ion to that of oxygen ($R_X : R_O = 0.214$) indicates that 4 coordination will be stable. The sulfur to oxygen bond in such an ionic group is very strong (e.v. = $1\frac{1}{2}$; see Fig. 9.1*b*) and covalent in its proper-

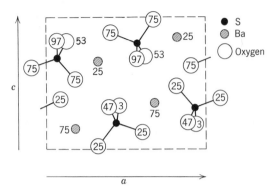

FIG. 9.20. The structure of barite, $BaSO_4$, as projected on (010). The dashed lines outline the unit cell. The oxygen atoms, which would normally superimpose, have been displaced from their positions in this projection.

ties and produces tightly bound groups that are not capable of sharing oxygens. These anionic $(SO_4)^{-2}$ groups are the fundamental structural units of the sulfate minerals.

The most important and common of the anhydrous sulfates are members of the barite group (space group *Pnma*), with large divalent cations coordinated with the sulfate ion. Part of the structure of *barite*, $BaSO_4$, is illustrated in Fig. 9.20. Each barium ion is coordinated to 12 oxygen ions belonging to seven different (SO_4) groups. The barite type structure is also found in manganates (with MnO_4 tetrahedral groups) and chromates (with CrO_4 tetrahedral groups) which contain large cations.

FIG. 9.21. The structure of anhydrite, $CaSO_4$.

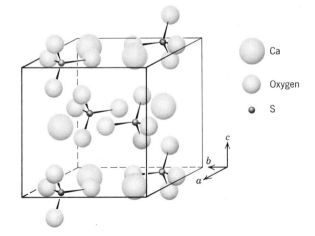

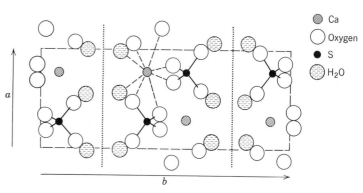

FIG. 9.22. The structure of gypsum, $CaSO_4 \cdot 2H_2O$, projected on (001). Perfect (010) cleavage is indicated by dotted lines. Dashed lines outline unit cell. The 8 coordination (six oxygens and two water molecules) is indicated for one Ca^{2+}.

Anhydrite, $CaSO_4$, because of the smaller size of Ca^{2+} compared to Ba^{2+}, has a very different structure from that of barite (see Fig. 9.21). Each Ca^{2+} is coordinated to eight nearest neighbor oxygens from the tetrahedral (SO_4) groups.

Of the hydrous sulfates, *gypsum,* $CaSO_4 \cdot 2H_2O$, is the most important and abundant. The structure of gypsum is illustrated in Fig. 9.22. Gypsum is monoclinic with space group $C2/c$. The structure consists of layers parallel to {010} of $(SO_4)^{-2}$ groups strongly bonded to Ca^{2+}. Successive layers of this type are separated by sheets of H_2O molecules. The bonds between H_2O molecules in neighboring sheets are weak, which explains the excellent {010} cleavage in gypsum. Loss of the water molecules causes collapse of the structure and transformation into a metastable polymorph of anhydrite (γ $CaSO_4$) with a large decrease in specific volume and loss of the perfect cleavage.

A large number of minerals belong to this class but only a few of them are common. The class can be divided into (1) the anhydrous sulfates and (2) the hydrous and basic sulfates.

Anhydrous Sulfates and Chromate

Barite Group

Barite	$BaSO_4$
Celestite	$SrSO_4$
Anglesite	$PbSO_4$
Anhydrite	$CaSO_4$
Crocoite	$PbCrO_4$

Hydrous and Basic Sulfates

Gypsum	$CaSO_4 \cdot 2H_2O$
Antlerite	$Cu_3SO_4(OH)_4$
Alunite	$KAl_3(SO_4)_2(OH)_6$

Barite Group

The sulfates of Ba, Sr, and Pb form an isostructural group with space group *Pnma*. They have closely related crystal constants and similar habits. The members of the group are: *barite, celestite,* and *anglesite.*

BARITE—$BaSO_4$

Crystallography. Orthorhombic; $2/m\,2/m\,2/m$. Crystals usually tabular on {001}; often diamond-shaped. Both first- and second-order prisms are usually present, either beveling the corners of the diamond-shaped crystals or the edges of the tables and forming rectangular prismatic crystals elongated parallel to either a or b (Fig. 9.23). Crystals may be very complex. Frequently in divergent groups of tabular crystals forming "crested barite" or "barite roses." Also coarsely laminated; granular, earthy.

Angles: $m(\bar{2}10) \wedge m'(210) = 101°38'$, $c(001) \wedge o(011) = 52°43'$, $c(001) \wedge d(101) = 38°52'$.

Pnma; $a = 8.87$, $b = 5.45$, $c = 7.14$ Å; $a:b:c = 1.627:1:1.310$; $Z = 4$. d's: 3.90(6), 3.44(10), 3.32(7), 3.10(10), 2.12(8).

FIG. 9.23. Barite crystals.

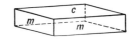

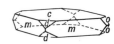

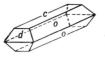

Physical Properties. *Cleavage* {001} perfect, {210} less perfect. **H** 3–3½. **G** 4.5 (heavy for a non-metallic mineral). *Luster* vitreous; on some specimens pearly on base. *Color* colorless, white, and light shades of blue, yellow, red. Transparent to translucent. *Optics:* (+); $\alpha = 1.636$, $\beta = 1.637$, $\gamma = 1.648$; 2V = 37°. $X = c$, $Y = b$. $r > v$.

Composition and Structure. BaO 65.7, SO_3 34.3% for pure barite. Sr substitutes for Ba and a complete solid solution series extends to celestite, but most material is near one end or the other of the series. A small amount of Pb may substitute for Ba. The structure of barite is illustrated in Fig. 9.20 and discussed on page 316.

Diagnostic Features. Recognized by its high specific gravity and characteristic cleavage and crystals. Fusible at 4, giving yellowish-green flame (Ba). Fused with reducing mixture gives a residue which, when moistened, produces a dark stain of silver sulfide on a clean silver surface.

Occurrence. Barite is a common mineral of wide distribution. It occurs usually as a gangue mineral in hydrothermal veins, associated with ores of silver, lead, copper, cobalt, manganese, and antimony. It is found in veins in limestone with calcite, or as residual masses in clay overlying limestone. Also in sandstone with copper ores. In places acts as a cement in sandstone. Deposited occasionally as a sinter by waters from hot springs.

Notable localities for the occurrence of barite crystals are in Westmoreland, Cornwall, Cumberland, and Derbyshire, England; Felsobanya and other localities, Rumania; Saxony; and Bohemia. In the United States at Cheshire, Connecticut; Dekalb, New York; and Fort Wallace, New Mexico. Massive barite, occurring usually as veins, nests, and irregular bodies in limestones, has been quarried in the United States in Georgia, Tennessee, Missouri, and Arkansas. At El Portal, California, at the entrance to Yosemite Park, barite is found in a vein with witherite.

Use. More than 80% of the barite produced is used in oil and gas-well drilling as "heavy mud" to aid in the support of drill rods, and to help prevent blowing out of gas. Barite is the chief source of Ba for chemicals. A major use of barium is in *lithopone*,

a combination of barium sulfide and zinc sulfate that form an intimate mixture of zinc sulfide and barium sulfate. Lithopone is used in the paint industry and to a lesser extent in floor coverings and textiles. Precipitated barium sulfate, "blanc fixe," is employed as a filler in paper and cloth, in cosmetics, as a paint pigment, and for barium meals in radiology.

Name. From the Greek word meaning *heavy*, in allusion to its high specific gravity.

✳CELESTITE—$SrSO_4$

Crystallography. Orthorhombic; $2/m\,2/m\,2/m$. Crystals closely resemble those of barite. Commonly tabular parallel to {001} or prismatic parallel to *a* or *b* with prominent development of the first- and second-order prisms. Crystals elongated parallel to *a* are frequently terminated by nearly equal development of faces of $d\{101\}$ and $m\{210\}$ (Fig. 9.24). Also radiating fibrous; granular.

Angles: $b(010) \wedge m(210) = 52°1'$, $c(001) \wedge o(011) = 52°3'$, $c(001) \wedge d(101) = 39°24'$, $c(001) \wedge l(102) = 22°19'$.

Pnma; $a = 8.38$, $b = 5.37$, $c = 6.85$ Å; $a:b:c = 1.561:1:1.276$; $Z = 4$. *d's:* 3.30(10), 3.18(6), 2.97(10), 2.73(6), 2.04(6).

Physical Properties. *Cleavage* {001} perfect, {210} good. **H** 3–3½. **G** 3.95–3.97. *Luster* vitreous to pearly. *Color* colorless, white, often faintly blue or red. Transparent to translucent. *Optics:* (+); $\alpha = 1.622$, $\beta = 1.624$, $\gamma = 1.631$; 2V = 50. $X = c$, $Y = b$. $r < v$.

Composition and Structure. SrO 56.4, SO_3 43.6% for pure celestite. Ba substitutes for Sr and a complete solid solution series exists between celestite and barite. At ordinary temperatures only a lim-

FIG. 9.24. Celestite crystals.

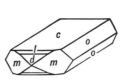
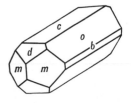

ited series exists between anhydrite, $CaSO_4$, and $SrSO_4$. Celestite is isostructural with barite (see Fig. 9.20).

Diagnostic Features. Closely resembles barite but is differentiated by lower specific gravity and crimson flame test (Sr). Fuses at $3\frac{1}{2}$–4. Usually decrepitates when touched with the blowpipe flame. Fused with sodium carbonate gives a residue which, when moistened, produces on a clean silver surface a dark stain of silver sulfide.

Occurrence. Celestite is found usually disseminated through limestone or sandstone, or in nests and lining cavities in such rocks. Associated with calcite, dolomite, gypsum, halite, sulfur, fluorite. Also found as a gangue mineral in lead veins.

Notable localities for its occurrence are with the sulfur deposits of Silicy; at Bex, Switzerland; Yate, Gloucestershire, England; and Herrengrund, Slovakia. Found in the United States at Clay Center, Ohio; Put-in-Bay, Lake Erie; Mineral County, West Virginia; Lampasas, Texas; and Inyo County, California.

Use. Use in the preparation of strontium nitrate for fireworks and tracer bullets, and other strontium salts used in the refining of beet sugar.

Name. Derived from *cœlestis,* in allusion to the faint blue color of the first specimens described.

✴ANGLESITE—PbSO$_4$

Crystallography. Orthorhombic; $2/m\,2/m\,2/m$. Frequently in crystals with habit often similar to that of barite but more varied. Crystals may be prismatic parallel to any one of the crystal axes and frequently show many forms, with a complex development (Fig. 9.25). Also massive, granular to compact. Frequently earthy, in concentric layers that may have an unaltered core of galena.

Angles: $b(010) \wedge m(210) = 51°52'$, $c(001) \wedge d(101) = 39°24'$, $c(001) \wedge o(011) = 52°13'$, $c(001) \wedge z(211) = 64°25'$.

Pnma; $a = 8.47$, $b = 5.39$, $c = 6.94$ Å; $a:b:c = 1.571:1:1.288$. $Z = 4$. $d's$: 4.26(9), 3.81(6), 3.33(9), 3.22(7), 3.00(10).

Physical Properties. *Cleavage* {001} good, {210} imperfect. Fracture conchoidal. **H** 3.0. **G**

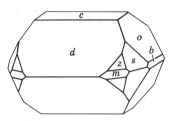

FIG. 9.25. Anglesite.

6.2–6.4 (unusually high). *Luster* adamantine when crystalline, dull when earthy. *Color* colorless, white, gray, pale shades of yellow. May be colored dark gray by impurities. Transparent to translucent. *Optics:* (+); $\alpha = 1.877$, $\beta = 1.883$, $\gamma = 1.894$; $2V = 75°$; $X = c$, $Y = b$. $r < v$.

Composition and Structure. PbO 73.6, SO_3 26.4%. Anglesite is isostructural with barite (see Fig. 9.20 and page 316).

Diagnostic Features. Recognized by its high specific gravity, its adamantine luster, and frequently by its association with galena. Distinguished from cerussite in that it will not effervesce in nitric acid. Fusible at $1\frac{1}{2}$. On charcoal with sodium carbonate reduces to a lead globule with yellow to white coating of lead oxide; the residue, when moistened, produces on clean silver surface a dark stain of silver sulfide.

Occurrence. Anglesite is a common supergene mineral found in the oxidized portions of lead deposits. It is formed through the oxidation of galena, either directly, as is shown by concentric layers of anglesite surrounding a core of galena, or by solution and subsequent deposition and recrystallization. Anglesite is commonly associated with galena, cerussite, sphalerite, smithsonite, hemimorphite, and iron oxides.

Notable localities for its occurrence are Monte Poni, Sardinia; Island of Anglesey, Wales; Derbyshire; and Leadhills, Scotland. From Sidi-Amor-ben-Salem, Tunis; Tsumeb, South-West Africa; Broken Hill, New South Wales; and Dundas, Tasmania. It is found in crystals embedded in sulfur at Los Lamentos, Chihuahua, Mexico. Occurs in the United States at Phoenixville, Pennsylvania; Tintic district, Utah; and Coeur l'Alene district, Idaho.

Use. A minor ore of lead.

Name. Named from the original locality on the Island of Anglesey.

ANHYDRITE—CaSO₄

Crystallography. Orthorhombic. $2/m\,2/m\,2/m$. Crystals rare; when observed are thick tabular on {010}, {100}, or {001} (see Fig. 9.26), also prismatic parallel to b. Usually massive or in crystalline masses resembling an isometric mineral with cubic cleavage. Also fibrous, granular, massive.

Angles: $b(010) \wedge m(110) = 45°1'$, $c(001) \wedge r(011) = 41°45'$, $c(001) \wedge o(111) = 51°38'$, $c(001) \wedge n(211) = 63°24'$.

Amma; $a = 6.95$, $b = 6.96$, $c = 6.21$ Å; $a:b:c = 0.998:1:0.892$; $Z = 4$. $d's$: 3.50(10), 2.85(3), 2.33(2), 2.08(2), 1.869(2).

Physical Properties. *Cleavage* {010} perfect, {100} nearly perfect, {001} good. **H** $3-3\frac{1}{2}$. **G** 2.89–2.98. *Luster* vitreous to pearly on cleavage. *Color* colorless to bluish or violet. Also may be white or tinged with rose, brown, or red. *Optics:* (+); $\alpha = 1.570$, $\beta = 1.575$, $\gamma = 1.614$; $2V = 44°$; $X = b$, $Y = a$; $r < v$.

Composition and Structure. CaO 41.2, SO₃ 58.8%. The structure illustrated in Fig. 9.21 is very different from the barite type. In anhydrite Ca is in 8 coordination with oxygen from SO₄ groups, whereas in barite Ca is in 12 coordination with oxygen. A metastable polymorph of anhydrite (γ CaSO₄) is hexagonal and is formed as the result of slow dehydration of gypsum (see Fig. 9.28).

Diagnostic Features. Fusible at 3. When moistened with HCl and ignited it gives an orange-red flame of Ca. When fused with reducing mixture it gives a residue that, when moistened with water, darkens silver. Anhydrite is characterized by its three cleavages at right angles. It is distinguished from calcite by its higher specific gravity and from gypsum by its greater hardness. Some massive varieties are difficult to recognize, and one should test for the sulfate radical.

Alteration. Anhydrite by hydration changes to gypsum with an increase in volume, and in places large masses have been thus altered.

Occurrence. Anhydrite occurs in much the same manner as gypsum and is often associated with that mineral but is not nearly as common. Found in beds associated with salt deposits in the cap rock of salt domes, and in limestones. Found in some amygdaloidal cavities in basalt.

Notable foreign localities are: Wieliczka, Poland; Aussee, Styria; Strassfurt, Prussia; Berchtesgaden, Bavaria; Hall near Innsbruck, Tyrol; and Bex, Switzerland. In the United States found in Lockport, New York; West Paterson, New Jersey; Nashville, Tennessee; New Mexico; and Texas. Found in large beds in Nova Scotia.

Use. Ground anhydrite is used as a soil conditioner and to a minor extent as a setting retardant in Portland cement. In Great Britain and Germany it has been used as a source of sulfur for the production of sulfuric acid.

Name. Anhydrite is from the Greek meaning *without water,* in contrast to the more common hydrous calcium sulfate, gypsum.

Crocoite—PbCrO₄

Crystallography. Monoclinic; $2/m$. Commonly in slender prismatic crystals, vertically striated, and columnar aggregates. Also granular.

$P2_1/n$; $a = 7.11$, $b = 7.41$, $c = 6.81$ Å, $\beta = 102°33'$; $a:b:c = 0.960:1:0.919$; $Z = 4$. $d's$: 4.96(3), 2.85(3), 2.33(2), 2.08(2), 1.869(2).

Physical Properties. *Cleavage* {110} imperfect. **H** $2\frac{1}{2}-3$. **G** 5.9–6.1. *Luster* adamantine. *Color* bright hyacinth-red. *Streak* orange-yellow. Translucent.

Composition and Structure. PbO 68.9, CrO₃ 31.1%. Crocoite is isostructural with *monazite,*

FIG. 9.26. Anhydrite.

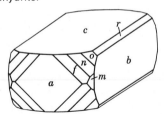

$CePO_4$, with Pb in 9 coordination with oxygen, linking six CrO_4 tetrahedra.

Diagnostic Features. Crocoite is characterized by its color, high luster, and high specific gravity. It may be confused with wulfenite, $PbMoO_4$, but can be distinguished from it by its redder color, lower specific gravity, and crystal form. Fusible at $1\frac{1}{2}$. Fused with sodium carbonate on charcoal gives a lead globule. Gives a green (Cr) borax bead in the oxidizing flame.

Occurrence. Crocoite is a rare mineral found in the oxidized zones of lead deposits in those regions where lead veins have traversed rocks containing chromite. Associated with pyromorphite, cerussite, and wulfenite. Notable localities are: Dundas, Tasmania; Beresovsk near Svedlovsk, Ural Mountains; and Rezbanya, Rumania. In the United States is found in small quantities in the Vulture district, Arizona.

Use. Not abundant enough to be of commercial value, but of historic interest, because the element chromium was first discovered in crocoite.

Name. From the Greek meaning *saffron,* in allusion to the color.

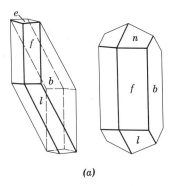

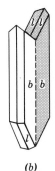

(b)

FIG. 9.27 (a) Gypsum crystals. (b) Gypsum twin.

✸ GYPSUM—CaSO₄·2H₂O

Crystallography. Monoclinic; $2/m$. Crystals are of simple habit (Fig. 9.27a); tabular on {010}; diamond-shaped, with edges beveled by {120} and {$\bar{1}11$}. Other forms rare. Twinning common, on {100} (Fig. 9.27b), often resulting in swallowtail twins.

Angles: $b(010) \wedge f(120) = 55°45'$, $b(010) \wedge l(\bar{1}11) = 71°54'$, $b(010) \wedge n(011) = 69°20'$.

$A2/n$; $a = 5.68$, $b = 15.18$, $c = 6.29$ Å, $\beta = 113°50'$; $a:b:c = 0.374:1:0.414$; $Z = 4$. d's: 7.56(10), 4.27(5), 3.06(6), 2.87(2), 2.68(3).

Physical Properties. *Cleavage* {010} perfect yielding thin folia; {100}, with conchoidal surface; {011}, with fibrous fracture. **H** 2. **G** 2.32. *Luster* usually vitreous; also pearly and silky. *Color* colorless, white, gray; various shades of yellow, red, brown, from impurities. Transparent to translucent.

Satin spar is a fibrous gypsum with silky luster. *Alabaster* is the fine-grained massive variety.

Selenite is a variety that yields broad colorless and transparent cleavage folia. *Optics:* (+); $\alpha = 1.520$, $\beta = 1.523$, $\gamma = 1.530$; $2V = 58°$. $Y = b$, $X \wedge c = -37°$; $r > v$.

Composition and Structure. CaO 32.6, SO_3 46.5, H_2O 20.9%. As the result of dehydration of gypsum, $CaSO_4·2H_2O$, several phases may form: γ $CaSO_4$, when all H_2O is lost and a metastable phase $CaSO_4·\frac{1}{2}H_2O$. During the dehydration process the first $1\frac{1}{2}$ molecules of H_2O in gypsum are lost relatively continuously between 0° and about 65°C (see Fig. 9.28) perhaps with only slight changes in the gypsum structure. At about 70°C the remaining $\frac{1}{2}H_2O$ molecule in $CaSO_4·\frac{1}{2}H_2O$ is still retained relatively strongly, but at about 95°C this is lost and the structure transforms to that of a polymorph of anhydrite. The structure of gypsum is given in Fig. 9.22 and discussed on page 317.

Diagnostic Features. Characterized by its softness and its three unequal cleavages. Fusible at

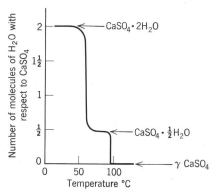

FIG. 9.28. Dehydration curve of gypsum showing formation of metastable $CaSO_4 \cdot \frac{1}{2}H_2O$ at about 65°C. At about 100°C the γ polymorph of $CaSO_4$ is formed.

3. Soluble in hot, dilute HCl and solution with barium chloride gives white precipitate of barium sulfate. In the closed tube turns white and yields much water. Its solubility in acid and the presence of much water distinguish it from anhydrite.

Occurrence. Gypsum is a common mineral widely distributed in sedimentary rocks, often as thick beds. It frequently occurs interstratified with limestones and shales and is usually found as a layer underlying beds of rock salt, having been deposited there as one of the first minerals to crystallize on the evaporation of salt waters. May recrystallize in veins, forming *satin spar*. Occurs also as lenticular bodies or scattered crystals in clays and shales. Frequently formed by the alteration of anhydrite and under these circumstances may show folding because of increased volume. Found in volcanic regions, especially where limestones have been acted upon by sulfur vapors. Also common as a gangue mineral in metallic veins. Associated with many different minerals, the more common being halite, anhydrite, dolomite, calcite, sulfur, pyrite, and quartz.

Gypsum is the most common sulfate, and extensive deposits are found in many localities throughout the world. The principal world producers are the United States, Canada, France, England, and the U.S.S.R. In the United States commercial deposits are found in many states, but the chief producers are located in New York, Michigan, Iowa,

Texas, and California. Gypsum is found in large deposits in Arizona and New Mexico in the form of windblown sand.

Use. Gypsum is used chiefly for the production of plaster of Paris. In the manufacture of this material, the gypsum is ground and then heated until about 75% of the water has been driven off, producing the substance $CaSO_4 \cdot \frac{1}{2}H_2O$. This material when mixed with water, slowly absorbs the water, crystallizes, and thus hardens or "sets." Plaster of Paris is used extensively for "staff," the material from which temporary exposition buildings are built, for gypsum lath, wallboard, and for molds and casts of all kinds. Gypsum is employed in making adamant plaster for interior use. Serves as a soil conditioner "land plaster," for fertilizer. Uncalcined gypsum is used as a retarder in Portland cement. Satin spar and alabaster are cut and polished for various ornamental purposes but are restricted in their uses because of their softness.

Name. From the Greek name for the mineral, applied especially to the calcined mineral.

Antlerite—Cu$_3$SO$_4$(OH)$_4$

Crystallography. Orthorhombic; $2/m\,2/m\,2/m$. Crystals usually tabular on {010} (Fig. 9.29). May be in slender prismatic crystals vertically striated. Also in parallel aggregates, reniform, massive.

Angles: $m(110) \wedge m(\bar{1}10) = 111°3'$, $b(010) \wedge o(011) = 63°19'$, $b(010) \wedge f(130) = 25°54'$, $m(110) \wedge r(111) = 48°23'$.

Pnam; $a = 8.24$, $b = 11.99$, $c = 6.03$ Å; $a:b:c = 0.687:1:0.503$; $Z = 4$. $d's$: 4.86(10), 3.60(8), 3.40(3), 2.68(6), 2.57(6).

Physical Properties. *Cleavage* {010} perfect. **H** $3\frac{1}{2}$–4. **G** 3.9±. *Luster* vitreous. *Color* emerald to blackish-green. *Streak* pale green. Transparent to translucent. *Optics:* (+); $\alpha = 1.726$, $\beta = 1.738$, $\gamma = 1.789$; $2V = 53°$; $X = b$, $Y = a$; pleochroism X yellow-green, Y blue-green, Z green.

Composition and Structure. CuO 67.3, SO$_3$ 22.5, H$_2$O 10.2%. The structure of antlerite contains Cu in two types of octahedral coordination: CuO(OH)$_5$ and CuO$_3$(OH)$_3$. These octahedra are linked with the SO$_4$ tetrahedra.

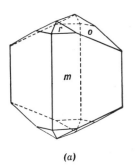 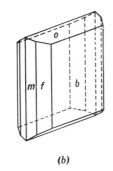

(a)   (b)

FIG. 9.29. Antlerite crystals.

Diagnostic Features. Fusible at $3\frac{1}{2}$. Yields a copper globule when fused with sodium carbonate on charcoal. HCl solution with barium chloride gives a white precipitate of barium sulfate. Yields water in the closed tube, and at a high temperature, yields sulfuric acid. Antlerite is characterized by its green color, {010} cleavage, and association. Lack of effervescence in HCl distinguishes it from malachite. It is commonly associated with atacamite and brochantite and one must use optical properties to distinguish it from these minerals.

Occurrence. Antlerite is found in the oxidized portions of copper veins, especially in arid regions. It was formerly considered a rare mineral, but in 1925 it was recognized as the chief ore mineral at Chuquicamata, Chile, the world's largest copper mine. It may form directly as a secondary mineral on chalcocite, or the copper may go into solution and later be deposited as antlerite, filling cracks. In the United States it is found at Bisbee, Arizona, and near Black Mountain, Nevada. Also found at Kennecott, Alaska.

Use. An ore of copper.

Name. From the Antler mine, Arizona, from which locality it was originally described.

Similar Species. *Brochantite*, $Cu_4SO_4(OH)_6$, is similar in all its properties to antlerite, but, although more widespread it is nowhere abundant. Until 1925 it was considered to be the chief ore mineral at Chuquicamata, Chile. Another rare hydrous copper sulfate is *chalcanthite*, $CuSO_4 \cdot 5H_2O$, which occurs abundantly at Chuquicamata and other arid areas in Chile. *Epsomite*, $MgSO_4 \cdot 7H_2O$, occurs as an efflorescence on the rocks in wine workings and on the walls of caves. *Melanterite*, $FeSO_4 \cdot 7H_2O$, is also a secondary mineral found as an efflorescence on mine timbers.

Alunite—KAl$_3$(SO$_4$)$_2$(OH)$_6$

Crystallography. Hexagonal-*R*; 3*m*. Crystals usually a combination of positive and negative trigonal pyramids resembling rhombohedrons with nearly cubic angles (90°50'). May be tabular on {0001}. Commonly massive or disseminated.

$R3m$; $a = 6.97$, $c = 17.38$ Å; $a:c = 1:2.494$; $Z = 3$. $d's$: 4.94(5), 2.98(10), 2.29(5), 1.89(6), 1.74(5).

Physical Properties. *Cleavage* {0001} imperfect. **H** 4. **G** 2.6–2.8. *Luster* vitreous to pearly in crystals, earthy in massive material. *Color* white, gray, or reddish. Transparent to translucent. *Optics:* (+); $\omega = 1.572$, $\epsilon = 1.592$.

Composition and Structure. K_2O 11.4, Al_2O_3 37.0, SO_3 38.6, H_2O 13.0%. Na may replace K at least up to Na:K = 7:4. When Na exceeds K in the mineral it is called *natroalunite*, $(Na,K)Al_3(SO_4)_2(OH)_6$. In the structure of alunite K and Al are in 6 coordination with respect to O and (OH). The SO_4 groups share some of their oxygens with the K-containing octahedra.

Diagnostic Features. Infusible and decrepitates before the blowpipe, giving a potassium flame. Heated with cobalt nitrate solution turns a fine blue color. In the closed tube gives acid water. Soluble in sulfuric acid. Alunite is usually massive and difficult to distinguish from rocks such as limestone and dolomite, and other massive minerals as anhydrite and granular magnesite. A positive test for acid water will serve to distinguish alunite from similar-appearing minerals.

Occurrence. Alunite also called *alumstone*, is usually formed by sulfuric acid solutions acting on rocks rich in potash feldspar, and in some places large masses have thus been formed. Found in smaller amounts about volcanic fumaroles. In the United States it is found at Red Mountain in the San Juan district, Colorado; Goldfield, Nevada; and Marysvale, Utah.

Use. In the production of alum. At Marysvale, Utah, alunite has been minded and treated in such a way as to recover potassium and aluminum.

Name. From the Latin meaning alum.

Similar Species. *Jarosite*, $KFe_3(SO_4)_2(OH)_6$, the Fe^{3+} analogue of alunite, is a secondary mineral found as crusts and coatings on ferruginous ores.

TUNGSTATES AND MOLYBDATES

W^{6+} and Mo^{6+} (radii of both = 0.62 Å) are considerably larger than S^{6+} and P^{5+}. Hence, when these ions enter into anisodesmic ionic groups with oxygen, the four coordinated oxygen ions do not occupy the apices of a regular tetrahedron, as is the case in the sulfates and phosphates, but form a somewhat flattened grouping of square outline. Although W (at. wt. 184) has a much greater atomic weight than Mo(96), both belong to the same family of the periodic table and, because of the lanthanide contraction, have the same ionic radii. As a result, each may freely substitute for the other as the coordinating cation in the warped tetrahedral oxygen groupings. In nature, however, the operation of the processes of geochemical differentiation often separates these elements, and it is not uncommon to find primary tungstates almost wholly free of Mo and vice versa. In secondary minerals, the two elements are more commonly in solid solution with each other.

The minerals of this chemical class fall mainly into two isostructural groups. The wolframite group contains compounds with fairly small divalent cations such as Fe^{2+}, Mn^{2+}, Mg, Ni, and Co in 6 coordination with (MoO_4). Complete solid solution occurs between Fe^{2+} and Mn^{2+} in the minerals of this group.

The scheelite group contains compounds of larger ions such as Ca^{2+} and Pb^{2+} in 8 coordination with $(WO_4)^{-2}$ and $(MoO_4)^{-2}$ groups. The structure of scheelite (see Fig. 9.30) is close to that of anhydrite and zircon, $ZrSiO_4$, but differs from these in the manner of linking of the CaO_8 polyhedra. The (WO_4) tetrahedra are somewhat flattened along the c axis and join edges with CaO_8. W and Mo

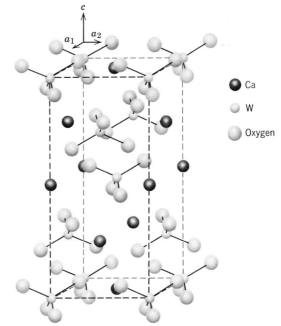

FIG. 9.30. Structure of scheelite, $CaWO_4$.

may substitute for each other forming partial series between *scheelite*, $CaWO_4$, and *powellite*, $CaMoO_4$; and *stolzite*, $PbWO_4$, and *wulfenite*, $PbMoO_4$. The substitution of Ca and Pb for one another forms partial series between scheelite and stolzite and between powellite and wulfenite. The tungstates and molybdates comprise a group of nine minerals of which we will discuss only three: wolframite, scheelite, and wulfenite.

WOLFRAMITE—(Fe, Mn)WO$_4$

Crystallography. Monoclinic; $2/m$. Crystals commonly tabular parallel to {100} (Fig. 9.31) giving bladed habit, with faces striated parallel to c. In bladed, lamellar, or columnar aggregates. Massive granular.

Angles: $m(110) \wedge m'(1\bar{1}0) = 79°5'$, $f(011) \wedge f'(0\bar{1}1) = 81°51'$, $a(100) \wedge t(102) = 61°57'$, $a(100) \wedge y(\bar{1}02) = 117°20'$.

$P2/c$; $a = 4.79$, $b = 5.74$, $c = 4.99$ Å; $\beta = 90°28'$; $a:b:c = 0.835:1:0.869$; $Z = 2$. $d's$:

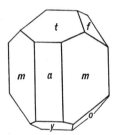

FIG. 9.31. Wolframite.

4.76(5), 3.74(5), 2.95(10), 2.48(6), 1.716(5). Cell parameters are slightly less for *ferberite,* $FeWO_4$, and greater for *huebnerite,* $MnWO_4$.

Physical Properties. *Cleavage* {010} perfect. **H** $4-4\frac{1}{2}$. **G** 7.0–7.5, higher with higher Fe content. *Luster* submetallic to resinous. *Color* black in ferberite to brown in huebnerite. *Streak* from nearly black to brown.

Composition and Structure. Fe^{2+} and Mn^{2+} substitute for each other in all proportions and a complete solid solution series exists between *ferberite,* $FeWO_4$, and *huebnerite,* $MnWO_4$. The percentage of WO_3 is 76.3 in ferberite and 76.6 in huebnerite. The structure of wolframite consists of distorted tetrahedral (WO_4) groups and octahedral $(Fe,Mn)O_6$ groups. From the interatomic distances around W, however, one may also conclude that W is in distorted octahedral coordination.

Diagnostic Features. The dark color, one direction of perfect cleavage, and high specific gravity serve to distinguish wolframite from other minerals. Fusible at 3–4 to a magnetic globule. Insoluble in acids. Fused with sodium carbonate, dissolves in HCl; tin added and solution boiled gives a blue color (W). In the oxidizing flame with sodium carbonate gives bluish-green bead (Mn).

Occurrence. Wolframite is a comparatively rare mineral found usually in pegmatites and high-temperature quartz veins associated with granites. More rarely in sulfide veins. Minerals commonly associated include cassiterite, scheelite, bismuth, quartz, pyrite, galena, sphalerite, and arsenopyrite. In some veins wolframite may be the only metallic mineral present.

Found in fine crystals from Schlaggenwald and

Zinnwald, Bohemia, and in the various tin districts of Saxony and Cornwall. Important producing countries are China, the U.S.S.R., Korea, Thailand, Bolivia, and Australia. Wolframite occurs in the United States in the Black Hills, South Dakota. Ferberite has been mined extensively in Boulder County, Colorado. Huebnerite is found near Silverton, Colorado; Mammoth district, Nevada; and Black Hills, South Dakota. More than 90% of the world's estimated tungsten resources are outside the United States, with almost 60% in southeastern China.

Use. Chief ore of tungsten. Tungsten is used as hardening metal in the manufacture of high-speed tool steel, valves, springs, chisels, files, etc. Its high melting point (3410°C) requires a special chemical process for reduction of the metal which is produced in the form of a powder. By powder-metallurgy, pure metal products such as lamp filaments are fabricated. A large use for tungsten is in the manufacture of carbides, harder than any natural abrasives (other than diamond), that are used for cutting tools, rock bits, and hard facings. Sodium tungstate is used in fireproofing cloth and as a mordant in dyeing.

Name. Wolframite is derived from an old word of German origin.

SCHEELITE—$CaWO_4$

Crystallography. Tetragonal; $4/m$. Crystals usually simple dipyramids of the second order (Fig. 9.32). The first-order dipyramid, {112}, closely resembles the octahedron in angles. Also massive granular.

Angles: $p(101) \wedge p'(\overline{1}01) = 130°33', \beta(013) \wedge \beta'(0\overline{1}3) = 71°48'$.

$I\ 4/a$; $a = 5.25$, $c = 11.40$ Å; $a:c = 1:2.171$; $Z = 4$. $d's$: 4.77(7), 3.11(10), 1.94(8), 1.596(9), 1.558(7).

Physical Properties. *Cleavage* {101}, distinct. **H** $4\frac{1}{2}$–5. **G** 5.9–6.1 (unusually high for a non-metallic mineral). *Luster* vitreous to adamantine. *Color* white, yellow, green, brown. Translucent; some specimens transparent. Most scheelite will fluoresce with bluish white color in short ultraviolet radiation. *Optics:* (+); $\omega = 1.920$, $\epsilon = 1.934$.

FIG. 9.32. Scheelite.

Composition and Structure. CaO 19.4, WO_3 80.6%. Mo can substitute for W and a partial series extends to *powellite,* $CaMoO_4$. The structure of scheelite (Fig. 9.30) consists of flattened (WO_4) tetrahedra and CaO_8 polyhedra.

Diagnostic Features. Difficultly fusible (5). Decomposed by boiling in HCl, leaving a yellow residue of tungstic oxide which, when tin is added to the solution and boiling continued, turns first blue and then brown. Recognized by its high specific gravity, crystal form, and fluorescence in short ultraviolet light. The test for tungsten may be necessary for positive identification.

Occurrence. Scheelite is found in granite pegmatites, contact metamorphic deposits, and high-temperature hydrothermal veins associated with granitic rocks. Associated with cassiterite, topaz, fluorite, apatite, molybdenite, and wolframite. Found in gold-silver-quartz veins as in the Porcupine-Ramore area of Ontario, Canada. In the Hollinger mine of this area gold, silver and scheelite are minded. Occurs with tin in deposits of Bohemia, Saxony, and Cornwall; and in quantity in New South Wales and Queensland. In the United States scheelite is mined near Mill City and Mina, Nevada; near Atolia, San Bernardino County, California; and in lesser amounts in Arizona, Utah, and Colorado.

Use. An ore of tungsten. Wolframite furnishes most of the world's supply of tungsten, but scheelite is more important in the United States.

Name. After K. W. Scheele (1742–1786), the discoverer of tungsten.

Wulfenite—PbMoO$_4$

Crystallography. Tetragonal; 4 or 4/m. Crystals usually square, tabular in habit with prominent {001}. Some crystals very thin. More rarely pyramidal (Fig. 9.33). Some crystals may be twinned on {001} giving them a dipyramidal habit.

Angles: $c(001) \wedge n(101) = 65°52'$, $c(001) \wedge s(103) = 36°38'$.

$I4_1/a$; $a = 5.42$, $c = 12.10$ Å; $a:c = 2.233$; $Z = 4$. $d's$: 3.24(10), 3.03(2), 2.72(2), 2.02(3), 1.653(3).

Physical Properties. *Cleavage* {011} distinct. **H** 3. **G** 6.8±. *Luster* vitreous to adamantine. *Color* yellow, orange, red, gray, white. *Streak* white. Transparent to subtranslucent. Piezoelectricity suggests symmetry 4 rather than 4/m as determined from structure. *Optics:* (−); $\omega = 2.404$, $\epsilon = 2.283$.

Composition and Structure. PbO 60.8, MoO_3 39.2%. Ca may substitute for Pb indicating at least a partial series to *powellite,* $Ca(Mo,W)O_4$. Wulfenite is isostructural with scheelite (see Fig. 9.30).

Diagnostic Features. Wulfenite is characterized by its square tabular crystals, orange to yellow color, high luster, and association with other lead minerals. Distinguished from crocoite by test for Mo. Fusible at 2. Gives a lead globule when fused with sodium carbonate on charcoal. With salt of phosphorus in the reducing flame it gives a green bead; in the oxidizing flame, yellowish-green when hot to almost colorless when cold.

Occurrence. Wulfenite is found in the oxidized portions of lead veins with other secondary lead minerals, especially cerussite, vanadinite, and pyromorphite. Found in the United States at Phoenixville, Pennsylvania, and in a number of places

FIG. 9.33. Wulfenite crystals.

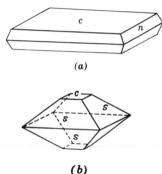

(a)

(b)

in Utah, Nevada, Arizona, and New Mexico. Found in beautiful crystals at Red Cloud, Arizona, and Schwarzenbach, Yugoslavia.

Use. A minor source of molybdenum. Molybdenite, MoS_2, is the chief ore.

Name. After X. F. Wulfen, Austrian mineralogist.

PHOSPHATES, ARSENATES, AND VANADATES

P^{5+} is only slightly larger than S^{6+} and, hence, like sulfur, forms a tetrahedral anionic $(PO_4)^{-3}$ group with oxygen (see Fig. 9.1). All phosphates contain this phosphate anionic complex as the fundamental building unit. Similar tetrahedral units, $(AsO_4)^{-3}$ and $(VO_4)^{-3}$ occur in arsenates and vanadates. P^{5+}, As^{5+}, and V^{5+} may substitute for each other in the anionic groups. This type of substitution is best shown in the pyromorphite series of the apatite group. *Pyromorphite*, $Pb_5(PO_4)_3Cl$, *mimetite*, $Pb_5(AsO_4)_3Cl$, and *vanadinite*, $Pb_5(CO_4)_3Cl$, are isostructural, and all gradations of composition between the end members exist.

The structure of *apatite*, $Ca_5(PO_4)_3(OH, F, Cl)$,

which is the most important and abundant phosphate, is illustrated in Fig. 9.34. The oxygens of the (PO_4) groups are linked to Ca, with $\frac{2}{5}$ of the calcium surrounded by six closest oxygens in the form of trigonal prisms, and $\frac{3}{5}$ of the calcium by five oxygens and one F. Each fluorine (or Cl or OH) lies in a triangle with three calciums. Apatite shows extensive solid solution with respect to anions as well as cations. (PO_4) may be substituted for by (AsO_4) or (VO_4) as noted above, but also in part by tetrahedral (CO_3OH) groups, giving rise to *carbonate-apatite*, $Ca_5F(PO_4,CO_3OH)_3$. Small amounts of (SiO_4) and (SO_4) may also be present in substitution for (PO_4); these types of substitution must be coupled with other cation substitutions in apatite in order to retain the electrical neutrality of the structure (see page 161). F may be replaced by (OH) or Cl producing *hydroxylapatite*, $Ca_5(PO_4)_3(OH)$ and *chlorapatite*, $Ca_5(PO_4)_3Cl$. Mn^{2+} and Sr^{2+} may substitute for Ca. These varied ionic substitutions are typical of the phosphates which generally have rather complicated structures.

This mineral class, composed mostly of phosphates, is very large but most of its members are so rare that they need not be mentioned here. Of the list of phosphates, arsenates, and vanadates given below only apatite can be considered as common.

FIG. 9.34. Structure of fluorapatite, $Ca_5(PO_4)_3F$, projected on the (0001) plane. The dashed parallelogram outlines the base of the unit cell. The tetrahedral (PO_4) groups, triangular coordination of F to Ca, and examples of the two types of coordination about Ca are shown.

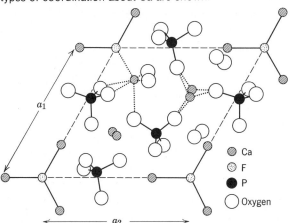

PHOSPHATES, ARSENATES AND VANADATES

Triphyllite—	Li(Fe, Mn)PO$_4$
Lithiophyllite	Li(Mn, Fe)PO$_4$
Monazite	(Ce, La, Y, Th)PO$_4$
Apatite Group	
Apatite	Ca$_5$(PO$_4$)$_3$(F, Cl, OH)
Pyromorphite	Pb$_5$(PO$_4$)$_3$Cl
Vanadinite	Pb$_5$(VO$_4$)$_3$Cl
Erythrite	Co$_3$(AsO$_4$)$_2$·8H$_2$O
Amblygonite	LiAlPO$_4$F
Lazulite—	(Mg, Fe)Al$_2$(PO$_4$)$_2$(OH)$_2$
Scorzalite	(Fe, Mg)Al$_2$(PO$_4$)$_2$(OH)$_2$
Wavellite	Al$_3$(PO$_4$)$_2$(OH)$_3$·5H$_2$O
Turquoise	CuAl$_6$(PO$_4$)$_4$(OH)$_8$·4H$_2$O
Autunite	Ca(UO$_2$)$_2$(PO$_4$)$_2$·10-12H$_2$O
Carnotite	K$_2$(UO$_2$)$_2$(VO$_4$)$_2$·3H$_2$O

Triphylite—Li(Fe, Mn)PO₄

Lithiophilite—Li(Mn, Fe)PO₄

Crystallography. Orthorhombic; $2/m2/m2/m$. Crystals rare. Commonly in cleavable masses. Also compact.

Pmcn. For triphylite: $a = 6.01$, $b = 4.86$, $c = 10.36$ Å; $a:b:c = 1.284:1:2.214$; $Z = 4$. d's: 4.29(8), 3.51(9), 3.03(9), 2.54(10), 1.75(5).

Physical Properties. *Cleavage* {001} nearly perfect, {010} imperfect. **H** 4½–5. **G** 3.42–3.56 increasing with Fe content. *Luster* vitreous to resinous. *Color* bluish gray in triphylite to salmon-pink or clove-brown in lithiophilite. May be stained black by manganese oxide. Translucent. *Optics:* (+); $\alpha = 1.669$–1.694, $\beta = 1.673$–1.695, $\gamma = 1.682$–1.700; 2V = 0°–55°, X = c, Y = a, Z = b. Indices increase with Fe content.

Composition and Structure. A complete Fe^{2+}-Mn^{2+} series exists between two essentially pure end members. In the structure of members of this series Li and (Mn, Fe) are in 6 coordination. These octahedra are linked along their edges into zigzag chains which are connected by (PO_4) tetrahedra.

Diagnostic Features. Fusible at 2½, giving red flame (Li). Triphylite becomes magnetic on heating in the reducing flame. Some Mn is usually present and therefore it gives a manganese bead test. Characterized by two cleavages at right angles, resinous luster, and association.

Occurrence. Triphylite and lithiophilite are pegmatite minerals associated with other phosphates, spodumene, and beryl. Notable localities are in Bavaria and Finland. Lithiophilite is found at several localities in Argentina. In the United States triphylite is found at Huntington, Massachusetts; Peru, Maine; Grafton, North Grafton, and Newport, New Hampshire; and the Black Hills, South Dakota. Lithiophilite is found at Branchville and Portland, Connecticut.

Name. Triphylite, from the Greek words meaning *three* and *family* in allusion to containing three cations. Lithiophilite from the Greek words meaning *lithium* and *friend*.

Monazite—(Ce, La, Y, Th)PO₄

Crystallography. Monoclinic; $2/m$. Crystals rare and usually small, often flattened on {100}, or elongated on b. Usually in granular masses, frequently as sand.

$P2_1/n$; $a = 6.79$, $b = 7.01$, $c = 6.46$ Å; $\beta = 103°38'$; $a:b:c = 0.969:1:0.921$; $Z = 4$. d's: 4.17(3), 3.30(5), 3.09(10), 2.99(2), 2.87(7).

Physical Properties. *Cleavage* {100} poor. Parting {001}. **H** 5–5½. **G** 4.6–5.4. *Luster* resinous. *Color* yellowish to reddish-brown. Translucent. *Optics:* (+); $\alpha = 1.785$–1.800, $\beta = 1.787$–1.801, $\gamma = 1.840$–1.850; 2V = 10°–20°; X = b, Z ∧ c = 2°–6°.

Composition and Structure. A phosphate of the rare-earth metals essentially (Ce,La,Y,Th)PO_4. Th content ranges from a few to 20% ThO_2. Si is often present up to several percent SiO_2. The Si has been ascribed to admixture of *thorite*, $ThSiO_4$, but may be in part due to substitution of (SiO_4) for (PO_4). In the structure of monazite the rare earths are in 9 coordination with oxygen, linking six PO_4 tetrahedra. It is isostructural with *crocoite*, $PbCrO_4$ (see page 320).

Diagnostic Features. Infusible. Insoluble in HCl. Decomposed by heating with concentrated sulfuric acid; solution after dilution with water and filtering gives, with ammonium oxalate, a precipitate of the oxalates of the rare earths. Large specimens may be distinguished from zircon by crystal form and inferior hardness, and from sphene by crystal form and higher specific gravity. On doubtful specimens it is usually well to make the chemical phosphate test.

Occurrence. Monazite is a comparatively rare mineral occurring as an accessory in granites, gneisses, aplites, and pegmatites, and as rolled grains in the sands derived from the decomposition of such rocks. It is concentrated in sands because of its resistance to chemical attack and its high specific gravity, and is thus associated with other resistant and heavy minerals such as magnetite, ilmenite, rutile, and zircon.

The bulk of the world's supply of monazite comes from beach sands in Brazil, India, and Australia. A dikelike body of massive granular monazite is mined near VanRhynsdorp, Cape Province, South Africa. Found in the United States in North Carolina, both in gneisses and in the stream sands; and in the beach sands of Florida.

Use. Monazite is the chief source of thorium oxide, which it contains in amounts varying between 1 and 20%; commercial monazite usually contains between 3 and 9%. Thorium oxide is used in the manufacture of mantles for incandescent gas lights.

Thorium is a radioactive element and is receiving considerable attention as a source of atomic energy. The natural isotope of thorium, Th^{232}, can be converted by neutron bombardment, first to Th^{233}, and then to U^{233}, a fissionable isotope.

Name. The name *monazite* is derived from a Greek word meaning *to be solitary,* in allusion to the rarity of the mineral.

Apatite Group

APATITE—$Ca_5(PO_4)_3(F, Cl, OH)$

Crystallography. Hexagonal; $6/m$. Commonly occurs in crystals of long prismatic habit; some short prismatic or tabular. Usually terminated by prominent pyramid of first order and frequently a basal plane. Some crystals show faces of a hexagonal dipyramid (μ, Fig. 9.35c) which reveals the true symmetry. Also in massive granular to compact masses.

Angles: $m(10\bar{1}0) \wedge x(10\bar{1}1) = 49°41'$, $c(0001) \wedge s(11\bar{2}1) = 55°46'$.

$P6_3/m$; $a = 9.39$, $c = 6.89$ Å; $a:c = 1:0.734$; $Z = 2$. $d's$: 2.80(10), 2.77(4), 2.70(6), 1.84(6), 1.745(3).

Physical Properties. *Cleavage* {0001} poor. **H** 5 (can just be scratched by a knife). **G** 3.15–3.20. *Luster* vitreous to subresinous. *Color* usually some shade of green or brown; also blue, violet, colorless. Transparent to translucent. *Optics:* $(-)$; $\omega = 1.633$, $\epsilon = 1.630$ (fluorapatite).

Composition and Structure. $Ca_5(PO_4)_3F$, *fluorapatite* is most common; more rarely $Ca_5(PO_4)_3Cl$, *chlorapatite,* and $Ca_5(PO_4)_3(OH)$, *hydroxylapatite.* F, Cl, and OH can substitute for each other, giving complete series. (CO_3, OH) may substitute for (PO_4) giving *carbonate-apatite.* The (PO_4) group can be partially replaced by (SO_4) as well as (SiO_4). The P^{5+} to S^{6+} replacement is compensated by the coupled substitution of Ca^{2+} by Na^+. Furthermore the P^{5+} to S^{6+} replacement may be balanced by substitution of Si^{4+} for P^{5+}. Mn and Sr can substitute in part for Ca (see page 327). The structure of apatite is illustrated in Fig. 9.34.

Collophane. The name collophane has been given to the massive, cryptocrystalline types of apatite that constitute the bulk of phosphate rock and fossil bone. X-ray study shows that collophane is essentially apatite and does not warrant designation as a separate species. In its physical appearance, collophane is usually dense and massive with a concretionary or colloform structure. It is usually impure and contains small amounts of calcium carbonate.

Diagnostic Features. Difficultly fusible ($5-5\frac{1}{2}$). Soluble in acids and gives the phosphate test with ammonium molybdate. Apatite is usually recognized by its crystals, color, and hardness. Distinguished from beryl by the prominent pyramidal terminations of its crystals and by its being softer than a knife blade.

Occurrence. Apatite is widely disseminated as an accessory constituent in all classes of rocks—igneous, sedimentary, and metamorphic. It is also found in pegmatites and other veins, probably of hy-

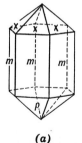 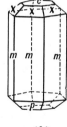 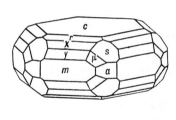

FIG. 9.35. Apatite crystals.　　　*(a)*　　　　　　*(b)*　　　　　　*(c)*

drothermal origin. Found in titaniferous magnetite bodies. Occasionally concentrated into large
deposits or veins associated with alkalic rocks.
Phosphate materials of bones and teeth are members of the apatite group.

Apatite occurs in large amounts along the
southern coast of Norway, between Langesund and
Arendal, where it is found in veins and pockets associated with gabbro. It is distributed through the magnetite iron ore at Kiruna, Sweden and Tahawas,
New York. In Ontario and Quebec, Canada, large
apatite crystals in crystalline limestone were formerly minded. Unusually fine crystals have come
from Renfrew County, Ontario. The world's largest
deposit of apatite is located on the Kola Peninsula,
near Kirovsk, U.S.S.R. Here apatite in granular aggregates intimately associated with nepheline and
sphene, is found in a great lens between two types
of alkalic rocks.

Finely crystallized apatite occurs at various lo-
calities in the Tyrol; in Switzerland; and Jumilla,
Spain. In the United States at Auburn, Maine; St.
Lawrence County, New York; Alexander County,
North Carolina; and San Diego County, California.

The variety collophane is an important constituent of the rock *phosphorite* or *phosphate rock*.
Bone is calcium phosphate, and large bodies of
phosphorite are derived from the accumulation of
animal remains as well as chemical precipitation
from sea water. Commercial deposits of phosphorite
are found in northern France, Belgium, Spain, and
especially in northern Africa in Tunisia, Algeria, and
Morocco. In the United States, high-grade phosphate deposits are found in Tennessee and in
Wyoming and Idaho. Deposits of "pebble" phosphate are found at intervals all along the Atlantic
coast from North Carolina to Florida. The most productive deposits in the United States are in Florida.

Use. Crystallized apatite has been used extensively as a source of phosphate for fertilizer but
today only the deposits on the Kola Peninsula are of
importance and phosphorite deposits supply most of
the phosphate for fertilizer. The calcium phosphate
is treated with sulfuric acid and changed to super-
phosphate to render it more soluble in the dilute
acids that exist in the soil. Transparent varieties of

apatite of fine color are occasionally used for gems.
The mineral is too soft, however, to allow its extensive use for this purpose.

Name. From the Greek word *to deceive*, because the gem varieties were confused with other
minerals.

Pyromorphite—$Pb_5(PO_4)_3Cl$

Crystallography. Hexagonal; $6/m$. Crystals usually
prismatic with basal plane (Fig. 9.36). Rarely shows
pyramid truncations. Often in rounded barrel-
shaped forms. Sometimes cavernous, the crystals
being hollow prisms. Also in parallel groups. Frequently globular, reniform, fibrous, and granular.

Angles: $c(0001) \wedge x(10\bar{1}1) = 40°22'$.

$P6_3m$; $a = 9.97$, $c = 7.32$ Å; $a:c = 1:0.734$;
$Z = 2$. $d's$: 4.31(6), 4.09(9), 2.95(10), 2.05(8),
1.94(7).

Physical Properties. H $3\frac{1}{2}$–4. **G** 7.04. *Luster*
resinous to adamantine. *Color* usually various
shades of green, brown, yellow; rarely orange-
yellow, gray, white. Subtransparent to translucent.
Optics: $(-)$, $\omega = 2.058$, $\epsilon = 2.048$.

Composition and Structure. For pure
$Pb_5(PO_4)_3Cl$, PbO 82.2, P_2O_5 15.7, Cl 2.6%. (AsO_4)
substitutes for (PO_4) and a complete series extends
to *mimetite,* $Pb_5(AsO_4)_3Cl$. Ca may substitute in part
for Pb. Isostructural with apatite (see Fig. 9.34).

Diagnostic Features. Fusible at 2. Gives a
lead globule with sodium carbonate. When fused
alone on charcoal gives a globule which on cooling

FIG. 9.36. Pyromorphite.

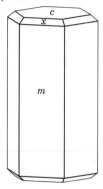

appears to show crystal forms. Gives phosphate test with ammonium molybdate. Pyromorphite is characterized by its crystal form, high luster, and high specific gravity.

Occurrence. Pyromorphite is a supergene mineral found in the oxidized portions of lead veins, associated with other oxidized lead and zinc minerals.

Notable localities for its occurrence are the lead mines of Poullaouen and Huelgoat, Brittany; at Ems in Nassau; at Zschopau, Saxony; Přibram, Bohemia; Beresovsk, Ural Mountains; in Cumberland, England and at Leadhills, Scotland. In the United States found at Phoenixville, Pennsylvania; Davidson County, North Carolina, and Idaho.

Use. A subordinate ore of lead.

Name. Derived from two Greek words meaning *fire* and *form*, in allusion to the apparent crystalline form it assumes on cooling from fusion.

Similar Species. *Mimetite*, $Pb_5(AsO_4)_3Cl$, is isostructural with pyromorphite and is similar in appearance, occurrence and most of its physical and chemical properties.

Vanadinite—$Pb_5(VO_4)_3Cl$

Crystallography. Hexagonal; $6/m$. Most commonly occurs in prismatic crystals with $\{10\bar{1}0\}$ and $\{0001\}$. May have small pyramidal faces, rarely the hexagonal dipyramid. In rounded crystals; in some cases cavernous. Also in globular forms. As incrustations.

$P6_3/m$; $a = 10.33$, $c = 7.35$ Å; $a:c = 1:0.711$; $Z = 2$. $d's$: 4.47(3), 4.22(4), 3.38(6), 3.07(9), 2.99(10).

Physical Properties. H 3. G 6.9. *Luster* resinous to adamantine. *Color* ruby-red, orange-red, brown, yellow. Transparent to translucent. *Optics:* (−); $\omega = 2.25–2.42$, $\epsilon = 2.20–2.35$. Indices lowered by substitution of As or P for V.

Composition. PbO 78.7, V_2O_5 19.4, Cl 2.5%. PO_4 and AsO_4 may substitute in small amounts for VO_4. In the variety *endlichite*, intermediate between vanadinite and mimetite, the proportion of V_2O_5 to As_2O_5 is nearly 1:1. Small amounts of Ca, Zn, and Cu substitute for Pb. Vanadinite is isostructural with apatite (see Fig. 9.34).

Diagnostic Features. Characterized by crystal form, high luster, and high specific gravity; distinguished from pyromorphite and mimetite by color. Fusible at $1\frac{1}{2}$. Gives globule of Pb when fused with sodium carbonate on charcoal. Gives an amber color in the oxidizing flame to salt of phosphorous bead (V). Dilute nitric acid solution gives with silver nitrate a white precipitate of silver chloride.

Occurrence. Vanadinite is a rare secondary mineral found in the oxidized portion of lead veins associated with other secondary lead minerals. Found in fine crystals near Oudjda, Morocco, and Grootfontein, South-West Africa. In the United States it occurs in various districts in Arizona and New Mexico.

Use. Source of vanadium and minor ore of lead. Vanadium is obtained chiefly from other ores, such as *patronite*, VS_4; the vanadate *carnotite*, $K_2(UO_2)_2(VO_4)_2 \cdot 3H_2O$; and a vanadium mica, *roscoelite*, $KV_2(AlSi_3O_{10})(OH)_2$. Vanadium is used chiefly as a steel-hardening metal. Metavanadic acid, HVO_3, is a yellow pigment, known as vanadium bronze. Vanadium oxide is a mordant in dyeing.

Name. In allusion to the composition.

Erythrite—$Co_3(AsO_4)_2 \cdot 8H_2O$

Crystallography. Monoclinic; $2/m$. Crystals prismatic and vertically striated. Usually as crusts in globular and reinform shapes. Also pulverulent and earthy.

$C2/m$; $a = 10.20$, $b = 13.37$, $c = 4.74$ Å; $\beta = 105°$. $a:b:c = 0.763:1:0.355$; $Z = 2$. $d's$: 6.65(10), 3.34(1), 3.22(1), 2.70(1), 2.32(1).

Physical Properties. *Cleavage* $\{010\}$ perfect. H $1\frac{1}{2}–2\frac{1}{2}$. G 3.06. *Luster* adamantine to vitreous, pearly on cleavage. *Color* crimson to pink. Translucent. *Optics:* (−); $\alpha = 1.626$, $\beta = 1.661$, $\gamma = 1.699$; $2V = 90°\pm$; $X = b$, $Z \wedge c = 31°$; $r > v$. Pleochroism X pink, Y violet, Z red.

Composition and Structure. CoO 37.5, As_2O_5 38.4, H_2O 24.1%. Ni substitutes for Co to form a complete series to *annabergite*, $Ni_3(AsO_4)_2 \cdot 8H_2O$. Annabergite, or *nickel bloom*, is light green in color. The structure of erythrite is of a layer type with strong bonds between (AsO_4) tetrahedra and $Co(O, H_2O)$ octahedra that are linked by common vertices. The layers parallel to (010) are held together by weak residual bonds.

Diagnostic Features. The association of erythrite with other cobalt minerals and its pink color are usually sufficient to distinguish it from all other minerals. Fusible at 2 to a gray bead. When heated on charcoal gives arsenical odor. Imparts a deep blue to the borax bead (Co). Soluble in HCl giving a red solution.

Occurrence. Erythrite is a rare secondary mineral. In pink crusts known as *cobalt bloom* it occurs as an alteration product of cobalt arsenides. It is rarely present in large amounts and usually forms as crusts or fine aggregates filling cracks. Notable localities are at Schneeberg, Saxony, and Cobalt, Ontario.

Use. Although erythrite has no economic importance, it is used by the prospector as a guide to other cobalt minerals and associated native silver.

Name. From the Greek word meaning *red*.

Similar Species. *Vivianite*, $Fe_3(PO_4)_2 \cdot 8H_2O$, a rare mineral which is a weathering product of primary Fe-Mn phosphates in pegmatites.

AMBLYGONITE—LiAlFPO$_4$

Crystallography. Triclinic: $\bar{1}$. Usually occurs in coarse, cleavable masses. Crystals are rare, equant, and usually rough when large. Frequently twinned on $\{\bar{1}\bar{1}1\}$.

Angles: $(001) \wedge (100) = 90°13'$. $(110) \wedge (100) = 44°51'$, $(100) \wedge (0\bar{1}1) = 75°22'$.

$P\bar{1}$; $a = 5.19$, $b = 7.12$, $c = 5.04$ Å; $\alpha = 112°02'$, $\beta = 97°50'$, $\gamma = 68°8'$; $a:b:c = 0.729:1:0.708$; $Z = 2$. $d's$: 4.64(10), 3.15(10), 2.93(10), 2.39(5), 2.11(4).

Physical Properties. *Cleavage* $\{100\}$ perfect, $\{110\}$ good, $\{0\bar{1}1\}$ distinct. **H** 6. **G** 3.0–3.1. *Luster* vitreous, pearly on $\{100\}$ cleavage. *Color* white to pale green or blue, rarely yellow. Translucent. *Optics:* usually $(-)$; $\alpha = 1.58–1.60$, $\beta = 1.59–1.62$, $\gamma = 1.60–1.63$; $2V = 50–90°$; $r > v$. Indices increase with increase in (OH) in substitution for F.

Composition and Structure. Li_2O 10.1, Al_2O_3 34.4, F 12.9, P_2O_5 47.9%. Na substitutes for Li; (OH) substitutes for F and probably forms a complete series. When OH $>$ F the mineral is known as *montebrasite*. In the structure of amblygonite AlO_6 octahedra and PO_4 tetrahedra are linked by vertices, Li is in 5 coordination and lies between PO_4 tetrahedra and the nearest Al octahedra. The structure is fairly compact as reflected in the relatively high density.

Diagnostic Features. Fusible at 2 with intumescence, giving a red flame (Li). Insoluble in acids. Gives test for phosphate with ammonium molybdate solution. Cleavage fragments may be confused with feldspar, but are distinguished by cleavage angles, easy fusibility, and flame test.

Occurrence. Amblygonite is a rare mineral found in granite pegmatites with spodumene, tourmaline, lepidolite, and apatite. Found at Montebras, France. In the United States it occurs at Hebron, Paris, Auburn, and Peru, Maine; Pala, California; and Black Hills, South Dakota.

Use. A source of lithium.

Name. From the two Greek words meaning *blunt* and *angle*, in allusion to the angle between the cleavages.

Lazulite—(Mg, Fe)Al$_2$(PO$_4$)$_2$(OH)$_2$; Scorzalite—(Fe, Mg)Al$_2$(PO$_4$)$_2$(OH)$_2$

Crystallography. Monoclinic; $2/m$. Crystals showing steep fourth-order prisms rare. Usually massive, granular to compact.

$P2_1/c$; $a = 7.12$, $b = 7.26$, $c = 7.24$ Å; $\beta = 118°55'$; $a:b:c = 0.981:1:0.997$; $Z = 2$. $d's$: 6.15(8), 3.23(8), 3.20(7), 3.14(10), 3.07(10).

Physical Properties. *Cleavage* $\{110\}$ indistinct. **H** 5–5$\frac{1}{2}$. **G** 3.0–3.1. *Luster* vitreous. *Color* azure-blue. Translucent. *Optics:* $(-)$; $\alpha = 1.604–1.639$, $\beta = 1.626–1.670$, $\gamma = 1.637 – 1.680$, $2V = 60°$; $Y = b$, $X \wedge c = 10°$; $r < v$. Absorption $X < Y < Z$. Indices increase with increasing Fe^{2+} content.

Composition and Structure. Probably a complete solid solution series exists from lazulite to scorzalite with the substitution of Fe^{2+} for Mg. In the structure (Mg, Fe)(O, OH)$_6$ octahedra are linked by edges and faces with Al(O, OH)$_6$ octahedra to form groups. These groups are joined to each other and to PO$_4$ tetrahedra.

Diagnostic Features. Infusible. Before the blowpipe it swells, loses its color, and falls to pieces. In the closed tube it whitens and yields water. Insoluble. Gives phosphate test with ammonium molybdate. If crystals are lacking, lazulite is difficult to distinguish from other blue minerals without optical, chemical, or blowpipe tests.

Occurrence. The members of the lazulite-scorzalite series are rare minerals found in high-grade, quartz-rich, metamorphic rocks and in pegmatites. They are usually associated with kyanite, andalusite, corundum, rutile, sillimanite, and garnet. Notable localities are Salzburg, Austria; Krieglach, Styria; and Horrsjoberg, Sweden. In the United States they are found with corundum on Crowder's Mountain, Gaston County, North Carolina; with rutile on Graves Mountain, Lincoln County, Georgia; and with andalusite in the White Mountains, Inyo County, California.

Use. A minor gem stone.

Name. Lazulite derived from an Arabic word meaning *heaven,* in allusion to the color of the mineral. Scorzalite after E. P. Scorza, Brazilian mineralogist.

Wavellite—Al$_3$(PO$_4$)$_2$(OH)$_3$·5H$_2$O

Crystallography. Orthorhombic; $2/m 2/m 2/m$. Single crystals rare. Usually in radiating spherulitic and globular aggregates.

Pcmn; $a = 9.62$, $b = 17.34$, $c = 6.99$ Å; $a:b:c = 0.555:1:0.403$; $Z = 4$. $d's$: 8.39(10), 5.64(6), 3.44(8), 3.20(8), 2.56(8).

Physical Properties. *Cleavage* {110} and {101} good. **H** 3½–4. **G** 2.36. *Luster* vitreous. *Color* white, yellow, green, and brown. Translucent. *Optics:* (+); $\alpha = 1.525$, $\beta = 1.535$, $\gamma = 1.550$; 2V = 70°. X = b, Y = a; r > v.

Composition and Structure. Al$_2$O$_3$ 38.0, P$_2$O$_5$ 35.2, H$_2$O 26.8%. F may substitute for OH. Details of the structure are uncertain. Al(O, OH)$_6$ octahedra are linked by common vertices to PO$_4$ tetrahedra.

Diagnostic Features. Almost invariably in radiating globular aggregates. Infusible, but on heating swells and splits into fine particles. Insoluble. Yields much water in the closed tube. When moistened with cobalt nitrate and then ignited it assumes a blue color (Al). Gives phosphate test.

Occurrence. Wavellite is a secondary mineral found in small amounts in crevices in aluminous, low-grade metamorphic rocks and in limonite and phosphorite deposits. Although it occurs in many localities, it only rarely is found in quantity. It is abundant in the tin veins of Llallagua, Bolivia. In the United States wavellite occurs in a number of localities in Pennsylvania and Arkansas.

Name. After Dr. William Wavel, who discovered the mineral.

Turquoise—CuAl$_6$(PO$_4$)$_4$(OH)$_8$·4H$_2$O

Crystallography. Triclinic; $\bar{1}$. Rarely in minute crystals, usually cryptocrystalline. Massive compact, reniform, stalactitic. In thin seams, incrustations, and disseminated grains.

$P\bar{1}$; $a = 7.48$ $b = 9.95$, $c = 7.69$; $\alpha = 111°39'$, $\beta = 115°23'$, $\gamma = 69°26'$; $a:b:c = 0.752:1:0.773$; $Z = 1$. $d's$: 6.17(7), 4.80(6), 3.68(10), 3.44(7), 3.28(7).

Physical Properties. *Cleavage* {001} perfect, {010} good (rarely seen). **H** 6. **G** 2.6–2.8. *Luster* waxlike. *Color* blue, bluish-green, green. Transmits light on thin edges. *Optics:* (+); $\alpha = 1.61$, $\beta = 1.62$, $\gamma = 1.65$; 2V = 40°; r < v strong.

Composition and Structure. Fe^{3+} substitutes for Al and a complete series exists between turquoise and *chalcosiderite,* CuFe$_6^{3+}$(PO$_4$)$_4$(OH)$_8$·4H$_2$O. The structure consists of PO$_4$ tetrahedra and (Al, Fe^{3+}) octahedra linked by common vertices. Fairly large holes in the structure contain Cu which is coordinated to 4(OH) and 2 H$_2$O molecules.

Diagnostic Features. Turquoise can be easily recognized by its color. It is harder than chrysocolla, the only common mineral which it resembles. Infusible. When moistened with HCl and heated gives a blue copper chloride flame. Soluble in HCl after ignition and solution gives test for phosphate. In the closed tube it turns dark and gives water.

Occurrence. Turquoise is a secondary mineral usually found in the form of small veins and stringers traversing more or less decomposed volcanic rocks in arid regions. The famous Persian deposits are found in trachyte near Nishapur in the province of Khorasan. In the United States it is found in a much altered trachytic rock in the Los Cerillos Mountains, near Santa Fe, and elsewhere in New Mexico. Turquoise has also been found in Arizona, Nevada, and California. Small crystals have been found in Virginia.

Use. As a gemstone. It is always cut in round or oval forms. Much cut turquoise is veined with the various gangue materials, and such stones are sold under the name of *turquoise matrix*.

Name. Turquoise is French and means *Turkish,* the original stones having come into Europe from the Persian locality through Turkey.

Similar Species. *Variscite,* $Al(PO_4)\cdot 2H_2O$, is a massive, bluish-green mineral somewhat resembling turquoise. It has been found in nodules in a large deposit at Fairfield, Utah.

Autunite—$Ca(UO_2)_2(PO_4)_2\cdot 10\text{–}12H_2O$

Crystallography. Tetragonal; $4/m\,2/m\,2/m$. Crystals tabular on $\{001\}$; subparallel growths are common; also foliated and scaly aggregates.

$I4/mmm$; $a = 7.00$, $c = 20.67$ Å; $a:c = 1:2.953$; $Z = 2$. $d's$: 10.33(10), 4.96(8), 3.59(7), 3.49(7), 3.33(7).

Physical Properties. *Cleavage* $\{001\}$ perfect. **H** $2\text{–}2\frac{1}{2}$. **G** 3.1–3.2. *Luster* vitreous, pearly on $\{001\}$. *Color* lemon yellow to pale green. *Streak* yellow. In ultraviolet light fluoresces strongly yellow-green. *Optics:* $(-)$; $\omega = 1.577$, $\epsilon = 1.553$. Pleochroism E pale yellow, O dark yellow.

Composition and Structure. Small amounts of Ba and Mg may subsitute for Ca. The H_2O content apparently ranges from 10 to 12 H_2O. On drying and slight heating autunite passes reversibly to *meta-autunite* I (a tetragonal phase with $6\frac{1}{2}\text{–}2\frac{1}{2}$ H_2O). On continued heating to about 80°C this passes irreversibly to *meta-autunite* II (an orthorhombic phase with 0–6 H_2O). Neither meta-I nor the meta-II hydrate occurs as a primary phase in na-

ture. The structure of autunite consists of (PO_4) tetrahedra and $(UO_2)O_4$ polyhedra which are joined into tetragonal corrugated layers of composition $UO_2(PO_4)$ parallel to $\{001\}$. These layers are held together by weak residual bonds to H_2O molecules.

Diagnostic Features. Autunite is characterized by yellow-green tetragonal plates and strong fluorescence in ultraviolet light. Fusible at 2–3. Soluble in acids. A soda bead with dissolved autunite fluoresces in ultraviolet light.

Occurrence. Autunite is a secondary mineral found chiefly in the zone of oxidation and weathering derived from the alteration of uraninite or other uranium minerals. Notable localities are near Autun, France; Sabugal and Vizeu, Portugal; Johanngeorgenstadt district and Falkenstein, Germany; Cornwall, England; and Katanga district of Zaire. In the United States autunite is found in many pegmatites, notably at the Ruggles mine, Grafton Center, New Hampshire; Black Hills, South Dakota; and Spruce Pine, Mitchell County, North Carolina. The finest specimens have come from the Daybreak mine near Spokane, Washington.

Use. An ore of uranium (see uraninite, page 275).

Name. From Autun, France.

Similar Species. *Torbernite,* $Cu(UO_2)_2(PO_4)_2\cdot 8\text{–}12H_2O$, is isostructural with autunite and has similar properties, but there is no evidence of a solid solution series. Color green, nonfluorescent. Associated with autunite.

Carnotite—$K_2(UO_2)_2(VO_4)_2\cdot 3H_2O$

Crystallography. Monoclinic; $2/m$. Only rarely in imperfect microscopic crystals flattened on $\{001\}$ or elongated on b. Usually found as a powder or as loosely coherent aggregates; disseminated.

$P2_1a$; $a = 10.47$, $b = 8.41$, $c = 6.91$ Å; $\beta = 103°40'$. $a:b:c = 1.245:1:0.822$; $Z = 2$. $d's$: 6.56(10), 4.25(3), 3.53(5), 3.25(3), 3.12(7).

Physical Properties. *Cleavage* $\{001\}$ perfect. *Hardness* unknown, but soft. **G** 4.7–5. *Luster* dull or earthy. *Color* bright yellow to greenish yellow. *Optics:* $(-)$; $\alpha = 1.75$, $\beta = 1.93$, $\gamma = 1.95$; $2V = 40°\pm$, $Y = b$, $X = c$. $r < v$. Indices increase with loss of water.

Composition and Structure. The water content of carnotite is partly zeolitic and varies with the humidity at ordinary temperatures; the $3H_2O$ is for fully hydrated material. Small amounts of Ca, Ba, Mg, Fe, and Na have been reported. The structure of carnotite consists of a layer pattern that is the result of strong bonds between the (VO_4) groups and the $UO_2(O_5)$ polyhedra. These layers, of composition $(UO_2)_2(VO_4)_2$, are held together by hydroxyl-hydrogen bonds to water molecules and also by K, Ca, and Ba between the layers.

Diagnostic Features. Carnotite is characterized by its yellow color, its pulverant nature, and its occurrence. Unlike many secondary uranium minerals, carnotite will not fluoresce in ultraviolet light but a soda bead with dissolved carnotite will fluoresce (test for uranium). Infusible. Soluble in acids.

Occurrence. Carnotite is of secondary origin, and its formation is usually ascribed to the action of meteoric waters on preexisting uranium and vanadium minerals. It has a strong pigmenting power and when present in a sandstone in amounts even less than 1% will color the rock yellow. It is found principally in the plateau region of southwestern Colorado and in adjoining districts of Utah where it occurs disseminated in a cross-bedded sandstone. Concentrations of relatively pure carnotite are found around petrified tree trunks.

Use. Carnotite is an ore of vanadium and, in the United States, a principal ore of uranium.

Name. After Marie-Adolphe Carnot (1839–1920), French mining engineer and chemist.

Similar Species. *Tyuyamunite,* $Ca(UO_2)_2(VO_4)_2 \cdot 5-8\frac{1}{2}H_2O$, is the calcium analogue of carnotite and similar in physical properties except for a slightly more greenish color and yellow-green fluorescence. It is found in almost all carnotite deposits. Named from Tyuya Muyum, southeastern Turkistan, U.S.S.R., where it is mined as a uranium ore.

References and Suggested Reading

Deer, W. A., Howie, R. A., and J. Zussman, 1962, *Rock-Forming Minerals,* vol. 5, Non-silicates. John Wiley & Sons, New York, 371 pp.

Palache, C., Berman, H., and C. Frondel, 1951, *The System of Mineralogy,* 7th ed., vol. II. John Wiley & Sons, 1124 pp.

Strunz, H., *Mineralogische Tabellen,* 1970, 5th ed., Akademische Verlagsgesellschaft, Leipzig, 621 pp.

10 SYSTEMATIC MINERALOGY

PART IV: SILICATES

The silicate mineral class is of greater importance than any other, for about 25% of the known minerals and nearly 40% of the common ones are silicates. With a few minor exceptions all the igneous rock-forming minerals are silicates, and they thus constitute well over 90% of the earth's crust.

When the average weight percentages of the eight most common elements in the earth's crust are recalculated on the basis of atomic percent (see Table 4.2), we find that out of every 100 atoms 62.5 are O, 21.2 are Si, and 6.5 are Al. Fe, Mg, Ca, Na, and K each accounts for about two to three more atoms. With the possible exception of Ti, all other elements are present in insignificant amounts in the upper levels of the earth's crust (see Table 4.1). When we recalculate the atomic percentages of the eight most abundant elements in terms of volume percentages (see Table 4.2, last column) we find that the earth's crust can be regarded as a packing of oxygen ions, with interstitial metal ions, such as Si^{4+}, Al^{3+}, Fe^{2+}, Ca^{2+}, Na^+, K^+, and so forth.

The dominant minerals of the crust are thus shown to be silicates, oxides, and other oxygen compounds such as carbonates. Of the different assemblages of silicate minerals that characterize igneous, sedimentary, and metamorphic rocks, ore veins, pegmatites, weathered rocks, and soils, each has the potential to tell us something of the environment in which it was formed.

We have a further deep and compelling reason to study the silicates. The soil from which our food is ultimately drawn is made up in large part of silicates. The brick, stone, concrete, and glass used in the construction of our buildings are either silicates or largely derived from silicates. Even with the coming of the space age, we need not fear obsolescence of our studies of the silicates, but rather an enlargement of their scope, because we now know that the moon and all the planets of our solar system have rocky crusts made of silicates and oxides much like those of Earth.

The radius ratio of Si^{4+} (radius = 0.39 Å) to that

of O^{-2} (radius = 1.40 Å) is 0.278. This radius ratio indicates that 4 coordination is the stable state of Si-O groupings. The fundamental unit on which the structure of all silicates is based consists of four O^{-2} at the apices of a regular tetrahedron surrounding and coordinated by one Si^{4+} at the center (Figs. 9.1*d* and 10.1). The powerful bond that unites the oxygen and silicon ions is literally the cement that holds the earth's crust together. This bond may be estimated by use of Pauling's electronegativity concept (page 137) as 50% ionic and 50% covalent. That is, although the bond arises in part from the attraction of oppositely charged ions, it also involves sharing of electrons and interpenetration of the electronic superstructures of the ions involved. The bond is strongly localized in the vicinity of these shared electrons.

Although electron sharing is present in the Si-O bond, the total bonding energy of Si^{4+} is still distributed equally among its four closest oxygen neighbors. Hence, the strength of any single Si-O bond is equal to just one-half the total bonding energy available in the oxygen ion. Each O^{-2} has, therefore, the potentiality of bonding to another silicon ion and entering into another tetrahedral grouping, thus uniting the tetrahedral groups through the shared (or *bridging*) oxygen. Such linking of tetrahedra is often referred to as *polymerization,* a term borrowed from organic chemistry, and the capacity for polymerization is the origin of the great variety of silicate structures. In no case, however, are three or even two oxygens shared between two adjacent tetrahedra in nature. Such sharing would place two highly charged Si^{4+} ions close together and the repulsion between them would render the structure unstable.

The sharing of oxygens may involve one, two, three, or all four of the oxygen ions in the tetrahedron, giving rise to a diversity of structural configurations. Figure 10.2 illustrates the various ways in which SiO_4 tetrahedra can be combined. Silicates with independent tetrahedral SiO_4 groups are known as *orthosilicates* or *nesosilicates*. Silicates in which two SiO_4 groups are linked, giving rise to Si_2O_7 groups, are classed as *sorosilicates*. If more than two tetrahedra are linked, closed ringlike struc-

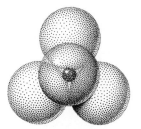

FIG. 10.1. Close packing representation of SiO_4 tetrahedron.

tures are formed of a general composition Si_xO_{3x}. Four-fold rings have composition Si_4O_{12}. This group of *ring silicates* is also known as *cyclosilicates*. Tetrahedra may be joined to form infinite single chains with a unit composition SiO_3. Infinite double chains give a ratio of Si : O = 4 : 11. Both types of *chain silicates* are also known as *inosilicates*. When three of the oxygens of a tetrahedron are shared between adjoining tetrahedra infinitely extending flat sheets are formed of unit composition Si_2O_5. Such *sheet silicates* are also referred to as *phyllosilicates*. When all four oxygens of a SiO_4 tetrahedron are shared by adjoining tetrahedra a three-dimensional network of unit composition SiO_2 is obtained. These *framework silicates* are also known as *tectosilicates*.

Next to O and Si, the most important constituent of the crust is Al. Al^{3+} has a radius of 0.51 Å and thus the radius ratio Al : O = 0.364, corresponding to 4 coordination with oxygen. However, the radius ratio is sufficiently close to the upper limit for 4 coordination so that 6 coordination is also possible. It is this capacity for playing a double role in silicate minerals that gives Al^{3+} its outstanding significance in the crystal chemistry of the silicates. When Al coordinates four O's arranged at the apices of a regular tetrahedron, the resultant grouping occupies approximately the same space as a silicon-oxygen tetrahedron and may link with silicon tetrahedra in polymerized groupings. On the other hand, Al^{3+} in 6 coordination may serve to link the tetrahedral groupings through simple ionic bonds, weaker than those that unite the ions in the tetrahedra. It is thus possible to have Al in silicate structures both in the tetrahedral sites, substituting for Si, and in the octahedral sites with 6 coordination, involved in solid

Class	Arrangement of SiO₄ tetrahedra (central Si⁴⁺ not shown)	Unit composition	Mineral example
Nesosilicates	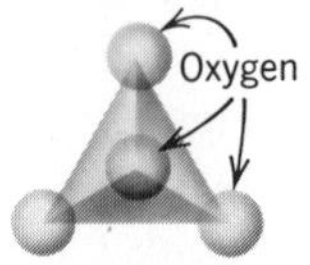	$(SiO_4)^{-4}$	Olivine, $(Mg, Fe)SiO_4$
Sorosilicates	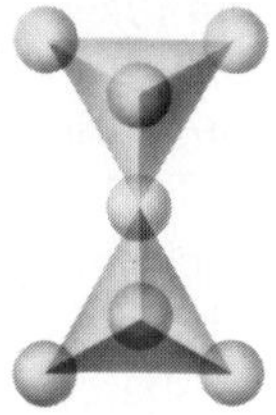	$(Si_2O_7)^{-6}$	Hemimorphite, $Zn_4Si_2O_7(OH) \cdot H_2O$
Cyclosilicates	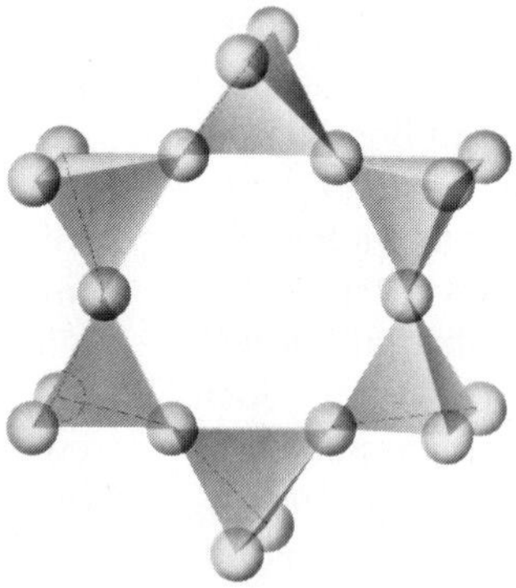	$(Si_6O_{18})^{-12}$	Beryl, $Be_3Al_2Si_6O_{18}$
Inosilicates (single chain)	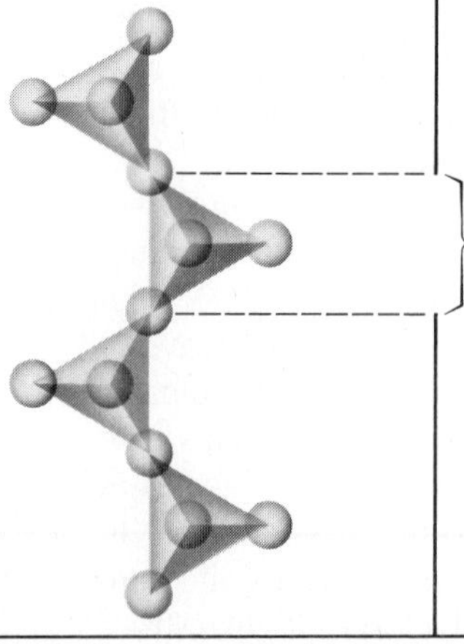	$(SiO_3)^{-2}$	Pyroxene e.g. Enstatite, $MgSiO_3$

FIG. 10.2. Silicate classification.

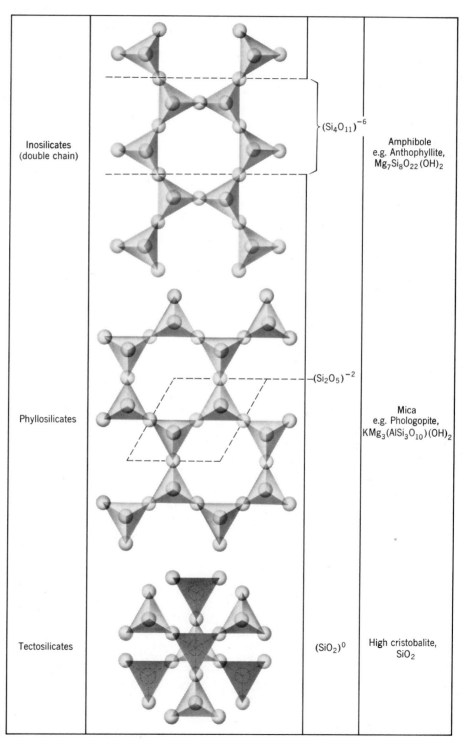

Inosilicates (double chain) — $(Si_4O_{11})^{-6}$ — Amphibole e.g. Anthophyllite, $Mg_7Si_8O_{22}(OH)_2$

Phyllosilicates — $(Si_2O_5)^{-2}$ — Mica e.g. Phologopite, $KMg_3(AlSi_3O_{10})(OH)_2$

Tectosilicates — $(SiO_2)^0$ — High cristobalite, SiO_2

FIG. 10.2 (*Continued*)

solution relations with elements such as Mg and Fe^{2+}.

Mg, Fe^{2+}, Fe^{3+}, Mn^{2+}, Al^{3+}, and Ti^{4+} all tend to occur in silicate structures in 6 coordination with oxygen (see Table 10.1). Although divalent, trivalent, and tetravalent ions are included here, all have about the same space requirements and about the same radius ratio relations with oxygen, and, hence, tend to occupy the same type of atomic site. Solid solution relations between ions of different charge are possible through the mechanism of coupled substitution (see page 161).

The larger and more weakly charged cations, Ca^{2+} and Na^+, of ionic radii 0.99 Å and 0.97 Å, respectively, generally enter sites having 8 coordination with respect to oxygen. Although the sizes of these two ions are very similar, their charges are different. This charge difference is compensated for by coupled substitution such as in the plagioclase feldspars where $Na^+ + Si^{4+}$ substitute for $Ca^{2+} + Al^{3+}$ in order to keep the overall structure neutral.

The largest ions common in silicate structures are those of K, Rb, Ba, and the rarer alkalis and alkali earths. These ions generally do not enter readily into Na or Ca sites and are found in high-coordination-number sites of unique type. Hence, solid solution between these ions and the other common ions is limited.

Ionic substitution is generally common and extensive between elements whose symbols lie between a pair of horizontal lines in Table 10.1, but it is rare and difficult between elements separated by a horizontal line. The usual role played by the commonest elements permits us to write a general formula for all silicates:

$$X_m Y_n (Z_p O_q) W_r$$

where X represents large, weakly charged cations in 8 or higher coordination with oxygen; Y represents medium sized, two to four valent ions in 6 coordination; Z represents small, highly charged ions in tetrahedral coordination; O is oxygen and W represents additional anionic groups such as $(OH)^-$ or anions such as Cl^- or F^-. The ratio $p:q$ depends on the degree of polymerization of the silicate framework, and the other subscript variables, m, n, and r, depend on the need for electrical neutrality. Any common silicate may be expressed by suitable substitution in this general formula.

The subsequent treatment of individual silicates will be on the basis of subclasses that reflect their internal structure (neso-, soro-, cyclosilicates, etc.) and their chemical composition. Such a scheme is the basis of *Mineralogische Tabellen,* 1970, 5th ed., by Hugo Strunz.

NESOSILICATES

In the nesosilicates the SiO_4 tetrahedra are isolated (Fig. 10.2) and bound to each other only by ionic bonds from interstitial cations. Their structures depend chiefly on the size and charge of the interstitial cations. The atomic packing of the nesosilicate structures is generally dense, causing the minerals of this group to have relatively high specific gravity and hardness. Because the SiO_4 tetrahedra are independent and not linked into chains or sheets, for example, the crystal habit of the nesosilicates is generally equidimensional and pronounced cleavage

Table 10.1
COORDINATION OF COMMON ELEMENTS IN SILICATES[a]

	Coordination Number	Ion	Ionic Radius (Å)	$R_X : R_O$
Z	4	Si^{4+}	0.39	0.278
	4	Al^{3+}	0.51	0.364
Y	6	Al^{3+}	0.51	0.364
	6	Fe^{3+}	0.64	0.457
	6	Mg^{2+}	0.66	0.471
	6	Ti^{4+}	0.68	0.486
	6	Fe^{2+}	0.74	0.529
	6	Mn^{2+}	0.80	0.571
X	8	Na^+	0.97	0.693
	8	Ca^{2+}	0.99	0.707
X	8–12	K^+	1.33	0.950
	8–12	Ba^{2+}	1.34	0.957
	8–12	Rb^+	1.47	1.050

[a] See Table 4.3 for a complete listing of ionic radii.

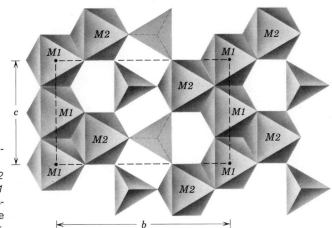

FIG. 10.3. Portion of the idealized structure of olivine projected on (100). *M1* and *M2* are octahedral sites. The *M1* octahedron is somewhat distorted in the real structure whereas *M2* is regular.

directions are absent. Although Al^{3+} substitutes commonly and easily in the Si position of silicates, the amount of Al substitution in SiO_4 tetrahedra in nesosilicates is generally low.

Very common members, especially in high-temperature igneous rocks, of the nesosilicate group are *forsterite*, Mg_2SiO_4, and *fayalite*, Fe_2SiO_4, end members of the $(Mg, Fe)_2SiO_4$ olivine series. The structure of olivine, which is shown in Fig. 10.3, can be viewed as consisting of layers parallel to {100}. These layers consist of octahedra cross-linked by independent SiO_4 tetrahedra. The octahedrally coordinated sites are known as *M1* and *M2* with *M1* somewhat distorted and *M2* relatively regular. In the $(Mg, Fe)_2SiO_4$ olivines Mg and Fe^{2+} occupy the *M1* and *M2* sites without any specific preference for either site. In the calcic olivines, however (e.g., *monticellite*, $CaMgSiO_4$), Ca enters into the *M1* site and Mg into *M2*.

Garnets are another group of very common nesosilicate minerals, especially in metamorphic rocks. Their structural formula may be represented as $A_3B_2(SiO_4)_3$, where A and B refer respectively to 8 and 6 coordinated cationic sites. The A sites are occupied by rather large divalent cations, whereas the B sites house smaller trivalent cations. Because of these size considerations in the filling of the A sites we may expect a fairly well-defined division of the garnets in those with Ca and those with easily interchangeable divalent ions such as Mg^{2+}, Fe^{2+},

Mn^{2+}. Likewise, because of the limited substitution possible in the B sites, we may expect a separation of garnets into Al^{3+}, Fe^{3+}, and Cr^{3+} bearing. These two trends are well marked and have given rise to a grouping of garnets into two series: *pyralspite* (Ca absent in A; B = Al) and *ugrandite* (A = Ca).

Pyralspite		**Ugrandite**	
*Py*rope	$Mg_3Al_2Si_3O_{12}$	*U*varovite	$Ca_3Cr_2Si_3O_{12}$
*Al*mandite	$Fe_3Al_2Si_3O_{12}$	*Gr*ossularite	$Ca_3Al_2Si_3O_{12}$
*Sp*essartite	$Mn_3Al_2Si_3O_{12}$	*And*radite	$Ca_3Fe^{3+}Si_3O_{12}$

The above classification serves as an excellent mnemonic aid for the names and formulas. Another grouping on the basis of the ions in the B site, yields three unequal groups:

Aluminum garnets	**Ferri-garnets**	**Chrome-garnet**
Pyrope	Andradite	Uvarovite
Almandite		
Spessartite		
Grossularite		

Hydroxyl, as tetrahedral $(OH)_4$ groups, may substitute to a limited extent for SiO_4 tetrahedra in hydrogarnets such as *hydrogrossularite*, $Ca_3Al_2Si_2O_8(SiO_4)_{1-m}(OH)_{4m}$, with m ranging from 0 to 1. Ti^{4+} may enter into the B sites concomitant with replacement of Ca by Na in the A sites, producing the black *melanite*.

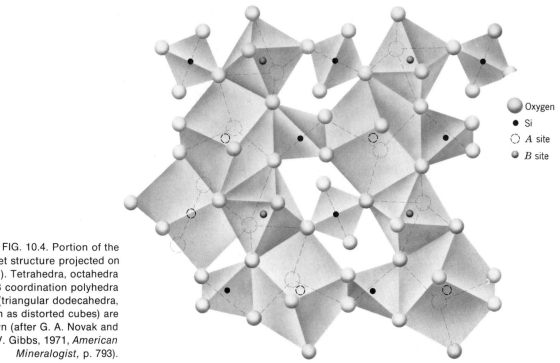

FIG. 10.4. Portion of the garnet structure projected on (001). Tetrahedra, octahedra and 8 coordination polyhedra (triangular dodecahedra, drawn as distorted cubes) are shown (after G. A. Novak and G. V. Gibbs, 1971, *American Mineralogist,* p. 793).

FIG. 10.5 (*a*) Projection of the sillimanite structure showing octahedral chains parallel to the *c* axis (redrawn after C. W. Burnham, 1963, *Zeitschrift für Kristallographie,* p. 146). (*b*) Projection of the andalusite structure showing octahedral chains parallel to the *c* axis, and the presence of AlO$_5$ polyhedra between SiO$_4$ tetrahedra (redrawn after C. W. Burnham and M. J. Buerger, 1961, *Zeitschrift für Kristallographie,* p. 287).

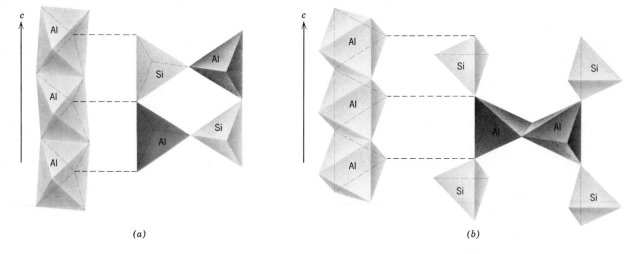

(*a*)   (*b*)

The structure of garnet, which is illustrated in Fig. 10.4, consists of alternating SiO_4 tetrahedra and BO_6 octahedra that share corners to form a continuous three-dimensional network. The A sites are surrounded by 8 oxygens in irregular coordination polyhedra.

The aluminosilicates of the nesosilicate group, kyanite, sillimanite, and andalusite are commonly found in medium- to high-grade metamorphic rocks of Al-rich bulk composition. All three minerals are polymorphs of Al_2SiO_5, which may be stated structurally as $Al^{[4-6]}Al^{[6]}SiO_5$ (digits in square brackets indicate coordination number). Chains of edge-sharing octahedra, parallel to the c axis, are characteristic of all three structures (see Fig. 10.5). In *kyanite*, $Al^{[6]}Al^{[6]}SiO_5$, the triclinic polymorph, with space group $P\bar{1}$, all of the Al is octahedrally coordinated. It occurs as octahedral chains parallel to c, and as isolated Al octahedra. In *sillimanite*, $Al^{[4]}Al^{[6]}SiO_5$, an orthorhombic polymorph, with space group *Pbnm*, the octahedrally coordinated Al is found in octahedral chains (see Fig. 10.5a) and adjacent tetrahedral chains consist of alternating tetrahedral AlO_4 and SiO_4 groups. In *andalusite*, $Al^{[5]}Al^{[6]}SiO_5$, another orthorhombic polymorph, with space group *Pnnm*, half the Al is found in octahedral chains and the other half occurs in 5 coordinated polyhedra (see Fig. 10.5b) which are linked by SiO_4 tetrahedra.

The following minerals in the nesosilicate group will be discussed in detail:

NESOSILICATES

Phenacite Group

Phenacite	Be_2SiO_4
Willemite	Zn_2SiO_4

Olivine Group

Forsterite	Mg_2SiO_4
Fayalite	Fe_2SiO_4

Garnet Group $A_3B_2(SiO_4)_3$

Pyrope	Mg_3Al_2	Uvarovite	Ca_3Cr_2
Almandite	Fe_3Al_2	Grossularite	Ca_3Al_2
Spessartite	Mn_3Al_2	Andradite	$Ca_3Fe_2^{3+}$

Zircon Group

Zircon	$ZrSiO_4$

Al_2SiO_5 Group

Andalusite	
Sillimanite	Al_2SiO_5
Kyanite	

NESOSILICATES (Continued)

Topaz	$Al_2SiO_4(F, OH)_2$
Staurolite	$FeAl_9O_6(SiO_4)_4(O, OH)_2$
Humite Group	
Chondrodite	$Mg_5(SiO_4)_2(OH, F)_2$
Datolite	$CaB(SiO_4)(OH)$
Sphene	$CaTiO(SiO_4)$
Chloritoid	$(Fe, Mg)_2Al_4O_2(SiO_4)_2(OH)_4$

Phenacite Group

Phenacite—Be_2SiO_4

Crystallography. Hexagonal-R;$\bar{3}$. Crystals usually flat rhombohedral or short prismatic. Often with complex development. Frequently twinned on $\{10\bar{1}0\}$.

$R3$; $a = 12.45$, $c = 8.23$ Å; $a:c = 1:0.661$; $Z = 18$. $d's$: 3.58(6), 2.51(8), 2.35(6), 2.18(8), 1.258(10).

Physical Properties. *Cleavage* $\{11\bar{2}0\}$ imperfect. **H** $7\frac{1}{2}$–8. **G** 2.97–3.00. *Luster* vitreous. *Color* colorless, white. Transparent to translucent. *Optics:* (+); $\omega = 1.654$, $\epsilon = 1.670$.

Composition and Structure. BeO 45.6, SiO_2 54.5%. The structure consists of SiO_4 and BeO_4 tetrahedra with each oxygen linked to 2 Be and one Si at the corners of an equilateral triangle.

Diagnostic Features. Characterized by its crystal form and great hardness. Infusible. Fused with sodium carbonate yields a white enamel.

Occurrence. Phenacite is a rare pegmatite mineral associated with topaz, chrysoberyl, beryl, and apatite. Fine crystals are found at the emerald mines in the Ural Mountains, U.S.S.R., and in Minas Gerais, Brazil. In the United States found at Mount Antero, Colorado.

Use. Occasionally cut as a gem stone.

Name. From the Greek meaning *a deceiver*, in allusion to its resemblance to quartz.

WILLEMITE—Zn_2SiO_4

Crystallography. Hexagonal-R;3. In hexagonal prisms with rhombohedral terminations. Usually massive to granular. Rarely in crystals.

$R\bar{3}$; $a = 13.96$, $c = 9.34$ Å; $a:c = 0.669$; $Z = 18$. $d's$: 2.84(8), 2.63(9), 2.32(8), 1.849(8), 1.423(10).

Physical Properties. *Cleavage* {0001} good. **H** $5\frac{1}{2}$. **G** 3.9–4.2. *Luster* vitreous to resinous. *Color* yellow-green, flesh-red, and brown; white when pure. Transparent to translucent. Most willemite from Franklin, New Jersey, fluoresces under ultraviolet light. *Optics:* (+); $\omega = 1.691$, $\epsilon = 1.719$.

Composition and Structure. ZnO 73.0, SiO$_2$ 27.0%. Mn^{2+} often replaces a considerable part of the Zn (manganiferous variety called *troostite*); Fe^{2+} may also be present in small amount. Willemite is isostructural with phenacite (see above), with SiO$_4$ and ZnO$_4$ tetrahedra. Because Zn^{2+} (radius = 0.74 Å) is much larger than Be^{2+} (radius = 0.35 Å) the structure of willemite is much expanded over that of phenacite.

Diagnostic Features. Willemite from Franklin, New Jersey, can usually be recognized by its association with franklinite and zincite. Pure willemite infusible. When heated on charcoal with cobalt nitrate assay turns blue. Troostite will give reddish-violet color to the borax bead in the oxidizing flame (Mn). Distinguished from hemimorphite by absence of water.

Occurrence. Willemite is found in crystalline limestone and may be the result of metamorphism of earlier hemimorphite or smithsonite. It is also found sparingly as a secondary mineral in the oxidized zone of zinc deposits.

Found at Altenberg, near Moresnet, Belgium; Algeria; Zambia; South-West Africa; and Greenland. The most important locality is in the United States at Franklin, New Jersey, where willemite occurs associated with franklinite and zincite and as grains imbedded in calcite. It has also been found at the Merritt mine, New Mexico, and Tiger, Arizona.

Use. A valuable zinc ore at Franklin, New Jersey.

Name. In honor of the King of the Netherlands, William I.

Olivine Group

The composition of the majority of olivines can be represented in the system CaO-MgO-FeO-SiO$_2$ (Fig.

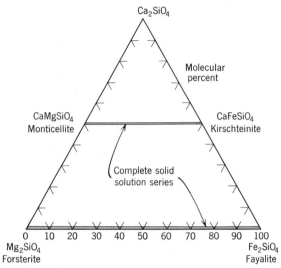

FIG. 10.6. Olivine compositions in the system Ca$_2$SiO$_4$-Mg$_2$SiO$_4$-Fe$_2$SiO$_4$.

10.6). The most common series in this system is from *forsterite*, Mg$_2$SiO$_4$, to *fayalite*, Fe$_2$SiO$_4$. Relatively rare olivines occur also along the *monticellite*, CaMgSiO$_4$, to *kirschteinite*, CaFe^{2+}SiO$_4$, join. Very little, if any, solid solution exists between these two series. Mn^{2+} may substitute for Fe^{2+}, forming a relatively rare series between fayalite and *tephroite*, Mn$_2$SiO$_4$. The structure of olivine, with very similar *M1* and *M2* octahedral sites, and independent SiO$_4$ tetrahedra is shown in Fig. 10.3.

Members of the forsterite-fayalite series are common as primary crystallization products in Fe- and Mg-rich, silica-poor melts (magmas). *Dunites* and *peridotites* are pure olivine and olivine plus pyroxene rocks, respectively. Olivine concentrations in igneous rocks may result from the accumulation of olivine crystals, under the influence of gravity, during the cooling stages of a magma. Members of the forsterite-fayalite series are highly refractory, as can be seen from Fig. 10.7 (forsterite melting point = 1890°C and fayalite melting point = 1205°C). This diagram represents a complete solid solution series without a maximum or minimum (see also Fig. 11.7). When a melt with composition X (50 weight percent Fe$_2$SiO$_4$) is cooled to the liquidus curve, olivine crystals of composition X_1, will start to form. The liquid, as a result of the

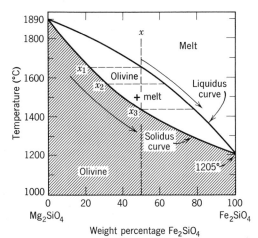

FIG. 10.7. Temperature-composition diagram for the system Mg_2SiO_4-Fe_2SiO_4 at atmospheric pressure.

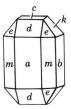

FIG. 10.8. Olivine.

Mg-rich olivine crystallization will become more Fe-rich, as shown by the upper arrow. This, in turn, will cause a more Fe-rich olivine to crystallize as shown by X_2, until finally all liquid is used up. At this point, under equilibrium conditions, the final composition to crystallize out is of composition X_3, which is the same as the original X. Chemical zoning in olivines with Mg-rich cores and more Fe-rich rims may be found in high-temperature igneous rocks. Such rimming is the result of nonequilibrium crystallization in the direction of the arrows in Fig. 10.7.

OLIVINE—$(Mg, Fe)_2SiO_4$

Crystallography. Orthorhombic; $2/m2/m2/m$. Crystals are usually a combination of the three prisms, the three pinacoids, and a dipyramid. Often flattened parallel to either {100} or {010} (see Fig. 10.8). Usually appears as embedded grains or in granular masses.

Angles: $b(010) \wedge m(110) = 55°1'$, $b(010) \wedge s(120) = 32°58'$, $k(021) \wedge k'(0\bar{2}1) = 99°6'$, $d(101) \wedge d'(\bar{1}01) = 103°6'$, $c(001) \wedge e(111) = 54°11'$.

Pmcn; $Z = 4$. Mg_2SiO_4: $a = 4.76$, $b = 10.20$, $c = 5.98$; $a:b:c = 0.467:1:0.586$. Fe_2SiO_4: $a = 4.82$, $b = 10.48$, $c = 6.11$; $a:b:c = 0.460:1:0.583$. $d's$ for common olivine: 2.49(10), 2.41(8), 2.24(7), 1.734(8), 1.498(7).

Physical Properties. *Fracture* conchoidal. **H** $6\frac{1}{2}$–7. **G** 3.27–4.37, increasing with increase in Fe content (see Fig. 10.9). *Luster* vitreous. *Color* pale yellow green to olive green in forsterite, darker, brownish green with increasing Fe^{2+}. Transparent to translucent. *Optics:* forsterite (+), others (−). For $Mg_{1.6}Fe_{0.4}SiO_4$: $\alpha = 1.674$, $\beta = 1.692$, $\gamma = 1.712$, $2V = 87°$; $X = b$, $Z = a$.

Composition and Structure. A complete solid solution exists from *forsterite,* Mg_2SiO_4, to *fayalite,* Fe_2SiO_4. The more common olivines are richer in Mg than in Fe^{2+}. An example of the recalculation of an olivine analysis is given in Table 4.13. The structure of olivine is given in Fig. 10.3 and discussed on page 341. Under very high pressures (50 to 100 kilobars) the olivine structure transforms to a spinel structure with Si in tetrahedral sites and Mg and Fe^{2+} in the octahedral sites of the spinel structure (see Spinel and Fig. 8.5). The spinel structure is slightly more dense than the olivine structure. Olivine is thought to be abundant in the mantle and probably occurs as the spinel type at such great depths.

Diagnostic Features. Distinguished usually by its glassy luster, conchoidal fracture, green color, and granular nature. Infusible. Rather slowly soluble in HCl and yields gelatinous silica upon evaporation.

Occurrence. Olivine is a rather common rock-forming mineral, varying in amount from that of an accessory to that of a main constituent. It is found principally in the dark-colored igneous rocks such as gabbro, peridotite, and basalt (see Table 11.2). In these rock types it coexists with plagioclase and pyroxenes. The rock, *dunite,* is made up almost wholly of olivine. Forsterite is not stable in the presence of free SiO_2 and will react with it to form pyroxene, according to $Mg_2SiO_4 + SiO_2 \rightarrow$

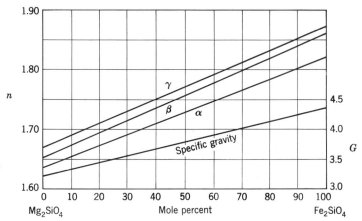

FIG. 10.9. Olivine. Variation of refractive indices and specific gravity with composition.

$2MgSiO_3$. Found as glassy grains in stony and stony-iron meteorites. Occasionally in crystalline dolomitic limestones where it formed by the reaction: $2CaMg(CO_3)_2 + SiO_2 \rightarrow Mg_2SiO_4 + 2CO_2$. Associated often with pyroxene, plagioclase, magnetite, corundum, chromite, and serpentine.

The transparent gem variety is known as *peridot*. It was used as a gem in ancient times in the East, but the exact locality for the stones is not known. At present peridot is found in Burma, and in rounded grains associated with pyrope garnet in the surface gravels of Arizona and New Mexico, but the best quality material comes from St. John's Island in the Red Sea. Crystals of olivine are found in the lavas of Vesuvius. Larger crystals, altered to serpentine, come from Snarum, Norway. Olivine occurs in granular masses in volcanic bombs in the Eifel district, Germany, and in Arizona. Dunite rocks are found at Dun Mountain, New Zealand, and with the corundum deposits of North Carolina.

Alteration. Very readily altered to *serpentine* minerals such as *antigorite*, $Mg_6Si_4O_{10}(OH)_8$. Magnesite, $MgCO_3$ and iron oxides form at the same time as a result of the alteration. Olivines in metamorphosed basic igneous rocks commonly show *coronas,* which are concentric rims consisting of pyroxene and amphibole. Such coronas are the result of the instability of the high-temperature olivine in a lower temperature, H_2O-containing environment.

Use. As the clear green variety, *peridot,* it has some use as a gem. Olivine is mined as refractory sand for the casting industry.

Name. Olivine derives its name from the usual olive-green color. *Chrysolite* is a synonym for olivine. Peridot is an old name for the species.

Similar Species. Other rarer members of the olivine group are: *monticellite,* $CaMgSiO_4$, a high-temperature contact metamorphic mineral in siliceous dolomitic limestones; *tephroite,* Mn_2SiO_4 and *larsenite,* $PbZnSiO_4$.

Garnet Group

The garnet group includes a series of isostructural subspecies with space group *Ia3d;* they crystallize in the hexoctahedral class of the isometric system and are similar in crystal habit. The structural arrangement (see Fig. 10.4) is such that the atomic population of the {100} and {111} families of planes is much depleted. As a result the cube and octahedron, common on most isometric hexoctahedral crystals, are rarely found on garnets.

Crystallography. Isometric; $4/m\overline{3}2/m$. Common forms are dodecahedron *d* (Fig. 10.10*a*) and trapezohedron, *n* (Fig. 10.10*b*), often in combination (Figs. 10.10*c, d,* and 10.11). Hexoctahedrons are observed occasionally (Fig. 10.10*e*). Other forms are rare. Usually distinctly crystallized; also appears in rounded grains; massive granular, coarse or fine.

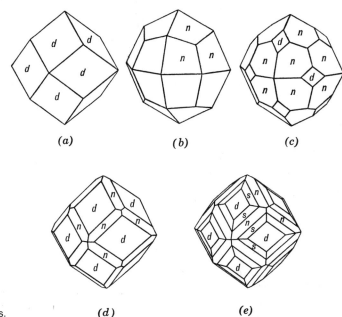

FIG. 10.10. Garnet crystals.

Ia3d; cell edge (see Table 10.2); Z = 8. *d's* for pyrope: 2.89(8), 2.58(9), 1.598(9), 1.542(10), 1.070(8).

Physical Properties. H $6\frac{1}{2}$–$7\frac{1}{2}$. **G** 3.5–4.3, varying with composition (see Table 10.2). *Luster* vitreous to resinous. *Color* varying with composition (see below); most commonly red, also brown, yellow, white, green, black. *Streak* white. Transparent to translucent.

Composition and Structure. Garnet compositions can be expressed by the general structural formula $A_3B_2(SiO_4)_3$ where the A site houses Ca, Mg, Fe^{2+} or Mn^{2+} and the B site incorporates Al, Fe^{3+}, and Cr^{3+} (see Fig. 10.4 and page 341). The formulas of the main subspecies are given in Table 10.2, with the refractive indices, specific gravity, and unit cell edges for the end member compositions. There is extensive substitution among the *pyralspite* group and also among the *ugrandite* group but relatively little solid solution between these two major categories. Rarely the pure end members are found. Hydrous garnets such as *hydrogrossularite,* may contain up to 8.5% H_2O. This water, in the form of $(OH)_4^{-4}$ groups, substitutes probably for (SiO_4) tetra-

Table 10.2 **CHEMICAL COMPOSITIONS AND PHYSICAL PROPERTIES OF GARNETS**					
	Subspecies	Composition	*n*	G	Unit Cell Edge (Å)
	Pyrope	$Mg_3Al_2Si_3O_{12}$	1.714	3.58	11.46
	Almandite	$Fe_3Al_2Si_3O_{12}$	1.830	4.32	11.53
	Spessartite	$Mn_3Al_2Si_3O_{12}$	1.800	4.19	11.62
	Grossularite	$Ca_3Al_2Si_3O_{12}$	1.734	3.59	11.85
	Andradite	$Ca_3Fe_2Si_3O_{12}$	1.887	3.86	12.05
	Uvarovite	$Ca_3Cr_2Si_3O_{12}$	1.868	3.90	12.00
	Hydrogrossularite	$Ca_3Al_2Si_2O_8(SiO_4)_{1-m}$	1.734 to	3.59 to	11.85 to
		$(OH)_{4m}$; $m = 0$–1	1.675	3.13	12.16

hedra in the structure, assuming the replacement $Si^{4+} \rightleftarrows 4H^+$.

Experimental studies at high pressures and temperatures show that the garnet structure is stable in the conditions of the earth's mantle. It appears that pyrope type compositions are most likely in the P and T regime of the earth's mantle.

Pyrope, Mg_3Al_2. Some Ca and Fe^{2+} usually present. Color deep red to nearly black. Often transparent and then used as a gem. Name derived from Greek, meaning *firelike. Rhodolite* is the name given to a pale rose-red or purple garnet, corresponding in composition to two parts of pyrope and one almandite.

Almandite, Fe_3Al_2. Fe^{3+} may replace Al, and Mg, Fe^{2+}. Color fine deep red, transparent in precious garnet; brownish red, translucent in common garnet. Name derived from Alabanda in Asia Minor where in ancient times garnets were cut and polished.

Spessartite, Mn_3Al_2. Fe^{2+} usually replaces some Mn^{2+} and Fe^{3+} some Al. Color brownish to red.

Grossularite, Ca_3Al_2 (*essonite, cinnamon stone*). Often contains Fe^{2+} replacing Ca, and Fe^{3+} replacing Al. Color, white, green, yellow, cinnamon-brown, pale red. Name derived from the botanical name for gooseberry, in allusion to the light green color of the original grossularite.

Andradite, $Ca_3Fe_2^{3+}$. Common garnet in part. Al may replace Fe^{3+}; Fe^{2+}, Mn^{2+} and Mg may replace Ca. Color various shades of yellow, green, brown to black. *Demantoid* is a green variety with a brilliant luster, used as a gem. Named after the Portuguese mineralogist, d'Andrada.

Uvarovite, $Ca_3Cr_2^{3+}$. Color emerald-green. Named after Count Uvarov.

Diagnostic Features. Garnets are usually recognized by their characteristic isometric crystals, their hardness, and their color. Specific gravity, refractive index, and unit cell dimension taken together serve to distinguish members of the group. With the exception of uvarovite, all garnets fuse at $3-3\frac{1}{2}$; uvarovite is almost infusible. The iron garnets, almandite, and andradite fuse to magnetic globules. Spessartite when fused with sodium carbonate gives a bluish-green bead (Mn). Uvarovite gives a green color to salt of phosphorous bead (Cr).

Occurrence. Garnet is a common and widely distributed mineral, occurring abundantly in some metamorphic rocks (see Fig. 10.11) and as an accessory constituent in some igneous rocks. Its most characteristic occurrence is in mica schists, hornblende schists, and gneisses. It is frequently used as an index mineral in the delineation of isograds in metamorphic rocks (see Chapter 11). Found in pegmatite dikes, more rarely in granitic rocks. *Pyrope*

FIG. 10.11. Garnet crystals in chlorite schist.

occurs in ultrabasic rocks such as peridotites or kimberlites and in serpentines derived from them. The garnets, which in eclogites coexist with pyroxenes and kyanite, vary in composition from pyrope to almandite. *Almandite* is the common garnet in metamorphic rocks, resulting from the regional metamorphism of argillaceous sediments. It is also a widespread detrital garnet in sedimentary rocks. *Spessartite* occurs in skarn deposits, and Mn-rich assemblages containing rhodonite, Mn-oxides, and so forth. *Grossularite* is found chiefly as a product of contact or regional metamorphism of impure limestones. *Andradite* occurs in geological environments similar to that of grossularite. It may be the result of metamorphism of impure siliceous limestone, by the reaction:

$$3CaCO_3 + SiO_2 + Fe_2O_3 \rightarrow$$

calcite   quartz   hematite

$$Ca_3Fe_2Si_3O_{12} + CO_2$$

andradite

Melanite, a black variety of andradite, occurs in alkaline igneous rocks. *Uvarovite* is the rarest of this group of garnets and is found in serpentine associated with chromite. The best known locality is Outokumpu, Finland.

Pyrope of gem quality is found associated with clear grains of olivine (peridot) in the surface sands near Fort Defiance, close to the Utah-Arizona line. A locality near Meronitz, Bohemia, is famous for pyrope gems. Almandite, of gem quality, is found in northern India, Sri Lanka, and Brazil. Fine crystals, although for the most part too opaque for cutting, are found in a mica schist on the Stikine River, Alaska. Grossularite is used only a little in jewelry, but essonite or cinnamon stones of good size and color are found in Sri Lanka.

Alteration. Garnet often alters to other minerals, particularly talc, serpentine, and chlorite.

Use. Chiefly as a rather inexpensive gemstone. A green andradite, known as *demantoid,* comes from the Ural Mountains, U.S.S.R., and yields fine gems known as *Uralian emeralds.* At Gore Mountain, New York, large crystals of almandite in an amphibolite are mined. The unusual angular fractures and high hardness of these garnets make them de-

sirable for a variety of abrasive purposes including garnet paper.

Name. *Garnet* is derived from the Latin *granatus,* meaning like a grain.

ZIRCON—ZrSiO₄

Crystallography. Tetragonal; $4/m2/m2/m$. Crystals usually show a simple combination of $a\{010\}$ and $p\{011\}$, but $m\{110\}$ and a ditetragonal dipyramid also observed (Fig. 10.12). Usually in crystals; also in irregular grains.

Angles: $a(100) \wedge p(101) = 47°50'$, $p(101) \wedge p'(\bar{1}01) = 84°20'$, $p(101) \wedge p'(011) = 56°41'$.

$I4/amd$; $a = 6.60$, c 5.98 Å; $a:c = 1:0.906$; $Z = 4$. $d's$: 4.41(7), 3.29(10), 2.52(8), 1.710(9).

Physical Properties. *Cleavage* $\{010\}$ poor. **H** $7\frac{1}{2}$. **G** 4.68. *Luster* adamantine. *Color* commonly some shade of brown; also colorless, gray, green, red. *Streak* uncolored. Usually translucent; in some cases transparent. *Optics:* (+), $\omega = 1.923–1.960$, $\epsilon = 1.968–2.015$. Metamict zircon, $n = 1.78$.

Composition and Structure. For ZrSiO₄, ZrO₂ 67.2, SiO₂ 22.8%. Zircon always contains some hafnium. Although the amount is usually small (1 to 4%) analyses have reported up to 24% HfO₂. The structure of zircon is shown in Fig. 10.13. Zr is in 8 coordination with oxygen in the form of distorted cubelike polyhedra. The eight oxygens surrounding each Zr belong to six different SiO₄ tetrahedra. Although the structure of zircon is resistant to normal chemical attack, it is often in a metamict state. This is caused by the structural damage from Th and U, which are present in small amounts in

FIG. 10.12. Zircon crystals.

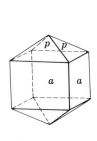
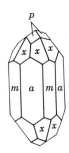
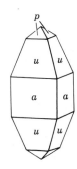

c

FIG. 10.13. Structure of zircon, ZrSiO₄, consisting of independent SiO₄ tetrahedra and distorted ZrO₈ cubes (redrawn from: *Polyhedral Structure Models* by T. Zoltai, University Minnesota).

many zircons. The "self irradiation" from the decay of U and Th produces as a final result an isotropic glass with a reduction in density of about 16% and a lowering of refractive index.

Diagnostic Features. Usually recognized by its characteristic crystals, color, luster, hardness, and high specific gravity. Infusible; insoluble.

Occurrence. Zircon is a common and widely distributed accessory mineral in all types of igneous rocks. It is especially frequent in the more silicic types such as granite, granodiorite, syenite, and monzonite, and very common in nepheline syenite. Found also in crystalline limestone, gneisses, and schists. Because zircon is a stable chemical compound, it is a common accessory mineral in many sediments. Its characteristics in the heavy mineral fraction of sandstones are often useful in the evaluation of provenance. Zircon is frequently found as rounded grains in stream and beach sands, often with gold. Zircon has been produced from beach sands in Australia, Brazil, and Florida.

Gem zircons are found in the stream sands at Matura, Sri Lanka, and in the gold gravels in the Ural Mountains and Australia. In large crystals from the Malagasy Republic. Found in the nepheline syenites of Norway. Found in the United States at Litchfield, Maine, and in Orange and St. Lawrence counties, New York; in considerable quantity in the sands of Henderson and Buncombe counties, North Carolina. Large crystals have been found in Renfrew County, Ontario, Canada.

Use. When transparent it serves as a gemstone. It is colorless in some specimens, but more often of a brownish or red-orange color. Blue is not a natural color for zircon but is produced by heat treatment. The colorless, yellowish, or smoky stones are called *jargon,* because although resembling diamond they have little value. Serves as the source of zirconium oxide, which is one of the most refractory substances known. Platinum, which fuses at 1755°C, can be melted in crucibles of zirconium oxide.

Overshadowing its other uses since 1945 is the use of zircon as the source of metallic zirconium. Pure zirconium metal is used in the construction of nuclear reactors. Its low neutron-absorption cross-section, coupled with retention of strength at high temperatures and excellent corrosion resistance, makes it a most desirable metal for this purpose.

Name. The name is very old and is believed to be derived from Persian words *zar,* gold, and *gun,* color.

Similar Species. *Thorite,* ThSiO₄, is isostructural with zircon. It is usually reddish-brown to black, hydrated, and radioactive.

Al₂SiO₅ Group

The three polymorphs of Al₂SiO₅ are: *andalusite,* Al$^{[5]}$Al$^{[6]}$SiO₅ (space group *Pnnm*), *sillimanite,* Al$^{[4]}$Al$^{[6]}$SiO₅ (space group *Pbnm*), and *kyanite,* Al$^{[6]}$Al$^{[6]}$SiO₅ (space group *P$\bar{1}$*). The structures of the three polymorphs have been discussed on page 343 and the andalusite and sillimanite structures are shown in Fig. 10.5. All three minerals may be found in metamorphosed aluminous rocks, such as pelitic schists. The stability relations of the three polymorphs have been determined experimentally as shown in Fig. 10.14. A knowledge of these stability fields is of great value in the study of regionally and contact metamorphosed terranes. Rocks of appropriate compositions tend to form sillimanite in high-temperature, regionally metamorphosed areas.

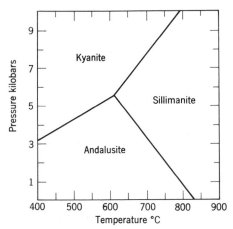

FIG. 10.14. Experimentally determined stability fields for the polymorphs of Al$_2$SiO$_5$.

Andalusite is frequently found in contact metamorphosed aureoles, and kyanite is located in metamorphic areas that have undergone considerable pressure. All three polymorphs are fairly easily recognized, in hand specimen, in medium- to coarse-grained rocks. The ease of recognition of these minerals and their presence in pelitic schists has led to their use as index minerals in the defining of metamorphic zones, as a function of temperature and pressure. Sillimanite can be used to define metamorphic zones that have been subjected to temperature of about 620°C and up (see Fig. 10.14). Kyanite is diagnostic of areas that have been subjected to a minimum pressure of about 3 kilobars (see Fig. 10.14). The metamorphic zones are named after the index mineral present, for example, sillimanite zone, kyanite zone. The boundaries between different zones are referred to as *isograds* (see page 470). It should be noted that kyanite, which has **G** = 3.55 to 3.66 (as compared to **G** = 3.23 for sillimanite and **G** = 3.16–3.20 for andalusite) has the densest structure of the three polymorphs and its stability field is located in the direction of increasing pressure (Fig. 10.14).

ANDALUSITE—Al$_2$SiO$_5$

Crystallography. Orthorhombic; $2/m2/m2/m$. Usually occurs in coarse, nearly square prisms terminated by {001}.

Angles: $(110) \wedge (1\bar{1}0) = 89°12'$.

Pnnm; $a = 7.78$, $b = 7.92$, $c = 5.57$ Å; $a:b:c = 0.982:1:0.703$; $Z = 4$. $d's$: 4.53(10), 3.96(8), 2.76(9), 2.17(10), 1.46(10).

Physical Properties. **H** $7\frac{1}{2}$. **G** 3.16–3.20. *Luster* vitreous. *Color* flesh-red, reddish-brown, olive-green. The variety *chiastolite* has dark-colored carbonaceous inclusions arranged in a regular manner forming a cruciform design (Fig. 10.15). Transparent to translucent. *Optics:* (−), $\alpha = 1.632$, $\beta = 1.638$, $\gamma = 1.643$; $2V = 85°$; $X = c$, $Z = a$. In some crystals strong pleochroism; X red, Y and Z green to colorless; $r > v$.

Composition and Structure. Al$_2$O$_3$ 63.2, SiO$_2$ 36.8%. The structure consists of chains of AlO$_6$ octahedra parallel to c cross-linked by SiO$_4$ tetrahedra and AlO$_5$ polyhedra (see Fig. 10.5b).

Diagnostic Features. Characterized by the nearly square prism and hardness. Chiastolite is readily recognized by the symmetrically arranged inclusions. Infusible. Insoluble. When fine powder is made into a paste with cobalt nitrate and intensely ignited it turns blue (Al).

Alteration. Pseudomorphs of fine-grained muscovite (sericite) after andalusite are common.

Occurrence. Andalusite is formed typically in the contact aureoles of igneous intrusions in argillaceous rocks. Here it commonly coexists with cordierite. Andalusite can be found in association with kyanite, or sillimanite, or both in regionally metamorphosed terrane. Such occurrences may reflect

FIG. 10.15. Successive cross-sections through a chiastolite crystal.

variations in P and T during metamorphism, as well as the sluggishness of the reactions in the system Al_2SiO_5.

Notable localities are in Andalusia, Spain; the Austrian Tyrol; in water-worn pebbles from Minas Gerais, Brazil. Crystals of chiastolite are found at Bimbowrie, South Australia. In the United States found in the White Mountains near Laws, California; at Standish, Maine; and Delaware County, Pennsylvania. Chiastolite is found at Westford, Lancaster, and Sterling, Massachusetts.

Use. Andalusite has been mined in large quantities in California for use in the manufacture of spark plugs and other porcelains of a highly refractory nature. When clear and transparent it may serve as a gemstone.

Name. From Andalusia, a province of Spain.

SILLIMANITE—Al_2SiO_5

Crystallography. Orthorhombic; $2/m2/m2/m$. Occurs in long, slender crystals without distinct terminations; often in parallel groups; frequently fibrous and called *fibrolite*.

$Pbnm$; $a = 7.44$; $b = 7.60$; $c = 5.75$ Å; $a:b:c = 0.979:1:0.757$; $Z = 4$. d's: 3.32(10), 2.49(7), 2.16(8), 1.677(7), 1.579(7).

Physical Properties. *Cleavage* {010} perfect. **H** 6–7. **G** 3.23. *Luster* vitreous. *Color* brown, pale green, white. Transparent to translucent. *Optics:* (+), $\alpha = 1.657$, $\beta = 1.658$, $\gamma = 1.677$; 2V = 20°, $X = b$, $Z = c$; $r > v$.

Composition and Structure. Oxide components as in andalusite. The structure consists of AlO_6 chains parallel to the c axis and linking SiO_4 and AlO_4 tetrahedra which alternate along the c direction (see Fig. 10.5a).

Diagnostic Features. Characterized by slender crystals with one direction of cleavage. Infusible. Insoluble. The finely ground mineral turns blue when heated with cobalt nitrate solution.

Occurrence. Sillimanite occurs as a constituent of high-temperature metamorphosed argillaceous rocks. In contact metamorphosed rocks it may occur in sillimanite-cordierite gneisses or sillimanite-biotite hornfels. In regionally metamor-

phosed rocks it is found, for example, in quartz-muscovite-biotite-oligoclase-almandite-sillimanite schists (see also Table 11.9). In silica-poor rocks it may be associated with corundum.

Notable localities for its occurrence are Maldau, Bohemia; Fassa, Austrian Tyrol; Bodenmais, Bavaria; Freiberg, Saxony; and waterworn masses in diamantiferous sands of Minas Gerais, Brazil. In the United States found at Worcester, Massachusetts; at Norwich and Willimantic, Connecticut; and New Hampshire.

Name. In honor of Benjamin Silliman (1779–1864), professor of chemistry at Yale University.

Similar Species. *Mullite,* a nonstoichiometric compound of approximate composition $Al_6Si_3O_{15}$, is rare as a mineral but common in artificial Al_2O_3-SiO_2 systems at high temperature. *Dumortierite,* $Al_7O_3(BO_3)(SiO_4)_3$, has been used in the manufacture of high-grade porcelain.

KYANITE—Al_2SiO_5

Crystallography. Triclinic; $\bar{1}$. Usually in long, tabular crystals, rarely terminated. In bladed aggregates.

$P\bar{1}$; $a = 7.10$, $b = 7.74$, $c = 5.57$ Å; $\alpha = 90°6'$, $\beta = 101°2'$, $\gamma = 105°45'$; $a:b:c = 0.899:1:0.709$; $Z = 4$. d's: 3.18(10), 2.52(4), 2.35(4), 1.93(5), 1.37(8).

Physical Properties. *Cleavage* {100} perfect. **H** 5 parallel to length of crystals, 7 at right angles to this direction. **G** 3.55–3.66. *Luster* vitreous to pearly. *Color* usually blue, often of darker shade toward the center of the crystal. Also, in some cases, white, gray, or green. Color may be in irregular streaks and patches. *Optics:* (−); $\alpha = 1.712$, $\beta = 1.720$, $\gamma = 1.728$; 2V = 82°; $r > v$.

Composition and Structure. Oxide components as in andalusite. The structure consists of AlO_6 octahedral chains parallel to c and AlO_6 octahedra and SiO_4 tetrahedra between the chains.

Diagnostic Features. Characterized by its bladed crystals, good cleavage, blue color, and different hardness in different directions. Infusible. Insoluble. A fragment moistened with cobalt nitrate solution and ignited assumes a blue color.

FIG. 10.16. Bladed kyanite crystals and prismatic anda-lusite (dark) in mica schist, St. Gothard, Switzerland.

Occurrence. Kyanite is typically a result of regional metamorphism of pelitic rocks and is often associated with garnet, staurolite, and corundum. It also occurs in some eclogites (rocks with omphacite-type pyroxenes associated with pyrope-rich garnets) and in garnet-omphacite-kyanite occurrences in kimberlite pipes. Both of these rock types reflect high to very high pressures of origin. Crystals of exceptional quality are found at St. Gothard, Switzerland (Fig. 10.16); in the Austrian Tyrol; and at Pontivy and Morbihan, France. Commercial deposits are located in India, Kenya, and in the United States in North Carolina and Georgia.

Use. Kyanite is used as is andalusite in the manufacture of spark plugs and other highly refractory porcelains.

Name. Derived from a Greek word meaning *blue*.

TOPAZ—$Al_2SiO_4(F, OH)_2$

Crystallography. Orthorhombic, $2/m2/m2/m$. Commonly in prismatic crystals terminated by dipyramids, first- and second-order prisms and basal pinacoid (Fig. 10.17). Vertical prism faces frequently striated. Usually in crystals but also in crystalline masses; granular, coarse, or fine.

Angles: $m(110) \wedge m'(\bar{1}10) = 55°43'$, $l(120) \wedge l'(1\bar{2}0) = 93°08'$, $c(001) \wedge f(011) = 43°39'$, $c(001) \wedge o(111) = 63°54'$.

Pbnm; $a = 4.65$, $b = 8.80$, $c = 8.40$ Å; $a:b:c = 0.528:1:0.955$; $Z = 4$. *d's:* 3.20(9), 2.96(10), 2.07(9), 1.65(9), 1.403(10).

Physical Properties. *Cleavage* {001} perfect. **H** 8. **G** 3.4–3.6. *Luster* vitreous. *Color* colorless, yellow, pink, wine-yellow, bluish, greenish. Transparent to translucent. *Optics:* (+); $\alpha = 1.606–1.629$, $\beta = 1.609–1.631$, $\gamma = 1.616–1.638$; $2V = 48°–68°$; $X = a$, $Y = b$; $r > v$.

Composition and Structure. $Al_2SiO_4(OH)_2$ contains Al_2O_3 56.6, SiO_2 33.4 and H_2O 10.0%. Most of the $(OH)^-$ is generally replaced by F^-, the maximum fluorine content being 20.7%. The structure of topaz consists of chains parallel to the c axis of AlO_4F_2 octahedra which are cross-linked by independent SiO_4 tetrahedra. This arrangement is morphologically expressed by the prismatic [001] habit. The perfect {001} cleavage of topaz passes through the structure without breaking Si-O bonds, only Al-O and Al-F bonds are broken. The structure is relatively dense because it is based on closest packing of oxygens and fluorine; this packing scheme is neither cubic nor hexagonal, but more complex, *ABAC*.

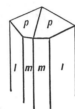

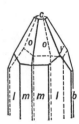
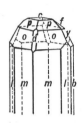

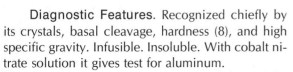

FIG. 10.17. Topaz crystals.

Diagnostic Features. Recognized chiefly by its crystals, basal cleavage, hardness (8), and high specific gravity. Infusible. Insoluble. With cobalt nitrate solution it gives test for aluminum.

Occurrence. Topaz is a mineral formed by fluorine-bearing vapors given off during the last stages of the solidification of siliceous igneous rocks. Found in cavities in rhyolite lavas and granite; a characteristic mineral in pegmatites, especially in those carrying tin. Associated with tourmaline, cassiterite, apatite, and fluorite; also with beryl, quartz, mica, and feldspar. Found in some localities as rolled pebbles in stream sands.

Notable localities for its occurrence are in the U.S.S.R. in the Nerchinsk district of Siberia in large wine-yellow crystals, and in Mursinsk, Ural Mountains, in pale blue crystals; in Saxony from various tin localities; Omi and Mino provinces, Japan; and San Luis Potosí, Mexico. Minas Gerais, Brazil, has long been the principal source of yellow gem-quality topaz. In the 1940s several well-formed, large crystals of colorless topaz were found there; the largest weighs 596 pounds. In the United States found at Pikes Peak, near Florissant and Nathrop, Colorado; Thomas Range, Utah; Streeter, Texas; San Diego County, California; Stoneham and Topsham, Maine; Amelia, Virginia; and Jefferson, South Carolina.

Use. As a gemstone. Frequently sold as "precious topaz" to distinguish it from citrine quartz, commonly called topaz. The color of the stones varies, being colorless, wine-yellow, golden brown, pale blue, and pink. The pink color is usually artificial, being produced by gently heating the dark yellow stones.

Name. Derived from Topazion, the name of an island in the Red Sea, but originally probably applied to some other species.

Similar Species. *Danburite,* $Ca(B_2Si_2O_8)$, a member of the tectosilicates, shows crystal form and physical properties that are very similar to those of topaz. It can be distinguished from topaz by testing for boron.

STAUROLITE—$Fe_2^{2+}Al_9O_6(SiO_4)_4(O,OH)_2$

Crystallography. Monoclinic; $2/m$ (pseudoorthorhombic). Prismatic crystals with common forms $\{110\}$, $\{010\}$, $\{001\}$, and $\{101\}$ (Fig. 10.18a). Cruciform twins very common, of two types: (1) with twin plane $\{031\}$ in which the two individuals cross at nearly 90° (Fig. 10.18b); (2) with twin plane $\{231\}$ in which they cross at nearly 60° (Fig. 10.18c). In some cases both types are combined in one twin group. Usually in crystals; rarely massive.

Angles: $m(110) \wedge m'(1\bar{1}0) = 50°40'$, $c(001) \wedge r(201) = 55°16'$.

$C2/m; a = 7.83$, $b = 16.62$, $c = 5.65$ Å, $\beta = 90°$; $a:b:c = 0.471:1:0.340$; $Z = 4$. $d's$: 3.01(8), 2.38(10), 1.974(9), 1.516(5), 1.396(10).

Physical Properties. H $7-7\frac{1}{2}$. G 3.65–3.75. *Luster* resinous to vitreous when fresh; dull to earthy when altered or impure. *Color* red-brown to brownish-black. Translucent. *Optics:* (+), $\alpha = 1.739-1.747$, $\beta = 1.745-1.753$, $\gamma = 1.752-1.761$; $2V = 82°-88°$; $X = b$ (colorless), $Y = a$ (pale yellow), $Z = c$ (deep yellow); $r > v$.

Composition and Structure. FeO 16.7, Al_2O_3 53.3, SiO_2 27.9, H_2O 2.0% for the pure Fe end member; however, Mg, replacing Fe^{2+}, and Fe^{3+}, replacing Al, are generally both present in amounts of a few percent. The structure closely resembles that of kyanite with layers of $4Al_2SiO_5$ composition (AlO_6 octahedra in chains parallel to the c axis) alternating with layers of $Fe_2AlO_3(OH)$ composition. The ideal formula, on the basis of 24(O, OH),

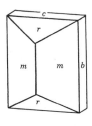 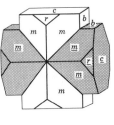 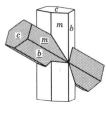

FIG. 10.18. Staurolite.　　*(a)*　　　　　　*(b)*　　　　　　*(c)*

should contain one OH, however, this OH content is generally exceeded.

Diagnostic Features. Recognized by its characteristic crystals and twins. Distinguished from andalusite by its obtuse prism. Infusible. Insoluble. On intense ignition in the closed tube it yields a little water.

Occurrence. Staurolite is formed during regional metamorphism of aluminum-rich rocks and is found in schists and gneisses. Often associated with almandite garnet and kyanite in high-grade metamorphic rocks. May grow on kyanite in parallel orientation. In somewhat lower grade metamorphic rocks it may occur with chloritoid. It is commonly used as an index mineral in medium grade metamorphic rocks (see Chapter 11).

Notable localities are Monte Campione, Switzerland; Goldenstein, Moravia; Aschaffenburg, Bavaria; and in large twin crystals in Brittany and Scotland. In the United States found at Windham, Maine; Franconia and Lisbon, New Hampshire; Chesterfield, Massachusetts; also in North Carolina, Georgia, Tennessee, Virginia, New Mexico, and Montana.

Use. Occasionally a transparent stone from Brazil is cut as a gem. In North Carolina the right angle twins are sold as amulets under the name "fairy stone," but most of the crosses offered for sale are imitations carved from a fine-grained rock and dyed.

Name. Derived from a Greek word meaning *cross*, in allusion to its cruciform twins.

Humite Group

The humite group includes the following four members:

Norbergite	$Mg_3(SiO_4)(F,OH)_2$
Chondrodite	$Mg_5(SiO_4)_2(F,OH)_2$
Humite	$Mg_7(SiO_4)_3(F,OH)_2$
Clinohumite	$Mg_9(SiO_4)_4(F,OH)_2$

All four species have similar structures, chemistry, and physical properties. The structure of the members of the humite group is closely related to that of olivine with layers parallel to {100} which have the atomic arrangement of olivine, and alternating layers of $Mg(OH,F)_2$ composition. For example, the norbergite composition can be rewritten to express this layering as $Mg_2SiO_4 \cdot Mg(OH,F)_2$. The replacement of F by OH is extensive but hydroxyl end members are unknown. The members of this group are rather restricted in their occurrence and are found mainly in metamorphosed and metasomatised limestones and dolomites, and skarn associated with ore deposits. Here we will discuss in detail only chondrodite, the most common member of the humite group.

Chondrodite—$Mg_5(SiO_4)_2(F,OH)_2$

Crystallography. Monoclinic; $2/m$. Crystals are frequently complex with many forms. Usually in isolated grains. Also massive.

$P2_1/c$; $a = 7.89$, $b = 4.74$, $c = 10.29$, $\beta = 109°$; $a:b:c = 1.665:1:2.171$; $Z = 2$. $d's$: 3.02(5), 2.76(4), 2.51(5), 2.26(10), 1.740(7).

Physical Properties. H $6–6\frac{1}{2}$. G 3.1–3.2. *Luster* vitreous to resinous. *Color* light yellow to red. Translucent. *Optics:* (+); $\alpha = 1.592–1.615$, $\beta = 1.602–1.627$, $\gamma = 1.621–1.646$; $2V = 71°–85°$; $Z = b$, $X \wedge c = 25°$; $r > v$.

Composition and Structure. The composition of chondrodite can be rewritten as $2Mg_2SiO_4 \cdot Mg(OH,F)_2$ to reflect the olivine-type and brucite-type layering in the structure. The two main substitutions are Mg by Fe^{2+} and F by OH. The Fe substitution is limited to about 6 weight percent FeO.

Diagnostic Features. Characterized by its light yellow to red color and its mineral associations in crystalline limestone. The members of the humite group cannot be distinguished from one another without optical tests. Infusible. Yields water in the closed tube.

Occurrence. Chondrodite occurs most commonly in metamorphosed dolomitic limestones. The mineral association including phlogopite, spinel, pyrrhotite, and graphite is highly characteristic. In skarn deposits it is found with wollastonite, forsterite, and monticellite. Noteworthy localities of chondrodite are Monte Somma, Italy; Paragas, Finland; and Kafveltorp, Sweden. In the United States it is found abundantly at the Tilly Foster magnetite deposit near Brewster, New York.

Name. Chondrodite is from the Greek meaning *a grain;* alluding to its occurrence as isolated grains. Humite is named in honor of Sir Abraham Hume.

DATOLITE—CaB(SiO₄)(OH)

Crystallography. Monoclinic; $2/m$. Crystals usually nearly equidimensional in the three axial directions and often complex in development (Fig. 10.19). Usually in crystals. Also coarse to fine granular. Compact and massive, resembling unglazed porcelain.

Angles: $m(110) \wedge m'(1\bar{1}0) = 64°47'$, $n(111) \wedge n'(1\bar{1}1) = 59°5'$, $M(011) \wedge M'(0\bar{1}1) = 103°23'$, $e(\bar{1}12) \wedge e'(\bar{1}\bar{1}2) = 48°20'$.

FIG. 10.19. Datolite.

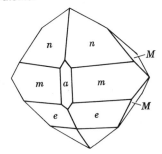

$P2_1/a$; $a = 4.83$, $b = 7.64$, $c = 9.66$, $\beta = 90°9'$; $a:b:c = 0.632:1:1.264$; $Z = 4$. $d's$: 3.76(5), 3.40(3), 3.11(10), 2.86(7), 2.19(6).

Physical Properties. **H** $5-5\frac{1}{2}$. **G** 2.8–3.0. *Luster* vitreous. *Color* white, often with faint greenish tinge. Transparent to translucent. *Optics:* $(-)$; $\alpha = 1.624$, $\beta = 1.652$, $\gamma = 1.688$; $2V = 74°$; $Y = b, Z = c; r > v$.

Composition and Structure. CaO 35.0, B_2O_3 21.8, SiO_2 37.6, H_2O 5.6%. The structure can be regarded as consisting of layers parallel to {100} of independent SiO_4 and $B(O,OH)_4$ tetrahedra. The layers are bonded by Ca in 8 coordination (6 oxygens and 2 OH groups). Despite the layerlike structure datolite has no conspicuous cleavage.

Diagnostic Features. Characterized by its glassy luster, pale green color, and its crystals with many faces. Fuses at $2-2\frac{1}{2}$ to a clear glass and colors the flame green (B). It is thus distinguished from quartz.

Occurrence. Datolite is a secondary mineral found usually in cavities in basalt lavas and similar rocks. Associated with zeolites, prehnite, apophyllite, and calcite. Notable foreign localities are Andreasberg, Harz Mountains; in Italy near Bologna; from Seiser Alpe and Theiso, Trentino; and Arendal, Norway. In the United States it occurs in the trap rocks of Massachusetts, Connecticut, and New Jersey, particularly at Westfield, Massachusetts, and Bergen Hill, New Jersey. Found associated with the copper deposits of Lake Superior.

Name. Derived from a Greek word meaning *to divide,* in allusion to the granular character of a massive variety.

SPHENE (*titanite*)—CaTiO(SiO₄)

Crystallography. Monoclinic; $2/m$. Wedge-shaped crystals common resulting from a combination of {001}, {110}, and {111} (Fig. 10.20). May be lamellar or massive.

Angles: $m(110) \wedge m(1\bar{1}0) = 66°29'$, $p(111) \wedge p'(1\bar{1}1)$ 43°29', $c(001) \wedge p(111) = 38°16'$.

$C2/c$, $a = 6.56$, $b = 8.72$, $c = 7.44$ Å, $\beta = 119°43'$; $a:b:c = 0.752:1:0.853$; $Z = 4$. $d's$: 3.23(10), 2.99(9), 2.60(9), 2.27(3), 2.06(4).

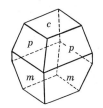 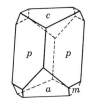

FIG. 10.20. Sphene crystals.

Physical Properties. *Cleavage* {110} distinct. Parting on {100} may be present. **H** $5-5\frac{1}{2}$. **G** 3.4–3.55. *Luster* resinous to adamantine. *Color* gray, brown, green, yellow, black. Transparent to translucent. *Optics:* (+); $\alpha = 1.900$, $\beta = 1.907$, $\gamma = 2.034$; $2V = 27°$; $Y = b$, $Z \wedge c = 51°$; $r > v$.

Composition and Structure. CaO 28.6, TiO_2 40.8, SiO_2 30.6%. Small amounts of rare earths, Fe, Al, Mn, Mg, and Zr may be present. The structure of sphene consists of independent SiO_4 tetrahedra and of CaO_7 and TiO_6 polyhedra. The polyhedra share edges and corners with each other. The SiO_4 tetrahedra share one edge with CaO_7 polyhedra and TiO_6 octahedra share four edges with CaO_7 polyhedra.

Diagnostic Features. Characterized by its wedge-shaped crystals and high luster. Hardness is less than that of staurolite and greater than that of sphalerite. Fusible at 4.

Occurrence. Sphene in small crystals is a common accessory mineral in granites, granodiorites, diorites, syenites, and nepheline syenites. Found in crystals of considerable size in metamorphic gneisses, chlorite schists, and crystalline limestone. Also found with iron ores, pyroxene, amphibole, scapolite, zircon, apatite, feldspar, and quartz.

The most notable occurrence of sphene is on the Kola Peninsula, U.S.S.R., where it is associated with apatite and nepheline syenite. It is mined there as a source of titanium. It is found in crystals at Tavetsch, Binnental, and St. Gothard, Switzerland; Zillertal, Tyrol; Ala, Piedmont; Vesuvius; and Arendal, Norway. In the United States in Diana, Rossie, Fine, Pitcairn, Edenville, and Brewster, New York; and Riverside, California. Also in various places in Ontario and Quebec, Canada.

Use. As a source of titanium oxide for use as a paint pigment. A minor gemstone.

Name. *Sphene* comes from a Greek word meaning *wedge*, in allusion to the wedge-shaped crystals.

Similar Species. *Benitoite*, $BaTiSi_3O_9$, a cyclosilicate occurs in association with *neptunite*, $KNa_2Li(Fe, Mn)_2Ti_2O(Si_4O_{11})_2$, a complex inosilicate, in San Benito, California. *Astrophyllite, aenigmatite, lamprophyllite, ramsayite,* and *fersmannite* are rare Ti-bearing silicates found associated with alkalic rocks.

Chloritoid—$(Fe, Mg)Al_4O_2(SiO_4)_2(OH)_4$

Crystallography. Generally monoclinic, $2/m$; may also be triclinic. Seldom in distinct tabular crystals, usually coarsely foliated, massive. Also in thin scales or small plates.

$C2/c$; $a = 9.52$, $b = 5.47$, $c = 18.19$; $\beta = 101°39'$; $a:b:c = 1.740:1:3.325$; $Z = 4$. *d's:* 4.50(10), 4.45(10), 3.25(6), 2.97(8), 2.70(7), 2.46(9).

Physical Properties. *Cleavage* {001} good (less so than in micas), producing brittle flakes. **H** $6\frac{1}{2}$. (much harder than chlorite). **G** 3.5–3.8. *Luster* pearly. *Color* dark green, greenish grey, often grass-green in very thin plates. *Streak* colorless. *Optics:* (+); $\alpha = 1.713-1.730$, $\beta = 1.719-1.734$, $\gamma = 1.723-1.740$; $2V = 36°-60°$; $X = b$, $Z \wedge c = 2-30°$. Pleochroism strong: Z yellow green, Y indigo blue, X olive green.

Composition and Structure. For the Fe^{2+} end member, SiO_2 23.8, Al_2O_3 40.5, FeO 28.5, and H_2O 7.2%. All chloritoids show some Mg replacing Fe^{2+} but this substitution is limited. Fe^{3+} substitutes to some extent for Al. The structure of chloritoid can be described as consisting of two close-packed octahedral layers; one of a brucite-type sheet with composition $(Fe, Mg)_2AlO_2(OH)_4$ and another of a corundum-type composition, Al_2O_3. These sheets alternate in a direction parallel to the c axis. Independent SiO_4 tetrahedra link the brucite- and corundum-like sheets together. The SiO_4 tetrahedra, therefore, do not occur in continuous sheets, as in the micas and chlorite. Synthetic studies show that chloritoid gives way to staurolite at elevated temperature, according to the reaction: chloritoid + andalusite → staurolite + quartz + fluid.

Diagnostic Features. Chloritoid is often difficult to distinguish from chlorite, with which it is generally associated. Optical study is necessary for unambiguous identification.

Occurrence. Chloritoid is a relatively common constituent of low- to medium-grade regionally metamorphosed iron-rich pelitic rocks. It generally occurs as porphyroblasts in association with muscovite, chlorite, staurolite, garnet and kyanite. The original chloritoid was described from Kosoibrod, in the Ural Mountains. It is most commonly found in greenschist facies rocks; it is not uncommon in Vermont, for example, and northern Michigan.

Name. Chloritoid is named from its superficial resemblance to chlorite.

SOROSILICATES

The sorosilicates are characterized by isolated, double tetrahedral groups formed by two SiO_4 tetrahedra sharing a single apical oxygen (Fig. 10.21). The resulting ratio of silicon to oxygen is 2:7. Over 70 minerals are known in this group but most of them are rare. We will discuss only the following six species, of which members of the *epidote group* and *idocrase* are most important.

SOROSILICATES

Hemimorphite	$Zn_4(Si_2O_7)(OH)_2 \cdot H_2O$
Lawsonite	$CaAl_2(Si_2O_7)(OH)_2 \cdot H_2O$
Epidote Group	
Clinozoisite	$Ca_2Al_3O(SiO_4)(Si_2O_7)(OH)$
Epidote	$Ca(Fe^{3+}, Al)Al_2O(SiO_4)(Si_2O_7)(OH)$
Allanite	$X_2Y_3O(SiO_4)(Si_2O_7)(OH)$
Idocrase	$Ca_{10}(Mg, Fe)_2Al_4(SiO_4)_5(Si_2O_7)_2(OH)_4$

The structure of epidote (Fig. 10.22) contains both independent SiO_4 tetrahedra as well as Si_2O_7 groups. Chains of AlO_6 and $AlO_4(OH)_2$ octahedra, which share edges, run parallel to the *b* axis. (Similar octahedral chains are also present in the three polymorphs of Al_2SiO_5 and in staurolite.) An addi-

FIG. 10.21. Close packing representation of Si_2O_7 group.

tional octahedral position occurs outside the chains; this site is occupied by Al in clinozoisite and by Fe^{3+} and Al in epidote. The chains are linked by independent SiO_4 and Si_2O_7 groups. Ca is in irregular 8 coordination with oxygen. The site in which Ca^{2+} is housed may in part be filled by Na^+. The octahedral site outside the chains may house, in addition to Al, and Fe^{3+}, Mn^{3+} and more rarely Mn^{2+}.

All members of the epidote group are isostructural and form monoclinic crystals characteristically elongate on *b*. Orthorhombic *zoisite* has a structure that may be derived from that of its monoclinic polymorph, *clinozoisite*, by a twinlike doubling of the cell along the *a* axis, such that *a* of zoisite = 2 *a* sin β of clinozoisite (or epidote) (see Fig. 10.23). The structure of clinozoisite is the same as that of epidote, with all octahedral positions occupied by Al. *Allanite* may be derived from the epidote structure by replacing some of the Ca^{2+} by rare earths and adjusting the electrostatic charge balance by replacing some of the Fe^{3+} by Fe^{2+}. As a result of the very similar structures of the members of the epidote group, the various ionic substitutions provide the major variables. These are:

	Ions in Ca Site	Ions in Al Site Outside Chains
Clinozoisite	Ca	Al
Epidote	Ca	Fe^{3+}, Al
Piemontite	Ca	Mn^{3+}, Fe^{3+}, Al
Allanite	Ca, Ce, La, Na	Fe^{3+}, Fe^{2+}, Mg, Al

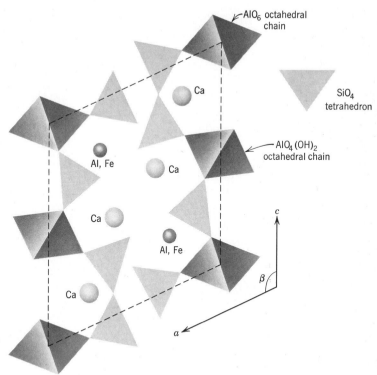

FIG. 10.22. Schematic representation of the structure of epidote, projected on (010). Dashed lines outline unit cell.

HEMIMORPHITE—$Zn_4(Si_2O_7)(OH)_2 \cdot H_2O$

Crystallography. Orthorhombic; $mm2$. Crystals usually tabular parallel to {010}. They show prism faces and are terminated above usually by a combination of domes and pedion, and below by a pyramid (Fig. 10.24), forming polar crystals. Crystals often divergent, giving rounded groups with slight reentrant notches between the individual crystals, forming knuckle or coxcomb masses. Also mammillary, stalactitic, massive, and granular.

Angles: $b(010) \wedge m(110) = 51°55'$, $c(001) \wedge i(031) = 55°06'$, $c(001) \wedge t(301) = 61°20'$.

$Imm2$, $a = 8.38$, $b = 10.72$, $c = 5.12$ Å; $a:b:c = 0.782:1:0.478$. $Z = 2$. $d's$: 6.60(9), 5.36(6), 3.30(8), 3.10(10), 2.56(5).

Physical Properties. *Cleavage* {110} perfect.

FIG. 10.23. Relation between unit cells of monoclinic epidote and orthorhombic zoisite, as projected on (010). The monoclinic unit cell outlined in solid can be related to its dashed equivalent by a mirror reflection (twinning). The orthorhombic unit cell is shown by shading.

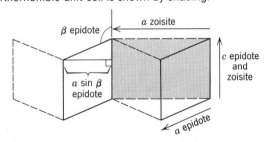

FIG. 10.24. Hemimorphite.

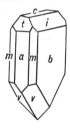

H $4\frac{1}{2}$–5. G 3.4–3.5. *Luster* vitreous. *Color,* white, in some cases with faint bluish or greenish shade; also yellow to brown. Transparent to translucent. Strongly pyroelectric and piezoelectric. *Optics:* (+); $\alpha = 1.614$, $\beta = 1.617$, $\gamma = 1.636$; 2V = 46°; $X = b$, $Z = c$; $r > v$.

Composition and Structure. ZnO 67.5, SiO_2 25.0, H_2O 7.5%. Small amounts of Fe and Al may be present. The structure contains Si_2O_7 groups linked by $ZnO_3(OH)$ tetrahedra. The tetrahedra in the Si_2O_7 groups have their bases parallel to (001) and their apices all point the same way along the c direction. This orientation causes the polar character of the structure. Each (OH) group is bound to two Zn^{2+} ions. H_2O molecules lie in holes between the tetrahedra. The H_2O molecules are lost continuously by heating hemimorphite up to 500°C. At this temperature all H_2O molecules are driven off and only (OH) groups are retained. These can be driven off only at much higher temperature with destruction of the crystal structure.

Diagnostic Features. Characterized by the grouping of crystals. Resembles prehnite but has a higher specific gravity. Fusible with difficulty at 5. When fused on charcoal with sodium carbonate it gives a nonvolatile coating of zinc oxide (yellow when hot, white when cold). When fused with cobalt nitrate on charcoal it turns blue. Gives water in the closed tube.

Occurrence. Hemimorphite is a secondary mineral found in the oxidized portions of zinc deposits, associated with smithsonite, sphalerite, cerussite, anglesite, and galena.

Notable localities for its occurrence are at Moresnet, Belgium; Aix la Chapelle, Germany; Carinthia, Rumania; Sardinia; Cumberland and Derbyshire, England; Algeria; and Chihuahua, Mexico. In the United States it is found at Sterling Hill, Ogdensburg, New Jersey; Friedensville, Pennsylvania; Wythe County, Virginia; with the zinc deposits of southwestern Missouri; Leadville, Colorado; Organ Mountains, New Mexico; and Elkhorn Mountains, Montana.

Use. An ore of zinc.

Name. From the hemimorphic character of the crystals. The mineral was formerly called *calamine*.

Lawsonite—$CaAl_2(Si_2O_7)(OH)_2 \cdot H_2O$

Crystallography. Orthorhombic; 222. Usually in tabular or prismatic crystals. Frequently twinned polysynthetically on {110}.

$C222_1$; $a = 8.79$, $b = 5.84$, $c = 13.12$ Å; $a:b:c = 1.505:1:2.247$; $Z = 4$. *d's:* 4.17(5), 3.65(6), 2.72(10), 2.62(7), 2.13(7).

Physical Properties. *Cleavage* {010} and {110} good. **H** 8. **G** 3.09. *Color* colorless, pale blue to bluish gray. *Luster* vitreous to greasy. Translucent. *Optics:* (+); $\alpha = 1.665$, $\beta = 1.674$, $\gamma = 1.684$; 2V = 84°; $X = a$, $Z = c$; $r > v$.

Composition and Structure. The composition of lawsonite is the same as that of anorthite, $CaAl_2Si_2O_8$, +H_2O. The structure consists of (AlO, OH) octahedra linked by Si_2O_7 groups. Ca^{2+} and H_2O molecules are located between these polyhedra.

Diagnostic Features. Lawsonite is characterized by its high hardness. It fuses to a blebby glass but once fused is not again fusible. On intense ignition in the closed tube it yields water.

Occurrence. Lawsonite is a typical mineral of the glaucophane schist facies associated with chlorite, epidote, sphene, glaucophane, garnet, and quartz. The type locality is on the Tiburon Peninsula, San Francisco Bay, California. It is a common constituent of gneisses and schists formed under low temperature and high pressure.

Name. In honor of Professor Andrew Lawson of the University of California.

Similar Species. *Ilvaite,* $CaFe_2^{2+}Fe^{3+}O$ $(Si_2O_7)(OH)$, is related to lawsonite with a similar although not identical structure.

Epidote Group

The structure and crystal chemistry of the epidote group are discussed on page 358 (see also Figs. 10.22 and 10.23).

CLINOZOISITE—$Ca_2Al_3O(SiO_4)(Si_2O_7)(OH)$
EPIDOTE—$Ca_2(Al, Fe)Al_2O(SiO_4)(Si_2O_7)(OH)$

Crystallography. Monoclinic; 2/m. Crystals are usually elongated parallel to b with a prominent

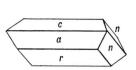 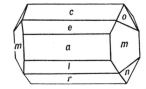

FIG. 10.25. Epidote crystals.

development of the faces of the [010] zone, giving them a prismatic aspect. Striated parallel to b (Fig. 10.25). Twinning on {100} common. Usually coarse to fine granular; also fibrous.

Angles: $c(001) \wedge a(100) = 64°36'$, $n(\bar{1}11) \wedge n'(11\bar{1}) = 70°29'$, $a(100) \wedge m(110) = 55°0'$, $c(001) \wedge o(011) = 58°28'$.

$P2_1/m$; $a = 8.98$, $b = 5.64$, $c = 10.22$ Å, $\beta = 115°24'$, $a:b:c = 1.592:1:1.812$; $Z = 2$. $d's$: 5.02(4), 2.90(10), 2.86(6), 2.53(6), 2.40(7).

Physical Properties. *Cleavage* {001} perfect and {100} imperfect. **H** 6–7. **G** 3.25–3.45. *Luster* vitreous. *Color* epidote: pistachio-green to yellowish-green to black; clinozoisite: pale green to gray. Transparent to translucent. *Optics.* Refractive indices and birefringence increase with iron content. Clinozoisite: (+): $\alpha = 1.670$–1.715, $\beta = 1.674$–1.725, $\gamma = 1.690$–1.734; $2V = 14°$–$90°$; $Y = b$, $X \wedge c\ -2°$ to $-7°$; $r < v$. Epidote: (−), $\alpha = 1.715$–1.751, $\beta = 1.725$–1.784, $\gamma = 1.734$–1.797; $2V = 64°$–$90°$; $Y = b$, $X \wedge c = 1$ to $-5°$; $r > v$. Absorption $Y > Z > X$. Transparent crystals may show strong absorption in ordinary light.

Composition and Structure. A complete solid solution series extends from clinozoisite (Al:Fe^{3+} = 3:0) to epidote (Al:Fe^{3+} = 2:1). *Piemontite*, $Ca_2Mn^{3+}Al_2O(SiO_4)(Si_2O_7)(OH)$, is isostructural with epidote and clinozoisite but contains mainly Mn^{3+} instead of Fe^{3+} or Al^{3+} in the Al site outside the chains of the epidote structure. The structure of epidote has been discussed on page 358 and is illustrated in Fig. 10.22.

Diagnostic Features. Fuses at 3–4 with intumescence to a black slag; once formed, this slag is not again fusible. On intense ignition in the closed tube yields a little water. Epidote is characterized by its peculiar green color and one perfect cleavage.

The slag formed on fusing, which is again infusible, is diagnostic.

Occurrence. Epidote forms under conditions of regional metamorphism of the epidote-amphibolite facies (for discussion of metamorphic facies see Chapter 11). Characteristic associations of actinolite-albite-epidote-chlorite occur in the upper part of the greenschist facies. Epidote forms also during retrograde metamorphism and forms as a reaction product of plagioclase, pyroxene, and amphibole. Epidote is common in metamorphosed limestones with calcium-rich garnets, diopside, idocrase, and calcite. *Epidotization* is a low-temperature metasomatism and is found in veins and joint fillings in some granitic rocks.

Epidote is a widespread mineral. Notable localities for its occurrence in fine crystals are Knappenwand, Untersulzbachthal, Salzburg, Austria; Bourg d'Oisans, Isère, France; and the Ala Valley and Traversella, Piemonte, Italy. In the United States found at Haddam, Connecticut; Riverside, California; and on Prince of Wales Island, Alaska (Fig. 10.26).

Name. Epidote from the Greek meaning *increase*, because the base of the vertical prism has one side longer than the other. Zoisite was named after Baron von Zois and piemontite after the locality, Piemonte, Italy.

Similar Species. *Zoisite* is an orthorhombic polymorph (space group *Pnmc*) of clinozoisite. It is similar in appearance and occurrence to clinozoisite but is less common. In 1967 gem quality, blue colored crystals were found in Tanzania. This variety is known as *tanzanite*.

Allanite—
(Ca, Ce)$_2$(Fe^{2+}, Fe^{3+})Al$_2$O(SiO$_4$)(Si$_2$O$_7$)(OH)

Crystallography. Monoclinic; $2/m$. Habit of crystals similar to epidote. Commonly massive and in embedded grains.

$P2_1/m$; $a = 8.98$, $b = 5.75$, $c = 10.23$ Å, $\beta = 115°0'$; $a:b:c = 1.562:1:1.779$; $Z = 2$. $d's$: 3.57(6), 2.94(10), 2.74(8), 2.65(6), 2.14(4).

Physical Properties. H $5\frac{1}{2}$–6. **G** 3.5–4.2. *Luster* submetallic to pitchy and resinous. *Color* brown to pitch-black. Often coated with a yellow-

FIG. 10.26. Epidote with quartz crystals, Prince of Wales Island, Alaska.

brown alteration product. Subtranslucent, will transmit light on thin edges. Slightly radioactive. *Optics:* (−), usually, with 2V = 40°–90°; when (+), 2V = 60°–90°, α = 1.690–1.791, β = 1.700–1.815, γ = 1.706–1.828; $Y = b$, $X \wedge c$ 1°–40°. Metamict allanites are isotropic with n = 1.54–1.72.

Composition and Structure. Of variable composition with Ce, La, Th, and Na in partial substitution for Ca, and Fe^{2+}, Fe^{3+}, Mn^{3+}, and Mg in partial substitution for some of the Al. The structure of well-crystallized allanite is the same as that of epidote (see Fig. 10.22). Allanite is commonly found in a metamict state as the result of "self irradiation" by radioactive constituents in the original mineral. Total destruction of the structure leads to a glassy product that adsorbs considerable H_2O.

Diagnostic Features. Characterized by its black color, pitchy luster, and association with granitic rocks. Fuses at $2\frac{1}{2}$ with intumescence to a black magnetic glass.

Occurrence. Allanite occurs as a minor accessory constituent in many igneous rocks, such as granite, syenite, diorite, and pegmatites. Frequently associated with epidote.

Notable localities are at Miask, Ural Mountains, U.S.S.R.; Greenland; Falun, Ytterby, and Sheppsholm, Sweden; and the Malagasy Republic. In the United States allanite is found at Moriah, Monroe, and Edenville, New York; Franklin, New Jersey; Amelia Court House, Virginia; and Barringer Hill, Texas.

Name. In honor of Thomas Allan, who first observed the mineral. *Orthite* is sometimes used as a synonym.

IDOCRASE (*vesuvianite*)— $Ca_{10}(Mg, Fe)_2Al_4(SiO_4)_5(Si_2O_7)_2(OH)_4$

Crystallography. Tetragonal; $4/m2/m2/m$. Crystals prismatic, often vertically striated. Common forms are {110}, {010}, and {001} (Fig. 10.27). Frequently found in crystals, but striated columnar aggregates more common. Also granular, massive.

Angles: $c(001) \wedge p(011) = 37°08'$, $(001) \wedge (112) = 28°11'$.

$P4nnc$; $a = 15.66$, $c = 11.85$ Å; $a:c = 1:0.757$; $Z = 4$. $d's$: 2.95(4), 2.75(10), 2.59(8), 2.45(5), 1.621(6).

Physical Properties. *Cleavage* {010} poor. **H** $6\frac{1}{2}$. **G** 3.35–3.45. *Luster* vitreous to resinous. *Color* usually green or brown; also yellow, blue, red. Subtransparent to translucent. *Streak* white. *Optics:* (−); ω = 1.703–1.752, ϵ = 1.700–1.746.

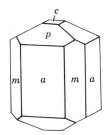

FIG. 10.27. Idocrase.

Composition and Structure. There is some substitution of Na for Ca; Mn^{2+} for Mg; Fe^{3+} and Ti for Al; and F for (OH). B and Be have been reported in some varieties. The structure of idocrase appears to be closely related to that of grossularite garnet. Some parts of the structure are common to both minerals. Isolated SiO_4 tetrahedra as well as Si_2O_7 groups occur. Three-fourths of the Ca is in 8 coordination and one-fourth in 6 coordination with oxygen.

Diagnostic Features. Brown tetragonal prisms and striated columnar masses are characteristic of idocrase. Fuses at 3 with intumescence to a greenish or brownish glass. Gives water in the closed tube.

Occurrence. Idocrase is usually formed as the result of contact metamorphism of impure limestones. Associated with other contact minerals, such as grossularite and andradite garnet, wollastonite, and diopside. Originally discovered in the ancient ejections of Vesuvius and in the dolomitic blocks of Monte Somma. Thus *vesuvianite* is sometimes used as a synonym.

Important localities are Zermatt, Switzerland; Ala, Piemonte; Monzoni, Trentino; Vesuvius; Christiansand, Norway; Achmatovsk, Ural Mountains, and River Vilui, Siberia, U.S.S.R.; and Morelos and Chiaspas, Mexico. In the United States it is found at Auburn and Sanford, Maine; near Amity, New York; and Franklin, New Jersey. Found at many contact metamorphic deposits in the western United States. A compact green variety resembling jade found in Siskiyou, Fresno, and Tulare counties, California, is known as *californite*. In Quebec, Canada, found at Litchfield, Pontiac County; and at Templeton, Ottawa County.

Use. The green variety californite is used to a small extent as a semiprecious stone sometimes sold as jade.

Name. From two Greek words meaning *form* and a *mixture* because the crystal forms present appear to be a combination of those found on different minerals.

CYCLOSILICATES

The cyclosilicates contain rings of linked SiO_4 tetrahedra having a ratio of $Si:O = 1:3$. Three possible closed cyclic configurations of this kind may exist as shown in Fig. 10.28. The simplest is the Si_3O_9 ring, represented among minerals only by the rare titano-silicate *benitoite*, $BaTiSi_3O_9$. The Si_4O_{12} ring occurs together with BO_3 triangles and (OH) groups in the complex structure of the triclinic mineral *axinite*. The Si_6O_{18} ring, however, is the basic framework of the structures of the common and important minerals, *beryl*, $Be_3Al_2Si_6O_{18}$, and *tourmaline*. In the structure of beryl (Figs. 10.29a and b) Si_2O_6 rings are arranged in layers parallel to {0001}. Sheets of Be and Al ions lie between the layers of rings. Be in 4 coordination and Al in 6 coordination tie the layers together horizontally and vertically. The silicon-oxygen rings are so arranged as to be nonpolar; that is, a mirror plane can be pictured passing through the tetrahedra in the plane of the ring (Fig. 10.29b). Rings are positioned one above the other in the basal sheets so that the central holes correspond, forming prominent channels parallel to the c axis. In these channels a wide variety of ions, neutral atoms, and molecules can be housed. $(OH)^-$, H_2O, F, He, and ions of Rb, Cs, Na, and K are housed in beryl in this way. With monovalent alkalies such as Na^+ in the channels, the overall charge of the structure is neutralized by the substitution 2 $alkalis^{+1} \leftrightarrows Be^{2+}$, or by 3 $alkalis^{+1} + 1Al^{3+} \leftrightarrows 3Be^{2+}$.

Cordierite, $(Mg,Fe)_2Al_4Si_5O_{18} \cdot nH_2O$, has a high-temperature polymorph, *indialite*, which is isostructural with beryl in which the Al is randomly distributed in the $(Si,Al)_6O_{18}$ rings. The common, lower temperature form, *cordierite* is orthorhombic (pseudohexagonal) in which two of the tetrahedra in

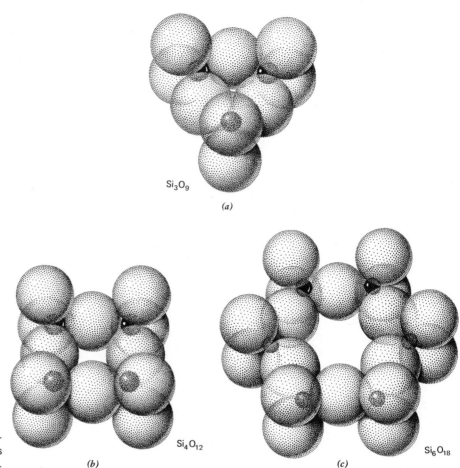

Si_3O_9

(a)

Si_4O_{12}

(b)

Si_6O_{18}

(c)

FIG. 10.28. Close packing representation of ring structures in the cyclosilicates.

the 6-fold ring are occupied by Al, producing an ordered structure (see Fig. 10.30). Al occupies the Be sites and Mg and Fe^{2+} the octahedral Al sites of the beryl structure. H_2O molecules may reside in the channels of the structure. It has been pointed out by G. V. Gibbs (*American Mineralogist,* 1966, pp. 1068–1088) that cordierite should be classified with the tectosilicates rather than the cyclosilicates, because Al- and Si-tetrahedra are in perfect alternation in all directions of the structure except for two SiO_4 tetrahedra that share a common oxygen in the six-membered ring.

The structure of *tourmaline,* with a complex chemical composition, had long remained a mystery. It has been shown to be built on Si_6O_{18} rings along the center of which Na^+ and $(OH)^-$ alternate (see Fig. 10.31). The Si_6O_{18} rings in tourmaline are polar; that is, the net strength of bonds to one side of the ring is not the same as the strength of bonds extending to the other, looking first in one direction, then in the other along the c axis. Interlayered with the rings are sheets of triangular BO_3 groups. (Li, Mg, Al)$O_4(OH)_2$ octahedral groups link the Si_6O_{18} rings and BO_3 groups together. The columns of Si_6O_{18} rings are linked to each other by (Al, Fe, Mn)$O_5(OH)$ groups. The composition of tourmaline is very complex, but the following generalizations apply: Ca may substitute for Na, Al may be replaced by Fe^{3+}, Mn^{3+} and Mg may be replaced by Fe^{2+}, Mn^{2+}, Al^{3+}, and Li^+. The varieties are determined by

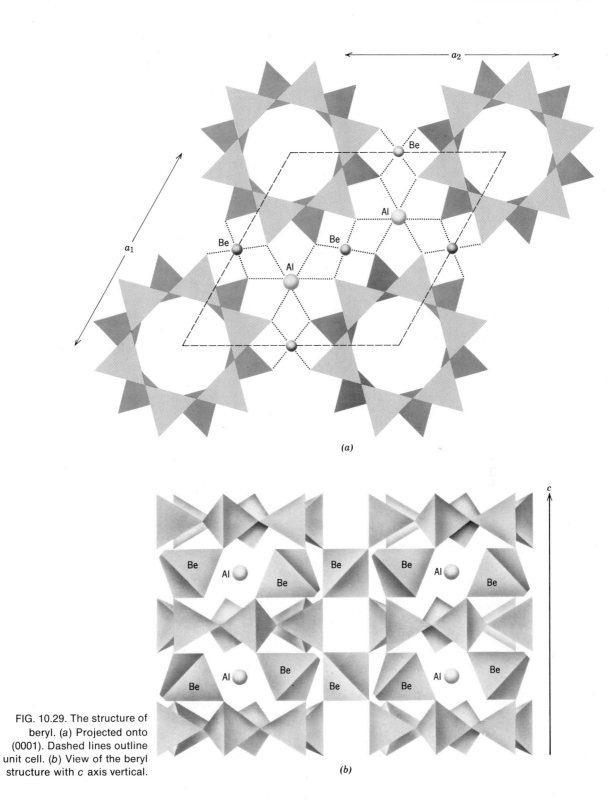

FIG. 10.29. The structure of beryl. (a) Projected onto (0001). Dashed lines outline unit cell. (b) View of the beryl structure with c axis vertical.

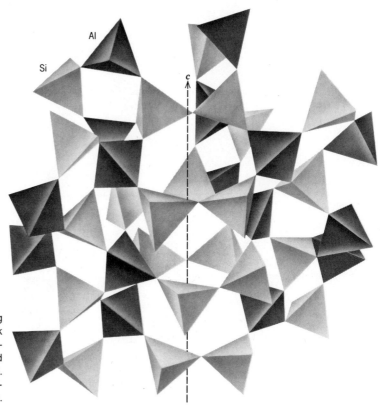

FIG. 10.30. Idealized drawing of the $(Al_4Si_5O_{18})$ framework in low cordierite. The octahedral coordination of Mg and Fe^{2+} is not shown (after G. V. Gibbs, *American Mineralogist*, 1966, p. 1081).

the relative proportions of the cations, and ionic substitution follows the usual pattern, with extensive mutual substitution of Mg by Fe^{2+} and Mn^{2+}, and Na^+ by Ca^{2+}, with concomitant coupled substitution to maintain electrical neutrality. We will discuss in detail the following cyclosilicates:

Axinite	$(Ca, Fe^{2+}, Mn)_3Al_2(BO_3)(Si_4O_{12})(OH)$
Beryl	$Be_3Al_2(Si_6O_{18})$
Cordierite	$(Mg, Fe)_2Al_4Si_5O_{18} \cdot nH_2O$
Tourmaline	$(Na, Ca)(Li, Mg, Al)(Al, Fe, Mn)_6$- $(BO_3)_3(Si_6O_{18})(OH)_4$

FIG. 10.31. Part of the structure of tourmaline projected on (0001).

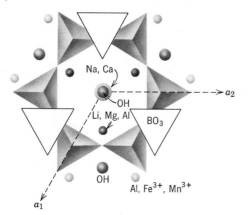

Axinite—$(Ca, Fe^{2+}, Mn)_3Al_2(BO_3)(Si_4O_{12})(OH)$

Crystallography. Triclinic; $\bar{1}$. Crystals usually thin with sharp edges but varied in habit (Fig. 10.32). Frequently in crystals and crystalline aggregates; also massive, lamellar to granular.

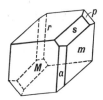

FIG. 10.32. Axinite.

$P\bar{1}$; $a = 7.15$, $b = 9.16$, $c = 8.96$ Å, $\alpha = 88°04'$, $\beta = 81°36'$, $\gamma = 77°42'$; $a:b:c = 0.779:1:0.978$. $Z = 2$. $d's$: 6.30(7), 3.46(8), 3.28(6), 3.16(9), 2.81(10).

Physical Properties. *Cleavage* {100} distinct. **H** $6\frac{1}{2}$–7. **G** 3.27–3.35. *Luster* vitreous. *Color* clove-brown, violet, gray, green, yellow. Transparent to translucent. Pyroelectric. *Optics:* (−); $\alpha = 1.674$–1.693, $\beta = 1.681$–1.701, $\gamma = 1.684$–1.704; $2V = 63°80'$; $r < v$.

Composition and Structure. A considerable range exists in composition with varying amounts of Ca, Mn, and Fe. Some Mg may be present. The structure of axinite contains Si_4O_{12} rings and BO_3 groups linked by Fe^{2+}, Al, and Ca ions.

Diagnostic Features. Characterized by the triclinic crystals with very acute angles. Fusible at $2\frac{1}{2}$–3 with intumescence. When mixed with boron flux (1 part powdered fluorite and 2 parts potassium bisulfate) and heated on platinum wire it gives a green flame (B). Gives water in the closed tube.

Occurrence. Axinite occurs in cavities in granite, and in the contact zones surrounding granitic intrusions. Notable localities for its occurrence are Bourg d'Oisans, Isère, France; various points in Switzerland; St. Just, Cornwall; and Obira, Japan. In the United States at Luning, Nevada, and a yellow manganous variety at Franklin, New Jersey.

Name. Derived from a Greek word meaning axe, in allusion to the wedgelike shape of the crystals.

BERYL—$Be_3Al_2(Si_6O_{18})$

Crystallography. Hexagonal; $6/m2/m2/m$. Strong prismatic habit. Frequently vertically striated and grooved. Cesium beryl frequently flattened on {0001}. Forms usually present consist only of {10$\bar{1}$0} and {0001} (Fig. 10.33a). Pyramidal forms are rare (Fig. 10.33b). Crystals frequently of considerable size with rough faces. At Albany, Maine, a tapering crystal 27 feet long weighed over 25 tons.

Angles: $c(0001) \wedge p(10\bar{1}2) = 29°57'$, $c(0001) \wedge s(11\bar{2}2) = 44°56'$.

$P6/mcc$; $a = 9.23$, $c = 9.19$ Å; $a:c = 1:0.996$; $Z = 2$. $d's$: 7.98(9), 4.60(5), 3.99(5), 3.25(10), 2.87(10).

Physical Properties. *Cleavage* {0001} imperfect. **H** $7\frac{1}{2}$–8. **G** 2.65–2.8. *Luster* vitreous. *Color* commonly bluish green or light yellow, may be deep emerald-green, gold-yellow, pink, white, or colorless. Transparent to translucent. Frequently the larger, coarser crystals show a mottled appearance due to the alternation of clear transparent spots with cloudy portions.

Color serves as the basis for several variety names of gem beryl. *Aquamarine* is the pale greenish blue transparent variety. *Morganite, or rose beryl*, is pale pink to deep rose. *Emerald* is the deep green transparent beryl. *Golden beryl* is a clear golden-yellow variety. *Optics:* (−); $\omega = 1.566$–1.608, $\epsilon = 1.562$–1.600.

Composition and Structure. BeO 14.0, Al_2O_3 19.0, SiO_2 67.0 are the theoretical percentages of the oxides in the formula. However, the presence of alkalis (Na and K) and Li may considerably reduce the percentage of BeO. Small and variable amounts of zeolitic H_2O are present. The structure of beryl is illustrated in Fig. 10.29 and discussed on page 363.

Diagnostic Features. Recognized usually by its hexagonal crystal form and color. Distinguished

FIG. 10.33. Beryl crystals.

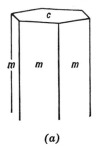

(a)

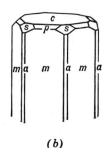

(b)

from apatite by greater hardness and from quartz by higher specific gravity. Fuses with difficulty at $5–5\frac{1}{2}$ to an enamel. Insoluble in acids.

Occurrence. Beryl, although containing the rare element Be, is rather common and widely distributed. It occurs usually in granitic rocks, or in pegmatites. It is also found in mica schists and associated with tin ores. The world's finest emeralds are found in Columbia in a dark bituminous limestone; the most notable localities are Muzo, Cosquez, and El Chivor. Emeralds of good quality are found in mica schists in the Transvaal, South Africa; Sandawana, Rhodesia, and near Sverdlovsk, U.S.S.R. Rather pale emeralds have been found in small amount in Alexander County, North Carolina, associated with the green variety of spodumene, *hiddenite*. Beryl of the lighter aquamarine color is much more common and is found in gem quality in many countries.

For many years Brazil has led in production not only of gem beryls other than emerald but also in the production of common beryl used as beryllium ore. India, the U.S.S.R., and the United States are also major producers of common beryl. In the United States gem beryls, chiefly aquamarine, have been found in various places in Maine, New Hampshire, Massachusetts, Connecticut, North Carolina, and Colorado. Rose-colored beryl has been found in San Diego County, California, associated with pink tourmaline and the pink spodumene, *kunzite*.

Use. Used as a gemstone of various colors. The emerald ranks as the most valuable of stones and may have a much greater value than the diamond. Beryl is also the major source of beryllium, a light metal similar to aluminum in many of its properties. One of its chief uses is as an alloy with copper. One and one-half percent of beryllium in copper greatly increases the hardness, tensile strength, and fatigue resistance.

Name. The name beryl is of ancient origin, derived from the Greek word which was applied to green gemstones.

Similar Species. *Euclase,* $BeAl(SiO_4)(OH)$, and *gadolinite,* $YFe^{2+}Be_2(SiO_4)_2O_2$, are rare beryllium silicates.

Cordierite—$(Mg, Fe)_2Al_4Si_5O_{18}\cdot nH_2O$

Mg Fe Al Silicate no specimen

Crystallography. Orthorhombic: $2/m2/m2/m$. Crystals are usually short prismatic, pseudohexagonal twins twinned on {110}. Also found as imbedded grains and massive.

Cccm; $a = 17.13$, $b = 9.80$, $c = 9.35$ Å; $a:b:c = 1.748:1:0.954$; $Z = 4$. *d's:* 8.54(10), 4.06(8), 3.43(8), 3.13(7), 3.03(8).

Physical Properties. *Cleavage* {010} poor. **H** $7–7\frac{1}{2}$. **G** 2.60–2.66. *Luster* vitreous. *Color* various shades of blue to bluish grey. Transparent to translucent. *Optics:* usually (−), may be (+). Indices increasing with Fe content. $\alpha = 1.522–1.558$, $\beta = 1.532–1.568$, $\gamma = 1.527–1.573$. $2V = 0°–90°$. $X = c$, $Y = a$; $r < v$. Pleochroism: cordierite is sometimes called *dichroite* because of pleochroism. Fe rich varieties: X colorless, Y and Z violet.

Composition and Structure. Although some substitution of Mg by Fe^{2+} occurs, most cordierites are magnesium-rich. Mn may replace part of the Mg. The Al content of cordierite shows little variation. Most analyses show appreciable but variable H_2O which is probably located in the large channels parallel to c. Small amounts of Na and K may be similarly housed. The structure of the low-temperature form, also known as low cordierite, is shown in Fig. 10.30 and is discussed on page 363. A high-temperature polymorph, *indialite*, with random distribution of Al in the $(AlSi)_6O_{18}$ ring, is isostructural with beryl and has space group $P6/mcc$.

Diagnostic Features. Cordierite resembles quartz and is distinguished from it with difficulty. Unlike quartz, it is fusible on thin edges. Distinguished from corundum by lower hardness. Pleochroism is characteristic if observed.

Alteration. Commonly altered to some form of mica, chlorite, or talc and is then various shades of grayish green.

Occurrence. Cordierite is a common constituent of contact and regionally metamorphosed argillaceous rocks. It is especially common in hornfels produced by contact metamorphism of pelitic rocks. Common assemblages are sillimanite-cordierite-spinel, and cordierite-spinel-plagioclase-orthopyroxene. Cordierite is also found in regionally meta-morphosed cordierite-garnet-sillimanite gneisses. Cordierite-anthophyllite coexistences have been described from several localities. It occurs also in norites resulting from the incorporation of argillaceous material in gabbroic magmas. It is found in some granites and pegmatites. Gem material has come from Sri Lanka.

Use. Transparent cordierite of good color has been used as a gem known by jewelers as *saphir d'eau* or *dichroite*.

Name. After the French geologist P. L. A. Cordier (1777–1861). *Iolite* is sometimes used as a synonym.

Na Mg Fe Al Boron

TOURMALINE—(Na, Ca)(Li, Mg, Al)- *Silicate* (Al, Fe, Mn)$_6$(BO$_3$)$_3$(Si$_6$O$_{18}$)(OH)$_4$

Crystallography. Hexagonal-*R*; 3*m*. Usually in prismatic crystals with a prominent trigonal prism and subordinate second-order hexagonal prism vertically striated. The prism faces may round into each other giving the crystals a cross-section like a spherical triangle. When doubly terminated, crystals usually show different forms at the opposite ends of the vertical axis (Fig. 10.34). May be massive compact; also coarse to fine columnar, either radiating or parallel.

Angles: $r(10\bar{1}1) \land r'(\bar{1}101) = 46°52'$, $-m(01\bar{1}0) \land o(02\bar{2}1) = 44°3'$, $r(10\bar{1}1) \land m(10\bar{1}0) = 62°40'$.

$R3m$; $a = 15.95$, $c = 7.24$ Å; $a:c = 1:0.448$; $Z = 3$. $d's$: 4.24(7), 4.00(7), 3.51(7), 2.98(9), 2.58(10).

Physical Properties. **H** 7–7$\frac{1}{2}$. **G** 3.0–3.25.

Luster vitreous to resinous. *Color* varied, depending on the composition. *Fracture* conchoidal.

Black, Fe-bearing tourmaline (*shorl*) is most common; brown tourmaline (*dravite*) contains Mg. The rarer Li-bearing varieties (*elbaite*) are light colored in fine shades of green (*verdelite*), yellow, red-pink (*rubellite*), and blue (*indicolite*). Rarely white or colorless *achroite*. A single crystal may show several different colors arranged either in concentric envelopes about the *c* axis or in layers transverse to the length. Strongly pyroelectric and piezoelectric. *Optics:* (−); $\omega = 1.635–1.675$, $\epsilon = 1.610–1.650$. Some varieties strongly pleochroic, $O > E$.

Composition and Structure. A complex silicate of B and Al (see Fig. 10.31) with the following substitutions: Ca for Na along the centers of the ring channels; Mg and Al for Li in 6 coordination between Si$_6$O$_{18}$ rings and BO$_3$ groups; Fe^{3+} and Mn^{3+} for Al in polyhedra that link the Si$_6$O$_{18}$ rings. The structure is discussed on page 364.

Diagnostic Features. Usually recognized by the characteristic rounded triangular cross-section of the crystals and conchoidal fracture. Distinguished from hornblende by absence of cleavage. Fusibility varies with composition: Mg-rich varieties fusible at 3; Fe-rich varieties fusible with difficulty; Li-rich varieties infusible. Fused with boron flux gives momentary green flame of boron.

Occurrence. The most common and characteristic occurrence of tourmaline is in granite pegmatites and in the rocks immediately surrounding them. It is found also as an accessory mineral in igneous rocks and metamorphic rocks. Most pegmatitic tourmaline is black and is associated with the

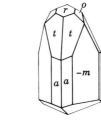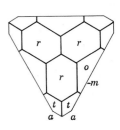

FIG. 10.34. Tourmaline.

FIG. 10.35. Tourmaline crystals with quartz and cleavelandite, Pala, California.

common pegmatite minerals, microcline, albite, quartz, and muscovite. Pegmatites are also the home of the light colored lithium-bearing tourmalines frequently associated with lepidolite, beryl, apatite, fluorite, and rarer minerals (see Fig. 10.35). The brown Mg rich tourmaline is found in crystalline limestones.

Famous localities for the occurrence of the gem tourmalines are the Island of Elba; the state of Minas Gerais, Brazil; Ural Mountains near Sverdlovsk; and the Malagasy Republic. In the United States found at Paris and Auburn, Maine; Chesterfield, Massachusetts; Haddam Neck, Connecticut; and Mesa Grande, Pala, Rincon, and Ramona in San Diego County, California. Brown crystals are found near Gouverneur, New York, and fine black crystals at Pierrepont, New York.

Use. Tourmaline forms one of the most beautiful of the semiprecious gemstones. The color of the stones varies, the principal shades being olive-green, pink to red, and blue. Sometimes a stone is so

cut as to show different colors in different parts. The green colored stones are usually known by the mineral name, tourmaline, or as *Brazilian emeralds*. The red or pink stones are known as *rubellite,* and the rarer dark blue stones are called *indicolite*.

Because of its strong piezoelectric property, tourmaline is used in the manufacture of pressure gauges to measure transient blast pressures.

Name. *Tourmaline* comes from *turamali,* a name given to the early gems from Sri Lanka.

INOSILICATES

SiO_4 tetrahedra may link into chains by sharing oxygens (Fig. 10.2). Such simple chains may be joined side by side by further sharing of oxygens in alternate tetrahedra to form bands or double chains (Fig. 10.2). In the simple chain structure, two of the four oxygens in each SiO_4 tetrahedron are shared giving a ratio of $Si:O = 1:3$. In the band structure half of the tetrahedra share three oxygens, the other half share two oxygens yielding a ratio of $Si:O = 4:11$.

Included in the inosilicates are two important rock-forming groups of minerals: the pyroxenes as single chain members and the amphiboles as double chain members. Many similarities exist between the two groups in crystallographic, physical, and chemical properties. Although most pyroxenes and amphiboles are monoclinic both groups have orthorhombic members. Also in both the repeat distance along the chains, that is, the c dimension of the unit cell, is approximately 5.2 Å. The a cell dimensions are also analogous but, because of the double chain, the b dimension of amphiboles is roughly twice that of the corresponding pyroxenes.

The same cations are present in both groups; but the amphiboles are characterized by the presence of (OH), which is lacking in pyroxenes. Although the color, luster, and hardness of analogous species are similar, the (OH) in amphiboles gives them, in general, slightly lower specific gravity and refractive indices than their pyroxene counterparts. Furthermore, the crystals have somewhat different habits. Pyroxenes commonly occur in stout prisms whereas amphiboles tend to form more elongated crystals, often acicular. Their cleavages are distinctly different and can be directly related to the underlying chain structure (see Figs. 10.38*b* and 10.49*b*).

Pyroxenes crystallize at higher temperatures than their amphibole analogues and hence are generally formed early in a cooling igneous melt and occur also in high-temperature metamorphic rocks rich in Mg and Fe (see Fig. 11.9). If water is present in the melt or as a metamorphic fluid, the early-formed pyroxene may react with the liquid at lower temperatures to form amphibole. Under prograde metamorphic conditions amphiboles commonly react to form pyroxenes, and under retrograde metamorphic conditions pyroxenes commonly give way to amphiboles (see Fig. 10.41).

Pyroxene Group   Fe Mg Al Silicate

The chemical composition of pyroxenes can be expressed by a general formula as XYZ_2O_6, where X represents Na^+, Ca^{2+}, Mn^{2+}, Fe^{2+}, Mg^{2+}, and Li^+ in the $M2$ crystallographic site; Y represents Mn^{2+}, Fe^{2+}, Mg^{2+}, Fe^{3+}, Al^{3+}, Cr^{3+}, and Ti^{4+} in the $M1$ site; and Z represents Si^{4+} and Al^{3+} in the tetrahedral sites of the chain. (It should be noted that the X cations in general are larger than the Y cations, in accordance with the cation size requirements of the sites $M2$ and $M1$.) The pyroxenes can be divided into several groups the most common of which can be represented in a part of the chemical system $CaSiO_3$ (wollastonite, a pyroxenoid)–$MgSiO_3$(enstatite)–$FeSiO_3$(ferrosilite). The trapezium part of this system includes members of the common series *diopside*, $CaMgSi_2O_6$-*hedenbergite*, $CaFeSi_2O_6$, and of the series *enstatite-orthoferrosilite* (see Fig. 10.36). The chemical compositions of pyroxenes in this trapezium are frequently stated in terms of molecular percentages of Wo (for wollastonite), En (for enstatite), and Fs (for ferrosilite). Table 4.14 illustrates the recalculation of a pyroxene analysis in terms of Wo-En-Fs end member components. Compositionally *augite* is closely related to members of the diopside-hedenbergite series but with some substitution of, for example, Na for Ca in $M2$, Al for Mg(or Fe^{2+}) in $M1$, and Al for Si. *Pigeonite* represents a field of Mg-Fe solid solutions with a Ca-content somewhat larger than in the *enstatite-orthoferrosilite* series which represents the compositional field of the orthopyroxenes. Sodium-containing pyroxenes are *aegirine*, $NaFe^{3+}Si_2O_6$, and *jadeite*, $NaAlSi_2O_6$. Aegirine and augite represent a complete solid solution series as shown by members of intermediate composition, *aegirine-augite*. *Omphacite* represents a solid solution series between augite and jadeite. *Spodumene*, $LiAlSi_2O_6$, is a relatively rare pyroxene found in Li-rich pegmatites.

The pyroxene structure is based on single SiO_3 chains that run parallel to the c axis. Figure 10.37 (see also Fig. 10.2) illustrates this tetrahedral chain as well as the double octahedral chain to which it is bonded. The structure contains two types of cation sites, labeled $M1$ and $M2$. The $M1$ site is a relatively regular octahedron, but, especially in the monoclinic pyroxenes, the $M2$ site is an irregular polyhedron of 8 coordination (in orthorhombic pyroxenes with Mg in the $M2$ site this polyhedron is closer to

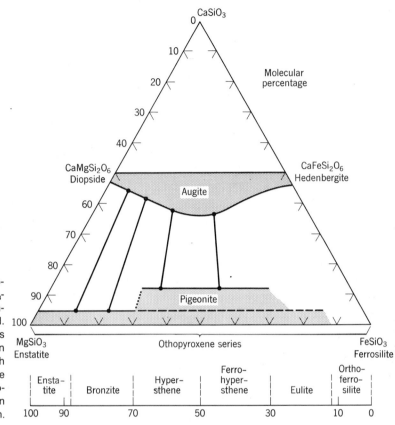

FIG. 10.36. Pyroxene compositions in the system CaSiO₃-MgSiO₃-FeSiO₃. General compositional fields are outlined. Representative tielines across the miscibility gap between augite and more Mg-Fe-rich pyroxenes are shown. The nomenclature of the orthopyroxene species is shown by the bar diagram.

a regular octahedron). Figure 10.38a shows the pyroxene structure and the distribution of cation sites as seen in a direction parallel to the c axis. The cations in the M1 sites are all coordinated by oxygens of two opposing SiO₃ chains and as such produce a tetrahedral-octahedral-tetrahedral ("t-o-t") strip. The coordination of cations in the M2 position, however, is such that several of these t-o-t strips are cross-linked. These t-o-t strips are often schematically represented as in Fig. 10.38b; this in turn shows the relationship of the t-o-t strips to the cleavage angles in pyroxenes.

The majority of pyroxenes can be assigned to one of three space groups; two monoclinic (C2/c and P2₁/c) and one orthorhombic (Pbca). The C2/c structure is found in most of the common clinopyroxenes such as diopside, $CaMgSi_2O_6$, jadeite,

$NaAlSi_2O_6$, and augite. This structure is illustrated in Fig. 10.37. The M1 site is generally occupied by cations that are smaller than those of the M2 site. For example, in the diopside-hedenbergite series, $CaMgSi_2O_6$-$CaFeSi_2O_6$, the M1 site is occupied by Mg and Fe^{2+} in random distribution, whereas the M2 is occupied solely by the larger Ca^{2+} ion, in 8 coordination. The M2 site, however, can also house Mn^{2+}, Fe^{2+}, Mg, or Li, in which case the coordination is 6-fold. The P2₁/c space group is found in pigeonite (see Fig. 10.36) which may be represented as $Ca_{0.25}(Mg, Fe)_{1.75}Si_2O_6$ in which the M2 site is occupied by all of the Ca in the formula as well as additional Mg and Fe so as to make the composition of the site $[Ca_{0.25}(Fe + Mg)_{0.75}]$. The M2 site, as present in the clinopyroxenes with space group C2/c, is too large to accommodate these smaller cations, so that

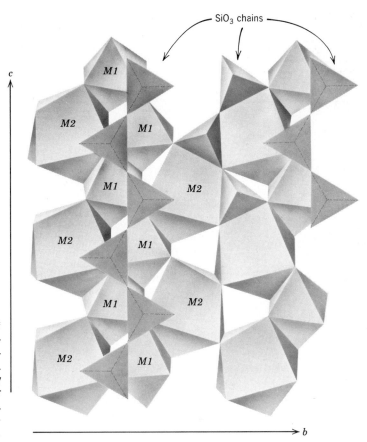

SiO₃ chains

FIG. 10.37. The structure of jadeite, NaAlSi₂O₆, a monoclinic pyroxene in an approximate projection onto (100). The $M2$ site is occupied by Na^+, the $M1$ site by Al^{3+} (after C. T. Prewitt and C. W. Burnham, *American Mineralogist,* 1966, p. 966).

FIG. 10.38 (a) Schematic projection of the monoclinic pyroxene structure on a plane perpendicular to the c axis. (b) Control of cleavage angles by *t-o-t* strips (also referred to as "I beams") in the pyroxene structure, as compared with naturally occurring pyroxene cleavage.

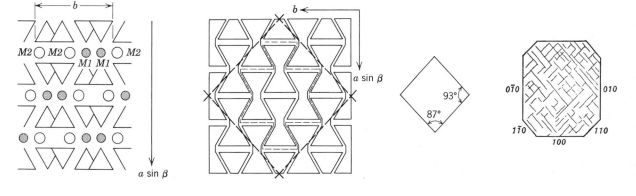

(a)

(b)

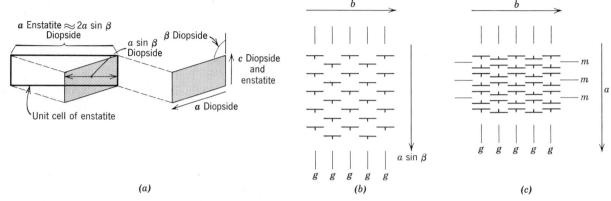

FIG. 10.39 (a) Relationship between unit cells of clinopyroxene (e.g., diopside) and orthopyroxene (e.g., enstatite) as projected on (010). The monoclinic unit cell outlined by shading can be related to the larger orthorhombic unit cell by a mirror reflection. (b) Schematic representation of possible monoclinic pyroxene (looking down the c axis) with the t-o-t strips (or "I beams") represented by strike and dip symbols; g's indicate glide planes (with a translation component parallel to c). (c) Schematic representation of a possible orthorhombic pyroxene; m's represent mirrors. (b and c after J. B. Thompson, Jr., Harvard University, pers. comm.)

M2 is somewhat smaller in pigeonite (hence, this site has irregular 7 coordination) due to a slight shift in the pigeonite structure as compared with the more regular diopside type structure. The orthorhombic Pbca structure is that found in the orthopyroxene series which contains virtually no Ca. The Mg and Fe²⁺ ions are distributed among M1 and M2 with the larger cation (Fe²⁺) showing strong preference for the somewhat larger and distorted M2 site. The coordination of both the M1 and M2 sites in orthopyroxenes is 6-fold. The unit cells of the orthorhombic pyroxenes are related to the monoclinic unit cells by a twinlike mirror across {100} accompanied by an approximate doubling of the a cell dimension (e.g., a of enstatite ≈ 2 a sin β of diopside; see Fig. 10.39a). Figures 10.39b and c show schematically the development of monoclinic and orthorhombic pyroxene structures built up from t-o-t strips. The compositions represented by the orthopyroxene series may in rare occurrences be found in a monoclinic form, known as the clinoenstatite-clinohypersthene series.

We will discuss in detail the following common pyroxenes:

PYROXENES

Enstatite-orthoferrosilite series

Enstatite	$MgSiO_3$
Hypersthene	$(Mg, Fe)SiO_3$
Pigeonite ~	$Ca_{0.25}(Mg, Fe)_{1.75}Si_2O_6$

Diopside-hedenbergite series

Diopside	$CaMgSi_2O_6$
Hedenbergite	$CaFeSi_2O_6$
Augite	$XY(Z_2O_6)$

Sodium pyroxene group

Jadeite	$NaAlSi_2O_6$
Aegirine	$NaFe^{3+}Si_2O_6$
Spodumene	$LiAlSi_2O_6$

ENSTATITE—MgSiO₃
HYPERSTHENE—(Mg, Fe)SiO₃

Crystallography. Orthorhombic; $2/m2/m2/m$. Prismatic habit, crystals rare. Usually massive, fibrous, or lamellar.

Angles: $(210) \wedge (2\bar{1}0) = 91°44'$.

$Pbca$; $a = 18.22, b = 8.81, c = 5.21Å$; $a:b:c = 2.068:1:0.590$; $Z = 8$. d's: 3.17(10), 2.94(4), 2.87(9), 2.53(4), 2.49(5).

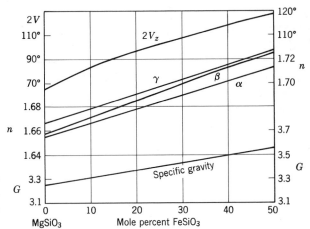

FIG. 10.40. Enstatite-hypersthene. Variation of refractive index, 2V, and specific gravity with composition.

Physical Properties. *Cleavage* {210} good. Because of the doubling of the *a* dimension in orthorhombic pyroxenes, the cleavage form is {210} rather than {110} as in monoclinic pyroxenes. Frequently good parting on {100}, less common on {001}. **H** $5\frac{1}{2}$–6. **G** 3.2–3.6 increasing with Fe content. *Luster* vitreous to pearly on cleavage surfaces; *bronzite* has submetallic, bronzelike luster. *Color* grayish, yellowish, or greenish-white to olive-green and brown. Translucent. *Optics:* enstatite (+); bronzite and hypersthene (−). $\alpha = 1.650$–1.715, $\beta = 1.653$–1.728, $\gamma = 1.658$–1.731, for enstatite-hypersthene. 2V = 35°–50°; X = b, Z = c. Indices increase with Fe content (see Fig. 10.40); in orthoferrosilite $\beta = 1.785$.

Composition and Structure. Fe^{2+} may substitute for Mg in all proportions up to nearly 90% $FeSiO_3$. However, in the more common orthopyroxenes the ratio of Fe:Mg rarely exceeds 1:1. Pure enstatite contains MgO 40.0, SiO_2 60.0%. The maximum amount of CaO in orthopyroxenes generally does not exceed 1.5 weight percent. The nomenclature for the orthopyroxenes is shown in Fig. 10.36; chemical compositions are generally expressed in terms of molecular percentages, for example, $En_{40}Fs_{60}$. The pure end member $FeSiO_3$, *orthoferrosilite*, is rarely found in nature because in most geologically observed pressure and temperature ranges the compositionally equivalent assemblage Fe_2SiO_4 (fayalite) + SiO_2 is more stable; all

other varieties of the orthopyroxene series are found. The structure of members of the orthopyroxene series can be considered to consist of the monoclinic *t-o-t* strips twinned along {100} so as to essentially double the *a* dimension of orthopyroxenes as compared to the *a* of clinopyroxenes. In the *Pbca* structure of orthopyroxenes Fe^{2+} shows a strong preference for the *M2* crystallographic site. Compositions between $MgSiO_3$-$FeSiO_3$ may also occur as members of the monoclinic series, *clinoenstatite-clinoferrosilite*, with space group $P2_1/c$. Experimental results of the stability fields of enstatite versus clinoenstatite are controversial. The common occurrence of orthopyroxenes versus the rare occurrence of clinopyroxenes in the series $MgSiO_3$-$FeSiO_3$ may indicate that the orthorhombic series is generally more stable, and at lower temperatures, than the monoclinic series.

Diagnostic Features. Usually recognized by its color, cleavage and unusual luster. Varieties high in iron are black and difficult to distinguish from augite without optical tests. Almost infusible; thin edges will become slightly rounded. Fuses more easily with increasing amounts of iron.

Occurrence. Mg-rich orthopyroxene is a common constituent of peridotites, gabbros, norites, and basalts and is commonly associated with Ca-clinopyroxenes (e.g., augite), olivine, and plagioclase. It may be the major constituent of pyroxenites. Orthopyroxenes may also be found in metamorphic rocks, some types of which are of high *T* and high *P* origin, such as in the granulite facies. Iron-rich members of the orthopyroxene series (e.g., eulite) are common in metamorphosed iron-formation in association with grunerite. In all such occurrences orthopyroxenes commonly coexist with clinopyroxenes because of a large miscibility gap between the two groups (see Fig. 10.36). Orthopyroxenes frequently show exsolved lamellae of a Ca-rich clinopyroxene. Enstatite as well as clinoenstatite and clinohypersthene occur in both iron and stony meteorites. In prograde metamorphic rocks orthopyroxenes form commonly at the expense of Mg-Fe amphibole (e.g., anthophyllite) and in retrograde metamorphic rocks orthopyroxenes may give way to Mg-Fe amphiboles (see Fig. 10.41).

Enstatite is found in the United States at the Tilly Foster Mine, Brewster, New York, and at Edwards, St. Lawrence County, New York; at Texas, Pennsylvania; Bare Hills, near Baltimore, Maryland; and Webster, North Carolina. Hypersthene occurs in New York in the norites of the Cortland region, on the Hudson River, and in the Adirondack region. Fe-rich orthopyroxenes are common constituents of the metamorphosed Lake Superior and Labrador Trough iron-formations.

FIG. 10.41. Schematic *P-T* diagram for the stability fields of anthophyllite, $Mg_7Si_8O_{22}(OH)_2$, and reaction products, enstatite, $MgSiO_3$, + quartz + H_2O (after H. J. Greenwood, 1963, *Jour. Petrology*, v. 4, p. 325).

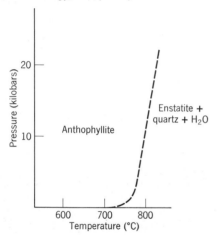

Name. Enstatite is named from the Greek word meaning *opponent* because of its refractory nature. Hypersthene is named from two Greek words meaning *very* and *strong* because its hardness is greater than that of hornblende.

Pigeonite ~ $Ca_{0.25}(Mg, Fe)_{1.75}Si_2O_6$

Crystallography. Monoclinic; $2/m$. Very rarely as well-formed phenocrysts with a prismatic habit parallel to *c*.

$P2_1/c$; $a = 9.71$, $b = 8.96$, $c = 5.25$, $\beta = 108°33'$; $a:b:c = 1.083:1:0.585$; $Z = 8$. *d's*: 3.21(8), 3.02(10), 2.908(8), 2.904(10), 2.578(6).

Physical Properties. *Cleavage* {110} good; parting on {100} may be present. **H** 6. **G** 3.30–3.46. *Color* brown, greenish brown, to black. *Optics:* (+); $\alpha = 1.682–1.722$, $\beta = 1.684–1.722$, $\gamma = 1.704–1.752$, increasing with Fe^{2+}. Two orientations occur: (1) with $Y = b$ and (2) more common, $X = b$. $Z \wedge c = 37°–44°$, $2V = 0–30°$.

Composition and Structure. Pigeonites are calcium-poor monoclinic pyroxenes which contain between 5 to 15 molecular percent of the $CaSiO_3$ component (see Fig. 10.36; field just above orthopyroxene base). The crystal structure of pigeonite is similar to that of diopside with all of the Ca and additional Fe and Mg in the *M2* site, and the remaining Mg and Fe in *M1*. Fe shows a strong preference for the *M2* sites. Pigeonite is stable at high temperatures in igneous rocks and inverts commonly at lower temperatures to orthopyroxene with augite-type exsolution lamellae. A possible stability diagram is given in Fig. 10.42.

Diagnostic Features. Can be distinguished from other pyroxenes only by optical or X-ray techniques. Augite $2V > 39°$; pigeonite $2V < 32°$.

Occurrence. Pigeonite is common in high-temperature, rapidly cooled lavas and in some intrusives such as diabases. It is present as phenocrysts in some volcanic rocks, but is not known from metamorphic rocks. If pigeonite formed in an igneous rock that cooled slowly, it may have exsolved augite lamellae on {001} and may subsequently have inverted to orthopyroxene (see Fig. 10.42) through a reconstructive transformation. At even lower tem-

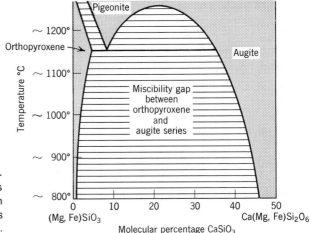

FIG. 10.42. Schematic *T*-composition section across the Wo-En-Fs diagram shown in Fig. 10.36. The section is at about En_{65} Fs_{35}.

peratures the orthopyroxene may have developed augite exsolution along {100} planes.

Name. After the locality, Pigeon Cove, Minnesota.

DIOPSIDE—$CaMgSi_2O_6$
HEDENBERGITE—$CaFeSi_2O_6$
AUGITE—$(Ca, Na)(Mg, Fe, Al)(Si, Al)_2O_6$

Diopside and hedenbergite form a complete solid solution series with physical and optical properties varying linearly with composition. Augite is a clinopyroxene in which some Na substitutes for Ca, some Al substitutes for both Mg (or Fe) and Si, and in which Fe and Mg contents are higher than in diopside or hedenbergite (see Fig. 10.36). Although the crystal constants vary slightly from one member to another, a single description suffices for all.

Crystallography. Monoclinic; $2/m$. In prismatic crystals showing square or eight-sided cross-section (Fig. 10.43). Also granular massive, columnar, and lamellar. Frequently twinned polysynthetically on {001}; less commonly twinned on {100}.

Angles: $m(110) \wedge m'(1\bar{1}0) = 92°50'$, $c(001) \wedge p(111) = 33°50'$, $s(\bar{1}11) \wedge s'(\bar{1}\bar{1}1) = 59°11'$, $c(001) \wedge a(100) = 74°10'$.

$C2/c$; $a = 9.73$, $b = 8.91$, $c = 5.25$ Å; $\beta = 105°50'$; $a:b:c = 1.092:1:0.589$; $Z = 4$. $d's$: 3.23(8), 2.98(10), 2.94(7), 2.53(4), 1.748(4).

Physical Properties. *Cleavage* {110}, at 87° and 93°, imperfect. Frequently parting on {001}, and less commonly on {100} in the variety *diallage*. **H** 5–6. **G** 3.2–3.3. *Luster* vitreous. *Color* white to light green in diopside; deepens with increase of Fe. Augite is black. Transparent to translucent. *Optics:* (+); $\alpha = 1.66–1.75$, $\beta = 1.67–1.73$, $\gamma = 1.69–1.75$. $2V = 55–65°$; $Y = b$, $Z \wedge c = 39–48°$; $r > v$. Darker members show pleochroism; X pale green, Y yellow-green, Z dark green.

Composition and Structure. In the diopside-hedenbergite series Mg and Fe^{2+} substitute for each other in all proportions. In the majority of analyses of members of this series the Al_2O_3 content varies between 1 and 3 weight percent. In augite, in addition to varying Mg and Fe^{2+} contents, Al substitutes for both Mg(or Fe^{2+}) and Si. Mn, Fe^{3+}, Ti, and

FIG. 10.43. Augite crystals.

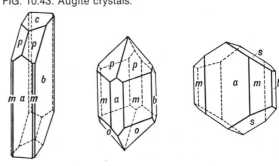

Na may also be present. A complete series exists toward *aegirine-augite,* (Na, Ca) (Fe^{3+}, Fe^{2+}, Mg, Al)Si$_2$O$_6$, by the replacement Ca(Mg, Fe^{2+}) $\leftrightarrows$ NaFe^{3+}. Compositional zoning is commonly found in igneous augites with the cores rich in the augite component and the rims tending toward aegirine-augite. Chemical analyses are generally recalculated in terms of molecular percentages of wollastonite (Wo), enstatite (En), and ferrosilite (Fs) and are expressed as Wo$_x$En$_y$Fs$_z$, where x, y, and z are molecular percentages (see Table 4.14). The structures of diopside, hedenbergite, and augite are all based on space group $C2/c$. Their structure is discussed on page 371, and a monoclinic structure is illustrated in Fig. 10.37. The Ca ions in the *M2* site are in 8 coordination, whereas the ions in *M1* are in 6 coordination.

Diagnostic Features. Characterized by crystal form and imperfect prismatic cleavage at 87° and 93°. Fusible at 4 to a green glass. Insoluble in acids. With fluxes gives tests for Ca, Mg, and Fe.

Occurrence. *Diopside* and *hedenbergite* are common in metamorphic rocks. Diopside, in association with forsterite and calcite, infrequently with monticellite, is the result of thermal metamorphism of siliceous, Mg-rich limestones, or dolomites. For example:

$$CaMg(CO_3)_2 + 2SiO_2 \rightarrow CaMgSi_2O_6 + 2CO_2$$
$$\text{dolomite} \qquad \text{quartz} \qquad \text{diopside}$$

Other associations include tremolite, scapolite, idocrase, garnet, and sphene. Hedenbergite occurs in more Fe-rich metamorphic rocks. Diopside and hedenbergite are also known as products of igneous crystallization. Early formed Ca-rich clinopyroxenes may be very close to diopside in composition whereas the latest stages of crystallization may be represented by hedenbergite compositions, due to enrichment of the residual magma in Fe. The Skaergaard intrusion, east Greenland, contains late, fine grained hedenbergite interstitial to earlier formed coarser grained diopside and augite grains.

Fine crystals have been found in the Ural Mountains, U.S.S.R.; Austrian Tyrol; Binnenthal, Switzerland; and Piemonte, Italy. At Nordmark, Sweden, fine crystals range between diopside and hedenbergite. In the United States they are found at Canaan, Connecticut; and DeKalb Junction and Gouverneur, New York.

Augite is the most common pyroxene and an important rock-forming mineral. It is found chiefly in the dark-colored igneous rocks, such as basaltic lavas and intrusives, gabbros, peridotites, and andesites. Chemically zoned augites are common in quickly cooled rocks such as the lunar basalts. The clinopyroxene crystallization history is very well documented for many lunar basalts and gabbros as well as for the Skaergaard intrusion, East Greenland, for example; early formed crystals are more magnesian than later pyroxene grains. Fine crystals of augite have been found in the lavas of Vesuvius; at Val di Fassa, Trentino, Italy; and Bilin, Bohemia.

Use. Transparent varieties of diopside have been cut and used as gemstones.

Name. Diopside is from two Greek words meaning *double* and *appearance,* because the vertical prism zone can apparently be oriented in two ways. Hedenbergite is named after M. A. Ludwig Hedenberg, the Swedish chemist who discovered and described it. Augite comes from a Greek word meaning *luster.* The name pyroxene, *stranger to fire,* is a misnomer and was given to the mineral because it was thought that it did not occur in igneous rocks.

Jadeite—NaAlSi$_2$O$_6$

Crystallography. Monoclinic; $2/m$. Rarely in isolated crystals. Usually fibrous in compact massive aggregates.

$C2/c$; $a = 9.50$, $b = 8.61$, $c = 5.24$ Å; $\beta = 107°26'$; $a:b:c = 1.103:1:0.609$; $Z = 4$. $d's$: 3.27(3), 3.10(3), 2.92(8), 2.83(10), 2.42(9).

Physical Properties. *Cleavage* {110} at angles of 87° and 93°. Extremely tough and difficult to break. **H** 6$\frac{1}{2}$–7. **G** 3.3–3.5. *Color* apple-green to emerald green, white. May be white with spots of green. *Luster* vitreous, pearly on cleavage surfaces. *Optics:* (+); $\alpha = 1.654$, $\beta = 1.659$, $\gamma = 1.667$; 2V = 70°; $X = b$, $Z \wedge c = 34°$; $r > v$.

Composition and Structure. Na$_2$O 15.4, Al$_2$O$_3$ 25.2, SiO$_2$ 59.4 for pure end member. There is no replacement of Si by Al in jadeite and Fe^{3+} substitution for Al is very limited. Jadeite has a composition that is intermediate between that of nepheline, $NaAlSiO_4$, and albite, $NaAlSi_3O_8$, but does not form under the normal crystallization conditions of these two minerals. However under high pressures (10 to 25 kilobars) and elevated temperatures (between 600 and 1000°C) jadeite forms by:

$$NaAlSiO_4 + NaAlSi_3O_8 \leftrightarrows 2NaAlSi_2O_6$$

nepheline   albite   jadeite

Similarly jadeite forms at high pressures at the expense of albite alone according to the reaction:

$$NaAlSi_3O_8 \leftrightarrows NaAlSi_2O_6 + SiO_2$$

(see Fig. 10.44)

The structure of jadeite is shown in Fig. 10.37.

Diagnostic Features. Fuses at $2\frac{1}{2}$ to a trans-

pact fibers. Distinguished from nephrite by its ease of fusion. On polished surfaces nephrite has an oily luster, jadeite is vitreous.

Occurrence. Jadeite is found only in metamorphic rocks. Laboratory experiments have shown that high pressure and only relatively low temperatures are necessary for the formation of jadeite. Such occurrences are found near the margins of the continental crust as in the Alps, California, and Japan. In the Fransiscan Formation of California, jadeitic pyroxene is associated with glaucophane, aragonite, muscovite, lawsonite, and quartz.

FIG. 10.44. Experimentally determined stability fields of albite and jadeite + quartz. Reaction curve is only approximately located as shown by dashes.

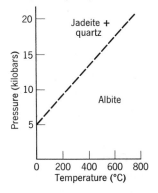

Use. Jadeite has long been highly prized in the Orient, especially in China, where it is worked into ornaments and utensils of great variety and beauty. It was also used by primative people for various weapons and implements.

Name. The term *jade* includes both jadeite and the amphibole, nephrite.

Similar Species. *Omphacite* is a bright green variety of augite-type composition rich in $NaAlSi_2O_6$. It occurs in ecologites (main constituents omphacite + garnet), which are generally considered to be the result of high pressure and high temperature of metamorphism. This is supported by the high density of both omphacite and garnet.

Aegirine (*acmite*)—$NaFe^{3+}Si_2O_6$

Crystallography. Monoclinic; $2/m$. Crystals slender prismatic with steep terminations. Often in fibrous aggregates. Faces often imperfect.

$C2/c$; $a = 9.66$, $b = 8.79$, $c = 5.26$ Å; $\beta = 107°20'$, $a:b:c = 1.099:1:0.598$; $Z = 4$. $d's$: 6.5(4), 4.43(4), 2.99(10), 2.91(4), 2.54(6).

Physical Properties. *Cleavage* {110} imperfect at angles of 87° and 93°. **H** $6-6\frac{1}{2}$. **G** 3.40–3.55. *Luster* vitreous. *Color* brown or green. Translucent. *Optics* for aegirine: $(-)$; $\alpha = 1.776$, $\beta = 1.819$, $\gamma = 1.836$; $2V = 60°$; $Y = b$, $Z \wedge c = 8°$; $r > v$. Aegirine-augite: $(+)$ with lower indices and pleochroism in greens and brown.

Composition and Structure. Although the term *acmite* has been used to describe $NaFe^{3+}Si_2O_6$, both *aegirine* and *acmite* are used for pyroxenes of this composition. Aegirines show a wide range of composition but in most species the replacement is according to $NaFe^{3+} \rightleftarrows Ca(Mg, Fe^{2+})$ which causes a complete series to augite, with an intermediate member known as *aegirine-augite*. Compositional zoning is very common in aegirine and aegirine-augite. Earlier crystallized material is richer in augite (in cores) and rims tend to be enriched in the aegirine component. The structure of aegirine is similar to that of other $C2/c$ pyroxenes (see Fig. 10.37).

Diagnostic Features. Fusible at 3, giving yellow sodium flame. Fused globule slightly magnetic. With fluxes gives tests for iron. The slender prismatic crystals, brown to green color, and association are characteristic. However, it is not easily distinguished without optical tests.

Occurrence. Aegirine is a comparatively rare rock-forming mineral found chiefly in igneous rocks rich in Na and poor in SiO_2 such as nepheline syenite and phonolite. Associated with orthoclase, feldspathoids, augite, and soda-rich amphiboles. It is also found in some metamorphic rocks associated with glaucophane or riebeckite. It occurs in the nepheline syenites and related rocks of Norway; southern Greenland; Kola Peninsula, the U.S.S.R. In the United States it is found in fine crystals at Magnet Cove, Arkansas. Found in Montana at Libby and in the Highwood and Bear Paw Mountains.

Name. From *Aegir,* the Icelandic god of the sea.

Spodumene—LiAlSi$_2$O$_6$

Crystallography. Monoclinic; $2/m$. Crystals prismatic, frequently flattened on {100}. Deeply striated vertically. Crystals usually coarse with roughened faces; some very large. Occurs also in cleavable masses. Twinning on {100} common.

$C2/c$; $a = 9.52$, $b = 8.32$, $c = 5.25$ Å; $\beta = 110°28'$, $a:b:c = 1.144:1:0.631$; $Z = 4$. $d's$: 4.38(5), 4.21(6), 2.93(10), 2.80(8), 2.45(6).

Physical Properties. *Cleavage* {110} perfect at angles of 87° and 93°. Usually a well-developed parting on {100}. **H** $6\frac{1}{2}$–7. **G** 3.15–3.20. *Luster* vitreous. *Color* white, gray, pink, yellow, green. Transparent to translucent. *Optics:* (+); $\alpha = 1.660$, $\beta = 1.666$, $\gamma = 1.676$; $2V = 58°$; $Y = b$, $Z \wedge c = 24°$. $r < v$. Absorption: $X > Y > Z$. The clear lilac-colored variety is called *kunzite,* and the clear emerald-green variety *hiddenite.*

Composition and Structure. Li_2O 8.0, Al_2O_3 27.4, SiO_2 64.6%. A small amount of Na usually substitutes for Li. The structure of spodumene is the same as that of other $C2/c$ pyroxenes. The cell volume of spodumene is smaller than that of diopside, for example, because the larger Ca and Mg ions are substituted for by smaller Li and Al. This reduction in ionic sizes causes a somewhat closer packing of the SiO_3 chains.

Diagnostic Features. Fusible at $3\frac{1}{2}$, throwing out fine branches and then fusing to a clear glass. Gives crimson flame (Li). Insoluble. Characterized by its prismatic cleavage and {100} parting. The angle formed by one cleavage direction and the {100} parting resembles the cleavage angle of tremolite. A careful angular measurement or a lithium flame is necessary to distinguish them.

Alteration. Spodumene very easily alters to other species, becoming dull. The alteration products include clay minerals, albite, *eucrypite,* $LiAlSiO_4$, muscovite, microcline.

Occurrence. Spodumene is a comparatively rare mineral found almost exclusively in lithium-rich pegmatites. It occasionally forms very large crystals. Occurs in Goshen, Chesterfield, Huntington, and Sterling, Massachusetts; Branchville, Connecticut; Newry, Maine; and Dixon, New Mexico. At the Etta mine, Black Hills, South Dakota, crystals measuring as much as 40 feet in length and weighing many tons have been found. Spodumene is mined as a source of lithium in the Black Hills, South Dakota; Kings Mountain district, North Carolina; and Gunnison County, Colorado. Hiddenite occurs with emerald beryl at Stony Point, Alexander County, North Carolina. Kunzite is found with pink beryl at Pala, San Diego County, California, and at various localities in the Malagasy Republic and Brazil.

Use. As a source of lithium. Lithium's largest use is in grease to which it is added to help it retain its lubricating properties over a wide range of temperatures. It is used in ceramics, storage batteries, air conditioning, and as a welding flux. The gem varieties of spodumene, hiddenite, and kunzite furnish very beautiful gemstones but are limited in their occurrence.

Names. *Spodumene* comes from a Greek word meaning *ash colored. Hiddenite* is named for W. E. Hidden, *kunzite,* for G. F. Kunz.

Similar Species. *Eucryptite,* $LiAlSiO_4$, is a common alteration product of spodumene.

Pyroxenoid Group

There are a number of silicate minerals that have, as do the pyroxenes, a ratio of Si:O = 1:3, but with

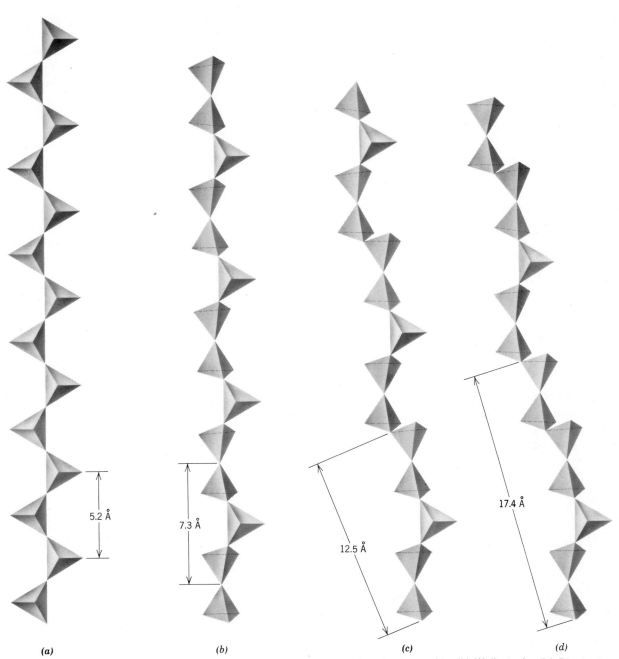

FIG. 10.45. SiO$_3$ chains in pyroxene (a) and pyroxenoids. (b) Wollastonite. (c) Rhodonite. (d) Pyroxmangite (after F. Liebau, 1959, *Acta Crystallographica,* v. 12, p. 180).

structures which are not identical to those of the pyroxenes. Both pyroxene and pyroxenoid structures contain octahedrally coordinated cations between SiO_3 chains but in pyroxenoids the geometry of the chains is not of the simple, infinitely extending type with a repeat distance of about 5.2 Å along the direction of the chain (see Fig. 10.45). In wollastonite, $CaSiO_3$, the smallest repeat of the chain consists of three twisted tetrahedra with a repeat distance of 7.3 Å and in rhondonite, $MnSiO_3$, the unit repeat is built of five twisted tetrahedra with a repeat distance of 12.5 Å. Because of the lower symmetry of the chains (as compared with the pyroxene chain) the structures of pyroxenoids are triclinic. The chain structure in the pyroxenoids is expressed by their generally splintery cleavages and sometimes fibrous habit. We will discuss in detail the following three members of the pyroxenoid group:

Wollastonite	$CaSiO_3$
Rhodonite	$MnSiO_3$
Pectolite	$Ca_2NaH(SiO_3)_3$

WOLLASTONITE—CaSiO₃

Crystallography. Triclinic; $\bar{1}$. Rarely in tabular crystals with either {001} or {100} prominent. Commonly massive, cleavable to fibrous; also compact. *Pseudowollastonite*, $CaSiO_3$, is a polymorphic form stable above 1120°C; it is triclinic, pseudohexagonal, with different properties from wollastonite.

$P\bar{1}$; $a = 7.94$, $b = 7.32$, $c = 7.07$; $\alpha = 90°2'$, $\beta = 95°22'$, $\gamma = 103°26'$; $a:b:c = 1.084:1:0.966$, $Z = 6$. $d's$: 3.83(8), 3.52(8), 3.31(8), 2.97(10), 2.47(6).

Physical Properties. *Cleavage* {100} and {001} perfect, {$\bar{1}$01} good giving splintery fragments elongated on b. **H** 5–5½. **G** 2.8–2.9. *Luster* vitreous, pearly on cleavage surfaces. May be silky when fibrous. *Color* colorless, white, or gray. Translucent. *Optics*: (−); $\alpha = 1.620$, $\beta = 1.632$, $\gamma = 1.634$; 2V = 40°; Y near b, X $\wedge$ c = 32°.

Composition and Structure. CaO 48.3, SiO_2 51.7% for pure $CaSiO_3$. Most analyses are very close to the pure end member in composition, although considerable amounts of Fe and Mn, and

lesser Mg, may replace Ca. The structure of wollastonite consists of infinite chains, parallel to the b axis, with a unit repeat of three twisted tetrahedra (see Fig. 10.45b). Ca is in irregular octahedral coordination and links the SiO_3 chains. *Pseudowollastonite*, stable above 1120°C, has space group $P\bar{1}$ but a much larger unit cell ($Z = 24$ as compared to $Z = 6$ for wollastonite). The basic structure of pseudowollastonite is very similar to that of wollastonite.

Diagnostic Features. Fusible at 4 to a white, almost glassy globule. Decomposed by HCl, with the separation of silica but without the formation of a jelly. Characterized by its two perfect cleavages of about 84°. It resembles tremolite but is distinguished from it by the cleavage angle and solubility in acid.

Occurrence. Wollastonite occurs chiefly as a contact metamorphic mineral in crystalline limestones, and forms by the reaction: $CaCO_3 + SiO_2 \rightarrow CaSiO_3 + CO_2$. It is associated with calcite, diopside, andradite, grossularite, tremolite, plagioclase feldspar, idocrase, and epidote. During progressive metamorphism of siliceous dolomites the following approximate sequence of mineral formation is often found, beginning with the lowest temperature product: talc-tremolite-diopside-forsterite-wollastonite-periclase-monticellite (see Fig. 11.16).

In places it may be so plentiful as to constitute the chief mineral of the rock mass. Such wollastonite rocks are found in the Black Forest; in Brittany; Willsboro, New York; in California; and Mexico. Crystals of the mineral are found at Csiklova in Rumania; Harz Mountains, Germany; and Chiapas, Mexico. In the United States found in New York at Diana, Lewis County, and also in Orange and St. Lawrence counties. In California at Crestmore, Riverside County.

Use. Wollastonite is mined in those places where it constitutes a major portion of the rock mass and is used in the manufacture of tile.

Name. In honor of the English chemist, W. H. Wollaston (1766–1828).

RHODONITE—MnSiO₃

Crystallography. Triclinic; $\bar{1}$. Crystals commonly tabular parallel to {001} (Fig. 10.46); often rough

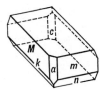

FIG. 10.46. Rhodonite.

with rounded edges. Commonly massive, cleavable to compact; in embedded grains.

Angles: $m(110) \wedge M(1\bar{1}0) = 92°29'$, $a(100) \wedge c(001) = 72°37'$, $c(001) \wedge n(\bar{2}\bar{2}1) = 73°52'$. These angles are given for the traditional morphological orientation with the intersection of the cleavages defining c, as in the pyroxenes. The cell parameters refer to a more recent orientation.

$P\bar{1}$; $a = 7.79$, $b = 12.47$, $c = 6.75$ Å; $\alpha = 85°10'$, $\beta = 94°4'$, $\gamma = 111°29'$; $a:b:c = 0.625:1:0.541$; $Z = 10$. $d's$: 4.78(4), 3.15(5), 3.09(3), 2.98(8), 2.93(9), 2.76(10).

Physical Properties. Cleavage {110} and {1$\bar{1}$0} perfect. H 5½–6. G 3.4–3.7. *Luster* vitreous. *Color* rose-red, pink, brown; frequently with black exterior of manganese oxide. Transparent to translucent. *Optics:* (+); $\alpha = 1.716$–1.733, $\beta = 1.720$–1.737, $\gamma = 1.728$–1.747; 2V = 60°–75°, $r < v$.

Composition and Structure. Rhodonite is never pure $MnSiO_3$ but always contains some Ca, with a maximum $CaSiO_3$ content of about 20 molecular percent. Fe^{2+} may replace Mn up to as much as 14 weight percent FeO. Zn may be present and Zn-rich varieties are known as *fowlerite*. The structure of rhodonite consists of SiO_3 chains, parallel to the c axis, with a unit repeat of five twisted tetrahedra (see Fig. 10.45c). Layers of cations alternate with the chains. The structure is similar to that of wollastonite and pyroxmangite, (Mn, Fe)SiO_3 (see Fig. 10.45d).

Diagnostic Features. Fusible at 3 to a nearly black glass. Insoluble in HCl. Gives a blue-green color to the sodium carbonate bead. Characterized by its pink color and near 90° cleavages. Distinguished from rhodochrosite by its greater hardness and insolubility.

Occurrence. Rhodonite occurs in manganese deposits and manganese-rich iron formations, as a result of metamorphic and commonly associated metasomatic activity. It may form from rhodochrosite by the reaction: $MnCO_3 + SiO_2 \rightarrow MnSiO_3 + CO_2$.

Rhodonite is found at Långban, Sweden, with other manganese minerals and iron ore; in large masses near Sverdlovsk in the Ural Mountains, U.S.S.R.; and from Broken Hill, New South Wales. In the United States rhodonite and fowlerite occur in good-sized crystals in crystalline limestone with franklinite, willemite, zincite, and so forth, at Franklin, New Jersey.

Use. Some rhodonite is polished for use as an ornamental stone. This material is obtained chiefly from the Ural Mountains, U.S.S.R., and Australia.

Name. Derived from the Greek word for a *rose*, in allusion to the color.

Similar Species. *Pyroxmangite,* (Mn, Fe)SiO_3, is structurally very similar to rhodonite, but with a unit repeat of seven tetrahedra in the SiO_3 chain (see Fig. 10.45d). *Pyroxferroite,* $Ca_{0.15}Fe_{0.85}SiO_3$, isostructural with pyroxmangite, is a relatively common mineral in lunar lavas. *Bustamite,* (Mn, Ca, Fe)SiO_3, is very similar in structure to wollastonite. *Tephroite,* Mn_2SiO_4, a red to gray mineral associated with rhodonite, is isostructural with olivine.

Pectolite—$Ca_2NaH(SiO_3)_3$

Crystallography. Triclinic; $\bar{1}$. Crystals elongated parallel to the b axis. Usually in aggregates of acicular crystals. Frequently radiating, with fibrous appearance. In compact masses.

$P\bar{1}$; $a = 7.99$, $b = 7.04$, $c = 7.02$ Å; $\alpha = 90°31'$, $\beta = 95°11'$, $\gamma = 102°28'$; $a:b:c = 1.135:1:0.997$; $Z = 2$. $d's$: 3.28(7), 3.08(9), 2.89(10), 2.31(7), 2.28(7).

Physical Properties. *Cleavage* {001} and {100} perfect. H 5. G 2.8±. *Luster* vitreous to silky. *Color* colorless, white, or gray. Transparent. *Optics:* (+); $\alpha = 1.595$, $\beta = 1.604$, $\gamma = 1.633$; 2V = 60°; $Z \approx b$, $X \wedge c = 19°$.

Composition and Structure. CaO 33.8, Na_2O 9.3, SiO_2 54.2, H_2O 2.7%. In some pectolite

Mn^{2+} substitutes for Ca. The structure of pectolite contains SiO_3 chains, parallel to the b axis, the unit repeat of which consists of three twisted tetrahedra (see Fig. 10.45b) similar to that found in wollastonite. The Ca ions are in octahedral coordination and Na is present in a very distorted octahedral coordination.

Diagnostic Features. Fuses quietly at $2\frac{1}{2}$–3 to a glass coloring the flame yellow (Na). Decomposed by HCl, with the separation of silica but without the formation of a jelly. Water in the closed tube. Characterized by two directions of perfect cleavage, yielding sharp, acicular fragments that will puncture the skin if not handled carefully. Resembles wollastonite but gives test for water and sodium. Distinguished from similar-appearing zeolites by absence of aluminum.

Occurrence. Pectolite is a secondary mineral similar in its occurrence to the zeolites. Found lining cavities in basalt, associated with various zeolites, prehnite, calcite, and so forth. Found at Bergen Hill and West Paterson, New Jersey.

Name. From the Greek word meaning *compact,* in allusion to its habits.

Amphibole Group

The chemical composition of members of the amphibole group can be expressed by the general formula $W_{0-1}X_2Y_5Z_8O_{22}$ (OH, F)$_2$ where W represents Na and K in the A site, X denotes Ca^{2+}, Na^+, Mn^{2+}, Fe^{2+}, Mg^{2+}, and Li^+ in the $M4$ sites, Y represents Mn^{2+}, Fe^{2+}, Mg^{2+}, Fe^{3+}, Al^{3+}, and Ti^{4+} in the $M1$, $M2$, and $M3$ sites and Z refers to Si^{4+} and Al^{3+} in the tetrahedral sites. Essentially complete ionic substitution may take place between Na and Ca and among Mg, Fe^{2+}, and Mn^{2+}. There is limited substitution between Fe^{3+} and Al and between Ti and other Y-type ions; and partial substitution of Al for Si in the tetrahedral sites of the double chains. Partial substitution of F and O for OH in the hydroxyl sites is also common. Several common series of the amphibole group can be represented compositionally in the chemical system $Mg_7Si_8O_{22}(OH)_2$ (anthophyllite)–$Fe_7Si_8O_{22}(OH)_2$ (grunerite)–"$Ca_7Si_8O_{22}$(OH)$_2$," a hypothetical end member composition, as in Fig. 10.47. This illustration is very similar to the Wo-En-Fs diagram for the pyroxene group (Fig. 10.36). A complete series exists from *tremolite*, $Ca_2Mg_5Si_8O_{22}(OH)_2$ to *ferroactinolite*, $Ca_2Fe_5Si_8O_{22}(OH)_2$. The commonly occurring *actinolite* composition is a generally Mg-rich member of the tremolite-ferroactinolite series. The compositional range from $Mg_7Si_8O_{22}(OH)_2$ to about $Fe_2Mg_5Si_8O_{22}(OH)_2$ is represented by the orthorhombic species, *anthophyllite*. The *cummingtonite-grunerite* series is monoclinic and extends from about $Fe_2Mg_5Si_8O_{22}(OH)_2$ to $Fe_7Si_8O_{22}(OH)_2$. Miscibility gaps are present between anthophyllite and the tremolite-actinolite series, as well as between the

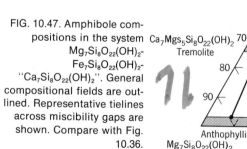
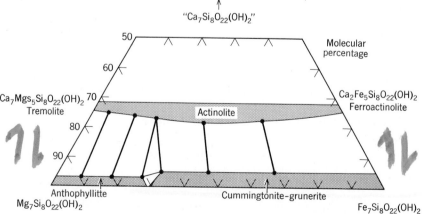

FIG. 10.47. Amphibole compositions in the system $Mg_7Si_8O_{22}(OH)_2$–$Fe_7Si_8O_{22}(OH)_2$–"$Ca_7Si_8O_{22}(OH)_2$". General compositional fields are outlined. Representative tielines across miscibility gaps are shown. Compare with Fig. 10.36.

members of the cummingtonite-grunerite series and the calcic amphiboles. These gaps are reflected in the common occurrence of anthophyllite-tremolite and grunerite-actinolite pairs, as well as exsolution textures between members of the Mg-Fe and the Ca-containing amphiboles of Fig. 10.47. No Ca-amphibole compositions are possible above the 2/7th line (representing two Ca out of a total of 7 $X + Y$ cations) because Ca can be housed only in the two $M4$ sites of the amphibole structure. *Hornblende* may be regarded as a tremolite-ferroactinolite type composition with additional partial substitution of Na in the A and $M4$ sites, Mn, Fe^{3+}, and Ti^{4+} for Y cations, and Al for Si in the tetrahedral sites, leading to a very complex general formula. Sodium containing amphiboles are represented by members of the *glaucophane*, $Na_2Mg_3Al_2Si_8O_{22}(OH)_2$,–*riebeckite*, $Na_2Fe_3^{2+}Fe_2^{3+}Si_8O_{22}(OH)_2$, series. *Arfvedsonite*, $NaNa_2Fe_4^{2+}Fe^{3+}Si_8O_{22}(OH)_2$, contains additional Na in the A site of the structure. Table 10.3 compares the composition of some of the common members of the pyroxene and amphibole groups, and Table 4.15 illustrates the recalculation of an amphibole analysis. The amphibole structure is based on double Si_4O_{11} chains that run parallel to the c axis. Figure 10.48 (see also Fig. 10.2) illustrates this chain as

well as the octahedral strip to which it is bonded. The structure contains several cation sites, labeled A, $M4$, $M3$, $M2$, $M1$, as well as the tetrahedral sites in the chains. The A site has 10 to 12 coordination with oxygen and (OH) and houses mainly Na, and at times small amounts of K. The $M4$ site has 6 to 8 coordination and houses X type cations (see Table 10.3). The $M1$, $M2$, and $M3$ octahedra accommodate Y type cations and share edges to form octahedral bands parallel to c. $M1$ and $M3$ are coordinated by four oxygens and two (OH, F) groups whereas $M2$ is coordinated by six oxygens. Figure 10.49a shows the monoclinic amphibole structure and the distribution of cation sites as seen in a direction parallel to the c axis. The t-o-t strips (Fig. 10.49b) are approximately twice as wide (in the b direction) as the equivalent t-o-t strips in pyroxenes (see Fig. 10.37b) because of the doubling of the chain width in amphiboles. This wider geometry causes the typical 56° and 124° cleavage angles as shown in Fig. 10.49b.

The majority of amphiboles can be assigned to one of three space groups; two monoclinic ($C2/m$ and $P2_1/m$) and one orthorhombic ($Pnma$). The $C2/m$ structure is found in all of the common clinoamphiboles such as tremolite, $Ca_2Mg_5Si_8O_{22}(OH)_2$, and hornblende. This type of structure is illustrated

Table 10.3

IONS IN COMMON PYROXENES AND AMPHIBOLES

PYROXENES			AMPHIBOLES			
Atomic Sites		**Name**		**Atomic Sites**		**Name**
$M2$	$M1$		A	$M4$	$(M1 + M2 + M3)$	
Mg	Mg	Enstatite	—	Mg	Mg	Anthophyllite
		Bronzite	—	Fe	Mg	Cummingtonite
Fe	Mg	Hypersthene				
		Clinohypersthene	—	Fe	Fe	Grunerite
Ca	Mg	Diopside	—	Ca	Mg	Tremolite
Ca	Fe	Hedenbergite	—	Ca	Fe	Ferroactinolite
Ca	Mn	Johannsenite				
Ca, Na	Mg, Fe Mn, Al Fe^{3+}, Ti	Augite	—	Ca, Na	Mg, Fe^{2+}, Mn Al, Fe^{3+}, Ti	Hornblende
Na	Al	Jadeite	—	Na	Mg, Al	Glaucophane
Na	Fe^{3+}	Aegirine	—	Na	Fe^{2+}, Fe^{3+}	Riebeckite
			Na	Na	Fe^{2+}, Fe^{3+}	Afrvedsonite
Li	Al	Spodumene	—	Li	Mg, Fe^{3+} Al, Fe^{2+}	Holmquistite

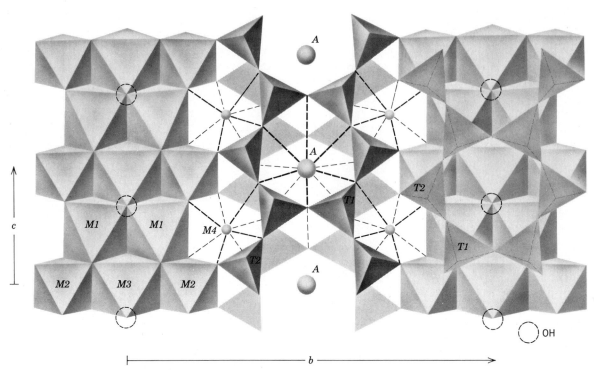

FIG. 10.48. Crystal structure of a monoclinic (C2/m) amphibole projected down the a axis. (OH) groups are located in the center of the large holes of the rings in the chains. M1, M2, and M3 sites house Y cations and are 6 coordinated. The M4 site houses larger X cations in 6 to 8 coordination. The ion in the A site, between the backs of double chains is 10 to 12 coordinated (after J. J. Papike et al., 1969, Mineralogical Society of America, Special Paper no. 2, p. 120).

FIG. 10.49 (a) Schematic projection of the monoclinic amphibole structure on a plane perpendicular to the c axis (after Colville et al., 1966, American Mineralogist, v. 51, p. 1739). Compare with Fig. 10.38a. (b) Control of cleavage angles by t-o-t strips (also referred to as "I beams") in the amphibole structure, as compared with naturally occurring cleavage angles.

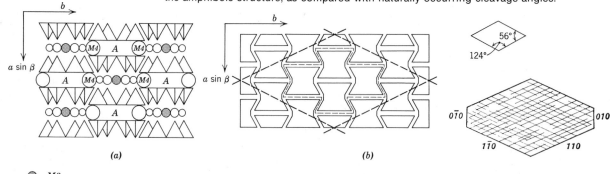

in Fig. 10.48. The $P2_1/m$ space group is found in some Mg-rich cummingtonites, close to $(Mg, Fe)_2Mg_5Si_8O_{22}(OH)_2$ in composition and results because the $M4$ site is somewhat collapsed to house the relatively small Mg and Fe^{2+} ions. The orthorhombic $Pnma$ structure is found in members of the anthophyllite, $Mg_7Si_8O_{22}(OH)_2$ to $Fe_2Mg_5Si_8O_{22}(OH)_2$, series and in gedrite, an Al- and Na-containing anthophyllite, and in holmquistite. The presence of cations of small size in $M4$, $M3$, $M2$, and $M1$ leads to an orthorhombic rather than a monoclinic structure. The unit cells of orthorhombic amphiboles are related to the monoclinic unit cells by a twinlike mirror across (100) accompanied by an approximate doubling of the a cell dimension (e.g., a of anthophyllite ≈ 2 a sin β of tremolite; see Figs. 10.39b and c for pyroxene unit cell relations).

The presence of (OH) groups in the structure of amphiboles causes a decrease in their thermal stabilities as compared to the more refractory pyroxenes. This causes amphiboles to decompose to anhydrous minerals (often pyroxenes) at elevated temperatures below the melting point (see Figs. 10.41 and 10.50).

FIG. 10.50. Comparison of the thermal stability limits of tremolite and ferroactinolite (after W. G. Ernst, 1968, *Amphiboles*, Springer-Verlag, p. 58).

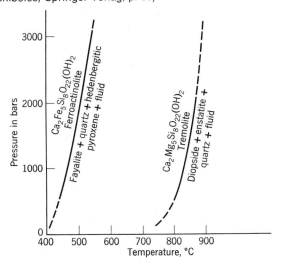

We will discuss in detail the following common amphiboles:

AMPHIBOLES

Anthophyllite	$(Mg, Fe)_7Si_8O_{22}(OH)_2$
Cummingtonite series	
Cummingtonite	$Fe_2Mg_5Si_8O_{22}(OH)_2$
Grunerite	$Fe_7Si_8O_{22}(OH)_2$
Tremolite series	
Tremolite	$Ca_2Mg_5Si_8O_{22}(OH)_2$
Actinolite	$Ca_2(Mg, Fe)_5Si_8O_{22}(OH)_2$
Hornblende	$X_{2-3}Y_5Z_8O_{22}(OH)_2$
Sodium amphibole group	
Glaucophane	$Na_2Mg_3Al_2Si_8O_{22}(OH)_2$
Riebeckite	$Na_2Fe_3^{2+}Fe_2^{3+}Si_8O_{22}(OH)_2$

Anthophyllite—$(Mg, Fe)_7Si_8O_{22}(OH)_2$

Crystallography. Orthorhombic; $2/m2/m2/m$. Rarely in distinct crystals. Commonly lamellar or fibrous.

$Pnma$; $a = 18.56$, $b = 18.08$, $c = 5.28$ Å; $a:b:c = 1.027:1:0.292$; $Z = 4$. $d's$: 8.26(6), 3.65(4), 3.24(6), 3.05(10), 2.84(4).

Physical Properties. *Cleavage* {210} perfect; $(210) \wedge (2\bar{1}0) = 55°$. **H** $5\frac{1}{2}$–6. **G** 2.85–3.2. *Luster* vitreous. *Color* gray to various shades of green and brown and beige. *Optics:* (−); $\alpha = 1.60$–1.69, $\beta = 1.61$–1.71, $\gamma = 1.62$–1.72. 2V = 70°–100°; $X = a$, $Y = b$. Absorption $Z > Y$ and X. Indices increase with Fe content.

Composition and Structure. Anthophyllite is part of a solid solution series from $Mg_7Si_8O_{22}(OH)_2$ to approximately $Fe_2Mg_5Si_8O_{22}(OH)_2$; at higher Fe contents the monoclinic cummingtonite structure results. *Gedrite* is an Al- and Na-containing variety of anthophyllite, with an end member composition approximating $Na_{0.5}(Mg, Fe^{2+})_2(Mg, Fe^{2+})_{3.5}(Al, Fe^{3+})_{1.5}Si_6Al_2O_{22}(OH)_2$, with Na in the A site of the structure. At moderate temperatures a miscibility gap exists between anthophyllite and gedrite as shown by coexisting anthophyllite and gedrite grains. The structures of anthophyllite and gedrite are similar, both with orthorhombic space group $Pnma$.

Diagnostic Features. Characterized by its clove-brown color but unless in crystals cannot be distinguished from other amphiboles such as cummingtonite or grunerite without optical or X-ray tests. Fusible at 5 to a black magnetic enamel. Insoluble in acids. Yields water in the closed tube.

Occurrence. Anthophyllite is a metamorphic product of Mg-rich rocks such as ultrabasic igneous rocks and impure dolomitic shales. It is common in cordierite-bearing gneisses and schists. It may also form as a retrograde product rimming relict orthopyroxenes and olivine (see Fig. 10.41). Occurs at Kongsberg, Norway; in many localities in southern Greenland. In the United States it is found at several localities in Pennsylvania, in southwestern New Hampshire and central Massachusetts, in the Gravelly Range and Tobacco Root Mountains of southwestern Montana and at Franklin, North Carolina.

Name. From the Latin *anthophyllum,* meaning *clove,* in allusion to the clove-brown color.

CUMMINGTONITE—(Mg, Fe)$_7$Si$_8$O$_{22}$(OH)$_2$
GRUNERITE—Fe$_7$Si$_8$O$_{22}$(OH)$_2$

Crystallography. Monoclinic; $2/m$. Rarely in distinct crystals. Commonly fibrous or lamellar, often radiated.

$C2/m$; for grunerite, $a = 9.59, b = 18.44, c = 5.34$, $\beta = 102°0'$; cell lengths decrease with increasing Mg (see Fig. 3.19). $a:b:c = 0.520:1:0.289$; $Z = 2$. $d's$: 9.21(5), 8.33(10), 3.07(8), 2.76(9), 2.51(6).

Physical Properties. *Cleavage* {110} perfect. **H** $5\frac{1}{2}$–6. **G** 3.1–3.6. *Luster* silky; fibrous. *Color* various shades of light brown. Translucent; will transmit light on thin edges. *Optics:* (−) for grunerite; (+) for cummingtonite; $\alpha = 1.65$–1.69; $\beta = 1.67$–1.71; $\gamma = 1.69$–1.73; 2V large; $Y = b$, $Z \wedge c = 13$–20°. $r < v$ for cummingtonite; $r > v$ for grunerite. Essentially nonpleochroic.

Composition and Structure. The cummingtonite-grunerite series extends from approximately Fe$_2$Mg$_5$Si$_8$O$_{22}$(OH)$_2$ to the end member Fe$_7$Si$_8$O$_{22}$(OH)$_2$. Members with Mg < Fe (atomic percentage) are referred to as cummingtonite; those with Fe > Mg as grunerite (in the literature this division is often taken at 30 atomic percent Fe). Mn^{2+} can be present to as much as

Mn$_2$Mg$_5$Si$_8$O$_{22}$(OH)$_2$, with probably most of the Mn in the *M4* structure site. Al$_2$O$_3$ and CaO range up to a maximum of 0.4 and 0.9 weight percent, respectively. The structure of the majority of members of the cummingtonite-grunerite series is $C2/m$ like that of tremolite; however, some Mg-rich cummingtonites have $P2_1/m$ symmetry.

Diagnostic Features. Characterized by light brown color and needlelike, often radiating habit. May be impossible to distinguish from anthophyllite or gedrite without optical or X-ray tests.

Occurrence. Cummingtonite is a constituent of regionally metamorphosed rocks and occurs in amphibolites. It commonly coexists with hornblende or actinolite (see tielines in Fig. 10.47). Mg-rich cummingtonite can also coexist with anthophyllite, on account of a small miscibility gap between anthophyllite and the Mg-rich part of the cummingtonite-grunerite series. Cummingtonite phenocrysts have been reported in some igneous rocks such as dacites. Mn-rich varieties occur in metamorphosed manganese-rich units. Grunerites are characteristic of metamorphosed iron-formations in the Lake Superior region and the Labrador Trough. Upon prograde metamorphism cummingtonite and grunerite give way to members of the orthopyroxene or olivine series.

Use. *Amosite,* an asbestiform variety with ash-gray color, and long flexible fibers is used as asbestos and mined for this purpose in South Africa.

Names. Cummingtonite is named after Cummington, Massachusetts and grunerite after Grüner, nineteenth century.

TREMOLITE—Ca$_2$Mg$_5$Si$_8$O$_{22}$(OH)$_2$
ACTINOLITE—Ca$_2$(Mg, Fe)$_5$Si$_8$O$_{22}$(OH)$_2$

Crystallography. Monoclinic; $2/m$. Crystals usually prismatic. The termination of the crystals is almost always formed by the two faces of a low first-order prism (similar to hornblende, see Fig. 10.51). Tremolite is often bladed and frequently in radiating columnar aggregates. In some cases in silky fibers. Coarse to fine granular. Compact. Angles close to those of hornblende.

$C2/m$; $a = 9.84, b = 18.05, c = 5.28$; $\beta =$

104°42'; $a:b:c = 0.545:1:0.293$; $Z = 2$. $d's$: 8.38(10), 3.27(8), 3.12(10), 2.81(5), 2.71(9).

Physical Properties. *Cleavage* {110} *perfect*; $(110) \wedge (1\bar{1}0) = 56°$ often yielding a splintery surface. **H** 5–6. **G** 3.0–3.3. *Luster* vitreous; often with silky sheen in the prism zone. *Color* varying from white to green in actinolite. The color deepens and the specific gravity increases with increase in Fe content. Transparent to translucent. A felted aggregate of tremolite fibers goes under the name of *mountain leather* or *mountain cork*. A tough, compact variety that supplies much of the material known as *jade* is called *nephrite* (see Jadeite). *Optics:* $(-)$; $\alpha = 1.56–1.63$, $\beta = 1.61–1.65$, $\gamma = 1.63–1.66$; $2V = 80°$; $Y = b$, $Z \wedge c = 15°$. Indices increase with iron content.

Composition and Structure. A complete solid solution exists from tremolite to ferroactinolite, $Ca_2Fe_5Si_8O_{22}(OH)_2$. The name actinolite is used for intermediate members. In high-temperature metamorphic and in igneous occurrences a complete series exists from the tremolite-ferroactinolite series to hornblende as a result of a wide range of Al substitution for Si, and concomitant cation replacements elsewhere in the structure. The names tremolite and ferroactinolite are generally used to refer to the Al-poor members of this large compositional range. In lower temperature occurrences tremolite (or actinolite) may coexist with hornblende as the result of a miscibility gap between tremolite (or actinolite) and hornblende. The structures of the members of the tremolite-ferroactinolite series are of the $C2/m$ type (see Figs. 10.48, 10.49, and page 385).

Diagnostic Features. Characterized by slender prisms and good prismatic cleavage. Distinguished from pyroxenes by the cleavage angle and from hornblende by lighter color. Fusible at 4 to a clear glass.

Occurrence. Tremolite is most frequently found in metamorphosed dolomitic limestones where it forms according to: $5CaMg(CO_3)_2 + 8SiO_2 + H_2O \rightarrow Ca_2Mg_5Si_8O_{22}(OH)_2 + 3CaCO_3 + 7CO_2$. At higher temperatures tremolite is unstable and yields to diopside (see Fig. 10.50). Actinolite is a characteristic mineral of the greenschist facies of metamorphism. It occurs also in glaucophane schists in coexistence with quartz, epidote, glaucophane, pumpellyite, and lawsonite.

Notable localities for crystals of tremolite are: Ticino, Switzerland; in the Tyrol; and in Piemonte, Italy. In the United States from Russell, Gouverneur, Amity, Pierrepont, DeKalb, and Edwards, New York. Crystals of actinolite are found at Greiner, Zillerthal, Tyrol. Tremolite and actinolite frequently are fibrous; this is the material to which the name *asbestos* was originally given. The asbestiform varieties have been found in metamorphic rocks in various states along the Applachian Mountains. The original jade was probably nephrite that came from the Khotan and Yarkand Mountains of Chinese Turkestan. Nephrite has been found in New Zealand, Mexico, and Central America; and, in the United States, near Lander, Wyoming.

Use. The fibrous varieties are used as asbestos, but fibrous serpentine, *chrysotile,* is more widely used and usually a better grade. The compact variety, *nephrite,* is used as ornamental material.

Names. Tremolite is derived from the Tremola Valley near St. Gothard, Switzerland. Actinolite comes from two Greek words meaning *a ray* and *stone,* in allusion to its frequently somewhat radiated habit.

HORNBLENDE—
$(Ca, Na)_{2-3}(Mg, Fe, Al)_5Si_6(Si, Al)_2O_{22}(OH)_2$
Crystallography. Monoclinic; $2/m$. Crystals prismatic, usually terminated by {011}. The vertical prism zone shows, in addition to the prism faces, usually {010}, and more rarely {100} (see Fig. 10.51). May be columnar or fibrous; coarse to fine grained.

Angles: $m(110) \wedge m'(1\bar{1}0) = 55°49'$, $r(011) \wedge r'(0\bar{1}1) = 31°32'$, $b(010) \wedge e(130) = 32°11'$, $c(001) \wedge a(100) = 74°16'$.

$C2/m$; $a = 9.87$, $b = 18.01$, $c = 5.33$ Å; $\beta = 105°44'$. $a:b:c = 0.548:1:0.296$; $Z = 2$. $d's$: 3.38(9), 3.29(7), 3.09(10), 2.70(10), 2.59(7).

Physical Properties. *Cleavage* {110} *perfect.* **H** 5–6. **G** 3.0–3.4. *Luster* vitreous; fibrous; fibrous varieties often silky. *Color* various shades of dark green to black. Translucent; will transmit light on

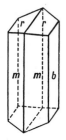

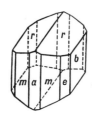

FIG. 10.51. Hornblende crystals.

thin edges. *Optics:* (−); α = 1.61–1.71, β = 1.62–1.72, γ = 1.63–1.73; 2V = 30°–90°; Y = b, Z ∧ c = 15°–34°. r > v usually; may be r < v. Pleochroism X yellow-green, Y olive green, Z deep green.

Composition and Structure. What passes under the name of hornblende is in reality a complex and large compositional range with variations in the ratios Ca/Na, Mg/Fe^{2+}, Al/Fe^{3+}, Al/Si and OH/F. A generalized formula for a "common" hornblende is given above and an example of the recalculation of an amphibole analysis is given in Table 4.15. At elevated temperatures hornblende shows a complete series toward tremolite-ferroactinolite, but at lower temperaturers a miscibility gap exists. In some volcanic rocks oxyhornblendes are found with considerable Fe^{3+}/Fe^{2+} ratios and low OH contents; an example composition is NaCa$_2$Fe$_4^{2+}$Fe^{3+}(AlSi$_7$)O$_{23}$(OH). The structure of hornblende is similar to that of C2/m tremolite (see Figs. 10.48 and 10.49). On account of the miscibility gap between Ca-amphiboles and Mg-Fe amphiboles, hornblende often contains grunerite exsolution lamellae (see Fig. 4.28).

Diagnostic Features. Crystal form and cleavage angles serve to distinguish hornblende from dark pyroxenes. Usually distinguished from other amphiboles by its dark color. Fusible at 4. Yields water in the closed tube.

Occurrence. Hornblende is an important and widely distributed rock-forming mineral, occurring in both igneous and metamorphic rocks; it is particularly characteristic of some medium-grade metamorphic rocks known as amphibolites in which hornblende and associated plagioclase are the major constituents. It characteristically alters from pyroxene both during the late magmatic stages of crystallization of igneous rocks and during metamorphism. Hornblende is a common constituent of syenites and diorites.

Name. From an old German word for any dark prismatic mineral occurring in ores but containing no recoverable metal.

Similar Species. Because of the large chemical variation in the hornblende series, many names have been proposed for members on the basis of chemical composition and physical and optical properties. The most common names on the basis of chemical composition are: *edenite, pargasite, hastingsite,* and *tschermakite.*

Glaucophane—Na$_2$Mg$_3$Al$_2$Si$_8$O$_{22}$(OH)$_2$
Riebeckite—Na$_2$Fe$_3^{2+}$Fe$_2^{3+}$Si$_8$O$_{22}$(OH)$_2$

Crystallography. Monoclinic; 2/m. In slender acicular crystals; frequently aggregated; sometimes asbestiform.

C2/m; a = 9.58, b = 17.80, c = 5.30 Å; β = 103°48'; a:b:c = 0.538:1:0.298; Z = 2. d's: 8.42(10), 4.52(5), 3.43(6), 3.09(8), 2.72(10).

Physical Properties. *Cleavage* {110} perfect. **H** 6. **G** 3.1–3.4. *Luster* vitreous. *Color* blue to lavender-blue to black, darker with increasing iron content. *Streak* white to light blue. Translucent. *Optics:* (−); α = 1.61–1.70, β = 1.62–1.71, γ = 1.63–1.72; 2V = 40°–90°; Y = b, Z ∧ c = 8°. Pleochroism in blue X < Y < Z.

Composition and Structure. Very few glaucophanes are close to the end member composition, Na$_2$Mg$_3$Al$_2$Si$_8$O$_{22}$(OH)$_2$, on account of some substitution by Fe^{2+} for Mg and Fe^{3+} for Al. Riebeckite analyses do not conform well with the end member formulation because the sum of X + Y cations is frequently larger than 7 with Na entering the normally vacant A site. A partial series exists between glaucophane and riebeckite, with intermediate compositions known as *crossite.* The structures of both sodic amphiboles are similar to that of C2/m tremolite (see Fig. 10.48).

Diagnostic Features. Glaucophane and riebeckite are characterized by their generally fibrous habit and blue color. Fusible at 3.

Occurrence. *Glaucophane* is found only in metamorphic rocks such as schists, eclogite, and marble. The occurrences of glaucophane reflect low-temperature, relatively high-pressure metamorphic conditions, in association with jadeite, lawsonite, and aragonite. It is a major constituent of glaucophane schists in the Fransiscan Formation of California. It is also found in metamorphic rocks from Mexico, Japan, Taiwan, Indonesia, and eastern Australia. *Riebeckite* occurs most commonly in igneous rocks such as granites, syenites, nepheline syenites, and related pegmatites. It is a conspicuous mineral in the granite of Quincy, Massachusetts. It is present in some schists of regional metamorphic origin. The asbestiform variety of riebeckite, *crocidolite,* occurs in South Africa, in a metamorphic iron-formation and is mined on a large scale. A similar occurrence is in the iron-formation of the Hamersley Range, Western Australia.

Use. Crocidolite makes up about 4% of the total world production of asbestos. In many places in South Africa there has been a pseudomorphous replacement of crocidolite by quartz. The preservation of the fibrous texture makes it an attractive ornamental material with a chatoyancy. This is widely used for jewelry under the name of *tiger eye.*

Name. Glaucophane is from the two Greek words meaning *bluish* and *to appear.* Riebeckite is in honor or E. Riebeck.

Similar Species. *Arfvedsonite,* $Na_3Fe_4^{2+}Fe^{3+}$-$Si_8O_{22}(OH)_2$, is a deep green amphibole similar to riebeckite in occurrence. It is commonly associated with aegirine, or aegirine-augite. In the structure of arfvedsonite the *A* site is completely or almost completely filled by Na.

PHYLLOSILICATES

As the derivation of the name of this important group implies (Greek: *phyllon,* leaf), most of its many members have a platy or flaky habit and one prominent cleavage. They are generally soft, of relatively low specific gravity and may show flexibility or even elasticity of the cleavage lamellae. All these characterizing peculiarities arise from the dominance in the structure of the indefinitely extended

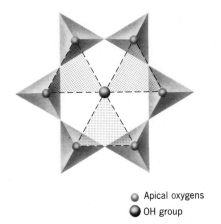

● Apical oxygens
◑ OH group

FIG. 10.52. Undistorted, hexagonal ring in Si_2O_5 sheet showing location of apical oxygens and (OH) group. In an ideal sheet structure the size of the triangles (outlined by shading) is the same as the size of triangular faces of XO_6 octahedra.

sheet of SiO_4 tetrahedra. In this sheet (see Fig. 10.2) three of the four oxygens in each SiO_4 tetrahedron are shared with neighboring tetrahedra, leading to a ratio of $Si:O = 2:5$. Each sheet, if undistorted, has 6-fold symmetry.

Most of the members of the phyllosilicates are hydroxyl bearing with the (OH) group located in the center of the 6-fold rings of tetrahedra, at the same height as the unshared apical oxygens in the SiO_4 tetrahedra (see Fig. 10.52). When ions, external to the Si_2O_5 sheet, are bonded to the sheets, they are coordinated to two oxygens and one OH, as shown in Fig. 10.52. The size of the triangle between 2 oxygens and one (OH) is just about the same as (but not identical to, see page 398) the triangular face of an XO_6 octahedron (with *X* commonly Mg or Al). This means that it is possible to bond to a regular net of apical oxygens and (OH) groups of $(Si_2O_5OH)^{-3}$ composition a sheet of regular octahedra, in which each octahedron is tilted onto one of its triangular sides (Fig. 10.53). When such tetrahedral and octahedral sheet are joined one obtains the general geometry of the *antigorite* or *kaolinite* structures (Fig. 10.54).

The cations in the octahedral layer may be divalent or trivalent. When the cations are divalent, for example, Mg or Fe^{2+}, the layer will have the

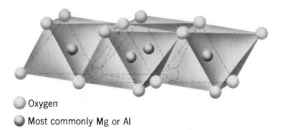

○ Oxygen

● Most commonly Mg or Al

FIG. 10.53. Infinitely extending sheet of XO_6 octahedra. All octahedra lie on triangular faces.

geometry of that of brucite (Fig. 8.19) in which each cation site is occupied. In such a layer six bonds, each with e.v. $= \frac{2}{6} = \frac{1}{3}$, originate from the Mg^{2+} ion. Three such bonds radiate from each oxygen, or (OH) group, thus neutralizing one-half the charge of the oxygen but all of the OH. A layer in which each oxygen or (OH) group is surrounded by three cations, as in the brucite, $Mg(OH)_2$, structure, is known as *trioctahedral*. When the cations in the octahedral layer are trivalent, charge balance is maintained when one out of every three cations sites is unoccupied. This is the case in the gibbsite, $Al(OH)_3$, structure (Fig. 8.2 and see page 282). Such a layer structure in which each oxygen or (OH) group is surrounded only by two cations, is known as *dioctahedral*. On the basis of the chemistry and geometry of the octahedral layers, the phyllosilicates are divided into two major groups: *trioctahedral* and *dioctahedral*.

It was pointed out above that the geometries of an average Si_2O_5 sheet and a layer of XO_6 octahedra are generally compatible such that one can be bonded to the other. Let us look at this in terms of the chemistries of the octahedral and tetrahedral layers. Brucite, $Mg(OH)_2$, consists of two (OH) sheets between which Mg is coordinated in octahedra. We can state this symbolically as $Mg_3 \frac{(OH)_3}{(OH)_3}$. If we replace two of the (OH) groups on one side of a brucite layer by two apical oxygens of an Si_2O_5 sheet, we obtain $Mg_3 \frac{Si_2O_5(OH)}{(OH)_3}$. This means that the other side of the brucite structure is not attached to an Si_2O_5 sheet as in Fig. 10.54. This is known as the *antigorite*, $Mg_3Si_2O_5(OH)_4$, structure. The equivalent structure with a dioctahedral sheet is *kaolinite*, $Al_2Si_2O_5(OH)_4$. In short, the antigorite and kaolinite structures are built of one tetrahedral ("t") and one octahedral ("o") sheet giving "t-o" layers (see Fig. 10.57). Such *t-o* layers are electrically neutral and are bonded to one another by weak van der Waals' bonds. As noted above, in the antigorite and kaolinite structures only one side of the octahedral sheet is coordinated to a tetrahedral sheet. However, we can derive further members of the phyllosilicate group by joining tetrahedral sheets on both sides of the *o* sheet. This leads to *t-o-t* type sandwiches (Fig. 10.55) as in *talc*, $Mg_3Si_4O_{10}(OH)_2$, and *pyrophyllite*, $Al_2Si_4O_{10}(OH)_2$. We may again begin with brucite, $Mg_3 \frac{(OH)_3}{(OH)_3}$ and replace 2(OH)

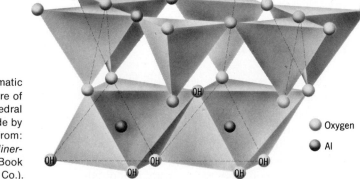

FIG. 10.54. Diagrammatic sketch of the structure of kaolinite with a tetrahedral sheet bonded on one side by an octahedral layer (from: R. E. Grim, 1968, *Clay Mineralogy*, McGraw-Hill Book Co.).

○ Oxygen

● Al

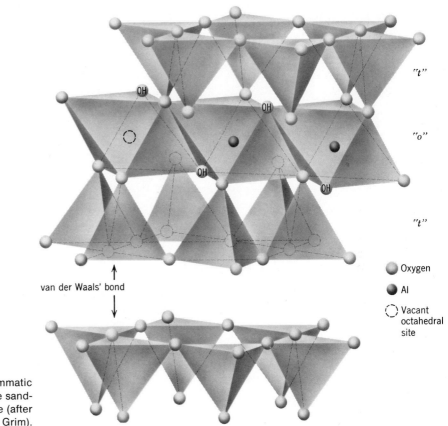

FIG. 10.55. Diagrammatic sketch of a *t-o-t* type sandwich as in pyrophyllite (after Grim).

● Oxygen

● Al

○ Vacant octahedral site

van der Waals' bond

groups in both the upper and lower (OH) layers by two oxygens of Si_2O_5 sheets, giving rise to $Mg_3 \frac{Si_2O_5(OH)}{Si_2O_5(OH)}$ or $Mg_3Si_4O_{10}(OH)_2$, talc. Similarly, gibbsite, $Al_2 \frac{(OH)_3}{(OH)_3}$, would result in $Al_2 \frac{Si_2O_5(OH)}{Si_2O_5(OH)}$ by replacement of 2(OH) groups on both sides of the gibbsite layer by oxygens of Si_2O_5 sheets; this leads to pyrophyllite, $Al_2Si_4O_{10}(OH)_2$ (see Figs. 10.55 and 10.57). The *t-o-t* sandwiches are electrically neutral and form stable structural units that are joined to each other by van der Waals' bonds. Because this is a very weak bond, we may expect such structures to have excellent cleavage, easy gliding, and greasy feel, as indeed found in the minerals talc and pyrophyllite.

We can carry the process of evolution of phyllosilicate structures one step farther by substituting Al for some of the Si in the tetrahedral sites of the Si_2O_5 sheets. Because Al is trivalent, whereas Si is tetravalent, each substitution of this kind causes a free electrical charge to appear on the surface of the *t-o-t* sandwich. If Al substitutes for every fourth Si, a charge of significant magnitude is produced to bind univalent cations in regular 12 coordination in between *t-o-t* sandwiches (see Fig. 10.57). By virtue of these sandwich-cation-sandwich bonds, the structure is more firmly held together, the ease of gliding is diminished, the hardness is increased, and slippery feel is lost. The resulting mineral structures are those of the true *micas* (see Figs. 10.56 and 10.57). In the trioctahedral group of micas the uni-

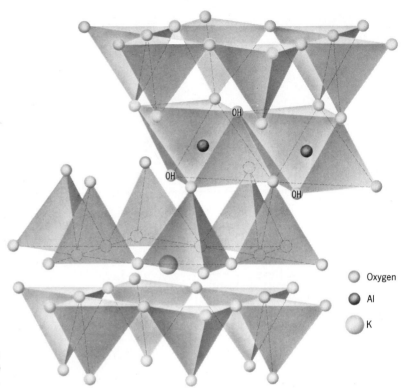

FIG. 10.56. Diagrammatic sketch of the muscovite structure (after Grim).

valent cation between the *t-o-t* sandwiches is K; in the dioctahedral group the cation can be K, as in muscovite, or Na, as in paragonite. It is easy to remember the formulas of the micas if it is recalled that *one* of the aluminums belongs in the tetrahedral sheet and the formulas are written accordingly. Thus,

Trioctahedral $KMg_3(AlSi_3O_{10})(OH)_2$ Phlogopite
Dioctahedral $KAl_2(AlSi_3O_{10})(OH)_2$ Muscovite

If half the Si in the tetrahedral sites of the Si_2O_5 sheets is replaced by Al, two charges per *t-o-t* sandwich become available for binding an interlayer cation. Such ions as Ca^{2+}, and to a smaller extent Ba^{2+}, may enter in between two sandwiches of the mica structure. These interlayer cations are held by ionic bonds so strong that the quality of the cleavage is diminished, the hardness increased, the flexibility of the layers is almost wholly lost, and the density is increased. The resulting minerals are the *brittle micas*, typified by

Trioctahedral $CaMg_3(Al_2Si_2O_{10})(OH)_2$ Xanthophyllite
Dioctahedral $CaAl_2(Al_2Si_2O_{10})(OH)_2$ Margarite

There is little solid solution between members of the dioctahedral and trioctahedral groups, although there may be extensive and substantially complete ionic substitution of Fe^{2+} for Mg, of Fe^{3+} for Al, of Ca for Na in appropriate sites. Ba may replace K, Cr may substitute for Al, and F may replace OH to a limited extent. Mn, Ti, and Cs are rarer constituents of some micas. The lithium micas are structurally distinct from muscovite and biotite because of the small lithium ion.

Additional members of the phyllosilicate group

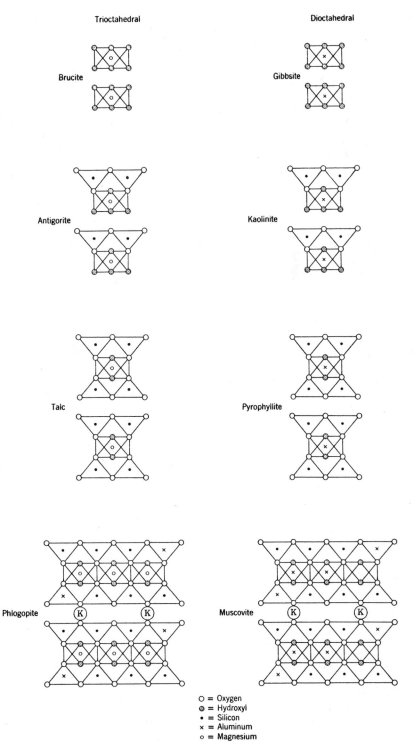

FIG. 10.57. Schematic development of some of the phyllosilicate structures.

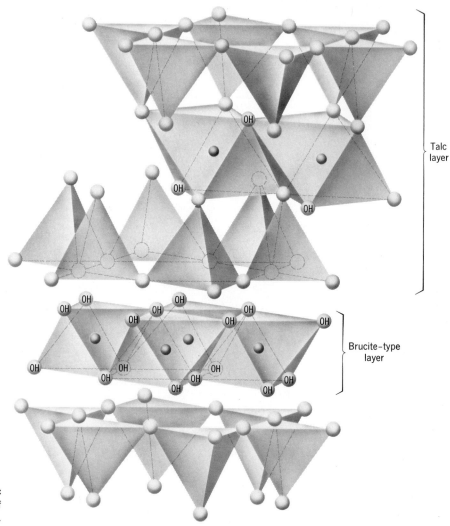

FIG. 10.58. Diagrammatic sketch of the structure of chlorite (after Grim).

can be developed. The important group of *chlorites* may be viewed as consisting of two layers of talc (or pyrophyllite) separated by a brucite (or gibbsite) type octahedral layer (see Fig. 10.58). This leads to a formula such as $Mg_3Si_4O_{10}(OH)_2 \cdot Mg_3(OH)_6$. However, in most chlorites Al, Fe^{2+}, and Fe^{3+} substitute for Mg in octahedral sites both in the talc and brucite sheets, and Al substitutes for Si in the tetrahedral sites. A more general chlorite formula would be:

$$(Mg, Fe^{2+}, Fe^{3+}, Al)_3(Al, Si)_4O_{10}(OH)_2 \cdot (Mg, Fe^{2+}, Fe^{3+}, Al)_3(OH)_6$$

The various members of the chlorite group differ from one another in amounts of substitution and the manner in which the successive octahedral and tetrahedral layers are stacked along the c axis.

The *vermiculite* minerals may be derived from the talc structure by interlamination of water molecules in definite sheets 4.98 Å thick, which is about

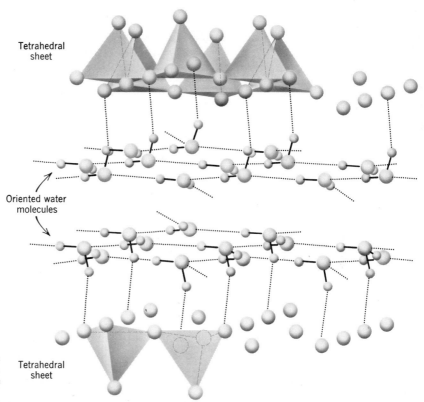

Tetrahedral
sheet

Oriented water
molecules

FIG. 10.59. Diagrammatic
sketch of the vermiculite
structure, showing layers of
water (after Grim).

Tetrahedral
sheet

the thickness of two water molecules (see Fig. 10.59). An example of a specific vermiculite formula would be $Mg_3(Si,Al)_4(OH)_2 \cdot 4.5H_2O[Mg]_{0.35}$ in which [Mg] represents exchangeable ions in the structure. The presence of exchangeable ions located between layers of H_2O molecules, and the ability of the structure to retain variable amounts of H_2O, are of great importance in agriculture. When the vermiculite structure is saturated with H_2O the basal spacing is approximately 14.8 Å. This water can be gradually extracted as shown by a discontinuous collapse sequence along the c axis, leading to a basal spacing of about 9.0 Å for a vermiculite without interlayer water. The structure of the *montmorillonite* (or *smectite*) group may be derived from the pyrophyllite structure by insertion of sheets of molecular water containing exchangeable cations between the pyrophllite *t-o-t* sandwiches, leading to a structure that is essentially identical to that of ver-

miculite. The members of the vermiculite and montmorillonite groups exhibit an unrivalled capacity for swelling when wetted because of their ability to incorporate large amounts of interlayer water.

If occasional random substitution of aluminum for silicon in the tetrahedral sites of pyrophyllite sheets takes place, there may not be enough of an aggregate charge on the *t-o-t* sandwiches to produce an ordered mica structure with all possible interlayer cation sites filled. Locally, however, occasional cation sites may be occupied, leading to properties intermediate between those of clays and micas. Introduction of some molecular water may further complicate this picture. K-rich minerals of this type, intermediate between montmorillonite clays and the true micas, are referred to as *illite* or *hydromica*.

In our previous discussion of the derivation of various types of structures among the phyllosilicates

FIG. 10.60 (a) Diagrammatic sketch of the antigorite structure as viewed along the *b* axis, showing curved *t* and *o* layers. (*b*) Highly schematic representation of the possible curvature in the chrysotile structure.

▭ Brucite layer
▲▲ Tetrahedral layer

(*a*)

(*b*)

(see Fig. 10.57) we suggested that there is a good geometric fit between an octahedral brucitelike layer and a tetrahedral Si_2O_5 sheet. In detail, however, one finds that there is considerable mismatch between a brucite type layer and an undistorted Si_2O_5 sheet with hexagonal rings. The misfit is due to the fact that the edges of an $Mg(OH)_6$ octahedron in the brucite layer are somewhat greater than the distances between apical oxygens in the Si_2O_5 or (Si, $Al)_2O_5$ layer; this means that the geometry shown in Fig. 10.52 is an oversimplification. In the case of the *serpentine minerals, antigorite* and *chrysotile,* both $Mg_3Si_2O_5(OH)_4$, this misfit is compensated for by a bending (stretching of the distance between apical oxygens) of the tetrahedral layer so as to provide a better fit with the adjoining octahedral brucite layer (see Fig. 10.60a). In the platy variety, *antigorite,* the bending is not continuous but occurs as corrugations. In the fibrous variety, *chrysotile,* the mismatch is resolved by a continuous bending of the structure into cylindrical tubes (see Fig. 10.60b). High magnification electron micrographs (see Fig. 10.61) have shown the existence of such a tubelike structure for the layers which occur in essentially parallel sheets in the kaolinite group (see Fig. 10.57).

In our discussion of the phyllosilicates we have pointed out the existence of hexagonal rings in the undistorted Si_2O_5 sheet. In order to join the *t* sheet to a brucite *o* layer, or to an *o-t* sequence in order to produce *t-o* or *t-o-t* type sandwiches, the octahedral sheet is staggered with respect to the tetrahedral sheets (see any of the phyllosilicate illustrations). This stagger reduces the symmetry within each sandwich to monoclinic (see Fig. 10.62), even though the tetrahedral sheets contain hexagonal

FIG. 10.61. Electron micrograph showing tubular fibers of chrysotile from Globe, Arizona. Magnification 35,000 × (after Nagy and Faust, 1956, *American Mineralogist*, v. 41, p. 822).

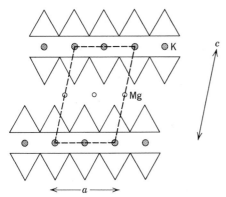

FIG. 10.62. Schematic projection along the b axis of the mica structure. Note stagger of t and o sheets with respect to each other. Unit cell is outlined by dashed lines.

holes. Most of the phyllosilicates, therefore, have monoclinic structures, some are triclinic, and a few orthorhombic or trigonal.

Because of the hexagonal symmetry about the (OH) group in the $Si_2O_5(OH)$ tetrahedral sheets, there are three alternate directions (see Fig. 10.63a) in which $Si_2O_5(OH)$ sheets can be stacked. These three directions can be represented by three vectors (1, 2, and 3; and negative directions $\bar{1}$, $\bar{2}$, and $\bar{3}$) at 120° to each other. If the stacking of $Si_2O_5(OH)$ sheets is always in the same direction (see Fig. 10.63b), the resulting structure will have monoclinic symmetry, with space group $C2/m$; this is referred to as the 1M (M = monoclinic) *polytype* with stacking sequence [1]. If the stacking sequence of

FIG. 10.63. Schematic illustration of some possible stacking polymorphs (*polytypes*) in mica. (*a*) Three vectorial directions for possible location of the (OH) group in an $Si_2O_5(OH)$ sheet which is stacked above or below the hexagonal ring shown. (*b*) Stacking of $Si_2O_5(OH)$ sheets in the same direction. (*c*) Stacking in 2 directions at 120° to each other. (*d*) Stacking in three directions at 120° to each other. Here, sheet 3 should lie directly above sheet 0 but is offset slightly for illustrative purposes.

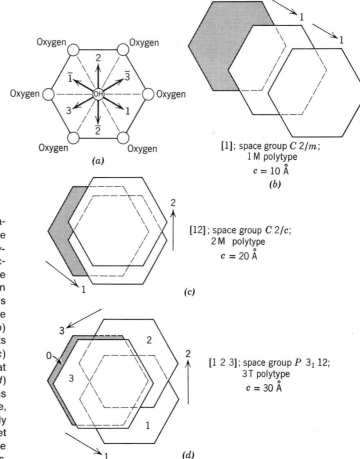

$Si_2O_5(OH)$ sheets consists of alternating 1 and $\bar{1}$ directions (in opposing directions along the same vector) a structure results which is best described as orthorhombic with space group $Ccm2_1$. This stacking sequence can be expressed as [1 $\bar{1}$] and the polytype is known as 2O(O = orthorhombic). If the stacking sequence consists of two vectors at 120° to each other, that is, [1 2], another monoclinic polytype results; this is known as $2M_1$ with space group $C2/c$ (see Fig. 10.63c). When all three vectors come into play, such as in [1 2 3] a trigonal polytype, 3T(T = trigonal) results which is compatible with either of two enantiomorphic space groups: $P3_112$ or $P3_212$. Additional *polytype* stacking sequences are possible but most micas that commonly show *polytypism* belong to 1M($C2/m$), $2M_1$($C2/c$) or 3T($P3_112$) polytypes.

The great importance of the phyllosilicates lies in part in the fact that the products of rock weathering, and hence the constituents of soils, are mostly of this structural type. The release and retention of plant foods, the stockpiling of water in the soil from wet to dry seasons, and the accessibility of the soil to atmospheric gases and organisms depend in large part on the properties of the sheet silicates.

Geologically, the phyllosilicates are of great significance. The micas are the chief minerals of schists and are widespread in igneous rocks. They form at lower temperatures than amphiboles or pyroxenes and frequently are formed as replacements of earlier minerals as a result of hydrothermal alteration. We will discuss in detail the following common phyllosilicates:

PHYLLOSILICATES

Serpentine Group

Antigorite }	
Chrysotile }	$Mg_3Si_2O_5(OH)_4$

Clay Mineral Group

Kaolinite	$Al_2Si_2O_5(OH)_4$
Talc	$Mg_3Si_4O_{10}(OH)_2$
Pyrophyllite	$Al_2Si_4O_{10}(OH)_2$

Mica Group

Muscovite	$KAl_2(AlSi_3O_{10})(OH)_2$
Phlogopite	$KMg_3(AlSi_3O_{10})(OH)_2$
Biotite	$K(Mg, Fe)_3(AlSi_3O_{10})(OH)_2$

Lepidolite	$K(Li, Al)_{2-3}(AlSi_3O_{10})(OH)_2$
Margarite	$CaAl_2(Al_2Si_2O_{10})(OH)_2$

Chlorite Group

Chlorite	$(Mg, Fe)_3(Si, Al)_4O_{10}(OH)_2$- $(Mg, Fe)_3(OH)_6$

and three minerals closely related to the above phyllosilicate groups:

Apophyllite	$KCa_4(Si_4O_{10})_2F \cdot 8H_2O$
Prehnite	$Ca_2Al(AlSi_3O_{10})(OH)_2$
Chrysocolla	$Cu_4H_4Si_4O_{10}(OH)_8$

Serpentine Group

ANTIGORITE and CHRYSOTILE— $Mg_3Si_2O_5(OH)_4$

Crystallography. Monoclinic; $2/m$. Crystals, except as pseudomorphs, unknown. Serpentine occurs in two distinct forms: (1) a platy variety, *antigorite*, which conforms in its properties to those of the common phyllosilicates, and (2) a fibrous variety, *chrysotile* (Fig. 10.64). Orthorhombic polymorphs of both antigorite and chrysotile are known.

Cm, $C2$, or $C2/m$. For antigorite: $a = 5.30$, $b = 9.20$, $c = 7.46$ Å; $\beta = 91°24'$; $a:b:c = 0.567:1:0.811$; $Z = 2$. d's: 7.30(10), 3.63(8), 2.52(2), 2.42(2), 2.19(1). For chrysotile: $a = 5.34$, $b = 9.25$, $c = 14.65$; $\beta = 93°16'$; $a:b:c = 0.577:1:1.584$; $Z = 4$.

Physical Properties. **H** 3–5, usually 4. **G** 2.5–2.6. *Luster* greasy, waxlike in the massive varieties, silky when fibrous. *Color* often variegated, showing mottling in lighter and darker shades of green. Translucent. *Optics:* (−); Chrysotile: $\alpha = 1.532–1.549$ (across fiber), $\gamma = 1.545–1.556$ (parallel to fiber length). Antigorite: nearly isotropic, $n = 1.55–1.56$.

Composition and Structure. Close to $Mg_3Si_2O_5(OH)_4$ with MgO 43.0, SiO_2 44.1, and H_2O 12.9%. Fe and Ni may substitute to some extent for Mg and Al for some Si. The idealized structure of antigorite is shown in Fig. 10.57 in terms of trioctahedral *t-o* layers, in analogy with the dioctahedral kaolinite structure. Because of a misfit between the *t*

FIG. 10.64. Veins of chrysotile asbestos in serpentine, Thetford, Quebec.

and *o* layers the actual antigorite structure consists of finite and corrugated layers parallel to {001} (see Fig. 10.60*a* and page 398). In chrysotile this same mismatch is responsible for the curvature of the *t-o* layers forming cylindrical tubes (see Fig. 10.60*b* and page 398).

Diagnostic Features. Infusible. Gives water in the closed tube. Recognized by its variegated green color and its greasy luster or by its fibrous nature. Distinguished from the fibrous amphibole by the presence of a large amount of water.

Occurrence. Serpentine is a common mineral and widely distributed; usually as an alteration of magnesium silicates, especially olivine, pyroxene, and amphibole. From forsterite it may form by the reaction:

$$2Mg_2SiO_4 + 3H_2O \rightarrow$$
$$Mg_3Si_2O_5(OH)_4 + Mg(OH)_2$$

Frequently associated with magnesite, chromite, and magnetite. Found in both igneous and meta-morphic rocks, frequently in disseminated particles, in places in such quantity as to make up practically the entire rock mass. Serpentine, as a rock name, is applied to such rock masses made up mostly of the variety antigorite. Large deposits of the fibrous variety, chrysotile, are located in the province of Quebec, Canada; in the Ural Mountains, U.S.S.R.; and in Rhodesia. In the United States chrysotile is found in Vermont, New York, New Jersey, and in Arizona near Globe from Sierra Ancha, and in the Grand Canyon.

Use. The variety *chrysotile* is the chief source of asbestos. The use of asbestos depends on its fibrous, flexible nature, which allows it to be made into felt and woven into cloth and other fabrics, and upon its incombustibility and slow conductivity of heat. Asbestos products, therefore, are used for fire-proofing and as an insulation material against heat and electricity. Massive serpentine, which is translucent and of a light to dark green color, is often used as an ornamental stone and may be valuable as building material. Mixed with white marble and showing beautiful variegated coloring, it is called *verde antique* marble.

Name. Serpentine refers to the green, serpent-like cloudings of the massive variety. Antigorite after Antigorio in Italy and chrysotile from the Greek words for *golden* and *fiber*.

Similar Species. *Greenalite*, (Fe, Mg)$_3$Si$_2$O$_5$(OH)$_4$, the Fe-rich analogue of antigorite, is fairly common in unmetamorphosed Precambrian iron-formations. *Garnierite*, (Ni, Mg)$_3$Si$_2$O$_5$(OH)$_4$, is an apple-green Ni-bearing serpentine formed as an alteration product of Ni-rich peridotites. It is mined as a nickel ore in New Caledonia, U.S.S.R.; and in Australia. In the United States it is found at Riddle, Oregon. *Sepiolite* (meerschaum), a hydrous magnesium silicate, is a claylike, secondary mineral associated with serpentine. It is used in the manufacture of meerschaum pipes.

Clay Mineral Group

Clay is a rock term, and like most rocks it is made up of a number of different minerals in varying proportions. Clay also carries the implication of very small particle size. Usually the term clay is used in reference to fine-grained, earthy material that becomes plastic when mixed with a small amount of water. With the use of X-ray techniques, clays have been shown to consist mainly of a group of crystalline substances known as the *clay minerals*. They are all essentially hydrous aluminum silicates. In some, Mg or Fe substitute in part for Al and alkalies or alkaline earths may be present as essential constituents. Although a clay may be made up of a single clay mineral, there are usually several mixed with other minerals such as feldspar, quartz, carbonates, and micas.

KAOLINITE—Al$_2$Si$_2$O$_5$(OH)$_4$

Crystallography. Triclinic; $\bar{1}$. In very minute, thin, rhombic or hexagonal plates. Usually in claylike masses, either compact or friable.

$P1$; $a = 5.14$. $b = 8.93$, $c = 7.37$ Å; $\alpha = 91°48'$; $\beta = 104°30'$, $\gamma = 90°$; $a:b:c = 0.576:1:0.825$. $Z = 2$. $d's$: 7.15(10), 3.57(10), 2.55(8), 2.49(9), 2.33(10).

Physical Properties. *Cleavage* {001} perfect.

H 2. **G** 2.6. *Luster* usually dull earthy; crystal plates pearly. *Color* white. Often variously colored by impurities. Usually unctuous and plastic. *Optics*: (−); $\alpha = 1.553–1.565$, $\beta = 1.559–1.569$, $\gamma = 1.560–1.570$; $2V = 24°–50°$. $r > v$.

Composition and Structure. Kaolinite shows little compositional variation; for Al$_2$Si$_2$O$_5$(OH)$_4$, Al$_2$O$_3$ 39.5, SiO$_2$ 46.5, H$_2$O 14.0%. The dioctahedral structure of kaolinite is shown schematically in Figs. 10.54 and 10.57. It consists of an Si$_2$O$_5$ sheet bonded to a gibbsite type layer. The clay minerals, *dickite* and *nacrite* are chemically identical to kaolinite but show a *t-o* stacking different from that in kaolinite.

Diagnostic Features. Infusible. Insoluble. Assumes a blue color when moistened with cobalt nitrate and ignited (Al). Gives water in the closed tube. Recognized usually by its claylike character, but without X-ray tests it is impossible to distinguish from the other clay minerals of similar composition which collectively make up *kaolin*.

Occurrence. Kaolinite is a common mineral, the chief constituent of kaolin or clay. It is always a secondary mineral formed by weathering or hydrothermal alteration of aluminum silicates, particularly feldspar. It is found mixed with feldspar in rocks that are undergoing alteration; in places it forms entire deposits where such alteration has been carried to completion. As one of the common products of the decomposition of rocks it is found in soils, and transported by water it is deposited, mixed with quartz and other materials, in lakes, etc., in the form of beds of clay.

Use. Clay is one of the most important of the natural industrial substances. It is available in every country of the world and is commercially produced in nearly every state in the United States. Many and varied products are made from it which include common brick, paving brick, drain tile, and sewer pipe. The commercial users of clay recognize many different kinds having slightly different properties, each of which is best suited for a particular purpose. High-grade clay, which is known as *china clay* or *kaolin*, has many uses in addition to the manufacture of china and pottery. Its largest is as a filler in paper, but it is also used in the rubber industry and in the manufacture of refractories.

The chief value of clay for ceramic products lies in the fact that when wet it can be easily molded into any desired shape, and when it is heated, part of the combined water is driven off, producing a hard, durable substance.

Name. Kaolinite is derived from *kaolin,* which is a corruption of the Chinese *kauling,* meaning high ridge, the name of a hill near Jauchu Fa, where the material is obtained.

Similar Species. *Dickite* and *nacrite* are of the same composition as kaolinite but differ somewhat in their structure; they are less important constituents of clay deposits.

Anauxite is also assigned to the kaolinite group but has a higher Si/Al ratio than kaolinite. *Halloysite* has two forms: one with kaolinite composition, $Al_2Si_2O_5(OH)_4$, the other with composition $Al_2Si_2O_5(OH)_4 \cdot 2H_2O$. The second type dehydrates to the first with loss of interlayed water molecules.

The *montmorillonite* group comprises a number of clay minerals composed of *t-o-t* silicate layers of both dioctahedral and trioctahedral type. The outstanding characteristic of members of this group is their capacity to absorb water molecules between the sheets, thus producing marked expansion of the structure (see Fig. 10.59 for the similar structure of vermiculite). The dioctahedral members are *montmorillonite, beidellite,* and *nontronite;* the trioctahedral members are *hectorite* and *saponite.*

Montmorillonite is the dominant clay mineral in *bentonite,* altered volcanic ash. Bentonite has the unusual property of expanding several times its original volume when placed in water. This property gives rise to interesting industrial uses. Most important is as a drilling mud in which the montmorillonite is used to give the fluid a viscosity several times that of water. It is also used for stopping leakage in soil, rocks, and dams.

Illite is a general term for the micalike clay minerals. The illites differ from the micas in having less substitution of Al for Si, in containing more water, and in having K partly replaced by Ca and Mg. Illite is the chief constituent in many shales.

TALC—$Mg_3Si_4O_{10}(OH)_2$

Crystallography. Monoclinic; $2/m$. Crystals rare. Usually tabular with rhombic or hexagonal outline.

Foliated and in radiating foliated groups. When compact and massive known as *steatite* or *soapstone.*

$C2/c$; $a = 5.27$, $b = 9.12$, $c = 18.85$ Å; $\beta = 100°0'$; $a:b:c = 0.578:1:2.067$; $Z = 4$. $d's$: 9.34(10), 4.66(9), 3.12(10), 2.48(7), 1.870(4).

Physical Properties. *Cleavage* {001} perfect. Thin folia somewhat flexible but not elastic. Sectile. **H** 1 (will make a mark on cloth). **G** 2.7–2.8. *Luster* pearly to greasy. *Color* apple-green, gray, white, or silver-white; in soapstone often dark gray or green. Translucent. Greasy feel. *Optics:* (−); $\alpha = 1.539$, $\beta = 1.589$, $\gamma = 1.589$; $2V = 6°–30°$. $Z = b$, $X \perp$ {001}; $r > v$.

Composition and Structure. There is little variation in the chemical composition of most talc; pure talc contains MgO 31.7, SiO_2 63.5, H_2O 4.8%. Small amounts of Al or Ti may substitute for Si and Fe may replace some of the Mg. The Fe-rich analogue, *minnesotaite,* $Fe_3Si_4O_{10}(OH)_2$, is common in low-grade metamorphic Precambrian iron deposits. It is probable that a complete solid solution exists between talc and minnesotaite. The trioctahedral structure of talc is similar to that of dioctahedral pyrophyllite (Fig. 10.55) and consists of essentially neutral *t-o-t* sandwiches held together by weak residual bonds.

Diagnostic Features. Fusible with difficulty (5). Unattacked by acids. Yields water in the closed tube when heated intensely. Characterized by its micaceous habit, cleavage, softness, and greasy feel. Distinguished from pyrophyllite by moistening a fragment with cobalt nitrate and heating intensely; talc will assume a pale violet color, pyrophyllite a blue color.

Occurrence. Talc is a secondary mineral formed by the alteration of magnesium silicates, such as olivine, pyroxenes, and amphiboles, and may be found as pseudomorphs after these minerals. Characteristically in low-grade metamorphic rocks, where, in massive form, *soapstone,* it may make up nearly the entire rock mass. It may also occur as a prominent constituent in schistose rocks, as in talc schist.

In the United States many talc or soapstone quarries are located along the line of the Appalachian Mountains from Vermont to Georgia. The major producing states are California, North Carolina, Texas, and Georgia.

Use. As slabs of the rock soapstone, talc is used for laboratory table tops and electric switchboards. Most of the talc and soapstone produced is used in powdered form as an ingredient in paint, ceramics, rubber, insecticides, roofing, paper, and foundry facings. The most familiar use is in talcum powder.

Name. The name talc is of ancient and doubtful origin, probably derived from the Arabic, *talk*.

Similar Species. *Minnesotaite,* $Fe_3Si_4O_{10}$-$(OH)_2$, may be isostructural with talc.

Pyrophyllite—$Al_2Si_4O_{10}(OH)_2$

Crystallography. Monoclinic; $2/m$. Not in distinct crystals. Foliated, in some cases in radiating lamellar aggregates. Also granular to compact. Identical with talc in appearance.

$C2/c$; $a = 5.15$, $b = 8.92$, $c = 18.59$ Å; $\beta = 99°55'$; $a:b:c = 0.577:1:2.084$; $Z = 4$. d's: 9.21(6), 4.58(5), 4.40(2), 3.08(10), 2.44(2).

Physical Properties. *Cleavage* {001} perfect. Folia somewhat flexible but not elastic. **H** 1–2 (will make a mark on cloth). **G** 2.8. *Luster* pearly to greasy. *Color* white, apple-green, gray, brown. Translucent, will transmit light on thin edges. *Optics:* (−); $\alpha = 1.552$, $\beta = 1.588$, $\gamma = 1.600$; $2V = 57°$; $X \perp$ {001}; $r > v$.

Composition and Structure. Pyrophyllite shows little deviation from the ideal formula; Al_2O_3 28.3, SiO_2 66.7, H_2O 5.0%. The dioctahedral structure of pyrophyllite consists of essentially neutral *t-o-t* sandwiches (Figs. 10.55 and 10.57) held together by weak residual bonds.

Diagnostic Features. Characterized chiefly by its micaceous habit, cleavage, and greasy feel. When moistened with cobalt nitrate and ignited it assumes a blue color (Al); talc under the same conditions becomes pale violet. Infusible, but, on heating, radiated varieties exfoliate in a fanlike manner. At high temperature it yields water in the closed tube.

Occurrence. Pyrophyllite is a comparatively rare mineral. Found in metamorphic rocks; fre-

quently with kyanite. Occurs in considerable amount in Guilford and Orange Counties, North Carolina.

Use. Quarried in North Carolina and used for the same purpose as talc but does not command as high a price as the best grades of talc. A considerable part of the so-called *agalmatolite*, from which the Chinese carve small images, is this species.

Name. From the Greek meaning *fire* and *a leaf,* because it exfoliates on heating.

Mica Group

The micas, composed of *t-o-t* sandwiches with interlayer cations and little or no exchangeable water, crystallize in the monoclinic system but with their crystallographic angle β close to 90°, so that their monoclinic symmetry is not easily seen. The crystals are usually tabular with prominent basal planes and have either a diamond-shaped or hexagonal outline with angles of approximately 60° and 120°. Crystals, as a rule, therefore, appear to be either orthorhombic or hexagonal. They are characterized by a highly perfect {001} cleavage. A blow with a somewhat dull-pointed instrument on a cleavage plate develops in all the species a six-rayed *percussion figure* (Fig. 10.65) two lines of which are nearly parallel to the prism edges and the third, which is most strongly developed, is parallel to the mirror plane {010}. There is limited ionic substitution between various members of the group and two members may crystallize together in parallel position in the same crystal plate with the cleavage extending through both.

FIG. 10.65. Percussion figure in mica.

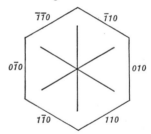

MUSCOVITE—KAl$_2$(AlSi$_3$O$_{10}$)(OH)$_2$

Crystallography. Monoclinic; $2/m$. Distinct crystals rare; usually tabular with prominent {001}. The presence of prism faces {110} at angles of nearly 60° gives some plates a diamond-shaped outline, making them simulate orthorhombic symmetry (Fig. 10.66). If {010} is also present, the crystals have a hexagonal appearance. The prism faces are roughened by horizontal striations and frequently taper. Foliated in large to small sheets; in scales that are, in some cases, aggregated into plumose or globular forms. Also cryptocrystalline and compact massive.

$C2/c$; $a = 5.19$, $b = 9.04$, $c = 20.08$ Å; $\beta = 95°30'$; $a:b:c = 0.574:1:2.221$; $Z = 4$. $d's$: 9.95(10), 3.37(10), 2.66(8), 2.45(8), 2.18(8).

Physical Properties. *Cleavage* {001} perfect allowing the mineral to be split into very thin sheets; folia flexible and elastic. **H** 2–2½. **G** 2.76–2.88. *Luster* vitreous to silky or pearly. *Color* colorless and transparent in thin sheets. In thicker blocks translucent, with light shades of yellow, brown, green, red. Some crystals allow more light to pass parallel to the cleavage than perpendicular to it. *Optics:* (−); $\alpha = 1.560–1.572$, $\beta = 1.593–1.611$, $\gamma = 1.599–1.615$; $2V = 30°–47°$; $Z = Y$, $X \wedge c = 0–5°$; $r > v$.

Composition and Structure. Essentially KAl$_2$(AlSi$_3$O$_{10}$)(OH)$_2$. Little solid solution occurs among the dioctahedral mica group or between members of the dioctahedral and trioctahedral groups. Minor substitutions may be Na, Rb, Cs for K; Mg, Fe^{2+}, Fe^{3+}, Li, Mn, Ti, Cr for Al; F for OH. *Paragonite,* NaAl$_2$(AlSi$_3$O$_{10}$)(OH)$_2$, which is isostructural with muscovite is practically indistinguishable from it without X-ray methods. At low to moderate temperatures a large miscibility gap exists between muscovite and paragonite. The structure of muscovite is illustrated in Figs. 10.56 and 10.57. It consists of t sheets of composition (Si, Al)$_2$O$_5$ linked to octahedral gibbsite-type sheets to form t-o-t sandwiches. These sandwiches have a net negative charge which is balanced by K or Na ions (in paragonite) between them (see discussion on page 393). Although the muscovite structure may show several polytypic forms due to various ways of stacking succeeding Si$_2$O$_5$ sheets (see page 399 and Fig. 10.63) the most commonly observed polytype is $2M_1$ with space group $C2/c$.

Diagnostic Features. Characterized by its highly perfect cleavage and light color. Distinguished from phlogopite by not being decomposed in sulfuric acid and from lepidolite by not giving a crimson flame. Gives water in the closed tube.

Occurrence. Muscovite is a widespread and common rock forming mineral. Characteristic of granites and granite pegmatites. In pegmatites, muscovite associated with quartz and feldspar may be in large crystals, called *books,* which in some localities are several feet across. It is also very common in metamorphic rocks, forming the chief constituent of certain mica schists. In the chlorite zone of metamorphism, muscovite is a characteristic constituent of albite-chlorite-muscovite schists. In some schistose rocks it occurs as fibrous aggregates of minute scales with a silky luster and is known as *sericite.* The development of sericite from feldspar and other minerals, such as topaz, kyanite, spodumene, and andalusite is a common feature of retrograde metamorphism. Sericite also forms as an alteration of the wall rock of hydrothermal ore veins. As *illite* it is a constituent of some shales, soils and recent sediments.

Several notable localities for muscovite are found in the Alps; also Mourne Mountains, Ireland;

FIG. 10.66. Muscovite. Diamond-shaped crystals.

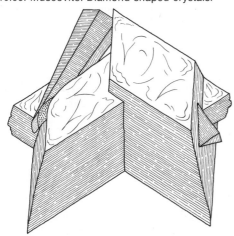

Cornwall, England; Norway; and Sweden. Large and important deposits occur in India. The most productive pegmatites in the United States are in New Hampshire, North Carolina, and in the Black Hills of South Dakota. Of less importance are the deposits in Colorado, Alabama, Maine, and Virginia. Crystals measuring 7 to 9 feet across have been mined in Mattawan Township, Ontario, Canada.

Use. Because of its high dielectric and heat-resisting properties, *sheet mica,* single cleavage plates, is used as an insulating material in the manufracture of electrical apparatus. The *isinglass* used in furnace and stove doors is sheet mica. Many small parts used for electrical insulation are built up of thin sheets of mica cemented together. They may thus be pressed into shape before the cement hardens. India is the largest supplier of mica used in this way. Ground mica is used in many ways: in the manufacture of wallpapers to give them a shiny luster; as a lubricant when mixed with oils; a filler; and as a fireproofing material.

Name. *Muscovite* was so called from the popular name of the mineral, *Muscovy-glass,* because of its use as a substitute for glass in Old Russia (Muscovy). *Mica* was probably derived from the Latin *micare,* meaning *to shine.*

Phlogopite—$KMg_3(AlSi_3O_{10})(OH)_2$

Crystallography. Monoclinic; $2/m$. Usually in six-sided plates or in tapering prismatic crystals. Crystals frequently large and coarse. Found also in foliated masses.

$C2/m$; $a = 5.33, b = 9.23, c = 10.26$ Å; $\beta = 100°12'$; $a:b:c = 0.577:1:1.112$; $Z = 2$. $d's$: 10.13(10), 3.53(4), 3.36(10), 3.28(4), 2.62(10).

Physical Properties. *Cleavage* {001} perfect. Folia flexible and elastic. **H** $2\frac{1}{2}$–3. **G** 2.86. *Luster* vitreous to pearly. *Color* yellowish-brown, green, white, often with copperlike reflections from the cleavage surface. Transparent in thin sheets. When viewed in transmitted light, some phlogopite shows asterism because of tiny oriented inclusions of rutile. *Optics:* (−); $\alpha = 1.53–1.59$, $\beta = 1.56–1.64$, $\gamma = 1.56–1.64$; $2V = 0°–15°$; $Y = b$, $X \wedge c = 0°–5'$; $r < v$.

Composition and Structure. Small amount of Na and lesser amounts of Rb, Cs, and Ba may substitute for K. Fe^{2+} substitutes for Mg forming a series toward biotite. F may substitute in part for OH. Only very limited solid solution occurs between muscovite and the phlogopite-biotite series. The trioctahedral structure of phlogopite is schematically illustrated in Fig. 10.57. It consists of t-o-t sandwiches bonded by interlayer K^+ ions; it is very similar to the muscovite structure except for the difference in the octahedral layer. It occurs most commonly in the 1M (space group $C2/m$) polytype, but $2M_1$ and 3T polytypes occur occasionally (see Fig. 10.63).

Diagnostic Features. Characterized by its micaceous cleavage and yellowish-brown color. Distinguished from muscovite by its decomposition in sulfuric acid and from biotite by its lighter color. It is, however, impossible to draw a sharp distinction between biotite and phlogopite. Fusible at $4\frac{1}{2}$–5. Yields water in the closed tube.

Occurrence. Phlogopite is found in metamorphosed Mg-rich limestones, dolomites, and ultrabasic rocks. It is a common mineral in kimberlite. Notable localities are in Finland; Sweden; Campolungo, Switzerland; Sri Lanka; and the Malagasy Republic. In the United States found chiefly in Jefferson and St. Lawrence Counties, New York. Found abundantly in Canada in Ontario at North and South Burgess, and in various other localities in Ontario and Quebec.

Use. Same as for muscovite; chiefly as electrical insulator.

Name. Name from a Greek word meaning *fire-like,* in allusion to its color.

BIOTITE—$K(Mg, Fe)_3(AlSi_3O_{10})(OH)_2$

Crystallography. Monoclinic; $2/m$. Rarely in tabular or short prismatic crystals with prominent {001} and pseudohexagonal outline. Usually in irregular foliated masses; often in disseminated scales or in scale aggregates.

$C2/m$; $a = 5.31, b = 9.23, c = 10.18$ Å; $\beta = 99°18'$; $a:b:c = 0.575:1:1.103$; $Z = 2$. $d's$: 10.1(10), 3.37(10), 2.66(8), 2.54(8), 2.18(8).

Physical Properties. *Cleavage* {001} perfect. Folia flexible and elastic. **H** $2\frac{1}{2}$–3. **G** 2.8–3.2. *Luster* splendent. *Color* usually dark green, brown to black. More rarely light yellow. Thin sheets usually have a smoky color (differing from the almost colorless muscovite). *Optics:* $(-)$; $\alpha = 1.57-1.63$, $\beta = 1.61-1.70$; $\gamma = 1.61-1.70$; $2V = 0°-25°$; $Y = b$, $Z \wedge a = 0°-9°$; $r < v$.

Composition and Structure. The composition is similar to phlogopite but with considerable substitution of Fe^{2+} for Mg. There is also substitution by Fe^{3+} and Al for Mg and by Al for Si. In addition Na, Ca, Ba, Rb, and Cs may substitute for K. A complete series exists between phlogopite and biotite. The trioctahedral biotite structure is the same as that of phlogopite (see Fig. 10.57). The 1M polytype with space group $C2/m$ is most common but $2M_1$ and 3T also occur (see Fig. 10.63).

Diagnostic Features. Characterized by its micaceous cleavage and dark color. Fusible with difficulty at 5. Decomposed by boiling concentrated sulfuric acid, giving a milky solution. Gives water in the closed tube.

Occurrence. Biotite forms under a very great variety of geological environments. It is found in igneous rocks varying from granite pegmatites, to granites, to diorites, to gabbros and peridotites. It occurs also in felsite lavas and porphyries. In metamorphic rocks it is formed over a wide range of temperature and pressure conditions, and it occurs in regionally as well as contact metamorphosed rocks. The crystallization of biotite in argillaceous rocks is taken as the onset of the temperature and pressure conditions of the biotite zone. Typical associations at this metamorphic grade are: biotite-chlorite and biotite-muscovite. Occurs in fine crystals in blocks included in the lavas of Vesuvius.

Name. In honor of the French physicist, J. B. Biot.

Similar Species. *Glauconite,* similar in composition to biotite, is an authigenic mineral found in green pellets in sedimentary rocks. *Vermiculite* is a platy mineral formed chiefly as an alteration of biotite. The structure is essentially that of talc with interlayered water molecules (see Fig. 10.59). The name comes from the Latin word meaning to breed worms, for when heated, the mineral expands into wormlike forms. In the expanded condition it is used extensively in heat and sound insulating. Vermiculite is mined at Libby, Montana, and Macon, North Carolina.

LEPIDOLITE—K(Li, Al)$_{2-3}$(AlSi$_3$O$_{10}$)(O,OH, F)$_2$

Crystallography. Monoclinic; $2/m$. Crystals usually in small plates or prisms with hexagonal outline. Commonly in coarse- to fine-grained scaly aggregates.

$C2/m$; $a = 5.21$, $b = 8.97$, $c = 20.16$ Å; $\beta = 100°48'$; $a:b:c = 0.581:1:1.124$; $Z = 4$. d's: 10.0(6), 5.00(5), 4.50(5), 2.58(10), 1.989(8).

Physical Properties. *Cleavage* {001} perfect. **H** $2\frac{1}{2}-4$. **G** 2.8–2.9. *Luster* pearly. *Color* pink and lilac to grayish white. Translucent. *Optics:* $(-)$; $\alpha = 1.53-1.55$, $\beta = 1.55-1.59$, $\gamma = 1.55-1.59$; $2V = 0°-60°$; $Y = b$, $Z \wedge a = 0°-7°$; $r > v$.

Composition and Structure. The composition of lepidolite varies depending chiefly on the relative amounts of Al and Li in octahedral coordination. Analyses show a range of Li_2O from 3.3 to 7 weight percent. In addition Na, Rb, and Cs may substitute for K. It is possible that there is a continuous chemical series from dioctahedral muscovite to trioctahedral lepidolite, in which all of the octahedral sites are occupied. Lepidolite can occur as one of three polytypes (1M, $2M_2$, and 3T) (see Fig. 10.63).

Diagnostic Features. Fusible at 2, giving a crimson flame (Li). Insoluble in acids. Gives acid water in the closed tube. Characterized chiefly by its micaceous cleavage and usually by its lilac to pink color. Muscovite may be pink, and lepidolite white, and therefore a flame test should be made to distinguish them.

Occurrence. Lepidolite is a comparatively rare mineral, found in pegmatites, usually associated with other lithium-bearing minerals such as pink and green tourmaline, amblygonite, and spodumene. Often intergrown with muscovite in parallel position. Notable foreign localities for its occurrence are the Ural Mountains, U.S.S.R.; Isle of Elba; Rozna, Moravia; Bikita, Rhodesia; South-West Africa and the Malagasy Republic. In the United States it is found in several localities in Maine; near Middletown, Connecticut; Pala, California; Dixon, New Mexico; and Black Hills, South Dakota.

Use. A source of lithium. Used in the manufacture of heat-resistant glass.

Name. Derived from a Greek word meaning *scale*.

Margarite—CaAl₂(Al₂Si₂O₁₀)(OH)₂

Crystallography. Monoclinic; 2/m. Seldom in distinct crystals. Usually in foliated aggregates with micaceous habit.

$C2/c$; $a = 5.13$, $b = 8.92$, $c = 19.50$ Å; $\beta = 100°48'$; $a:b:c = 0.575:1:2.186$; $Z = 4$. $d's$: 4.40(8), 3.39(8), 3.20(9), 2.51(10), 2.42(8).

Physical Properties. *Cleavage* {001} perfect. **H** $3\frac{1}{2}$–5 (harder than the true micas). **G** 3.0–3.1. *Luster* vitreous to pearly. *Color* pink, white, and gray. Translucent. Folia somewhat brittle; because of this brittleness margarite is known as a *brittle mica*. *Optics:* (−); $\alpha = 1.632$–1.638; $\beta = 1.643$–1.648; $\gamma = 1.645$–1.650; $2V = 40°$–67°; $Z = b$, $Y \wedge a = 7°$.

Composition and Structure. Most analyses are close to the above end member composition with CaO 14.0, Al_2O_3 51.3, SiO_2 30.2 and H_2O 4.5%. A small amount of Na may replace Ca. The dioctahedral structure of margarite is very similar to that of muscovite. In margarite, however, the tetrahedral layer has the composition $(Si_2Al_2)O_{10}$ instead of $(Si_3Al)O_{10}$ as in muscovite. Because of the greater electrical charge on the $(Si_2Al_2)O_{10}$ sheet the structure can be balanced by incorporated divalent Ca^{2+} ions instead of monovalent K^+. The bond strength between the layers is therefore greater; this is expressed in the brittle nature of margarite.

Diagnostic Features. Characterized by its micaceous cleavage, brittleness, and association with corundum. Fuses at 4–4½, turning white. Slowly and incompletely decomposed by boiling HCl.

Occurrence. Margarite occurs usually with corundum and diaspore and apparently as an alteration product. It is found in this way with the emery deposits of Asia Minor and on the islands of Naxos and Nicaria, of the Grecian Archipelago. In the United States associated with emery at Chester, Massachusetts; Chester County, Pennsylvania; and with corundum deposits in North Carolina.

Name. From the Greek meaning *pearl*.

Similar Species. Other *brittle micas* are *clintonite* and *xanthophyllite* which may be regarded as the Ca analogues of phlogopite.

Chlorite Group

A number of minerals are included in the chlorite group all of which have similar chemical, crystallographic, and physical properties. Without quantitative chemical analyses or careful study of the optical and X-ray properties, it is extremely difficult to distinguish between the members. The following is a composite description of the principal members of the group: *clinochlore, penninite,* and *prochlorite*.

CHLORITE—
(Mg, Fe)₃(Si, Al)₄O₁₀(OH)₂·(Mg, Fe)₃(OH)₆

Crystallography. Monoclinic; 2/m, some polymorphs of chlorite are triclinic. In pseudohexagonal tabular crystals, with prominent {001}. Similar in habit to crystals of the mica group, but distinct crystals rare. Usually foliated massive or in aggregates of minute scales; also in finely disseminated particles.

Cell parameters vary with composition. For clinochlore: $C2/m$; $a = 5.2$–5.3, $b = 9.2$–9.3, $c = 28.6$ Å; $\beta = 96°50'$; $Z = 4$. $d's$: 3.53(10), 1.998(9), 1.564(9), 1.535(10), 1.393(10).

Physical Properties. *Cleavage* {001} perfect. Folia flexible but not elastic. **H** 2–2½. **G** 2.6–3.3. *Luster* vitreous to pearly. *Color* green of various shades. Rarely yellow, white, rose-red. Transparent to translucent. *Optics:* Most (+), some (−); in all Bxa nearly ⊥ {001}. $\alpha = 1.57$–1.66, $\beta = 1.57$–1.67, $\gamma = 1.57$–1.67; $2V = 20°$–60°. Pleochroism in green (+), X, $Y > Z$; (−) $X < Y$, Z. The indices increase with increasing iron content.

Composition and Structure. A hypothetical chlorite composition made of talc and brucite layers would be $Mg_3(Si_4O_{10})(OH)_2 + Mg_3(OH)_6$. In most chlorites there is considerable deviation from this composition with Fe^{2+}, Fe^{3+}, and Al substituting for Mg in both the talc and brucite layers, and Al substituting for Si in the tetrahedral sites. The great range in composition is reflected in variations in physical, optical, and X-ray properties. A diagrammatic representation of the chlorite structure is given in Fig. 10.58.

Diagnostic Features. Characterized by its green color, micaceous habit and cleavage, and by the fact that the folia are not elastic. Decomposed by boiling concentrated sulfuric acid, giving a milky solution. Water in the closed tube at high temperature.

Occurrence. Chlorite is a common mineral in metamorphic rocks and it is the diagnostic mineral of the greenschist facies. In pelitic schists it occurs in quartz-albite-chlorite-sericite-garnet assemblages. It is also commonly found with actinolite and epidote. Chlorite is also a common constituent of igneous rocks where it has formed as an alteration of Mg-Fe silicates such as pyroxenes, amphiboles, biotite, garnet, and idocrase. The green color of many igneous rocks is due to the chlorite to which the ferromagnesian silicates have altered; and the green color of many schists and slates is due to finely disseminated particles of the mineral.

Name. Chlorite is derived from a Greek word meaning *green,* in allusion to the common color of the mineral.

Apophyllite—$KCa_4(Si_4O_{10})_2F \cdot 8H_2O$

Crystallography. Tetragonal; $4/m2/m2/m$. Usually in crystals showing a combination of {110}, {011}, and {001} (Fig. 10.67). Crystals may resemble a combination of cube and octahedron, but are shown to be tetragonal by difference in luster between faces of prism and base.

Angles: $c(001) \wedge p(101) = 60°17'$, $c(001) \wedge e(114) = 23°39'$.

$P4mnc$; $a = 9.02$, $c = 15.8$ Å; $a:c = 1:1.1752$; $Z = 2$. d's: 4.52(2), 3.94(10), 3.57(1), 2.98(7), 2.48(3).

Physical Properties. *Cleavage* {001} perfect. **H** $4\frac{1}{2}$–5. **G** 2.3–2.4. *Luster* of base pearly, other faces vitreous. *Color* colorless, white, or grayish; may show pale shades of green, yellow, rose. Transparent to translucent. *Optics:* (+); $\omega = 1.537$, $\epsilon = 1.535$. May be optically (−).

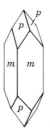 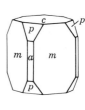

FIG. 10.67. Apophyllite.

Composition and Structure. The structure of apophyllite differs from that of the other phyllosilicates in that the sheets are composed of four-fold and eight-fold rings. These sheets are linked to each other by Ca, K, and F ions. Water, which is bonded to the Si_2O_5 sheet by hydrogen-bonding, is lost at about 250°C. This indicates that the water is more strongly held than adsorbed water, but less strongly than structural (OH) groups.

Diagnostic Features. Recognized usually by its crystals, color, luster, and basal cleavage. Fusible at 2 with swelling; forms a white vesicular enamel. Colors the flame pale violet (K). Yields much water in the closed tube.

Occurrence. Apophyllite occurs as a secondary mineral lining cavities in basalt and related rocks, associated with zeolites, calcite, datolite, and pectolite.

It is found in fine crystals at Andreasberg, Harz Mountains; Aussig, Bohemia; on the Seiser Alpe in Trentino, Italy; near Bombay, India; Faeroe Islands; Iceland; Greenland; and Guanajuato, Mexico. In the United States at Bergen Hill and Paterson, New Jersey; and Lake Superior copper district. Found in fine crystals in Nova Scotia.

Name. Apophyllite is derived from two Greek words meaning *from* and a *leaf,* because of its tendency to exfoliate when ignited.

PREHNITE—$Ca_2Al(AlSi_3O_{10})(OH)_2$

Crystallography. Orthorhombic; $2mm$. Distinct crystals are rare, commonly tabular parallel to {001}. Usually reniform, stalactitic, and in rounded groups of tabular crystals.

$P2cm$; $a = 4.65$, $b = 5.48$, $c = 18.49$ Å;

$a:b:c = 0.849:1:3.374$; $Z = 2$. $d's$: 3.48(9), 3.28(6), 3.08(10), 2.55(10), 1.77(7).

Physical Properties. H $6-6\frac{1}{2}$. **G** 2.8–2.95. *Luster* vitreous. *Color* usually light green, passing into white. Translucent. *Optics:* (+); $\alpha = 1.616$, $\beta = 1.626$, $\gamma = 1.649$; $2V = 66°$; $X = a$, $Z = c$; $r > v$.

Composition and Structure. CaO 27.1, Al_2O_3 24.8, SiO_2 43.7, H_2O 4.4%. Some Fe^{3+} may replace Al. The structure of prehnite contains layers of Al and Si tetrahedra parallel to {001}. Ca is in 7 coordination and lies between Si tetrahedra in adjoining layers.

Diagnostic Features. Characterized by its green color and crystalline aggregates forming reniform surfaces. Resembles hemimorphite but is of lower specific gravity and fuses easily ($2\frac{1}{2}$). Ease of fusion also distinguishes it from quartz and beryl. Yields water in the closed tube.

Occurrence. Prehnite occurs as a secondary mineral lining cavities in basalt and related rocks. Associated with zeolites, datolite, pectolite, and calcite. In the United States, it occurs at Farmington, Connecticut; Paterson and Bergen Hill, New Jersey; Westfield, Massachusetts; and Lake Superior copper district. Found in good crystals at Coopersburg, Pennsylvania.

Name. In honor of Colonel Prehn, who brought the mineral from the Cape of Good Hope.

CHRYSOCOLLA ≈ $Cu_4H_4Si_4O_{10}(OH)_8$

Crystallography and Structure. Generally amorphous and thus in the strict sense cannot be considered a mineral. Partially crystalline material shows the presence of Si_4O_{10} layers in a very much defect structure. Massive compact; in some cases earthy. Individual specimens inhomogeneous.

Physical Properties. Fracture conchoidal. **H** 2–4. **G** 2.0–2.4. *Luster* vitreous to earthy. *Color* green to greenish blue; brown to black when impure. Refractive index 1.40±.

Composition. Chrysocolla is a hydrogel or gelatinous precipitate and shows a wide range in composition. Chemical analyses report: CuO 32.4–42.2, SiO_2 37.9–42.5, H_2O 12.2–18.8%. In addition Al_2O_3 and Fe_2O_3 are usually present in small amounts.

Diagnostic Features. Characterized by its green or blue color and conchoidal fracture. Distinguished from turquoise by inferior hardness. Infusible. Gives a copper globule when fused with sodium carbonate on charcoal. In the closed tube darkens and gives water.

Occurrence. Chrysocolla forms in the oxidized zones of copper deposits associated with malachite, azurite, cuprite, or native copper, for example. It is commonly found in the oxidized portions of porphyry copper ores. Found in the copper districts of Arizona and New Mexico and at Chuquicamata, Chile.

Use. A minor ore of copper.

Name. Chrysocolla, derived from two Greek words meaning *gold* and *glue,* which was the name of a similar-appearing material used to solder gold.

Similar Species. *Dioptase,* $Cu_6(Si_6O_{18})\cdot 6H_2O$, is a rhombohedral cyclosilicate occurring in well-defined green rhombohedral crystals. *Plancheite,* $Cu_8(Si_4O_{11})_2(OH)_2\cdot H_2O$, and *shattuckite,* $Cu_5(SiO_3)_4(OH)_2$, have inosilicate structures.

TECTOSILICATES

Nearly three-quarters of the rocky crust of the earth is made up of minerals built about a three-dimensional framework of linked SiO_4 tetrahedra. These minerals belong to the *tectosilicate* class in which all the oxygen ions in each SiO_4 tetrahedron are shared with neighboring tetrahedra. This results in a stable, strongly bonded structure in which the ratio of Si:O is 1:2 (Fig. 10.2). We will discuss in detail the following groups and species:

TECTOSILICATES

SiO_2 Group
Quartz	
Tridymite	SiO_2
Cristobalite	
Opal	$SiO_2\cdot nH_2O$

Feldspar Group
K-Feldspars
Microcline	
Orthoclase	$KAlSi_3O_8$
Sanidine	

Plagioclase Feldspars
Albite $NaAlSi_3O_8$
Anorthite $CaAl_2Si_2O_8$
Danburite $CaB_2Si_2O_8$

Feldspathoid Group
Leucite $KAlSi_2O_6$
Nepheline $(Na, K)AlSiO_4$
Sodalite $Na_8(AlSiO_4)_6Cl_2$
Lazurite $(Na, Ca)_8(AlSiO_4)_6(SO_4, S, Cl)_2$
Petalite $LiAlSi_4O_{10}$

Scapolite Series
Marialite $Na_4(AlSi_3O_8)_3(Cl_2, CO_3, SO_4)$
Meionite $Ca_4(Al_2Si_2O_8)_3(Cl_2, CO_3, SO_4)$
Analcime $NaAlSi_2O_6 \cdot H_2O$

Zeolite Group
Natrolite $Na_2Al_2Si_3O_{10} \cdot 2H_2O$
Chabazite $CaAl_2Si_4O_{12} \cdot 6H_2O$
Heulandite $CaAl_2Si_7O_{18} \cdot 6H_2O$
Stilbite $CaAl_2Si_7O_{18} \cdot 7H_2O$

SiO_2 Group

An SiO_2 framework that does not contain other structural units is electrically neutral. There are at least nine different ways in which such a framework can be built. These modes of geometrical arrangement correspond to nine known polymorphs of SiO_2 one of which is synthetic (see Table 10.4). Each of these polymorphs has its own space group, cell dimensions, characteristic morphology, and lattice energy. Which polymorph is stable is determined chiefly by energy considerations; the higher-temperature forms with greater lattice energy possess the more expanded structures which are reflected in lower specific gravity and refractive index. In addition to the nine polymorphs of SiO_2, there are two related and essentially amorphous substances, *lechatelierite,* a high silica glass of variable composition and *opal,* $SiO_2 \cdot nH_2O$, with a locally ordered structure of silica spheres and highly variable H_2O content.

The principal naturally occurring SiO_2 polymorphs fall into three structural categories: *low quartz,* with the lowest symmetry and the most compact structure; *low tridymite,* with higher symmetry and a more open structure; and *low cristobalite,* with the highest symmetry and the most expanded structure of the three polymorphs. These polymorphs are related to each other by reconstructive transformations, a process that requires considerable energy. The sluggishness and energy requirements of reconstructive transformations allow the phases to exist metastably for long periods of time. The temperatures of reconstructive inversions vary widely, depending chiefly on the rate and direction of temperature change. Each of the three above structure types has also a high-low inversion, as shown by the existence of high and low quartz, high and low tridymite, and high and low cristobalite (see Table 10.4). These transformations are displacive; they take place quickly and reversibly at a fairly constant and sharply defined temperature of inversion and may be repeated over and over again without physical disintegration of the crystal. The

Table 10.4
POLYMORPHS OF SiO_2

Name	Symmetry	Space Group	Specific Gravity	Refractive Index (mean)
Stishovite[a]	Tetragonal	$P4_2/mnm$	4.35	1.81
Coesite	Monoclinic	$C2/c$	3.01	1.59
Low (α) quartz	Hexagonal	$P3_221$ (or $P3_121$)	2.65	1.55
High (β) quartz	Hexagonal	$P6_222$ (or $P6_422$)	2.53	1.54
Keatite (synth.)	Tetragonal	$P4_12_12$ (or $P4_32_12$)	2.50	1.52
Low (α) tridymite	Monoclinic or Orthorhombic	$C2/c$ (or Cc) $C222_1$	2.26	1.47
High (β) tridymite	Hexagonal	$P6_3/mmc$	2.22	1.47
Low (α) cristobalite	Tetragonal	$P4_12_12$ (or $P4_32_12$)	2.32	1.48
High (β) cristobalite	Isometric	$Fd3m$	2.20	1.48

[a] Only polymorph with Si in octahedral coordination with oxygen.

nearly instant high → low inversions take place with the release of a fairly constant amount of energy; very near the same temperature the low → high inversions take place with the absorption of energy (see Table 10.5). The structures of several SiO_2 polymorphs are illustrated in Fig. 10.68 (see Fig. 4.21 for high and low quartz structures).

The low-temperature form of each of the displacive polymorphic pairs has a lower symmetry than the higher temperature form (see Table 10.4) but this symmetry change is less than in the reconstructive transformations. The effect of increased pressure is to raise all inversion temperatures and for any temperature to favor the crystallization of the polymorph most economical of space (Fig. 10.69).

The most dense of the silica polymorphs are *coesite* and *stishovite*. *Coesite* was synthesized in 1953 and *stishovite*, which is isostructural with rutile, TiO_2, in 1961 and it was not until later that they were found in nature. They have both been identified in very small amounts at Meteor Crater, Arizona. Their formation is attributed to the high pressure and high temperature resulting from the impact of a meteorite. *Coesite* has recently been found in a sanidine-coesite eclogite from a kimberlite pipe in South Africa. *Keatite* has not been found in nature.

QUARTZ—SiO₂

Crystallography. Quartz, hexagonal-R; 32. High-quartz, hexagonal; 622. Crystals commonly prismatic, with prism faces horizontally striated. Terminated usually by a combination of positive and negative rhombohedrons, which often are so equally developed as to give the effect of a hexagonal dipyramid (Fig. 10.70a). In some crystals one rhombohedron predominates or occurs alone (Fig.

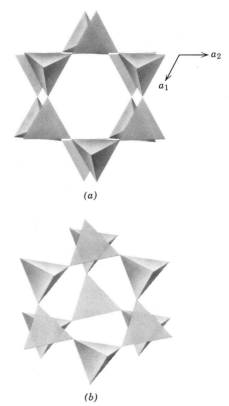

(a)

(b)

FIG. 10.68. Structures of some polymorphs of SiO_2 (see Fig. 4.21 for illustration of low and high quartz structures). (a) Tetrahedral layers in high tridymite projected onto (0001). (b) Portion of the high cristobalite structure projected onto (111).

10.70b). The prism faces may be wanting, and the combination of the two rhombohedrons gives what appears to be a hexagonal dipyramid (a *quartzoid*) (Fig. 10.70c). Some crystals are malformed, but the recognition of the prism faces by their horizontal striations assists in the orientation. The trigonal trapezohedral faces x are occasionally observed and

Table 10.5

INVERSION TEMPERATURES FOR DISPLACIVE TRANSFORMATIONS IN SOME SiO₂ POLYMORPHS	High T Polymorph	Minimum Crystallization T for Stable Form at 1 Atmosphere P	Inversion to Low T Form at 1 Atmosphere P
	High cristobalite	1470°C	~268°C
	High tridymite	870°C	~120–140°C
	High quartz	574°C	573°C

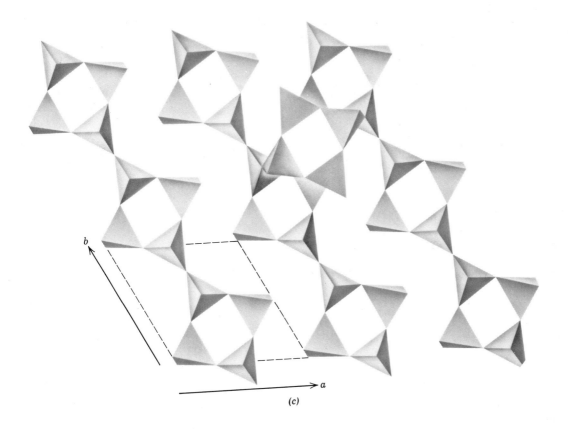

(c)

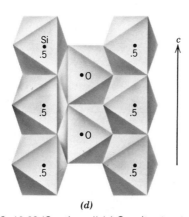

(d)

FIG. 10.68 (*Continued*) (c) Coesite structure showing four-membered tetrahedral rings that lie parallel to (001). (d) Structure of stishovite, with Si in octahedral coordination with oxygen, projected on (100) (a, b, c, and d after J. J. Papike and M. Cameron, 1976, *Reviews of Geophysics and Space Physics,* v. 14, pp. 37–80).

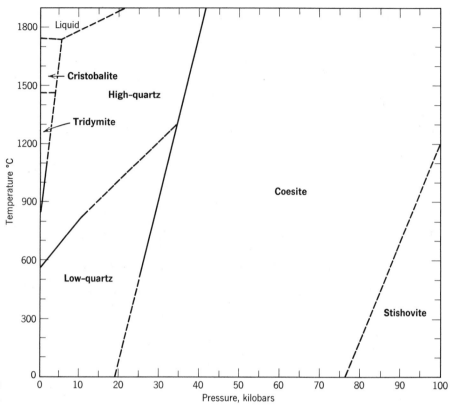

FIG. 10.69. Stability relations of the SiO₂ polymorphs.

reveal the true symmetry. They occur at the upper right of alternate prism faces in right-hand quartz and to the upper left of alternate prism faces in left-hand quartz (Figs. 10.70*d* and e). The right- and left-hand trigonal trapezohedrons are enantimorphous forms and reflect the arrangement of the SiO_4 tetrahedra, either in the form of a right- or left-hand screw (see space groups). In the absence of *x* faces, the "hand" can be recognized by observing whether plane polarized light passing parallel to *c* is rotated to the left or right.

Crystals may be elongated in tapering and sharply pointed forms, and some appear twisted or bent. Equant crystals with apparent 6-fold symmetry (Figs. 10.70*a*,*c*) are characteristic of high-temperature quartz, but the same habit is found in quartz that crystallized as the low-temperature form. Most quartz is twinned according to one or both of two laws (see Figs. 2.174 and 2.184). These are *Dauphiné,* c the twin axis; and *Brazil,* {11$\overline{2}$0} the

FIG. 10.70. Quartz crystals.

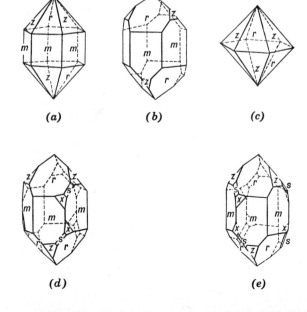

twin plane. Both types are penetration twins and external evidence of them is rarely seen.

The size of crystals varies from individuals weighing several tons to finely crystalline coatings, forming "drusy" surfaces. Also common in massive forms of great variety. From coarse to fine grained crystalline to flintlike or cryptocrystalline, giving rise to many variety names (see below). May form in concretionary masses.

Angles: $m(10\bar{1}0) \wedge r(10\bar{1}1) = 38°13'$, $r(10\bar{1}1) \wedge z(01\bar{1}1) = 46°16'$, $m(10\bar{1}0) \wedge s(11\bar{2}1) = 37°58'$, $m(10\bar{1}0) \wedge x(51\bar{6}1) = 12°1'$.

$P3_221$ or $P3_121$, $a = 4.91$, $c = 5.41$ Å; $a:c = 1:1.102$; $Z = 3$. $d's$: 4.26(8), 3.34(10), 1.818(6), 1.541(4), 1.081(5) (see Fig. 3.21).

Physical Properties. H 7. G 2.65. Fracture conchoidal. *Luster* vitreous, in some specimens greasy, splendent. *Color* usually colorless or white, but frequently colored by impurities and may then be any color. Transparent to translucent. Strongly piezoelectric and pyroelectric. *Optics:* (+); $\omega = 1.544$, $\epsilon = 1.553$.

Composition and Structure. Of all the minerals, quartz is most nearly a pure chemical compound with constant physical properties. Si 46.7, O 53.3%. The structure of low (α) quartz with one of two enantiomorphic space groups, $P3_121$ or $P3_221$, is discussed on page 151 and illustrated in Fig. 4.21a. At 573°C, and at atmospheric pressure, this structure transforms instantaneously to high (β) quartz, with hexagonal symmetry and one of two enantiomorphic space groups, $P6_222$ or $P6_422$. (This structure is shown in Fig. 4.21b.) The displacive transformation from low to high quartz involves only minor atomic adjustments without breaking of Si-O bonds. Upon cooling high quartz, through the inversion point at 573°C, Dauphiné twinning may be produced (see page 152).

Diagnostic Features. Characterized by its glassy luster, conchoidal fracture, and crystal form. Distinguished from calcite by its high hardness and from white varieties of beryl by its inferior hardness. Infusible. Insoluble except in hydrofluoric acid.

Varieties. A great many different forms of quartz exist, to which varietal names have been given. The more important varieties, with a brief description of each, follow.

Coarsely Crystalline Varieties

Rock Crystal. Colorless quartz, commonly in distinct crystals.

Amethyst. Quartz colored various shades of violet, often in crystals. The color apparently results from small amounts of Fe^{3+}.

Rose Quartz. Coarsely crystalline but usually without crystal form, color a rose-red or pink. Often fades on exposure to light. Small amounts of Ti^{4+} appear to be the coloring agent.

Smoky Quartz; Cairngorm Stone. Frequently in crystals; smoky yellow to brown to almost black. Named *cairngorm* for the locality of Cairngorm in Scotland. Spectrographic analyses of smoky quartz show no dominant impurity and are similar to those of colorless quartz. The dark color is attributed to free silicon formed by exposure to radioactive material.

Citrine. Light yellow resembling topaz in color.

Milky Quartz. Milky white owing to minute fluid inclusions. Some specimens have a greasy luster.

Quartz may contain parallel fibrous inclusions which give the mineral a chatoyancy. When stones are cut *en cabochon* they are called *quartz cat's*

FIG. 10.71. Rutilated quartz, Brazil.

eye. *Tiger eye* is a yellow fibrous quartz pseudomorphic after the fibrous amphibole crocidolite. It is also chatoyant.

With Inclusions. Many other minerals occur as inclusions in quartz and thus give rise to variety names. *Rutilated quartz* has fine needles of rutile penetrating it (Fig. 10.71). Tourmaline and other minerals are found in quartz in the same way. *Aventurine* quartz includes brilliant scales of colored minerals such as hematite (red) or chromium mica (green) and is used as a gem material. Liquids and gases may occur as inclusions; both liquid and gaseous carbon dioxide exist in some quartz.

Microcrystalline Varieties
Depending on their structure, the microcrystalline varieties of quartz may be divided into two types: *fibrous* and *granular*. It is difficult to distinguish between them macroscopically.

A. Fibrous Varieties
Chalcedony is the general term applied to fibrous varieties. More specifically it is a brown to gray, translucent variety, with a waxy luster, often mammillary and in other imitative shapes. Chalcedony has been deposited from aqueous solutions and is frequently found lining or filling cavities in rocks. Color and banding give rise to the following varieties:

Carnelian, a red chalcedony which grades into brown *sard*.

Chrysoprase is an apple-green chalcedony colored by nickel oxide.

Agate is a variety with alternating layers of chalcedony having different colors and porosity. The colors are usually in delicate, fine parallel bands which are commonly curved, in some specimens concentric (Fig. 10.72). Most agate used for commercial purposes is colored by artificial means. Some agates have the different colors not arranged in bands but irregularly distributed. *Moss agate* is a variety in which the variation in color is caused by visible impurities, commonly manganese oxide in mosslike patterns.

Wood that has been petrified by replacement

FIG. 10.72. Agate cut and polished, Brazil.

by clouded agate is known as *silicified* or *agatized wood.*

Onyx, like agate, is a layered chalcedony, with layers arranged in parallel planes. *Sardonyx* is an onyx with sard alternating with white or black layers.

Heliotrope or *bloodstone,* is a green chalcedony with small red spots of jasper in it.

B. Granular Varieties
Flint and *chert* resemble each other and there is no sharp distinction between them. Dark, siliceous nodules, usually found in chalk, are called flint; whereas lighter colored bedded deposits are called chert.

Jasper is a granular microcrystalline quartz with dull luster usually colored red by included hematite.

Prase has a dull green color; otherwise it is similar to jasper, and occurs with it.

Occurrence. Quartz is a common and abundant mineral occurring in a great variety of geological environments. It is present in many igneous and metamorphic rocks and is a major constituent of granite pegmatites. It is the most common gangue mineral in hydrothermal metal-bearing veins and in many veins is essentially the only mineral present.

In the form of flint and chert, quartz is deposited on the sea floor contemporaneously with the enclosing rock; or solutions carrying silica may replace limestone to form chert horizons. On the breakdown of quartz-bearing rocks, the quartz, because of its mechanical and chemical stability, persists as detrital grains to accumulate as sand. Sandstone and its metamorphic equivalent, quartzite, may be composed mainly of quartz.

Rock crystal is found widely distributed, some of the more notable localities being the Alps; Minas Gerais, Brazil; the Malagasy Republic; and Japan. The best quartz crystals from the United States are found at Hot Springs, Arkansas, and Little Falls and Ellenville, New York. Important occurrences of amethyst are in the Ural Mountains, Czechoslovakia, Tyrol, Zambia, and Brazil. Found at Thunder Bay on the north shore of Lake Superior. In the United States found in Delaware and Chester Counties, Pennsylvania; Oxford County, Maine; Black Hills, South Dakota; and Wyoming. Smoky quartz is found in large and fine crystals in Switzerland; and in the United States at Pikes Peak, Colorado; Alexander County, North Carolina; and Oxford County, Maine.

The chief source of agates at present is in southern Brazil and northern Uruguay. Most of these agates are cut at Idar-Oberstein, Germany, itself a famous agate locality. In the United States agate is found in numerous places, notably in Oregon and Wyoming. The chalk cliffs at Dover, England, are famous for the flint nodules that weather from them. Similar nodules are found on the French coast of the English channel and on islands off the coast of Denmark. Massive quartz, occurring in veins or with feldspar in pegmatite dikes, is mined in Connecticut, New York, Maryland, and Wisconsin for its various commercial uses.

Use. Quartz has many and varied uses. It is widely used as gemstones or ornamental material, as amethyst, rose quartz, cairngorm, tiger eye, aventurine, carnelian, agate, and onyx. As sand, quartz is used in mortar, in concrete, as a flux, as an abrasive, and in the manufacture of glass and silica brick. In powdered form it is used in porcelain, paints, sandpaper, scouring soaps, and as a wood filler. In the form of quartzite and sandstone it is used as a building stone and for paving purposes.

Quartz has many uses in scientific equipment. Because of its transparency in both the infrared and ultraviolet portions of the spectrum, quartz is made into lenses and prisms for optical instruments. The *optical activity* of quartz (the ability to rotate the plane of polarization of light) is utilized in the manufacture of an instrument to produce monochromatic light of differing wavelengths. Quartz wedges, cut from transparent crystals, are used as an accessory to the polarizing microscope. Because of its piezoelectric property, quartz has specialized uses. It is cut into small oriented plates and used as radio oscillators to permit both transmission and reception on a fixed frequency. The tiny quartz plate used in digital quartz watches serves the same function as quartz oscillators used to control radio frequencies. This property also renders it useful in the measurement of instantaneous high pressures such as result from firing a gun or atomic explosion.

Artificial. Since 1947 much of the quartz used for optical and piezoelectrical purposes has been manufactured by hydrothermal methods. More recently yellow, brown, blue and violet quartz has been synthesized for use as gem material.

Name. The name quartz is a German word of ancient derivation.

Similar Species. *Lechatelierite*, SiO_2, is fused silica or silica glass. Found in fulgurites, tubes of fused sand formed by lightning, and in cavities in some lavas. Lechatelierite is also found at Meteor Crater, Arizona, where sandstone has been fused by the heat generated by the impact of a meteorite.

Tridymite—SiO_2

Crystallography. Low (α) tridymite: monoclinic or orthorhombic; $2/m$, m, or 222. High (β) tridymite: hexagonal; $6/m\ 2/m\ 2/m$. Crystals are small and commonly twinned and at room temperature are pseudomorphs after high tridymite.

Low tridymite, for $C2/c$ or Cc; $a = 18.54$, $b = 5.01$, $c = 25.79$ Å; $\beta = 117°40'$; $a:b:c = 3.700:1:5.148$; $Z = 48$; for $C222_1$; $a = 8.74$, $b = 5.04$, $c = 8.24$; $a:b:c = 1.734:1:1.635$;

$Z = 8$. *d's:* (low tridymite): 4.30(10), 4.08(9), 3.81(9), 2.96(6), 2.47(6). High tridymite, $P6_3/mmc$; $a = 5.04$, $c = 8.24$ Å; $a:c = 1.635$; $Z = 4$.

Physical Properties. H 7. **G** 2.26. *Luster* vitreous. *Color* colorless to white. Transparent to translucent. *Optics:* (+); $\alpha = 1.468–1.479$, $\beta = 1.470–1.480$, $\gamma = 1.475–1.483$; $2V = 40–90°$.

Composition and Structure. Ideally SiO_2. However, small amounts of Na and Al may be in solid solution. The crystal structure of low (α) tridymite has not been investigated in detail but it is undoubtedly closely related to that of high tridymite. The high (β) tridymite structure consists of sheets of tetrahedra that lie parallel to {0001}; the tetrahedra within each sheet share corners to form six-membered rings (see Fig. 10.68a) and in these rings tetrahedra alternately point up or down providing linkage between the sheets. Tridymite is the stable form of SiO_2 at temperatures between 870° and 1470°, at atmospheric pressure (see Fig. 10.69). At higher temperatures it transforms to cristobalite, at lower temperatures to high quartz. These transformations are reconstructive and extremely sluggish.

Diagnostic Features. It is impossible to identify tridymite by macroscopic means, but under the microscope its crystalline outline and refractive index distinguish it from the other silica minerals.

Occurrence. Tridymite occurs commonly in certain siliceous volcanic rocks such as rhyolite, obsidian, and andesite, and for this reason may be considered an abundant mineral. Commonly associated with sanidine and cristobalite. It is found in large amounts in the lavas of the San Juan district of Colorado. It is also found in stony meteorites and the lunar basalts.

Name. From the Greek meaning *threefold,* in allusion to its common occurrence in trillings.

Cristobalite—SiO_2

Crystallography. Low (α) cristobalite: tetragonal; 422. High (β) cristobalite; isometric; $4/m\overline{3}2/m$. Crystals small octahedrons; this form is retained on inversion from high to low cristobalite. Also in spherical aggregates.

Low cristobalite, $P4_12_12$ (or $P4_32_12$): $a = 4.97$, $c = 6.93$ Å; $a:c = 1:1.394$; $Z = 4$. *d's* (low cristobalite): 4.05(10), 2.84(1), 2.48(2), 1.929(1),

1.870(1). High cristobalite, $Fd3m$: $a = 7.13$ Å; $Z = 8$.

Physical Properties. H $6\frac{1}{2}$. **G** 2.32. *Luster* vitreous. Colorless. Translucent. *Optics:* (+); $\omega = 1.484$, $\epsilon = 1.487$.

Composition and Structure. Ideally SiO_2, but most natural material contains some Na and Al in solid solution. The low cristobalite structure is tetragonal whereas high cristobalite is isometric. In high cristobalite six-membered tetrahedral rings are stacked parallel to {111}; see Fig. 10.68b. High cristobalite is stable from 1470° to the melting point, 1728°C, at atmospheric pressure (see Fig. 10.69). The transformation at 1470° to tridymite is of the reconstructive type.

Diagnostic Features. Infusible, but when heated to 200°C inverts to high cristobalite and becomes nearly transparent; on cooling, again inverts and assumes its initial white, translucent appearance. The occurrence in small lava cavities as spherical aggregates and its behavior when heated are characteristic, but it cannot be determined with certainty without optical or X-ray measurements.

Occurrence. Cristobalite is present in many siliceous volcanic rocks, both as the lining of cavities and as an important constituent in the fine-grained groundmass. It is, therefore, an abundant mineral. Associated with tridymite in the lavas of the San Juan district, Colorado.

Name. From the Cerro San Cristobal near Pachuca, Mexico.

OPAL—$SiO_2 \cdot nH_2O$

Crystallography. Generally amorphous. Massive; often botryoidal, stalactitic. Although X-ray studies indicate that much opal is essentially amorphous, precious opals contain silica spheres in an ordered packing (see Structure).

Physical Properties. *Fracture* conchoidal. **H** 5–6. **G** 2.0–2.25. *Luster* vitreous; often somewhat resinous. *Color* colorless, white, pale shades of yellow, red, brown, green, gray, and blue. The darker colors result from impurities. Often has a milky or "opalescent" effect and may show a fine play of colors. Transparent to translucent. Some opal, especially hyalite, shows a greenish-yellow

FIG. 10.73. Scanning electron micrograph of an opal with chalky appearance showing hexagonal packing of silica spheres (diameter of spheres is approximately 3000 Å). Because of the weak bonding between the spheres they are completely intact; in typical precious opal samples many of the spheres are cleaved (courtesy P. J. Darragh, A. J. Gaskin, and J. V. Sanders; see reference at end of chapter).

fluorescence in ultraviolet light. *Optics:* Refractive index 1.44–1.46.

Composition and Structure. $SiO_2 \cdot n H_2O$. The water content, usually between 4 and 9% may be as high as 20%. The specific gravity and refractive index decrease with increasing water content.

Although opal is essentially amorphous, it has been shown to have an ordered structure. It is not a crystal structure with atoms in a regular three-dimensional array but is made up of closely packed spheres of silica in hexagonal and/or cubic closest packing (see Figs. 4.6 and 10.73). Air or water occupy the voids between the spheres. In common opal the domains of equal size spheres with uniform packing are small or nonexistent but in precious opal large domains are made up of regularly packed spheres of the same size. The sphere diameters vary from one opal to another and range from 1500 Å to 3000 Å. When white light passes through the essentially colorless opal and strikes planes of voids between spheres, certain wavelengths are diffracted and flash out of the stone as nearly pure spectral colors. This phenomenon has been described as

analogous to the diffraction of X-rays by crystals. In X-ray diffraction the interplanar spacings (*d*) are of the same order of magnitude as the wavelengths of X-rays. In precious opal the spacings, determined by sphere diameters, are far greater but so are the wavelengths of visible light (4000–7000 Å). The different wavelengths that satisy the Bragg equation are diffracted with change in the angle of incident light (θ). Because light is refracted when it enters precious opal, the equation must include the refractive index, μ, (=1.45) and the equation is written $n\lambda = 2d\mu \sin \theta$.

Diagnostic Features. Distinguished from microcrystalline varieties of quartz by lesser hardness and specific gravity and by the presence of water. Soluble in hot strongly alkaline solutions. Upon intense ignition gives water in the closed tube.

Varieties. Precious opal is characterized by a brillant internal play of colors that may be red, orange, green or blue. The body color is white, milky-blue, yellow, or black (*black opal*). *Fire opal* is a variety with intense orange to red reflections. A logical explanation of the play of colors in precious opal has been proposed only recently (see discussion of Structure).

Common Opal. Milk-white, yellow, green, red, etc., without internal reflections.

Hyalite. Clear and colorless opal with a globular or botryoidal surface.

Geyserite or *Siliceous Sinter.* Opal deposited by hot springs and geysers. Found about the geysers in Yellowstone National Park.

Wood Opal. Fossil wood with opal as the petrifying material.

Diatomite. Fine-grained deposits, resembling chalk in appearance. Formed by sinking from near the surface and the accumulation on the sea floor of the siliceous tests of diatoms. Also known as *diatomaceous earth* or *infusorial earth.*

Occurrence. Opal may be deposited by hot springs at shallow depths by meteoric waters or low-temperature hypogene solutions. It is found lining and filling cavities in rocks and may replace wood buried in volcanic tuff. The largest accumlations of opal are as siliceous tests of silica-secreting organisms.

Precious opals are found at Caernowitza, Hungary; in Querétaro, Mexico; Queensland and New South Wales, Australia. Black opal has been found in the United States in Nevada and Idaho. Diatomite is mined in several western states, principally at Lompoc, California.

Artificial. Recently Pierre Gilson in Switzerland has synthesized precious opal that is nearly identical to natural material in chemical and physical properties including a beautiful play of colors.

Use. As a gem. Opal is usually cut in round shapes, *en cabochon.* Stones of large size and exceptional quality are very highly prized. Diatomite is used extensively as an abrasive, filler, filtration powder, and in insulation products.

Name. The name opal originated in the Sanskrit, *upala,* meaning stone or precious stone.

Feldspar Group

The compositions of the majority of common feldspars can be expressed in terms of the system $KAlSi_3O_8$ (orthoclase; Or)-$NaAlSi_3O_8$ (Albite; Ab)-$CaAl_2Si_2O_6$ (anorthite; An). The members of the series between $KAlSi_3O_8$ and $NaAlSi_3O_8$ are known as the *alkali feldspars* and the members in the series between $NaAlSi_3O_8$ and $CaAl_2Si_2O_8$ as the *plagioclase feldspars.* Members of both of these feldspar groups are given specific names as shown in Fig. 10.74. The chemical compositions of feldspars in this ternary system are generally expressed in terms of molecular percentages or Or, Ab, and An; e.g., $Or_{20}Ab_{75}An_5$. Barium feldspars such as *celsian,* $BaAl_2Si_2O_8$, and *hyalophane,* $(K, Ba)(Al, Si)_2Si_2O_8$, are relatively rare. All feldspars show good cleavages in two directions which make an angle of 90°, or close to 90°, with each other. Their hardness is about 6 and specific gravity ranges from 2.55 to 2.76 (excluding the Ba feldspars).

The unambiguous characterization of a feldspar requires a knowledge not only of the chemical composition but also of the structural state of the species. The structural state, which refers to the Al and Si distribution in tetrahedral sites of the framework structure, is a function of the crystallization temperature and subsequent thermal history of a feldspar. In general, feldspars that cooled rapidly after crystallization at high temperature show a disordered Al-Si distribution (high structural state). Those that cooled slowly from high temperatures or those that crystallized at low temperatures generally show an ordered Al-Si distribution (low structural state).

Structure

The feldspar structure, similar to the structures of the various polymorphs of SiO_2, consists of an infinite network of SiO_4 as well as AlO_4 tetrahedra. The feldspar structure can be considered a derivative of the SiO_2 structures, by incorporation of Al into the tetrahedral network, and concomitant housing of Na^+ (or K^+ or Ca^{2+}) in available voids. When only one Si^{4+} (per feldspar formula unit) is substituted for by Al^{3+}, the structure can be neutralized by incorporation of one K^+ or one Na^+. Similarly when two Si^{4+} (per feldspar formula unit) are substituted for by Al^{3+}, the electrostatic charge of the network can be balanced by a divalent cation such as Ca^{2+}.

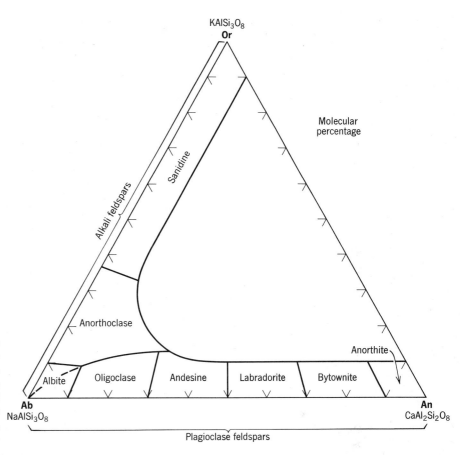

FIG. 10.74. Nomenclature for the plagioclase feldspar series and the high-temperature alkali feldspars (after Deer, Howie, and Zussman, 1963, *Rock Forming Minerals*, v. 4, Longman Group Ltd., p. 2).

We may state this as follows: Si_4O_8 framework $\rightarrow$ $Na(AlSi_3O_8) \rightarrow Ca(Al_2Si_2O_8)$. In the plagioclase structures the amount of tetrahedral Al varies in proportion to the relative amounts of Ca^{2+} and Na^+ so as to maintain electrical neutrality; the more Ca^{2+}, the greater the amount of Al^{3+}.

The general architecture of the feldspar structure can be illustrated with the aid of an illustration of the high-temperature polymorph of $KAlSi_3O_8$, *sanidine,* with space group $C2/m$ (see Fig. 10.75). In this structure the Al-Si distribution is completely disordered, meaning that the Al and Si ions are randomly distributed among the two crystallographically distinct tetrahedral sites, $T1$ and $T2$. The K^+ ions, bonded to 10 nearest oxygens in large interstices, occupy special positions on mirror planes perpendicular to the b axis. The Si-Al tetrahedral

framework consists of four-membered rings of tetrahedra that are linked into chains (of a double crankshaft type) parallel to the a axis (see Fig. 10.76). The square, blocky outline of these chains, imparted by the four-membered rings, finds outward expression in the right-angled cleavage and pseudotetragonal habit characteristic of feldspar. The structure of *microcline,* a low-temperature polymorph of $KAlSi_3O_8$, has triclinic symmetry (space group $C\bar{1}$) and lacks the mirror planes and rotation axes of sandine as shown in Fig. 10.75. In other words, its structure is less symmetric and the K^+ ions are no longer located in special positions. The Al-Si distribution is completely ordered in what is known as low-temperature or maximum microcline (maximum refers to maximum triclinity which results from the complete order). The tetrahedra that contain Al in

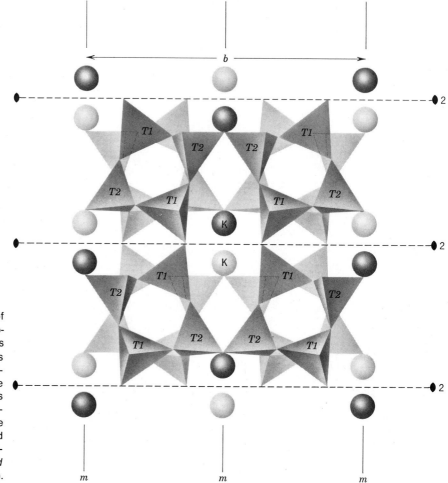

FIG. 10.75. The structure of high sanidine, KAlSi$_3$O$_8$, projected on (201). Mirror planes (*m*) and 2-fold rotation axes (2) are shown. Other symmetry elements such as glide planes and 2-fold screw axes are also present but their location is not shown here (after J. J. Papike and M. Cameron, 1976, *Reviews of Geophysics and Space Physics*, v. 14, p. 66).

this structure can be unambiguously located whereas in sandine the Al-Si distribution is completely random. *Orthoclase* represents a polymorphic form of KAlSi$_3$O$_8$ in which the Al-Si distribution is between the total randomness as found in sanidine, and the total order as seen in microcline. Orthoclase, with space group *C2/m*, crystallizes at intermediate temperatures (see Fig. 10.77). Although unambigous distinction between the three structurally different K-feldspars, sandine, orthoclase, and microcline, is based on careful measurements of unit cell dimensions and/or optical parameters, such as 2V and extinction angle *b* $\wedge$ *Z*, we will use the terms orthoclase and microcline more broadly. In the following descriptions the terms

will be based on generally recognizable characteristics in handspecimen. It must be remembered however that the definitions of the three types of K-feldspar as used in the recent literature are based on parameters which can be obtained only by X-ray and optical techniques (see J. V. Smith, 1974, *Feldspar Minerals,* Springer Verlag, New York). Such measurements allow for the definition of maximum-microcline, high-, intermediate, and low orthoclase, and provide the investigator with information about the state of order or disorder of the Al-Si distribution in the feldspar structure.

Microcline is particularly characteristic of deep-seated rocks and pegmatites, orthoclase of intrusive rocks formed at intermediate temperatures and sanidine of extrusive, high-temperature lavas.

The general structure of members of the plagioclase series is very similar to that of microcline. The Na end member, *albite* is generally triclinic (space group $C\bar{1}$) with a low albite form that shows a highly ordered Al-Si distribution and a high albite form with a highly disordered Al-Si distribution. A monoclinic variety of albite occurs at very high temperature and is known as *monalbite*. The Ca end member, *anorthite*, is also triclinic with space group $P\bar{1}$ at room temperature, and perfect Al-Si ordering in the structure. At elevated temperatures the structure of anorthite becomes body-centered with space group $I\bar{1}$. The general stability fields of the various forms of feldspar are shown in Figs. 10.77 and 10.79.

Composition

The alkali feldspar series ($NaAlSi_3O_8$ to $KAlSi_3O_8$) shows complete solid solution only at high temperatures (see Fig. 10.77). For example, members of the sanidine–high albite series are stable at elevated

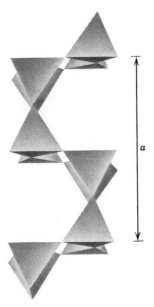

FIG. 10.76. The four-membered rings in Fig. 10.74 are linked to form chains that run parallel to the *a* axis (after J. J. Papike and M. Cameron, 1976, *Reviews of Geophysics and Space Physics*, v. 14, p. 67).

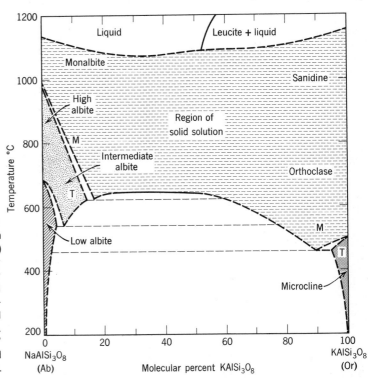

FIG. 10.77. Schematic phase diagram for the system $NaAlSi_3O_8$(Ab)-$KAlSi_3O_8$(Or) showing a large miscibility gap at temperatures below approximately 600°C. M and T mean monoclinic and triclinic, respectively. Modified after J. V. Smith (1974). Smith uses ''low sanidine'' for part of the region marked as orthoclase.

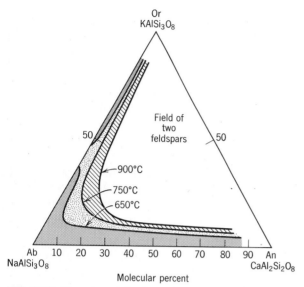

FIG. 10.78. Experimentally determined extent of solid solution in the system Or-Ab-An at P_{H_2O} = 1 kilobar (after P. Ribbe, 1975, *Feldspar Mineralogy*, see references at end of chapter).

temperatures but at lower temperature two separate phases, low albite and microcline, become stable. As can be seen from Fig. 10.77 the compositional ranges of low albite and microcline are very small. If a homogeneous feldspar, of composition Or_{50} Ab_{50},

in which the Na^+ and K^+ ions are randomly distributed, is allowed to cool slowly segregation of Na^+ and K^+ ions will result because the size requirements of the surrounding structure become more stringent. The Na^+ will diffuse to form Na-rich regions and the K^+ will segregate into K-rich regions in the structure, causing the originally homogeneous feldspar to become an inhomogeneous intergrowth. The separation most commonly results in thin layers of albite in a host crystal of K feldspar. Such intergrowths are known as *perthites,* and are the result of *exsolution* (see Fig. 4.29 and discussion on page 167). In the alkali feldspar series the orientation of the exsolution lamellae is roughly parallel to {100}. When these intergrowths are visible to the naked eye they are known as *macroperthite,* when visible only by optical microscope they are referred to as *microperthite* and when detectable only by X-ray or electron microscope techniques they are called *cryptoperthite*. More rarely the host mineral is a plagioclase feldspar and the lamellae are of $KAlSi_3O_8$ composition; this is called *antiperthite*.

Only very partial solid solution occurs between $KAlSi_3O_8$ and $CaAl_2Si_2O_8$ (see Fig. 10.78). Essentially complete solid solution, however, exists at elevated temperatures in the plagioclase series ($NaAlSi_3O_8$ to $CaAl_2Si_2O_8$) (see Fig. 10.79). The gen-

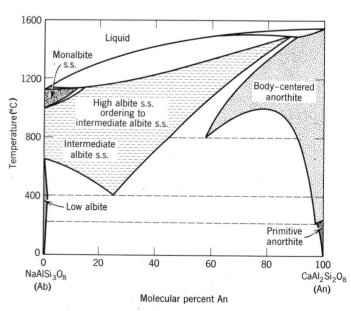

FIG. 10.79. Possible phase diagram for plagioclase feldspars showing various miscibility regions at lower temperatures. Simplified after J. V. Smith (1974). S.S. means solid solution.

eral formula of a feldspar in this series may be written as: $Na_{1-x}Ca_x(Si_{3-x}Al_x)O_8$, where x ranges from 0 to 1. The structural interpretation of the region of essentially complete solid solution is complicated because of the varying ratio of Al/Si from albite, $NaAlSi_3O_8$, to anorthite, $CaAl_2Si_2O_6$. Three types of exsolution textures found in the plagioclase series are not visible to the naked eye, but may be detected because of iridescence. *Peristerite* intergrowths occur in the range An_2 to An_{15}. *Bøggild intergrowths* occur in some plagioclase with composition between An_{47} and An_{58}; their presence is indicated by the play of colors in labradorite. A third intergrowth occurs in the An_{60} to An_{85} region and is known as *Huttenlocher intergrowths*. These discontinuities in the solid solution series between Ab and An are on a very fine scale, however, and most properties, such as specific gravity or refractive index, show a generally linear change with chemical composition. Thus determination of a suitable property with sufficient precision permits a close approximation of the chemical composition in the plagioclase series (see Fig. 10.89).

K-Feldspars

MICROCLINE—KAlSi$_3$O$_8$

Crystallography. Triclinic; $\bar{1}$. The habit, crystal forms, and interfacial angles are similar to those of orthoclase, and microcline may be twinned according to the same laws as orthoclase; *Carlsbad* twins are common, but *Baveno* and *Manebach* twins are rare (see Fig. 10.80). It is also twinned according to the *albite law,* with {010} the twin plane, and the *pericline law,* with b the twin axis. These two types of twinning are usually present in microcline and on {001} the lamellae cross at nearly 90° giving a characteristic "tartan" structure (see Fig. 2.175). Microcline is found in cleavable masses, in crystals, and as a rock constituent in irregular grains. Microcline probably forms the largest known crystals. In a pegmatite in Karelia, U.S.S.R., a mass weighing over 2000 tons showed the continuity of a single crystal. Also in pegmatites microcline may be intimately intergrown with quartz, forming *graphic granite* (Fig. 10.81).

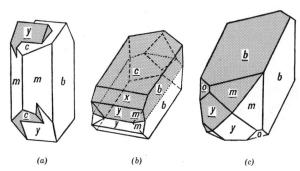

FIG. 10.80. Twinning in feldspar. (a) Carlsbad twin. (b) Manebach twin. (c) Baveno twin.

Microcline frequently has irregular and discontinuous bands crossing {001} and {010} that result from the exsolution of albite. The intergrowth as a whole is called *perthite,* or, if very fine, *microperthite* (see page 168 and Fig. 10.82).

$C\bar{1}$; for low microcline $a = 8.59$, $b = 12.97$, $c = 7.22$; $\alpha = 90°41'$, $\beta = 115°59'$, $\gamma = 87°39'$; $a:b:c = 0.662:1:0.557$; $Z = 4$. $d's$: 4.21(5), 3.83(3), 3.48(3), 3.37(5), 3.29(5), 3.25(10).

Physical Properties. *Cleavage* {001} perfect, {010} good at an angle of 89°30'. **H** 6. **G** 2.54–2.57. *Luster* vitreous. *Color* white to pale yellow, more rarely red or green. Green microcline is known as *Amazonstone.* Translucent to transparent. *Optics:* (−); $\alpha = 1.522$, $\beta = 1.526$, $\gamma = 1.530$; $2V = 83°$; $r > v$.

Composition and Structure. For $KAlSi_3O_8$, K_2O 16.9, Al_2O_3 18.4, SiO_2 64.7%; there is some substitution of Na for K (see Fig. 10.77). The structure of microcline is triclinic and therefore less symmetric than that of sanidine, which is shown in Fig. 10.75. Maximum low microcline shows perfect Al-Si order in the tetrahedral framework. With increasing disorder (increasing T) the terms low-, intermediate-, and high-microcline are applied. At considerably higher T the microcline structure transforms to orthoclase (see Fig. 10.77).

Diagnostic Features. Distinguished from orthoclase only by determining the presence of microcline ("tartan") twinning, which can rarely be determined without the aid of the microscope. With only minor exceptions, deep green feldspar is microcline.

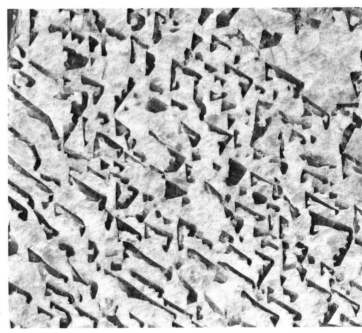

FIG. 10.81. Graphic granite, Hybla, Ontario. Quartz dark, microcline light.

Occurrence. Microcline is a prominent constituent of igneous rocks such as granites and syenites that cooled slowly and at considerable depth. In sedimentary rocks it is present in arkose and conglomerate. In metamorphic rocks in

FIG. 10.82. Microperthite as seen in polarized light under the microscope.

0.1 mm

gneisses. Microcline is the common K-feldspar of pegmatites and is quarried extensively in North Carolina, South Dakota, Colorado, Virginia, Wyoming, Maine, and Connecticut. *Amazonstone*, a green variety of microcline, is found in the Ural Mountains and in various places in Norway and the Malagasy Republic. In the United States it is found at Pikes Peak, Colorado, and Amelia Court House, Virginia (see Fig. 10.83).

Use. Feldspar is used chiefly in the manufacture of porcelain. It is ground very fine and mixed with kaolin or clay, and quartz. When heated to high temperature the feldspar fuses and acts as a cement to bind the material together. Fused feldspar also furnishes the major part of the glaze on porcelain ware. A small amount of feldspar is used in the manufacture of glass to contribute alumina to the batch. Amazonstone is polished and used as an ornamental material.

Name. *Microcline* is derived from two Greek words meaning *little* and *inclined*, referring to the slight variation of the cleavage angle from 90°. *Feldspar* is derived from the German word *feld*, field.

FIG. 10.83. Microcline and smoky quartz, Crystal Peak, Colorado.

ORTHOCLASE—KAISi₃O₈

Crystallography. Monoclinic; 2/m. Crystals are usually short prismatic elongated parallel to a, or elongated parallel to c and flattened on {010}. The prominent forms are b{010}, c{001}, and m{110}, often with smaller second- and fourth-order prisms (Fig. 10.84). Frequently twinned according to the following laws: *Carlsbad* a penetration twin with c as twin axis; *Baveno* with {021} as twin and composition plane; *Manebach* with {001} as twin and composition plane (see Fig. 10.80). Commonly in crystals or in coarsely cleavable to granular masses; more rarely fine grained, massive, and cryptocrystalline. Most abundantly in rocks as formless grains.

FIG. 10.84. Orthoclase and sanidine.

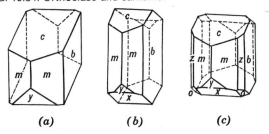

(a) (b) (c)

Angles: $m(110) \wedge m'(1\bar{1}0) = 61°13'$, $z(130) \wedge z'(\bar{1}30) = 58°48'$, $c(001) \wedge y(\bar{2}01) = 80°18'$, $c(001) \wedge x(\bar{1}01) = 50°17'$.

$C2/m$; $a = 8.56$, $b = 12.96$, $c = 7.21$ Å; $\beta = 115°50'$; $a:b:c = 0.663:1:0.559$. $Z = 4$. $d's$: 4.22(6), 3.77(7), 3.46(5), 3.31(10), 3.23(8).

Physical Properties. *Cleavage* {001} perfect, {010} good, {110} imperfect. **H** 6. **G** 2.57. *Luster* vitreous. *Color* colorless, white, gray, flesh-red, rarely yellow or green. *Streak* white. *Optics:* (−); $\alpha = 1.518$, $\beta = 1.524$, $\gamma = 1.526$; $2V = 10–70°$; $Z = b$, $X \wedge a = 5°$; $r > v$.

Composition and Structure. Most analyses contain small amounts of Na, but a complete solid solution series is possible from orthoclase toward intermediate albite (see Fig. 10.77). Orthoclase represents a K-feldspar structure that crystallizes at intermediate temperature and has a partially ordered Al-Si distribution. At high temperatures the Al-Si distribution in the structure becomes less ordered and is known as sanidine.

Diagnostic Features. Is usually recognized by its color, hardness, and cleavage. Distinguished from the other feldspars by its right-angle cleavage

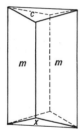

FIG. 10.85. Adularia.

and the lack of twin striations on the best cleavage surface. Difficultly fusible (5). Insoluble in acids.

Occurrence. Orthoclase is a major constituent of granites, granodiorites, and syenites, which have cooled at moderate depth and at reasonably fast rates. In more slowly cooled granites and syenites microcline will be the characteristic K-feldspar.

Name. The name *orthoclase* refers to the right-angle cleavage possessed by the mineral.

Similar Species. *Adularia,* KAlSi$_3$O$_8$, is a colorless, translucent to transparent K-feldspar which is commonly found in pseudo-orthorhombic crystals (Fig. 10.85). It occurs mainly in low-temperature veins in gneisses and schists. Some adularia shows an opalescent play of colors and is called *moonstone.*

Sanidine—(K,Na)AlSi$_3$O$_8$

Crystallography. Monoclinic; $2/m$. Crystals are often tabular parallel to {010}; also elongated on a with square cross-section as in Fig. 10.84a. Carlsbad twins common.

$C2/m$; for high sanidine, $a = 8.56$, $b = 13.03$, $c = 7.17$; $\beta = 115°59'$; $a:b:c = 0.657:1:0.550$. $Z = 4$. *d's*: 4.22(6), 3.78(8), 3.31(10), 3.278(6), 3.225(8).

Physical Properties. Cleavage {001} perfect, {010} good. **H** = 6. **G** = 2.56–2.62. *Luster* vitreous. *Color* colorless and commonly transparent. *Streak* white. *Optics:* (−); $\alpha = 1.518$–1.525, $\beta = 1.523$–1.530, $\gamma = 1.525$–1.531. Occurs in two orientations with optic plane parallel with {010} and 2V = 0–60°, and with optic plane normal to {010} and 2V = 0–25°.

Composition and Structure. A complete solid solution exists at high temperature between sanidine and high albite; part of the intermediate region is known as *anorthoclase* (see Fig. 10.74). The structure of sanidine shows a disordered (random) distribution of Al and Si in the tetrahedral framework (see Fig. 10.75). The Al and Si distribution in orthoclase is more ordered.

Diagnostic Features. Can be characterized with confidence only by optical or X-ray techniques. Optic orientation and 2V differ from orthoclase. Microcline and plagioclase show different types of twinning.

Occurrence. As phenocrysts (see Fig. 10.86) in extrusive igneous rocks such as rhyolites and trachytes. Sanidine is characteristic of rocks that cooled quickly from an initial high temperature of eruption. Most sanidines are cryptoperthitic.

Name. Sanidine is derived from the Greek *sanis,* tablet, and *idos,* appearance, in allusion to its typical tabular habit.

FIG. 10.86. Polished slab of basalt with sanidine phenocrysts.

Plagioclase Feldspar Series

The plagioclase feldspars form, at elevated temperatures, an essentially complete solid solution series from pure *albite*, $NaAlSi_3O_8$, to pure *anorthite*, $CaAl_2Si_2O_8$ (see Fig. 10.79). Minor discontinuities in this series, as shown by the existence of peristerite and other intergrowths (see page 425) can be detected only by careful electron optical and X-ray studies. The nomenclature for the series, as based on six arbitrary divisions, is given in Fig. 10.74. If species names are not given, the composition is expressed by $Ab_{20}An_{80}$, for example, which may be shortened to Ab_{20}. Most of the properties of the various species in this series vary in a uniform manner with the change in chemical composition. For this reason the series can be more easily understood if one comprehensive description is given, rather than six individual descriptions, and the dissimilarities between members indicated. The distinction between high- and low-temperature forms of the series can be made only by X-ray and optical means.

ALBITE—$NaAlSi_3O_8$
ANORTHITE—$CaAl_2Si_2O_8$

Crystallography. Triclinic; $\bar{1}$. Crystals commonly tabular parallel to {010}; occasionally elongated on *b* (Fig. 10.87). In anorthite crystals may be prismatic elongated on *c*.

Crystals frequently twinned according to the various laws governing the twins of orthoclase, that is, *Carlsbad, Baveno,* and *Manebach*. In addition, they are nearly always twinned according to the *albite law* and less frequently according to the *pericline law.* Albite twinning with twin plane {010} is commonly polysynthetic and because (010) $\wedge$ (001) $\approx$ 86°, {001} either as a crystal face

FIG. 10.87. Albite crystals.

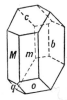

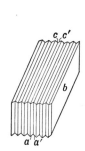

FIG. 10.88. Albite twinning (see also Fig. 2.171).

or cleavage is crossed by parallel groovings or striations (Figs. 10.88 and 2.171). Often these striations are so fine as to be invisible to the unaided eye, but on some specimens they are coarse and easily seen. The presence of the striations on the basal cleavage is one of the best proofs that a feldspar belongs to the plagioclase series. Pericline twinning with *b* as the twin axis is also polysynthetic. Striations resulting from it can be seen on {010}. Their direction on {010} is not constant but varies with the composition.

Distinct crystals are rare. Usually in twinned, cleavable masses; as irregular grains in igneous rocks.

Angles: c(001) $\wedge$ b(010) = 86°24', albite; 85°50', anorthite. m(110) $\wedge$ M(1$\bar{1}$0) = 59°14', albite; 59°29', anorthite. The corresponding angles for feldspars of intermediate composition lie between those of albite and anorthite.

For albite, C$\bar{1}$: $a = 8.14$, $b = 12.8$, $c = 7.16$; $\alpha = 94°20'$, $\beta = 116°34'$, $\gamma = 87°39'$. $a:b:c = 0.636:1:0.559$; $Z = 4$. For anorthite, $P\bar{1}$: $a = 8.18$, $b = 12.88$, $c = 14.17$; $\alpha = 93°10'$, $\beta = 115°51'$, $\gamma = 91°13'$; $a:b:c = 0.635:1:1.100$; $Z = 8$. $d's$(Ab-An): 4.02–4.03(7), 3.77–3.74(5), 3.66–3.61(6), 3.21(7)–3.20(10), 3.18(10)–3.16(7).

Physical Properties. Cleavage {001} perfect, {010} good. **H** 6. **G** 2.62 in albite to 2.76 in anorthite (Fig. 10.89). *Color* colorless, white, gray; less frequently greenish, yellowish, flesh-red. *Luster* vitreous to pearly. Transparent to translucent. A beautiful play of colors is frequently seen, especially in labradorite and andesine. *Optics:* Albite: (+);

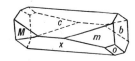

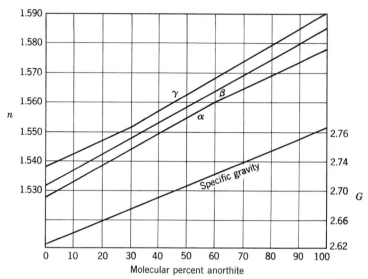

FIG. 10.89. Variation of refractive indices and specific gravity with composition in the plagioclase feldspar series.

$\alpha = 1.527$, $\beta = 1.532$, $\gamma = 1.538$; 2V = 74°. Anorthite: (−); $\alpha = 1.577$, $\beta = 1.585$, $\gamma = 1.590$, 2V = 77° (see Fig. 10.89).

Composition and Structure. An essentially complete solid solution series extends from $NaAlSi_3O_8$ to $CaAl_2Si_2O_8$ at elevated temperatures (see Fig. 10.79). Considerable K may be present especially toward the albite end of the series (see Fig. 10.78). Very fine scale discontinuities in the series, which can be detected only by electron optical and X-ray means, are reflected in the occurrence of very fine lamellar intergrowths: peristerite, Bøggild and Huttenlocher intergrowths (see page 425). The sturcture of albite is triclinic ($C\bar{1}$) at low to moderate temperatures, and shows a highly ordered Al-Si distribution; low albite is structurally analogous to low microcline. High albite is also triclinic and has a highly disordered Al-Si distribution. At temperatures above 920°C a monoclinic variety of albite, known as *monalbite*, with a totally disordered Al-Si distribution occurs. Anorthite has space group $P\bar{1}$ at low temperatures. At elevated temperatures the structure of anorthite inverts to $I\bar{1}$ (see Fig. 10.79).

Diagnostic Features. The plagioclase feldspars can be distinguished from other feldspars by the presence of albite twin striations {001}. They can be placed accurately in their proper places in the series only by quantitative chemical analyses or optical tests, but they can be roughly distinguished from one another by specific gravity. Fusible at 4–4½ to a colorless glass. Albite is insoluble in HCl, but anorthite is decomposed by HCl. Between these extremes the intermediate members show a greater solubility the greater the amount of calcium. A strong yellow flame (Na) is given by members rich in sodium.

Occurrence. The plagioclase feldspars, as rock-forming minerals, are even more widely distributed and more abundant than the potash feldspars. They are found in igneous, metamorphic, and, more rarely, sedimentary rocks.

The classification of igneous rocks is based largely on the kind and amount of feldspar present (see Table 11.2). As a rule, the greater the percentage of SiO_2 in a rock the fewer the dark minerals, the greater amount of potash feldspar, and the more sodic the plagioclase; and conversely, the lower the percentage of SiO_2 the greater the percentage of dark minerals and the more calcic the plagioclase.

In an igneous crystallization sequence the more refractory minerals crystallize before the less refractory ones. The An end member of the plagioclase series has a much higher melting point than the Ab end member (see Fig. 10.79). For this reason plagioclase feldspars that crystallize early from a magma are generally more Ca-rich than those that crystallize later. Continuous chemical zoning occurs commonly in phenocrysts with Ca-rich centers and more Na-rich rims.

Albite is included with orthoclase and microcline in what are known as the *alkali* feldspars, all of which have a similar occurrence. They are commonly found in granites, syenites, rhyolites, and trachytes. Albite is common in pegmatites where it may replace earlier microcline; also in small crystals as the platy variety, *cleavelandite*. Some albite and oligoclase shows an opalescent play of colors and is known as *moonstone*. The name albite is derived from the Latin *albus,* meaning *white,* in allusion to the color.

Oligoclase is characteristic of granodiorites and monzonites. In some localities, notably at Tvedestrand, Norway, it contains inclusions of hematite, which give the mineral a golden shimmer and sparkle. Such feldspar is called *aventurine* oligoclase, or *sunstone.* The name is derived from two Greek words meaning *little* and *fracture,* because it was believed to have less perfect cleavage than albite.

Andesine is rarely found except as grains in andesites and diorites. Named from the Andes Mountains where it is the chief feldspar in the andesite lavas.

Labradorite is the common feldspar in gabbros and basalts and in anorthosite it is the only important constituent. Found on the coast of Labrador in large cleavable masses which show a fine iridescent play of colors. The name is derived from this locality.

Bytownite is rarely found except as grains in gabbros. Named from Bytown, Canada (now the city of Ottawa).

Anorthite is rarer than the more sodic plagioclase. Found in rocks rich in dark minerals and in druses of ejected volcanic blocks and in granular limestones of contact metamorphic deposits. Name derived from the Greek word meaning *oblique,* because its crystals are triclinic.

Use. Plagioclase feldspars are less widely used than potash feldspars. Albite, or *soda spar,* as it is called commercially, is used in ceramics in a manner similar to microcline. Labradorite that shows a play of colors is polished and used as an ornamental stone. Those varieties that show opalescence are cut and sold under the name of *moonstone.*

Name. The name plagioclase is derived from the Greek meaning *oblique,* an allusion to the oblique angle between the cleavages. (See under "Occurrence" for names of specific species.)

Feldspathoid Group

The feldspathoids are anhydrous framework silicates that are chemically similar to the feldspars. The chief chemical difference between feldspathoids and feldspars lies in the SiO_2 content. The feldspathoids contain only about two-thirds as much silica as alkali feldspar and hence tend to form from melts rich in alkalis (Na and K) and poor in SiO_2 (see Fig. 11.11).

Examples of the aluminosilicate frameworks in feldspathoids are given in Figs. 10.90 and 10.91. The structure of *leucite,* $KAlSi_2O_6$, has tetragonal symmetry (space group $I4_1/a$) at low to intermediate temperatures; at approximately 605°C it inverts to a

FIG. 10.90. Portion of the leucite structure projected down the *c* axis (after J. J. Papike and M. Cameron, 1976, *Reviews of Geophysics and Space Physics,* v. 14, p. 74).

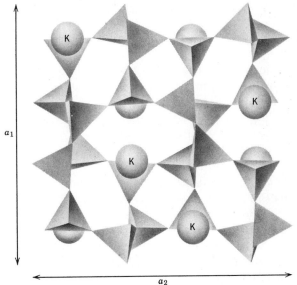

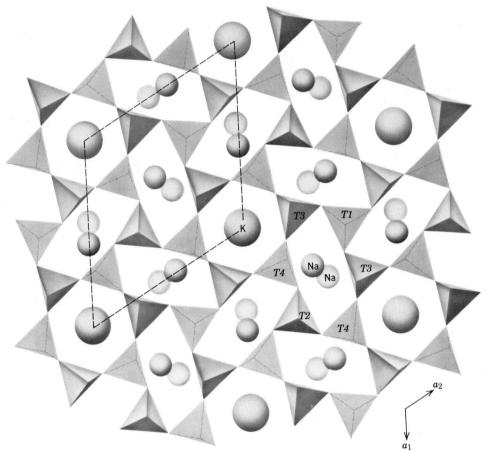

FIG. 10.91. Projection of the nepheline structure onto (0001). Dashed lines outline unit cell. *T1, T2, T3* and *T4* refer to crystallographically distinct tetrahedral sites (after J. J. Papike and M. Cameron, 1976, *Review of Geophysics and Space Physics*, v. 14, p. 56).

cubic structure with space group *Ia3d*. In the low-temperature form the Al and Si tetrahedra share corners to form four- and six-membered rings. The K^+ ions are in 12-coordination with oxygen in large cavities in the structure. The structure of *nepheline*, $(Na, K)AlSiO_4$ (space group $P6_3$), can be considered as a derivative of the high tridymite structure with Al replacing Si in half the tetrahedra and Na and K in interstitial voids. The Si and Al are ordered in specific tetrahedral sites of the structure; the *T1* and *T4* sites are Al-rich, the *T2* and *T3* sites Si-rich.

Some of the members of the feldspathoid group contain unusual anions. Sodalite contains Cl and carnotite incorporates CO_3. Noselite houses SO_4 and lazurite SO_4, S, and Cl ions. These large anionic groups and anions are located in large interstices of the structure.

LEUCITE—$KAlSi_2O_6$

Crystallography. Tetragonal; $4/m$ below 605°C. Isometric, $4/m\overline{3}2/m$ above 605°C. Usually in trapezohedral crystals which at low temperature are pseudomorphs of the high-temperature form (Fig. 10.92).

Below 605°C: $I4_1/a$; $a = 13.04$, $c = 13.85$; $a:c = 1:1.062$; $Z = 16$. Above 605°C: $Ia3d$; $a = 13.43$ Å; $Z = 16$. d's (low T form): 5.39(8), 3.44(9), 2.27(10), 2.92(7), 2.84(7).

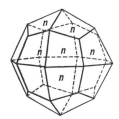

FIG. 10.92. Leucite.

Physical Properties. **H** $5\frac{1}{2}$–6. **G** 2.47. *Luster* vitreous to dull. *Color* white to gray. Translucent. *Optics:* (+); $\omega = 1.508$, $\epsilon = 1.509$.

Composition and Structure. Most leucites are close to $KAlSi_2O_6$ with K_2O 21.5, Al_2O_3 23.5, SiO_2 55.0%. Small amounts of Na may replace K. The structure of the low-temperature polymorph is shown in Fig. 10.90.

Diagnostic Features. Infusible. Characterized by its trapezohedral form and infusibility. Leucite is usually embedded in a fine-grained matrix (see Fig. 10.93), whereas analcime is usually in cavities in free-growing crystals. Moreover, analcime is fusible and yields water in the closed tube.

Occurrence. Although leucite is a rather rare mineral it is abundant in certain recent lavas; rarely observed in deep-seated rocks. It is found only in silica-deficient rocks and thus never in rocks containing quartz. Its most notable occurrence is as phenocrysts in the lavas of Mount Vesuvius. In the United States it is found in rocks of the Leucite Hills, Wyoming, and in certain of the rocks in the Highwood and Bear Paw Mountains, Montana. Pseudomorphs of a mixture of nepheline, orthoclase, and analcime after leucite, *pseudoleucite,* are found in syenites of Arkansas, Montana, and Brazil.

Name. From a Greek word meaning *white.*

Similar Species. *Pollucite,* $CsAlSi_2O_6 \cdot H_2O$, isostructural with leucite, is a rare isometric mineral usually occurring in pegmatites.

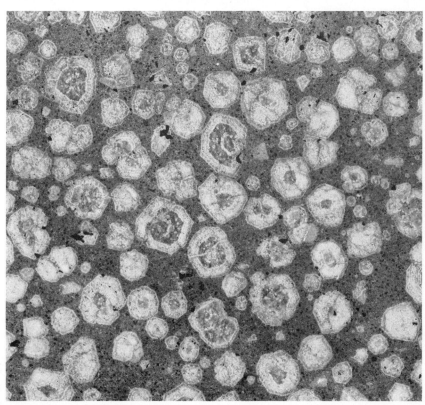

FIG. 10.93. Polished slab of basalt showing large leucite phenocrysts.

NEPHELINE—(Na, K)AlSiO$_4$

Crystallography. Hexagonal; 6. Rarely in small prismatic crystals with base. Almost invariably massive, compact, and in embedded grains.

$P6_3$; $a = 10.01$, $c = 8.41$ Å; $a:c = 1:0.840$; $Z = 8$. d's: 4.18(7), 3.27(7), 3.00(10), 2.88(7), 2.34(6).

Physical Properties. Cleavage $\{10\overline{1}0\}$ distinct. **H** $5\frac{1}{2}$–6. **G** 2.60–2.65. *Luster* vitreous in clear crystals; greasy in the massive variety and hence called *eleolite. Color* colorless, white, or yellowish. In the massive variety gray, greenish, and reddish. Transparent to translucent. *Optics:* (−); $\omega = 1.529$–1.546, $\epsilon = 1.526$–1.542.

Composition and Structure. The end member composition NaAlSiO$_4$ contains Na$_2$O 21.8, Al$_2$O$_3$ 35.9, SiO$_2$ 42.3%. The amount of K in natural nephelines may range from 3 to 12 weight percent K$_2$O. A large miscibility gap exists at low to moderate temperatures between NaAlSiO$_4$ and *kalsilite,* KAlSiO$_4$. However, at temperatures above 1070°C a complete solid solution series exists between K and Na end members. The structure of nepheline is illustrated in Fig. 10.91.

Diagnostic Features. Fusible at 4 to a colorless glass giving a strong yellow flame of sodium. Readily soluble in HCl and on evaporation yields a silica jelly. Characterized in massive varieties by its greasy luster. Distinguished from quartz by inferior hardness and from feldspar by gelatinizing in acid.

Occurrence. Nepheline is a rock-forming mineral found in silica-deficient intrusive and extrusive rocks. Crystals are present in the lavas of Mount Vesuvius. The largest known mass of intrusive nepheline rocks is on the Kola Peninsula, U.S.S.R., where, locally, nepheline is associated with apatite. Extensive masses of nepheline rocks are found in Norway and South Africa. In the United States nepheline, both massive and in crystals, is found at Litchfield, Maine, associated with cancrinite. Found near Magnet Cove, Arkansas, and Beemerville, New Jersey. Common in the syenites of the Bancroft region of Ontario, Canada, where the associated pegmatites contain large masses of nearly pure nepheline.

Use. Iron-free nepheline, because of its high Al$_2$O$_3$ content, has been used in place of feldspar in the glass industry. Most nepheline for this purpose comes from Ontario. Nepheline produced as a byproduct of apatite mining on the Kola Peninsula is used by the Russians in several industries including ceramics, leather, textile, wood, rubber, and oil.

Name. *Nepheline* is derived from a Greek word meaning a *cloud,* because when immersed in acid the mineral becomes cloudy. *Eleolite* is derived from the Greek word for *oil,* in allusion to its greasy luster.

Similar Species. *Cancrinite,* Na$_6$Ca(CO$_3$)(AlSiO$_4$)$_6$·2H$_2$O, is a rare mineral similar to nepheline in occurrence and associations.

SODALITE—Na$_8$(AlSiO$_4$)$_6$Cl$_2$

Crystallography. Isometric; $\overline{4}3m$. Crystals rare, usually dodecahedrons. Commonly massive, in embedded grains.

$P\overline{4}3n$; $a = 8.83$–8.91 Å; $Z = 2$. d's: 6.33(8), 3.64(10), 2.58(5), 2.10(2), 1.750(3).

Physical Properties. *Cleavage* $\{011\}$ poor. **H** $5\frac{1}{2}$–6. **G** 2.15–2.3. *Luster* vitreous. *Color* usually blue, also white, gray, green. Transparent to translucent. *Optics:* refractive index 1.483.

Composition and Structure. Na$_2$O 25.6, Al$_2$O$_3$ 31.6, SiO$_2$ 37.2, Cl 7.3%. Only small substitution of K for Na. In the structure of sodalite the AlSiO$_4$ framework is linked such as to produce cagelike, cubo-octahedral cavities in which the Cl is housed.

Diagnostic Features. Fusible at $3\frac{1}{2}$–4, to a colorless glass, giving a strong yellow flame (Na). Soluble in HCl and gives gelantinous silica upon evaporation. Nitric acid solution with silver nitrate gives white precipitate of silver chloride. In a salt of phosphorus bead with copper oxide gives azureblue copper chloride flame. Usually distinguished by its blue color, and told from lazurite by the different occurrence and absence of associated pyrite. If color is not blue, a positive test for chlorine is necessary to distinguish it from analcime, leucite, and haüynite.

Occurrence. Sodalite is a comparatively rare

rock-forming mineral associated with nepheline, cancrinite, and other feldspathoids in nepheline syenites, trachytes, phonolites, and so forth. Found in transparent crystals in the lavas of Vesuvius. The massive blue variety is found at Litchfield, Maine; Bancroft, Ontario; and Ice River, British Columbia.

Name. Named in allusion to its sodium content.

Similar Species. Other rare feldspathoids are *haüynite*, $(Na, Ca)_{4-8}(AlSiO_4)_6(SO_4)_{1-2}$ and *noselite*, $Na_8(AlSiO_4)_6SO_4$.

Lazurite—$(Na, Ca)_8(AlSiO_4)_6(SO_4, S, Cl)_2$

Crystallography. Isometric; $\bar{4}3m$. Crystals rare, usually dodecahedral. Commonly massive, compact.

$P\bar{4}3m$; $a = 9.08$; $Z = 1$. $d's$: 3.74(10), 2.99(10), 2.53(9), 1.545(9), 1.422(7).

Physical Properties. Cleavage {011} imperfect. **H** $5-5\frac{1}{2}$. **G** 2.4–2.45. *Luster* vitreous. *Color* deep azure-blue, greenish blue. Translucent. *Optics:* refractive index 1.50±.

Composition and Structure. Small amounts of Rb, Cs, Sr and Ba may substitute for Na. There is considerable variation in the amounts of SO_4, S, and Cl. Lazurite appears isostructural with sodalite. The large cagelike cavities in the structure house the Cl and S anions and the (SO_4) anionic groups.

Diagnostic Features. Characterized by its blue color and the presence of associated pyrite. Fusible at $3\frac{1}{2}$, giving strong yellow flame (Na). Soluble in HCl with slight evolution of hydrogen sulfide gas.

Occurrence. Lazurite is a rare mineral, occurring usually in crystalline limestones as a product of contact metamorphism. *Lapis lazuli* is a mixture of lazurite with small amounts of calcite, pyroxene, and other silicates, and commonly contains small disseminated particles of pyrite. The best quality of lapis lazuli comes from northeastern Afghanistan. Also found at Lake Baikal, Siberia, and in Chile.

Use. Lapis lazuli is highly prized as an ornamental stone, for carvings etc. As a powder it was formerly used as the paint pigment ultramarine. Now ultramarine is produced artificially.

Name. Lazurite is an obsolete synonym for azurite, and hence the mineral is named because of its color resemblance to azurite.

Petalite—$Li(AlSi_4O_{10})$

Crystallography. Monoclinic; $2/m$. Crystals rare, flattened on {010} or elongated on [100]. Usually massive or in foliated cleavable masses.

$P2/a$; $a = 11.76$, $b = 5.14$, $c = 7.62$ Å; $\beta = 112°24'$; $a:b:c = 2.288:1:1.482$; $Z = 2$. $d's$: 3.73(10), 3.67(9), 3.52(3), 2.99(1), 2.57(2).

Physical Properties. *Cleavage* {001} perfect, {201} good. Brittle. **H** $6-6\frac{1}{2}$. **G** 2.4. *Luster* vitreous, pearly on {001}. *Color* colorless, white, gray. Transparent to translucent. *Optics:* (+); $\alpha = 1.505$, $\beta = 1.511$, $\gamma = 1.518$; $2V = 83°$; $Z = b$. $X \wedge a = 2-8°$. $r > v$.

Composition and Structure. LiO 4.9, Al_2O_3 16.7, SiO_2 78.4%. The structure of petalite consists of a framework of AlO_4 and SiO_4 tetrahedra that contains Si_4O_{10} layers linked by AlO_4 tetrahedra. The Li is in tetrahedral coordination with oxygen.

Diagnostic Features. Characterized by its platy habit. Distinguished from spodumene by its cleavage and lesser specific gravity. Fusible at 5 giving the red lithium flame. When gently heated it emits a blue phosphorescent light. Insoluble.

Occurrence. Petalite is found in pegmatites where it is associated with other lithium-bearing minerals such as spodumene, tourmaline, and lepidolite. Petalite was considered a rare mineral until it was found at the Varutrask pegmatite, Sweden. More recently it has been found abundantly in Bikita, Rhodesia, and in South-West Africa. At these localities, associated with lepidolite and eucryptite, it is mined extensively.

Use. An important ore of lithium. (See spodumene, page 380.)

Name. From the Greek, *leaf*, alluding to the cleavage.

Scapolite Series

The *scapolites* are metamorphic minerals with compositions suggestive of the feldspars. There is a complete solid solution series between *marialite,* $3NaAlSi_3O_8 \cdot NaCl$, and *meionite,* $3CaAl_2Si_2O_8 \cdot CaSO_4$ or $CaCO_3$. When the formulas are written in this way it is clear that the composition of meionite is the equivalent of three formula weights of albite, $NaAlSi_3O_8$, plus one formula weight $NaCl$, and that the marialite composition is equivalent to three formula weights of anorthite, $CaAl_2Si_2O_8$, plus one formula weight of $CaCO_3$ or $CaSO_4$. In this series there is complete substitution of Ca for Na with charge compensation effected as in the feldspars by concomitant substitution of Al for Si. There is also complete substitution of CO_3, SO_4, and Cl_2 for each other. The name *wernerite* has been used for members intermediate in composition between marialite and meionite.

The structure of scapolite consists of a framework of SiO_4 and AlO_4 tetrahedra with large cavities containing the (Ca, Na) ions and (CO_3, Cl_2, SO_4) anionic groups (see Fig. 10.94).

Marialite–Meionite

Crystallography. Tetragonal; $4/m$. Crystals usually prismatic. Prominent forms are prisms {010} and {110} and dipyramid {011}; rarely shows the faces of the tetragonal dipyramid z (Fig. 10.95). Crystals are usually coarse, or with faint fibrous appearance.

Angles: $a(100) \wedge p(101) = 58°12'$, $p(101) \wedge p'(011) = 43°45'$.

$I4/m$; $a = 12.2$, $c = 7.6$ Å; $a:c = 1:0.623$; $Z = 2$. a varies slightly with composition. $d's$: (marialite), 4.24(7), 3.78(9), 3.44(10), 3.03(10), 2.68(9).

Physical Properties. *Cleavage* {100} and {110} imperfect but distinct. **H** 5–6. **G** 2.55–2.74. The specific gravity and refractive indices increase with increasing Ca content. *Luster* vitreous when fresh. *Color* white, gray, pale green; more rarely yellow, bluish, or reddish. Transparent to translucent. *Optics:* (−); $\omega = 1.55–1.60$, $\epsilon = 1.54–1.56$.

Composition and Structure. The extent of

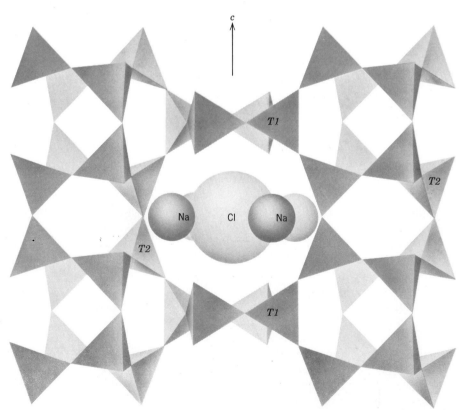

FIG. 10.94. The structure of scapolite projected on (100) (after J. J. Papike and M. Cameron, 1976, *Reviews of Geophysics and Space Physics,* v. 14, p. 75).

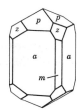

FIG. 10.95. Scapolite.

solid solution between *marialite,* $Na_4(AlSi_3O_8)_3$ (Cl_2, CO_3, SO_4), and *meionite,* $Ca_4(Al_2Si_2O_8)_3(Cl_2, CO_3, SO_4)$, is discussed on p. 436. The structure, illustrated in Fig. 10.94, contains two crystallographically distinct tetrahedra, *T1* and *T2*. In marialite compositions Al is confined to the *T2* tetrahedra (46% Al, 54% Si) and only Si occupies the *T1* sites. In meionite compositions Al is housed in both tetrahedral sites.

Diagnostic Features. Fusible at 3 with intumescence to a white blebby glass and colors the flame yellow. Characterized by its crystals with square cross-section and four cleavage directions at 45°. When massive, resembles feldspar but has a characteristic fibrous appearance on the cleavage surfaces. It is also more readily fusible. Its fusibility with intumescence distinguishes it from pyroxene.

Occurrence. Scapolite occurs in crystalline schists, gneisses, amphibolites, and granulite facies rocks. In many cases it is derived by alteration from plagioclase feldspars. It is also characteristic in crystalline limestones as a contact metamorphic mineral. Associated with diopside, amphibole, garnet, apatite, sphene, and zircon.

Crystals of gem quality with a yellow color occur in the Malagasy Republic. In the United States found in various places in Massachusetts, notably at

FIG. 10.96. Analcime crystals.

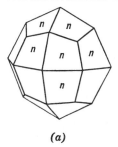

(a)

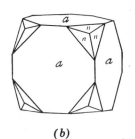

(b)

Bolton; Orange, Lewis, and St. Lawrence Counties, New York. Also found at various points in Ontario, Canada.

Name. From the Greek meaning a *shaft,* in allusion to the prismatic habit of the crystals.

ANALCIME—$NaAlSi_2O_6 \cdot H_2O$*
Crystallography. Isometric; $4/m\overline{3}2/m$. Usually in trapezohedrons (Fig. 10.96a). Cubes with trapezohedral truncations also known (Fig. 10.96b). Usually in crystals, also massive granular.

$Ia3d$; $a = 13.71$ Å; $Z = 16$. $d's$: 5.61(8), 4.85(4), 3.43(10), 2.93(7), 2.51(5).

Physical Properties. H $5-5\frac{1}{2}$. **G** 2.27. *Luster* vitreous. *Color* colorless or white. Transparent to translucent. *Optics:* refractive index 1.48–1.49.

Composition and Structure. The chemical composition of most analcimes is fairly constant with minor amounts of K or Ca substituting for Na; also some Al substitution for Si. $NaAlSi_2O_6 \cdot H_2O$ contains Na_2O 14.1, Al_2O_3 23.2, SiO_2 54.5, H_2O 8.2%. The structure is made of a framework of AlO_4 and SiO_4 tetrahedra with four- and six-membered tetrahedral rings. This framework contains continuous channels along the 3-fold axes of the structure which are occupied by H_2O molecules. The octahedrally coordinated Na is housed in somewhat smaller voids in the structure.

Diagnostic Features. Fusible at $3\frac{1}{2}$, to a clear glass, coloring the flame yellow (Na). Gives water in the closed tube. Usually recognized by its free-growing crystals and its vitreous luster. Crystals resemble leucite but distinguished from it by ease of fusibility, presence of water, and occurrence. (Leucite is always embedded in rock matrix.)

Occurrence. Analcime occurs as a primary mineral in some igneous rocks and is also the product of hydrothermal action in the filling of basaltic cavities. It is an original constituent of analcime basalts and may occur in alkaline rocks such as aegirine-analcime-nepheline syenites. It is also found in vesicles of igneous rocks in association with prehnite, calcite, and zeolites. Fine crystals

* Analcime may be classified as a member of the zeolite group, however, its structure, chemistry, and occurrence are very similar to those of the feldspathoids.

are found in the Cyclopean Islands near Sicily; in the Val di Fassa and on the Seiser Alpe, Trentino, Italy; in Victoria, Australia; and Kerguelen Island in the Indian Ocean. In the United States found at Bergen Hill, New Jersey; in the Lake Superior copper district; and at Table Mountain, near Golden, Colorado. Also found at Cape Blomidon, Nova Scotia.

Name. Derived from a Greek word meaning *weak,* in allusion to its weak electric property when heated or rubbed.

Zeolite Group

The zeolites form a large group of hydrous silicates and show close similarities in composition, association, and mode of occurrence. They are framework aluminosilicates with Na and Ca, and highly variable amounts of H_2O in the voids of the framework. They average between $3\frac{1}{2}$ and $5\frac{1}{2}$ in hardness and between 2.0 and 2.4 in specific gravity. Many of them fuse readily with marked intumescence (a swelling up), hence the name *zeolite,* from two Greek words meaning *to boil* and *stone.* Traditionally, the zeolites have been thought of as well-crystallized minerals found in cavities and veins in basic igneous rocks. Recently, large deposits of zeolites have been found in the western United States and in Tanzania as alterations of volcanic tuff and volcanic glass. The occurrence of zeolites in various rock types is used to define the low-grade regional metamorphic zone known as the "zeolite facies."

The zeolites, like the feldspars and feldspathoids, are built of frameworks of AlO_4 and SiO_4 tetrahedra; the zeolite frameworks, however, are very open with large interconnecting spaces or channels. The zeolites can be divided in those with a fibrous habit and an underlying chain structure (Fig. 10.97); those with a platy habit and an underlying sheet structure; and those with an equant habit and an underlying framework structure (Fig. 10.98).

Much of the interest in zeolites derives from the presence of the spacious channels. When a zeolite is heated, the water in the channelways is given off easily and continuously as the temperature rises, leaving the structure intact. This is in sharp contrast to other hydrated compounds, as gypsum, $CaSO_4\cdot$

FIG. 10.97. Chains of AlO_4 and SiO_4 tetrahedra as found in natrolite and other fibrous zeolites.

$2H_2O$, in which the water molecules play a structural role and complete dehydration produces structural collapse. After essentially complete dehydration of a zeolite, the channels may be filled again with water or with ammonia, mercury vapor, iodine vapor, or a variety of substances. This process is selective and depends on the particular zeolite structure and the size of the molecules, and, hence, zeolites are used as "molecular sieves."

Zeolites have a further useful property that derives from their structure. Water may pass easily through the channelways, and, in the process, ions in solution may be exchanged for ions in the structure. This process is called "base exchange" or "cation exchange," and by its agency zeolites or synthetic compounds with zeolitic structure are used for water softening. The zeolite used has a composition about $Na_2Al_2Si_3O_{10}\cdot 2H_2O$ (like natrolite). "Hard" water, that is, water containing many calcium ions in solution, is passed through a tank filled with zeolite grains. The Ca^{2+} ions replace the Na^+ ions in the zeolite, forming $CaAl_2Si_3O_{10}\cdot 2H_2O$,

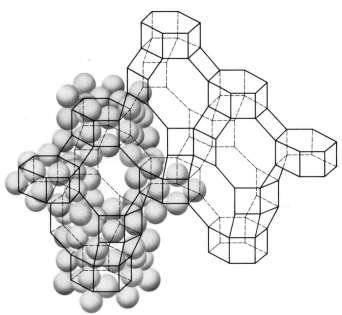

FIG. 10.98. Schematic representation of the structure of chabazite. Si and Al (not shown) occupy the corners of the framework outlined by the lines. The balls represent oxygen in close packing. Each framework unit contains a cavity which is connected to adjacent cavities by channels. The channel diameter in chabazite is 3.9 Å (after D. W. Breck and J. V. Smith, 1959, Scientific American, v. 200, no. 1, p. 88).

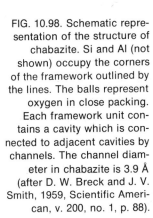

contributing Na^+ ions to the solution. Water containing sodium does not form scum and is said to be "soft." When the zeolite in the tank has become saturated with calcium, a strong NaCl brine is passed through the tank. The high concentration of sodium ion forces the reaction to go in the reverse direction, and the $Na_2Al_2Si_3O_{10} \cdot 2H_2O$ is reconstituted, calcium going into solution. By base exchange many ions, including silver, may be substituted for the alkali-metal cation in the zeolite structure.

NATROLITE—$Na_2Al_2Si_3O_{10} \cdot 2H_2O$

Crystallography. Orthorhombic; $mm2$. Prismatic, often acicular with prism zone vertically striated. Usually in radiating crystal groups; also fibrous, massive, granular, or compact.

$Fdd2$; $a = 18.35$, $b = 18.70$, $c = 6.61$; $a:b:c = 0.981:1:0.353$; $Z = 8$. $d's$: 6.6(9), 5.89(8), 3.19(9), 2.86(10), 2.42(6).

Physical Properties. *Cleavage* {110} perfect. **H** $5-5\frac{1}{2}$. **G** 2.25. *Luster* vitreous. *Color* colorless or white, rarely tinted yellow to red. Transparent to translucent. *Optics:* (+); $\alpha = 1.480$, $\beta = 1.482$, $\gamma = 1.493$; $2V = 63°$; $X = a$; $Y = b$; $r < v$.

Composition and Structure. Na_2O 16.3, Al_2O_3 26.8, SiO_2 47.4, H_2O 9.5%. Some K and Ca may replace Na. Natrolite is one of the group of fibrous zeolites. Its structure consists of an Si-Al-O framework in which chains parallel to the c axis are prominent (see Fig. 10.97). The chains are linked laterally by the sharing of some oxygens of opposing tetrahedra. Na is coordinated between the chains to six oxygens, four of which are tetrahedral oxygens and two of which are from water molecules. The perfect {110} cleavage in natrolite is due to the relatively small number of bonds between chains as compared to the large number of bonds within the chainlike structure.

Diagnostic Features. Fusible at $2\frac{1}{2}$ to a clear glass, giving a yellow sodium flame. Yields water in the closed tube. Soluble in HCl and gelatinizes upon evaporation.

Occurrence. Natrolite is characteristically found lining cavities in basalt associated with other zeolites and calcite. Notable localities are Aussig and Salesel, Bohemia; Puy-de-Dôme, France; and Val di Fassa, Trentino, Italy. In the United States found at Bergen Hill, New Jersey. Also found in various places in Nova Scotia.

Name. From the Latin *natrium,* meaning *sodium,* in allusion to its composition.

Similar Species. *Scolecite,* $CaAl_2Si_3O_{10} \cdot 3H_2O$, is another fibrous zeolite, similar in structure to natrolite but monoclinic in symmetry.

CHABAZITE—$Ca_2Al_2Si_4O_{12} \cdot 6H_2O$

Crystallography. Hexagonal-R; $\bar{3}2/m$. Usually in rhombohedral crystals, {1011}, with nearly cubic angles. May show several different rhombohedrons (Fig. 10.99). Often in penetration twins.

Angles: $r(10\bar{1}1) \wedge r'(\bar{1}101) = 85°14'$, $e(01\bar{1}2) \wedge r(10\bar{1}1) = 42°37'$.

$R\bar{3}m$; $a = 13.78$, $c = 14.97$ Å; $a:c = 1:1.086$; $Z = 6$. $d's$: 9.35(5), 5.03(3), 4.33(7), 3.87(3), 2.925(10).

Physical Properties. *Cleavage* {10$\bar{1}$1} poor. **H** 4–5. **G** 2.05–2.15. *Luster* vitreous. *Color* white, yellow, pink, red. Transparent to translucent. *Optics:* (+); $\omega = 1.482$, $\epsilon = 1.480$.

Composition and Structure. The ideal composition is $Ca_2Al_2Si_4O_{12} \cdot 6H_2O$ but there is considerable replacement of Ca by Na and K as well as (Na, K)Si for CaAl. The structure of chabazite consists of an Al-Si-O framework with large cagelike openings bounded by rings of tetrahedra (see Fig. 10.98). The cages are connected to each other by channels which allow for the diffusion of molecules through the structure of a size comparable to the diameter (3.9 Å) of the channels. Argon (3.84 Å in diameter) is rapidly absorbed by the chabazite structure, but *iso*-butane (5.6 Å in diameter) cannot enter the structure. In this way chabazite can act as a molecular sieve.

FIG. 10.99. Chabazite.

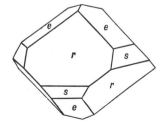

FIG. 10.100. Chabazite in trap rock cavity.

Diagnostic Features. Recognized usually by its rhombohedral crystals, and distinguished from calcite by its poorer cleavage and lack of effervescence in HCl. Fuses with swelling at 3 to a blebby glass. Decomposed by HCl with the separation of silica but without the formation of a jelly. Gives much water in the closed tube.

Occurrence. Chabazite is found, usually with other zeolites, lining cavities in basalt. Notable localities are the Faeroe Islands; the Giant's Causeway, Ireland; Aussig, Bohemia; Seiser Alpe, Trentino, Italy; and Oberstein, Germany. In the United States found at West Paterson, New Jersey, and Goble Station, Oregon. Also found in Nova Scotia and there known as *acadialite* (see Fig. 10.100).

Name. Chabazite is derived from a Greek word which was an ancient name for *a stone.*

HEULANDITE—$CaAl_2Si_7O_{18} \cdot 6H_2O$

Crystallography. Monoclinic; $2/m$ but crystals often simulate orthorhombic symmetry (Fig. 10.101); often diamond-shaped with {010} prominent.

$C2/m$; $a = 17.71$, $b = 17.84$, $c = 7.46$ Å; $\beta = 116°20'$; $a:b:c = 0.993:1:0.418$; $Z = 4$. $d's$: 8.9(7), 5.10(4), 3.97(10), 3.42(5), 2.97(7).

Physical Properties. *Cleavage* {010} perfect. **H** $3\frac{1}{2}$–4. **G** 2.18–2.2. *Luster* vitreous, pearly on {010}. *Color* colorless, white, yellow, red. Trans-

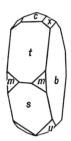

FIG. 10.101. Heulandite.

FIG. 10.102. Stilbite.

parent to translucent. *Optics:* (+); α = 1.482, β = 1.485, γ = 1.489; 2V = 50°; X = b. Y $\wedge$ c = 35°. $r > v$.

Composition and Structure. There is considerable variation in the Si/Al ratio with concomitant variation in the proportions of Ca and Na. The structure of heulandite consists of a very open Si-Al-O framework in which two-thirds of the $(Si,Al)O_4$ tetrahedra are linked to form networks of six-membered rings parallel to {010}; this is responsible for the perfect {010} cleavage. The framework contains several sets of channels which house the water molecules as well as Ca.

Diagnostic Features. Fusible at 3 with intumescence to a white glass. Yields water in the closed tube. Characterized by its crystal form and one direction of perfect cleavage with pearly luster.

Occurrence. Heulandite is usually found in cavities of basic igneous rocks associated with other zeolites and calcite. Found in notable quality in Iceland; the Faeroe Islands; Andreasberg, Harz Mountains; Tyrol, Austria; and India, near Bombay. In the United States found at West Paterson, New Jersey. Also found in Nova Scotia.

Name. In honor of the English mineral collector, H. Heuland.

STILBITE—$CaAl_2Si_7O_{18} \cdot 7H_2O$

Crystallography. Monoclinic; $2/m$. Crystals usually tabular on {010} or in sheaflike aggregates (Figs. 10.102 and 10.103). They may also form cruciform penetration twins.

$C2/m$; a = 13.63, b = 18.17, c = 11.31 Å; β = 129°10'; $a:b:c$ = 0.750:1:0.622; Z = 4. $d's$: 9.1(9), 4.68(7), 4.08(10), 3.41(5), 3.03(7).

Physical Properties. *Cleavage* {010} perfect. **H** $3\frac{1}{2}$–4. **G** 2.1–2.2. *Luster* vitreous; pearly on {010}. *Color* white, more rarely yellow, brown, red. Translucent. *Optics:* (−); α = 1.494, β = 1.498, γ = 1.500; 2V = 33°; Y = b, X $\wedge$ a = 5°.

Composition and Structure. Na and K are usually present substituting for Ca by the coupled substitution mechanisms Na^+Si^{4+} $Ca^{2+}Al^{3+}$ and $(Na, K)^+Al^{3+}$ Si^{4+}. The structure of stilbite is simi-

FIG. 10.103. Stilbite, Nova Scotia.

lar to that of heulandite with sheets of six-membered $(Si,Al)O_4$ tetrahedra parallel to {010}. This accounts for the tabular habit and excellent {010} cleavage.

Diagnostic Features. Characterized chiefly by its cleavage, pearly luster on the cleavage face, and common sheaflike groups of crystals. Fuses with intumescence at 3 to a white enamel. Decomposed by HCl with the separation of silica but without the formation of a jelly. Yields water in the closed tube.

Occurrence. Stilbite is found in cavities in basalts and related rocks associated with other zeolites and calcite. Notable localities are Poonah, India; Isle of Skye; Faeroe Islands; Kilpatrick, Scotland; and Iceland. In the United States it is found in northeastern New Jersey. Also found in Nova Scotia.

Name. Derived from a Greek word meaning *luster,* in allusion to the pearly luster.

Similar Species. Other zeolites of lesser importance than those described are:

Phillipsite, $KCa(Al_3Si_5O_{16}) \cdot 6H_2O$, monoclinic.
Harmotome, $Ba(Al_2Si_6O_{16}) \cdot 6H_2O$, monoclinic.
Thomsonite, $NaCa_2(Al_5Si_5O_{20}) \cdot 6H_2O$, orthorhombic.
Gmelinite, $(Na_2, Ca)(Al_2Si_4O_{12}) \cdot 6H_2O$, hexagonal.
Laumontite, $Ca(Al_2Si_4O_{12}) \cdot 4H_2O$, monoclinic.
Scolecite, $Ca(Al_2Si_3O_{10}) \cdot 3H_2O$, monoclinic.

References and Suggested Reading

Bates, R. L., 1960, *Geology of the Industrial Rocks and Minerals.* Harper and Row, New York, 441 pp.

Breck, D. W., 1974, *Zeolite Molecular Sieves.* John Wiley & Sons, New York, 771 pp.

Darragh, P. J., Gaskin, A. J., and Sanders, J. V., 1976, Opals. *Scientific American,* v. 234, no. 4, pp. 84–95.

Deer, W. A., Howie, R. A., and J. Zussman, 1962, *Rock-Forming Minerals,* v. 1, 2, 3, and 4 (silicates). John Wiley & Sons, New York.

Ernst, W. G., 1968, *Amphiboles, Crystal Chemistry, Phase Relations and Occurrence.* Springer-Verlag, New York, 125 pp.

Frondel, C., 1962, *The System of Mineralogy,* v. 3, Silica Minerals. John Wiley & Sons, New York, 334 pp.

Gillson, J. L., ed., 1960, *Industrial Minerals and Rocks,* 3rd ed. American Institute of Mining, Metallurgical and Petroleum Engineers, New York, 934 pp.

Grim, Ralph E., 1968, *Clay Mineralogy,* 2nd ed. McGraw-Hill Book Co., New York, 596 pp.

Liebau, F., 1972, *Silicon,* v. II-3, section 14A in *Handbook of Geochemistry.* Springer-Verlag, New York.

Mineralogical Society of America, 1975, *Feldspar Mineralogy* (Short Course Notes), P. H. Ribbe, ed. Southern Printing Co., Blacksburg, Virginia.

Papike, J. J. and Maryellen Cameron, 1976, Crystal chemistry of silicate minerals of geophysical interest. *Reviews of Geophysics and Space Physics,* v. 14, pp. 37–80.

Smith, J. V., 1974, *Feldspar Minerals,* v. 1 and 2. Springer-Verlag, New York.

Strunz, H., 1970, *Mineralogische Tabellen,* 5th ed. Akademische Verlagsgesellschaft, Geest und Portig K. G., Leipzig, 621 pp.

Zussman, J., 1968, The crystal chemistry of pyroxenes and amphiboles, 1. Pyroxenes. *Earth Science Reviews,* v. 4, pp. 39–67.

U.S. Bureau of Mines, *Minerals Yearbook.* Annual publication by the Government Printing Office, Washington, D.C.

11
MINERAL ASSOCIATIONS: AN INTRODUCTION TO PETROLOGY

In the previous four chapters we have discussed specific mineral species and mineral groups. Under the heading of **occurrence** we have mentioned typical assemblages or rock types in which a specific mineral is found, and we have often illustrated the stability of a mineral or mineral group with pressure-temperature (*P-T*) or temperature-composition (*T-X*) diagrams (see also page 163). In our previous treatment we have not, however, addressed ourselves to the systematics of mineral occurrences as reflected in naturally occurring rock types or in experimental studies of mineral and rock compositions. Because rocks are generally made up of a variety of minerals, the study of rocks, known as *petrology*, concerns itself to a large degree with the identification of the individual minerals in a rock, their textures, abundances and grain size; this type of information is fundamental in understanding the origin of a rock. A petrologist must have a strong background in mineralogy and mineral identification, but in addition he must be conversant with the

theories of origin of rocks and with experimental studies that elucidate their possible origins. Within the context of this book we cannot cover the many aspects of petrology that are treated extensively in books on the subject (see references at end of chapter); we will, however, attempt to provide a link between the study of minerals and the study of rocks. In order to provide this link with field studies and with experimental and theoretical studies, it is necessary to introduce some aspects of phase equilibria.

PHASE EQUILIBRIA

The field of *phase equilibria* involves the application of physical chemistry to the study of the origin of minerals and rocks. A *phase* is a homogeneous substance with a well-defined set of physical and chemical properties. As such the term phase can be used interchangeably with the term mineral only if the mineral is essentially a pure substance (i.e., exhibits no compositional variation). For example, low quartz (SiO_2) is a low-temperature phase in the chemical system Si-O_2 (or SiO_2); kyanite (Al_2SiO_5) is a high-pressure phase in the chemical system Al_2O_3-SiO_2 (or Al_2SiO_5). When a mineral exhibits solid solution, as in the complete solid solution series between forsterite and fayalite, we speak of a *phase region*. A phase may be solid, liquid, or gaseous as in the case of H_2O with three distinct phases ice, water, and steam. In Fig. 4.30 we have illustrated the *stability regions* (or *fields*) for the various polymorphs of ice and for liquid H_2O in terms of the variables of pressure and temperature. The results of investigations in the system H_2O at lower pressures are given in Fig. 11.1. Along the various curves in this diagram two phases can coexist stably (*in equilibrium*, see below). For example, ice and water can coexist along the *P-T* curve (freezing point curve) on the left side of the diagram (*a-t*). At the point where all three curves meet, three phases, ice, water, and steam can coexist; this is known as the *triple point, t* (a similar triple point is seen in the *P-T* diagram for Al_2SiO_5, Fig. 10.14). Along the curve *t-c* water and

steam can coexist but with increasing *P* and *T* (going in the upper right direction along the curve) the water phase becomes less dense (expands due to increasing temperature) and the steam phase more compressed (due to increasing *P*). At point *c* (*critical point*) the two phases become identical, hence indistinguishable. In *P-T* space, to the upper right of *c* we no longer speak of water or steam but we speak of a *supercritical* aqueous fluid.

The concept of *equilibrium* is related to time. If, for example, water and ice coexist together in constant amounts indefinitely, that is, no water is forming at the expense of ice, or vice versa (as along curve *a-c* in Fig. 11.1), we can say that under these specific conditions water and ice are *in equilibrium*. In rocks, where the constituent minerals have coexisted since their formation, perhaps several million years ago, one cannot always conclude unambiguously whether the mineral constituents are in equilibrium with each other or not. If no *reaction rims* are observed between minerals that touch each other in the rock, we may assume that the minerals were in equilibrium at the time of formation; this assumption must generally be supported by further chemical information about element distribution between the physically touching minerals. If, however, megascopically or microscopically visible rims exist between minerals we may tend to conclude that some of the minerals were *not* in equilibrium with each other. For example, garnet may be separated by a chlorite rim from a coexisting biotite. Here we would conclude that the garnet and biotite were *not* in equilibrium with each other, as they are separated by a reaction product, chlorite. In experimental petrology, the experimentalist will conclude that the phases under investigation are in equilibrium when no further change takes place between them during a certain time interval. This may range from a few hours, to several months, or even years, depending on the speed or sluggishness of the reactions studied, and on the patience of the investigator.

In Fig. 11.1 we have illustrated stability fields, in terms of bordering equilibrium curves, for phases in a system that can be described chemically by one compound component, H_2O. Phases in a system are

FIG. 11.1. Pressure-temperature diagram for H_2O. The stability relations of dense forms of ice, at high pressures, are shown in Fig. 4.30.

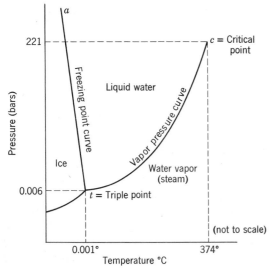

described by independent chemical species known as *components*; a minimum number of chemical variables is generally chosen. For example, in the case of H_2O we chose the compound component H_2O, but we could have defined the chemical system in terms of H_2 and O_2. In the system Al_2SiO_5 (andalusite-sillimanite-kyanite) Al_2SiO_5 is generally chosen as the compound component although three elements, Al, Si, and O, or two oxide components, Al_2O_3 and SiO_2, could have been selected to define the system chemically. If one is interested in the stability relations of wollastonite, $CaSiO_3$, one might choose to represent one's findings in terms of the compound component, $CaSiO_3$. If, however, one wishes to know the stability fields of pyroxenes in the system $CaO-MgO-FeO-SiO_2$, one generally chooses the three components $CaSiO_3-MgSiO_3-FeSiO_3$ to define the system chemically (see Fig. 10.36).

Phase diagrams

Let us now look at the appearance of some additional examples of *phase* (or *stability*) *diagrams* which outline the stability fields of minerals. At low temperatures and pressures, such as prevail under atmospheric conditions, it is often useful to express the stability fields of phases (or minerals) in terms of Eh (oxidation potential) and pH (the negative logarithm of the hydrogen-ion concentration; it allows for the definition of acidic conditions with pH = 0–7, for basic conditions with pH = 7–14, and neutral solutions at pH = 7). The construction of such diagrams on the basis of thermodynamic parameters is outlined in Garrels and Christ (1965, see reference list). Figure 11.2 illustrates the stability fields of some iron oxides and iron sulfides as calculated for atmospheric conditions (25°C, 1 atmosphere pressure). It shows at a glance that hematite is stable in an oxidizing region (high Eh); magnetite, on the other hand, is stable under more reducing conditions, lower Eh. The two sulfides, pyrite and pyrrhotite, occur under reducing conditions and pH values between 4 and 9. Along the lines that separate the various mineral fields, two of the minerals can coexist in equilibrium. For example, hematite

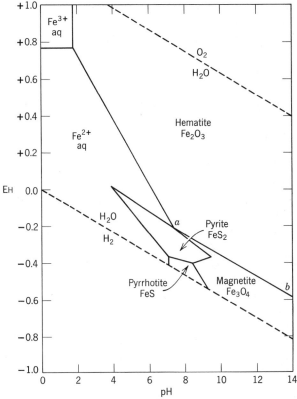

FIG. 11.2. Stability of iron oxides and iron sulfides in water at 25°C and 1 atmosphere total pressure, with an activity of total dissolved sulfur = 10^{-6}. The pyrrhotite field is calculated on the basis of the formula FeS in this diagram; more correctly it would be $Fe_{1-x}S$ (adapted from Garrels and Christ, 1965, *Solutions, Minerals, and Equilibria*, Freeman, Cooper and Company, Fig. 7.20).

and magnetite can coexist under conditions of variable pH and Eh (line *ab*, Fig. 11.2); indeed hematite and magnetite often occur together in sedimentary iron-formations. Diagrams of this type are especially useful in evaluating some of the physical chemical parameters that prevail during conditions of atmospheric weathering and of chemical sedimentation and diagenesis of water-laid sediments (at essentially atmospheric pressure conditions and temperatures ranging from 25°C to about 100°C). On the basis of the frequent coexistence of hematite and magnetite in Proterozoic banded iron-formations we may be able to evaluate to some extent parameters such as Eh and pH of the original Proterozoic sedimentary basin.

In the evaluation of minerals and rocks that have undergone metamorphic conditions, P-T and T-X diagrams are frequently used. Examples of such diagrams are Figs. 10.14, 10.41, 10.44, and 10.50. An example of a composite P-T diagram with reaction curves for several minerals is shown in Fig. 11.3. The reaction curves for Al_2SiO_5 and for muscovite + quartz = K-feldspar + sillimanite + H_2O are especially relevant to metamorphic rocks with high Al_2O_3 contents relative to other components such as CaO, MgO, and FeO. Shales with abundant clay minerals have relatively high Al_2O_3 contents and during metamorphism mineral reactions take place in the shale, as well as recrystallization. The metamorphic equivalents of shales, known collectively as pelitic schists may contain abundant sillimanite, muscovite, and quartz. This type of schist, on the basis of the reaction curves, and stability fields of minerals in Fig. 11.3 may have equilibrated under P-T conditions as outlined by the shaded area in Fig. 11.3. In metamorphic rocks that may have undergone very high pressures such as kimberlites and eclogites, the originally present al-

bite will, during metamorphism, have reacted to form jadeite + quartz, and any original calcite may have transformed to aragonite. The presence of jadeite + quartz, for example, would indicate a minimum pressure during metamorphism of about 10 kilobars at 300°C (see upper left curve in Fig. 11.3).

It should be noted that stability diagrams such as Figs. 11.2 and 11.3 are obtained by theoretical calculations or laboratory experimentation. In the case of the experimental study of the stability fields of the three polymorphs of Al_2SiO_5 very high purity minerals may be used for the experiments although most commonly the starting materials are high purity chemicals. For example, gem grade and inclusion-free kyanite may be found at increasing T to react to form sillimanite (see Fig. 11.3). The position of the curve separating the kyanite and sillimanite fields is based on the first evidence of sillimanite forming at the expense of kyanite with increasing T and on evidence of the reverse reaction, namely sillimanite giving way to kyanite at decreasing T and increasing P. The determination of the beginning of such reactions is generally based on a combination of X-ray powder diffraction and optical microscopic techniques. Because of uncertainties in various aspects of the experimental techniques in locating specific reactions the reaction curves, defined in P and T values, may represent relatively broad reaction zones although they are commonly shown as narrow lines or curves. The most difficult problem is relating the experimental, and chemically relatively simple system, to the naturally occurring and more complex system.

The value of stability diagrams is especially great in the evaluation of parameters such as temperature and pressure during metamorphism. In our subsequent discussion of metamorphic rocks it will be clear that such rocks can be assigned to fairly specific ranges of temperature or pressure of origin on the basis of the mineral assemblage.

In the evaluation of the origin of igneous rocks it is very instructive to study experimentally the melting of igneous rock compositions as well as the crystallization sequence of minerals from a melt. Diagrams based on studies involving a liquid phase, *melt,* are in general known as *liquidus* diagrams.

FIG. 11.3. Reaction curves for some common metamorphic minerals. For significance of shaded area, see text.

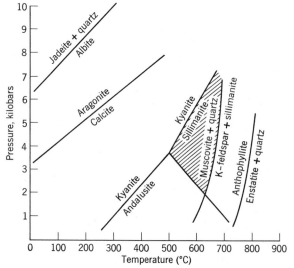

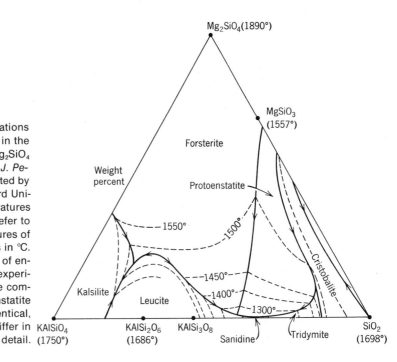

FIG. 11.4. Melting relations among minerals in the system SiO$_2$-KAlSiO$_4$-Mg$_2$SiO$_4$ (after W. C. Luth, 1967, *J. Petrology*, p. 373, reprinted by permission of the Oxford University Press). Temperatures next to compositions refer to the melting temperatures of those compositions in °C. Protoenstatite instead of enstatite is produced in experiments of this type. The compositions of protoenstatite and enstatite are identical, but their structures differ in detail.

An example of such a diagram is given in Fig. 11.4, in terms of three compound components SiO$_2$, KAlSiO$_4$, and Mg$_2$SiO$_4$. The contours in this diagram represent melting temperatures and are known as *isotherms*; the surface defined by these isotherms is the *liquidus surface*. The arrows along the boundaries of the various phase fields indicate crystallization paths, with decreasing temperature.

The temperature-composition (*T-X*) sections through such liquidus diagrams are strongly influenced by the extent of solid solution (compositional variation in the minerals). We will here provide a few examples of *T-X* sections at constant pressure (isobaric sections) of some two-component systems. Figure 11.5*a* is a *T-X* section showing a *eutectic* relationship. Phases *A* and *B* are pure substances and as such there is no solid solution between them. For composition *A* the melting temperature is *T$_A$*; similarly for composition *B* the melting temperature is *T$_B$*. The addition of some of composition *B* to a melt of *A* lowers the temperature of the liquid that can coexist with *A*, along the curve between *T$_A$* and e (*eutectic point*). Similarly the

melting temperature of the liquid that can coexist with *B* is lowered by the addition of some of *A* as shown by the curve between *T$_B$* and e. The lowest temperature at which crystals and melt are in equilibrium is *T$_C$*, the temperature of the eutectic. Let us now look at the crystallization sequence of a melt of composition *M*. Upon lowering the temperature of the melt at *M* (original temperature = *T$_B$*) the melt will start to crystallize some of the pure substance *B* at *T$_1$*. The crystallization of *B* from the melt will continue along the *liquidus curve* (*T$_B$* to e), continually increasing the content of substance *A* in the melt. At the eutectic point (e) substance *A* will join *B* as a crystallization product. At this point there is no further change in composition of the melt because both *A* and *B* crystallize from the melt in the same proportions as are present in the melt. With continued crystallization the melt will disappear and the final crystalline products will be *B* and *A* in the proportions represented by the original bulk composition *M*. Fig. 11.5*b* is an illustration of a eutectic melting relationship in the system NaAlSi$_3$O$_8$-SiO$_2$.

A different type of *T-X* section through a liq-

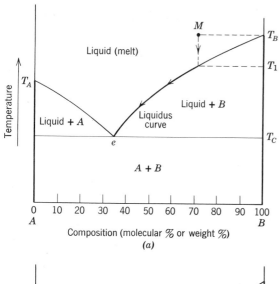

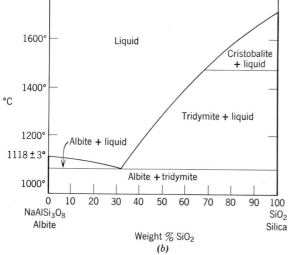

FIG. 11.5. (a) Schematic temperature-composition section showing eutectic crystallization of components A and B; both components are pure substances (no solid solution between them). (b) The system NaAlSi$_3$O$_8$(albite)-SiO$_2$ (after J. F. Schairer and N. L. Bowen, 1956, *Amer. Jour. Sci.*, v. 254, p. 161).

uidus diagram is shown in Fig. 11.6a. This diagram illustrates a *peritectic* as well as *eutectic* phenomenon in the crystallization sequence. This diagram, in addition to phases A and B (which lack solid solution between them), shows the presence of a third phase of constant composition, C. Phases of composition A and B melt to a liquid at T_A and T_B, respec-

tively, but phase C melts *incongruently*, which means that it reacts and decomposes to form another solid plus liquid, in this case B and melt of composition M_C at temperature T_C. If we look at the crystallization history of a melt of composition M (originally at temperature T_B) it can be cooled until it reaches the liquidus curve at T_1 at which point phase B begins to crystallize. With continued loss of heat the melt composition will move toward A along the liquidus curve with continued precipitation of phase B until point p (*peritectic point*) is reached. At this point all of the earlier formed phase B will react with the melt to form phase C. With continued decrease in temperature in the melt, the temperature of the product phase C will fall until it finally reaches point e (eutectic). At this point C is joined by crystallization products of composition A. Figure 11.6b illustrates these types of crystallization phenomena in the system Mg$_2$SiO$_4$-SiO$_2$.

A very different type of *T-X* section through a liquidus diagram is seen when complete solid solution exists between the two end member compositions A and B, as in Fig. 11.7. This diagram shows that pure end member A melts at T_A and pure end member B at T_B, but intermediate compositions (AB), which consist of a single phase (part of the solid solution series), melt at temperatures intermediate between T_A and T_B. The upper curve in the *T-X* section (the *liquidus curve*) represents the locus of melt compositions in equilibrium with crystalline products, the lower curve, known as the *solidus curve*, is the locus for the composition of crystalline phases in equilibrium with the melt. A melt of composition M at temperature T_B will start to crystallize out a member of the solid solution series AB with specific composition xA,yB at temperature T_1. These crystals are enriched in the B component with the respect to the melt composition M, and their growth will deplete the melt in component B. The melt composition will, as a result of this depletion, move along the liquidus curve toward A as indicated by the upper arrow. As a result of the continual lowering of the temperature the solid phase of original composition xA,yB will react with the melt in the direction of the lower arrow. As such both the melt and crystalline products will increase in content of

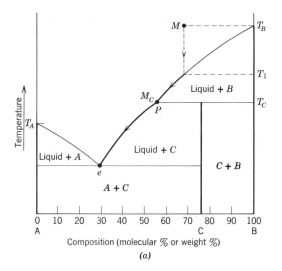

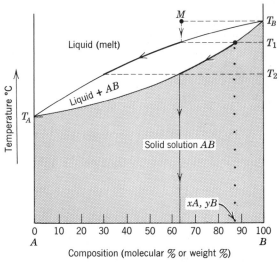

FIG. 11.7. Schematic temperature-composition section for a chemical system which shows a complete solution series between two end member components.

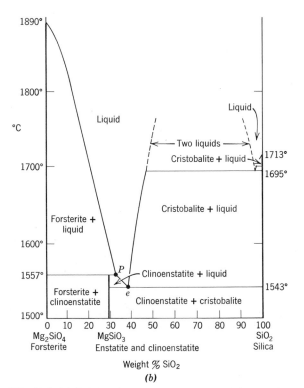

FIG. 11.6. (a) Schematic temperature-composition section showing peritectic as well as eutectic crystallization involving three pure phases, A, B, and C. (b) The system Mg_2SiO_4(forsterite)-SiO_2 (after N. L. Bowen and O. Andersen, modified by J. W. Greig, *Phase Diagrams for Ceramists,* Amer. Ceramic Soc., 1964, p. 112).

A with decreasing T. Finally at T_2 the crystallized products have a composition which is that of the original melt M and the amount of melt in equilibrium with the crystals will reach zero. With continued lowering of the temperature the composition of the crystalline product will remain constant at the bulk composition M of the original melt. This type of melting and crystallization behavior is found in the olivine system between forsterite, Mg_2SiO_4 and fayalite, Fe_2SiO_4 (see Fig. 10.7). A similar but not identical T-X section is also seen in the plagioclase feldspar system (Fig. 10.79). Because of the occurrence of monalbite and the discontinuity between body-centered anorthitic compositions and high albite solid solution, the T-X section for the plagioclase feldspars is more complex than the schematic representation of Fig. 11.7.

The above four figures (Figs. 11.4 to 11.7) illustrate some specific crystallization sequences which act as constraints in the crystallization history of igneous melts.

Assemblage and phase rule

One of the first steps in the study of a rock is to list the minerals of which it is composed. A simple igneous granite may consist of orthoclase, albite, quartz, and biotite. The texture of this specific granite may indicate that all four minerals formed as crystallization products at about the same elevated temperature. A petrologist, therefore, might conclude that orthoclase-albite-quartz-biotite is the *mineral assemblage* (also referred to as the *mineral paragenesis*) of this granite. Although the term *assemblage* (or *paragenesis*) is often loosely used to include all the minerals that compose a rock, it should be restricted to those minerals in a rock which appear to have been *in equilibrium* (the *equilibrium mineral assemblage*). The above granite may be found to contain vugs in outcrop which are lined by several clay minerals, bauxite as well as some limonite. Clearly the minerals in the vugs represent relatively low-temperature alteration products of the original much higher temperature granite. The high-temperature granitic assemblage consists of orthoclase-albite-quartz-biotite, and a separate, later formed, low-temperature assemblage consists of clays-bauxite-limonite. In short, an assemblage consists of minerals that formed under the same, or very similar, conditions of pressure and temperature. In practice all minerals that coexist (physically touch each other) and show no alter-

ation or rimming relations are considered to constitute the assemblage. In a distinctly banded rock the mineral assemblages in the various bands are often very different from each other because the bands may represent major differences in bulk chemistries.

Let us illustrate the concept of assemblage with a rock that has undergone metamorphism. A shale, which at low temperatures may have consisted of the assemblage sericite-kaolin-dolomite-quartz-feldspar has reacted at elevated temperatures of metamorphism to produce garnet, sillimanite, biotite, and feldspar. Garnet-sillimanite-biotite-feldspar is therefore the high-temperature metamorphic assemblage.

The evaluation of whether all minerals in a rock are in *equilibrium* with each other is often not straightforward. If the texture does not show reaction rims and alterations, it is possible that the minerals are in equilibrium, however, further detailed chemical tests are often needed to define the equilibrium assemblage without ambiguity.

The mineral assemblages in a specific chemical system are commonly represented in graphical form. Because almost all rock types or mineral associations are part of chemical systems with many elements, *mineral assemblage diagrams* are generally

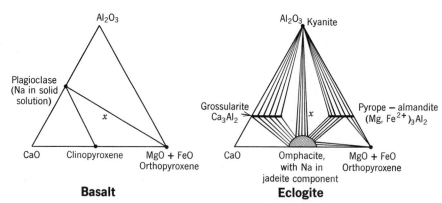

Basalt   **Eclogite**

FIG. 11.8. Coexisting minerals in the system Al_2O_3-CaO-(MgO + FeO)-SiO_2-Na_2O. In *basalt* a common assemblage is plagioclase-clinopyroxene-orthopyroxene, as shown by the triangle of connecting tielines (the *x* represents a possible bulk composition of basalt with this mineral assemblage). In *eclogite*, of the same bulk composition, marked by *x*, the assemblage kyanite-omphacite-pyrope (as a component in almandite) occurs. Composition bars and omphacite field in the eclogite diagram outline approximate extent of solid solution. Eclogitic rocks may form in the mantle of the earth from basaltic compositions.

triangular with individual chemical components or combinations of components of the system at each corner (see Figs. 7.16, 8.10, 9.5). Figure 11.8 illustrates two assemblage diagrams in which two oxide components have been grouped together in the right-hand corner. This allows for a more complete representation of the mineral compositions in this instance. Coexisting minerals are connected by *tielines* (see also page 166). In a triangular diagram the number of chemical components and hence the number of coexisting minerals that can be represented is small and therefore tetrahedral diagrams may be used to allow for a fourth chemical component in the graphical representation. The chemistry of most rocks and mineral associations is so complex that a graphical representation of mineral assemblages is not feasible, and algebraic expressions must be used instead.

The number of minerals that may coexist in equilibrium in an assemblage is not unlimited; indeed this number is restricted by what is known as the *phase rule* which states

$$f = c - p + 2$$

where f = the variance, or the number of degrees of freedom in a chemical system, c = the number of components, and p = the number of phases. In the case of three components and three phases in equilibrium (these would outline a triangle in a three-component phase diagram), f turns out to be 2. This means that two parameters such as P and T may be varied, and that a system with three phases and three components has two degrees of freedom.

In graphical assemblage representations, we group minerals of similar origin in terms of constant P and T. In other words we attempt to fix the pressure at which the assemblage formed as well as the temperature. This leads to assemblage diagrams that are *isobaric* (= constant pressure) and *isothermal* (= constant temperature). In Fig. 11.8 the basalt assemblage may have crystallized at a constant and relatively low P (e.g., 1–2 kilobars) and a temperature of somewhere between 1000 and 1200°C. The chemically equivalent eclogite assemblage formed at a much higher pressure, possibly over a range from 12 to 30 kilobars, and a temperature range of 400° to 800°C. In other words the two assemblage diagrams represent very different P and T conditions for identical bulk chemical compositions. In assemblage diagrams obtained from experimental studies the temperatures and pressures can be closely controlled and isobaric and isothermal phase diagrams can be obtained for much narrower ranges in P and T than is generally possible for natural rock or mineral systems.

ROCK TYPES

All rocks can be classified in terms of three major groups: I. igneous, II. sedimentary, and III. metamorphic.

I. Igneous Rocks

Igneous rocks comprise approximately 95 percent of the upper 10 miles of the earth's crust, but their great abundance is hidden on the earth's surface by a relatively thin but widespread layer of sedimentary and metamorphic rocks. Igneous rocks have crystallized from a silicate melt, known as a *magma*. The high temperatures (about 900 to 1600°C) necessary to generate magmas, as concluded from experimental liquidus diagrams obtained in the laboratory and from the similarly high temperatures measured in the boiling magmas of active volcanoes suggest that their source lies deep in the earth's crust. A magma is principally made up of O, Si, Al, Fe, Ca, Mg, Na, and K but it also contains considerable amounts of H_2O and CO_2 as well as lesser gaseous components such as H_2S, HCl, CH_4, and CO. As a magma cools there is generally a definite order of crystallization of the various mineral constituents. In a magma consisting mainly of O, Si, Mg, and Fe, for example, the mineral with the highest melting point, Mg-rich olivine, Mg_2SiO_4, would crystallize first followed by more Fe-rich olivines approaching fayalite, Fe_2SiO_4 (see Fig. 10.7). In a magma of the appropriate composition for pyroxene crystallization, the Mg-rich ortho- and clino-pyroxenes (see Fig. 10.36), such as enstatite and diopsidic augite, would crystallize before hedenbergite and Fe-rich orthopyroxenes that have lower relative melting points. In magmas rich in the plagioclase component (see Fig. 10.79) the An-rich plagioclases will crystallize before the Ab-rich members. All of the above crystallization sequences are the result of *continuous reactions* that take place under equilibrium conditions between the melt and the precipitated crystals as a function of falling temperature. If chemical equilibrium is not maintained between melt and crystals during cooling, the resulting crystals may show compositional zoning. This is especially common in the plagioclase feldspars and in members of the pyroxene series. Igenous rocks, however, also illustrate *discontinuous reactions* that occur at fairly definite temperatures and do not take place over a range of temperatures as do the continuous reactions. In Fig. 11.6*b* we have shown the peritectic reaction of early formed forsterite transforming to enstatite + melt at lower temperatures. In igneous rocks of appropriate bulk composition it is not uncommon to find early-formed olivine crystals rimmed by later orthopyroxenes. Such reactions (see Fig. 11.6*b*) force the composition of the melt toward the eutectic point at which crystallization of an SiO_2 phase (e.g., cristobalite or tridymite in high-temperature volcanic rocks) occurs.

Early-formed crystals produced in a cooling melt may separate from the liquid by settling out because of gravitational forces, or may be removed from the melt by tectonic deformation. Therefore,

the crystals may not remain in equilibrium with the melt from which they crystallized and systematic changes will occur in the bulk composition of the remaining (residual) magma.

The above continuous and discontinuous reactions as well as separation of magma and crystals lead to what is known as *magmatic differentiation*. This concept was first developed by N. L. Bowen, an American petrologist, on the basis of his studies of the textures and mineralogical make-up of many igneous rock types and his correlative experimental studies. Mg-Fe-rich igneous silicates constitute a series of mineral groups that are related to each other by discontinuous reactions. For example, early-formed olivine may be rimmed by pyroxene; amphiboles may form at the expense of the pyroxene rims, and biotite may form as a reaction product of the earlier crystallized amphiboles. On the other hand, members of the plagioclase feldspar series represent a continuous reaction series in which An-rich plagioclase crystallizes early from the melt, enriching the residual melt in alkalies (Na and K). Figure 11.9 illustrates schematically what is known as *Bowen's reaction series*. As a magma of basaltic composition cools, olivine and An-rich plagioclase may crystallize first. If these minerals remain in contact with the magma they will tend to form pyroxene and more Ab-rich plagioclase and the resulting rock will be a gabbro or a basalt. If, however, the bulk of the early olivine and An-rich

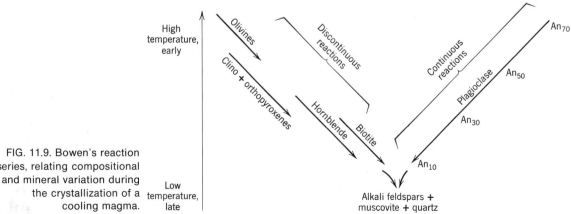

FIG. 11.9. Bowen's reaction series, relating compositional and mineral variation during the crystallization of a cooling magma.

plagioclase is removed, as by crystal settling, the bulk composition of the remaining melt will tend to become enriched in Si, Al, Fe^{2+}, alkalies and H_2O and CO_2. Such a melt could produce a mineral assemblage consisting mainly of amphiboles, some mica, alkali feldspar, and SiO_2. It should be noted in Fig. 11.9 that amphiboles, micas, alkali feldspars and quartz are relatively low-temperature, late-stage crystallization products. The various processes that cause magmatic differentiation can produce a large variety of igneous rocks from a common parent magma.

General Occurrence and Texture

There are two major types of igneous rocks, *extrusive* and *intrusive*. The first group includes those igneous rocks that reached the earth's surface in a molten or partly molten state. Modern volcanoes produce lava flows that pour from a vent or fracture in the earth's crust. Such extrusive or *volcanic* rocks tend to cool and crystallize rapidly with the result that their grain size is generally small. If the cooling has been so rapid as to prevent the formation of even small crystals of the mineral constituents, the resulting rock may be a *glass*. Ordinarily the mineral constituents of fine grained extrusive rocks can be determined only by microscopic examination of thin sections of the rocks. *Intrusive* or *plutonic* rocks are the result of crystallization from a magma that did not reach the earth's surface. Magmatic intrusions that are discordant with the surrounding country rock are referred to as *batholiths* or *plutons* depending on their size and shape; when the intrusive is tabular and concordant it is known as a *sill*. A magma that is deeply buried in the earth's crust generally cools slowly and the mineral constituents crystallizing from it have time to grow to considerable size giving the rock a medium to coarse grained texture. The mineral grains in such rocks can generally be identified with the naked eye. When a magma intrudes in *dikes* (discordant tabular bodies) the textures are usually finer grained than those of plutonic rocks but coarser than those of volcanic rocks; these rocks of intermediate grain size are known as *hypabyssal*.

Some igneous rocks show distinct crystals of some minerals embedded in a much finer grained matrix. The larger crystals are *phenocrysts,* and the finer grained material is the *groundmass* (see Fig. 10.86). Such rocks are known as *porphyries.* The phenocrysts may vary in size from crystals an inch or more across down to very small individuals. The groundmass may also be composed of fairly coarse grained material, or its grains may be microscopic. The difference in size between the phenocrysts and the particles of the groundmass is the distinguishing feature of a porphyry. The porphyritic texture develops when some of the crystals grow to considerable size before the main mass of the magma consolidates into the finer and uniform grained material. Any of the types of igneous rocks described below may have a porphyritic variety, such as *granite porphyry, diorite porphyry, rhyolite porphyry.* Porphyritic varieties are more common in volcanic rocks, especially in the more siliceous types.

Chemical Composition

The chemical bulk compositions of igneous rocks exhibit a fairly limited range. The largest oxide component, SiO_2, ranges from about 40 to 75 weight percent in common igneous rock types (see Table 11.1 and Fig. 11.10). Al_2O_3 ranges generally from about 10 to 20 weight percent (except for peridotite and dunite, see analyses Table 11.1) and each of the other major components generally does not exceed 10 weight percent (except MgO in peridotite and dunite, see Table 11.1).

When the magma is fairly low in SiO_2 the resulting rocks will contain mainly relatively silica-poor minerals such as olivine, pyroxene, hornblende, or biotite and little or no free SiO_2 (i.e., quartz, cristobalite, tridymite; see Fig. 11.10). These rocks, which tend to be dark because of their high percentage of ferromagnesian minerals, are known as *mafic* rock types. When the melt is poor in SiO_2 (*subsiliceous,* or silica undersaturated) and high in alkalies and Al_2O_3 (as in a nepheline syenite composition, see Table 11.1) the resultant crystallization products will contain SiO_2-poor minerals such as feldspathoids and will lack free SiO_2 as quartz (see Figs. 11.10 and 11.11). Crystallization of a melt high

Table 11.1

AVERAGE CHEMICAL COMPOSITIONS OF SOME IGNEOUS ROCKS[a]

Oxide	Nepheline syenite	Syenite	Granite	Tonalite	Diorite	Gabbro	Peridotite	Dunite
SiO_2	54.83	59.41	72.08	66.15	51.86	48.36	43.54	40.16
TiO_2	0.39	0.83	0.37	0.62	1.50	1.32	0.81	0.20
Al_2O_3	22.63	17.12	13.86	15.56	16.40	16.84	3.99	0.84
Fe_2O_3	1.56	2.19	0.86	1.36	2.73	2.55	2.51	1.88
FeO	3.45	2.83	1.67	3.42	6.97	7.92	9.84	11.87
MnO	trace	0.08	0.06	0.08	0.18	0.18	0.21	0.21
MgO	trace	2.02	0.52	1.94	6.12	8.06	34.02	43.16
CaO	1.94	4.06	1.33	4.65	8.40	11.07	3.46	0.75
Na_2O	10.63	3.92	3.08	3.90	3.36	2.26	0.56	0.31
K_2O	4.16	6.53	5.46	1.42	1.33	0.56	0.25	0.14
H_2O	0.18	0.63	0.53	0.69	0.80	0.64	0.76	0.44
P_2O_5	—	0.38	0.18	0.21	0.35	0.24	0.05	0.04
Total	99.77	100.00	100.00	100.00	100.00	100.00	100.00	100.00

[a] All analyses except the nepheline syenite from S. R. Nockolds, 1954, *Geol. Soc. Amer. Bul.*, v. 65, pp. 1007–1032.

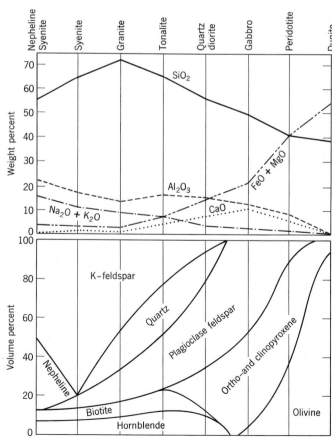

FIG. 11.10. Relationships of variation in chemical and mineral compositions in igneous rocks.

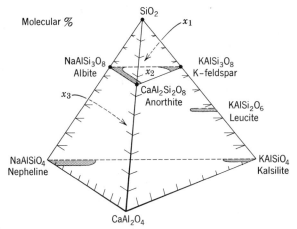

FIG. 11.11. Feldspar and feldspathoid compositions in the system SiO_2(quartz)-$NaAlSiO_4$(nepheline)-$KAlSiO_4$(kalsilite)-$CaAl_2O_4$. Bars and fields indicate average extent of solid solution. A melt of composition x_1 (*silica oversaturated*) would produce a quartz-feldspar assemblage; a melt of x_2 composition (*silica saturated*) and lying in the feldspar plane would produce only feldspars and a melt of composition x_3 (*silica undersaturated*) would produce feldspar-feldspathoid assemblages.

in SiO_2 (silica oversaturated) results in rocks with abundant quartz and alkali feldspars, with or without muscovite, and only minor amounts of ferromagnesian minerals. Such rock types are referred to as *felsic* (alkali feldspar-rich) or *silicic,* and are lighter in overall color than mafic rocks. In general, the darker the rock the greater the abundance of ferromagnesian minerals and the lighter the rock the greater the abundance of quartz, or feldspars, or feldspathoids.

Classification
Because of considerable variation in magmas both in chemistry and conditions of crystallization, igneous rock types show a wide variation in mineralogy and texture. There is a complete gradation from one rock type into another, so the names of igneous rocks, and the boundaries between types, are largely arbitrary (see Figs. 11.10 and 11.12).

Many schemes have been proposed for the classification of igneous rocks, but the most practical for the introductory student is based on mineralogy and texture. In general, four criteria are to be considered in classifying an igneous rock. (1) The relative amount of silica; quartz (or tridymite, or cristobalite) indicates an excess of silica; feldspathoids indicate a deficiency of silica. (2) The kinds of feldspar (alkali feldspar versus plagioclase) and the relative amount of each kind. (3) The relative amount and type of dark minerals. (4) The texture or size of the grains. Is the rock coarse or fine grained; that is, is it plutonic or volcanic?

It is obvious that the exact determination of the kind of feldspar or a correct estimate of the amount of each kind is impossible in the field or in the hand specimen. It is also impossible in many fine grained rocks to recognize individual minerals. Such precise work must be left for the laboratory and carried out by the microscopic examination of thin sections of rocks. Nevertheless, it is important that the basis for the general classification be understood in order that a simplified field classification may have more meaning.

Three major divisions may be made on the basis of the silica content (see Fig. 11.11). (1) Quartz present in amounts greater than 5% (silica oversaturated). (2) No quartz and no feldspathoids present (silica saturated). (3) Feldspathoids in amounts greater than 5% (silica undersaturated). The above divisions made on the basis of silica content are further subdivided according to the kind and amount (or the absence) of feldspar. Most of the rocks thus classified have a coarse and fine grained variety, which receive different names. Figure 11.12 illustrates the classification of the principal plutonic and volcanic rock types. Table 11.2 gives examples of the principal rock types according to such a classification. Although these rock names are the most important, more than 600 have been proposed to indicate specific types.

Mineralogical Composition
Many minerals are found in the igneous rocks, but those that can be called rock-forming minerals are comparatively few. Table 11.3, which lists the major mineral constituents of igneous rocks, is divided into two parts: (1) the common rock-forming minerals of igneous rocks, and (2) the accessory

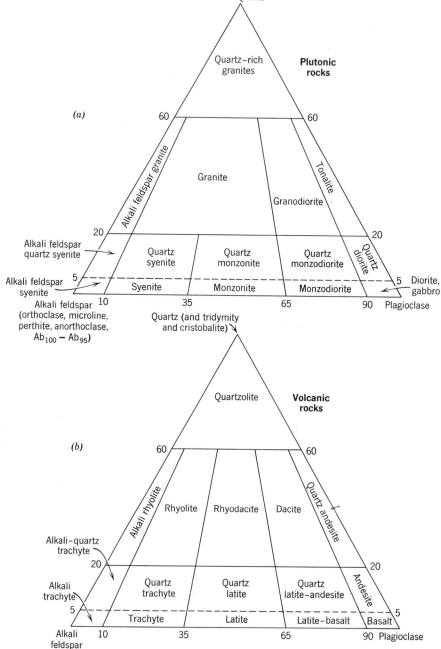

FIG. 11.12. General classification and nomenclature of some common plutonic rock types (*a*) and some common volcanic rock types (*b*). This classification is based on the relative percentages of quartz, alkali feldspar, and plagioclase, measured in volume percent (adapted from Subcommission on the Systematics of Igneous Rocks, *Geotimes,* 1973, v. 18, no. 10, pp. 26–30 and Hyndman, D. W., 1972, *Petrology of Igneous and Metamorphic Rocks,* McGraw-Hill Book Company, p. 35).

Table 11.2

SIMPLIFIED CLASSIFICATION OF THE IGNEOUS ROCKS

Feldspar	Quartz > 5%		No Quartz; No Feldspathoids		Nepheline or Leucite > 5%	
	Coarse	Fine	Coarse	Fine	Coarse	Fine
K-feldspar[a] > plagioclase	Granite	Rhyolite	Syenite	Trachyte	Nepheline syenite Leucite syenite	Phonolite Leucite phonolite
Plagioclase > K-feldspar	Granodiorite	Dacite	Monzonite	Latite	Nepheline monzonite	
Plagioclase (oligoclase or andesine)	Tonalite	Quartz andesite	Monzodiorite	Latite basalt	Nepheline diorite	
Plagioclase (labradorite to anorthite)	Quartz diorite	Andesite	Gabbro	Basalt	Nepheline gabbro	Tephrite (− olivine) Basanite (+ olivine)
No feldspar			Peridotite (olivine dominant) Pyroxenite (pyroxene dominant) Hornblendite (hornblende dominant)		Ijolite	Nephelinite (− olivine) Nepheline basalt (+ olivine)

[a] K-feldspar includes orthoclase, microcline and microperthite; in high T volcanic rocks it can be sanidine or anorthoclase.

minerals of igneous rocks. See Table 11.4 for a listing of common mineral assemblages in some plutonic rocks.

Plutonic Rocks

Granite-Granodiorite. Granite is a granular rock of light color and even texture consisting chiefly of feldspar and quartz. Usually both K-feldspar and oligoclase (or albite) are present; K-feldspar may be flesh-colored or red, whereas oligoclase (or albite) is commonly white and can be recognized by the presence of albite twinning striations. The quartz can be recognized by its glassy luster and lack of cleavage. Granites usually carry a small amount (about 10%) of mica or hornblende.

The mica is commonly biotite, but muscovite may also be present. The minor accessory minerals are zircon, sphene, apatite, magnetite, ilmenite. Figure 11.13 illustrates the liquidus diagram for the hydrous granite system, $NaAlSi_3O_8$-$KAlSi_3O_8$-SiO_2-H_2O, with a temperature minimum of about 770°C in the central part of a low-temperature trough between the liquidus fields of K-Na feldspar solid solutions and high quartz. When the mineral compositions of hundreds of granites are expressed in terms of percentages of albite, K-feldspar, and quartz they can be plotted on Fig. 11.12. The distribution of the resulting data points represents the residual melt compositions from which the granites formed during their crystallization history. These

Table 11.3
MINERALOGY OF THE IGNEOUS ROCKS

Common Rock-Forming Minerals	Common Accessory Minerals
1. Quartz, tridymite, cristobalite	1. Zircon
2. Feldspars	2. Sphene
Orthoclase	3. Magnetite
Microcline	4. Ilmenite
Sanidine	5. Hematite
Plagioclase	6. Apatite
3. Nepheline	7. Pyrite
4. Sodalite	8. Rutile
5. Leucite	9. Corundum
6. Micas	10. Garnet
Muscovite	
Biotite	
Phlogopite	
7. Pyroxenes	
Augite	
Hypersthene	
Aegirine	
8. Amphiboles	
Hornblende	
Arfvedsonite	
Riebeckite	
9. Olivine	

melt compositions coincide broadly with the experimentally determined region of minimum temperature, between 860° and 770°C. Many granites, therefore, are concluded to be the result of crystallization from melts at relatively low temperatures.

There is a complete series grading from granite, with feldspar almost entirely the potash varieties, to granodiorite with feldspar mostly plagioclase and only slightly more than 5% K-feldspar. The boundary between the two types is arbitrarily set. Granites are those rocks in which K-feldspar generally exceeds plagioclase; granodiorites are those in which plagioclase exceeds K-feldspar. In most instances it so happens that as the plagioclase increases in amount the percentage of dark minerals also increases, and thus, in general, granodiorites are darker than granites. However, in the field or in a hand specimen, it is usually impossible to distinguish between the two rock types with certainty.

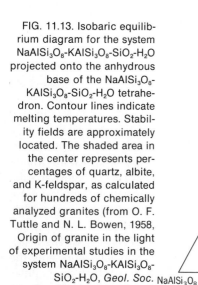

FIG. 11.13. Isobaric equilibrium diagram for the system $NaAlSi_3O_8$-$KAlSi_3O_8$-SiO_2-H_2O projected onto the anhydrous base of the $NaAlSi_3O_8$-$KAlSi_3O_8$-SiO_2-H_2O tetrahedron. Contour lines indicate melting temperatures. Stability fields are approximately located. The shaded area in the center represents percentages of quartz, albite, and K-feldspar, as calculated for hundreds of chemically analyzed granites (from O. F. Tuttle and N. L. Bowen, 1958, Origin of granite in the light of experimental studies in the system $NaAlSi_3O_8$-$KAlSi_3O_8$-SiO_2-H_2O, *Geol. Soc. America*, Memoir 74, 153 pp.)

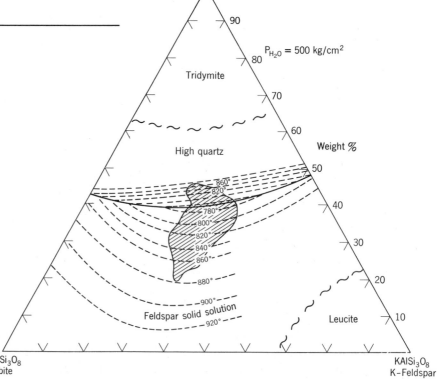

Syenite-Monzonite. A syenite is a granular rock of light color and even texture composed essentially of K-feldspar and oligoclase, with lesser amounts of hornblende, biotite, and pyroxene. It thus resembles granite but differs from granite in that it contains less than 5% quartz. Accessory minerals are apatite, sphene, zircon, and magnetite.

A series exists between syenite and monzonite, with increasing plagioclase content in the monzonite (see Fig. 11.12). Monzonites are usually darker than syenites, for an increase in dark minerals frequently accompanies an increase in plagioclase. However, without microscopic aid it is rarely possible to distinguish between the two types.

Nepheline is present in some syenites; if the amount exceeds 5%, the rock is called a *nepheline syenite*. The nepheline has a greasy luster and may be mistaken for quartz, but can be distinguished by its hardness ($5\frac{1}{2}$–6). Some nepheline syenites may contain sodalite; others, corundum.

Syenites in which leucite is present in amounts greater than 5% are called *leucite syenites*. The leucite can be recognized by its trapezohedral form. Such rocks are extremely rare.

Tonalite-Quartz Diorite. A tonalite is composed essentially of plagioclase feldspar and quartz with only minor amounts of K-feldspar (less than 5%). The plagioclase is oligoclase or andesine. Dark minerals, especially biotite and hornblende, are plentiful; pyroxene is more rarely present. Apatite, sphene, magnetite are common accessory minerals. Although not essential to the classification, dark

Table 11.4

TYPICAL MINERAL ASSEMBLAGES OF SOME PLUTONIC ROCKS

	Major Constituents (>10%)	Minor Constituents (each <10%)
Granite	mic(often perthitic) + Ab (or olig) + qtz	bio,musc,hbld,mg,zirc,ap,sph
Granodiorite	and + K-spar + qtz + hbld	bio,sph,ap,mag
Syenite	mic + Ab (or olig)	qtz,bio,hbld,sph,ap,zirc
Nepheline syenite	orth + neph	Ab,aeg,arf,sod,ap,zirc,sph
Monzonite	plag + orth + aug	hbld,qtz,bio,mag,ap,sph
Diorite	olig (or and) + hbld	bio,orth,qtz,hbld,pyx,ap,zirc,mag
Gabbro	plag(An_{50}-An_{90}) + aug + hyp + oliv	bio,mag,sp,ilm,hbld
Norite	lab(or byt) + bronz(or hyp) + some oliv	aug,hbld,bio,ap,mag
Anorthosite	lab($\sim$90%)	aug,hbld,bio
Peridotite	oliv(forsteritic) + enst	hbld,chrom,plag
Dunite	oliv(forsteritic) $\sim$90%	chrom,mag,ilm,pyrrh

Abbreviations: Ab = albite, aeg = aegirine, and = andesine, ap = apatite, arf = arfvedsonite, aug = augite, bio = biotite, bronz = bronzite, byt = bytownite, chrom = chromite, enst = enstatite, hbld = hornblende, hyp = hypersthene, ilm = ilmenite, K-spar = K-feldspar, lab = labradorite, mag = magnetite, mic = microcline, musc = muscovite, neph = nepheline, olig = oligoclase, oliv = olivine, orth = orthoclase, plag = plagioclase, pyrrh = pyrrhotite, pyx = pyroxene, qtz = quartz, sph = sphene, sp = spinel, sod = sodalite, zirc = zircon.

minerals are usually abundant, and thus, in general, tonalites are darker in color than granites and granodiorites.

As the plagioclase becomes more An-rich and with a lessening of the amount of quartz tonalite grades into a *quartz-diorite* and finally into a *diorite*.

Diorite-Gabbro. A diorite is a granular rock characterized by plagioclase feldspar (oligoclase to andesine) but lacking quartz and K-feldspar in appreciable amounts. Hornblende is the principal dark mineral, but biotite is usually present. Pyroxenes are rare. Magnetite, ilmenite, apatite, and, less commonly, sphene and zircon are accessory minerals. Normally dark minerals are present in sufficient amount to give the rock a dark appearance.

If the plagioclase is more calcic in composition than andesine (labradorite to anorthite) the rock is a gabbro. Although the distinction is made on this criterion alone, it so happens that rocks carrying labradorite or more calcic plagioclase usually have pyroxene as the chief dark constituent, whereas the diorites with more sodic plagioclase usually have amphiboles as dark minerals. Olivine is also present in most gabbros. The association of pyroxene-olivine-An-rich plagioclase is diagnostic of the relatively high temperatures of crystallization of mafic rock types. In Fig. 11.4 forsterite and protoenstatite are shown crystallizing together from 1557°C to about 1300°C (along the crystallization path between the two fields). These temperatures are much higher than those determined for granites, for example, in Fig. 11.13. See also Fig. 11.9 which expresses these relative temperature differences qualitatively.

The name *norite* is given to a gabbro in which the pyroxene is essentially hypersthene; it is usually impossible to make this distinction without microscopic aid. A type of igneous rock known as *anorthosite* is composed almost entirely of plagioclase feldspar and may therefore be light in color.

If amounts of nepheline in diorites and gabbros exceed 5% the rocks are called respectively *nepheline diorite* and *nepheline gabbro*. These rocks are rare and unimportant.

The term *diabase* is sometimes used to indicate a fine grained gabbro characterized by a certain texture. This "diabasic" texture is shown microscopically to have augite filling the interstices of tabular plagioclase crystals.

Peridotite. A peridotite is a granular rock composed of dark minerals; feldspar is negligible (less than 5%). The dark minerals are chiefly pyroxene and olivine in varying proportions, but hornblende may be present. If the rock is composed almost wholly of pyroxene it is called a *pyroxenite;* if it is composed almost wholly of olivine it is called a *dunite*. The name *hornblendite* is given to a rare type of rock composed almost wholly of hornblende. Magnetite, chromite, ilmenite, and garnet are frequently associated with peridotites. Platinum is associated with chromite in some peridotites, usually dunites, whereas diamond is found in a variety of altered peridotite known as *kimberlite*.

The olivine in peridotites is usually altered in whole or in part to the mineral serpentine. If the entire rock is thus altered the name *serpentinite* is given to it (see also page 473).

Volcanic Rocks

Because of their fine grained texture it is much more difficult to distinguish between the different types of volcanic rocks than between their plutonic equivalents. In the field only an approximate classification, based chiefly on whether the rock is light or dark in color, can be made. The term *felsite* is thus used to include the dense, fine grained rocks of all colors except dark gray, dark green, or black. Felsite thus embraces the following types described below: rhyolite, trachyte, quartz latite, latite, dacite, and andesite. The experienced petrologist may be able, by the aid of a hand lens, to discern differences in texture or mineral composition that enable him to classify these rocks, but to the untrained observer they frequently appear much the same.

Fine grained rocks that are a very dark green or black are called *traps*. This term is applied to dark, fine grained rocks of indefinite mineral composition irrespective of whether they have been intruded as dikes or extruded as lava flows. It so happens that most rocks thus classified as traps in the field or hand specimen are basalts and satisfy the more rigorous classification based on microscopic examination.

Rhyolite is a dense fine grained rock, the volcanic equivalent of a granite. It is thus composed essentially of alkali feldspar and quartz, but much of the silica may be present as tridymite or cristobalite. Phenocrysts of quartz, sanidine, and oligoclase are common. Dark minerals are never abundant, but dark brown biotite is most common. Augite and hornblende are found in some rhyolites.

Rhyolites may be very uniform in appearance or may show a flow structure, giving a banded or streaked appearance to the rock. The groundmass may be partly or wholly glassy. When the rock is completely glassy and compact, it is known as *obsidian* and is usually black. Similar glassy rocks of a brown, pitchy appearance are called *pitchstones*. *Pumice* is rhyolite glass in which expanding gas bubbles have distended the magma to form a highly vesicular material. In pumice, therefore, cavities are so numerous as to make up the bulk of the rock and give it an apparent low specific gravity.

Trachyte is the volcanic equivalent of syenite. It is thus composed chiefly of alkali feldspar with some dark minerals but lacks quartz. Small amounts of tridymite and cristobalite are often found in gas cavities. Phenocrysts of sanidine are frequently present and characteristically show Carlsbad twinning; phenocrysts of oligoclase, biotite, hornblende, and pyroxene are less common. Olivine may be present.

Banding or streaking, due to flow, is common in the trachytes. Unlike the rhyolites, glass is seldom found in the groundmass, and there are thus few glassy or vesicular types. As a result of flow the tabular feldspar frequently shows a subparallel orientation which is so common in trachytes that it is called *trachytic texture.*

Phonolite is the volcanic equivalent of nepheline syenite and is thus poorer in silica than trachyte. This is expressed mineralogically by the presence of feldspathoids. Orthoclase or sanidine is the common feldspar; albite is rarely present. Nepheline occurs in the groundmass as minute hexagonal crystals and can be observed only by microscopic aid. Sodalite and other feldspathoids may be present, usually altered to zeolites. Leucite is present in *leucite phonolite.* It is in well-formed crystals which range from microscopic sizes to half

an inch in diameter. Aegirine is the common dark mineral and normally occurs as phenocrysts, but biotite may be abundant in the leucite-rich rocks. The phonolites are completely crystalline, and there are thus no glassy varieties.

Latite and **quartz latite** are the volcanic equivalents of monzonite and quartz monzonite, respectively. They, therefore, contain about equal amounts of plagioclase and alkali feldspar. The dark minerals are chiefly biotite and hornblende. The distinction between them rests on the amount of quartz present; quartz latites would contain more than 5% quartz, latites less than 5% quartz. Both of these rock types occur infrequently.

Dacite is the dense volcanic equivalent of granodiorite. It contains plagioclase feldspar and quartz, both of which may occur as phenocrysts. The dark mineral is usually hornblende, but biotite is found in some varieties. Some glass may be present in the groundmass, but glassy equivalents of dacites are rare.

Andesite is the volcanic equivalent of quartz diorite and thus is composed chiefly of oligoclase or andesine feldspar. K-feldspar and quartz are absent or present in amounts of less than 10%. Hornblende, biotite, augite, or hypersthene may be present, frequently as phenocrysts. Andesites are usually named according to the dark mineral present, such as *hornblende andesite* or *hypersthene andesite.* In some andesites the groundmass is partly glassy and in rarer types completely so.

Andesites are abundant in certain localities, notably in the Andes Mountains of South America, from which locality the rock receives its name.

Basalt is a dark-colored, fine grained rock, the volcanic equivalent of gabbro. Labradorite feldspar is the chief constituent of the groundmass, whereas more calcic plagioclase (bytownite or anorthite) may be present as phenocrysts. Augite and olivine are usually present; the augite is frequently found both as phenocrysts and in the groundmass, but olivine, as a rule, is only in phenocrysts. Brown hornblende and brown biotite are present in some basalts.

The groundmass of some basalts contains small amounts of interstitial glass and in rare instances is wholly glassy. Gas cavities near the top of basalt flows may be abundant enough to make the rock vesicular.

The presence of nepheline or leucite in basalt gives rise to the rare rock types *tephrite* and *leucite tephrite.*

Basalts are the most abundant of the volcanic rocks and form extensive lava flows in many regions; the most noted are the Columbia River flows in the western United States and the Deccan "traps" of western India. The oceanic basins are underlain by extensive flows of basaltic composition. Many of the great volcanoes, such as form the Hawaiian Islands, are built up of basaltic material. In addition to forming extrusive rock masses, basalt is widely found forming many small dikes and other intrusives.

Fragmental Igneous Rocks

During periods of igneous activity volcanoes eject much fragmental material which accumulates and forms the *fragmental igneous rocks,* or *pyroclastic rocks.* The ejecta vary greatly in size. Rock composed of finer particles of *volcanic ash* and *volcanic dust* is called *tuff;* that composed of coarser volcanic bombs is called *agglomerate,* or *volcanic breccia.* Such rocks are frequently waterlaid and bedded, and thus form a transition between the igneous and sedimentary rocks.

Pegmatites

Pegmatites are extremely coarse grained bodies that are commonly closely related genetically and in space to large masses of plutonic rocks. They may be found as veins or dikes traversing the granular igneous rock but more commonly extend out from it into the surrounding country rock. Granites more frequently than any other rock have pegmatites genetically associated with them; consequently, unless modified by other terms, pegmatite refers to *granite pegmatite.* The minerals in most pegmatites, therefore, are the common minerals in granite (quartz, feldspar, mica) but of extremely large size. Crystals of these minerals measuring a foot across are common, and in some localities they reach gigantic sizes. Probably the largest crystals ever found were of feldspar in pegmatites in Karelia, U.S.S.R., where material weighing thousands of tons was mined from single crystals. Quartz crystals weighing thousands of pounds and mica crystals over 10 feet across have been found. A common characteristic of pegmatites is the simultaneous and interpenetrating crystallization of quartz and K-feldspar (usually microcline) to form *graphic granite* (see Fig. 10.81).

Although most pegmatites are composed entirely of the minerals found abundantly in granite, those of greatest interest contain other, and rarer, minerals. In these pegmatites there has apparently been a definite sequence in deposition. The earliest minerals are microcline and quartz, with smaller amounts of garnet and black tourmaline. These are followed by and partly replaced by albite, lepidolite, gem tourmaline, beryl, spodumene, amblygonite, topaz, apatite, and fluorite. A host of rarer minerals such as triphylite, columbite, monazite, molybdenite, and uranium minerals may be present. In places some of the above minerals are abundant and form large crystals that are mined for their rare constituent elements. Thus spodumene crystals over 40 feet long have been found in the Black Hills of South Dakota, and beryl crystals from Albany, Maine, have measured as much as 27 feet long and 6 feet in diameter.

The formation of pegmatite dikes is believed to be directly connected with the crystallization of the larger masses of associated plutonic rock. The process of crystallization brings about a concentration of the volatile constituents in the residual melt of the magma. The presence of these volatiles (H_2O, B, F, Cl, and P) decreases the viscosity and thus facilitates crystallization. Such an end product of magmatic crystallization is also enriched in the rare elements originally disseminated through the magma. When this residual liquid is injected into the cooler surrounding rock, it crystallizes from the borders inward frequently giving a zonal distribution of minerals with massive quartz at the center.

Nepheline syenite pegmatites have been found in a number of localities. They are commonly rich in unusual constituents and contain numerous zirconium, titanium, and rare-earth minerals.

II. Sedimentary Rocks

The sedimentary rock cover over the continents of the earth's crust is extensive, but the total contribution of sedimentary rocks to the upper 10 miles is estimated to be only about 5%. As such, the sedimentary sequences we see represent only a thin veneer over a crust consisting mainly of igneous and metamorphic rocks.

The materials of which sedimentary rocks are composed have been derived from the weathering of some previously existing rock mass. They are transported and deposited in areas of accumulation by the action of water or, less frequently, by glacial or wind action. Such loose deposits are converted into rocks by the processes of *lithification* and *diagenesis* which include compaction and cementation of loose materials. Weathering includes both mechanical disintegration and chemical decomposition and thus the end products consist of sedimentary materials which can be divided into two categories: *clastic* (or *detrital*) sediments, which are the products of mechanical accumulation of individual grains, and *chemical* (and biochemical) sediments, which are the product of chemical precipitation from solutions. These two categories are not exclusive because most clastic rocks carry some chemical sediment and most chemical sediments contain detrital material. Mechanically deposited loose sediments include gravel and sand which, upon lithification, form conglomerates and sandstones. Such rocks contain inert minerals such as quartz, zircon, rutile, and magnetite. Chemically deposited sediments are represented by carbonate and evaporite sequences and sedimentary iron-formations. Such chemical sediments are the result of dissolution of materials from an earlier sequence of rocks and subsequent transportation of dissolved chemical species into a sea or lake; in such a basin the precipitation of the chemical species from solution may be the result of inorganic or organic processes.

All sedimentary rocks are, in general, characterized by a parallel arrangement of their mineral constituents forming layers or beds distinguished from each other by differences in thickness, size of grain, or color. In all the coarser grained sedimentary rocks there is material that acts as a cement and surrounds the individual mineral particles, binding them together. This cement is usually silica, calcium carbonate, or iron oxide, which formed as chemical precipitates.

Chemical Composition

The range in chemical compositions of igneous rocks (see Table 11.1) is generally relatively small

Table 11.5

CHEMICAL COMPOSITION OF AVERAGE IGNEOUS ROCK COMPARED WITH AVERAGE COMPOSITIONS OF SOME SEDIMENTARY ROCK TYPES[a]	Average Continental Igneous Rock	Average Sandstone	Average Shale	Average Limestone
SiO_2	59.14	78.33	58.10	5.19
TiO_2	1.05	0.25	0.65	0.06
Al_2O_3	15.34	4.77	15.40	0.81
Fe_2O_3	3.08	1.07	4.02	0.54
FeO	3.80	0.30	2.45	—
MgO	3.49	1.16	2.44	7.89
CaO	5.08	5.50	3.11	42.57
Na_2O	3.84	0.45	1.30	0.05
K_2O	3.13	1.31	3.24	0.33
H_2O	1.15	1.63	5.00	0.77
P_2O_5	0.30	0.08	0.17	0.04
CO_2	0.10	5.03	2.63	41.54
SO_3	—	0.07	0.64	0.05
C(elemental)	—	—	0.80	—
Total	99.50	99.95	99.95	99.84

[a] After F. W. Clarke, 1924, Data of Geochemistry, *U. S. Geol. Surv. Bul.* 770.

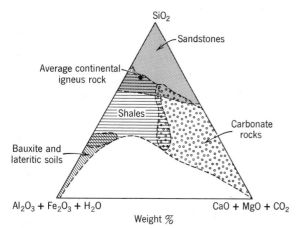

FIG. 11.14. Compositional variations of common sediments (after B. Mason, 1966, *Principles of Geochemistry*, p. 154).

because the crystallization sequence of a magma is governed by physicochemical and chemical principles that control the sequence of products crystallizing from the melt. Sedimentary rocks, on the other hand, present a much larger compositional range (see Table 11.5). For example, a sandstone consisting essentially of quartz grains may contain as much as 99% SiO_2; $FeO + Fe_2O_3$ may be as high as 58% in banded iron-formations rich in hematite and magnetite; CaO may reach 55% in pure limestones. This large range in compositions which is illustrated in Fig. 11.14, is caused by the weathering cycle which tends to produce mechanical sediments which are compositionally very distinct from chemical sediments.

Mineralogic Composition

The minerals of sedimentary rocks can be divided into two major groups: minerals that are resistant to the mechanical and chemical breakdown of the weathering cycle and minerals that are newly formed from the products of chemical weathering. The relative stabilities of minerals to weathering are shown in Table 11.6. This listing means that quartz is one of the chemically and mechanically most resistant minerals, whereas olivine is easily altered chemically. The positions of minerals listed between quartz and olivine represent intermediate stabilities ("survival rates") during the weathering

process. It is of interest to note that this sequence is very similar to Bowen's reaction series in reverse (see Fig. 11.9). This similarity indicates that the minerals formed at the lowest temperature in the crystallization of a melt are also the most stable at ordinary temperatures and pressures (atmospheric conditions).

Detrital sedimentary rocks consist mainly of the most resistant rock-forming minerals, quartz, K-feldspar, mica and lesser plagioclase, as well as small amounts of garnet, zircon, and spinel (magnetite). The detrital rock types may be regarded as accidental, mechanical mixtures of genetically unrelated resistant minerals. For example, a feldspar-rich sandstone may contain orthoclase and microcline, as well as several members of the plagioclase series. Such a random variety of feldspar compositions is not found in igneous assemblages because of physical-chemical controls in the crystallization sequence.

Sedimentary rocks that result from the inorganic or organic precipitation of minerals can be interpreted in large part in terms of chemical and physico-chemical principles that apply at low temperatures (25°C) and atmospheric pressure (see, for example, Fig. 11.2). The chemical sedimentary assemblages, therefore, are not random or accidental but reflect the concentrations of ions in solution, as well as conditions such as temperature, pressure, and salinity of the sedimentary basin. For

Table 11.6
RELATIVE STABILITIES OF SOME ROCK-FORMING MINERALS IN THE WEATHERING CYCLE

High stability		Quartz
		Muscovite
		K-feldspar
		Biotite
Increasing stability	Albite intermediate plagioclases Anorthite	Hornblende Augite
Low stability		Olivine

example, the sequence of minerals in evaporite beds can be related to the concentration of ions in solution in the brine from which the sequence precipitated (see p. 466). Examples of common chemical precipitates are calcite, aragonite, gypsum, and halite. In sedimentary iron-formations hematite, magnetite, siderite and ankerite as well as chert are considered products of chemical sedimentation.

In addition to the detrital minerals and chemical precipitates, sediments may contain highly variable amounts of layer silicates—mainly clay minerals (kaolinite, montmorillonite, illite) and chlorite. These layer silicates are the product of low-temperature decomposition of earlier primary silicates such as feldspars, olivines, and pyroxenes.

Rock Types

Conglomerate. Conglomerates may be considered consolidated gravels. They are composed of coarse pebbles, usually rounded by stream transportation. The individual pebbles may be entirely composed of quartz, or may be rock fragments that have not been decomposed. Fine conglomerates grade into coarse sandstones.

Sandstone. Beds of sand that have been consolidated into rock masses are called sandstones. The constituent grains are usually rounded and waterworn but may be more or less angular. The cement that binds the sand grains together may be silica, a carbonate (usually calcite), an iron oxide (hematite or goethite), or fine grained clays or chloritic material. The color of the rock depends in large measure on the character of the cement. The rocks that have silica or calcite as their binding material are light in color, usually pale yellow, buff, white to gray; those that contain an iron oxide are red to reddish brown. It is to be noted that when a sandstone breaks it is commonly the cement that is fractured, the individual grains remaining unbroken, so that the fresh surfaces of the rock may have a granular appearance and feeling. The chief mineral of sandstones is quartz and if the rock contains more than 90% quartz grains it is known as an *orthoquartzite*. If the rock contains notable amounts of feldspar it is termed an *arkose*. If the sandstone contains in addition to abundant quartz, considerable amounts of

feldspar grains, rock fragments, and fine grained silt or clay material, it is known as a *greywacke*. Greywackes may contain as much as 30% very fine grained clay or chlorite, or both. The finer grained greywackes grade into the shales.

Shale. The shales are very fine grained sedimentary rocks that have been formed by the consolidation of beds of mud, clay, or silt. They usually have a thinly laminated structure. Their color is commonly some tone of gray, although they may be white, yellow, brown, red, or green to black. They are composed chiefly of clay minerals with quartz and mica but are too fine grained to permit the recognition of their mineral constituents by the eye alone. By the introduction of quartz and an increase in the grain size they grade into greywackes and sandstones, and with the presence of calcite they grade into the limestones.

Limestone. Limestone composed essentially of calcite, $CaCO_3$, is the most abundant chemically precipitated sedimentary rock. Although the calcite may be precipitated directly from seawater, most limestone is the result of organic precipitation. Many organisms living in the sea extract calcium carbonate from the water to build hard protective shells. On the death of the organisms the hard calcareous parts accumulate on the sea floor. When marine life is abundant, great thicknesses of shells and other hard parts may build up, which, when consolidated, become limestone. There are several varieties of limestone. *Chalk* is a porous, fine grained limestone composed for the most part of foraminiferal shells. *Coquina,* found on the coast of Florida, consists of only partially consolidated shells and shell fragments. *Lithographic limestone* is an extremely fine grained rock from Solenhofen, Bavaria. Although limestones are dominantly calcite they may contain small amounts of other minerals. If impurities of clay are abundant, limestone grades toward shale and would be called an *argillaceous limestone*. When dolomite becomes an important constituent of a limestone, the rock grades toward the rock *dolomite*. *Oölitic limestone* is composed of small spherical concretions resembling fish roe, and is believed to have been chemically precipitated. The tiny concretions have a nucleus of a sand grain, shell fragment, or some foreign particle around which deposition has taken place. Oölitic sand, which is forming at the present time on the floor of the Great Salt Lake, would on consolidation, produce an oölitic limestone.

Travertine. This is a calcareous material deposited from spring waters (frequently thermal springs) under atmospheric conditions. If the deposit is porous it is known as *calcareous tufa*. Such deposits are prevalent in limestone regions where circulating ground water containing CO_2 has taken considerable calcium carbonate in solution. When the ground water reaches the surface as springs, some of the CO_2 is given off, resulting in the precipitation of some of the calcium carbonate as travertine.

Dolomite. The rock *dolomite* resembles limestone so closely in its appearance that it is usually impossible to distinguish between the two rock types without a chemical test. Moreover, dolomite as a rock name is not restricted to a material of the composition of the mineral dolomite but may have admixed calcite. Dolomites have generally not formed as original chemical precipitates but are the result of alteration of limestone in which part of the calcium is replaced by magnesium. This process of *dolomitization* is believed to have taken place either by the action of sea water shortly after original deposition, or by the action of circulating ground water after the rock has been consolidated and raised above sea level.

Magnesite. As a sedimentary rock *magnesite* is much more restricted in its distribution than dolomite. It has been formed by the almost complete replacement by magnesium of the calcium of an original limestone. Such magnesite rocks are found in Austria; in the Ural Mountains, U.S.S.R.; and in Stevens County, Washington.

Siliceous Sinter. In certain volcanic regions hot springs deposit an opaline material known as *siliceous sinter* or *geyserite*. The deposit apparently results from both evaporation and secretion of silica by algae.

Diatomite. Diatoms are minute one-celled organisms that live in both fresh and seawater and have the power of secreting tests of opaline mate-

rial. When the organisms die, their tiny shells accumulate to build up a chalklike deposit of *diatomaceous earth*.

Iron-formation. Sedimentary, banded iron-formations are common in rock sequences of late Precambrian age. They consist mainly of chert-magnetite, chert-hematite, and chert-hematite-magnetite. The sedimentary banding in these rocks is generally very well preserved and is considered to reflect the chemical precipitation of iron-oxides and chert in basins of Proterozoic age. Other minerals that are commonly present are siderite, ankerite, greenalite, and minnesotaite. Regions of vast Proterozoic iron formations are found in the Lake Superior region, in the Labrador Trough, Canada, in the Hamersley Range, Western Australia, and Brazil.

Evaporites. When a cut off body of seawater or the waters of saline lakes evaporate, the elements in solution (see Table 11.7) are precipitated in what are known as *evaporites*. More than 80 minerals (excluding clastic material) have been recorded in evaporites and most of these are chlorides, sulfates, carbonates, and borates; only about 11 rank as major constituents (see Table 11.7). On evaporation the general sequence of precipitation is: some calcite (when the original volume of seawater is reduced by evaporation to about one-half), gypsum (with the volume reduced to one-fifth of the original), halite (with the volume reduced to one-tenth of the original), and finally sulfates and chlorides of Mg and K. If all the salt in a 1000 foot (305 m) column of seawater were precipitated, it would form 0.5 feet (0.15 m) of calcium sulfate, 11.8 feet (3.6 m) of NaCl, and 2.6 feet (0.8 m) of K- and Mg-bearing salts, producing a total salt column of 15 feet (4.6 m) thick.

In natural deposits, however, the minerals that precipitate early in the sequence tend to show increased abundance. Hence, gypsum and anhydrite are by far the most abundant evaporite minerals and commonly form massive beds. The deposition of gypsum or anhydrite depends on the temperature and salinity of the brine; anhydrite is formed at higher salt concentrations and at higher temperatures than gypsum.

Halite forms about 95% of the chloride miner-

Table 11.7

MAJOR IONIC CONSTITUENTS OF SEAWATER AND MOST COMMON MINERALS IN EVAPORITE SEQUENCES[a]

	Normal Seawater (Ion Concentration as Parts per Million)	Common Minerals in Marine Evaporites
K^+	380	Halite—$NaCl$
Na^+	10,556	Sylvite—KCl
Ca^{2+}	400	Carnallite—$KMgCl_3.6H_2O$
Mg^{2+}	1,272	Anhydrite—$CaSO_4$
Cl^{-1}	18,980	Gypsum—$CaSO_4.2H_2O$
SO_4^{-2}	2,649	Langbeinite—$K_2Mg_2(SO_4)_3$
HCO_3^{-1}	140	Polyhalite—$K_2Ca_2Mg(SO_4)_4 \cdot 2H_2O$
Total	34,387	Kieserite—$MgSO_4 \cdot H_2O$
		Calcite—$CaCO_3$
		Magnesite—$MgCO_3$
		Dolomite—$CaMg(CO_3)_2$

[a] After F. H. Stewart, 1963, Marine Evaporites, U.S. Geol. Surv. Prof. Paper no. 440-Y.

als in an evaporite sequence; thick and extensive beds of *rock salt* overlie the gypsum-anhydrite zone of marine evaporites.

Deposits of the more soluble salts such as sylvite, carnallite, and polyhalite are rare for they are not deposited until nearly complete dryness has been reached. Nevertheless, in places large deposits, particularly of sylvite, have formed and are mined extensively as the chief source of potassium.

III. Metamorphic Rocks

Metamorphic rocks are derived from preexisting rocks (igneous, sedimentary, or metamorphic) by mineralogical, textural, and structural changes. Such changes may be the result of marked variations in temperature, pressure, and shearing stress at considerable depth in the earth's crust. Weathering effects, at atmospheric conditions, are not considered part of metamorphism, nor are chemical reactions involving partial melting as these are part of igneous processes. Metamorphic changes such as recrystallization and chemical reactions among mineral constituents take place essentially in the solid state although the solids may exchange chemical species with small amounts of a liquid phase consisting mainly of H_2O (as water, steam, or supercritical fluid, depending on the temperature and pressure at which the reactions took place). The general condi-

tions of formation of metamorphic rocks lie between those of sedimentary rocks that form at essentially atmospheric conditions of T and P and those of igneous rocks that are the result of crystallization from a melt at high T. Metamorphic rocks may be the result also of very large changes in pressure in conjunction with increasing metamorphic temperature; mineral assemblages in the earth's mantle have formed in response to very high confining pressures. Excluding gain or loss of H_2O many metamorphic reactions are generally considered to be essentially *isochemical;* this implies that during recrystallization and the processes of chemical reactions the bulk chemistry of the rocks has remained essentially constant. If this is not the case, and if additional elements have been introduced into the rock by circulating fluids, for example, it is said to have undergone *metasomatism.*

The most obvious textural feature of most metamorphic rocks (except those of contact metamorphic origin, see below) is the alignment of minerals along planar surfaces. For example, a shale that has undergone only slight metamorphic changes may show well-developed cleavage along planar surfaces, producing a *slate.* With increased temperature of metamorphism recrystallization of originally very fine grained minerals produces coarser grained *schists* which exhibit minerals aligned in parallel layers (known as *schistosity*). The coarsest grained metamorphic rocks that show distinct mineralogic banding are known as *gneisses* (exhibiting *gneissosity*).

In general, metamorphic rocks can be divided into two groups: (1) those formed by contact metamorphism and (2) those formed by regional metamorphism. *Contact metamorphic rocks* occur as concentric zones (*aureoles*) around hot igneous intrusive bodies. Such metamorphic rocks lack schistosity, and the relatively large temperature gradient from the intrusive contact to the unaffected country rock gives rise to zones which may differ greatly in mineral assemblages. Sandstones are converted to quartzites and shales are changed to *hornfels,* a fine grained dense rock. *Regional metamorphic rocks* are the result of increases in *T* or *P*, or both, on a regional scale (areas a few hundred to thousands of miles in extent) in response to mountain building, or deep burial of rocks. Contact and regionally metamorphosed rocks generally reflect an increase in temperature (progressive metamorphism) in their assemblages. However, when assemblages of high-temperature origin (e.g., igneous or high-temperature metamorphic rocks) fail to survive conditions of lower temperature metamorphism, the process is referred to as *retrograde* (or retrogressive) metamorphism.

Chemical Composition

As metamorphic rocks are the recrystallized and generally foliated (from the Latin *folia* meaning leaves) products of rocks of igneous or sedimentary bulk compositions, the general range of their chemical compositions is as large as that of igneous and sedimentary rocks combined. Although metamorphic reactions take place without addition of chemical species to the rock, some chemical components may be lost especially during the higher temperature ranges of metamorphism. For example, a shale, which is largely composed of hydrous minerals such as clays, may convert into a slate and subsequently into a schist at increasing metamorphic temperatures, during which processes the mineral assemblage becomes less and less hydrous, due to progressive *dehydration*. In other words, with an increase in metamorphic temperatures water is generally lost from the rock. Similarly, carbonate-rich sedimentary rocks while undergoing increasing temperatures of metamorphism will tend to loose CO_2 which is known as *decarbonation*. Such processes can be illustrated as follows:

$$Al_4Si_4O_{10}(OH)_8 \rightarrow Al_4Si_{4-x}O_{10} +$$

$$\text{kaolinite} \qquad \text{``meta-kaolinite''}$$

$$xSiO_2 + H_2O\nearrow \text{ and}$$

$$CaCO_3 + SiO_2 \rightarrow CaSiO_3 + CO_2\nearrow$$

$$\text{calcite} \quad \text{chert} \quad \text{wollastonite}$$

Although such losses of H_2O and CO_2 to a thin layer of fluid phase surrounding the mineral grains are common during the conditions of progressive metamorphism, additions of chemical species from the fluid phase to the rock are considered part of metasomatic processes.

Mineralogical Composition

The mineralogic composition of a rock that has not undergone metamorphic conditions will generally be very different from that of its metamorphosed equivalent. The extent of the differences is largely controlled by the marked changes of *T* and *P* during metamorphism; the relative degrees of metamorphism can be expressed in terms of *grades,* such as very-low, low, medium, and high grade (see Fig.

FIG. 11.15. Schematic *P-T* diagram outlining approximate fields for various metamorphic grades. The shaded area marked "diagenetic conditions" represents the general conditions of lithification of a sediment at low temperature (after H. G. F. Winkler, 1974, *Petrogenesis of Metamorphic Rocks*, Springer-Verlag, p. 5, with some modifications).

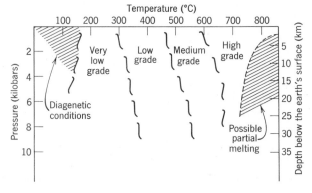

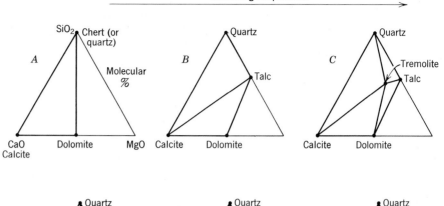

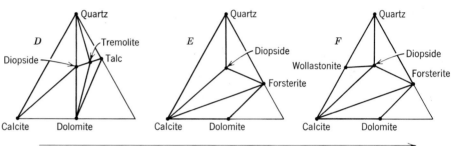

FIG. 11.16. Some of the assemblage changes as a result of increasing temperature in the system CaO-MgO-SiO$_2$-CO$_2$-H$_2$O. Triangle *A* represents the original, sedimentary assemblage in this system (the composition of a naturally occurring cherty, dolomitic limestone is only partly represented in this triangle because it also contains small amounts of FeO). Triangle *F* represents the highest temperature (contact metamorphic) assemblage, and the other triangles represent intermediate grades of metamorphism.

11.15). *Changes in metamorphic grade are reflected in changes in the mineral assemblages of the rocks.* For example (see Fig. 11.16), a cherty dolomitic limestone that has undergone very-low grade metamorphic conditions (*T* between about 150°C and 250°C; see Fig. 11.15) may show the formation of talc according to:

$$3CaMg(CO_3)_2 + 4SiO_2 + H_2O \rightarrow$$
dolomite chert

$$Mg_3Si_4O_{10}(OH)_2 + 3CaCO_3 + 3CO_2\nearrow$$
talc calcite

At low-grade metamorphic conditions (*T* between about 250° and 450°C) the assemblage on the right of the above equation may react as follows:

$$2Mg_3Si_4O_{10}(OH)_2 + 3CaCO_3 \rightarrow$$
talc calcite

$$Ca_2Mg_5Si_8O_{22}(OH)_2 + CaMg(CO_3)_2 + CO_2\nearrow + H_2O\nearrow$$
tremolite dolomite

At medium-grade metamorphic conditions (about 450° to 600°C) the tremolite may give way to diopside according to:

$$Ca_2Mg_5Si_8O_{22}(OH)_2 + 3CaCO_3 + 2SiO_2 \rightarrow$$
tremolite calcite quartz

$$5CaMgSi_2O_6 + 3CO_2\nearrow + H_2O\nearrow$$
diopside

At high-grade metamorphic conditions (above about 600°C) forsterite may form as follows:

$$CaMgSi_2O_6 + 3CaMg(CO_3)_2 \rightarrow$$
diopside dolomite

$$2Mg_2SiO_4 + 4CaCO_3 + 2CO_2\nearrow$$
forsterite calcite

At very high temperatures, such as in the immediate border zone of an igneous intrusive (contact metamorphism) wollastonite may be formed by the reaction:

$$CaCO_3 + SiO_2 \rightarrow CaSiO_3 + CO_2\nearrow$$

    calcite   quartz   wollastonite

The above types of reactions can be studied experimentally, providing us with quantitative information in terms of temperature and pressure ranges at which reactants give way to products (see Fig. 11.3). This provides us with reasonably close estimates of the conditions of metamorphism of rocks that are similar in bulk chemistry to those studied in the laboratory.

In the regional metamorphism of argillaceous (clay-rich) rocks such as shales one can similarly study the various reactions that must have taken place among the mineral constituents as a result mainly of increasing temperature. In the succession from shale (clay minerals + chlorite + muscovite + feldspar + quartz) to slate, to staurolite-almandite-biotite-muscovite-quartz schist and finally to sillimanite-almandite-cordierite-K-feldspar-quartz gneiss at the highest temperatures of regional metamorphism, many reactions have taken place among the mineral constituents. As in the case of the above reactions for carbonate rocks, the sequence ranges from fine grained, H_2O-rich mineral assemblages at low temperature, to coarse grained anhydrous assemblages at the highest temperatures (see Table 11.9).

In a systematic study of shales and their metamorphic equivalents in the Scottish Highlands, published in 1912, G. Barrow delineated various metamorphic zones on the basis of the occurrence of *index* minerals. At successively higher grades of metamorphism, he noted that the argillaceous rocks showed the development of the following index minerals: first chlorite, then biotite, next almandite, subsequently staurolite, then kyanite, and at the highest temperatures sillimanite. Similar progressions of minerals are found in other metamorphic terranes containing argillaceous rocks such as in the New England region. When areas of specific index mineral occurrence are outlined on a geological map, such as a chlorite-rich region or a biotite-rich region, the line that marks the first appearance of an index mineral is known as an *isograd*. The isograds reflect positions of similar metamorphic grade in terms of T and P.

In regions where very high pressures have been generated during metamorphism the final mineral assemblages will be very different, for a specific bulk rock composition, from those where high temperatures have been most prevalent. Argillaceous rocks metamorphosed at high temperature commonly contain sillimanite, whereas those subjected to high pressures (with some increase in temperature, see Fig. 10.14) contain kyanite. Similarly, high-temperature basalt assemblages occur as coarse grained eclogites (see Fig. 11.8) in the high pressure regions of the lower crust of the earth.

In addition to the subdivision of P and T fields in terms of very low, low, medium, and high grades of metamorphism (Fig. 11.15) metamorphic rocks are often classified in terms of metamorphic facies. A *metamorphic facies* is a set of metamorphic mineral assemblages, repeatedly associated in space and time, such that there is a constant and therefore predictable relation between mineral composition and chemical composition. A metamorphic facies, therefore, is defined not in terms of a single index mineral but by an association of mineral assemblages; some examples of metamorphic facies follow. *Zeolite facies* represents the lowest grade of metamorphism. The mineral assemblages include zeolites, chlorite, muscovite, and quartz. *Greenschist facies* is the low-grade metamorphic facies of many regionally metamorphosed terranes. The mineral assemblages may include chlorite, epidote, muscovite, albite, and quartz. *Amphibolite facies* occurs in medium- to high-grade metamorphic terranes. The mineral constituents include hornblende, plagioclase, and almandite. This facies occurs where staurolite and sillimanite-grade metamorphic conditions have prevailed. *Glaucophane-lawsonite schist* (or *blueschist*) *facies* is represented by relatively low temperatures but elevated pressures of metamorphism in young orogenic zones, such as California and Japan. Characteristic constituents are lawsonite, jadeite, albite, glaucophane, muscovite and garnet. *Granulite facies* reflects the maximum temperature conditions of regional metamorphism such as have been commonly attained in Archean terranes. Characteristic mineral constituents are plagioclase, hypersthene, garnet and diopside. *Eclogite facies* represents the most deep-seated con-

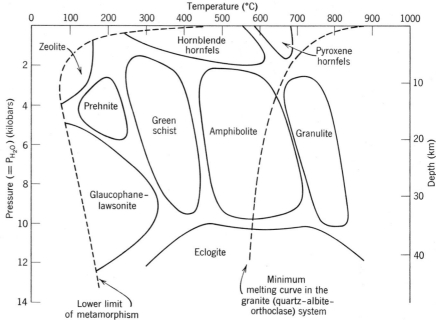

FIG. 11.17. Tentative scheme of metamorphic facies in relation to pressure (P_{H_2O}) and temperature. All boundaries are gradational (from F. J. Turner, *Metamorphic Petrology,* 1968, p. 366, with slight modifications; copyright 1968 by McGraw-Hill, Inc. Used with permission of McGraw-Hill Book Co.). Compare with Fig. 11.15.

Table 11.8

COMMON SILICATES OF METAMORPHIC ROCKS

Phyllosilicates	talc
	serpentine
	chlorite
	muscovite
	biotite
Inosilicates	anthophyllite
	cummingtonite-grunerite
	tremolite-actinolite
	hornblende
	glaucophane
	clinopyroxene with jadeite component (high pressure)
	diopside
	orthopyroxene
	wollastonite[a]
Tectosilicates	quartz
	plagioclase, except for very An-rich compositions
	microcline and orthoclase
Nesosilicates	garnet (pyrope at high pressures)
	epidote
	kyanite, sillimanite, andalusite
	idocrase[a]
	forsterite
	staurolite
	choritoid

[a] Especially common in contact metamorphic rocks.

ditions of metamorphism. Characteristic mineral constituents are pyrope-rich garnet and omphacite. Such assemblages are common in *kimberlite* pipes, many of which carry associated diamond. A diagram outlining the approximate fields of the various metamorphic facies in terms of *P* and *T* is given in Fig. 11.17.

Minerals that are especially common in metamorphic rocks are listed in Table 11.8. Examples of mineral assemblages in carbonate-rich and argillaceous rocks as a function of increasing metamorphic grade are given in Table 11.9.

Rock Types

Some of the most common metamorphic rocks are briefly described below.

Slate. Slates are exceedingly fine grained rocks that have a remarkable property known as *slaty cleavage* which permits them to be split into thin, broad sheets. Their color is commonly gray to black but may be green, yellow, brown, and red. Slates are usually the result of the metamorphism of shales. Their characteristic slaty cleavage may or may not be parallel to the bedding planes of the original shales. Slate is rather common in occurrence.

Marble. A marble is a metamorphosed limestone. It is a crystalline rock composed of grains of calcite or, more rarely, dolomite. The individual grains may be so small that they cannot be distinguished by the eye, and again they may be coarse and show clearly the characteristic calcite cleavage. Like limestone, a marble is characterized by its

softness and its effervescence with acids. When pure, marble is white, but various impurities may create a wide range of color. It is a rock that is found in many localities and may be in thick and extensive beds. Commercially *marble* is used to indicate any Ca-carbonate rock capable of taking a polish and thus includes some limestones.

Schist. Schists are metamorphic rocks that are distinguished by the presence of well-developed foliation or schistosity, along which the rock may be easily broken. A common example is a *mica schist,* which consists essentially of quartz and mica, usually muscovite or biotite, or both. Mica is the prominent mineral, occurring in irregular leaves and in foliated masses. The mica plates all lie with their cleavage planes parallel to each other and give to the rock a striking laminated appearance. Mica schists frequently carry characteristic accessory minerals, such as garnet, staurolite, kyanite, sillimanite, andalusite, epidote, and hornblende; thus the rock may be called *garnet-mica schist* or *staurolite-mica schist.* Other varieties of schists that are derived chiefly by the metamorphism of igneous rocks rich in ferromagnesian minerals are *talc schist, chlorite schist, hornblende schist* and *amphibolite.* They are characterized, as their names indicate, by the preponderance of some metamorphic ferromagnesian mineral.

Gneiss. When the word gneiss is used alone it refers to a coarsely foliated metamorphic rock. The banding is caused by the segregation of quartz and feldspar into layers alternating with layers of dark

Table 11.9

EXAMPLES OF METAMORPHIC MINERAL ASSEMBLAGES PRODUCED DURING PROGRADE METAMORPHISM IN CARBONATE-RICH ROCKS AND ARGILLACEOUS SHALE	Ca carbonate-rich rock (see Fig. 11.16)	Argillaceous shale
Very low grade	calcite-dolomite-talc and calcite-quartz-talc	muscovite-chlorite-quartz-feldspar
Low grade	calcite-dolomite-tremolite and calcite-tremolite-quartz	biotite-muscovite-chlorite-quartz-feldspar
Medium grade	calcite-dolomite-diopside and calcite-diopside-quartz	saturolite-garnet-biotite-muscovite-quartz-feldspar
High grade	calcite-dolomite-forsterite and calcite-diopside-quartz	sillimanite-garnet-biotite-muscovite-quartz-feldspar

minerals. Because the metamorphism of many igneous or sedimentary rocks may result in a gneiss, there are many varieties, with varied mineral associations. Thus they are given such names as *plagioclase-biotite gneiss, hornblende gneiss* or *pyroxene-garnet gneiss*. When it is certain that a gneiss is the result of metamorphism of an earlier formed igneous rock the igneous rock name is used in the metamorphic terminology such as *granite gneiss* or *syenite gneiss*. Granite gneisses, that is rocks derived from the metamorphism of granites, are common, especially in Archean terranes.

Quartzite. As its name indicates, a quartzite is a rock composed essentially of quartz. It has been derived from a sandstone by high-grade metamorphism. It is a common and widely distributed rock in which solution and redeposition of silica have yielded a compact rock of interlocking quartz grains. It is distinguished from a sandstone by noting the fracture, which in a quartzite passes through the grains but in a sandstone passes around them.

Serpentinite. A serpentinite is a rock composed essentially of the mineral serpentine, derived by retrograde metamorphism from high-temperature igneous rocks such as dunite or peridotite. The primary forsteritic olivine gives way, in a lower temperature (retrograde) and hydrous metamorphic environment, to serpentine and brucite according to:

$$4Mg_2SiO_4 + H_2O \rightarrow Mg_6Si_4O_{10}(OH)_8 +$$

forsterite   serpentine

$$2Mg(OH)_2$$

brucite

Serpentinites are compact, of a green to greenish yellow color, and may have a slightly greasy feel. Serpentinites may be a source for associated chromite and platinum, as in the Ural Mountains, or a source of nickel from associated garnierite, as in New Caledonia.

VEINS AND VEIN MINERALS

Although ore and mineral deposits can be of igneous, metamorphic, or sedimentary origin a large number of mineral deposits exist as tabular or lentic-

ular bodies known as *veins*. The veins have been formed by the filling with minerals of a preexisting fracture or fissure.

The shape and general physical character of a vein depends on the type of fissure in which its minerals have been deposited, and the type of fissure in turn depends on the character of the rock in which it occurs. In a firm, homogeneous rock, like a granite, fissures may be fairly regular and clean-cut. They are likely to be comparatively narrow with respect to their horizontal and vertical extent and reasonably straight in their course. On the other hand, if a rock that is easily fractured and splintered, like a slate or a schist, is subjected to a breaking stress, a zone of narrow and interlacing fissures is more likely to have formed than one straight crack. In an easily soluble rock like a limestone, a fissure will often be extremely irregular owing to the differential solution of its walls by the solutions that have moved through it.

A typical vein consists of minerals that have filled a fissure solidly from wall to wall and shows sharply defined boundaries. There are, however, many variations from this type. Frequently irregular openings termed *vugs* may occur along the center of the vein. It is from these vugs that many well-crystallized mineral specimens are obtained. Again, the walls of a vein may not be sharply defined. The mineralizing solutions that filled the fissure may have acted on the wall rocks and partially replaced them with the vein minerals. Consequently, there may be an almost complete gradation from the unaltered rock to the vein filling with no sharp line of division between. Some deposits have been largely formed by the deposition of vein minerals in the surrounding rocks (wall rocks) and are known as *replacement deposits*. There is every gradation possible, from a vein formed by filling of an open fissure with sharply defined walls to a replacement deposit with indefinite boundaries.

Vein mineralization is generally the result of what is known as deposition from *hydrothermal solutions*. This term refers to heated or hot magmatic emanations rich in water as well as heated aqueous solutions that have no demonstrable magmatic affiliations. An example of relatively high-temperature vein mineralization in and adjacent to granitic intrusions occurs in Cornwall, England. The mineral assemblage in such veins consists of quartz-mica-tourmaline-wolframite-stannite-molybdenite. In the Tintic district, Utah, veins peripheral to granitic intrusions in limestone consist of pyrite, enargite, tetrahedrite, and galena. In many mining districts, however, there is no apparent relationship between the ore veins and possible igneous activity.

Associated with the economically useful minerals (*ore minerals*) are minerals of no commercial value (*gangue*). Careful assemblage studies, involving both ore and gangue mineralizations, together with the study of very small inclusions of remaining hydrothermal fluid (*fluid inclusions*) in mineral grains allow for the division of hydrothermal ore deposits in terms of temperature of origin: low (50–150°C), intermediate (150–400°C) and high temperature (400–600°C).

The *primary* or *hypogene* ore minerals may alter near the surface to *secondary* or *supergene* minerals. The primary sulfide minerals such as pyrite, chalcopyrite, galena, and sphalerite are particularly subject to such alteration. Under the influences of low-temperature, oxygenated meteoric waters, sphalerite alters to hemimorphite and smithsonite; galena to anglesite and cerussite; copper sul-

fides such as chalcopyrite and bornite to covellite, chalcocite, native copper, cuprite, malachite, and azurite. The circulating meteoric waters dissolve primary minerals at near-surface conditions producing a barren and leached zone (see Fig. 11.18) and frequently redeposit part of the soluble species below and close to the groundwater table producing *secondary enrichment zones*. This process of *supergene enrichment* is especially important in ore formation because metals leached from the oxidized upper parts of mineral deposits may be redeposited at depth. As all such chemical processes take place at essentially atmospheric conditions

FIG. 11.19. Stability relations of some copper minerals commonly found in supergene enrichment deposits, in the system Cu-H_2O-O_2-S-CO_2 at 25°C and 1 atmosphere total pressure (adapted from J. Anderson in Garrels and Christ, 1965, *Solutions, Minerals, and Equilibria*, Freeman, Cooper and Company, Fig. 7.27b; see same book for construction of such diagrams).

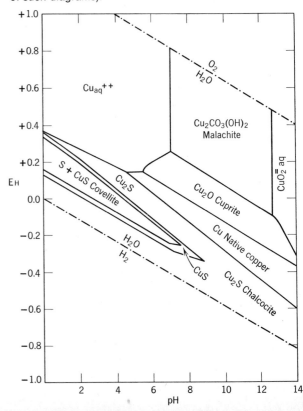

FIG. 11.18. Diagrammatic representation of weathering and enrichment of a vein rich in primary Cu sulfides.

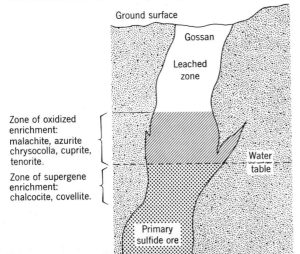

(about 25°C and one atmosphere pressure) Eh-pH stability diagrams for appropriate chemical systems can be applied to the assemblages in supergene deposits. Figure 11.19 shows the stability fields of malachite, cuprite, native copper, chalcocite, and covellite. Malachite and cuprite stability fields occupy the high Eh (highly oxygenated) part of the diagram; covellite and chalcocite are found in more reducing conditions, directly below the water table.

Because pyrite is generally a very abundant sulfide in veins as well as other mineral deposits, it forms upon oxidation insoluble compounds such as goethite or limonite which remain at the surface. The upper portion of a vein is thus frequently converted to a cellular rusty mass of goethite called *gossan* (see Fig. 11.18). The yellow-brown, rusty appearance of such outcrops is so characteristic as to be a guide to ore.

References and Suggested Reading

Blatt, H., Middleton, G., and Murray, R., 1972, *Origin of Sedimentary Rocks.* Prentice-Hall, Inc., Englewood Cliffs, N.J., 634 pp.

Carmichael, I. S. E., Turner, F. J., and Verhoogen, J., 1974, *Igneous Petrology.* McGraw-Hill Book Co., New York, 739 pp.

Garrels, R. M., and Christ, C. L., 1965, *Solutions, Minerals, and Equilibria.* Freeman, Cooper and Co., San Francisco. 450 pp.

Garrels, R. M., and MacKenzie, F. T., 1971, *Evolution of Sedimentary Rocks.* W. H. Norton and Co., New York, 397 pp.

Hyndman, D. W., 1972, *Petrology of Igneous and Metamorphic Rocks.* McGraw Hill Book Co., New York, 533 pp.

Jahns, R. H., 1952. The study of pegmatites, *Economic Geology,* 50th Ann. Vol., pp. 1025–1130.

Park, C. F., Jr. and MacDiarmid, R. A., 1975, *Ore Deposits,* 3rd ed. W. H. Freeman and Co., San Francisco, 530 pp.

Pettijohn, F. J., 1975, *Sedimentary Rocks,* 3rd ed. Harper and Row, Publishers, New York, 628 pp.

Stewart, F. H., 1963, *Marine Evaporites,* U.S. Geol. Survey, Prof. Paper 440-Y, 53 pp.

Turner, F. J., 1968, *Metamorphic Petrology, Mineralogical and Field Aspects.* McGraw-Hill Book Co., New York, 403 pp.

Winkler, H. G. F., 1974, *Petrogenesis of Metamorphic Rocks.* Springer-Verlag, New York, 320 pp.

12
DETERMINATIVE
TABLES

The determinative tables given in this chapter are based on the physical properties of minerals that can be easily and quickly determined. They should be used with the understanding that for many minerals there is a variation in physical properties from specimen to specimen. Color for some minerals is constant but for others is extremely variable. Hardness, although more definite, may vary slightly and, by change in the state of aggregation of a mineral, may appear to vary more widely. Cleavage may also be obscured in a fine grained aggregate of a mineral. In using the tables it is often impossible to differentiate between two or three similar species. However, by consulting the descriptions of these possible minerals in the sections on Systematic Mineralogy and the specific tests given there, a definite decision can usually be made.

In the determinative tables only the common minerals or those which, though rarer, are of economic importance have been included. The chances of having to determine a mineral that is not included in these tables are small, but it must be borne in mind

that there is such a possibility. The names of the minerals have been printed in three different styles of type, as **PYRITE**, CHALCOCITE, and Covellite, in order to indicate their relative importance and frequency of occurrence. Whenever there is ambiguity in placing a mineral, it has been included in the two or more possible divisions.

The general scheme of classification is outlined on page 477. The proper division in which to look for a mineral can be determined by means of the tests indicated there. The tables are divided into two main sections on the basis of luster: (1) metallic and submetallic and (2) nonmetallic. The metallic minerals are opaque and give black or dark-colored streaks; whereas, nonmetallic minerals are non-opaque and give either colorless or light-colored streaks. The tables are next subdivided according to hardness. These divisions are easily determined. For nonmetallic minerals they are: $<2\frac{1}{2}$, can be scratched by the fingernail; $>2\frac{1}{2}-<3$, cannot be scratched by fingernail, can be scratched by copper cent; $>3-<5\frac{1}{2}$, cannot be scratched by copper cent, can be scratched by a knife; $>5\frac{1}{2}-<7$, cannot be scratched by a knife, can be scratched by quartz; >7, cannot be scratched by quartz. Metallic minerals have fewer divisions of hardness; they are $<2\frac{1}{2}$; $>2\frac{1}{2}-<5\frac{1}{2}$; $>5\frac{1}{2}$.

Nonmetallic minerals are further subdivided according to whether they show a *prominent* cleavage. If the cleavage is not prominent and imperfect or obscure, the mineral is included with those that have no cleavage. Minerals in which cleavage may not be observed easily, because of certain conditions in the state of aggregation, have been included in both divisions.

The minerals that fall in a given division of the tables have been arranged according to various methods. In some cases, those that possess similar cleavages have been grouped together; frequently color determines the order, etc. The column farthest to the left will indicate the method of arrangement.

The figures given in the column headed **G** are specific gravity. This is an important diagnostic physical property but less easily determined than luster, hardness, color, and cleavage. For a discussion of specific gravity and methods for its accurate determination, see page 186. However, if a specimen is pure and of sufficient size, its approximate specific gravity can be determined by simply weighing it in the hand. Below is a list of common minerals over a wide range of specific gravity. By experimenting with specimens of these one can become expert in the approximate determination of specific gravity. Following the determinative tables is a list of the common minerals arranged according to increasing specific gravity.

Gypsum,	2.32	Topaz,	3.53	Cassiterite,	6.95
Orthoclase,	2.57	Corundum,	4.02	Galena,	7.50
Quartz,	2.65	Barite,	4.45	Cinnabar,	8.10
Calcite,	2.71	Pyrite,	5.02	Copper,	8.9
Fluorite,	3.18	Arsenopyrite,	6.07	Silver,	10.5

Many minerals are polymorphic. The crystal system given in the tables is that of the most commonly observed forms.

DETERMINATIVE TABLES

General Classification of the Tables

Luster—Metallic or Submetallic

I. Hardness: $<2\frac{1}{2}$. (Will leave a mark on paper.) Page 478

II. Hardness: $>2\frac{1}{2}$, $<5\frac{1}{2}$. (Can be scratched by knife; will not readily leave a mark on paper.) Page 479

III. Hardness: $>5\frac{1}{2}$. (Cannot be scratched by knife.) Page 483

Luster—Nonmetallic

I. Streak definitely colored. Page 485

II. Streak colorless.

  A. Hardness: $<2\frac{1}{2}$. (Can be scratched by fingernail.) Page 488

  B. Hardness: $>2\frac{1}{2}$, <3. (Cannot be scratched by fingernail; can be scratched by cent.)

    1. Cleavage prominent. Page 490

    2. Cleavage not prominent.

      a. A small splinter is fusible in the candle flame. Page 491

      b. Infusible in candle flame. Page 492

  C. Hardness: >3, $<5\frac{1}{2}$. (Cannot be scratched by cent; can be scratched by knife.)

    1. Cleavage prominent. Page 493

    2. Cleavage not prominent. Page 497

  D. Hardness: $>5\frac{1}{2}$, <7. (Cannot be scratched by knife; can be scratched by quartz.)

    1. Cleavage prominent. Page 500

    2. Cleavage not prominent. Page 503

  E. Hardness: >7. (Cannot be scratched by quartz.)

    1. Cleavage prominent. Page 506

    2. Cleavage not prominent. Page 507

Streak	Color	G.	H.	Remarks	Name, Composition, Crystal System
Black	Iron-black	4.7	1–2	Usually splintery or in radiating fibrous aggregates.	**PYROLUSITE** p. 273 MnO_2 Tetragonal
	Steel-gray to iron-black	2.23	$1–1\frac{1}{2}$	Cleavage perfect {0001}. May be in hexagonal-shaped plates. Greasy feel.	**GRAPHITE** p. 234 C Hexagonal
Black to greenish black	Blue-black	4.7	$1–1\frac{1}{2}$	Cleavage perfect {0001}. May be in hexagonal-shaped leaves. Greenish streak on glazed porcelain (graphite, black). Greasy feel.	MOLYBDENITE p. 253 MoS_2 Hexagonal
Gray-black	Blue-black to lead-gray	7.6	$2\frac{1}{2}$	Cleavage perfect cubic {100}. In cubic crystals. Massive granular.	**GALENA** p. 240 PbS Isometric
	Blue-black	4.5	2	Cleavage perfect {010}. Bladed with cross striations. Fuses in candle flame.	**STIBNITE** p. 249 Sb_2S_3 Orthorhombic
Bright red	Red to vermilion	8.1	$2–2\frac{1}{2}$	Cleavage perfect {10$\bar{1}$0}. Luster adamantine. Usually granular massive.	**CINNABAR** p. 247 HgS Rhombohedral
Red-brown	Red to vermilion	5.2	1+	Earthy. Frequently as pigment in rocks. Crystalline hematite is harder and black.	**HEMATITE** p. 269 Fe_2O_3 Rhombohedral
Black. May mark paper	Gray-black	7.3	$2–2\frac{1}{2}$	Usually massive or earthy. Easily sectile. Bright steel-gray on fresh surfaces; darkens on exposure.	ACANTHITE p. 236 Ag_2S Morph.: isometric
	Blue; may tarnish to blue-black	4.6	$1\frac{1}{2}–2$	Usually in platy masses or in thin 6-sided platy crystals. Moistened with water turns purple.	COVELLITE p. 247 CuS Hexagonal

Streak	Color	G.	H.	Remarks	Name, Composition, Crystal System
Black	Iron-black	4.7	1–2	Usually splintery or in radiating fibrous aggregates.	**PYROLUSITE** p. 273 MnO_2 Tetragonal
	Gray-black	4.4	3	Cleavage {110}. Usually in bladed masses showing cleavage. Associated with other copper minerals.	**ENARGITE** p. 259 Cu_3AsS_4 Orthorhombic
		5.5 to 6.0	2–3	Fuses easily in candle flame. Characteristically in fibrous, featherlike masses.	Jamesonite p. 260 $Pb_4FeSb_6S_{14}$ Monoclinic
		5.8 to 5.9	$2\frac{1}{2}$–3	Fuses easily in candle flame. In stout prismatic crystals; characteristically twinned with re-entrant angles.	Bournonite p. 259 $PbCuSbS_3$ Orthorhombic
	Pale copper-red. May be silver-white, pinkish	7.8	5–$5\frac{1}{2}$	Usually massive. May be coated with green nickel bloom. Associated with cobalt and nickel minerals.	NICCOLITE p. 245 NiAs Hexagonal
	Fresh surface brownish bronze; purple tarnish	5.1	3	Usually massive. Associated with other copper minerals; chiefly chalcocite and chalcopyrite.	**BORNITE** p. 239 Cu_5FeS_4 Morph.: isometric
	Brownish bronze	4.6	4	Small fragments magnetic. Usually massive. Often associated with chalcopyrite and pyrite.	**PYRRHOTITE** p. 245 $Fe_{1-x}S$ Morph.: hexagonal
		4.6 to 5.0	$3\frac{1}{2}$–4	Octahedral parting. Resembles pyrrhotite with which it usually is associated but nonmagnetic	Pentlandite p. 246 $(Fe, Ni)_9S_8$ Isometric
	Brass-yellow	4.1 to 4.3	$3\frac{1}{2}$–4	Usually massive, but may be in crystals resembling tetrahedrons. Associated with other copper minerals and pyrite.	**CHALCOPYRITE** p. 243 $CuFeS_2$ Tetragonal
	Brass-yellow. Slender crystals greenish	5.5	3–$3\frac{1}{2}$	Cleavage {10$\bar{1}$1}, rarely seen. Usually in radiating groups of hairlike crystals.	MILLERITE p. 246 NiS Rhombohedral

Streak	Color	G.	H.	Remarks	Name, Composition, Crystal System
Black	Blue; may tarnish to blue black	4.6	$1\frac{1}{2}$–2	Usually in platy masses or in thin 6-sided platy crystals. Moistened with water turns purple.	Covellite p. 247 CuS Hexagonal
Gray-black. Will mark paper	Lead-gray	4.6	2	Cleavage perfect {010}. Fuses easily in candle flame. In bladed crystal aggregates with cross striations.	**STIBNITE** p. 249 Sb_2S_3 Orthorhombic
		7.5	$2\frac{1}{2}$	Cleavage perfect {100}. In cubic crystals and granular masses. If held in the candle flame does not fuse but small globules of metallic lead collect on the surface.	**GALENA** p. 240 PbS Isometric
Black. May mark paper	Steel-gray on fresh surface. Tarnishes to dull gray	7.3	2–$2\frac{1}{2}$	Easily sectile. Usually massive or earthy. Rarely in cubic crystals.	Acanthite p. 236 Ag_2S Morph.: isometric
Usually black. May be brownish	Black	3.7 to 4.7	5–6	Massive botryoidal and stalactitic. Usually associated with pyrolusite.	PSILOMELANE p. 285 $(Ba, Mn)_3(O, OH)_6Mn_8O_{16}$ Appears amorphous
Black; may have brown tinge	Steel-gray May tarnish to dead black on exposure	4.7 to 5.0	3–$4\frac{1}{2}$	Massive or in tetrahedral crystals. Often associated with silver ores.	TETRAHEDRITE p. 258 $(Cu, Fe, Zn, Ag)_{12}Sb_4S_{13}$ Isometric
Gray-black		5.7	$2\frac{1}{2}$–3	Somewhat sectile. Usually compact massive. Associated with other copper minerals.	CHALCOCITE p. 237 Cu_2S Morph.: orthorhombic
Gray-black	Tin-white; tarnishes to dark gray	5.7	$3\frac{1}{2}$	Usually occurs in fibrous botryoidal masses. Heated in the candle flame gives off white fumes and yields a strong garlic odor.	Arsenic p. 227 As Rhombohedral
	Tin-white	8 to 8.2	2	Cleavage {010}. Fuses easily in candle flame. Often as thin coatings and in lath-shaped crystals.	Sylvanite p. 256 $(Au, Ag)Te_2$ Monoclinic

LUSTER: METALLIC OR SUBMETALLIC II. Hardness: $>2\frac{1}{2}$, $<5\frac{1}{2}$

(Can be scratched by a knife; will not readily leave a mark on paper.) (Continued)

Streak	Color	G.	H.	Remarks	Name, Composition, Crystal System
Gray-black	Tin-white to brass yellow	9.4	$2\frac{1}{2}$	Fuses easily in candle flame. In irregular masses or in thin deeply striated lath-shaped crystals. Told from sylvanite by its lack of cleavage.	Calaverite p. 255 $AuTe_2$ Monoclinic
Dark brown to black	Iron-black to brownish black	4.6	$5\frac{1}{2}$	Luster pitchy. May be accompanied by yellow or green oxidation products. Usually in masses in peridotites.	**CHROMITE** p. 279 $FeCr_2O_4$ Isometric
	Brown to black	7.0 to 7.5	5–$5\frac{1}{2}$	Cleavage perfect {010}. With greater amounts of Mn streak and color are darker.	WOLFRAMITE p. 324 $(Fe, Mn)WO_4$ Monoclinic
Dark brown to black	Steel-gray to iron-black	4.3	4	In radiating fibrous or crystalline masses. Distinct prismatic crystals often grouped in bundles. Associated with pyrolusite.	MANGANITE p. 284 $MnO(OH)$ Morph.: orthorhombic
Light to dark brown	Dark brown to coal-black. More rarely yellow or red	3.9 to 4.1	$3\frac{1}{2}$–4	Cleavage perfect {110} (6 directions). Usually cleavable granular; may be in tetrahedral crystals. The streak is always of a lighter color than the specimen.	**SPHALERITE** p. 242 ZnS Isometric
Red brown to Indian red	Dark brown to steel-gray to black	4.8 to 5.3	$5\frac{1}{2}$–$6\frac{1}{2}$	Usually harder than knife. Massive, radiating, reniform, micaceous.	**HEMATITE** p. 269 Fe_2O_3 Rhombohedral
	Deep red to black	5.85	$2\frac{1}{2}$	Cleavage {10$\bar{1}$1}. Fusible in candle flame. Shows dark ruby-red color in thin splinters. Associated with other silver minerals.	Pyrargyrite p. 257 Ag_3SbS_3 Rhombohedral
	Red-brown to deep red. Ruby-red if transparent	6.0	$3\frac{1}{2}$–4	Massive or in cubes or octahedrons. May be in very slender crystals. Associated with malachite, azurite, native copper.	CUPRITE p. 266 Cu_2O Isometric

LUSTER: METALLIC OR SUBMETALLIC
II. Hardness: $>2\frac{1}{2}$, $<5\frac{1}{2}$

(Can be scratched by a knife; will not readily leave a mark on paper.) (*Continued*)

Streak	Color	G.	H.	Remarks	Name, Composition, Crystal System
Bright red	Ruby-red	5.55	$2-2\frac{1}{2}$	Cleavage $\{10\bar{1}1\}$. Fusible in candle flame. Associated with pyrargyrite.	Proustite p. 257 Ag_3AsS_3 Rhombohedral
Yellow brown. Yellow ocher	Dark brown to black	4.37	$5-5\frac{1}{2}$	Cleavage $\{010\}$. In radiating fibers, mammillary and stalactitic forms. Rarely in crystals.	**GOETHITE** p. 286 $\alpha FeO\cdot OH$ Orthorhombic
Dark red. (Some varieties mark paper.)	Dark red to vermilion	8.10	$2\frac{1}{2}$	Cleavage $\{10\bar{1}1\}$. Usually granular or earthy. Commonly impure and dark red or brown. When pure, translucent or transparent and bright red.	**CINNABAR** p. 247 HgS Rhombohedral
Copper-red, shiny	Copper-red on fresh surface; black tarnish	8.9	$2\frac{1}{2}-3$	Malleable. Usually in irregular grains. May be in branching crystal group or in rude isometric crystals.	COPPER p. 223 Cu Isometric
Silver-white, shiny	Silver-white on fresh surface. Black tarnish	10.5	$2\frac{1}{2}-3$	Malleable. Usually in irregular grains. May be in wire, plates, branching crystal groups.	SILVER p. 222 Ag Isometric
Gray, shiny	White or steel-gray	14 to 19	$4-4\frac{1}{2}$	Malleable. Irregular grains or nuggets. Unusually hard for a metal. Rare.	Platinum p. 224 Pt Isometric
Silver-white, shiny	Silver-white with reddish tone	9.8	$2-2\frac{1}{2}$	Cleavage perfect $\{0001\}$. Sectile. Fusible in candle flame. When hammered, at first malleable but soon breaks into small pieces.	Bismuth p. 227 Bi Rhombohedral
Gold-yellow, shiny	Gold-yellow	15.0 to 19.3	$2\frac{1}{2}-3$	Malleable. Irregular grains, nuggets, leaves. Very heavy; specific gravity varies with silver content.	GOLD p. 221 Au Isometric

LUSTER: METALLIC OR SUBMETALLIC III. Hardness: >$5\frac{1}{2}$ (Cannot be scratched by a knife.)	Streak	Color	G.	H.	Remarks	Name, Composition, Crystal System
	Black	Silver- or tin-white	6.0 to 6.2	$5\frac{1}{2}$–6	Usually massive. Crystals pseudo-orthorhombic.	**ARSENOPYRITE** p. 254 FeAsS Monoclinic
			6.1 to 6.9	$5\frac{1}{2}$–6	Usually massive. Crystals pyritohedral. May be coated with pink nickel bloom.	Skutterudite-Nickel skutterudite p. 255 $(Co, Ni, Fe)As_3$- $(Ni, Co, Fe)As_3$ Isometric
			6.33	$5\frac{1}{2}$	Commonly in pyritohedral crystals with pinkish cast. Also massive.	Cobaltite-Gersdorffite p. 253 $(Co, Fe)AsS$-NiAsS Pseudoisometric
		Copper-red to silver-white with pink tone	7.5	5–$5\frac{1}{2}$	Usually massive. May be coated with green nickel bloom.	NICCOLITE p. 245 NiAs Hexagonal
		Pale brass-yellow	5.0	6–$6\frac{1}{2}$	Often in pyritohedrons or striated cubes. Massive granular. Most common sulfide.	**PYRITE** p. 251 FeS_2 Isometric
		Pale yellow to almost white	4.9	6–$6\frac{1}{2}$	Frequently in "cock's comb" crystal groups and radiating fibrous masses.	**MARCASITE** p. 252 FeS_2 Orthorhombic
		Black	5.18	6	Strongly magnetic. Crystals octahedral. May show octahedral parting.	**MAGNETITE** p. 278 Fe_3O_4 Isometric
	Dark brown to black	Black	9.0 to 9.7	$5\frac{1}{2}$	Luster pitchy. Massive granular, botryoidal crystals.	Uraninite p. 275 UO_2 Isometric
			4.7	$5\frac{1}{2}$–6	May be slightly magnetic. Often associated with magnetite. Massive granular; platy crystals; as sand.	**ILMENITE** p. 271 $FeTiO_3$ Rhombohedral
			3.7 to 4.7	5–6	Compact massive, stalactitic, botryoidal. Associated with other manganese minerals and told by greater hardness.	PSILOMELANE p. 285 $(Ba, Mn)_3(O, OH)_6Mn_8O_{16}$ Orthorhombic

	Streak	Color	G.	H.	Remarks	Name, Composition, Crystal System
LUSTER: METALLIC OR SUBMETALLIC III. Hardness: $>5\frac{1}{2}$ (Cannot be scratched by a knife.) (Continued)	Dark brown to black	Black	5.3 to 7.3	6	Luster black and shiny on fresh surface. May have slight bluish tarnish. Granular or in stout prismatic crystals.	Columbite-Tantalite p. 281 $(Fe, Mn)(Nb, Ta)_2O_6$ Orthorhombic
	Dark brown	Iron-brown to brownish-black	7.0 to 7.5	$5-5\frac{1}{2}$	Cleavage perfect {010}. With greater amounts of Mn streak and color are darker.	WOLFRAMITE p. 324 $(Fe, Mn)WO_4$ Monoclinic
			4.6	$5\frac{1}{2}$	Luster pitchy. Frequently accompanied by green oxidation products. Usually in granular masses in peridotites.	**CHROMITE** p. 279 $FeCr_2O_4$ Isometric
			5.15	6	Slightly magnetic. Granular or in octahedral crystals. Common only at Franklin, N. J., associated with zincite and willemite.	FRANKLINITE p. 279 $(Fe, Zn, Mn)(Fe, Mn)_2O_4$ Isometric
	Red-brown, Indian red	Dark brown to steel-gray to black	4.8 to 5.3	$5\frac{1}{2}-6\frac{1}{2}$	Radiating, reniform, massive, micaceous. Rarely in steel-black rhombohedral crystals. Some varieties softer.	**HEMATITE** p. 269 Fe_2O_3 Rhombohedral
	Pale brown	Brown to black	4.18 to 4.25	$6-6\frac{1}{2}$	In prismatic crystals vertically striated; often slender acicular. Crystals frequently twinned. Found in black sands.	**RUTILE** p. 272 TiO_2 Tetragonal
	Yellow-brown to yellow ocher	Dark brown to black	4.37	$5-5\frac{1}{2}$	Cleavage {010}. Radiating, colloform, stalactitic.	**GOETHITE** p. 286 $\alpha FeO \cdot OH$ Orthorhombic

Streak	Color	G.	H.	Remarks	Name, Composition, Crystal System
Dark red	Dark red to vermilion	8.10	$2\frac{1}{2}$	Cleavage $\{10\bar{1}0\}$. Usually granular or earthy. Commonly impure and dark red or brown. When pure, translucent or transparent and bright red.	**CINNABAR** p. 247 HgS Rhombohedral
	Red-brown. Ruby-red when transparent	6.0	$3\frac{1}{2}$–4	Massive or in cubes or octahedrons. May be in very slender crystals. Associated with malachite, azurite, native copper.	CUPRITE p. 266 Cu_2O Isometric
Red-brown, Indian brown	Dark brown to steel-gray to black	4.8 to 5.3	$5\frac{1}{2}$–$6\frac{1}{2}$	Radiating, reniform, massive, micaceous. Rarely in steel-black rhombohedral crystals. Some varieties softer.	**HEMATITE** p. 269 Fe_2O_3 Rhombohedral
	Deep red to black	5.8	$2\frac{1}{2}$	Cleavage $\{10\bar{1}1\}$. Fusible in candle flame. Shows dark ruby-red color in thin splinters. Associated with other silver minerals.	Pyrargyrite p. 257 Ag_3SbS_3 Rhombohedral
Bright red	Ruby-red	5.55	2–$2\frac{1}{2}$	Cleavage $\{10\bar{1}1\}$. Fusible in candle flame. Light "ruby silver." Associated with pyrargyrite.	Proustite p. 257 Ag_3AsS_3 Rhombohedral
Pink	Red to pink	2.95	$1\frac{1}{2}$–$2\frac{1}{2}$	Cleavage perfect $\{010\}$. Usually reniform or as pulverulent or earthy crusts. Found as coatings on cobalt minerals.	Erythrite (cobalt bloom) p. 331 $Co_3(AsO_4)_2\cdot8H_2O$ Monoclinic
Yellow-brown to yellow-ocher	Dark brown to black	4.4	5–$5\frac{1}{2}$	Cleavage $\{010\}$. In radiating fibers, mammillary and stalactitic forms. Rarely in crystals.	**GOETHITE** p. 286 $\alpha FeO\cdot OH$ Orthorhombic
Brown	Dark brown	7.0 to 7.5	5–$5\frac{1}{2}$	Cleavage perfect $\{010\}$. With greater amounts of manganese the streak and color are darker.	WOLFRAMITE p. 324 $(Fe, Mn)WO_4$ Monoclinic
	Light to dark brown	3.83 to 3.88	$3\frac{1}{2}$–4	In cleavable masses or in small curved rhombohedral crystals. Becomes magnetic after heating in candle flame.	**SIDERITE** p. 301 $FeCO_3$ Rhombohedral

Streak	Color	G.	H.	Remarks	Name, Composition, Crystal System
Light brown	Light to dark brown	3.9 to 4.1	$3\frac{1}{2}$–4	Cleavage perfect {011} (6 directions). Usually cleavable granular; may be in tetrahedral crystals. The streak is always of a lighter color than the specimen.	**SPHALERITE** p. 242 ZnS Isometric
	Brown to black	6.8 to 7.1	6–7	Occurs in twinned crystals. Fibrous, reniform and irregular masses; in rolled grains.	**CASSITERITE** p. 274 SnO_2 Tetragonal
	Reddish brown to black	4.18 to 4.25	6–$6\frac{1}{2}$	Crystals vertically striated; often acicular. Twinning common.	**RUTILE** p. 272 TiO_2 Tetragonal
Orange-yellow	Deep red to orange-yellow	5.68	4–$4\frac{1}{2}$	Cleavage {0001}. Found only at Franklin, N.J., associated with franklinite and willemite.	ZINCITE p. 266 ZnO Hexagonal
	Bright red	5.9 to 6.1	$2\frac{1}{2}$–3	Luster adamantine. In long slender crystals, often interlacing groups. Decrepitates in candle flame.	Crocoite p. 320 $PbCrO_4$ Monoclinic
	Deep red	3.48	$1\frac{1}{2}$–2	Frequently earthy. Associated with orpiment. Fusible in candle flame.	REALGAR p. 248 AsS Monoclinic
Pale yellow	Lemon-yellow	3.49	$1\frac{1}{2}$–2	Cleavage {010}. Luster resinous. Associated with realgar. Fusible in candle flame.	ORPIMENT p. 249 As_2S_3 Monoclinic
	Pale yellow	2.05 to 2.09	$1\frac{1}{2}$–$2\frac{1}{2}$	Burns with blue flame giving odor of SO_2. Crystallized, granular, earthy.	**SULFUR** p. 230 S Orthorhombic
Light green	Dark emerald-green	3.75 to 3.77	3–$3\frac{1}{2}$	One perfect cleavage {010}. In granular cleavable masses or small prismatic crystals.	Atacamite p. 293 $Cu_2Cl(OH)_3$ Orthorhombic

LUSTER: NONMETALLIC I. Streak definitely colored. (*Continued*)	Streak	Color	G.	H.	Remarks	Name, Composition, Crystal System
	Light green	Dark emerald-green	3.9±	3½–4	One good cleavage {010}. In small prismatic crystals or granular masses.	Antlerite p. 322 $Cu_3(SO_4)(OH)_4$ Orthorhombic
		Bright green	3.9 to 4.03	3½–4	Radiating fibrous, mammillary. Associated with azurite and may alter to it. Effervesces in cold acid.	**MALACHITE** p. 309 $Cu_2CO_3(OH)_2$ Monoclinic
	Light blue	Intense azure-blue	3.77	3½–4	In small crystals, often in groups. Radiating fibrous, usually as alteration from malachite. Effervesces in cold acid.	**AZURITE** p. 310 $Cu_3(CO_3)_2(OH)_2$ Monoclinic
	Very light blue	Light green to turquoise blue	2.0 to 2.4	2–4	Massive compact. Associated with oxidized copper minerals.	CHRYSOCOLLA p. 410 $\sim Cu_4H_4Si_4O_{10}(OH)_8$ Amorphous

LUSTER:
NONMETALLIC
II. Streak colorless

A. Hardness: $<2\frac{1}{2}$.
(Can be scratched
by fingernail.)

Cleavage, Fracture	Color	G.	H.	Remarks	Name, Composition, Crystal System
Pale brown, green, yellow, white	2.76 to 2.88	$2-2\frac{1}{2}$	In foliated masses and scales. Crystals tabular with hexagonal or diamond-shaped outline. Cleavage flakes elastic.	**MUSCOVITE** p. 405 $KAl_2(AlSi_3O_{10})(OH)_2$ Monoclinic
	Dark brown, green to black; may be yellow	2.95 to 3	$2\frac{1}{2}-3$	Usually in irregular foliated masses. Crystals have hexagonal outline, but rare. Cleavage flakes elastic.	**BIOTITE** p. 406 $K(Mg, Fe)_3(AlSi_3O_{10})(OH)_2$ Monoclinic
	Yellowish brown, green, white	2.86	$2\frac{1}{2}-3$	Often in 6-sided tabular crystals; in irregular foliated masses. May show copperlike reflection from cleavage.	PHLOGOPITE p. 406 $KMg_3(AlSi_3O_{10})(OH)_2$ Monoclinic
	Green of various shades	2.6 to 2.9	$2-2\frac{1}{2}$	Usually in irregular foliated masses. May be in compact masses of minute scales. Thin sheets flexible but not elastic.	**CHLORITE** p. 408 $(Mg, Fe)_3(Si, Al)_4O_{10}(OH)_2$-$(Mg, Fe)_3(OH)_6$ Monoclinic
	White, apple-green, gray. When impure, as in soapstone, dark gray, dark green to almost black	2.7 to 2.8	1	Greasy feel. Frequently distinctly foliated or micaceous. Cannot be positively identified by physical tests.	**TALC** p. 403 $Mg_3Si_4O_{10}(OH)_2$ Monoclinic
		2.8 to 2.9	1-2		Pyrophyllite p. 404 $Al_2Si_4O_{10}(OH)_2$ Monoclinic
	White, gray, green	2.39	$2\frac{1}{2}$	Pearly luster on cleavage face, elsewhere vitreous. Sectile. Commonly foliated massive, may be in broad tabular crystals. Thin sheets flexible but not elastic.	BRUCITE p. 283 $Mg(OH)_2$ Rhombohedral
{001} perfect; seldom seen. Fract. earthy	White; may be darker	2.6 to 2.63	$2-2\frac{1}{2}$	Compact, earthy. Breathed upon gives argillaceous odor. Will adhere to the dry tongue.	**KAOLINITE** p. 402 $Al_2Si_2O_5(OH)_4$ Triclinic

Perfect cleavage in one direction

The micas or related micaceous minerals, which possess such a perfect cleavage that they can be split into exceedingly thin sheets. They may occur as aggregates of minute scales, when the micaceous structure may not be readily apparent.

LUSTER:
NONMETALLIC
II. Streak colorless

A. Hardness: $<2\frac{1}{2}$
(Can be scratched
by fingernail.)
(*Continued*)

Cleavage, Fracture	Color	G.	H.	Remarks	Name, Composition, Crystal System
Cubic {100}	Colorless or white	1.99	2	Soluble in water, bitter taste. Resembles halite but softer. In granular cleavable masses or cubic crystals.	Sylvite p. 290 KCl Isometric
{010} perfect, {100} {011} good	Colorless, white, gray. May be colored by impurities	2.32	2	Occurs in crystals, broad cleavage flakes. May be compact massive without cleavage, or fibrous with silky luster.	**GYPSUM** p. 321 $CaSO_4 \cdot 2H_2O$ Monoclinic
Rhomb. {10$\bar{1}$1} poor		2.29	1–2	Occurs in saline crusts. Readily soluble in water; cooling and salty taste. Fusible in candle flame.	SODA NITER p. 311 $NaNO_3$ Rhombohedral
Prismatic {110} seldom seen. Fract. conchoidal	Colorless or white	2.09 to 2.14	2	Usually in crusts, silky tufts and delicate acicular crystals. Readily soluble in water; cooling and salty taste. Fusible in the candle flame.	Niter p. 311 KNO_3 Orthorhombic
Fract. uneven	Pearl-gray or colorless. Turns to pale brown on exposure to light	5.5±	2–3	Perfectly sectile. Translucent in thin plates. In irregular masses, rarely in crystals. Distinguished from other silver halides only by chemical tests.	Cerargyrite p. 291 AgCl Isometric
	Pale yellow	2.05 to 2.09	1$\frac{1}{2}$–2$\frac{1}{2}$	Burns with blue flame, giving odor of SO_2. Crystallized, granular, earthy.	**SULFUR** p. 230 S Orthorhombic
	Yellow, brown, gray, white	2.0 to 2.55	1–3	In rounded grains, often earthy and claylike. Usually harder than 2$\frac{1}{2}$.	**BAUXITE** p. 287 A mixture of Al hydroxides
Cleavage seldom seen	White	1.95	1	Usually in rounded masses of fine fibers and acicular crystals.	ULEXITE p. 315 $NaCaB_5O_6(OH) \cdot 5H_2O$ Triclinic

LUSTER: NONMETALLIC

II. Streak colorless

B. Hardness: $>2\frac{1}{2}$, <3. (Cannot be scratched by fingernail; can be scratched by cent.)

1. Cleavage prominent

Cleavage, Fracture	Color	G.	H.	Remarks	Name, Composition, Crystal System
{001}	Lilac, grayish white	2.8 to 3.0	$2\frac{1}{2}$–4	Crystals 6-sided prismatic. Usually in small irregular sheets and scales. A pegmatite mineral.	LEPIDOLITE p. 407 $K(Li, Al)_{2-3}(AlSi_3O_{10})(O, OH, F)_2$ Monoclinic
{001}	Pink, gray, white	3.0 to 3.1	$3\frac{1}{2}$–5	Usually in irregular foliated masses; folia brittle. Associated with emery.	MARGARITE p. 408 $CaAl_2(Al_2Si_2O_{10})(OH)_2$ Monoclinic
{010}	Colorless or white	4.3	$3\frac{1}{2}$	Usually massive, with radiating habit. Effervesces in cold acid.	**WITHERITE** p. 305 $BaCO_3$ Orthorhombic
Cleavage in 2 directions {001} {100}	Colorless or white to gray	1.95	3	Occurs in cleavable crystalline aggregates.	KERNITE p. 313 $Na_2B_4O_6(OH)_2 \cdot 3H_2O$ Monoclinic
Cubic {100}	Colorless, white, red, blue	2.1 to 2.3	$2\frac{1}{2}$	Common salt. Soluble in water, taste salty, fusible in candle flame. In granular masses or in cubic crystals.	**HALITE** p. 289 $NaCl$ Isometric
Cubic {100}	Colorless or white	1.99	2	Resembles halite, but distinguished from it by more bitter taste and lesser hardness.	Sylvite p. 290 KCl Isometric
{001} {010} {100}	Colorless, white, blue, gray, red	2.89 to 2.98	3–$3\frac{1}{2}$	Commonly in massive aggregates, not showing cleavage; then distinguished only by chemical tests.	**ANHYDRITE** p. 320 $CaSO_4$ Orthorhombic
Cleavage in 3 directions not at right angles. Rhombohedral {10$\bar{1}$1}	Colorless, white, and variously tinted	2.71	3	Effervesces in cold acid. Crystals show many forms. Occurs as limestone and marble. Clear varieties show strong double refraction.	**CALCITE** p. 297 $CaCO_3$ Rhombohedral

(Left margin notes:) Perfect cleavage in one direction. See also the minerals of the mica group, p. 404 which may be harder than the fingernail.

Cleavage in 3 directions at right angles.

II. Streak colorless

B. Hardness: $>2\frac{1}{2}$, <3. (Cannot be scratched by fingernail; can be scratched by cent.)

1. Cleavage prominent (*Continued*)

Cleavage, Fracture	Color	G.	H.	Remarks	Name, Composition, Crystal System
Cleavage in 3 directions not at right angles. Rhombohedral $\{10\bar{1}1\}$.	Colorless, white, pink	2.85	$3\frac{1}{2}$–4	Usually harder than 3. Often in curved rhombohedral crystals with pearly luster; as dolomitic limestone and marble. Powdered mineral will effervesce in cold acid.	**DOLOMITE** p. 308 $CaMg(CO_3)_2$ Rhombohedral
	Colorless, white, blue, yellow red	4.5	3–$3\frac{1}{2}$	Frequently in aggregates of platy crystals. Pearly luster on basal cleavage. Characterized by high specific gravity and thus distinguished from celestite.	**BARITE** p. 317 $BaSO_4$ Orthorhombic
Cleavage in 3 directions. Basal $\{001\}$ at right angles to prismatic $\{210\}$	Colorless, white, blue, red	3.95 to 3.97	3–$3\frac{1}{2}$	Similar to barite but lower specific gravity. A crimson strontium flame will distinguish it.	**CELESTITE** p. 318 $SrSO_4$ Orthorhombic
	Colorless or white. Gray brown when impure	6.2 to 6.4	3	Adamantine luster. Usually massive but may be in small tabular crystals. Alteration of galena. When massive may need test for SO_4 to distinguish from cerrusite $(PbCO_3)$.	**ANGLESITE** p. 319 $PbSO_4$ Orthorhombic

2. Cleavage not prominent

a. Small splinter is fusible in the candle flame.

	Color	G.	H.	Remarks	Name, Composition, Crystal System
	Colorless or white	1.7±	2–$2\frac{1}{2}$	Soluble in water. One good cleavage, seldom seen. In crusts and prismatic crystals. In candle flame swells and then fuses. Sweetish alkaline taste.	**BORAX** p. 313 $Na_2B_4O_5(OH)_4\cdot 8H_2O$ Monoclinic
		2.95 to 3.0	$2\frac{1}{2}$	Massive with peculiar translucent appearance. Fine powder becomes nearly invisible when placed in water. Ivigtut, Greenland, only important locality. Pseudocubic parting.	CRYOLITE p. 291 Na_3AlF_6 Monoclinic
		6.55	3–$3\frac{1}{2}$	Luster adamantine. In granular masses and platy crystals, usually associated with galena. Effervesces in cold nitric acid. Reduced in candle flame, producing small globules of lead.	**CERUSSITE** p. 306 $PbCO_3$ Orthorhombic

	Color	G.	H.	Remarks	Name, Composition, Crystal System
LUSTER: NONMETALLIC II. Streak colorless	Colorless or white	4.3	$3\frac{1}{2}$	Often in radiating masses; granular; rarely in pseudohexagonal crystals. Effervesces in cold acid.	**WITHERITE** p. 305 $BaCO_3$ Orthorhombic
B. Hardness: $>2\frac{1}{2}$, <3. (Cannot be scratched by fingernail; can be scratched by cent.)		2.6 to 2.63	$2-2\frac{1}{2}$	Usually compact, earthy. When breathed upon gives an argillaceous odor. The basis of most clays.	**KAOLINITE** p. 402 $Al_2Si_2O_5(OH)_4$ Triclinic
2. Cleavage not prominent. b. Infusible in candle flame.	Colorless, white, blue, gray, red	2.89 to 2.98	$3-3\frac{1}{2}$	Commonly in massive fine aggregates, not showing cleavage, and can be distinguished only by chemical tests.	**ANHYDRITE** p. 320 $CaSO_4$ Orthorhombic
	Yellow, brown, gray, white	2.0 to 2.55	$1-3$	Usually pisolitic; in rounded grains and earthy masses. Often impure.	**BAUXITE** p. 287 A mixture of aluminum hydroxides
	Ruby-red, brown, yellow	6.7 to 7.1	3	Luster resinous. In slender prismatic and cavernous crystals; in barrel-shaped forms.	Vanadinite p. 331 $Pb_5(VO_4)_3Cl$ Hexagonal
	Yellow, green, white, brown	2.33	$3\frac{1}{2}-4$	Characteristically in radiating hemispherical globular aggregates. Cleavage seldom seen.	Wavellite p. 333 $Al_3(PO_4)_2(OH)_3 \cdot 5H_2O$ Orthorhombic
	Olive- to blackish-green, yellow-green, white	2.2 to 2.65	$2-5$	Massive. Fibrous in the asbestos variety, chrysotile. Frequently mottled green in the massive variety.	**SERPENTINE** p. 401 $Mg_3Si_2O_5(OH)_4$ Monoclinic

LUSTER:
NONMETALLIC
II. Streak
colorless

C. Hardness: >3, <5½.
(Cannot be scratched
by cent; can be
scratched by knife.)

1. Cleavage prominent

Cleavage	Color	G.	H.	Remarks	Name, Composition, Crystal System
{100}	Blue, usually darker at center.	3.56 to 3.66	5–7	In bladed aggregates with cleavage parallel to length. Can be scratched by knife parallel to the length of the crystal but not in a direction at right angles to this.	**KYANITE** p. 352 Al_2SiO_5 Triclinic
{010}	White, yellow, brown, red	2.1 to 2.2	3½–4	Characteristically in sheaflike crystal aggregates. May be in flat tabular crystals. Luster pearly on cleavage face.	**STILBITE** p. 441 $CaAl_2Si_7O_{18}·7H_2O$ Monoclinic
{001}	Colorless, white, pale green, yellow, rose	2.3 to 2.4	4½–5	In prismatic crystals vertically striated. Crystals often resemble cubes truncated by the octahedron. Luster pearly on base, elsewhere vitreous.	APOPHYLLITE p. 409 $KCa_4(Si_4O_{10})_2F·8H_2O$ Tetragonal
{010}	White, yellow, red	2.18 to 2.20	3½–4	Luster pearly on cleavage, elsewhere vitreous. Crystals often tabular parallel to cleavage plane. Found in cavities in igneous rocks.	HEULANDITE p. 440 $CaAl_2Si_7O_{18}·6H_2O$ Monoclinic
{010}	Colorless, white	2.42	4–4½	In crystals and in cleavable aggregates. Decrepitates violently in the candle flame.	COLEMANITE p. 315 $CaB_3O_4(OH)_3·H_2O$ Monoclinic
{010}	Colorless, white	4.3	3½	Often in radiating crystal aggregates; granular. Rarely in pseudohexagonal crystals.	WITHERITE p. 305 $BaCO_3$ Orthorhombic
{010} {110} poor	Colorless, white	2.95	3½–4	Effervesces in cold acid. Frequently in radiating groups of acicular crystals; in pseudohexagonal twins.	**ARAGONITE** p. 304 $CaCO_3$ Orthorhombic
{001} {010}	Blueish gray, salmon to clove-brown	3.42 to 3.56	4½–5	Commonly cleavable, massive. Found in pegmatites with other lithium minerals.	Triphylite-lithiophilite p. 328 $Li(Fe, Mn)PO_4$ Orthorhombic
{001} {100}	Colorless, white, gray	2.8 to 2.9	5–5½	Usually cleavable massive to fibrous. Also compact. Associated with crystalline limestone.	WOLLASTONITE p. 382 $CaSiO_3$ Triclinic

One cleavage direction

Two cleavage directions

C. Hardness: >3, $<5\frac{1}{2}$.
(Cannot be scratched
by cent; can be
scratched by knife.)

1. Cleavage prominent
(*Continued*)

Cleavage	Color	G.	H.	Remarks	Name, Composition, Crystal System
{001} {100}	Colorless, white, gray	2.7 to 2.8	5	In radiating aggregates of sharp acicular crystals. Associated with zeolites in cavities in igneous rocks.	Pectolite p. 383 $Ca_2NaH(SiO_3)_3$ Triclinic
{110}	Colorless, white	2.25	5–5½	Slender prismatic crystals, prism faces vertically striated. Often in radiating groups. Found lining cavities in igneous rocks.	**NATROLITE** p. 439 $Na_2Al_2Si_3O_{10}\cdot2H_2O$ Orthorhombic
{110}	Colorless, white	3.7	3½–4	In prismatic crystals and pseudohexagonal twins. Also fibrous and massive. Effervesces in cold acid.	**STRONTIANITE** p. 306 $SrCO_3$ Orthorhombic
{110}	White, pale green, blue	3.4 to 3.5	4½–5	Often in radiating crystal groups. Also stalactitic, mammillary. Prismatic cleavage seldom seen.	**HEMIMORPHITE** p. 359 $Zn_4(Si_2O_7)(OH)_2\cdot H_2O$ Orthorhombic
{110}	Brown, gray, green, yellow	3.4 to 3.55	5–5½	Luster adamantine to resinous. In thin wedge-shaped crystals with sharp edges. Prismatic cleavage seldom seen.	SPHENE p. 356 $CaTiO(SiO_4)$ Monoclinic
Prismatic at angles of 56° and 124°	White, green, black	3.0 to 3.3	5–6	Crystals usually slender, fibrous, asbestiform. Tremolite (white, gray, violet), actinolite (green) common in metamorphic rocks. Hornblende and arfvedsonite (dark green to black) common in igneous and metamorphic rocks.	**AMPHIBOLE GROUP** p. 384 Essentially hydrous Ca-Mg-Fe silicates Monoclinic
	Gray, clove-brown, green	2.85 to 3.2	5½–6	An amphibole. Distinct crystals rare. Commonly in aggregates and fibrous masses	Anthophyllite p. 387 or grunerite p. 388 $(Mg, Fe)_7Si_8O_{22}(OH)_2$ Orthorhombic or monoclinic
Prismatic at nearly 90° angles	White, green, black	3.1 to 3.5	5–6	In stout prisms with rectangular cross section. Often in granular crystalline masses. Diopside (colorless, white, green), aegirine (brown, green), augite (dark green to black).	**PYROXENE GROUP** p. 371 Essentially Ca-Mg-Fe silicates Monoclinic

Two cleavage directions

**C. Hardness: >3, <5½.
(Cannot scratched by
cent; can be scratched
by knife.)**

**1. Cleavage prominent
(Continued)**

Cleavage	Color	G.	H.	Remarks	Name, Composition, Crystal System
Prismatic at nearly 90° angles	Rose-red, pink, brown	3.58 to 3.70	5½–6	Color diagnostic. Usually massive, cleavable to compact, in imbedded grains; in large rough crystals, with rounded edges.	RHODONITE p. 382 MnSiO₃ Triclinic
	Colorless, white, and variously tinted	2.72	3	Effervesces in cold acid. Crystals show many forms. Occurs as limestone and marble. Clear varieties show strong double refraction.	**CALCITE** p. 297 CaCO₃ Rhombohedral
	Colorless, white, pink	2.85	3½–4	Often in curved rhombohedral crystals with pearly luster. In coarse masses as dolomitic limestone and marble. Powdered mineral will effervesce in cold acid.	**DOLOMITE** p. 308 CaMg(CO₃)₂ Rhombohedral
	White, yellow, gray, brown	3.0 to 3.2	3½–5	Commonly in dense compact masses; also in fine to coarse cleavable masses. Effervesces in hot hydrochloric acid.	**MAGNESITE** p. 300 MgCO₃ Rhombohedral
	Light to dark brown	3.83 to 3.88	3½–4	In cleavable masses or in small curved rhombohedral crystals. Becomes magnetic after heating.	**SIDERITE** p. 301 FeCO₃ Rhombohedral
	Pink, rose-red, brown	3.45 to 3.6	3½–4½	In cleavable masses or in small rhombohedral crystals. Characterized by its color.	RHODOCROSITE p. 302 MnCO₃ Rhombohedral
	Brown, green, blue, pink, white	4.35 to 4.40	5	In rounded botryoidal aggregates and honeycombed masses. Effervesces in cold hydrochloric acid. Cleavage rarely seen.	SMITHSONITE p. 302 ZnCO₃ Rhombohedral
	White, yellow, flesh-red	2.05 to 2.15	4–5	In small rhombohedral crystals with nearly cubic angles. Found lining cavities in igneous rocks.	CHABAZITE p. 440 Ca₂Al₂Si₄O₁₂·6H₂O Rhombohedral
{001} {010} {100}	Colorless, white, blue, gray, red	2.89 to 2.98	3–3½	Commonly in massive fine aggregates, not showing cleavage, and can be distinguished only by chemical tests.	**ANHYDRITE** p. 320 CaSO₄ Orthorhombic

Three cleavage directions

Three directions not at right angles. Rhombohedral {10$\bar{1}$1}

LUSTER: NONMETALLIC
II. Streak colorless

C. Hardness: >3, <5½. (Cannot be scratched by cent; can be scratched by knife.)

1. Cleavage prominent (*Continued*)

Cleavage	Color	G.	H.	Remarks	Name, Composition, Crystal System
Three cleavage directions — {001} at rt. angles to {110}	Colorless, white, blue, yellow, red	4.5	3–3½	Frequently in aggregates of platy crystals. Pearly luster on basal cleavage. Characterized by high specific gravity and thus distinguished from celestite.	**BARITE** p. 317 $BaSO_4$ Orthorhombic
	Colorless, white, blue, red	3.95 to 3.97	3–3½	Similar to barite but lower specific gravity. A crimson strontium flame will distinguish it.	**CELESTITE** p. 318 $SrSO_4$ Orthorhombic
Four cleavage directions — {111} Octahedral	Colorless, violet, green, yellow, pink	3.18	4	In cubic crystals often in penetration twins. Characterized by cleavage.	**FLUORITE** p. 292 CaF_2 Isometric
{100} {110}	White, pink, gray, green, brown	2.65 to 2.74	5–6	In prismatic crystals, granular or massive. Commonly altered. Prismatic cleavage obscure.	SCAPOLITE p. 435 Essentially Na, Ca aluminum silicate Tetragonal
Six cleavage directions — Dodecahedral {011}	Yellow, brown, white	3.9 to 4.1	3½–4	Luster resinous. Small tetrahedral crystals rare. Usually in cleavable masses. If massive difficult to determine.	**SPHALERITE** p. 242 ZnS Isometric
	Blue, white, gray, green	2.15 to 2.3	5½–6	Massive or in embedded grains; rarely in crystals. A feldspathoid associated with nepheline, never with quartz.	SODALITE p. 434 $Na_8(AlSiO_4)_6Cl_2$ Isometric

LUSTER:
NONMETALLIC
II. Streak
colorless

C. Hardness: >3, <5½.
(Cannot be scratched
by cent; can be
scratched by knife.)

2. Cleavage not
prominent

Color	G.	H.	Remarks	Name, Composition, Crystal System
Colorless, white	2.25	5–5½	In slender prismatic crystals, prism faces vertically striated. Often in radiating groups. Found lining cavities in igneous rocks. Poor prismatic cleavage.	NATROLITE p. 439 $Na_2Al_2Si_3O_{10}\cdot 2H_2O$ Orthorhombic
	2.95	3½–4	Effervesces in cold acid. Falls to powder in the candle flame. Frequently in radiating groups of acicular crystals; in pseudohexagonal twins. Cleavage indistinct.	ARAGONITE p. 304 $CaCO_3$ Orthorhombic
	2.27	5–5½	Usually in trapezohedrons with vitreous luster. Found lining cavities in igneous rocks.	ANALCIME p. 437 $NaAlSi_2O_6\cdot H_2O$ Isometric
	3.7	3½–4	Occurs in prismatic crystals and pseudohexagonal twins. Also fibrous and massive. Effervesces in cold acid.	STRONTIANITE p. 306 $SrCO_3$ Orthorhombic
	3.0 to 3.2	3½–5	Commonly in dense compact masses showing no cleavage. Effervesces in hot hydrochloric acid.	MAGNESITE p. 300 $MgCO_3$ Rhombohedral
	4.3	3½	Often in radiating masses; granular; rarely in pseudohexagonal crystals. Effervesces in cold hydrochloric acid.	WITHERITE p. 305 $BaCO_3$ Orthorhombic
	2.7 to 2.8	5	Commonly fibrous in radiating aggregates of sharp acicular crystals. Associated with zeolites in cavities in igneous rocks.	Pectolite p. 383 $Ca_2NaH(SiO_3)_3$ Triclinic
Colorless, pale green, yellow	2.8 to 3.0	5–5½	Usually in crystals with many brilliant faces. Occurs with zeolites lining cavities in igneous rocks.	DATOLITE p. 356 $CaB(SiO_4)(OH)$ Monoclinic
White, pale green, blue	3.4 to 3.5	4½–5	Often in radiating crystal groups. Also stalactitic, mammillary. Prismatic cleavage seldom seen.	HEMIMORPHITE p. 359 $Zn_4(Si_2O_7)(OH)_2\cdot H_2O$ Orthorhombic

**LUSTER:
NONMETALLIC
II. Streak
colorless**

**C. Hardness: >3, <5½.
(Cannot be scratched
by cent; can be
scratched by knife.)**

**2. Cleavage not
prominent (*Continued*)**

Color	G.	H.	Remarks	Name, Composition, Crystal System
White, pink, gray, green, brown	2.65 to 2.74	5–6	In prismatic crystals, granular or massive. Commonly altered. Prismatic cleavage obscure.	SCAPOLITE p. 435 Essentially Na, Ca⁻ aluminum silicate Tetragonal
White, grayish, red	2.6 to 2.8	4	May be in rhombohedral crystals. Usually massive granular. Definitely determined only by chemical tests. Cleavage {0001}, poor.	ALUNITE p. 323 $KAl_3(SO_4)_2(OH)_6$ Rhombohedral
Colorless, white, yellow, red, brown	1.9 to 2.2	5–6	Conchoidal fracture. Precious opal shows internal play of colors. Specific gravity and hardness less than fine-grained quartz.	**OPAL** p. 418 $SiO_2 \cdot nH_2O$ Essentially amorphous
Brown, green, blue, pink, white	4.35 to 4.40	5	Usually in rounded botryoidal aggregates and in honeycombed masses. Effervesces in cold hydrochloric acid. Cleavage rarely seen.	**SMITHSONITE** p. 302 $ZnCO_3$ Rhombohedral
Brown, gray, green, yellow	3.4 to 3.55	5–5½	Adamantine to resinous luster. In thin wedge-shaped crystals with sharp edges. Prismatic cleavage seldom seen.	SPHENE p. 356 $CaTiO(SiO_4)$ Monoclinic
Colorless, white, yellow, red, brown	2.72	3	May be fibrous or fine granular, banded in Mexican onyx variety. Effervesces in cold hydrochloric acid.	**CALCITE** p. 297 $CaCO_3$ Rhombohedral
Yellowish to reddish brown	5.0 to 5.3	5–5½	In small crystals or as rolled grains. Found in pegmatites.	Monazite p. 328 $(Ce, La, Y, Th)PO_4$ Monoclinic
Light to dark brown	3.83 to 3.88	3½–4	Usually cleavable but may be in compact concretions in clay or shale—clay ironstone variety. Becomes magnetic on heating.	**SIDERITE** p. 301 $FeCO_3$ Rhombohedral
White, yellow, green, brown	5.9 to 6.1	4½–5	Luster vitreous to adamantine. Massive and in octahedral-like crystals. Frequently associated with quartz. Will fluoresce.	SCHEELITE p. 325 $CaWO_4$ Tetragonal

LUSTER:
NONMETALLIC
II. Streak
colorless

C. Hardness: >3, <5½.
(Cannot be scratched
by cent; can be
scratched by knife.)

2. Cleavage not
prominent (Continued)

Color	G.	H.	Remarks	Name, Composition, Crystal System
Yellow, orange, red, gray, green	6.8±	3	Luster adamantine. Usually in square tabular crystals. Also granular massive. Characterized by color and high specific gravity.	WULFENITE p. 326 $PbMoO_4$ Tetragonal
White, yellow, brown, gray	2.6 to 2.9	3–5	Occurs massive as rock phosphate. Difficult to identify without chemical tests.	**APATITE** p. 329 $Ca_5(PO_4)_3(F, Cl, OH)$ Appears amorphous
Yellow, brown, gray, white	2.0 to 2.55	1–3	Usually pisolitic; in rounded grains and earthy masses. Often impure.	**BAUXITE** p. 287 A mixture of aluminum hydroxides
Green, blue, violet, brown, colorless	3.15 to 3.20	5	Usually in hexagonal prisms with pyramid. Also massive. Poor basal cleavage.	**APATITE** p. 329 $Ca_5(PO_4)_3(F, Cl, OH)$ Hexagonal
Green, brown, yellow, gray	6.5 to 7.1	3½–4	In small hexagonal crystals, often curved and barrel-shaped. Crystals may be cavernous. Often globular and botryoidal.	Pyromorphite p. 330 $Pb_5(PO_4)_3Cl$ Hexagonal
Yellow, green, white, brown	2.33	3½–4	Characteristically in radiating hemispherical globular aggregates. Cleavage seldom seen.	Wavellite p. 333 $Al_3(PO_4)_2(OH)_3 \cdot 5H_2O$ Orthorhombic
Olive- to blackish green, yellow-green, white	2.2	2–5	Massive. Fibrous in the asbestos variety, chrysotile. Frequently mottled green in the massive variety.	**SERPENTINE** p. 400 $Mg_3Si_2O_5(OH)_4$ Monoclinic
Yellow-green, white, blue, gray, brown	3.9 to 4.2	5½	Massive and in disseminated grains. Rarely in hexagonal prisms. Associated with red zincite and black franklinite at Franklin, N.J. Will fluoresce.	WILLEMITE p. 343 Zn_2SiO_4 Rhombohedral
White, gray, blue, green	2.15 to 2.3	5½–6	Massive or in embedded grains; rarely in crystals. A feldspathoid associated with nepheline, never with quartz. Dodecahedral cleavage poor.	SODALITE p. 434 $Na_8(AlSiO_4)_6Cl_2$ Isometric
Deep azure-blue, greenish blue	2.4 to 2.45	5–5½	Usually massive. Associated with feldspathoids and pyrite. Poor dodecahedral cleavage.	Lazurite p. 435 $(Na, Ca)_8(AlSiO_4)_6(SO_4, S, Cl)_2$ Isometric

LUSTER:
NONMETALLIC
II. Streak
colorless

D. Hardness: $>5\frac{1}{2}$, <7.
(Cannot be scratched
by knife; can be
scratched by quartz.)

1. Cleavage prominent

Cleavage	Color	G.	H.	Remarks	Name, Composition, Crystal System
{010}	White, gray, pale lavender, yellow-green	3.35 to 3.45	$6\frac{1}{2}$–7	In thin tabular crystals. Luster pearly on cleavage face. Associated with emery, margarite, chlorite.	Diaspore p. 285 $\alpha AlO \cdot OH$ Orthorhombic
{010} perfect	Hair-brown, grayish green	3.23	6–7	In long, prismatic crystals. May be in parallel groups—columnar or fibrous. Found in schists.	SILLIMANITE p. 352 Al_2SiO_5 Orthorhombic
{001}	Yellowish to black-ish green	3.35 to 3.45	6–7	In prismatic crystals striated parallel to length. Found in metamorphic rocks and crystalline limestones.	**EPIDOTE** p. 360 $Ca_2(Al, Fe)Al_2O(SiO_4)$-$(Si_2O_7)(OH)$ Monoclinic
{100}	Blue. May be gray, or green	3.56 to 3.66	5–7	In bladed aggregates with cleavage parallel to length. H5 parallel to length of crystal.	**KYANITE** p. 352 Al_2SiO_5 Triclinic
{100} good {110} poor	White, pale green, or blue	3.0 to 3.1	6	Usually cleavable, resembling feldspar. Found in pegmatites associated with other lithium minerals.	AMBLYGONITE p. 332 $LiAlFPO_4$ Triclinic
{100} good {001}	Colorless, white, gray	2.8 to 2.9	5–$5\frac{1}{2}$	Usually cleavable massive to fibrous. Also compact. Associated with crystalline limestone.	WOLLASTONITE p. 382 $CaSiO_3$ Triclinic
{001} {100}	Grayish white, green, pink	3.25 to 3.37	6–$6\frac{1}{2}$	In prismatic crystals deeply striated. Also massive, columnar, compact. Luster pearly on cleavage, elsewhere vitreous.	Clinozoisite p. 360 $Ca_2Al_3O(SiO_4)(Si_2O_7)$-$(OH)$ Monoclinic
{110}	Colorless, white	2.25	5–$5\frac{1}{2}$	In slender prismatic crystals, prism faces vertically striated. Often in radiating groups lining cavities in volcanic rocks.	**NATROLITE** p. 439 $Na_2Al_2Si_3O_{10} \cdot 2H_2O$ Orthorhombic
{110}	Brown, gray, green, yellow	3.4 to 3.55	5–$5\frac{1}{2}$	Luster adamantine to resinous. In thin wedge-shaped crystals with sharp edges. Prismatic cleavage seldom seen.	SPHENE p. 356 $CaTiO(SiO_4)$ Monoclinic

One cleavage direction

Two cleavage directions

LUSTER: NONMETALLIC
II. Streak colorless

D. Hardness: $>5\frac{1}{2}$, <7. (Cannot be scratched by knife; can be scratched by quartz.)

1. Cleavage prominent (Continued)

Two cleavage directions at 90° or nearly 90°.

Cleavage	Color	G.	H.	Remarks	Name, Composition, Crystal System
{001} {010}	Colorless, white, gray, cream, red, green	2.54 to 2.56	6	In cleavable masses or in irregular grains as rock constituents. May be in crystals in pegmatite. Distinguished with certainty only with the microscope. Green amazonstone is microcline.	**ORTHOCLASE** p. 427 Monoclinic **MICROCLINE** p. 425 Triclinic $KAISi_3O_8$
{001} {010}	Colorless, white, gray, bluish. Often shows play of colors	2.62 (albite) to 2.76 (anorthite)	6	In cleavable masses or in irregular grains as a rock constituent. On the better cleavage can be seen a series of fine parallel striations due to albite twinning; these distinguish it from orthoclase.	**PLAGIOCLASE** p. 429 Various proportions of albite, $NaAlSi_3O_8$ and anorthite, $CaAl_2Si_2O_8$ Triclinic
{110}	White, gray, pink, green	3.15 to 3.20	6½–7	In flattened prismatic crystals, vertically striated. Also massive cleavable. Found in pegmatites. Frequently shows good {100} parting.	SPODUMENE p. 380 $LiAlSi_2O_6$ Monoclinic
{110}	White, green, black	3.1 to 3.5	5–6	In stout prisms with rectangular cross section. Often in granular crystalline masses. Diopside (colorless, white, green), aegirine (brown, green), augite (dark green to black). Characterized by cleavage.	**PYROXENE GROUP** p. 371 Essentially Ca-Mg-Fe silicates Monoclinic
{110}	Gray-brown, green, bronze-brown, black	3.2 to 3.5	5½	Crystals usually prismatic but rare. Commonly massive, fibrous, lamellar. Fe may replace Mg and mineral is darker.	ENSTATITE p. 374 $MgSiO_3$ Orthorhombic
{110}	Rose-red, pink, brown	3.58 to 3.70	5½–6	Color diagnostic. Usually massive; cleavable to compact, in embedded grains; in large rough crystals with rounded edges.	RHODONITE p. 382 $MnSiO_3$ Triclinic

LUSTER: NONMETALLIC
II. Streak colorless

D. Hardness: $>5\frac{1}{2}$, <7.
(Cannot be scratched by knife; can be scratched by quartz.)

1. Cleavage prominent (Continued)

Cleavage	Color	G.	H.	Remarks	Name, Composition, Crystal System
{110}	White, green, black	3.0 to 3.3	5–6	Crystals usually slender fibrous asbestiform. Tremolite (white, gray, violet) and actinolite (green) are common in metamorphic rocks. Hornblende and arfvedsonite (dark green to black) are common in igneous rocks. Characterized by cleavage angle.	**AMPHIBOLE GROUP** p. 384 Essentially hydrous Ca-Mg-Fe silicates Monoclinic
{110}	Gray, clove-brown, green	2.85 to 3.2	$5\frac{1}{2}$–6	An amphibole. Distinct crystals rare. Commonly in aggregates and fibrous massive.	Anthophyllite p. 387 and grunerite p. 388 $(Mg,Fe)_7(Si_8O_{22})(OH)_2$ Orthorhombic and monoclinic
{110}	Blue, gray, white, green	2.15 to 2.3	$5\frac{1}{2}$–6	Massive or embedded grains; rarely in crystals. A feldspathoid associated with nepheline, never with quartz.	SODALITE p. 434 $Na_8(AlSiO_4)_6Cl_2$ Isometric

Two cleavage directions at 54° and 126° angles (rows 1–2)

Six cleavage directions (row 3)

LUSTER:
NONMETALLIC
II. Streak
colorless

D. Hardness: $>5\frac{1}{2}$, <7.
(Cannot be scratched
by knife; can be
scratched by quartz.)

2. Cleavage not
prominent

Color	G.	H.	Remarks	Name, Composition, Crystal System
Colorless	2.26	7	Occurs as small crystals in cavities in volcanic rocks. Difficult to determine without optical aid.	Tridymite p. 417 SiO_2 Pseudohexagonal
Colorless or white	2.27	5–5$\frac{1}{2}$	Usually in trapezohedrons with vitreous luster. Found lining cavities in igneous rocks.	ANALCIME p. 437 $NaAlSi_2O_6 \cdot H_2O$ Isometric
	2.32	7	Occurs in spherical aggregations in volcanic rocks. Difficult to determine without optical aid.	Cristobalite p. 418 SiO_2 Pseudoisometric
Colorless, yellow, red, brown, green, gray, blue	1.9 to 2.2	5–6	Conchoidal fracture. Precious opal shows internal play of colors. Gravity and hardness less than fine-grained quartz.	**OPAL** p. 418 $SiO_2 \cdot nH_2O$ Essentially amorphous
Gray, white, colorless	2.45 to 2.50	5$\frac{1}{2}$–6	In trapezohedral crystals embedded in dark igneous rock. Does not line cavities as analcime does.	LEUCITE p. 432 $KAlSi_2O_6$ Pseudoisometric
Colorless, pale green, yellow	2.8 to 3.0	5–5$\frac{1}{2}$	Usually in crystals with many brilliant faces. Occurs with zeolites lining cavities in igneous rocks.	DATOLITE p. 356 $CaB(SiO_4)(OH)$ Monoclinic
Colorless, white, smoky. Variously colored.	2.65	7	Crystals usually show horizontally striated prism with rhombohedral terminations.	**QUARTZ** p. 412 SiO_2 Rhombohedral
Colorless, gray, greenish, reddish	2.55 to 2.65	5$\frac{1}{2}$–6	Greasy luster. A rock constituent, usually massive; rarely in hexagonal prisms. Poor prismatic cleavage. A feldspathoid.	NEPHELINE p. 434 $(Na, K)AlSiO_4$ Hexagonal
White, gray, light to dark green, brown	2.65 to 2.74	5–6	In prismatic crystals, granular or massive. Commonly altered and prismatic cleavage obscure.	SCAPOLITE p. 435 Essentially Na-Ca aluminum silicate Tetragonal

Color	G.	H.	Remarks	Name, Composition, Crystal System
Light yellow, brown, orange	3.1 to 3.2	$6-6\frac{1}{2}$	Occurs in disseminated crystals and grains. Commonly in crystalline lime-stones.	Chondrodite p. 355 $Mg_5(SiO_4)_2(F, OH)_2$ Monoclinic
Light brown, yellow, red, green	2.6	7	Chalcedony, waxy luster commonly colloform; as agate lining cavities. Dull luster, massive: jasper red, flint black, chert gray.	**MICROCRYSTALLINE QUARTZ** p. 416 SiO_2
Blue, blueish green, green	2.6 to 2.8	6	Usually appears amorphous in reni-form and stalactitic masses.	TURQUOISE p. 333 $CuAl_6(PO_4)_4(OH)_8 \cdot 4H_2O$ Triclinic
Apple-green, gray, white	2.8 to 2.95	$6-6\frac{1}{2}$	Reniform and stalactitic with crystal-line surface. In subparallel groups of tabular crystals.	PREHNITE p. 409 $Ca_2Al(AlSi_3O_{10})(OH)_2$ Orthorhombic
Yellow-green, white, blue, gray, brown	3.9 to 4.2	$5\frac{1}{2}$	Massive and in disseminated grains. Rarely in hexagonal prisms. Associ-ated with red zincite and black frank-linite at Franklin, N.J. Will fluoresce.	WILLEMITE p. 343 Zn_2SiO_4 Rhombohedral
Olive to grayish green, brown	3.27 to 4.37	$6\frac{1}{2}-7$	Usually in disseminated grains in basic igneous rocks. May be massive granular.	**OLIVINE** p. 345 $(Mg, Fe)_2SiO_4$ Orthorhombic
Black, green, brown, blue, red, pink, white	3.0 to 3.25	$7-7\frac{1}{2}$	In slender prismatic crystals with tri-angular cross section. Crystals may be in radiating groups. Found usually in pegmatites. Black most common, other colors associated with lithium minerals.	**TOURMALINE** p. 369 Complex boron sili-cate of Na-Ca-Al-Mg-Fe-Mn Rhombohedral
Green, brown, yellow, blue, red	3.35 to 4.45	$6\frac{1}{2}$	In square prismatic crystals vertically striated. Often columnar and granular massive. Found in crystalline lime-stones.	IDOCRASE p. 362 Complex hydrous Ca-Mg-Fe-Al silicate Tetragonal
Clove-brown, gray, green, yellow	3.27 to 3.35	$6\frac{1}{2}-7$	In wedge-shaped crystals with sharp edges. Also lamellar.	AXINITE p. 366 $(Ca, Fe, Mn)Al_2(BO_3)Si_4O_{12}(OH)$ Triclinic

LUSTER: NONMETALLIC
II. Streak colorless

D. Hardness: $>5\frac{1}{2}$, <7. (Cannot be scratched by knife; can be scratched by quartz.)

2. Cleavage not prominent (Continued)

Color	G.	H.	Remarks	Name, Composition, Crystal System
Red-brown to brownish black	3.65 to 3.75	$7-7\frac{1}{2}$	In prismatic crystals; commonly in cruciform penetration twins. Frequently altered on the surface and then soft. Found in schists.	**STAUROLITE** p. 354 $Fe_2Al_9O_6(SiO_4)_4$-$(O, OH)_2$ Pseudoorthorhombic
Reddish brown, flesh-red, olive-green	3.16 to 3.20	$7\frac{1}{2}$	Prismatic crystals with nearly square cross section. Cross section may show black cross (chiastolite). May be altered to mica and then soft. Found in schists.	**ANDALUSITE** p. 351 Al_2SiO_5 Orthorhombic
Brown, gray, green, yellow	3.4 to 3.55	$5-5\frac{1}{2}$	Luster adamantine to resinous. In thin wedge-shaped crystals with sharp edges. Prismatic cleavage seldom seen.	**SPHENE** p. 356 $CaTiO(SiO_4)$ Monoclinic
Yellowish to reddish brown	5.0 to 5.3	$5-5\frac{1}{2}$	In isolated crystals, granular. Commonly found in pegmatites.	Monazite p. 328 $(Ce, La, Y, Th)PO_4$ Monoclinic
Brown to black	6.8 to 7.1	$6-7$	Rarely in prismatic crystals, twinned. Fibrous, giving reniform surface. Rolled grains. Usually gives light brown streak.	**CASSITERITE** p. 274 SnO_2 Tetragonal
Reddish brown to black	4.18 to 4.25	$6-6\frac{1}{2}$	In prismatic crystals vertically striated; often slender acicular. Crystals frequently twinned. A constituent of black sands.	**RUTILE** p. 272 TiO_2 Tetragonal
Brown to pitch-black	3.5 to 4.2	$5\frac{1}{2}-6$	Crystals often tabular. Massive and in embedded grains. An accessory mineral in igneous rocks.	Allanite p. 361 Ce, La, Th epidote Monoclinic
Blue, rarely colorless	2.60 to 2.66	$7-7\frac{1}{2}$	In embedded grains and massive, resembling quartz. Commonly altered and foliated; then softer than a knife.	Cordierite p. 368 $(Mg, Fe)_2Al_4Si_5O_{18} \cdot nH_2O$ Orthorhombic (Pseudohexagonal)
Deep azure-blue, greenish blue	2.4 to 2.45	$5-5\frac{1}{2}$	Usually massive. Associated with feldspathoids and pyrite. Poor dodecahedral cleavage.	LAZURITE p. 435 $(Na, Ca)_8(AlSiO_4)_6$-$(SO_4, S, Cl)_2$ Isometric

LUSTER:
NONMETALLIC
II. Streak
colorless

D. Hardness: $>5\frac{1}{2}$, <7.
(Cannot be scratched
by knife; can be
scratched by quartz.)

2. Cleavage not
prominent (*Continued*)

Color	G.	H.	Remarks	Name, Composition, Crystal System
Azure-blue	3.0 to 3.1	$5-5\frac{1}{2}$	Usually in pyramidal crystals, which distinguishes it from massive lazurite. A rare mineral.	**LAZULITE** p. 332 $(Mg, Fe)Al_2(PO_4)_2(OH)_2$ Monoclinic
Blue, green, white, gray	2.15 to 2.3	$5\frac{1}{2}-6$	Massive or in embedded grains; rarely in crystals. A feldspathoid associated with nepheline, never with quartz. Poor dodecahedral cleavage.	SODALITE p. 434 $Na_8(AlSiO_4)_6Cl_2$ Isometric

LUSTER:
NONMETALLIC
II. Streak
colorless

E. Hardness: >7.
(Cannot be scratched
by quartz.)

1. Cleavage prominent

Cleavage		Color	G.	H.	Remarks	Name, Composition, Crystal System
One cleavage direction	{001}	Colorless, yellow, pink, bluish, greenish	3.4 to 3.6	8	Usually in crystals, also coarse to fine granular. Found in pegmatites.	**TOPAZ** p. 353 $Al_2SiO_4(F, OH)_2$ Orthorhombic
	{010}	Brown, gray, greenish gray	3.23	6–7	Commonly in long slender prismatic crystals. May be in parallel groups, columnar or fibrous. Found in schistose rocks.	**SILLIMANITE** p. 352 Al_2SiO_5 Orthorhombic
Two cleavage directions	{110}	White, gray, pink, green	3.15 to 3.20	$6\frac{1}{2}-7$	In flattened prismatic crystals, vertically striated, Also massive cleavable. Pink variety, kunzite; green, hiddenite. Found in pegmatites. Frequently shows good {100} parting.	SPODUMENE p. 380 $LiAlSi_2O_6$ Monoclinic
Three cleavage directions	{010} {110}	Colorless pale blue, gray	3.09	8	Commonly in tabular or prismatic crystals in schists.	Lawsonite p. 360 $CaAl_2(Si_2O_7)(OH)_2 \cdot H_2O$ Orthorhombic
Four cleavage directions	{111}	Colorless, yellow, red, blue, black	3.5	10	Adamantine luster. In octahedral crystals, frequently twinned. Faces may be curved.	Diamond p. 231 C Isometric
No cleavage. Rhombohedral and basal parting.		Colorless, gray, blue, red, yellow, brown, green	3.95 to 4.1	9	Luster adamantine to vitreous. Parting fragments may appear nearly cubic. In rude barrel-shaped crystals.	**CORUNDUM** p. 267 Al_2O_3 Rhombohedral

**LUSTER:
NONMETALLIC
II. Streak
colorless**

**E. Hardness: >7.
(Cannot be scratched
by quartz.)**

**2. Cleavage not
prominent**

Color	G.	H.	Remarks	Name, Composition, Crystal System
Colorless, white, smoky, variously colored	2.65	7	Crystals usually show horizontally striated prism with rhombohedral terminations.	**QUARTZ** p. 412 SiO_2 Rhombohedral
White, colorless	2.97 to 3.0	$7\frac{1}{2}$–8	In small rhombohedral crystals. A rare mineral.	Phenacite p. 343 Be_2SiO_4 Rhombohedral
White and almost any color	3.95 to 4.1	9	Luster adamantine to vitreous. Parting fragments may appear nearly cubic. In rude barrel-shaped crystals.	**CORUNDUM** p. 267 Al_2O_3 Rhombohedral
Red, black, blue, green, brown	3.6 to 4.0	8	In octahedrons; twinning common. Associated with crystalline lime-stones.	**SPINEL** p. 276 $MgAl_2O_4$ Isometric
Blueish green, yellow, pink, colorless	2.65 to 2.8	$7\frac{1}{2}$–8	Commonly in hexagonal prisms termi-nated by the base; pyramid faces are rare. Crystals large in places. Poor basal cleavage.	**BERYL** p. 367 $Be_3Al_2(Si_6O_{18})$ Hexagonal
Yellowish to emerald-green	3.65 to 3.8	$8\frac{1}{2}$	In tabular crystals frequently in pseudohexagonal twins. Found in pegmatites.	CHRYSOBERYL p. 280 $BeAl_2O_4$ Orthorhombic
Green, brown, blue, red, pink, black	3.0 to 3.25	7–$7\frac{1}{2}$	In slender prismatic crystals with tri-angular cross section. Found usually in pegmatites. Black most common, other colors associated with lithium minerals.	TOURMALINE p. 369 Complex boron silicate of Na-Ca-Al-Mg-Fe-Mn Rhombohedral
Green, gray, white	3.3 to 3.5	$6\frac{1}{2}$–7	Massive, closely compact. Poor pris-matic cleavage at nearly 90° angles. A pyroxene.	Jadeite p. 378 $NaAlSi_2O_6$ Monoclinic
Olive to grayish green, brown	3.27 to 4.37	$6\frac{1}{2}$–7	Usually in disseminated grains in basic igneous rocks. May be massive granular.	**OLIVINE** p. 345 $(Mg, Fe)_2SiO_4$ Orthorhombic

LUSTER: NONMETALLIC II. Streak colorless	Color	G.	H.	Remarks	Name, Composition, Crystal System
E. Hardness: >7. (Cannot be scratched by quartz.)	Green, brown, yellow, blue, red	3.35 to 3.45	$6\frac{1}{2}$	In square prismatic crystals vertically striated. Often columnar and granular massive. Found in crystalline limestones.	IDOCRASE p. 362 Complex hydrous Ca-Mg-Fe-Al silicate Tetragonal
2. Cleavage not prominent (Continued)	Dark green	4.55	$7\frac{1}{2}$–8	Usually in octahedrons characteristically striated. A zinc spinel.	Gahnite p. 277 $ZnAl_2O_4$ Isometric
	Reddish brown to black	6.8 to 7.1	6–7	Rarely in prismatic crystals twinned. Fibrous, giving reniform surface. Rolled grains. Usually gives light brown streak.	**CASSITERITE** p. 274 SnO_2 Tetragonal
	Reddish brown, flesh-red, olive-green	3.16 to 3.20	$7\frac{1}{2}$	Prismatic crystals with nearly square cross section. Cross section may show black cross (chiastolite). May be altered to mica and then soft. Found in schists.	**ANDALUSITE** p. 351 Al_2SiO_5 Orthorhombic
	Clove-brown, green, yellow, gray	3.27 to 3.35	$6\frac{1}{2}$–7	In wedge-shaped crystals with sharp-edges. Also lamellar.	AXINITE p. 366 $Ca_2(Fe, Mn)Al_2(BO_3)$-$(Si_4O_{12})(OH)$ Triclinic
	Red-brown to brownish black	3.65 to 3.75	7–$7\frac{1}{2}$	In prismatic crystals; commonly in cruciform penetration twins. Frequently altered on the surface and then soft. Found in schists.	**STAUROLITE** p. 354 $Fe_2Al_9O_6(SiO_4)_4(O, OH)_2$ Pseudo-orthorhombic
	Brown, red, gray, green, colorless	4.68	$7\frac{1}{2}$	Usually in small prisms truncated by the pyramid. An accessory mineral in igneous rocks. Found as rolled grains in sand.	**ZIRCON** p. 349 $ZrSiO_4$ Tetragonal
	Usually brown to red. Also yellow, green, pink	3.5 to 4.3	$6\frac{1}{2}$–$7\frac{1}{2}$	Usually in dodecahedrons or trapezohedrons or in combinations of the two. An accessory mineral in igneous rocks and pegmatites. Commonly in metamorphic rocks. As sand.	**GARNET** p. 346 $A_3B_2(SiO_4)_3$ Isometric

MINERALS ARRANGED ACCORDING TO INCREASING SPECIFIC GRAVITY

G.	Name	G.	Name	G.	Name
1.6	Carnallite	2.65–2.74	**Scapolite**	3.27–4.37	**Olivine**
1.7	**Borax**	2.65–2.8	**Beryl**	3.2–3.5	**Enstatite**
1.95	**Kernite**	2.72	**Calcite**		
1.96	**Ulexite**	2.6–2.9	**Chlorite**	**3.4–3.59**	
1.99	Sylvite	2.62–2.76	**Plagioclase**	3.27–4.27	Olivine
		2.6–2.9	**Collophane**	3.3–3.5	Jadeite
2.0–2.19		2.74	**Bytownite**	3.35–3.45	**Diaspore**
		2.7–2.8	Pectolite	3.35–3.45	**Epidote**
2.0–2.55	**Bauxite**	2.7–2.8	**Talc**	3.35–3.45	**Idocrase**
2.0–2.4	**Chrysocolla**	2.76	**Anorthite**	3.4–3.5	**Hemimorphite**
2.05–2.09	**Sulfur**			3.45	Arfvedsonite
2.05–2.15	**Chabazite**			3.40–3.55	Aegirine
1.9–2.2	**Opal**	**2.8–2.99**		3.4–3.55	**Sphene**
2.09–2.14	Niter			3.48	**Realgar**
2.1–2.2	**Stilbite**	2.6–2.9	**Collophane**	3.42–3.56	Triphylite
2.16	**Halite**	2.8–2.9	Pyrophyllite	3.49	**Orpiment**
2.18–2.20	**Heulandite**	2.8–2.9	**Wollastonite**	3.4–3.6	**Topaz**
		2.85	**Dolomite**	3.5	**Diamond**
		2.86	Phlogopite	3.45–3.60	**Rhodochrosite**
2.2–2.39		2.76–3.1	**Muscovite**	3.5–4.3	Garnet
		2.8–2.95	**Prehnite**		
2.0–2.4	**Chrysocolla**	2.8–3.0	**Datolite**	**3.6–3.79**	
2.2–2.65	**Serpentine**	2.8–3.0	**Lepidolite**		
2.23	**Graphite**	2.89–2.98	**Anhydrite**	3.27–4.37	Olivine
2.25	**Natrolite**	2.9–3.0	Boracite	3.5–4.2	Allanite
2.26	Tridymite	2.95	**Aragonite**	3.5–4.3	**Garnet**
2.27	**Analcime**	2.95	Erythrite	3.6–4.0	**Spinel**
2.29	Soda niter	2.8–3.2	**Biotite**	3.56–3.66	**Kyanite**
2.30	Cristobalite	2.95–3.0	**Cryolite**	3.58–3.70	**Rhodonite**
2.30	**Sodalite**	2.97–3.00	Phenacite	3.65–3.75	**Staurolite**
2.32	**Gypsum**			3.7	**Strontianite**
2.33	Wavellite	**3.0–3.19**		3.65–3.8	**Chrysoberyl**
2.3–2.4	**Apophyllite**			3.75–3.77	Atacamite
2.39	**Brucite**	2.97–3.02	Danburite	3.77	**Azurite**
		2.85–3.2	Anthophyllite		
2.4–2.59		3.0–3.1	**Amblygonite**	**3.8–3.99**	
		3.0–3.1	**Lazulite**		
2.0–2.55	**Bauxite**	3.0–3.2	**Magnesite**	3.7–4.7	**Psilomelane**
2.2–2.65	**Serpentine**	3.0–3.1	Margarite	3.6–4.0	**Spinel**
2.42	**Colemanite**	3.0–3.25	**Tourmaline**	3.6–4.0	Limonite
2.42	Petalite	3.0–3.3	**Tremolite**	3.83–3.88	**Siderite**
2.4–2.45	**Lazurite**	3.09	Lawsonite	3.5–4.2	Allanite
2.45–2.50	**Leucite**	3.1–3.2	Autunite	3.5–4.3	**Garnet**
2.2–2.8	**Garnierite**	3.1–3.2	Chondrodite	3.9	Antlerite
2.54–2.57	**Microcline**	3.15–3.20	**Apatite**	3.9–4.03	**Malachite**
2.57	**Orthoclase**	3.15–3.20	**Spodumene**	3.95–3.97	**Celestite**
		3.16–3.20	**Andalusite**		
2.6–2.79		3.18	**Fluorite**	**4.0–4.19**	
2.55–2.65	**Nepheline**	**3.2–3.39**		3.9–4.1	**Sphalerite**
2.6–2.63	**Kaolinite**			4.02	**Corundum**
2.62	**Albite**	3.1–3.3	Scorodite	3.9–4.2	**Willemite**
2.60–2.66	Cordierite	3.2	**Hornblende**		
2.65	**Oligoclase**	3.23	**Sillimanite**	**4.2–4.39**	
2.65	**Quartz**	3.2–3.3	**Diopside**		
2.69	**Andesine**	3.2–3.4	**Augite**	4.1–4.3	**Chalcopyrite**
2.6–2.8	Alunite	3.25–3.37	Clinozoisite	3.7–4.7	**Psilomelane**
2.6–2.8	Turquoise	3.26–3.36	Dumortierite	4.18–4.25	**Rutile**
2.71	**Labradorite**	3.27–3.35	Axinite		

MINERALS ARRANGED ACCORDING TO INCREASING SPECIFIC GRAVITY (*Continued*)

G.	Name	G.	Name	G.	Name
4.3	**Manganite**	5.06–5.08	**Bornite**	5.3–7.3	Columbite
4.3	**Witherite**	5.15	**Franklinite**	6.33	**Cobaltite**
4.37	Goethite	5.0–5.3	Monazite		
4.35–4.40	**Smithsonite**	5.18	**Magnetite**	**6.5–6.99**	
4.4–4.59		**5.2–5.39**		6.5	**Skutterudite**
				6.55	**Cerussite**
4.43–4.45	**Enargite**	**5.4–5.59**		6.78	Bismuthinite
4.5	**Barite**			6.5–7.1	Pyromorphite
4.55	Gahnite	5.5	Millerite	6.8	**Wulfenite**
4.52–4.62	**Stibnite**	5.5±	**Cerargyrite**	6.7–7.1	Vanadinite
		5.55	**Proustite**	6.8–7.1	**Cassiterite**
4.6–4.79					
		5.6–5.79		**7.0–7.49**	
3.7–4.7	**Psilomelane**				
4.6	**Chromite**	5.5–5.8	**Chalcocite**	7.0–7.5	**Wolframite**
4.58–4.65	**Pyrrhotite**	5.68	**Zincite**	7.3	**Acanthite**
4.7	**Ilmenite**	5.7	Arsenic		
4.75	**Pyrolusite**	5.5–6.0	Jamesonite	**7.5–7.99**	
4.6–4.76	Covellite	5.3–7.3	Columbite		
4.62–4.73	**Molybdenite**			7.4–7.6	**Galena**
4.68	**Zircon**	**5.8–5.99**		7.3–7.9	Iron
				7.78	**Niccolite**
4.8–4.99		5.8–5.9	Bournonite		
		5.85	**Pyrargyrite**	**>8.0**	
4.6–5.0	Pentlandite				
4.6–5.1	**Tetrahedrite-Tennantite**	**6.0–6.49**		8.0–8.2	Sylvanite
				8.10	**Cinnabar**
4.89	**Marcasite**	5.9–6.1	Crocoite	8.9	**Copper**
		5.9–6.1	**Scheelite**	9.0–9.7	Uraninite
5.0–5.19		6.0	**Cuprite**	9.35	Calaverite
		6.07	**Arsenopyrite**	9.8	Bismuth
5.02	**Pyrite**	6.0–6.2	Polybasite	10.5	**Silver**
4.8–5.3	**Hematite**	6.2–6.4	**Anglesite**	15.0–19.3	**Gold**
				14–19	**Platinum**

MINERAL INDEX

In this index the mineral name is followed by the commonly sought information: composition, crystal system (Xl Sys.), specific gravity (**G**), hardness (**H**), and index of refraction (*n*). For uniaxial crystals $n = \omega$, for biaxial crystals $n = \beta$. The refractive index is given as a single entry when the range is usually no greater than ± 0.01. Mineral names in bold face (e.g. **Arsenopyrite**) represent species for which complete descriptions are given in the chapters on Systematic Mineralogy (chapters 7 to 10); those in light face (e.g. Arfvedsonite) refer to minerals which are briefly treated or only mentioned in the text.

Name, Page	Composition	Xl Sys.	G	H	n	Remarks
Acadialite, 440	....					See chabazite
Acanthite, 236	Ag_2S	{Mon / Iso	7.3	$2-2\frac{1}{2}$	—	Low temp. Ag_2S
Achroite, 369						Colorless tourmaline
Acmite, 379	$NaFe^{3+}Si_2O_6$	Mon	3.5	$6-6\frac{1}{2}$	1.82	See aegirine
Actinolite, 388	$Ca_2(Mg, Fe)_5Si_8O_{22}(OH)_2$	Mon	3.1–3.3	5–6	1.64	Green amphibole
Adularia, 428	$KAlSi_3O_8$	Mon	....			Colorless, translucent K-feldspar
Aegirine, 379	$NaFe^{3+}Si_2O_6$	Mon	3.5	$6-6\frac{1}{2}$	1.82	Brown, green pyroxene
Aegirine-augite, 378, 379	$(Na, Ca)(Fe^{3+}, Fe^{2+}-Mg, Al)Si_2O_6$	Mon	3.4–3.5	6	1.71–1.78	Pyroxene
Aenigmatite, 357	$Na_2Fe_5^{2+}TiO_2(Si_2O_6)_3$	Tric	3.75	$5\frac{1}{2}$	1.80	In prismatic crystals, black
Agate, 416						Concentric layers of chalcedony
Alabandite, 242, 274	MnS	Iso	4.0	$3\frac{1}{2}-4$	—	Black
Alabaster, 321						See gypsum
Albite, 429	$NaAlSi_3O_8$	Tric	2.62	6	1.53	Na end member of plagioclase
Alexandrite, 280						Gem chrysoberyl
Alkali feldspar, 420						Na or K feldspar
Allanite, 361	$(Ca, Ce)_2(Fe^{2+}, Fe^{3+})Al_2O-(SiO_4)(Si_2O_7)(OH)$	Mon	3.5–4.2	$5\frac{1}{2}-6$	1.70–1.81	Brown-black, pitchy luster
Allemontite, 227	$AsSb$	Hex	5.8–6.2	3–4	—	One cleavage
Almandite, 347	$Fe_3Al_2Si_3O_{12}$	Iso	4.32	7	1.83	A garnet
Altaite, 242, 255	$PbTe$	Iso	8.16	3	—	Tin-white
Alumstone, 323						See alunite
Alunite, 323	$KAl_3(SO_4)_2(OH)_6$	Rho	2.6–2.8	4	1.57	Usually massive
Amalgam, 223	$Ag-Hg$					See silver
Amazonstone, 425						Green microcline
Amblygonite, 332	$LiAlFPO_4$	Tric	3.0–3.1	6	1.60	Fusible at 2
Amethyst, 415	SiO_2					Purple quartz
Amosite, 388						Asbestiform cummingtonite
Amphibole, 384						A mineral group
Analcime, 437	$NaAlSi_2O_6 \cdot H_2O$	Iso	2.27	$5-5\frac{1}{2}$	1.48–1.49	Usually in trapezohedrons
Anatase, 272	TiO_2	Tet	3.9	$5\frac{1}{2}-6$	2.6	Adamantine luster
Anauxite, 403		Mon	2.6	2	1.56	Si-rich kaolinite
Andalusite, 351	Al_2SiO_5	Orth	3.16–3.20	$7\frac{1}{2}$	1.64	Square cross-sections
Andesine, 421, 431	$Ab_{70}An_{30}-Ab_{50}An_{50}$	Tric	2.69	6	1.55	Plagioclase feldspar
Andradite, 347	$Ca_3Fe_2Si_3O_{12}$	Iso	3.86	7	1.89	A garnet
Anglesite, 319	$PbSO_4$	Orth	6.2–6.4	3	1.88	$Cl\{001\}\{210\}$
Anhydrite, 320	$CaSO_4$	Orth	2.89–2.98	$3-3\frac{1}{2}$	1.58	$Cl\{010\}\{100\}\{001\}$
Ankerite, 308	$CaFe(CO_3)_2$	Rho	2.95–3	$3\frac{1}{2}$	1.70–1.75	$Cl\{10\overline{1}1\}$
Annabergite, 331	$Ni_3(AsO_4)_2 \cdot 8H_2O$	Mon	3.0	$2\frac{1}{2}-3$	1.68	Nickel bloom. Green
Anorthite, 429	$CaAl_2Si_2O_8$	Tric	2.76	6	1.58	Ca end member of plagioclase
Anorthoclase, 428	$KAlSi_3O_8-NaAlSi_3O_8$	Tric	2.58	6	1.53	An alkali feldspar between K-spar and Ab
Anthophyllite, 387	$(Mg, Fe)_7Si_8O_{22}(OH)_2$	Orth	2.85–3.2	$5\frac{1}{2}-6$	1.61–1.71	An amphibole
Antigorite, 400	$Mg_3Si_2O_5(OH)_4$	Mon	2.5–2.6	~4	1.55	Platy serpentine
Antimony, 219	Sb	Rho	6.7	3	—	$Cl\{0001\}$
Antlerite, 222	$Cu_3SO_4(OH)_4$	Orth	3.9±	$3\frac{1}{2}-4$	1.74	Green
Apatite, 229	$Ca_5(PO_4)_3(F, Cl, OH)$	Hex	3.15–3.20	5	1.63	$Cl\{0001\}$ poor
Apophyllite, 409	$KCa_4(Si_4O_{10})_2F \cdot 8H_2O$	Tet	2.3–2.4	$4\frac{1}{2}-5$	1.54	$Cl\{001\}$
Aquamarine, 367						Greenish-blue gem beryl
Aragonite, 304	$CaCO_3$	Orth	2.95	$3\frac{1}{2}-4$	1.68	$Cl\{010\}\{110\}$
Arfvedsonite, 391	$Na_3Fe_4^{2+}Fe^{3+}Si_8O_{22}(OH)_2$	Mon	3.45	6	1.69	Deep green amphibole

Name, Page	Composition	Xl Sys.	G	H	n	Remarks
Argentite, 236	Ag_2S	....				Now known as acanthite
Arsenic, 227	As	Rho	5.7	$3\frac{1}{2}$	—	Cl$\{0001\}$
Arsenopyrite, 254	FeAsS	Mon	6.07	$5\frac{1}{2}$–6	—	Pseudo-orthorhombic
Asbestos, 389, 401		....				See amphibole and serpentine
Astrophyllite, 357	$(K, Na)_3(Fe, Mn)_7(Ti, Zr)_2$- $Si_8(O, OH)_{31}$	Tric	3.35	3	1.71	Micaceous cleavage
Atacamite, 293	$Cu_2Cl(OH)_3$	Orth	3.75–3.77	3–$3\frac{1}{2}$	1.86	Green. Cl$\{010\}$
Augite, 377	$(Ca, Na)(Mg, Fe, Al)(Si, Al)_2O_6$	Mon	3.2–3.4	5–6	1.67–1.73	Dark green to black pyroxene
Aurichalcite, 310	$(Zn, Cu)_5(CO_3)_2(OH)_3$	Orth	3.64	2	1.74	Green to blue
Autunite, 334	$Ca(UO_2)_2(PO_4)_2 \cdot 10$–$12\ H_2O$	Tet	3.1–3.2	2–$2\frac{1}{2}$	1.58	Yellow-green
Aventurine, 416, 431		....				Quartz or oligoclase
Axinite, 366	$(Ca, Fe, Mn)_3Al_2(BO_3)$- $(Si_4O_{12})(OH)$	Tric	3.27–3.35	$6\frac{1}{2}$–7	1.69	Crystal angles acute
Azurite, 310	$Cu_3(CO_3)_2(OH)_2$	Mon	3.77	$3\frac{1}{2}$–4	1.76	Always blue
Balas ruby, 277		....				Red gem spinel
Barite, 317	$BaSO_4$	Orth	4.5	3–$3\frac{1}{2}$	1.64	Cl$\{001\}\{210\}$
Bauxite, 287	mixture of diaspore, gibbsite, boehmite		2.0–2.55	1–3	—	An earthy rock
Beidellite, 403	$(Ca, Na)_{0.3}Al_2(OH)_2$- $(Al, Si)_4O_{10} \cdot (H_2O)_4$	Mon	2–3	1–2	....	Member of montmorillo- nite group
Benitoite, 357	$BaTiSi_3O_9$	Hex	3.6	$6\frac{1}{2}$	1.76	Blue
Bentonite, 403		....				Montmorillonite alteration of volcanic ash
Beryl, 367	$Be_3Al_2(Si_6O_{18})$	Hex	2.65–2.8	$7\frac{1}{2}$–8	1.57–1.61	Usually green
Biotite, 406	$K(Mg, Fe)_3(AlSi_3O_{10})(OH)_2$	Mon	2.8–3.2	$2\frac{1}{2}$–3	1.61–1.70	Black mica
Bismuth, 227	Bi	Rho	9.8	2–$2\frac{1}{2}$	—	Cl$\{0001\}$
Bismuthinite, 250	Bi_2S_3	Orth	6.78 ± 0.03	2	—	Cl$\{010\}$
Black-band ore, 301		....				See siderite
Black jack, 243		....				See sphalerite
Bloodstone, 416						Green and red chalcedony
Boehmite, 285	$\gamma AlO \cdot OH$	Orth	3.01–3.06	$3\frac{1}{2}$–4	1.65	In bauxite
Bog-iron ore, 287						See limonite
Boracite, 316	$Mg_3ClB_7O_{13}$	Orth	2.9–3.0	7	1.66	Pseudoisometric
Borax, 313	$Na_2B_4O_5(OH)_4 \cdot 8H_2O$	Mon	$1.7\pm$	2–$2\frac{1}{2}$	1.47	Cl$\{100\}$
Bornite, 239	Cu_5FeS_4	Tet / Iso	5.06–5.08	3	—	Purple-blue tarnish
Bort, 231	C	Iso	3.5	10	2.4	Variety of diamond
Boulangerite, 260	$Pb_5Sb_4S_{11}$	Orth	$6.0\pm$	$2\frac{1}{2}$–3	—	Cl$\{001\}\{010\}$
Bournonite, 259	$PbCuSbS_3$	Orth	5.8–5.9	$2\frac{1}{2}$–3	—	Fusible at 1
Bravoite, 251	$(Ni, Fe)S_2$	Iso	4.66	$5\frac{1}{2}$–6	—	Steel gray
Brazilian emerald, 369		....				Green gem tourmaline
Breithauptite, 246	NiSb	Hex	8.23	$5\frac{1}{2}$	—	Copper-red
Brittle mica, 408		....				A type of mica
Brochantite, 323	$Cu_4SO_4(OH)_6$	Mon	3.9	$3\frac{1}{2}$–4	1.78	Cl$\{010\}$. Green
Bromyrite, 291	AgBr	Iso	5.9	1–$1\frac{1}{2}$	2.25	Sectile
Bronzite, 375	$(Mg, Fe)SiO_3$	Orth	$3.3\pm$	$5\frac{1}{2}$	1.68	Orthopyroxene member
Brookite, 272	TiO_2	Orth	3.9–4.1	$5\frac{1}{2}$–6	2.6	Adamantine luster
Brucite, 283	$Mg(OH)_2$	Rho	2.39	$2\frac{1}{2}$	1.57	Cl$\{0001\}$
Bustamite, 383	$(Mn, Ca, Fe)SiO_3$	Tric	3.3–3.4	$5\frac{1}{2}$–$6\frac{1}{2}$	1.67–1.70	Pink to brownish pyroxenoid
Bytownite, 421, 431	$Ab_{30}An_{70}$-$Ab_{10}An_{90}$	Tric	2.74	6	1.57	Plagioclase feldspar
Cairngorm Stone, 415	SiO_2	....				Smoky quartz
Calamine, 303		....				See smithsonite, hemi- morphite
Calaverite, 255	$AuTe_2$	Mon	9.35	$2\frac{1}{2}$	—	Fusible at 1

Name, Page	Composition	XI Sys.	G	H	n	Remarks
Calcite, 297	$CaCO_3$	Rho	2.71	3	1.66	$Cl\{10\bar{1}1\}$
Californite, 363						See idocrase
Cancrinite, 434	$Na_6Ca(CO_3)(AlSiO_4)_6 \cdot 2H_2O$	Hex	2.45	5–6	1.52	Feldspathoid
Capillary pyrites, 246						See millerite
Carbonado, 231	C	Iso	3.5	10	2.4	Black, cryptocrystalline diamond
Carbonate-apatite, 329	$CaF(PO_4, CO_3, OH)_3$					See apatite
Carnallite, 290	$KMgCl_3 \cdot 6H_2O$	Orth	1.6	1	1.48	Deliquescent
Carnelian, 416						Red chalcedony
Carnotite, 334	$K_2(UO_2)_2(VO_4)_2 \cdot 3H_2O$	Mon	4.7–5	Soft	1.93	Yellow
Cassiterite, 274	SnO_2	Tet	6.8–7.1	6–7	2.00	Luster adamantine
Cat's eye, 280, 415						See chrysoberyl and quartz
Celestite, 318	$SrSO_4$	Orth	3.95–3.97	3–3½	1.62	$Cl\{001\}\{210\}$
Celsian, 420	$BaAl_2Si_2O$	Mon	3.37	6	1.59	Ba-feldspar
Cerargyrite, 291	AgCl	Iso	5.5±	2–3	2.07	Perfectly sectile
Cerianite, 276	$(Ce, Th)O_2$	Iso			>2.0	Very rare
Cerussite, 306	$PbCO_3$	Orth	6.55	3–3½	2.08	Efferv. in HNO_3
Chabazite, 440	$Ca_2Al_2Si_4O_{12} \cdot 6H_2O$	Rho	2.05–2.15	4–5	1.48	Cubelike crystals
Chalcanthite, 323	$CuSO_4 \cdot 5H_2O$	Tric	2.12–2.30	2½	1.54	Soluble in water
Chalcedony, 416						Microcryst. quartz
Chalcocite, 237	Cu_2S	Orth / Hex	5.5–5.8	2½–3	—	Imperfectly sectile
Chalcopyrite, 243	$CuFeS_2$	Tet	4.1–4.3	3½–4	—	Brittle, yellow
Chalcosiderite, 333	$CuFe_6(PO_4)_4(OH)_8 \cdot 4H_2O$	Tric	3.22	4½	1.84	Light green
Chalcotrichite, 266						Fibrous cuprite
Chalk, 299						See calcite
Chalybite, 302						See siderite
Chert, 416	SiO_2	—	2.65	7	1.54	Microcryst. quartz
Chiastolite, 351						Variety of andalusite
Chloanthite, 255	$(Ni, Co)As_{3-x}$	Iso	6.5	5½–6	—	See skutterudite
Chlorapatite, 329	$Ca_5(PO_4)_3Cl$					See apatite
Chlorite, 408	$(Mg, Fe)_3(Si, Al)_4O_{10}(OH)_2 \cdot (Mg, Fe)_3(OH)_6$	Mon / Tric	2.6–3.3	2–2½	1.57–1.67	Green, $Cl\{001\}$
Chloritoid, 357	$(Fe, Mg)Al_4O_2(SiO_4)_2(OH)_4$	Mon / Tric	3.5–3.8	6½	1.72–1.73	Appears similar to chlorite
Chondrodite, 355	$Mg_5(SiO_4)_2(F, OH)_2$	Mon	3.1–3.2	6–6½	1.60–1.63	Yellow-red
Chromite, 279	$FeCr_2O_4$	Iso	4.6	5½	2.16	Luster submetallic
Chrysoberyl, 280	$BeAl_2O_4$	Orth	3.65–3.8	8½	1.75	Crystals tabular
Chrysocolla, 410	$\sim Cu_4H_4Si_4O_{10}(OH)_8$	?	2.0–2.4	2–4	1.40±	Blueish green
Chrysolite, 346						See olivine
Chrysoprase, 416						Green chalcedony
Chrysotile, 400	$Mg_3Si_2O_5(OH)_4$	Mon	2.5–2.6	~4	1.55	Fibrous serpentine
Cinnabar, 247	HgS	Rho	8.10	2½	2.81	Red
Cinnamon stone, 348						See grossularite
Citrine, 415	SiO_2					Yellow quartz
Clay ironstone, 301						See siderite
Cleavelandite, 431						White, platy albite
Clinochlore, 408					1.58	See chlorite
Clinoenstatite, 375	$MgSiO_3$	Mon	3.19	6	1.66	In meteorites
Clinoferrosilite, 375	$FeSiO_3$	Mon				Clinopyroxene end member
Clinohumite, 355	$Mg_9(SiO_4)_4(F, OH)_2$	Mon	3.1–3.2	6	1.64	See chondrodite
Clinohypersthene, 376	$(Mg, Fe)SiO_3$	Mon	3.4–3.5	5–6	1.68–1.72	In meteorites
Clinozoisite, 360	$Ca_2Al_3O(SiO_4)Si_2O_7(OH)$	Mon	3.25–3.37	6–6½	1.67–1.72	Crystals striated
Clintonite, 408	$Ca(Mg, Al)_{3-2}Al_2Si_2O_{10}(OH)_2$	Mon	3–3.1	3½	1.65	Brittle mica
Cobalt bloom, 331	$Co_3(AsO_4)_2 \cdot 8H_2O$					See erythrite

Name, Page	Composition	Xl Sys.	G	H	n	Remarks
Cobaltite, 253	(Co, Fe)AsS	Orth	6.33	$5\frac{1}{2}$	—	In pseudoisometric pyritohedrons
Coesite, 412	SiO_2	Mon	3.01	7	1.59	High pressure SiO_2
Cogwheel ore, 259						See bournonite
Colemanite, 315	$CaB_3O_4(OH)_3 \cdot H_2O$	Mon	2.42	$4-4\frac{1}{2}$	1.59	Cl{010} perfect
Collophane, 329						See apatite
Columbite, 281	$(Fe, Mn)Nb_2O_6$	Orth	5.2–7.3	6	—	Iron-black, submetallic
Common salt, 289						See halite
Copper, 223	Cu	Iso	8.9	$2\frac{1}{2}$–3	—	Malleable
Copper glance, 237						See chalcocite
Copper pyrites, 243						See chalcopyrite
Cordierite, 368	$(Mg, Fe)_2Al_4Si_5O_{18} \cdot nH_2O$	Orth	2.60–2.66	$7-7\frac{1}{2}$	1.53–1.57	Light blueish gray
Corundum, 267	Al_2O_3	Rho	4.02	9	1.77	Rhomb. parting
Cotton balls, 315						See ulexite
Covellite, 247	CuS	Hex	4.6–4.76	$1\frac{1}{2}$–2	—	Blue
Cristobalite, 418	SiO_2	{Tet {Iso	2.32	$6\frac{1}{2}$	1.48	In volcanic rocks
Crocidolite, 391	$NaFe_3^{2+}Fe_2^{3+}Si_8O_{22}(OH)_2$	Mon	3.2–3.3	4	1.70	Blue amphibole asbestos
Crocoite, 320	$PbCrO_4$	Mon	5.9–6.1	$2\frac{1}{2}$–3	2.36	Orange-red
Crossite, 390						Amphibole between glaucophane and riebeckite
Cryolite, 291	Na_3AlF_6	Mon	2.95–3.0	$2\frac{1}{2}$	1.34	White
Cummingtonite, 388	$(Mg, Fe)_7Si_8O_{22}(OH)_2$	Mon	3.1–3.3	$5\frac{1}{2}$–6	1.66–1.68	Light beige, needlelike
Cuprite, 266	Cu_2O	Iso	6.1	$3\frac{1}{2}$–4	—	Ruby-red in transparent crystals
Cymophane, 280						Chatoyant chrysoberyl
Danburite, 354	$Ca(B_2Si_2O_8)$	Orth	2.97–3.02	7	1.63	In crystals
Datolite, 356	$CaB(SiO_4)(OH)$	Mon	2.8–3.0	$5-5\frac{1}{2}$	1.65	Usually in crystals
Demantoid, 348						Green andradite
Diallage, 377						Variety of diopside
Diamond, 231	C	Iso	3.51	10	2.42	Adamantine luster
Diaspore, 285	$\alpha AlO \cdot OH$	Orth	3.35–3.45	$6\frac{1}{2}$–7	1.72	Cl{010} perfect
Diatomaceous earth, 420						See opal
Diatomite, 420						See opal
Dichroite, 368						See cordierite
Dickite, 403	$Al_2Si_2O_5(OH)_4$	Mon	2.6	$2-2\frac{1}{2}$	1.56	Clay mineral
Digenite, 239	Cu_9S_5	Iso	5.6	$2\frac{1}{2}$–3	—	Like chalcocite
Diopside, 377	$CaMgSi_2O_6$	Mon	3.2	5–6	1.67	White to light green
Dioptase, 410	$Cu_6(Si_6O_{18}) \cdot 6H_2O$	Rho	3.3	5	1.65	Green
Dolomite, 308	$CaMg(CO_3)_2$	Rho	2.85	$3\frac{1}{2}$–4	1.68	Cl{10$\bar{1}$1}
Dravite, 369						Brown tourmaline
Dry-bone ore, 302						See smithsonite
Dumortierite, 352	$Al_7O_3(BO_3)(SiO_4)_3$	Orth	3.26–3.36	7	1.69	Radiating
Edenite, 390	$NaCa_2Mg_5AlSi_7O_{22}(OH)_2$	Mon	3.0	6	1.63	See hornblende
Elbaite, 369						See tourmaline
Electrum, 221						See gold
Eleolite, 434						See nepheline
Embolite, 291	Ag(Cl, Br)	Iso	5.3–5.4	$1-1\frac{1}{2}$	—	Sectile
Emerald, 367						Deep green gem beryl
Emery, 267						Corundum with magnetite
Enargite, 259	Cu_3AsS_4	Orth	4.45	3	—	Cl{110}
Endlichite, 331						See vanadinite
Enstatite, 374	$MgSiO_3$	Orth	3.2–3.5	$5\frac{1}{2}$	1.65	Cl{210}~90°
Epidote, 360	$Ca_2(Al, Fe)Al_2O(SiO_4)(Si_2O_7)(OH)$	Mon	3.35–3.45	6–7	1.72–1.78	Cl{001}, {100}, green

Name, Page	Composition	Xl Sys.	G	H	n	Remarks
Epsomite, 323	$MgSO_4 \cdot 7H_2O$	Orth	1.75	$2-2\frac{1}{2}$	1.46	Bitter taste
Epsom salt, 323						See epsomite
Erythrite, 331	$Co_3(AsO_4)_2 \cdot 8H_2O$	Mon	3.06	$1\frac{1}{2}-2\frac{1}{2}$	1.66	Pink; cobalt bloom
Essonite, 348						See grossularite
Euclase, 368	$BeAl(SiO_4)(OH)$	Mon	3.1	$7\frac{1}{2}$	1.66	Cl{010}
Eucryptite, 380	$LiAlSiO_4$	Hex	2.67		1.55	Spodumene alteration
Fahlore, 258						See tetrahedrite
Fairy stone, 355						See staurolite
Famatinite, 259	Cu_3SbS_4	Tet	4.52	$3\frac{1}{2}$	—	Gray
Fayalite, 345	Fe_2SiO_4	Orth	4.39	$6\frac{1}{2}$	1.86	An olivine
Feather ore, 260						See jamesonite
Feldspar, 420						A mineral group
Feldspathoid, 431						A mineral group
Ferberite, 325	$FeWO_4$	Mon	7.0-7.5	5	—	See wolframite
Fergusonite, 282	$(Y, Er, Ce, Fe)NbO_4$	Tet	5.8	$5\frac{1}{2}-6$	—	Brown-black
Ferrimolybdite, 253	$Fe_2(MoO_4)_3 \cdot 8H_2O$	Orth?	2.99	~1	1.73-1.79	Canary yellow, soft
Ferroactinolite, 384	$Ca_2Fe_5Si_8O_{22}(OH)_2$	Mon	3.2-3.3	5-6	1.68	Dark green amphibole
Fersmannite, 357	$Na_4Ca_4Ti_4(SiO_4)_3(O, OH, F)_3$	Mon	3.44	$5\frac{1}{2}$	1.93	Brown
Fibrolite, 352						See sillimanite
Flint, 416	SiO_2	—	2.65	7	1.54	Microcryst. quartz
Flos ferri, 305						See aragonite
Fluorapatite, 329	$Ca_5(PO_4)_3F$					See apatite
Fluorite, 292	CaF_2	Iso	3.18	4	1.43	Cl octahedral
Forsterite, 345	Mg_2SiO_4	Orth	3.2	$6\frac{1}{2}$	1.63	An olivine
Fowlerite, 383						Zn-bearing rhodonite
Franklinite, 279	$(Zn, Fe, Mn)(Fe, Mn)_2O_4$	Iso	5.15	6	—	At Franklin, N.J.
Freibergite, 258						Ag-bearing tetrahedrite
Gadolinite, 368	$YFeBe_2(SiO_4)_2O_2$	Mon	4.0-4.5	$6\frac{1}{2}-7$	1.79	Black
Gahnite, 277	$ZnAl_2O_4$	Iso	4.55	$7\frac{1}{2}-8$	1.80	Dark green octahedrons
Galaxite, 277	$MnAl_2O_4$	Iso	4.03	$7\frac{1}{2}-8$	1.92	Mn spinel
Galena, 240	PbS	Iso	7.4-7.6	$2\frac{1}{2}$	—	Cl cubic
Garnet, 346	$A_3B_2(SiO_4)_3$	Iso	3.5-4.3	$6\frac{1}{2}-7\frac{1}{2}$	1.71-1.88	In crystals
Garnierite, 402	$(Ni, Mg)_3Si_2O_5(OH)_4$	Mon	2.2-2.8	2-3	1.59	Green Ni serpentine
Gaylussite, 310	$Na_2Ca(CO_3)_2 \cdot 5H_2O$	Mon	1.99	2-3	1.52	Cl{110} perfect
Gedrite, 387	$\sim Na_{0.5}(Mg, Fe)_2(Mg, Fe)_{3.5}$ $(Al, Fe^{3+})_{1.5}Si_6Al_2O_{22}(OH)_2$	Orth				See anthophyllite
Geikielite, 271	$MgTiO_3$	Rho	4.05	$5\frac{1}{2}-6$	—	Cl{10$\bar{1}$1}
Geocronite, 260	$Pb_5(Sb, As)_2S_8$	Orth	6.4±	$2\frac{1}{2}$	—	Lead-gray to grayish blue
Gersdorffite, 254	$NiAsS$	Iso	5.9	$5\frac{1}{2}$	—	See cobaltite
Geyserite, 420						Opal in hot springs
Gibbsite, 284	$Al(OH)_3$	Mon	2.3-2.4	$2\frac{1}{2}-3\frac{1}{2}$	1.57	Basal Cl {001}
Glaucodot, 254	$(Co, Fe)AsS$	Orth	6.04	5	—	Tin-white
Glauconite, 407	$(K, Na, Ca)_{0.5-1}(Fe^{3+}, Al, Fe^{2+},$ $Mg)_2(Si, Al)_4O_{10}(OH)_2 \cdot nH_2O$	Mon	2.4±	2	1.62	In green sands
Glaucophane, 390	$Na_2Mg_3Al_2Si_8O_{22}(OH)_2$	Mon	3.1-3.3	$6-6\frac{1}{2}$	1.62-1.67	Blue to black amphibole
Gmelinite, 442	$(Na_2, Ca)(Al_2Si_4O_{12}) \cdot 6H_2O$	Rho	2.1±	$4\frac{1}{2}$	1.49	A zeolite
Goethite, 286	$\alpha FeO \cdot OH$	Orth	4.37	$5-5\frac{1}{2}$	2.39	Cl{010} perfect
Gold, 221	Au	Iso	15-19.3	$2\frac{1}{2}-3$	—	Yellow, soft
Graphite, 234	C	Hex	2.23	1-2	—	Black, platy
Gray copper, 258						See tetrahedrite
Greenalite, 402	$(Fe, Mg)_3Si_2O_5(OH)_4$	Mon	3.2		1.67	In iron-formations
Greenockite, 243	CdS	Hex	4.9	$3-3\frac{1}{2}$	—	Yellow-orange
Grossularite, 347	$Ca_3Al_2Si_3O_{12}$	Iso	3.59	$6\frac{1}{2}$	1.73	A garnet
Grunerite, 388	$Fe_7Si_8O_{22}(OH)_2$	Mon	3.6	6	1.71	Light brown, needlelike
Gypsum, 321	$CaSO_4 \cdot 2H_2O$	Mon	2.32	2	1.52	Cl{010}{100}{011}

Name, Page	Composition	Xl Sys.	G	H	n	Remarks
Halite, 289	$NaCl$	Iso	2.16	$2\frac{1}{2}$	1.54	Cl cubic. Salty
Halloysite, 403	$Al_2Si_2O_5(OH)_4$ and $Al_2Si_2O_5(OH)_4 \cdot 2H_2O$	Mon	2.0–2.2	1–2	1.54	Clay mineral
Harmotome, 442	$Ba(Al_2Si_6O_{16}) \cdot 6H_2O$	Mon	2.45	$4\frac{1}{2}$	1.51	A zeolite
Hastingsite, 390	$NaCa_2Fe_4(Al, Fe)Al_2Si_6$-$O_{22}(OH)_2$	Mon	3.2	6	1.66	See hornblende
Haüynite, 435	$(Na, Ca)_{4-8}(AlSiO_4)_6(SO_4)_{1-2}$	Iso	2.4–2.5	$5\frac{1}{2}$–6	1.50	A feldspathoid
Heavy spar, 317	....					See barite
Hectorite, 403	$(Mg, Li)_3Si_4O_{10}(OH)_2$-$Na_{0.3}(H_2O)_4$	Mon	2.5	1–$1\frac{1}{2}$	1.52	Li montmorillonite
Hedenbergite, 377	$CaFeSi_2O_6$	Mon	3.55	5–6	1.73	Clinopyroxene end member
Heliotrope, 416						Green and red chalcedony
Hematite, 269	Fe_2O_3	Rho	5.26	$5\frac{1}{2}$–$6\frac{1}{2}$	—	Red streak
Hemimorphite, 359	$Zn_4(Si_2O_7)(OH)_2 \cdot H_2O$	Orth	3.4–3.5	$4\frac{1}{2}$–5	1.62	Cl{110}
Hercynite, 277	$FeAl_2O_4$	Iso	4.39	$7\frac{1}{2}$–8	1.80	Fe spinel
Hessite, 255	Ag_2Te	⌠Iso ⌊Mon	8.4	$2\frac{1}{2}$–3	—	Steel-gray
Heulandite, 440	$CaAl_2Si_7O_{18} \cdot 6H_2O$	Mon	2.18–2.20	$3\frac{1}{2}$–4	1.48	Cl{010} perfect
Hiddenite, 380	....					Green gem spodumene
Hollandite, 285	$Ba_2Mn_8O_{16}$	⌠Tet ⌊Mon	4.9	6	—	Silver-gray to black
Holmquistite, 385	$Li_2(Mg, Fe)_3(Al, Fe^{3+})_2$-$Si_8O_{22}(OH)_2$	Orth	3.06–3.13	5–6	1.64–1.66	Blue to violet blue; near Li-pegmatites
Hornblende, 389	$(Ca, Na)_{2-3}(Mg, Fe, Al)_5$-$Si_6(Si, Al)_2O_{22}(OH)_2$	Mon	3.0–3.4	5–6	1.62–1.72	Green to black amphibole
Horn silver, 291	....					See cerargyrite
Huebnerite, 325	$MnWO_4$	Mon	7.0	5	—	See wolframite
Humite, 355	$Mg_7(SiO_4)_3(F, OH)_2$	Orth	3.1–3.2	6	1.64	See chondrodite
Hyacinth, 349	....					See zircon
Hyalite, 420	....					Colorless, globular opal
Hyalophane, 420	$(K, Ba)(Al, Si)_2Si_2O_8$	Mon	2.8	6	1.54	A feldspar
Hydroboracite, 316	$CaMgB_6O_8(OH)_6 \cdot 3H_2O$	Mon	2.17	2	1.53	Clear, colorless to white
Hydrogrossularite, 347	$Ca_3Al_2Si_2O_8(SiO_4)_{1-m}(OH)_{4m}$	Iso	3.13–3.59	6–7	1.67–1.73	Hydrous garnet
Hydroxylapatite, 329	$Ca_5(PO_4)_3(OH)$	....				See apatite
Hydrozincite, 303	$Zn_5(CO_3)_2(OH)_6$	Mon	3.6–3.8	2–$2\frac{1}{2}$	1.74	Secondary mineral
Hypersthene, 374	$(Mg, Fe)SiO_3$	Orth	3.4–3.5	5–6	1.68–1.73	Cl{210} ~90°
Iceland spar, 298	....					See calcite
Idocrase, 362	$Ca_{10}(Mg, Fe)_2Al_4(SiO_4)_5$-$(Si_2O_7)_2(OH)_4$	Tet	3.35–3.45	$6\frac{1}{2}$	1.70–1.75	Prismatic crystals
Illite, 403	....				1.57–1.61	Micalike clay minerals
Ilmenite, 271	$FeTiO_3$	Rho	4.7	$5\frac{1}{2}$–6	—	May be slightly magnetic
Ilvaite, 360	$CaFe_2^{2+}Fe^{3+}O(Si_2O_7)(OH)$	Orth	4.0	$5\frac{1}{2}$–6	1.91	Black
Indialite, 368	$(Mg, Fe)_2Al_4Si_5O_{18} \cdot nH_2O$	Hex	~2.6	7	....	High T form of cordierite
Indicolite, 369	....					Blue tourmaline
Inyoite, 316	$CaB_3O_3(OH)_5 \cdot 4H_2O$	Mon	1.88	2	1.51	Colorless, transparent
Iodobromite, 291	$Ag(Cl, Br, I)$	Iso	5.71	1–$1\frac{1}{2}$	2.20	Sectile
Iodyrite, 291	AgI	Hex	5.5–5.7	1–$1\frac{1}{2}$	2.18	Sectile
Iolite, 368	....					See cordierite
Iridium, 225	Ir	Iso	22.7	6–7	—	A platinum metal
Iridosmine, 225	Ir-Os	Rho	19.3–21.1	6–7	—	See platinum
Iron, 225	Fe	Iso	7.3–7.9	$4\frac{1}{2}$	—	Very rare
Iron pyrites, 251	....					See pyrite
Jacinth, 349	....					See zircon
Jacobsite, 266	$MnFe_2O_4$	Iso	5.1	$5\frac{1}{2}$–$6\frac{1}{4}$	2.3	A spinel
Jade, 379	....					See nephrite and jadeite

Name, Page	Composition	XI Sys.	G	H	n	Remarks
Jadeite, 378	$NaAlSi_2O_6$	Mon	3.3–3.5	$6\frac{1}{2}$–7	1.66	Green, compact
Jamesonite, 260	$Pb_4FeSb_6S_{14}$	Mon	5.5–6.0	2–3	—	Feather ore
Jargon, 350						See zircon
Jarosite, 324	$KFe_3(SO_4)_2(OH)_6$	Rho	3.2±	3	1.82	Yellow-brown
Jasper, 416						Red, microcryst. quartz
Johannsenite, 385	$CaMnSi_2O_6$	Mon	3.4–3.5	6	1.71–1.73	Mn analogue of diopside
Kainite, 291	$KMg(Cl, SO_4) \cdot 2^3/_4H_2O$	Mon	2.1	3	1.51	Cl{001}. Salty, bitter
Kalsilite, 434	$KAlSiO_4$	Hex	2.61	6	1.54	Isostructural with nepheline
Kamacite, 225	Fe-Ni	Iso	7.3–7.9	4	—	In meteorites
Kaolin, 402						Mixture of clay minerals
Kaolinite, 402	$Al_2Si_2O_5(OH)_4$	Tric	2.6	2	1.56	Earthy
Keatite, 411	SiO_2	Tet	2.50	7	1.52	Synthetic
Kernite, 313	$Na_2B_4O_6(OH)_2 \cdot 3H_2O$	Mon	1.95	3	1.47	Cl{001}{100}
K-feldspar, 425						See microcline,, ortho-clase, sanidine
Kieserite, 467	$MgSO_4 \cdot H_2O$	Mon	2.57	$3\frac{1}{2}$	1.53	Massive, granular. Whitish grey.
Krennerite, 255	$AuTe_2$	Orth	8.62	2–3	—	Basal Cl{001}
Kunzite, 380						Pink gem spodumene
Kutnahorite, 302, 308	$CaMn(CO_3)_2$	Rho	3.12	$3\frac{1}{2}$–4	1.74	Dolomite structure
Kyanite, 352	Al_2SiO_5	Tric	3.55–3.66	5–7	1.72	Blue, bladed
Labradorite, 421, 431	$Ab_{50}An_{50}$–$Ab_{30}An_{70}$	Tric	2.71	6	1.56	Plagioclase feldspar
Lamprophyllite, 357	$Na_3Sr_2Ti_3(Si_2O_7)_2(O, OH, F)_2$	Orth	3.45	4	1.75	Platy
Langbeinite, 467	$K_2Mg_2(SO_4)_3$	Iso	2.83	$3\frac{1}{2}$–4	1.53	Colorless; in evaporites
Lapis lazuli, 435						See lazurite
Larsenite, 346	$PbZnSiO_4$	Orth	5.9	3	1.95	An olivine
Laumontite, 442	$Ca(Al_2Si_4O_{12}) \cdot 4H_2O$	Mon	2.28	4	1.52	A zeolite
Lawsonite, 360	$CaAl_2(Si_2O_7)(OH)_2 \cdot H_2O$	Orth	3.09	8	1.67	In gneisses and schists
Lazulite, 332	$(Mg, Fe)Al_2(PO_4)_2(OH)_2$	Mon	3.0–3.1	5–$5\frac{1}{2}$	1.64	Blue
Lazurite, 435	$(Na, Ca)_8(AlSiO_4)_6$-$(SO_4, S, Cl)_2$	Iso	2.4–2.45	5–$5\frac{1}{2}$	1.50	Blue; associated with pyrite
Lechatelierite, 417	SiO_2	Amor	2.2	6–7	1.46	Fused silica
Lepidochrosite, 287	$\gamma FeO \cdot OH$	Orth	4.09	5	2.2	Red
Lepidolite, 407	$K(Li, Al)_{2-3}(AlSi_3O_{10})$-$(O, OH, F)_2$	Mon	2.8–2.9	$2\frac{1}{2}$–4	1.55–1.59	Pink, lilac, gray mica
Leucite, 433	$KAlSi_2O_6$	{Tet {Iso	2.47	$5\frac{1}{2}$–6	1.51	In trapezohedrons
Leucoxene, 273						Brownish alteration of Ti minerals
Limonite, 287	$FeO \cdot OH \cdot nH_2O$	Amor	3.6–4.0	5–$5\frac{1}{2}$	—	Streak yellow-brown
Linnaeite, 255	Co_3S_4	Iso	4.8	$4\frac{1}{2}$–$5\frac{1}{2}$	—	Steel-gray
Litharge, 242	PbO	Tet	9.14	2	2.66	Red
Lithiophilite, 328	$Li(Mn, Fe)PO_4$	Orth	3.5	5	1.67	Cl{001}{010}
Lodestone, 278						Natural magnet
Lonsdaelite, 232	C	Hex	3.3+	10	2.42	Hex. diamond
Luzonite, 259	Cu_3AsS_4	Tet	4.4	3–4	—	Low T form of enargite
Magnesiochromite, 280	$MgCr_2O_4$	Iso	4.2	$5\frac{1}{2}$	—	A spinel
Magnesioferrite, 279	$MgFe_2O_4$	Iso	4.5–4.6	$5\frac{1}{2}$–$6\frac{1}{2}$	—	A spinel
Magnesite, 300	$MgCO_3$	Rho	3.0–3.2	$3\frac{1}{2}$–5	1.70	Commonly massive
Magnetite, 278	Fe_3O_4	Iso	5.18	6	—	Strongly magnetic
Malachite, 309	$Cu_2CO_3(OH)_2$	Mon	3.9–4.03	$3\frac{1}{2}$–4	1.88	Green
Manganite, 284	$MnO(OH)$	Mon	4.3	4	—	Prismatic, pseudo-orthorhombic crystals
Manganotantalite, 281	$(Mn, Fe)Ta_2O_6$	Orth	6.6±	$4\frac{1}{2}$	—	See columbite

Name, Page	Composition	XI Sys.	G	H	n	Remarks
Marcasite, 252	FeS_2	Orth	4.89	$6-6\frac{1}{2}$	—	White iron pyrites
Margarite, 408	$CaAl_2(Al_2Si_2O_{10})(OH)_2$	Mon	3.0–3.1	$3\frac{1}{2}-5$	1.65	Brittle mica
Marialite, 436	$Na_4(AlSi_3O_8)_3(Cl_2, CO_3, SO_4)$	Tet	2.60±	$5\frac{1}{2}-6$	1.55	See scapolite
Martite, 269	....					See hematite
Meerschaum, 402	....					See sepiolite
Meionite, 436	$Ca_4(Al_2Si_2O_8)_3(Cl_2, CO_3, SO_4)$	Tet	2.69	$5\frac{1}{2}-6$	1.59	See scapolite
Melaconite, 266	....					See tenorite
Melanite, 349	$Ca_3Fe_2(SiO_4)_3$	Iso	3.7	7	1.94	Black andradite
Melanterite, 323	$FeSO_4 \cdot 7H_2O$	Mon	1.90	2	1.48	Green-blue
Meneghinite, 260	$CuPb_{13}Sb_7S_{24}$	Orth	6.36	$2\frac{1}{2}$	—	Cl{010}
Mercury, 219	Hg	....	13.6	0	—	Fluid. Quicksilver
Metacinnabar, 248	$Hg_{1-x}S$	Iso	7.65	3.0	—	Grayish black
Mica, 404						A mineral group
Microcline, 425	$KAlSi_3O_8$	Tric	2.54–2.57	6	1.53	Low T K-feldspar
Microlite, 282	$Ca_2Ta_2O_6(O, OH, F)$	Iso	5.48–5.56	$5\frac{1}{2}$	1.92–1.99	Streak yellow to brown
Microperthite, 424	....					K-spar—albite inter-growth
Millerite, 246	NiS	Rho	5.5 ± 0.2	$3-3\frac{1}{2}$	—	Capillary crystals
Mimetite, 331	$Pb_5(AsO_4)_3Cl$	Hex	7.0–7.2	$3\frac{1}{2}$	2.1–2.2	Pale yellow, yellow-brown
Minium, 242	Pb_3O_4	?	8.9–9.2	$2\frac{1}{2}$	2.42	Earthy, brownish red
Minnesotaite, 403	$Fe_3Si_4O_{10}(OH)_2$	Mon	3.01	1?	~1.60	In iron-formations
Molybdenite, 253	MoS_2	Hex	4.62–4.73	$1-1\frac{1}{2}$	—	Lead-gray, platy
Monalbite, 423	$NaAlSi_3O_8$	Mon	....			Monoclinic, high T albite
Monazite, 328	$(Ce, La, Y, Th)PO_4$	Mon	4.6–5.4	$5-5\frac{1}{2}$	1.79	Parting{001}
Montebrasite, 332	$(Li, Na)Al(PO_4)(OH, F)$	Tric	~2.98	$5\frac{1}{2}-6$	1.61	See amblygonite
Monticellite, 346	$CaMgSiO_4$	Orth	3.2	5	1.65	An olivine
Montmorillonite, 403	$(Al, Mg)_8(Si_4O_{10})_3 \cdot (OH)_{10} \cdot 12H_2O$	Mon	2.5	$1-1\frac{1}{2}$	1.50–1.64	A clay mineral
Moonstone, 428, 431	....					See adularia, albite and orthoclase
Morganite, 367	....					Rose gem beryl
Moss agate, 416	....					Agate with mosslike patterns
Mountain cork, 389	....					Felted tremolite
Mountain leather, 389	....					Felted tremolite
Mullite, 352	$\sim Al_6Si_2O_{15}$	Orth	3.23	6–7	1.67	Cl{010}
Muscovite, 405	$KAl_2(AlSi_3O_{10})(OH)_2$	Mon	2.76–2.88	$2-2\frac{1}{2}$	1.60	Cl{001} perfect
Nacrite, 403	$Al_2Si_2O_5(OH)_4$	Mon	2.6	$2-2\frac{1}{2}$	1.56	Clay mineral
Nagyagite, 255	$AuTe \cdot 6Pb(S, Te)$?	Rho	7.4	$1-1\frac{1}{2}$	—	Blackish lead-gray
Natroalunite, 323	$(Na, K)Al_3(SO_4)_2(OH)_6$	Rho	2.6–2.9	$3\frac{1}{2}-4$	1.57	White, grayish, massive
Natrolite, 439	$Na_2Al_2Si_3O_{10} \cdot 2H_2O$	Orth	2.25	$5-5\frac{1}{2}$	1.48	Cl{110} perfect
Nepheline, 434	$(Na, K)AlSiO_4$	Hex	2.60–2.65	$5\frac{1}{2}-6$	1.54	Greasy luster
Nephrite, 389	....					Tough, compact tremolite
Neptunite, 357	$KNa_2Li(Fe, Mn)_2 \cdot TiO_2(Si_4O_{11})_2$	Mon	3.23	5–6	1.70	Black
Niccolite, 245	NiAs	Hex	7.78	$5-5\frac{1}{2}$	—	Copper-red
Nickel bloom, 331	....					See annabergite
Nickel iron, 225	Fe-Ni	Iso	7.8–8.2	5	—	Kamacite, taenite
Nickel skutterudite, 255	$(Ni, Co)As_3$	Iso	6.5 ± 0.4	$5\frac{1}{2}-6$	—	Tin-white
Niter, 311	KNO_3	Orth	2.09–2.14	2	1.50	Cooling taste
Nontronite, 403	$Fe_2(Al, Si)_4O_{10}(OH)_2 \cdot Na_{0.3}(H_2O)_4$	Mon	2.5	$1-1\frac{1}{2}$	1.60	Clay mineral
Norbergite, 355	$Mg_3(SiO_4)(F, OH)_2$	Orth	3.1–3.2	6	1.57	See chondrodite
Noselite, 435	$Na_8(AlSiO_4)_6SO_4$	Iso	2.3±	6	1.50	A feldspathoid
Oligoclase, 421, 431	$Ab_{90}An_{10}-Ab_{70}An_{30}$	Tric	2.65	6	1.54	Plagioclase feldspar
Olivine, 344	$(Mg, Fe)_2SiO_4$	Orth	3.27–4.37	$6\frac{1}{2}-7$	1.69	Green rock mineral

Name, Page	Composition	Xl Sys.	G	H	n	Remarks
Omphacite, 379	$(Ca, Na)(Mg, Fe, Al)Si_2O_6$	Mon	3.2–3.4	5–6	1.67–1.70	Green pyx. in eclogite
Onyx, 416	....					Layered chalcedony
Onyx marble, 299	....					See calcite and aragonite
Opal, 418	$SiO_2 \cdot nH_2O$	Amor	2.0–2.25	5–6	1.44	Conchoidal fracture
Orpiment, 249	As_2S_3	Mon	3.49	$1\frac{1}{2}$–2	2.8	Cl{010}. Yellow
Orthite, 362	....					See allanite
Orthoclase, 427	$KAlSi_3O_8$	Mon	2.57	6	1.52	Medium T K-feldspar
Orthoferrosilite, 375	$FeSiO_3$	Orth	3.9	6	1.79	Orthopyroxene end member
Palladium, 225	Pd	Iso	11.9	$4\frac{1}{2}$–5	—	See platinum
Paragonite, 405	$NaAl_2(AlSi_3O_{10})(OH)_2$	Mon	2.85	2	1.60	Like muscovite
Pargasite, 390	$NaCa_2Fe_4(Al, Fe)$-$Al_2Si_6O_{22}(OH)_2$	Mon	3–3.5	$5\frac{1}{2}$	1.62	See hornblende
Patronite, 331	VS_4	Mon	....			Ore of vanadium
Pectolite, 383	$Ca_2NaH(SiO_3)_3$	Tric	2.8±	5	1.60	Crystals acicular
Penninite, 408	....				1.58	See chlorite
Pentlandite, 246	$(Fe, Ni)_9S_8$	Iso	4.6–5.0	$3\frac{1}{2}$–4	—	With pyrrhotite
Periclase, 267	MgO	Iso	3.56	$5\frac{1}{2}$	1.73	Cl{001}, cubic
Peridot, 346	....					Gem olivine
Peristerite, 425	....					Intergrowths between An_2 and An_{15}
Perovskite, 272	$CaTiO_3$	Orth	4.03	$5\frac{1}{2}$	2.38	Pseudo-isometric crystals
Perthite, 424	....					K-feldspar—albite intergrowth
Petalite, 435	$Li(AlSi_4O_{10})$	Mon	2.4	6–$6\frac{1}{2}$	1.51	Cl{001}{201}
Petzite, 255	$(Ag, Au)_2Te$	Iso	8.7–9.0	$2\frac{1}{2}$–3	—	Steel-gray to iron-black
Phenacite, 343	Be_2SiO_4	Rho	2.97–3.0	$7\frac{1}{2}$–8	1.65	In pegmatites
Phillipsite, 442	$KCa(Al_3Si_5O_{16}) \cdot 6H_2O$	Mon	2.2	$4\frac{1}{2}$–5	1.50	A zeolite
Phlogopite, 406	$KMg_3(AlSi_3O_{10})(OH)_2$	Mon	2.86	$2\frac{1}{2}$–3	1.56–1.64	Yellow-brown mica
Phosgenite, 307	$Pb_2CO_3Cl_2$	Tet	6.0–6.3	3	2.12	Adamantine luster
Phosphorite, 330	....					See apatite
Picotite, 277	....					Cr-rich, yellow-brown spinel
Piemontite, 361	$Ca_2MnAl_2O(SiO_4)(Si_2O_7)(OH)$	Mon	3.4	$6\frac{1}{2}$	1.75–1.81	Reddish brown
Pigeonite, 376	$\sim Ca_{0.25}(Mg, Fe)_{1.75}Si_2O_6$	Mon	3.30–3.46	6	1.64–1.72	Pyroxene
Pitchblende, 275	....					Massive UO_2
Plagioclase, 429	$Ab_{100}An_0$-Ab_0An_{100}	Tric	2.62–2.76	6	1.53–1.59	A feldspar series
Plagionite, 260	$Pb_5Sb_8S_{17}$	Mon	5.56	$2\frac{1}{2}$	—	Blackish lead-gray
Plancheite, 410	$Cu_8(Si_4O_{11})_2(OH)_2 \cdot H_2O$	Orth	3.3	5.5	1.66	Fibrous, mammillary, blue
Platinum, 224	Pt	Iso	14–19	4–$4\frac{1}{2}$	—	As grains in placers
Pleonaste, 277	....					Fe-rich, dark spinel
Polianite, 273	MnO_2	Tet	5.0	6–$6\frac{1}{2}$	—	Steel-gray
Pollucite, 433	$CsAlSi_2O_6 \cdot H_2O$	Iso	2.9	$6\frac{1}{2}$	1.52	Colorless; in pegmatites
Polybasite, 258	$Ag_{16}Sb_2S_{11}$	Mon	6.0–6.2	2–3	—	Pseudorhombohedral
Polyhalite, 291	$K_2Ca_2Mg(SO_4)_4 \cdot 2H_2O$	Tric	2.78	$2\frac{1}{2}$–3	1.56	Bitter taste
Potash feldspar, 425	$KAlSi_3O_8$	....				Microcline, orthoclase, sanidine
Powellite, 326	$CaMoO_4$	Tet	4.23	$3\frac{1}{2}$–4	1.97	Fluoresces yellow
Prase, 416	....					Dull green, microcrystalline quartz
Prehnite, 409	$Ca_2Al(AlSi_3O_{10})(OH)_2$	Orth	2.8–2.95	6–$6\frac{1}{2}$	1.63	Tabular crystals, green
Prochlorite, 408	....				1.60	See chlorite
Proustite, 257	Ag_3AsS_3	Rho	5.57	2–$2\frac{1}{2}$	3.09	Light ruby silver
Pseudoleucite, 433	....					See leucite
Pseudowollastonite, 382	$CaSiO_3$	Tric	....			High T form of wollastonite

Name, Page	Composition	Xl Sys.	G	H	n	Remarks
Psilomelane, 285	$(Ba, Mn)_3(O, OH)_6Mn_8O_{16}$	Orth	3.7–4.7	5–6	—	Botryoidal
Pyrargyrite, 257	Ag_3SbS_2	Rho	5.85	2–$2\frac{1}{2}$	3.08	Dark ruby silver
Pyrite, 251	FeS_2	Iso	5.02	6–$6\frac{1}{2}$	—	Crystals striated
Pyrochlore, 282	$(Ca, Na)_2(Nb, Ta)_2O_6$-(O, OH, F)	Iso	4.3±	5	—	Usually metamict
Pyrolusite, 273	MnO_2	Tet	4.75	1–2	—	Sooty
Pyromorphite, 330	$Pb_5(PO_4)_3Cl$	Hex	7.04	$3\frac{1}{2}$–4	2.06	Adamantine luster
Pyrope, 347	$Mg_3Al_2Si_3O_{12}$	Iso	3.58	7	1.71	A garnet
Pyrophanite, 271	$MnTiO_3$	Rho	4.54	5–6	2.48	$Cl\{02\bar{2}1\}$
Pyrophyllite, 404	$Al_2Si_4O_{10}(OH)_2$	Mon	2.8	1–2	1.59	Smooth feel, micaceous habit
Pyroxene, 371	....					Mineral group
Pyroxenoid, 380	....					Mineral group
Pyroxferroite, 383	$Ca_{0.15}Fe_{0.85}SiO_3$	Tric	3.7	....	1.75	Pyroxenoid; in lunar rocks
Pyroxmangite, 383	$(Mn, Fe)SiO_3$	Tric	3.6–3.8	$5\frac{1}{2}$–6	1.72–1.75	Metamorphic Mn-rich rocks
Pyrrhotite, 245	$Fe_{1-x}S$	{Mon Hex	4.58–4.65	4	—	Magnetic
Quartz, 412	SiO_2	{Rho Hex	2.65	7	1.54	Conchoidal fracture
Ramsayite, 357	$Na_2Ti_2Si_2O_9$	Orth	3.43	6	2.01	From Kola Peninsula
Realgar, 248	AsS	Mon	3.48	$1\frac{1}{2}$–2	2.60	$Cl\{010\}$. Red
Red ocher, 269	....					See hematite
Rhodochrosite, 302	$MnCO_3$	Rho	3.5–3.7	$3\frac{1}{2}$–4	1.82	$Cl\{10\bar{1}1\}$. Pink
Rhodolite, 348	....					See garnet
Rhodonite, 382	$MnSiO_3$	Tric	3.4–3.7	$5\frac{1}{2}$–6	1.73–1.75	Pink
Riebeckite, 390	$Na_2Fe_3^{2+}Fe_2^{3+}Si_8O_{22}(OH)_2$	Mon	3.4	5	1.66–1.71	Dark green to black amphibole
Ringwoodite, 265	$(Fe, Mg)_2SiO_4$	Iso	3.90	....	1.77	Meteoritic; spinel structure
Rock crystal, 415	SiO_2					Colorless quartz crystal
Rock salt, 289	....					See halite
Roscoelite, 331	$KV_2(AlSi_3O_{10})(OH)_2$	Mon	2.97	$2\frac{1}{2}$	1.69	Vanadium mica
Rubellite, 369	....					Red to pink tourmaline
Ruby, 267	....					Red gem corundum
Ruby copper, 266	....					See cuprite
Ruby silver, 257	....					See pyrargyrite and proustite
Ruby spinel, 277	$MgAl_2O_4$	....				Red gem spinel
Rutile, 272	TiO_2	Tet	4.18–4.25	6–$6\frac{1}{2}$	2.61	Adamantine luster
Saltpeter, 311	....					See niter
Sanidine, 428	$KAlSi_3O_8$	Mon	2.56–2.62	6	1.53	High T K-feldspar
Saponite, 403	$(Mg, Fe)_3(Al, Si)_4O_{10}(OH)_2$-$(\frac{1}{2}Ca, Na)_{0.3}(H_2O)_4$	Mon	2.5	1–$1\frac{1}{2}$	1.52	Clay mineral
Sapphire, 267	....					Blue gem corundum
Sard, 416	....					Brown chalcedony
Sardonyx, 416	....					Onyx with sard
Satin spar, 321	....					Fibrous gypsum
Scapolite, 435	$3NaAlSi_3O_8 \cdot NaCl$ to $3CaAl_2Si_2O_8 \cdot CaCO_3$	Tet	2.55–2.74	5–6	1.55–1.60	$Cl\{100\}\{110\}$
Scheelite, 325	$CaWO_4$	Tet	5.9–6.1	$4\frac{1}{2}$–5	1.92	$Cl\{101\}$
Schorl, 369	....					Black tourmaline
Scolecite, 440, 442	$CaAl_2Si_3O_{10} \cdot 3H_2O$	Mon	2.2±	5–$5\frac{1}{2}$	1.52	A zeolite
Scorzalite, 332	$(Fe, Mg)Al_2(PO_4)_2(OH)_2$	Mon	3.35	$5\frac{1}{2}$–6	1.67	Blue, see lazulite
Selenite, 321	....					See gypsum

Name, Page	Composition	Xl Sys.	G	H	n	Remarks
Semseyite, 260	$Pb_9Sb_8S_{21}$	Mon	5.8	$2\frac{1}{2}$	—	Cl{112}, gray to black
Sepiolite, 402	$Mg_4(OH)_2Si_6O_{15} \cdot H_2O + 4H_2O$	Orth	2.0	$2–2\frac{1}{2}$	1.52	Meerschaum
Sericite, 405						Fine grained muscovite
Serpentine, 400	$Mg_3Si_2O_5(OH)_4$	{Mon Orth	2.3–2.6	3–5	1.55	Green to yellow
Shattuckite, 410	$Cu_5(SiO_3)_4(OH)_2$	Orth	3.8	....	1.78	Blue
Siderite, 301	$FeCO_3$	Rho	3.96	$3\frac{1}{2}–4$	1.88	Cl{10$\bar{1}$1}
Sillimanite, 352	Al_2SiO_5	Orth	3.23	6–7	1.66	Cl{010} perfect
Silver, 222	Ag	Iso	10.5	$2\frac{1}{2}–3$	—	White, malleable
Silver glance, 236						See acanthite
Skutterudite, 255	$(Co, Ni)As_3$	Iso	6.5 ± 0.4	$5\frac{1}{2}–6$	—	Tin-white
Smaltite, 255	$(Co, Ni)As_{3-x}$	Iso	6.5 ± 0.4	$5\frac{1}{2}–6$	—	See skutterudite
Smithsonite, 302	$ZnCO_3$	Rho	4.30–4.45	$4–4\frac{1}{2}$	1.85	Reniform
Smoky quartz, 415	SiO_2					Brown to black
Soapstone, 403						See talc
Sodalite, 434	$Na_8(AlSiO_4)_6Cl_2$	Iso	2.15–2.30	$5\frac{1}{2}–6$	1.48	Usually blue
Soda niter, 311	$NaNO_3$	Rho	2.29	1–2	1.59	Cooling taste
Soda spar, 431						Albite
Specularite, 269						Platy, metallic hematite
Sperrylite, 225	$PtAs_2$	Iso	10.50	6–7	—	See platinum
Spessartite, 347	$Mn_3Al_2Si_3O_{12}$	Iso	4.19	7	1.80	A garnet
Sphalerite, 242	ZnS	Iso	3.9–4.1	$3\frac{1}{2}–4$	2.37	Cl{011} 6 directions
Sphene, 356	$CaTiO(SiO_4)$	Mon	3.40–3.55	$5–5\frac{1}{2}$	1.91	Wedge-shaped xls
Spinel, 276	$MgAl_2O_4$	Iso	3.5–4.1	8	1.72	In octahedrons
Spodumene, 380	$LiAlSi_2O_6$	Mon	3.15–3.20	$6\frac{1}{2}–7$	1.67	Cl{110}~90°. In pegmatites
Staetite, 403						See talc
Stannite, 244, 275	Cu_2FeSnS_4	Tet	4.3–4.5	4	—	Metallic, grey to black
Staurolite, 354	$Fe_2Al_9O_6(SiO_4)_4(O, OH)_2$	Mon	3.65–3.75	$7–7\frac{1}{2}$	1.75	Pseudo-orthorhombic; cruciform twins
Stephanite, 258	Ag_5SbS_4	Orth	6.2–6.3	$2–2\frac{1}{2}$	—	Pseudohexagonal
Stibnite, 249	Sb_2S_3	Orth	4.52–4.62	2.0	—	Cl{010} perfect
Stilbite, 441	$CaAl_2Si_7O_{18} \cdot 7H_2O$	Mon	2.1–2.2	$3\frac{1}{2}–4$	1.50	Sheaflike aggregates
Stishovite, 412	SiO_2	Tet	4.35	7	1.80	In meteorite craters
Stolzite, 324	$PbWO_4$	Tet	7.9–8.3	$2\frac{1}{2}–3$	2.27	Cl{001}{011}
Stromeyerite, 239	$(Ag, Cu)_2S$	Orth	6.2–6.3	$2\frac{1}{2}–3$	—	Steel-gray
Strontianite, 306	$SrCO_3$	Orth	3.7	$3\frac{1}{2}–4$	1.67	Efferv. in HCl
Sulfur, 230	S	Orth	2.05–2.09	$1\frac{1}{2}–2\frac{1}{2}$	2.04	Burns with blue flame
Sunstone, 431						See oligoclase
Sylvanite, 256	$(Au, Ag)Te_2$	Mon	8.0–8.2	$1\frac{1}{2}–2$	—	Cl{010} perfect
Sylvite, 290	KCl	Iso	1.99	2	1.49	Cl cubic. Bitter
Taenite, 225	Fe-Ni	Iso	7.8–8.2	5	—	In meteorites
Talc, 403	$Mg_3Si_4O_{10}(OH)_2$	Mon	2.7–2.8	1	1.59	Greasy feel, sectile
Tantalite, 281	$(Fe, Mn)Ta_2O_6$	Orth	6.5±	6	—	See columbite
Tanzanite, 361						Blue gem zoisite
Tennantite, 258	$Cu_{12}As_4S_{13}$	Iso	4.6–5.1	$3–4\frac{1}{2}$	—	In tetrahedrons
Tenorite, 266	CuO	Tric	6.5	3–4	—	Black
Tephroite, 346, 383	Mn_2SiO_4	Orth	4.1	6	1.70–1.80	An olivine
Tetrahedrite, 258	$Cu_{12}Sb_4S_{13}$	Iso	4.6–5.1	$3–4\frac{1}{2}$	—	In tetrahedrons
Thomsonite, 442	$NaCa_2(Al_5Si_5O_{20}) \cdot 6H_2O$	Orth	2.3	5	1.52	A zeolite
Thorianite, 276	ThO_2	Iso	9.7	$6\frac{1}{2}$	—	Dark gray, brownish black
Thorite, 328, 350	$ThSiO_4$	Tet	5.3	5	1.8	Brown to black, radioactive
Thulite, 361						Rose-red zoisite
Tiger eye, 390, 416						See riebeckite and quartz
Tin, 219	Sn	Tet	7.3	2	—	Very rare
Tincalconite, 313	$Na_2B_4O_5(OH)_4 \cdot 3H_2O$	Rho	1.88	1	1.46	Alteration of borax

Name, Page	Composition	Xl Sys.	G	H	n	Remarks
Tin stone, 274	....					See cassiterite
Titanic iron ore, 271	....					See ilmenite
Titanite, 356	....					See sphene
Topaz, 353	$Al_2SiO_4(F, OH)_2$	Orth	3.4–3.6	8	1.61–1.63	Cl{001} perfect
Torbernite, 334	$Cu(UO_2)_2(PO_4)_2 \cdot 8\text{-}12H_2O$	Tet	3.22	2–2½	1.59	Green
Tourmaline, 369	$(Na, Ca)(Li, Mg, Al)$-$(Al, Fe, Mn)_6(BO_3)_3$-$(Si_6O_{18})(OH)_4$	Rho	3.0–3.25	7–7½	1.64–1.68	Trigonal cross-sections and conchoidal fracture
Travertine, 299	....					See calcite
Tremolite, 388	$Ca_2Mg_5Si_8O_{22}(OH)_2$	Mon	3.0–3.2	5–6	1.61	Cl{110}, white to light green
Tridymite, 417	SiO_2	{Mon {Orth	2.26	7	1.47	In volcanic rocks
Triphylite, 328	$Li(Fe, Mn)PO_4$	Orth	3.42–3.56	4½–5	1.69	Cl{001}{010}
Troilite, 245	FeS	Hex	4.7	4	—	In meteorites
Trona, 310	$Na_3H(CO_3)_2 \cdot 2H_2O$	Mon	2.13	3	1.49	Alkaline taste
Troostite, 344						Manganiferous willemite
Tschermakite, 390	$Ca_2Mg_3(Al, Fe)_2$-$Al_2Si_6O_{22}(OH)_2$	....				See hornblende
Tufa, 299						See calcite
Turquoise, 333	$CuAl_6(PO_4)_4(OH)_8 \cdot 4H_2O$	Tric	2.6–2.8	6	1.62	Blue-green
Tyuyamunite, 335	$Ca(UO_2)_2(VO_4)_2 \cdot 5\text{-}8\tfrac{1}{2}H_2O$	Orth	3.7–4.3	2	1.86	Yellow
Ulexite, 315	$NaCaB_5O_6(OH)_6 \cdot 5H_2O$	Tric	1.96	1–2½	1.50	"Cotton balls"
Ulvöspinel, 278	Fe_2TiO_4	Iso	4.78	7½–8	—	Exsolution in magnetite
Uralian emerald, 349						Green andradite
Uraninite, 275	UO_2	Iso	7.5–9.7	5½	—	Pitchy luster
Uvarovite, 347	$Ca_3Cr_2Si_3O_{12}$	Iso	3.90	7½	1.87	Green garnet
Vanadinite, 331	$Pb_5(VO_4)_3Cl$	Hex	6.9	3	2.25–2.42	Adamantine luster; red to yellow
Variscite, 334	$Al(PO_4) \cdot 2H_2O$	Orth	2.57	3½–4½	1.58	Green, massive
Verde antique, 402	....					Marble and green serpentine
Verdelite, 369						Green tourmaline
Vermiculite, 407	$(Mg, Ca)_{0.3}(Mg, Fe, Al)_{3.0}$-$(Al, Si)_4O_{10}(OH)_4 \cdot 8H_2O$	Mon	2.4	1½	1.55–1.58	Altered biotite
Vesuvianite, 362	....					See idocrase
Vivianite, 332	$Fe_3(PO_4)_2 \cdot 8H_2O$	Mon	2.58–2.68	1½–2	1.60	Cl{010} perfect
Wad, 274						Manganese ore
Wavellite, 333	$Al_3(PO_4)_2(OH)_3 \cdot 5H_2O$	Orth	2.36	3½–4	1.54	Radiating aggregates
Wernerite, 436					1.55–1.60	See scapolite
White iron pyrites, 252	....					See marcasite
Willemite, 343	Zn_2SiO_4	Rho	3.9–4.2	5½	1.69	From Franklin, N.J.
Witherite, 305	$BaCO_3$	Orth	4.3	3½	1.68	Efferv. in HCl
Wolframite, 324	$(Fe, Mn)WO_4$	Mon	7.0–7.5	4–4½	—	Cl{010} perfect
Wollastonite, 382	$CaSiO_3$	Tric	2.8–2.9	5–5½	1.63	Cl{100}, {001} perfect
Wood opal, 420	....					Fossil wood with opal
Wood tin, 274	....					See cassiterite
Wulfenite, 326	$PbMoO_4$	Tet	6.8±	3	2.40	Orange-red
Wurtzite, 243	ZnS	Hex	3.98	4	2.35	See sphalerite
Xanthophyllite, 408	$Ca(Mg, Al)_{3\text{-}2}(Al_2Si_2O_{10})(OH)_2$	Mon	3–3.1	3½	1.65	Brittle mica
Zeolite, 438						A mineral group
Zinc blende, 242	....					See sphalerite
Zincite, 266	ZnO	Hex	5.68	4	2.01	At Franklin, N.J.
Zinc spinel, 277	....					See gahnite

Name, Page	Composition	Xl Sys.	G	H	n	Remarks
Zinkenite, 260	$Pb_6Sb_{14}S_{27}$	Hex	5.3	$3-3\frac{1}{2}$	—	Steel-gray
Zircon, 349	$ZrSiO_4$	Tet	4.68	$7\frac{1}{2}$	1.92–1.96	In small crystals
Zoisite, 361	$Ca_2Al_3O(SiO_4)(Si_2O_7)(OH)$	Orth	3.35	6	1.69	Gray, green-brown, metamorphic

INDEX